physiology and medicine

D0024637

ecology and population biology

ecology and population biology

fisheries and oceanography

fisheries and oceanography

genetics

physics

chemistry

general

St. Olaf College

JUL 20 1984

Science Library

mathematics for the biosciences

michael r. cullen

loyola marymount university

Prindle, Weber & Schmidt
Boston, Massachusetts

PWS PUBLISHERS

Prindle, Weber & Schmidt • ✣ • Willard Grant Press • **WG** • Duxbury Press • ♦
Statler Office Building • 20 Park Plaza • Boston, Massachusetts 02116

QH
323.5
.C84

To Papa, in loving memory.

© Copyright 1983 PWS Publishers

Library of Congress Cataloging in Publication Data

Cullen Michael R.
   Mathematics for the biosciences.

   Includes bibliographical references and index.
   1. Biomathematics. I. Title.
QH323.5.C84 1983   510   82-14185
ISBN 0-87150-352-2

ISBN 0-87150-352-2

PWS Publishers is a division of Wadsworth, Inc.

Cover print *Sky and Water* by M. C. Escher. © BEELDRECHT, Amsterdam/V.A.G.A, New York, Collection Haags Gemeentemuseum—The Hague, 1981.

Text and cover design by Susan London, under the direction of Betty O'Bryant. Text composition in Linotron Times Roman and Universe by Intercontinental Photocomposition Ltd. Technical art by Deborah Schneck. Text printed and bound by Maple-Vail Book Manufacturing Group, New York, U.S.A.

83 84 85 86 87 — 10 9 8 7 6 5 4 3 2 1

preface

mathematics for the biosciences presents a year course in mathematics for biology students with special emphasis on calculus and its applications. The book is purposely structured so that it can accomodate shorter courses including a one quarter or one semester introduction to calculus. The text has been developed and class tested over a four year period. Although the best prerequisite for the book is a course in algebra and trigonometry, much of the text can be covered with only high school algebra as background.

I have based the writing on several premises.

1. *The main job of the text is to teach mathematics not biology.* The applications have been carefully selected to reinforce the mathematics and provide motivation. Generally biological models are presented at the end of a section and only after the mathematical techniques have been practiced.

2. *The reader is a freshman or sophomore life science student.* This student is just beginning his or her study of the sciences and I do not assume familiarity with the technical jargon of biology, chemistry, or physics. Hence, before a particular model is presented, you will find a good deal of explanation of precisely what is being modeled.

3. *The biological applications can be emphasized or de-emphasized according to the tastes of the instructor and without interrupting the logical development of the mathematics.* There is a wealth of biological and physical applications throughout the text that the instructor can select from according to the time available. It is important that the student see honest and real applications of the mathematics developed. Accordingly many models important to the practicing biologist have been chosen from advanced texts and journals and adapted to this audience. In addition, applications from business and the social sciences are occasionally presented to stress analogies and mathematical similarities amongst many different disciplines.

4. *Formal proofs are avoided whenever possible in favor of arguments based on geometric and physical intuition.* Although all theorems are stated carefully, the major goal of the text is to teach the mathematical techniques through examples and applications. I have not emphasized rigorous proofs or the weakest hypotheses under which a given theorem is valid.

text structure   The book is divided into *eight parts* which can be used to construct a variety of courses.

Part I reviews the *elementary functions* and introduces the student to the *limit concept*. For students coming out of a pre-calculus or algebra and trigonometry course, the first five sections can be omitted or quickly reviewed. The course would then begin with section 6 on limits. Section 15 (Exponential and Logarithm Functions) and section 29 (Trigonometric Functions) are independent, self-contained chapters that can be added to Part I in order to give a complete review of the elementary functions. The examples and exercises emphasize those skills from algebra important for the study of calculus.

Part II presents a course in *differential calculus* with special attention given to curve sketching techniques, optimization problems, and applications of exponential and logarithm functions. There is sufficient material for a one quarter course. In addition, sections 30 and 31 (from Part III) on the calculus of trigonometric functions can be added to give a complete course in differential calculus.

Part III covers *integral calculus* and emphasizes applications of the integral rather than techniques of integration. The substitution method, integration by parts, and the partial fractions method are included, but the student is encouraged to use integral tables whenever possible. This leaves time to present applications to physiology, oceanography, and physics. Section 25 studies density functions in an oceanographic setting and can serve as a lead-in to section 26 on probability density functions. Section 27 presents numerical integration methods, which are applied in section 28 to a standard model for measuring cardiac output. Finally, sections 29–31 cover trigonometric functions and their applications.

Part IV is a short course in *differential equations and modeling* that stresses the separation of variables method, first and second order linear equations, and linear systems. Starting with section 32 on implicit differentiation and related rates, the dy/dt notation for derivative and its interpretation as a rate of change are emphasized. Sections 34–36 are optional sections that contain models from ecology and fishery science including the famous logistic model and the von Bertalanffy growth model. Sections 37 and 39 introduce the student to compartmental models in physiology and ecology.

Part V presents *multivariante calculus* including partial derivatives, curve fitting by least squares, and double integration in rectangular and polar coordinates.

Part VI introduces the student to methods for solving *difference equations* and to *discrete mathematical models*. Also included is a brief introduction to infinite series and power series. Section 51 on Taylor polynomials, however, does not depend on earlier sections and could be covered after Part II on differential calculus.

Part VII provides an introduction to *vectors and matrices* and gives applications to biomechanics, population biology, and systems ecology.

Finally, Part VIII is a short course in *discrete probability* that starts with the sample space concept and proceeds to conditional probability and Bayes Theorem. Counting techniques are deemphasized in favor of applications to simple games of chance, Mendelian inheritance, and problems of genetic counseling. The final section presents the famous Hardy-Weinberg model in population genetics and uses difference equations to construct selection models.

text format   All key results and techniques are in boxes to set them apart from the remainder of the discussion and for easy reference. Worked-out examples are arranged in order of difficulty and the special applications to the biosciences generally appear at the end of a section. Individual sections in the book are usually 7–10 pages in length and require two-hour lectures for full coverage.

An important part of the text is the exercise sections and you will find almost 2,500 exercises in the book. Exercises are graded and divided into three parts. Part A exercises are drill exercises based on the examples in the section. More difficult problems are marked with a star (★). Part B and C exercises present more involved biological applications, optional topics, or problems of a more theoretical nature. They were written to provide challenges for the better student.

There are some exercises and examples using trigonometry that occur prior to section 29 where trig functions are formally covered. These problems are marked with a triangle (▲) and can be skipped if algebra is the only prerequisite for the course. My own preference is to include these functions early in the course so that they may be used to illustrate the derivative rules. I postpone the derivation of d/dt (sin t) = cos t, for example, until later in the course. Alternately, section 29 can be covered after section 5 and section 30 after section 8.

role of the calculator   I have assumed that all students will have access to a standard scientific calculator. I use the calculator in the text to illustrate the limit concept, the definition of derivative, and the definition of integral. Many exercises are best done with the aid of a calculator. In addition, the student is urged to discover many of the properties of transcendental functions by experimenting with his or her calculator.

suggested course outlines   *Mathematics for the Biosciences* can be adapted to a large number of quarter and semester courses. Shown below are suggested course outlines:

one quarter course in differential calculus	sections 1–5 (as needed) sections 6 and 7, and part 2
one semester course in calculus	sections 1–5 (as needed) and sections 6–22
one quarter course in integral calculus	sections 18–25 and sections 27–31
one quarter course in finite mathematics	sections 47–49 and sections 53–61
year course in calculus	sections 1–5 (as needed) 6–7, part 2, sections 18–25, 27–31, 32, 33, 35, 37–39, and 40–43
year course in biomathematics (meeting 4 times/week)	the entire book excluding the optional sections

acknowledgements   It is a pleasure to acknowledge the many individuals who have had a positive effect on the development of the text. Very special thanks are due to Professors Richard Kronmal and Vincent Gallucci of the Biomathematics Group at University of Washington for their kindness and hospitality during the 1977–78 academic year. In addition, hundreds of students endured earlier versions of the text and many offered helpful suggestions for clarifying the presentation.

I am pleased to acknowledge the assistance of the following reviewers who had a substantial effect on the final form of the book: Susann Novalis, San Francisco State University; Robert A. Clark, Case Western Reserve University; H. Howard Frisinger, Colorado State University; Carolyn C. Volpe, San Diego Mesa College; T.L. Herdman, Virginia Polytechnic Institute and State University; Frederic T. Metcalf, University of California at Riverside; and Neal C. Raber, University of Akron.

Additional thanks are due to Mr. Andy Castanon, who read the entire manuscript and worked out many of the exercises; to my colleague Professor Michael Grady, for reviewing the probability material; and to the staff of PSW Publishers, for their faith in this project.

And finally, I must thank my wife Kathy for her patience, love, and encouragement.

Michael R. Cullen
February, 1983

table of contents

the derivative and its applications, 78

the integral and its applications, **197**

*Optional section.

an introduction to differential equations and modeling, 345

*Optional section.

an introduction to multivariate calculus, **435**

sequences, difference equations, and series, **523**

an introduction to vectors and matrices, **592**

*Optional section.

an introduction to discrete probability, **643**

*Optional section.

preparations: elementary functions and the limit concept

prologue   It is the dream of every scientist to capture in a nutshell (i.e., in a mathematical formula or **function**) the relationship among several quantities in nature. So far in the history of science those special formulas that have come to light have had one common feature: the basic building blocks in these formulas are the so-called *elementary functions*. You will find these functions (e.g., x^2, $\sqrt{x}$, x^y, log x, ln x, sin x, cos x) on any standard scientific calculator. It is the purpose of the first part of this text to review elementary functions by (1) recalling their definitions and basic properties, (2) carrying out computations with the aid of a calculator, and (3) indicating examples where the given functions might occur in nature. We will begin our study by defining the meaning of *numbers*.

the function concept

Shown in figure 1.1 is the real number line.

figure 1.1

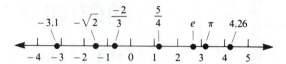

On this line are the **integers** $(0, \pm 1, \pm 2, \ldots)$ and **rational numbers** (fractions such as -3.1, $-\frac{2}{3}$, $\frac{5}{4}$, and 4.26). The real number line also includes the **irrational numbers**, whose decimal expansions are infinite and nonrepeating. Examples of irrational numbers are $-\sqrt{2} = -1.4142136\ldots$, $\pi = 3.1415926\ldots$, and $e = 2.71828183\ldots$, which play a key role in calculus.

We will let $\mathcal{R}$ denote the collection of all real numbers and will use capital letters $A, B, C, \ldots$ to denote special subsets of $\mathcal{R}$. Small letters $a, b, c, \ldots$ will denote particular real numbers. Hence, $Z = \{0, \pm 1, \pm 2, \ldots\}$ denotes the integers and $Q = \{p/q : p, q \in Z \text{ and } q \neq 0\}$ denotes the collection of rational numbers.

The word **function** is derived from the Latin word *functo*, meaning *to perform*. A function f takes a given real number x and "performs upon it," producing a second real number, $y = f(x)$.

example 1.1

inversion Let's take a look at the reciprocal button $\boxed{1/x}$ on your calculator and do a few computations (table 1.1).

table 1.1

Inputs	Outputs
2 ⟶	.5 = 1/2
1/3 ⟶	3
4/5 ⟶	1.25 = 5/4
.25897 ⟶	3.8614511

This function allows real numbers, to "stand on their heads" so to speak. In general, $x \to 1/x$ and we write $f(x) = 1/x$. Notice that $f(0)$ is not defined, but $f(x)$ does make sense for any $x \neq 0$. The collection of inputs or performers $\{x : x \neq 0\}$ is called the **domain** of the function. The collection of the final results of the performance (i.e., the outputs) is called the **range** of the function.   ▤

definition 1.1   Let A be a subset of $\mathcal{R}$. A *function f* from A into $\mathcal{R}$ is a rule that assigns to each a in A a single real number $f(a)$. The set A is called the **domain of definition** of the function f.

One good way of visualizing a function is with the input-output machine illustrated in figure 1.2. Think of as a button on a calculator. A real number x in the domain is entered, the "f button" is hit, and the final functional value $f(x)$ appears.

figure 1.2

input x ⟶ [f] ⟶ output $f(x)$

definition 1.2   The **range** of the function f is the collection of all outputs $f(a)$ for a in the domain A.

example 1.2   For $x \geq 0$, define $f(x) = x + 1$. Find the range of f.

solution 1.2   For $x = 0$, $f(0) = 1$. As x increases from 0 to $+\infty$, $f(x) = x + 1$ increases from 1 to $+\infty$. Thus, the range is the set of real numbers $y \geq 1$.   ▤

example 1.3   Find the largest possible domain of definition for each of the following functions.

$$\text{(a) } f(x) = \frac{x+1}{x^2 - 4} \qquad \text{(b) } g(x) = \sqrt{x-1} \qquad \text{(c) } h(x) = \sqrt{x^2 + 1}$$

solution 1.3   In all three cases, we must determine the restrictions on x that will guarantee that all computations in the rule can be performed. For $f(x)$, x must not equal -2 or $+2$, in order to avoid division by zero. Thus, f has largest domain $\{x: x \neq \pm 2\}$. For $g(x)$, note that $g(0) = \sqrt{-1}$ is not a real number. We must avoid taking the square root of a negative number. This will be guaranteed if $x \geq 1$. Finally, the largest domain for $h(x)$ is $\mathcal{R}$, since $x^2 + 1$ is always greater than 0.   ▤

In calculus, it is important to perform algebraic manipulations of the type given in the next example.

example 1.4   If $f(x) = x^2 - 2x$, compute:

$$\text{(a) } f(-2) \qquad \text{(b) } f(2a) \qquad \text{(c) } f(x+h) \qquad \text{(d) } \frac{f(x+h) - f(x)}{h}$$

solution 1.4

If general, if the input is Δ, then $f(\Delta) = (\Delta)^2 - 2(\Delta)$. Thus, for **(a)**, $f(-2) = (-2)^2 - 2(-2) = 4 + 4 = 8$. To work **(b)**, let $\Delta = 2a$. Then $f(2a) = (2a)^2 - 2(2a) = 4a^2 - 4a$. For **(c)**, let $\Delta = x + h$. Then $f(x + h) = (x + h)^2 - 2(x + h) = x^2 + 2xh + h^2 - 2x - 2h$. Finally, for **(d)**,

$$\frac{f(x+h) - f(x)}{h} = \frac{f(x+h) - (x^2 - 2x)}{h}$$

$$= \frac{2xh + h^2 - 2h}{h}, \quad \text{using (c)}$$

$$= 2x + h - 2 \qquad \blacksquare$$

The next example shows that a function need not be given by a single formula.

example 1.5

Define

$$f(x) = \begin{cases} x + 1, & x > 1 \\ x - 1, & 0 \leq x \leq 1 \\ 3, & x < 0 \end{cases} \qquad \text{and} \qquad g(x) = \begin{cases} x + 1, & x > 1 \\ x - 1, & 0 \leq x \leq 1 \\ 3, & x \leq 0 \end{cases}$$

Are f and g functions?

solution 1.5

For the rule f, a single real number has been assigned to each real number x. For example, $f(0) = 0 - 1 = -1$ and $f(-2) = 3$. For the rule g, we have not been careful enough. For $x = 0$, $g(0) = 0 - 1 = -1$, since $g(x) = x - 1$, for $0 \leq x \leq 1$. But we also have $g(0) = 3$ since $x = 0 \leq 0$. Thus, $g(0) = -1$ and $g(x) = 3$. Hence, the rule g does not define a function. $\qquad \blacksquare$

example 1.6

Define

$$f(x) = \begin{cases} x - 1, & x \geq 1 \\ 1 - x^2, & x \leq 1 \end{cases}$$

Is f a function?

solution 1.6

If $x > 1$ or if $x < 1$, a single real number is assigned to x. We can compute $f(1)$ in two ways: $f(1) = 1 - 1 = 0$ and $f(1) = 1 - 1^2 = 0$. Hence, the single real number 0 has been assigned to 1, and f is a function. $\qquad \blacksquare$

A second way to visualize a function is to draw two number lines, as shown in figure 1.3, which indicates that $f(-1) = 1$ and $f(0) = 2$.

figure 1.3

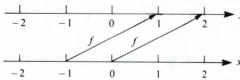

The third and most common way of picturing a function is to draw its *graph*, as illustrated in figure 1.4. This is the main tool we will use to obtain a *global view* of a function's behavior.

figure 1.4

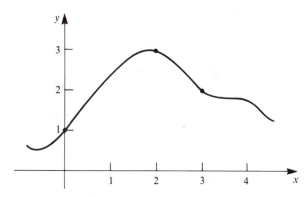

definition 1.3   The collection of all points $(x, f(x))$ in a Cartesian coordinate plane, where x varies over the domain of definition A, is called the **graph** of the function f.

In general, the domain of definition A consists of *infinitely many real numbers*, and so we cannot plot all the points $(x, f(x))$. Graphing then consists of two steps:

step 1. Plotting a finite number of points.
step 2. Knowing how to *connect* these points.

Let's illustrate this with three examples.

example 1.7   Sketch the graph of the function $f(x) = \frac{1}{2}x + 1$.

solution 1.7   Table 1.2 gives the coordinates of three points on the graph of $f(x)$. Actually, it is sufficient to plot only two points since the shape of the graph is well known to be a **line** (see section 3). A third point is plotted in figure 1.5 as a check.

table 1.2

x	$\frac{1}{2}x + 1$
0	1
2	2
-2	0

figure 1.5

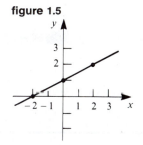

example 1.8   Sketch the graph of the function $f(x) = x^2 - 2x + 1$.

solution 1.8   We rewrite the function as $f(x) = x^2 - 2x + 1 = (x - 1)^2$. It is known that the graph of any quadratic function $y = ax^2 + bx + c$ is a **parabola** (see section 4). In our case, the vertex occurs at $x = 1$, and the parabola opens upward. Table 1.3 displays seven points, which are plotted and connected in figure 1.6.

table 1.3

x	$f(x) = (x - 1)^2$
0,2	1
1	0
−1,3	4
−2,4	9

figure 1.6

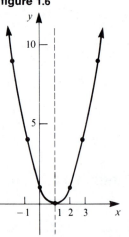

example 1.9   Sketch the graph of the function $f(x) = x^3 - 3x + 1$.

solution 1.9   There is no elementary way of determining the shape of the graph of f. Calculus is needed for this step! We will begin by determining specific points, displayed in table 1.4 and plotted as dots in figure 1.7. How do we connect the

table 1.4

x	−3	−2	−1	0	1	2	3
$f(x)$	−17	−1	3	1	−1	3	19

figure 1.7

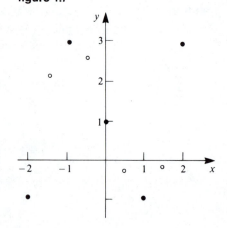

points? Although we could plot a few more points [$f(\pm 0.5)$ and $f(\pm 1.5)$ are shown as ∘], there are still many different ways of connecting these 9 points. Shown in figure 1.8 are points on the graph of f that were plotted with the aid of a computer graphics device. Although we may think that we know the shape, there is always the possibility that something unexpected happens between the points we have plotted.

figure 1.8

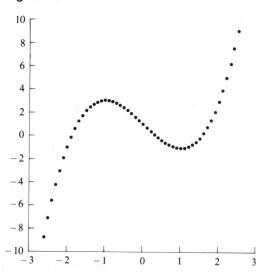

To determine the shape of a graph, it is important to know where a function is increasing (see figure 1.9**a**) and where it is decreasing (see figure 1.9**b**).

figure 1.9(a) increasing

figure 1.9(b) decreasing

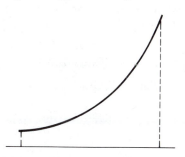
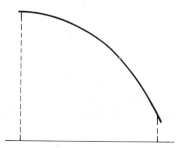

We will devise ways of determining regions of increase and decrease after we have studied the **derivative**—the first of the two main concepts in calculus.

A large number of drill exercises on the concept of a function are given in the following exercises.

exercises for section 1

Find the range of each of the following functions on the specified domain A.

1. $f(x) = x + 2$, for $x \geq 0$

2. $f(x) = x - 1$, for $x \leq 2$

3. $f(x) = x^2$, for $x \geq 0$

4. $f(x) = x^3$, for $x \geq 1$

5. $f(x) = x^2$, for $x \geq -1$

6. $f(x) = x^3$, for $x \geq -1$

7. $f(x) = \sqrt{x}$, for $x > 4$

8. $f(x) = \sqrt{x + 1}$, for $x > 0$

9. $f(x) = 1/x$, for $x > 2$

10. $f(x) = 1/x^2$, for $0 < x < 1$

Find the largest possible domain of definition (the *natural* domain) for each of the following functions.

11. $f(x) = \dfrac{x}{x - 1}$

12. $f(x) = \dfrac{x}{2x + 1}$

13. $f(x) = \dfrac{1}{x^2 - x}$

14. $f(x) = \dfrac{x^2}{x^2 - 9}$

15. $f(x) = \sqrt{x + 3}$

16. $f(x) = \sqrt{x - 4}$

17. $f(x) = \sqrt{3x + 4}$

18. $f(x) = \sqrt{x^2 + 4}$

19. $f(x) = 4 - \sqrt{x}$

20. $f(x) = x\sqrt{x + 1}$

For each of the following functions, compute and simplify (a) $f(2a)$, (b) $f(x + h)$, and (c) $[f(x + h) - f(x)]/h$.

21. $f(x) = x^2$

22. $f(x) = 2x$

23. $f(x) = 3x + 2$

24. $f(x) = x^2 + x$

25. $f(x) = 4$

26. $f(x) = 1/x$

★27. $f(x) = \sqrt{x}$ [*Hint*: Rationalize the numerators to do (c).]

★28. $f(x) = \sqrt{x + 1}$

★29. $f(x) = 1/(x + 1)$

★30. $f(x) = x/(x + 1)$

Which of the following rules are functions? Compute (if possible) $f(0)$, $f(1)$, and $f(2)$.

31. $f(x) = \begin{cases} x^2, & x \geq 1 \\ 1 - x^2, & x < 1 \end{cases}$

32. $f(x) = \begin{cases} x^2, & x \geq 1 \\ 2 - x, & x \leq 1 \end{cases}$

33. $f(x) = \begin{cases} x - 3, & x \geq 2 \\ 3x - 7, & x \leq 2 \end{cases}$

34. $f(x) = \begin{cases} x + 1, & x \geq 0 \\ x - 1, & x \leq 0 \end{cases}$

35. $f(x) = \begin{cases} 2x - 3, & x \leq 0 \\ 4x + 1, & 0 < x < 1 \\ 5x - 6, & x \geq 1 \end{cases}$

Sketch the graph of each of the following functions $y = f(x)$ over the specified domain.

36. $y = 4x + 3$, for all x **37.** $y = x^2$, for all x

38. $y = 1/x$, for $x > 0$ **39.** $y = 1/x^2$, for $x < 0$

40. $y = \sqrt{x}$, for $x \geq 0$ **41.** $y = x^2 - 4x + 4$, for all x

42. $y = x^3$, for all x **43.** $y = (x-1)^2 + 3$, for all x

★44. $y = \dfrac{x}{x+1}$, for $x \geq 0$ **45.** $y = \dfrac{1}{x^2+1}$, for all x

For each of the following functions f, determine the zeros—i.e., solve the equation $f(x) = 0$.

46. $3x - 4$ **47.** $x^2 + 3x$

48. $x^2 - x - 2$ **49.** $x^3 - x^2 - 2x$

50. $(x^2 - 4)/(x^2 - 1)$ **51.** $(x-1)(2x+3)(4x-5)$

52. $(x^2 - 1)(x^2 + 1)$ **53.** $x^2 - 3x - 3$ [*Hint*:

54. $x^2 + 2x + 4$ [*Hint*:   Use the quadratic formula.]

   Use the quadratic formula.]   **★55.** $x^3 + 6x^2 + 5x - 12$

part B

Complete the following input-output tables. In each case, determine the pattern of the outputs.

56.

input x	output $3x - 1$
1	
.1	
.01	
.001	
.0001	

57.

input x	output $\sqrt{x}$
4.5	
4.05	
4.005	
4.0005	

58.

input x	output $\dfrac{x^3 - 1}{x - 1}$
1.1	
1.01	
1.001	
1.0001	

59.

input x	output $x^{1/x}$
.1	
.01	
.001	
.0001	

60.

input x	output $\dfrac{\sqrt{x}-2}{x-4}$
3.9	
3.99	
3.999	
3.9999	

61.

input x	output $(\tfrac{1}{2})^x$
4	
5	
6	
7	
8	

62.

input x	output $\sqrt{x^2+x}-x$
10	
100	
1000	
10000	

63.

input x	output $\dfrac{\sqrt{x}-1}{x-1}$
.9	
.99	
.999	
.9999	

64. $x \to \boxed{x^2 - x} \to 0$
Find x.

65. $x \to \boxed{4x + 3} \to 7$
Find x.

66. $x \to \boxed{\dfrac{x+1}{x-1}} \to 4$
Find x.

67. $x \to \boxed{1/x} \to 4$
Find x.

68. $x \to \boxed{2^x} \to 4$
Find x.

69. $x \to \boxed{2^x} \to 1/4$
Find x.

70. If $f(x) = 1/x$, what is $f(f(x))$?

71. If an object is dropped from a 250-foot building, its distance (in feet) from the ground t seconds later is given by:

$$d(t) = 250 - 16t^2$$

(a) Where is the object 2 seconds after it has been dropped?
(b) When will the object hit the ground?

2 solving inequalities

The domains and ranges of the functions encountered in calculus are **intervals**, or, more generally, **unions of intervals**. We will now introduce a useful notation for these special sets.

Consider the sets A_1, A_2, A_3, and A_4, shown in figure 2.1. Our special notation for A_1 is $(-\infty, -3]$. Here we use the bracket "]" to indicate that $x = -3$ *is included* in the set. The parenthesis "(" indicates that the symbol of direction $-\infty$ is *not to be included* in A_1. Following this convention, we express sets A_2, A_3, and A_4 as

$$A_2 = [-2, -1] \qquad A_3 = (0, 1] \qquad A_4 = (2, +\infty)$$

figure 2.1

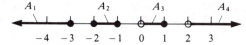

example 2.1

Find the domains of definition of the following functions and express the results using interval notation.

(a) $f(x) = \sqrt{x + 2}$ (b) $g(x) = 1/(\sqrt{x + 2})$ (c) $h(x) = \sqrt{x^2 - 1}$

solution 2.1

(a) In order to calculate the square root, we must require $x + 2 \geq 0$. Hence, $x \geq -2$, and the domain is then $A = [-2, +\infty)$.
(b) For g, we must require that $(x + 2)$ be nonnegative and that $\sqrt{x + 2}$ be nonzero. Hence, $x + 2 > 0$, and $x > -2$. The domain is then $(-2, +\infty)$.
(c) Finally, the domain of definition of $h(x)$ is the collection of real numbers satisfying $x^2 \geq 1$. Hence, $x \geq 1$ *or* $x \leq -1$, and $A = [1, +\infty)$ *or* $(-\infty, -1] = [1 + \infty) \cup (-\infty, -1]$. ◼

The reader may recall that for subsets A and B of the set of real numbers $\mathcal{R}$.

$$A \cup B = \{x : x \in A \text{ or } x \in B\} \qquad \text{and} \qquad A \cap B = \{x : x \in A \text{ and } x \in B\}.$$

example 2.2

Express each of the following sets in interval notation.

(a) $[0, 1] \cap (.5, +\infty)$ (b) $[0, 1] \cup (\tfrac{3}{4}, 2)$ (c) $(-\infty, 4) \cap [3, +\infty)$

solution 2.2

The easiest method of solution is to sketch the sets, as shown in figure 2.2.

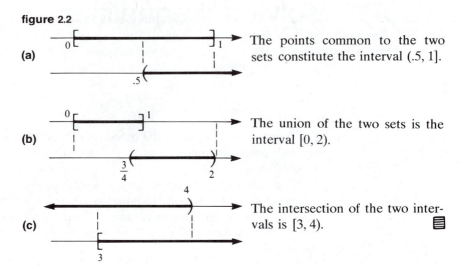

figure 2.2

(a) The points common to the two sets constitute the interval (.5, 1].

(b) The union of the two sets is the interval [0, 2).

(c) The intersection of the two intervals is [3, 4].

As example 2.1 shows, it is often necessary to solve inequalities in order to find the domain of definition of a function. In addition, after we have studied the derivative $f'(x)$, we will need to solve the inequalities $f'(x) > 0$ and $f'(x) < 0$. The remainder of this section is devoted to various methods for solving inequalities.

---

the Laws of Inequalities

Law 1. If $a < b$, then $a + c < b + c$ for any real number c. The direction (or sense) of the inequality has been *preserved*.

Law 2. If $a < b$, then
  (i) for $c > 0$, $ac < bc$.
  (ii) for $c < 0$, $ac > bc$. The direction of the inequality has been *reversed*.

Law 3. If $0 < a < b$, then $\dfrac{1}{a} > \dfrac{1}{b}$.

It is easy to convince yourself of the validity of these rules by considering numerical examples. [For example, $3 < 5$ and $(-2)3 = -6 > (-2)5 = -10$.] Now let's put the rules to work.

example 2.3   For $0 < x < y$, compare:

(a) x^2 and xy **(b)** $-x$ and $-y$ **(c)** x and $x + y$ **(d)** $\dfrac{1}{x}$ and $\dfrac{1}{x + y}$

solution 2.3

(a) Since $x>0$, we may multiply both sides of the inequality $x<y$ by x to obtain $x^2<xy$.
(b) Multiplying both sides of the inequality $x<y$ by (-1), we obtain $-x>-y$. The direction of the inequality has been reversed.
(c) Since $0<y$, we have $0+x<y+x$, or $x<x+y$.
(d) Since $x>0$, we may apply Law 3 to the inequality in (c) to obtain $1/x>1/(x+y)$. ▤

example 2.4

Solve the inequality $3-4x\le2$.

solution 2.4

Adding -3 to both sides yields the inequality $-4x\le2-3=-1$. We now multiply both sides by the negative number $-\frac14$ to obtain $x\ge\frac14$. We have reversed the direction of the inequality. Hence, the solution set is $[\frac14,+\infty)$. ▤

example 2.5

Find the domain of definition of the function $f(x)=\sqrt{x^2-5x+6}$.

solution 2.5

To find the domain, we must solve the inequality

$$x^2-5x+6=(x-2)(x-3)\ge0$$

There are two different ways this product can be nonnegative:
(i) $(x-2)\ge0$ and $(x-3)\ge0$. In order for both inequalities to hold, we must have $x\ge3$, i.e., $x\in[3,+\infty)$.
(ii) $(x-2)\le0$ and $(x-3)\le0$. Both inequalities will hold if $x\le2$, i.e., $x\in(-\infty,2]$.
Therefore, the solution to the inequality is $x\le2$ *or* $x\ge3$, and the domain of definition is $A=(-\infty,2]\cup[3,+\infty)$. ▤

example 2.6

Solve the inequality $x/(x-1)>2$.

solution 2.6

One is strongly tempted to multiply both sides of the inequality by $(x-1)$ in order to clear out the fraction, but Law 2 reminds us that we must know the *sign* of $(x-1)$ before doing so. Since x is unknown, we must divide our search for solutions into two directions.

case I $x>1$

Since $(x-1)>0$, $x>2(x-1)=2x-2$. Hence, $-x>-2$ or $x<2$. Hence, one part of the solution set is the interval $(1,2)$, as shown in case I.

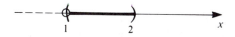

case II $x<1$

Since $(x-1)<0$, $x<2(x-1)=2x-2$. [*Note*: We have reversed the direction of the inequality.] Hence, $2<x$. But this is an impossible requirement for the case $x<1$. Hence, there are no solutions for case II.

The solution set is then $(1,2)$. ▤

As you can see from example 2.6, you must be careful! Here are two more inequalities that can be solved using the basic laws:

1. $(x-1)(x-3)(2x+3)<0$

2. $\dfrac{x^2-3x+2}{x(x-3)}>0$

There are many cases to consider and it is easy to make errors. Fortunately, there is a simple method for handling inequalities like (1) and (2) whose justification depends on concepts of calculus. The step-by-step approach is summarized in the accompanying box.

---

the Sign Chart Method for solving inequalities

Step 1. Set the expression $f(x)=0$ and solve. Call the solutions z_1, z_2, z_3, . . .

Step 2. Determine points not in the domain of definition of $f(x)$. Call these points u_1, u_2, u_3, . . .

Step 3. Mark off the points in Steps 1 and 2 (shown as ●) on a number line:

Step 4. Select *test points* t_i in the intervals formed and compute $f(t_i)$. Test points are labelled with ○ in the above figure.

Step 5. If $f(t_i)>0$, then $f(x)>0$ over the entire interval. Place a plus sign over the interval.

If $f(t_i)<0$, then $f(x)<0$ over the entire interval. Place a minus sign over the interval.

Step 6. The union of all "+" intervals is the solution to $f(x)>0$. The union of all "−" intervals is the solution to $f(x)<0$.

The Sign Chart Method is illustrated in our final three examples.

example 2.7   Solve the inequality $(x-1)(x-3)(2x+3)<0$, using the sign chart method.

solution 2.7   Let $f(x)=(x-1)(x-3)(2x+3)$. Then $f(x)=0$ for $x=1$, 3, and $-\frac{3}{2}$. These points are shown as dots in figure 2.3. Note that $f(x)$ is defined everywhere and so we can bypass Step 2. Now $f(-2)=(-3)(-5)(-1)<0$, $f(0)=$

figure 2.3

$(-1)(-3)(3) < 0$, $f(2) = (1)(-1)(7) < 0$, and $f(4) = (3)(1)(11) > 0$. The resulting sign chart for $f(x)$ is shown in figure 2.4. Thus, the solution to $f(x) < 0$ is $(-\infty, -\frac{3}{2}) \cup (1, 3)$.

figure 2.4

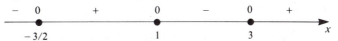

example 2.8   Solve the inequality $\dfrac{x^2 - 3x + 2}{x(x-3)} > 0$, using the sign chart method.

solution 2.8   Let $f(x) = \dfrac{x^2 - 3x + 2}{x(x-3)}$. Then $f(x) = 0$ when $x^2 - 3x + 2 = (x-1)(x-2) = 0$, i.e., when $x = 1, 2$. Furthermore, $f(x)$ is not defined at $x = 0, 3$. These points and selected test points are marked off in figure 2.5. The reader can check that

figure 2.5

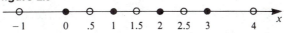

$f(-1) = 1.5 > 0$, $f(.5) = -.6 < 0$, $f(1.5) = \frac{1}{9} > 0$, $f(2.5) = -.6 < 0$, and $f(4) = 1.5 > 0$. Hence, the sign chart for $f(x)$ may be completed as shown in figure 2.6. We have placed u's over $x = 0$ and $x = 3$ as reminders that $f(x)$ is undefined there. The solution to $f(x) > 0$ is, therefore, $(-\infty, 0) \cup (1, 2) \cup (3, +\infty)$.

figure 2.6

example 2.9   Solve the inequality $x^2 > x - 1$.

solution 2.9   The inequality is equivalent to $x^2 - x + 1 > 0$. Let $f(x) = x^2 - x + 1$. The solutions to $f(x) = 0$ are $x = (1 \pm \sqrt{1-4})/2$. Hence, there are no real roots. Since $f(0) = 1$, we conclude $f(x) > 0$ for all real numbers. The solution set is then $\mathcal{R}$, as shown in figure 2.7. [*Note*: We concluded that $f(x) > 0$ for *all* x based on the single computation $f(0) = 1 > 0$. This giant step should be regarded with some suspicion!]

figure 2.7

We will practice this technique again when we study **continuity** in section 7. The sign chart method is valid for almost all functions you are likely to encounter, but it does require justification.

exercises for section 2

part A

Find the domains of each of the following functions and express the results using interval notation.

  1. $f(x) = \sqrt{x-8}$ **2.** $f(x) = 1/\sqrt{x-8}$

  3. $f(x) = \sqrt{x^2-4}$ **4.** $f(x) = 1/(x^2-4)$

  5. $f(x) = 1/\sqrt{x^2-4}$

Express each of the following sets using interval notation.

  6. $(0, 2) \cap (1, 2)$ **7.** $(3, +\infty) \cap (-\infty, 4)$

  8. $[-1, 2) \cup [2, 3)$ **9.** $(5, 6) \cap (4, 5]$

  10. $(1, 2) \cup [2, +\infty)$

Use the laws of inequalities to solve each of the following.

  11. $-2 + 3x > 1$ **12.** $-3x - 4 < 0$

  13. $x^2 - x > 0$ **14.** $x^2 - 4x + 3 \leq 0$

  15. $2x^2 - x < 3$ **16.** $1/x > \frac{1}{2}$

  17. $\dfrac{x+2}{3} > \dfrac{x-3}{2}$ **★18.** $\dfrac{2}{x-1} > \dfrac{x}{x+2}$

  ★ 19. $\dfrac{1}{x^2-1} > \dfrac{1}{x+1}$ **20.** $(x+1)^3 \geq 8$

Use the sign chart method to solve each of the following inequalities.

  21. $x(2x - 1) < 0$ **22.** $\dfrac{x^2}{x-1} > 0$

  23. $(x+1)(x-1)(x+2) < 0$ **24.** $\dfrac{x^2-4}{x-3} < 0$

  25. $x^2 + 3x + 3 > 0$ **26.** $x^2 + 5x + 8 < 0$

  27. $\dfrac{\sqrt{x}-1}{x+3} < 0$ **28.** $\dfrac{2^x - 2}{x-2} < 0$

  29. $6x^2 + 14x + 4 < 0$ [*Hint*: Factor!] **30.** $x(x-1)(x-2)(x-3) < 0$

  31. What is the largest domain of definition for the function $y = \sqrt{x(x-1)}$?

  32. For what values of x is $f(x) = \sqrt{\dfrac{x^2-5}{x^2-1}}$ defined?

linear functions

A **linear function** is a function of the form $f(x) = mx + b$, where m and b are fixed constants. As we shall soon see, the graph of $f(x)$ is a **straight line**. Let's now consider some special cases.

case I $y = f(x) = b$

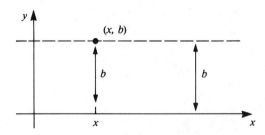

In this case, $m = 0$ and all points on the graph are of the form (x, b). Hence, the graph is a **horizontal straight line**, as shown above.

case II $y = f(x) = mx$

example 3.1 Sketch the graphs of $y = x$ and $y = \frac{1}{2}x$ on the same axes.

solution 3.1 Table 3.1 shows the coordinates of several points on the graphs of $y = x$ and $y = \frac{1}{2}x$. These points are plotted in figure 3.1. Notice that the line $y = \frac{1}{2}x$ is not

table 3.1

x	$y = x$	$y = \frac{1}{2}x$
-2	-2	-1
-1	-1	$-1/2$
0	0	0
1	1	$1/2$
2	2	1
3	3	$3/2$

figure 3.1

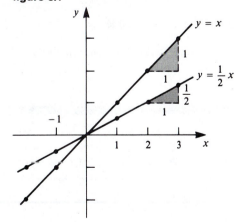

as steep as the line $y = x$. As shown in figure 3.1, given any point on the graph for a step of one unit to the right, one must move up $\frac{1}{2}$ unit to reach the graph of $y = \frac{1}{2}x$.

In general, the graph of $y = mx$ is a line through $(0, 0)$ and $(1, m)$. Note that $f(x + 1) = m(x + 1) = mx + m = f(x) + m$. Thus, for every unit step to the right, we go up or down m units to reach the graph of $y = mx$, depending upon whether $m > 0$ or $m < 0$. *The value of m is a measure of the steepness of the line, and is called the slope.*

figure 3.2

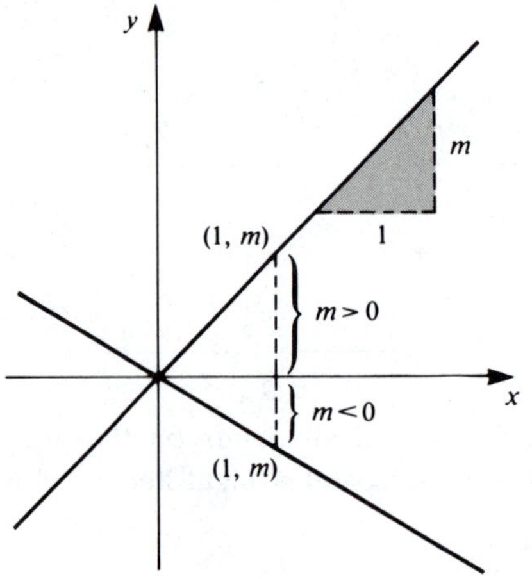

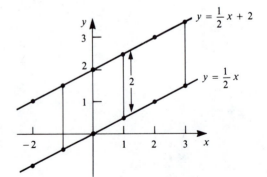

case III $y = f(x) = mx + b$

example 3.2   Sketch the graphs of $y = \frac{1}{2}x$ and $y = \frac{1}{2}x + 2$ on the same axes.

solution 3.2   Table 3.2 gives the coordinates of several points on the graphs of $y = \frac{1}{2}x$ and $y = \frac{1}{2}x + 2$. These points are plotted in figure 3.3.

table 3.2

x	$y = \frac{1}{2}x$	$y = \frac{1}{2}x + 2$
-3	$-3/2$	$-1/2$
-2	-1	1
-1	$-1/2$	$3/2$
0	0	2
1	$1/2$	$5/2$
2	1	3
3	$3/2$	$7/2$

figure 3.3

Notice in figure 3.3 that the graph of $y = \frac{1}{2}x$ has been shifted upward 2 units in the vertical direction.   ▤

In general, the graph of $y = mx + b$ is the graph of $y = mx$ translated vertically b units, as shown in figure 3.4. *The point $(0, b)$ is called the y-intercept.*

figure 3.4

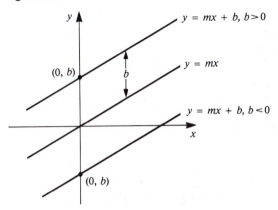

Note that if (x_1, y_1) and (x_2, y_2) are two points on the graph of $y = mx + b$, then $y_2 - y_1 = (mx_2 + b) - (mx_1 + b) = m(x_2 - x_1)$. Hence, the slope of the line may be computed as shown in the accompanying box.

the slope formula $m = \dfrac{y_2 - y_1}{x_2 - x_1}$

If we are given m and a point (x_1, y_1) on the graph, then any other point (x, y) on the graph must satisfy the point-slope formula, given in the box.

the point-slope formula $\dfrac{y - y_1}{x - x_1} = m$

example 3.3   Find the equation of linear function passing through the points $(1, -2)$ and $(4, 3)$. Determine where the line intersects the coordinate axes.

solution 3.3   Using the slope formula, we obtain $m = \dfrac{3 - (-2)}{4 - 1} = \dfrac{5}{3}$. Applying the point-slope formula we get

$$\frac{y-3}{x-4} = \frac{5}{3}$$

Hence, $3y - 9 = 5x - 20$ or $y = 5/3x - 11/3$. The y-intercept is $(0, -11/3)$ since $b = -11/3$. To find the x-intercept, set $y = 0$. It follows that $(5/3)x = 11/3$ or $x = 11/5 = 2.2$.   ▤

example 3.4   Find the equation of the straight line that passes through $(2, 3)$ and is parallel to the line $4x - 3y + 5 = 0$.

solution 3.4   The expression $4x - 3y + 5 = 0$ may be rewritten as $y = \frac{4}{3}x + \frac{5}{3}$. Hence, the slope is $m = \frac{4}{3}$. Since parallel lines have equal slopes, the problem reduces to finding the equation of the line that passes through $(2, 3)$ and has slope $\frac{4}{3}$. It follows from the point-slope formula that

$$\frac{y-3}{x-2} = \frac{4}{3}$$

Thus, $3y - 9 = 4x - 8$ *or* $y = \frac{4}{3}x + \frac{1}{3}$.   ▤

direct variation   What do we mean when we state that two variables y and x are *directly proportional*? First, we mean $y = f(x)$ for some function f. This idea is illustrated in figure 3.5. Secondly, whenever the input x is multiplied by a factor c the output is cy, c times the original output, as depicted in figure 3.6. This is expressed in function notation as $f(cx) = cf(x)$ for any constant c.

figure 3.5   **figure 3.6**

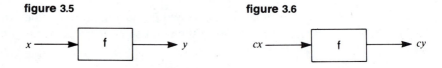

example 3.5   Show that if $f(x) = mx$, then y and x are directly proportional, i.e., the function f satisfies $f(cx) = cf(x)$.

solution 3.5 $f(cx) = m(cx) = c(mx) = cf(x)$ ▤

It can be shown that $f(x) = mx$ is the *only* function with the property illustrated in example 3.5. Thus, we may say that **y is directly proportional to x if and only if $y = mx$, for some constant m.** The constant m is called the **constant of proportionality**.

example 3.6

Hooke's Law   For many springs, the displacement x of the spring is directly proportional to the weight W on the spring (see figure 3.7). Hence, if the weight were doubled, the displacement x would also double. If a 10-pound weight stretches the spring 3 inches, find the formula relating W and x.

figure 3.7

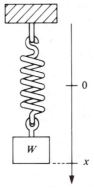

solution 3.6

Hooke's Law states that $W = mx$. Since $10 = (m)(3)$, we have $m = \frac{10}{3}$ (pounds/inch). Hence, $W = \frac{10}{3}x$.

example 3.7

A temperature of $0°$ Celsius corresponds to $32°$ Fahrenheit and $100°$C corresponds to $212°$F. Find the linear function relating Fahrenheit temperature F to the Centigrade temperature C.

solution 3.7

The points $(0, 32)$ and $(100, 212)$ lie on the graph of the function. Hence, the slope m is $(212 - 32)/(100 - 0) = 180/100 = 1.8$. Since $(0, 32)$ is the intercept, it follows that $b = 32$. Hence,
$$F = 1.8C + 32$$

Linear functions occur frequently in biological settings. Typically, they serve to summarize collected data that show a "linear pattern," as illustrated in figure 3.8. The formula $y = mx + b$ is certainly not an exact law, but it often

figure 3.8

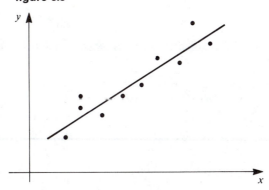

gives a nice tidy way of summarizing quantitative relationships obtained in an experiment. We will have more to say about such lines (called *regression lines*) later in the text. We now present two such examples from biology.

example 3.8   **the blue whale**   The blue whale is the largest mammal that has ever lived; the largest recorded length was 107 feet. What is the relationship between the length L (in feet) and the expected weight W (in British tons)? A rule of thumb adopted by whalers at the turn of the century was that each foot corresponded to a ton. Thus, an 80-foot specimen would weigh roughly 80 tons, according to this rule of thumb. A more precise relationship, adopted in the late 1960s by the International Whaling Commission, is

$$W = 3.51L - 192$$

Thus, an 80-foot specimen would have an expected weight of $3.51(80) - 192 = 88.8$ tons. [*Note*: The formula is used only for adult whales.]

example 3.9   In 1902, the pioneer statistician Karl Pearson reported the following relationship between a father's height x (in inches) and the *average* of his sons' heights, y:

$$y = \tfrac{1}{2}x + 34.8, \quad \text{for} \quad x \geq 59$$

Thus, if a 72-inch father were to have many sons, the average of their heights should be approximately $f(72) = 70.8$ inches. Pearson's linear function, illustrated in figure 3.9, conveniently summarizes the results of his statistical survey. No claims are made about the formula's applicability to other times or places.

figure 3.9   (Reproduced by permission of the Publishers, Charles Griffin & Company Ltd, of London and High Wycombe, from Smith *Biomathematics*, Vol. I.)

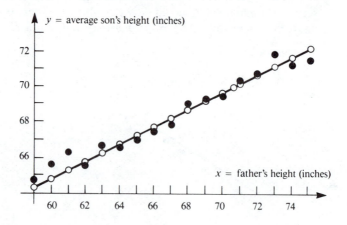

exercises for section 3

part A   Sketch the graphs of the following linear functions $y = f(x)$. For each function, determine the slope m, and the x- and y-intercepts.

1. $y = -1$

2. $y = 2$

3. $4x - y = 0$

4. $x = 2y$

5. $y = -2x + 1$

6. $y = 1/2x + 3$

7. $3x - 2y = 2$

8. $(y - 1)/(x - 1) = 4$

9. $y = (2x + 1)/3$

10. $x = 2 + y$

Find the equation of the line satisfying the following conditions.

11. The line passes through $(1, 2)$ and $(3, -2)$.

12. The line passes through $(0, 1)$ and $(-1, 2)$.

13. The line has slope 4 and passes through $(4, 2)$.

14. The line has slope -1 and passes through $(0, -3)$.

15. The line passes through $(2, 3)$ and is parallel to the line $y = 4x$.

16. The line passes through $(0, 0)$ and is parallel to the line $3x - 2y = 1$.

17. The line has slope -1 and passes through $(1, 2)$.

18. The line passes through $(0, 4)$ and $(4, 0)$.

★19. The equation of the line satisfies $f(x + 1) = 10 + f(x)$ and $f(0) = 5$.

★20. The equation of the line satisfies $f(x + 1) = 2 + f(x)$ and $f(0) = 10$.

21. If y and x are directly proportional and $y = 1$ when $x = 4$, express y as a function of x.

22. If y is directly proportional to x^2 and $y(2) = 3$, express y as a function of x.

23. For a spring obeying Hooke's Law, express W as a function of x if a 20-pound weight stretches the spring 5 inches.

24. A temperature of 0°C corresponds to 273K on the Kelvin scale, and a 1-degree increase in Centigrade temperature results in a 1-degree increase on the Kelvin scale. Express K as a function of C.

25. In January 1979, the average price of a gallon of regular gasoline was $.68. In January 1980, the average price had risen to $1.12. Fit a linear function to this information to predict the price in January 1981. What is the predicted

increase in price per month? [*Hint*: Let January 1979 correspond to $x = 0$.]*

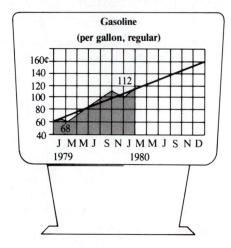

26. Assume that the amount C of fish food eaten in an aquarium is directly proportional to the amount A actually present in the aquarium. When 2 grams of fish food are in the aquarium, 1.5 grams are left the next day.

(a) Find a formula relating C and A.

(b) If 2 grams of food are in the aquarium and no food is added for four days, how much remains at the end of four days? [*Hint*: The answer is *not* 0.]

27. If the number C of school-age children in Los Angeles County is directly proportional to the population size N, find the formula relating C and N if there are 5,000 students in a city of 25,000 people.

28. Newborn blue whales measure about 24 feet, weigh 3 tons at birth, and are nursed for seven months. They weigh 23 tons and measure 53 feet after weaning.

(a) If t denotes time in months, find a linear function for L in terms of t that fits the given data.

(b) Express W as a linear function of t.

(c) According to these linear models for growth, what are the *daily* increases in length and weight?

29. Charles's Law for gases states that when pressure remains constant, the relationship between the volume V that the gas occupies and the temperature T in °C is given by

$$V = V_0 \left(\frac{1+T}{273} \right)$$

---

* Redrawn with permission from the Los Angeles Herald Examiner.

(a) What is the interpretation of V_0? Explain.

(b) What increase in temperature is needed to increase the volume from V_0 to $2V_0$?

30. The formula for the expected weight W (in long tons) of a humpback whale in terms of its length L (in feet) is given by

$$W = 1.70L - 42.8$$

(a) What is the expected weight of a 40-foot humpback whale?

(b) What is the expected length of a 35-ton humpback whale?

31. It costs a local whale-watching boat $100 for each afternoon excursion. If they charge $2.50 per ticket, how many must they sell to realize a $50 profit?

part B   Find the slope of the line joining $(x, f(x))$ and $(1, f(1))$ for each of the following functions.

32. $f(x) = x^2$ 33. $f(x) = 4x + 3$

34. $f(x) = \dfrac{1}{x}$ 35. $f(x) = 2x^2$

Express each of the following *hypotheses* in function form. Which of the assumptions do you believe are reasonable?

36. The loss in body heat is directly proportional to the surface area of the body.

37. The amount of material entering a cell per unit time is directly proportional to the area of the cell membrane.

38. The number of eggs laid in a stream is directly proportional to the number of spawners.

39. The momentum of a moving object is directly proportional to the velocity of the object.

40. The weight of an object is directly proportional to its volume.

41. The height to which a golf ball bounces is directly proportional to the height from which it is dropped.

42. Sketch the graph of the function $f(x) = \begin{cases} x - 3, & x \geq 2 \\ 3x - 7, & x < 2 \end{cases}$

43. Sketch the graph of the function $f(x) = \begin{cases} 2x - 3, & x \leq 0 \\ 4x + 1, & 0 < x < 1 \\ 5x - 6, & x \geq 1 \end{cases}$

 # power, polynomial, and rational functions

A **power function** is any function of the form $f(x) = x^k$, where $k \neq 0$ is a fixed real number. The simplest cases are the functions:

$$y = x, \; y = x^2, \; y = x^3, \; y = x^4, \ldots, \; y = x^n,$$

As we will see, the functions shown above are quite easy to graph.

example 4.1   Sketch the graphs of $y = x^2$ and $y = x^3$ on the same axes.

solution 4.1   Tables 4.1**(a)** and **(b)** supply points on the graphs of $y = x^2$ and $y = x^3$ respectively. These points are plotted in figure 4.1. For the function $f(x) = x^2$. $f(-x) = (-x)^2 = x^2 = f(x)$. Thus, $f(x) = x^2$ is an *even function* and its graph is

table 4.1(a)

x	x^2
0	0
±.5	.25
±1	1
±1.5	2.25
±2	4
±2.5	6.25

table 4.1(b)

x	x^3
0	0
.5	.125
−.5	−.125
1	1
−1	−1
1.5	3.375
−1.5	−3.375
2	8
−2	−8

figure 4.1

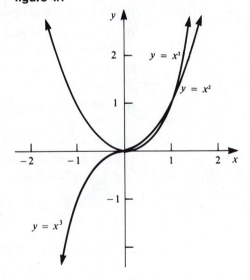

symmetric with respect to the y-axis. For $f(x) = x^3$, $f(-x) = (-x)^3 = (-1)^3 x^3 = -x^3 = -f(x)$. Thus $f(x) = x^3$ is an *odd function*.   ▤

We will now sketch the graph of $y = x^n$ for $x \geq 0$. For all values of n, $f(0) = 0$ and $f(1) = 1^n = 1$. Also, the function is increasing. The graph then has the general shape shown in figure 4.2. Notice that if n is even, then $f(-x) = f(x)$. Thus, f is an even function and the graph is symmetric about the y-axis.

figure 4.2

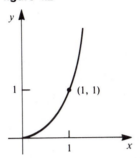

If n is odd, then $f(-x) = (-1)^n x^n = -x^n$, and f is odd. Thus, we must reflect the graph for $x \geq 0$ first through the y-axis and then through the x-axis, to obtain the graph for $x \leq 0$, as shown in figure 4.3. Of course, for the special case $n = 1$, we have the line $y = x$.

figure 4.3

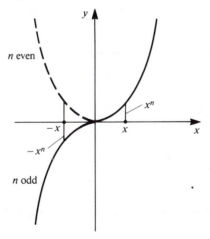

Using these simple power functions x, x^2, $x^3, \ldots$, we may form polynomial functions

$$p(x) = a_0 + a_1 x + a_2 x^2 + \cdots + a_n x^n$$

where $a_n \neq 0$. The fixed real numbers $a_0, a_1, \ldots, a_n$ are called **coefficients**, and n

is called the **degree of the polynomial**. For $n = 1$, we obtain the linear function $p(x) = a_0 + a_1 x$. The slope is $m = a_1$ and the y-intercept is $a_0 = p(0)$. For $n = 2$, we have the **quadratic function**

$$q(x) = a_2 x^2 + a_1 x + a_0$$

Figure 4.4 illustrates the graphs of general quadratic functions.

figure 4.4

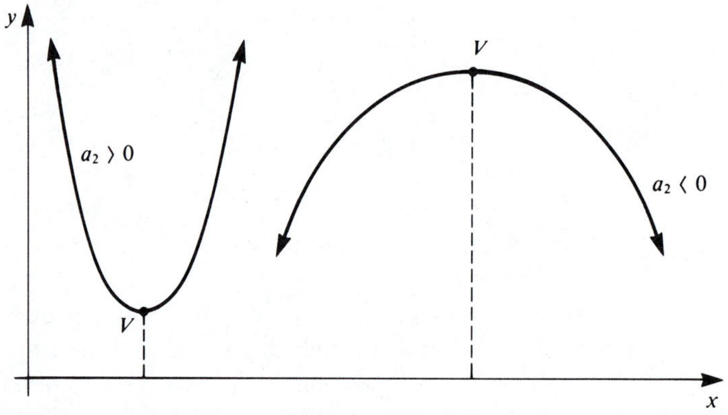

An important step in graphing $q(x)$ is to determine the vertex V. We will subsequently verify the formula in the accompanying box.

The vertex V of a quadratic function occurs when

$$x = -\frac{a_1}{2a_2}.$$

Hence, if $f(x) = -2x^2 + 3x + 4$, then the vertex V has x-coordinate $-3/2(-2) = 3/4$ and y-coordinate $f(3/4) = 5.125$.

example 4.2   Sketch the graphs of $y = x^2$, $y = \frac{1}{2}x^2$, and $y = -\frac{1}{2}x^2$ on the same axes.

solution 4.2   Table 4.2 supplies points on the graphs of the three functions. The graphs shown in figure 4.5 demonstrate that the effect of a_2 in the function $y = a_2 x^2$ is to adjust the width of the parabola and/or to change the direction in which the parabola opens.

table 4.2

x	x^2	$\frac{1}{2}x^2$	$-\frac{1}{2}x^2$
0	0	0	0
±.5	.25	.125	−.125
±1	1	.5	−.5
±2	4	2	−2
±3	9	4.5	−4.5

figure 4.5

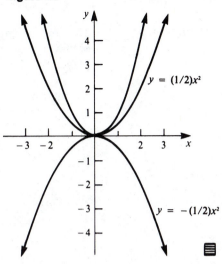

$y = (1/2)x^2$

$y = -(1/2)x^2$

example 4.3

Sketch the graphs of $y = (x - 4)^2$ and $y = (x - 4)^2 + 2$ on the same axes.

solution 4.3

Each graph has its vertex at $x = 4$ and table 4.3 gives the coordinates of several points around the vertex.

table 4.3

x	$(x - 4)^2$	$(x - 4)^2 + 2$
4	0	2
3,5	1	3
2,6	4	6

figure 4.6

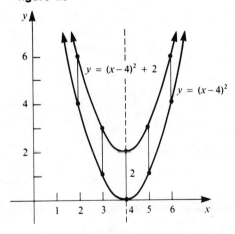

$y = (x - 4)^2 + 2$

$y = (x - 4)^2$

Thus, as can be seen in figure 4.6, the graph of $y = (x - 4)^2$ is just the graph of $y = x^4$ shifted four units to the right, and the graph of $y = (x - 4)^2 + 2$ is just the graph of $y = (x - 4)^2$ translated two units upward.

example 4.4   Sketch the graph of $y = x^2 - 4x + 3$.

solution 4.4   The first problem is to find the vertex. We may use the vertex formula to obtain $x = -(-4)/(2 \cdot 1) = 2$, and $y = f(2) = -1$. We then plot the points displayed in table 4.4 on either side of the vertex, as shown in figure 4.7.

table 4.4

x	$x^2 - 4x + 3$
2	-1
3	0
1	0
4	3
0	3

figure 4.7

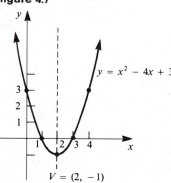

$V = (2, -1)$

A second method is to try to rewrite $y = x^2 - 4x + 3$ in the form $y = (x - a)^2 + b$, and then use the results of Example 4.3 to graph the function. We have

$$x^2 - 4x + 3 = x^2 - 4x + ? + 3 - ?$$

The algebraic "trick" we will use is called **completing the square**. In order to transform $x^2 - 4x$ into a perfect square, we must add $? = 4$ to obtain $(x - 2)^2$. Hence,

$$x^2 - 4x + 3 = (x^2 - 4x + 4) + (3 - 4) = (x - 2)^2 - 1$$

The vertex is then $(2, -1)$ and, as in Example 4.3, the final graph is the graph of $y = x^2$ shifted two units to the right and one unit downward. ▤

We can apply the algebraic trick of example 4.4 to derive the general formula for the vertex that was presented previously (page 28):

$$y = a_2 x^2 + a_1 x + a_0 = a_2 \left[x^2 + \frac{a_1}{a_2} x + \frac{a_0}{a_2} \right] = a_2 \left[\left(x^2 + \frac{a_1}{a_2} x + \frac{a_1^2}{4a_2^2} \right) + \frac{a_0}{a_2} - \frac{a_1^2}{4a_2^2} \right]$$

$$= a_2[(x - a)^2 + b], \quad \text{where } a = -\frac{a_1}{2a_2} \quad \text{and } b = \frac{a_0}{a_2} - \frac{a_1^2}{4a_2^2}$$

If $a_2 > 0$, then y is smallest when $(x - a)^2$ is smallest, i.e., when $x = a = -a_1/(2a_2)$.

Quadratic functions frequently occur in science because their graphs are **parabolas**—curves that occur naturally in the motion of objects like comets. A simple application describes the path of a ball thrown across a field (see figure 4.8).

figure 4.8

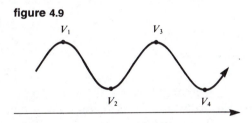

There is no elementary way to graph polynomials of degree greater than two, except to plot many points (see example 1.9). Calculus is needed to determine those points $V_1, V_2, \ldots$, where the graph changes direction (see figure 4.9).

figure 4.9

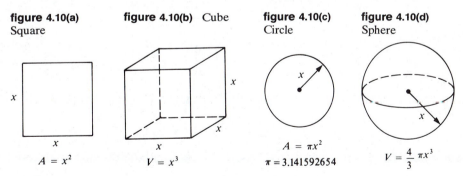

Simple polynomials of degree two and three arise regularly in formulas for area and volume. Four common formulas are presented in figure 4.10.

figure 4.10(a)   **figure 4.10(b)**   Cube   **figure 4.10(c)**   **figure 4.10(d)**
Square   Circle   Sphere

$A = x^2$ $V = x^3$ $A = \pi x^2$ $V = \dfrac{4}{3}\pi x^3$
                                $\pi = 3.141592654$

The next simple class of power functions $y = x^k$ are the functions:

$$y = x^{-1} = \frac{1}{x}, \qquad y = x^{-2} = \frac{1}{x^2}, \qquad \ldots, \qquad y = x^{-n} = \frac{1}{x^n}, \ldots$$

Note that we must *exclude* $x = 0$ from the domain of definition of each function.

example 4.5   Sketch the graphs of the functions $y = x^{-1}$ and $y = x^{-2}$ on the same axes.

solution 4.5   Note that if $f(x) = x^{-1} = 1/x$, then $f(-x) = 1/(-x) = -f(x)$. Thus, $f(x) = 1/x$ is an odd function. Since $f(x) = 1/(-x)^2 = 1/x^2$, $f(x) = 1/x^2$ is an even function. Note that as x increases, both $1/x$ and $1/x^2$ decrease, and the graphs approach the line $y = 0$.

table 4.5(a)

x	$1/x$
1	1.
2	.5
3	.33 ...
4	.25
10	.10
100	.01
.5	2
.25	4
.1	10
.01	100

table 4.5(b)

x	$1/x^2$
1	1.
2	.25
4	.0625
10	.01
100	.0001
.5	4
.25	16
.1	100
.01	10,000

We write $\lim_{x \to +\infty} 1/x = 0$ ("the limit of $1/x$ as x approaches $+\infty$ equals 0"). The line $y = 0$ is called a **horizontal asymptote**. From tables 4.5**(a)** and **(b)** and the graph of figure 4.11, we see that, as x approaches 0, the graphs approach the line $x = 0$. The line $x = 0$ is called a **vertical asymptote**.

figure 4.11

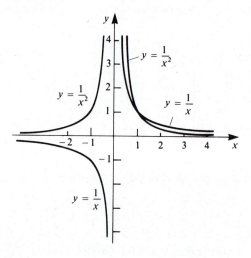

The general shape of the graph of $y = 1/x^n$ for $x > 0$ is illustrated in figure 4.12. Now, if $f(x) = 1/x^n$, then

$$f(-x) = 1/(-x)^n = \begin{cases} 1/x^n = f(x), \text{ for } n \text{ even} \\ -1/x^n = -f(x), \text{ for } n \text{ odd} \end{cases}$$

figure 4.12

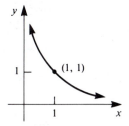

The function $f(x) = 1/x^n$ is even for n even, and odd for n odd. Therefore, we may complete the graph of $f(x)$ as shown in figure 4.13. Thus, the pattern shown in example 4.5 may be generalized. Note that $y = 0$ is a horizontal asymptote. Using the limit notation, we may conclude:

$$\underset{x \to +\infty}{\text{limit}} \frac{1}{x^n} = 0 \quad \text{and} \quad \underset{x \to -\infty}{\text{limit}} \frac{1}{x^n} = 0$$

figure 4.13

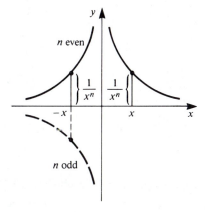

indirect variation   Suppose that two variables x and y are related by a function $y = f(x)$ and the following special property holds: When x is doubled, y is halved; when x is multiplied by $\frac{1}{3}$, y is tripled. In general, when x is multiplied by c, the corresponding output y is multiplied by $1/c$, as illustrated in figure 4.14. Expressing this in function notation, we say that y and x are **inversely proportional** if $f(cx) = f(x)/c$ for any nonzero constant c.

figure 4.14

example 4.6   Show that if $f(x) = k/x$ for some constant k, then y and x are inversely proportional, i.e., $f(cx) = f(x)/c$.

solution 4.6 $$f(cx) = \frac{k}{cx} = \left(\frac{k}{x}\right)\left(\frac{1}{c}\right) = \frac{f(x)}{c}$$

It can be shown that a function of the form $y = k/x$ is the *only* type of function with the special property described in example 4.6. Thus, we may say that **y is inversely proportional to x if and only if $y = k/x$, for some constant k.** The following are examples of indirect variation in the physical sciences.

example 4.7   **Boyle's Law**   A gas is held at constant temperature in the piston shown in figure 4.15. *Boyle's Law* for an ideal gas asserts that the pressure P within the container (in pounds per square inch, for example) is inversely proportional to the volume V (in cubic inches, for example). When $V = 60$, $P = 15$. Express V as a function of P. What is P when $V = 30$?

figure 4.15

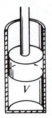

solution 4.7   If V and P are inversely proportional, then $P = k/V$ for some constant k. When $V = 60$, $P = 15 = k/60$. Hence, $k = 900$, and $P = 900/V$. Thus, when $V = 30$, $P = 900/30 = 30$. Consequently, halving the volume doubles the pressure. Of course, this is also a direct consequence of the definition of indirect variation: $f(cV) = f(V)/c$ with $c = 1/2$.

example 4.8   **an inverse square law**   Figure 4.16 illustrates an example of an *inverse square law*, one of the most commonly occurring laws in physics. As x decreases, the strength I of the light source on the screen increases, but I is *not* inversely proportional to x. Rather, I is inversely proportional to x^2, the square of the

figure 4.16

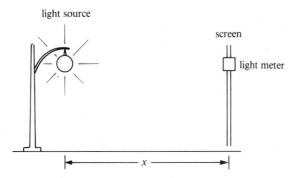

distance between the light source and the screen. Thus, $I = k/x^2$, and so

$$\frac{I_1}{I_2} = \frac{k/x_1^2}{k/x_2^2} = \left(\frac{x_2}{x_1}\right)^2$$

Hence, if $x_2 = .5x_1$, then $I_1/I_2 = .5^2 = 1/4$ or $I_2 = 4I_1$. The strength of the light source on the screen is four times as great when the distance between the light source and the screen is halved.

The power functions $y = x^{-n} = 1/x^n$ are the simplest examples of a much larger class of functions called rational functions. A *rational function $f(x)$* is a function of the form $[p(x)]/[q(x)]$, where $p(x)$ and $q(x)$ are polynomial functions. Thus, for example,

$$f_1(x) = \frac{x}{1+x}$$

$$f_2(x) = \frac{3x^4 - x + 1}{x^2 - 3x + 2}$$

$$f_3(x) = \frac{x}{x^2 - 4}$$

are all rational functions. Note that we must exclude the real numbers for which $q(x) = 0$ for the domain of definition. For f_3, the domain consists of all real numbers except $x = \pm 2$.

Until now, we have developed no technique for sketching the graphs of rational functions other than plotting points. Shown in figure 4.17 is a

figure 4.17   The Rational Function $x/(x^2 - 4)$

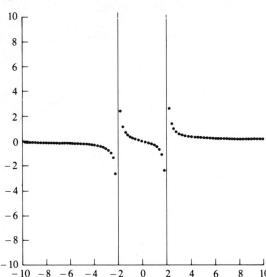

computer-generated graph of $f_3(x) = x/(x^2 - 4)$. Vertical asymptotes occur at $x = 2$ and $x = -2$. The line $y = 0$ is a horizontal asymptote.

We close this section with a rational function occurring in ecology.

example 4.9   **Holling's functional response curve**   Suppose we are studying the feeding habits of a predator (e.g., a fox). How does the number y of prey (e.g., rabbits) eaten over a prescribed period of time depend upon the density x of the prey? Surely as x increases, that is, as the prey become more abundant, $y = f(x)$ increases. Since the predator can consume only a certain number of prey, it should be the case that $f(x_1) \approx f(x_2)$ for large values of x_1 and x_2. The curve relating y and x should possess a horizontal asymptote. Holling (1959) has discovered a rational function* that works well in describing the feeding habits of invertebrate predators and some fish:

$$y = \frac{ax}{1 + abx} \text{ for } x \geq 0$$

The curve (called Holling's functional response curve) has the general shape shown in figure 4.18. Thus, the line $y = 1/b$ is a horizontal asymptote and so

$$\lim_{x \to +\infty} \frac{ax}{1 + abx} = \frac{1}{b}.$$

* Robert W. Poole, *An introduction to quantitative ecology* New York: McGraw-Hill, 1974, pp. 162–174.

figure 4.18

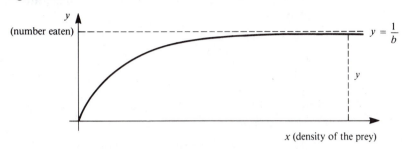

example **4.10**

Sketch the Holling's functional response curve

$$y = \frac{4x}{1 + 2x}$$

by plotting points.

solution **4.10**

Eleven points that lie on the functional response curve are displayed in table 4.6 and graphed in figure 4.19.

table **4.6**

x	0	.2	.4	.6	.8	1	2	3	4	5	6
$y = 4x/(1 + 2x)$	0	.57	.8$\overline{8}$	1.09	1.23	$\frac{4}{3}$	1.6	1.71	1.78	1.8	1.85

figure 4.19

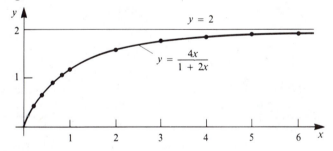

exercises for section 4

part **A**

Sketch the graphs of each of the following power functions.

1. $y = x^4$ **2.** $y = x^5$

3. $y = 1/x^3$ **4.** $y = 1/x^4$

Sketch the graphs of each of the following quadratic functions by first finding the vertex V.

5. $y = x^2 - 4x + 4$

6. $y = x^2 - 4x + 6$

7. $y = x^2 + x - 6$

8. $y = x^2 - 6x + 8$

9. $y = 3x^2 - 6x - 4$

10. $y = 5x^2 - 5x - 5$

11. If y is inversely proportional to x and $y = 4$ when $x = 2$, express y as a function of x.

12. If y is inversely proportional to x^2 and $y(2) = 4$, express y as a function of x.

13. Assume that the demand D for a product is inversely proportional to its selling price p. Presently, the price of the product is \$4 and the demand is 100 units per day. Predict the demand when the price is \$6.

14. Assume that the time T an animal spends searching for food is *inversely* proportional to the abundance A of the food, and that the energy E expended in searching is *directly* proportional to the time T. Write E in terms of A.

15. The rate R at which water flows through a pipe is inversely proportional to the length L of the pipe. When $L = 100$ feet, $R = 10$ cubic feet per minute. Find R when $L = 150$ feet.

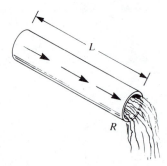

16. The force between two oppositely charged particles is inversely proportional to the square of the distance between them.

(a) What is the effect of decreasing the distance by a factor of .8?

(b) What is the effect of increasing the distance by a factor of 3?

17. The time it takes to build an apartment complex is assumed to be inversely proportional to the number of workers hired. If a crew of twenty workers completed a similar job in six months, how many workers should be hired to complete this job in four months?

Find the domains of definition of each of the following rational functions.

18. $f(x) = \dfrac{x^3 + x + 2}{x^2 - 1}$

19. $f(x) = \dfrac{x^3 + x + 2}{x^2 + 1}$

20. $f(x) = \dfrac{x^2 + 2x + 1}{x^2 + 3x + 2}$ **21.** $f(x) = \dfrac{x}{x-2} - \dfrac{x}{x-1}$

★**22.** $f(x) = \dfrac{x^7 + 1}{x^4 - 5x^2 + 4}$

Sketch the following functional response curves by plotting points and determine the horizontal asymptote for each curve.

23. $f(x) = \dfrac{3x}{1 + 5x}$ **24.** $f(x) = \dfrac{10x}{1 + 5x}$

25. $f(x) = \dfrac{3x}{1 + 10x}$ **26.** $f(x) = \dfrac{10x}{1 + 10x}$

★**27.** $f(x) = \dfrac{ax}{1 + ax}$ What is the effect of increasing a?

★**28.** $f(x) = \dfrac{x}{1 + bx}$ What is the effect of increasing b?

part B

29. Four hundred feet of fence are available to enclose a rectangular field. Find the length x and width y that will enclose the largest area. [*Hint*: $A = xy$ and $x + y = 200$.]

30. The Great Zambini is shot out of a cannon. The laws of physics predict that his height above the ground after t seconds is given by

$$h(t) = -16t^2 + 24t + 6$$

Find the maximum height he reaches.

31. Special proteins known as *enzymes* act as catalysts for a wide variety of chemical reactions in living things. The term *substrate* is used for the substance that is being acted upon. In 1913, Michaelis and Menten devised a formula relating the initial speed V with which the reaction begins to the original amount of substrate x:

The Michaelis-Menten Relation $V = \dfrac{ax}{x + b}$

(Typical units are moles/liter for x, and moles/liter/second for V.) This equation has been verified experimentally for a variety of enzyme-controlled reactions. There also exist theoretical derivations of the equations.*

* See S.I. Rubinow, *Introduction to mathematical biology*, New York: John Wiley (1975), pp. 47–51.

(a) Show that $V = a/(1 + b/x)$. Conclude that when x is very large, $V \approx a$. Thus, the line $V = a$ is a horizontal asymptote.

(b) Specify the conditions for which $V = a/2$.

(c) Sketch the graph of $V = 10x/(x + 5)$, by plotting points.

32. In 1938, Hill* devised an empirical formula relating the speed V with which a muscle shortens to the weight W that is being raised by the muscle:

Hill's Law $\qquad V = \dfrac{a(W_0 - W)}{W + b}$

Here a, b, and W_0 are constants characteristic of the given muscle. Typical units are grams for W and cm/sec for V.

(a) Compute $V(W_0)$. What is the interpretation of W_0?

(b) Compute $V(0)$.

(c) Graph $V = \dfrac{100(10 - W)}{W + 20}$ by plotting points.

---

* For more information on Hill's Law, see R.M. Alexander, *Animal mechanics*, Seattle: University of Washington Press (1968), pp. 107–9.

inverse functions

What is the rule that "unravels" what the first rule has done? The idea of **inverses** occurs frequently in our day-to-day activities. The operation of tying shoes is a very difficult task for a young child to learn. The inverse operation of untying is mastered from the very beginning!

In our mathematical setting, we are looking for a rule f^{-1} that unravels what f has done, restoring the original output $f(x)$ to its prior value x, as illustrated in figure 5.1.

figure 5.1

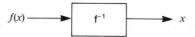

As a simple example, let $f(x) = 2x$. The function f sends 1 to 2, 2 to 4, 3 to 6, and so forth. To find f^{-1} we must devise a rule that sends 2 to 1, 4 to 2, 6 to 3, and so forth. Clearly, the desired formula is $f^{-1}(y) = y/2$, or $f^{-1}(x) = x/2$. Note then that the defining property for f^{-1} is:

$$f^{-1}(f(x)) = x$$

The inverse function f^{-1} has unraveled what f has done.

example 5.1　Determine the inverse function for $f(x) = 2x - 1$.

solution 5.1　The functional values for $f(x)$ displayed in table 5.1**(a)** lead to the values of $f^{-1}(x)$ shown in table 5.1**(b)**. What rule will send -1 to 0, 1 to 1, 3 to 2, and so forth? If we add 1 to the values of x in table 5.1**(b)**, we obtain $2f^{-1}(x)$. Hence, $x + 1 = 2f^{-1}(x)$ or $f^{-1}(x) = (x + 1)/2$.

table 5.1(a)

x	$2x - 1$
0	−1
1	1
2	3
−1	−3
−2	−5

table 5.1(b)

x	$f^{-1}(x)$	$x + 1$
−1	0	0
1	1	2
3	2	4
−3	−1	−2
−5	−2	−4

example 5.2

Determine whether $f(x) = x^2$ has an inverse function.

solution 5.2

Let's try the same method used in example 5.1. Tables 5.2**(a)** and **(b)** display selected functional values for $f(x)$ and $f^{-1}(x)$, respectively. We must devise a function that sends 0 to 0, 1 to 1, -1 to -1, and so forth. But $f^{-1}(1)$ cannot be *both* 1 and -1, because functions must be *single-valued*. Hence, for $f(x) = x^2$, *there is no inverse function*.

table 5.2(a)	
x	x^2
0	0
1	1
-1	1
2	4
-2	4

table 5.2(b)	
x	$f^{-1}(x)$
0	0
1	1
1	-1
4	2
4	-2

When does a function f have an inverse? A clue for the answer comes from the next example.

example 5.3

Fingerprinting is a useful technique in identifying criminals. What makes the technique so useful? To use mathematical language, let x represent a given individual, and let $f(x)$ stand for "the fingerprints of x," as depicted in figure 5.2. Having found fingerprints at the scene of a crime, we wish to identify the

figure 5.2

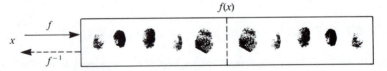

individual. This is theoretically possible because no two individuals have the same fingerprints. Thus, if $x_1 \neq x_2$, then $f(x_1) \neq f(x_2)$. It is this condition that is needed for a workable system—i.e., for the existence of f^{-1}. Contrast this to the totally ineffective system based on shoe size s, instead of fingerprints. It is impossible to go from "size 9D" back to a unique individual, because $s(x_1)$ may equal $s(x_2)$ even if $x_1 \neq x_2$.

definition 5.1

A function f with domain A and range B is called *one-to-one* if, whenever x_1 and x_2 are distinct real numbers in A, $f(x_1) \neq f(x_2)$. Thus, distinct inputs have distinct outputs.

Given a one-to-one function f with range B, we may define the inverse function on the set B as follows. Given y in B, let $f^{-1}(y)$ be that unique input x in A with $f(x) = y$. It follows that:

$$f(f^{-1}(y)) = y \qquad \text{for } y \text{ in } B$$
$$f^{-1}(f(x)) = x \qquad \text{for } x \text{ in } A$$

Two special types of functions we have mentiond are always one-to-one and therefore always have inverses. The general situations are graphed below.

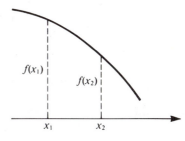

case I f is increasing   **case II** f is decreasing

In either case, $f(x_1) \neq f(x_2)$ whenever $x_1 \neq x_2$.

In example 5.1, we guessed the formula for f^{-1} by examining specific numerical values. In general, this strategy is not feasible. Fortunately, there is a mechanical method for finding f^{-1}, summarized in the accompanying box.

---

method for finding the inverse of a one-to-one function $y = f(x)$

Step 1. Interchange x and y in the formula $y = f(x)$ to obtain $x = f(y)$.

Step 2. Solve for y in terms of x.

Step 3. The resulting expression for y is $f^{-1}(x)$.

---

example 5.4

Find the inverse function for $f(x) = 3x - 2$, using the method outlined in the box. Sketch the graphs of $y = f(x)$ and $y = f^{-1}(x)$ on the same axes.

solution 5.4

Applying Step 1 to $y = 3x - 2$ yields $x = 3y - 2$. Hence, $y = (x + 2)/3 = (1/3)$ $x + 2/3$. The graphs of $y = 3x - 2$ and $y = (1/3)x + 2/3$ are shown in figure 5.3. [*Note*: The graphs are the reflections of one another in the line $y = x$.]

figure 5.3

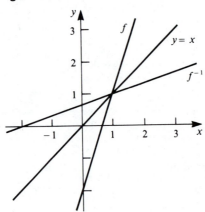

example 5.5   If $f(x) = (x + 1)/(x - 2)$, determine $f^{-1}(x)$.

solution 5.5   We begin with $y = (x + 1)/(x - 2)$. Interchanging y and x yields $x = (y + 1)/(y - 2)$. Hence,

$$xy - 2x = y + 1 \quad \text{or} \quad xy - y = 2x + 1$$

It follows that $y(x - 1) = 2x + 1$ or $y = (2x + 1)/(x - 1)$. Thus, $f^{-1}(x) = (2x + 1)/(x - 1)$.

The next example illustrates the special relationship between the graphs of f and f^{-1}.

example 5.6   Let $y = f(x)$ be the one-to-one function whose graph is shown in figure 5.4.

table 5.3

x	$f(x)$
-1	$1/2$
0	1
2	3
3	7
r	s

figure 5.4

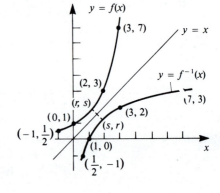

Although we do not have an explicit expression for f, we do know from table 5.3 that f^{-1} sends $\frac{1}{2}$ to -1, 1 to 0, 3 to 2, 7 to 3, and s to r. Hence, the points $(\frac{1}{2}, -1)$, $(1, 0)$, $(3, 2)$, $(7, 3)$, and (s, r) are on the graph of f^{-1}. As you can see from figure 5.4, *the graph of f^{-1} is the reflection of the graph of f in the line $y = x$*, exercise 64 will explore why this is true.   ▤

example 5.7   The function $f(x) = x^2$ is increasing for $x \geq 0$ and hence possesses an inverse. Since $f(f^{-1}(x)) = x$, $[f^{-1}(x)]^2 = x$. Thus, f^{-1} is the positive square root of x. We write:

$$f^{-1}(x) = \sqrt{x} = x^{1/2}$$

The graphs of $f(x) = x^2$ and $f^{-1}(x) = x^{1/2}$ are shown in figure 5.5.

figure 5.5

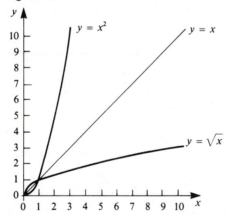

The result of example 5.7 may be generalized as shown in figure 5.6. The function $f(x) = x^n$ is increasing for $x \geq 0$. [*Note*: If n is odd, then x^n is

figure 5.6

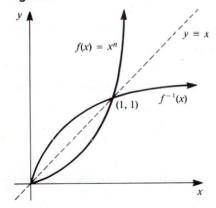

increasing for all x.] From the general equation $f(f^{-1}(x)) = x$, we obtain $[f^{-1}(x)]^n = x$. Thus, $f^{-1}(x)$ is the positive nth root of x, and we write:

$$f^{-1}(x) = \sqrt[n]{x} = x^{1/n}$$

We are now in a position to define $f(x) = x^k$ for any fraction (i.e., rational number) k:

definition 5.2   If $k = m/n$ where m, $n > 0$, then $x^k = (x^{1/n})^m = (\sqrt[n]{x})^m$. If $k \equiv -m/n$, then $x^k = 1/(x^{m/n})$.

For example, if $f(x) = x^{-1/2}$, then $f(4) = 1/4^{1/2} = 1/2$. If $g(x) = x^{3/4}$, then $g(16) = (\sqrt[4]{16})^3 = 2^3 = 8$.

Based on these definitions, the usual laws of exponents can be derived. They are summarized in the accompanying box.

---

laws of exponents

1. $x^{k_1} x^{k_2} = x^{k_1 + k_2}$
2. $(x^{k_1})^{k_2} = x^{k_1 k_2}$
3. $(xy)^k = x^k y^k$
4. $(x/y)^k = x^k/y^k$

These laws are illustrated in the next two examples.

example 5.8   If $f(x) = x^{1/4}$, show that $f(x) = \sqrt{\sqrt{x}}$. Then compute $f(1.519)$ using the $\boxed{\sqrt{}}$ button on your calculator.

solution 5.8   Using the second law of exponents, we obtain $x^{1/4} = (x^{1/2})^{1/2} = \sqrt{\sqrt{x}}$. To compute $f(1.519)$, enter 1.519 and hit the $\boxed{\sqrt{}}$ button twice to obtain 1.11016988.   ▤

example 5.9   If $f(x) = x^{-2/3}$, find $f(2)$ and $f(-8)$.

solution 5.9   To find $f(2)$ as a decimal, use the power button $\boxed{y^x}$ on your calculator. If your calculator is set up in "algebraic logic" (as are the Texas Instruments calculators), the steps are:

$$2 \;\boxed{y^x}\; \boxed{(}\; 2 \;\boxed{\div}\; 3 \;\boxed{\pm}\; \boxed{)}\; \boxed{=}$$

You should obtain .62996053 for your answer. If your calculator is equipped with "reverse Polish logic" (as are the Hewlett-Packard calculators), the steps are:

$$\boxed{\text{ch s}}\; 2 \;\boxed{\text{enter}}\; 3 \;\boxed{\div}\; 2 \;\boxed{y^x}$$

If you try to compute $f(-8)$ using your calculator, it may indicate an error message. *Your calculator is probably not designed to compute x^y when $x < 0$.* To determine if $f(-8)$ exists, we must refer to the original definitions of $x^{-m/n}$

$$f(-8) = (-8)^{-2/3} = \frac{1}{(\sqrt[3]{-8})^2} = \frac{1}{(-2)^2} = \frac{1}{4}$$

Power functions of the form $y = \alpha x^\beta$, where α and β are fixed constants, occur frequently in biology and ecology. Such applications include **allometric laws**—i.e., formulas relating physical dimensions of an organism, such as total length and weight, or various limb lengths or weights of organs. The next two examples will serve to illustrate the concept.

example 5.10   The formula for the weight (in metric tons) of sperm whales is given by $W = .000137L^{3.18}$, where L is the length in feet.
  (a) Find the expected weight of a 40-foot sperm whale.
  (b) What is the expected length of a 20-ton specimen?

solution 5.10   **(a)** When $L = 40$, $W = (.000137)40^{3.18} = 17.03$ metric tons.
  (b) We must first solve the given equation for L. Raising both sides to the power $1/3.18$ yields

$$W^{1/3.18} = [.000137L^{3.18}]^{1/3.18} = .06097225L$$

Hence,

$$L = 16.4009\,W^{1/3.18} = 16.4009\,W^{.314465}$$

Finally, when $W = 20$, $L = (16.4009)20^{.314465} = 42.07$ feet.

example 5.11   **the area-species curve**   A remarkable empirical relationship has been found relating S and A, where A is the area of an island in square miles, and if S is the number of species of a given taxon on the island:

$$S = cA^{.3}$$

The constant c depends upon the geographical location of the cluster of islands. If a 100-square-mile island contains 15 different species of reptile, how many different species can one expect to find on a much larger island with an area of 10,000 square miles?

solution 5.11   When $A = 100$, $S = 15$. Hence, $15 = c100^{.3} = 3.98107c$. Therefore, $c = 15/3.98107 = 3.7678296$. Hence, for this collection of islands, $S = 3.7678296\,A^{.3}$. When $A = 10,000$, $S = (3.7678296)10,000^{.3} = 59.716 \approx 60$ species of reptile.*

---

* For more information on the area-species curve, see Edward O. Wilson and W.H. Bossert, *A primer of population biology*. Sunderland, Mass.: Sinauer Associates, (1971), pp. 166–169.

How does one determine whether a given set of data follows a power law? If $y = \alpha x^\beta$, how are α and β determined? We will answer these questions in section 17 of the text.

The next two examples introduce logarithm and inverse trigonometric functions.

example 5.12

Sketch the graph of the exponential function $f(x) = 2^x$ and its inverse $f^{-1}(x)$ on the same axes. What is $f^{-1}(x)$?

solution 5.12

Tables 5.4(a) and (b) display points on the graphs of f and f^{-1}, respectively. We have reversed the columns of table 5.4(a) to obtain points on the graph of f^{-1} shown in figure 5.7.

table 5.4(a)

x	2^x	
-2	1/4	
-1	1/2	
0	1	$(2^0 = 1)$
1	2	
2	4	

figure 5.7

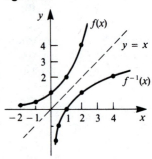

table 5.4(b)

x	$f^{-1}(x)$
1/4	-2
1/2	-1
1	0
2	1
4	2

From the relationship $f(f^{-1}(x)) = x$, we obtain:

$$2^{f^{-1}(x)} = x$$

Thus, **$f^{-1}(x)$ is the power to which 2 must be raised in order to obtain x.** For example, since $2^3 = 8$, we have $f^{-1}(8) = 3$. The more common notation is $f^{-1}(x) = \log_2 x$, the *logarithm of x to the base 2*. We write, for example, $\log_2 16 = 4$ and $\log_2 (1/2) = -1$.

On any scientific calculator, you will find the special exponential function $f(x) = e^x$, where $e \approx 2.718$. The button is usually shown as $\boxed{e^x}$ or $\boxed{\text{inv}}$ $\boxed{\text{ln}}$. The inverse function $f^{-1}(x) = \ln x$ is called the **natural logarithm function** and is designated $\boxed{\text{ln}}$.

The graphs of e^x and $\ln x$ are similar to those in example 5.12 and are shown in figure 5.8.

figure 5.8

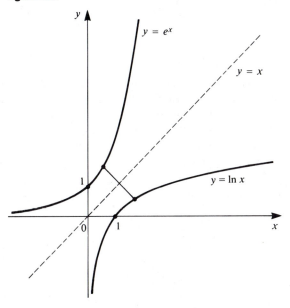

The reader may complete the study of exponential and logarithm functions, without loss of continuity, by proceeding to section 15.

By restricting the domain of definition, we may construct an inverse function for $f(x) = \sin x$, as demonstrated in the next example.

▲ **example 5.13**   **the inverse sine function**   Shown in figure 5.9 is the familiar oscillating graph of the function $f(x) = \sin x$. Make sure your calculator is in the *radian mode* and use the $\boxed{\sin}$ button to complete table 5.5:

table 5.5

x	0	$\pi/4$	1	$\pi/2$	2	π	$-\pi/2$	-1
$\sin x$	0		.8415			0	-1	

figure 5.9

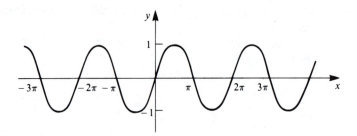

As you can see from either figure 5.9 or table 5.5, $f(x) = \sin x$ is *not* a one-to-one function. If, however, we confine our attention to the interval $[-\pi/2, \pi/2]$, then $f(x) = \sin x$ is increasing and hence one-to-one, as demonstrated in figure 5.10. Therefore, we may find its inverse function $f^{-1}(x)$. The more common notation for $f^{-1}(x)$ is $\sin^{-1} x$ or arcsin x. Since $f(f^{-1}(x)) = x$, we have $\sin(\sin^{-1} x) = x$. **Thus, $\sin^{-1} x$ is that unique real number in $[-\pi/2, \pi/2]$ whose sine is x.**

figure 5.10 $\sin x$ is 1–1 for $-\pi/2 \le x \le \pi/2$

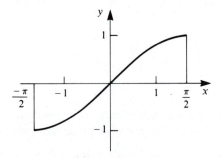

Make sure your calculator is in radian mode and use the $\boxed{\text{inv}}$ $\boxed{\text{sin}}$ buttons to complete table 5.6:

table 5.6

x	0	.2	.4	.6	.8	1	−.2	−.4	−.6	−.8
$\sin^{-1} x$	0		.4115		.9273			−.4115		

The graphs of $y = \sin x$ and $y = \sin^{-1} x$ are shown in figures 5.11**(a)** and **(b)**. Note that the domain of $\sin^{-1} x$ is $[-1, 1]$ and its range is $[-\pi/2, \pi/2]$.

figure 5.11(a)   **figure 5.11(b)**

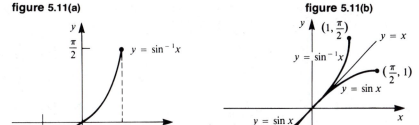

The reader may complete the study of trigonometric functions, without loss of continuity, by proceeding directly to section 29.

Table 5.7 provides a summary of the inverse function pairs that can be found on a standard scientific calculator.

table 5.7

f	f^{-1}	Remarks
$\boxed{x^2}$	$\boxed{\sqrt{x}}$	$x \geq 0$
$\boxed{x^n}$	$\boxed{x^{1/n}}$	$x \geq 0$ for n even
$\boxed{1/x}$	$\boxed{1/x}$	for $x \neq 0$
$\boxed{e^x}$	$\boxed{\ln}$	See section 15 for complete details.
$\boxed{\log}$	$\boxed{10^x}$	log denotes logarithm to the base 10. See section 15.
$\boxed{\sin}$	$\boxed{\sin^{-1}}$	See section 29 for complete details.
$\boxed{\cos}$	$\boxed{\cos^{-1}}$	See section 29.
$\boxed{\tan}$	$\boxed{\tan^{-1}}$	See section 30.

exercises for section 5

part A

By inspection, find the inverse function $f^{-1}(x)$ for each of the following elementary functions f.

1. $f(x) = 3x$

2. $f(x) = x + 1$

3. $f(x) = 1/x$

4. $f(x) = x/2$

5. $f(x) = 2x + 1$

6. $f(x) = 1/(x + 1)$

7. $f(x) = x^2$ for $x \geq 0$

★**8.** $f(x) = x^4$ for $x \leq 0$

9. $f(x) = \sqrt{x}$ for $x \geq 0$

10. $f(x) = \sqrt[3]{x}$

Show that the following functions are *not* one-to-one by producing, for each function, two inputs with the same output.

11. $f(x) = x^4$

12. $f(x) = \dfrac{1}{x^2}$

13. $f(x) = 2^{-x^2}$

14. $f(x) = \sqrt{x^2 + 1}$

15. $f(x) = |x + 1|$, where $| \; |$ denotes absolute value

★16. $f(x) = \dfrac{x + 1}{x^2 + 1}$

▲★17. $f(x) = \sin x$

▲★18. $f(x) = \cos x$

Using the special method for finding f^{-1} described on page 43, find $f^{-1}(x)$ for each of the following functions.

19. $f(x) = 4x - 3$

20. $f(x) = \dfrac{x - 1}{2}$

21. $f(x) = \dfrac{1}{x}$

22. $f(x) = \dfrac{1}{2x + 3}$

23. $f(x) = \dfrac{x - 1}{x + 2}$

24. $f(x) = \dfrac{2x + 3}{x - 4}$

25. $f(x) = \sqrt{x} + 1$, for $x \geq 0$

26. $f(x) = 1/\sqrt{x + 1}$, for $x > -1$

27. $f(x) = x^3 - 1$

28. $f(x) = x^2 + 1$, for $x > 0$

★29. $f(x) = x^4 + 1$, for $x < 0$

★30. $f(x) = (\tfrac{1}{2})^x$ [*Hint:* Use properties of $\log_{10}$.]

On the same set of axes, graph $y = f(x)$ and $y = f^{-1}(x)$ for each of the following functions in exercises 31–36.

31. $f(x) = 4x - 3$

32. $f(x) = \dfrac{x - 1}{2}$

33. $f(x) = x^3$

34. $f(x) = x^4$, for $x \geq 0$

35.

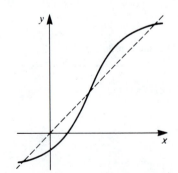

36.

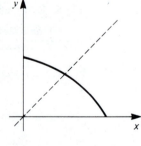

37. Use the laws of exponents to show how $\sqrt[8]{x}$ can be computed using only the $\boxed{\sqrt{}}$ button on the calculator.

Using your calculator, compute (if possible):

38. $\sqrt[4]{7}$ **39.** $\sqrt[8]{.7}$

40. $\sqrt[8]{4}$ **41.** $\sqrt[4]{-4}$

Use the $\boxed{y^x}$ button on your calculator to compute:

42. $5^{-2/3}$ **43.** $4^{1/3}$

44. $(2.1)^{3.21}$ **45.** $\sqrt[5]{2.1}$

46. $1/(\sqrt{3})^5$ **47.** $2^{-.4}$

Simplify the following exponential expressions *without* the aid of your calculator.

48. $\sqrt[3]{-8}$ **49.** $\sqrt{8}$

50. $(16)^{-3/4}$ **51.** $(-125)^{-2/3}$

52. $(-32)^{2/5}$ **53.** $4^{-1/2}$

54. For the sperm whale in example 5.10, find the expected weights given the lengths shown in the table.

L	20	25	30	40	45
W					

55. Refer to example 5.10. What is the expected length of a 35-ton sperm whale?

56. For the area-species curve of example 5.11, suppose that 10 species of reptile have been discovered on an island with area 1,000 square miles. How many reptile species does one expect to find on a nearby island of area 125 square miles?

57. According to the area-species curve of example 5.11, what is the effect of increasing the area of the island by a factor of 10?

58. On the British island of Montserrat in the West Indies, there are nine different species of amphibians and reptiles. The area of this small island is thirty-two square miles. Use the area-species curve to predict the number of such species on:

(a) Cuba (42,857 square miles)

(b) Puerto Rico (3,423 square miles)

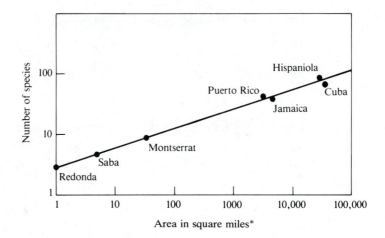

Complete the following tables in exercises 59–61 and then plot f and f^{-1} on the same axes. Make sure your calculator is in the radian mode.

59.

x	-2	-1	0	$\frac{1}{2}$	1	$\frac{3}{2}$	2
3^x							

60.

x	-2	-1	0	$\frac{1}{2}$	1	$\frac{3}{2}$	2	$\frac{5}{2}$	3
$(1/2)^x$									

▲61.

x	0	$.5$	1.0	$\pi/2$	2.0	2.5	3.0	π
$\cos x$								

part B

62. In 1916, DuBois and DuBois devised an empirically based formula for the surface area S of the human body (in square feet) in terms of weight W (in pounds) and height H (in inches):

$$S = .1091\,W^{.425}H^{.725}$$

(a) Estimate your own body surface area.

(b) Compare your own surface area to that of Kareem Abdul-Jabbar, who weighs 235 pounds and is 87 inches tall.

---

* The relationship between the number of species of West Indian amphibians and reptiles and the area of the islands. [Adapted from E.O. Wilson and W.H. Bossert, *A Primer of Population Biology*, Sunderland, Mass.: Sinauer Associates Inc., (1971).]

63. The length-weight relationship for the Pacific halibut is well described by
the formula

$$W = 10.375 \, L^3$$

where W is in kilograms and L is in meters.

(a) In 1973, the Pacific Halibut Commision adopted a 32-inch minimum size
limit. What is the minimum expected weight in pounds? [*Hint*: Use 1
inch = 2.54 centimeters and 1 kilogram = 2.2 pounds.]

(b) In 1973, Hart reported that a halibut caught off the Canadian coast
measured 267 cm, but did not report its weight. Estimate the weight.

(c) Although Pacific halibut weighing up to 800 pounds have been reported, the largest specimen that was thoroughly documented weighed 507
pounds. Estimate its length in feet.

64. Prove that the graph of f^{-1} is the reflection of the graph of f in the line
$y = x$ by completing the following:

(a) If (r, s) is on the graph of f, then (s, r) is on the graph of f^{-1}.

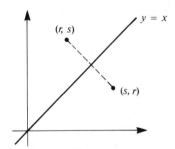

(b) The midpoint of the line segment joining (x_1, y_1) and (x_2, y_2) is given by
$M = \left(\dfrac{x_1 + x_2}{2}, \dfrac{y_1 + y_2}{2} \right)$. Show that the midpoint of the line segment
joining (r, s) to (s, r) is on the line $y = x$.

(c) Two lines are perpendicular if the product of their slopes is -1. Show
that the line segment joining (r, s) to (s, r) has slope -1, and is
therefore perpendicular to the line $y = x$.

65. An earthquake centered in the ocean can set up tidal waves that travel at
great speeds. The function $v = \sqrt{32x}$ gives the velocity in feet per second
at ocean depth x feet.

(a) What is the velocity at a depth of 100 feet?

(b) The average depth of the Pacific Ocean is 3.5 miles, or about 18,000
feet. Find the velocity near the bottom.

(c) How long would it take for a wave centered in the Pacific Ocean to
reach a point 2,000 miles away?

66. When the speed of an object nears the speed of light, $c = 186,000$ miles/second the theory of relativity predicts that its mass (m) increases according to the formula

$$m = f(v) = \frac{m_0}{\sqrt{1 - v^2/c^2}}$$

where m_0 is the mass at rest. At what speed does the mass double?

 # limits of functions

In this section we will use a pocket calculator to introduce the limit concept and learn to compute limits. Let's begin with a simple example.

example 6.1
Compute $\lim_{x\to 2} (x^2 - 4)$, the limit of the function $x^2 - 4$ as x approaches 2.

solution 6.1
In tables 6.1**(a)** and **(b)**, we select values of x closer and closer to 2. In table 6.1**(a)** we input values of x slightly larger than 2, and in table 6.1**(b)** we compute the function for values of x slightly smaller than 2. *Compute the functional values we have left out.* We *do not* place $x = 2$ in either table, since x is only approaching 2. Now let's examine the outputs. It appears that the outputs in each table are getting closer and closer to 0. Thus, we say:

$$\lim_{x\to 2} (x^2 - 4) = 0$$

table 6.1(a)

x	$x^2 - 4$
2.1	.41
2.05	
2.01	.0401
2.005	
2.001	.004001
2.0005	

table 6.1(b)

x	$x^2 - 4$
1.9	$-.39$
1.95	
1.99	-0.399
1.995	
1.999	$-.003999$
1.9995	

definition 6.1
If the outputs $f(x)$ become closer and closer to a single real number (L) as the inputs (x) get closer and closer to a (from either the right or left side), then we say

$$\lim_{x\to a} f(x) = L$$

The definition is illustrated by figure 6.1, in which the functional values are seen to be moving closer and closer to L. There are situations in which the limit L is not easily guessed, as the following examples demonstrate.

figure 6.1

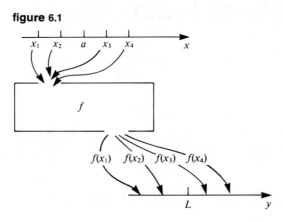

example 6.2

Compute $\displaystyle\lim_{x\to 1}\frac{x^6-1}{x-1}$.

solution 6.2

In this example, although $f(1)$ is not defined, we can still input values of x close to 1 into the function. Fill in the missing entries in tables 6.2**(a)** and **(b)**.

table 6.2(a)

x	$\dfrac{x^6-1}{x-1}$
1.2	9.92992
1.1	
1.02	6.3081
1.01	
1.001	
1.0002	6.003
1.0001	

table 6.2(b)

x	$\dfrac{x^6-1}{x-1}$
.9	4.6856
.95	
.99	5.8520
.995	
.999	5.9850
.9995	
.9999	5.9985

Examination of the tables suggests that $\displaystyle\lim_{x\to 1}\frac{x^6-1}{x-1}=6$.

example 6.3

Determine whether $\displaystyle\lim_{x\to 0}|x|/x$ exists, where $|x|$ denotes the absolute value of x.

solution 6.3

When x is slightly larger than 0, $|x|=x$. Hence, $|x|/x=x/x=1$. For x smaller than 0, $|x|=-x$ [e.g., $|-2|=-(-2)=2$]. These observations lead to tables 6.3**(a)** and **(b)**.

One may be tempted to claim there are two limits. However, our definition of limit requires that the outputs cluster around a *single real number*. Hence, we must conclude that $\lim_{x\to 0}|x|/x$ *does not exist.*

table 6.3(a)	
x	$|x|/x$
.1	1
.01	1
.001	1

table 6.3(b)	
x	$|x|/x$
−.1	−1
−.01	−1
−.001	−1

example 6.4   Compute $\displaystyle\lim_{x\to 3} \frac{x^2-9}{x-3}$.

solution 6.4   Although we could make tables as before, it is simpler to note that $x^2-9 = (x-3)(x+3)$. Hence, for $x\neq 3$, $(x^2-9)/(x-3) = x+3$. Thus, it is sufficient to determine $\lim_{x\to 3}(x+3)$. It is easy to see that the answer is 6. This commonly used technique, called the **algebraic approach**, consists of simplifying the expression for $f(x)$ before trying to compute the limit.

Next, suppose we have just determined that

$$\lim_{x\to a} f(x) = L \qquad \text{and} \qquad \lim_{x\to a} g(x) = M.$$

The Limit Principles stated in the accompanying box allow us to find the limits of new functions constructed from $f(x)$ and $g(x)$.

---

the limit principles

If $\lim_{x\to a} f(x) = L$ and $\lim_{x\to a} g(x) = M$, then

1. $\displaystyle\lim_{x\to a} [f(x) + g(x)] = L + M$

2. $\displaystyle\lim_{x\to a} [f(x) - g(x)] = L - M$

3. $\displaystyle\lim_{x\to a} [f(x)g(x)] = LM$

4. $\displaystyle\lim_{x\to a} [f(x)/g(x)] = L/M$, provided $M \neq 0$.

5. $\displaystyle\lim_{x\to a} \sqrt[n]{f(x)} = \sqrt[n]{L}$, provided $\sqrt[n]{L}$ is defined.

These principles are illustrated in the next two examples.

example 6.5   Suppose we have verified that $\lim_{x \to 4} (5x - 10) = 10$ and that $\lim_{x \to 4} \sqrt{x} = 2$. (Do so!) Compute:

(a) $\lim_{x \to 4} (5x - 10 - \sqrt{x})$ (b) $\lim_{x \to 4} \sqrt{5x - 10}$

(c) $\lim_{x \to 4} \dfrac{(5x - 10)}{\sqrt{x}}$ (d) $\lim_{x \to 4} \sqrt{x}(5x - 10)$.

solution 6.5   We will apply the limit principles with $f(x) = 5x - 10$ and $g(x) = \sqrt{x}$. Principle 2 applied to **(a)** yields $10 - 2 = 8$ as the desired limit. Using principles 5, 4, and 3, respectively, we obtain answers $\sqrt{10}$, $10/2 = 5$, and $2(10) = 20$, respectively, for **(b)**, **(c)**, and **(d)**. ▤

example 6.6   Determine $\lim_{x \to 2} \dfrac{x^2 - 4}{x - 2}$.

solution 6.6   In example 6.1 we showed that $\lim_{x \to 2}(x^2 - 4) = 0$, and it is clear that $\lim_{x \to 2} x - 2 = 0$. However, limit principle 4 does not apply since $M = 0$. *Instead of concluding that the limit does not exist,* we must attack the problem differently. Using the algebraic approach, we rewrite the function as:

$$\frac{x^2 - 4}{x - 2} = \frac{(x - 2)(x + 2)}{(x - 2)} = x + 2 \quad \text{for} \quad x \neq 2$$

Hence, we instead compute

$$\lim_{x \to 2} (x + 2) = 4$$ ▤

Notice that in example 6.5, all of the limits could have been found by simply computing the functional value $h(4)$, that is,

$$\lim_{x \to 4} h(x) = h(4)$$

It is important then to determine precisely when it will be true that $\lim_{x \to a} f(x) = f(a)$. For such functions, computing limits is easy and the rather tedious process of constructing input-output tables can be avoided. We will turn to this matter in the next section.

limits as $x \to \pm\infty$ Instead of letting x approach a *finite* real number, we can also let "x approach $+\infty$." By this we mean that x will take on larger and larger positive values. We can then input these values of x into our function machine and try to determine if the outputs $f(x)$ follow a pattern. In figure 6.2,

figure 6.2

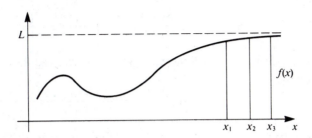

$f(x)$ is approaching the real number L and the graph of f approaches the line $y = L$. We write $\text{limit}_{x \to +\infty} f(x) = L$ and say that the line $y = L$ is a **horizontal asymptote** for the graph of f.

Likewise, we may let "x approach $-\infty$" and attempt to find $\text{limit}_{x \to -\infty} f(x)$. In figure 6.3, $\text{limit}_{x \to -\infty} f(x) = 2$.

figure 6.3

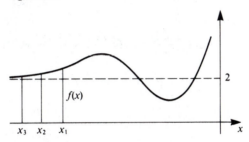

In section 4, we found that the line $y = 0$ was a horizontal asymptote for the power functions $y = 1/x^n$, for $n = 1, 2, 3, \ldots$. Thus,

$$\operatorname*{limit}_{x \to +\infty} \frac{1}{x^n} = 0 \qquad \text{and} \qquad \operatorname*{limit}_{x \to -\infty} \frac{1}{x^n} = 0$$

For x very large, x^n is even larger and hence, $1/x^n$ approaches 0.

On page 59 we stated the limit principles for obtaining $\text{limit}_{x \to a} f(x)$. *All of the limit principles are valid with "$x \to +\infty$" replacing "$x \to a$" or with "$x \to -\infty$" replacing "$x \to a$."* Let's apply these principles to compute these new types of limits.

example 6.7   Compute $\operatorname*{limit}_{x \to +\infty} \dfrac{2x - 1}{x + 1}$.

solution 6.7   It is not possible to apply the limit principles directly to $(2x - 1)/(x + 1)$.

Rather, by factoring out an x from the numerator and the denominator, we obtain

$$\frac{2x-1}{x+1} = \frac{x(2-1/x)}{x(1+1/x)} = \frac{2-(1/x)}{1+(1/x)}$$

Since we know that $\lim_{x\to+\infty} 1/x = 0$, we can use limit principles **(1)** and **(4)** to obtain

$$\lim_{x\to+\infty} \frac{2x-1}{x+1} = \frac{2-\lim(1/x)}{1+\lim(1/x)} = \frac{2-0}{1+0} = 2$$

The trick then is to rewrite $f(x)$ so that the terms of the form $1/x^n$ appear.

example 6.8   Compute each of the following limits.

$$\textbf{(a)}\ \lim_{x\to-\infty} \frac{x+1}{x^2+3} \qquad \textbf{(b)}\ \lim_{x\to+\infty} \frac{x^2-3x+2}{2x^2-x+4}$$

solution 6.8   For **(a)**, we first rewrite $\dfrac{x+1}{x^2+3}$ as $\dfrac{x(1+1/x)}{x^2(1+3/x^2)}$. Hence,

$$\lim_{x\to-\infty} \frac{x+1}{x^2+3} = \lim_{x\to-\infty} \left(\frac{1}{x}\right)\frac{1+(1/x)}{1+(3/x^2)} = (0)\frac{1+0}{1+0} = 0$$

For **(b)**, we factor out x^2 from the numerator and the denominator to obtain

$$\frac{1-(3/x)+(2/x^2)}{2-(1/x)+(4/x^2)}$$

Using the limit principles, we conclude that the limit is 1/2.

example 6.9   Compute

$$\lim_{x\to+\infty} \frac{x}{\sqrt{4x^2+1}} \qquad \text{and} \qquad \lim_{x\to-\infty} \frac{x}{\sqrt{4x^2+1}}$$

solution 6.9   We must take care in manipulating $\sqrt{4x^2+1}$.

$$\sqrt{4x^2+1} = \sqrt{x^2(4+1/x^2)} = \sqrt{x^2}\sqrt{4+1/x^2}$$

But $\sqrt{x^2} = \sqrt{|x|^2} = |x|$, since the positive square root must be taken. Hence,

$$\frac{x}{\sqrt{4x^2+1}} = \frac{x}{|x|}\frac{1}{\sqrt{4+1/x^2}} = \begin{cases} 1/\sqrt{4+1/x^2} & \text{for } x>0 \\ -1/\sqrt{4+1/x^2} & \text{for } x<0 \end{cases}$$

Applying the limit principles to each case, we obtain

$$\lim_{x\to+\infty} \frac{x}{\sqrt{4x^2+1}} = \frac{1}{2} \qquad \text{and} \qquad \lim_{x\to-\infty} \frac{x}{\sqrt{4x^2+1}} = -\frac{1}{2}$$

exercises for section 6

part A

Estimate each of the following limits *correct to three decimal places* by constructing appropriate input-output tables. Be sure to let x approach a from both the right and left.

1. $\lim_{x \to 3} (3x - 2)$

2. $\lim_{x \to 6} \sqrt{x}$

3. $\lim_{x \to .45} x^2$

4. $\lim_{x \to 1} (1/2)^x$

5. $\lim_{x \to 1} \dfrac{x^5 - 1}{x - 1}$

6. $\lim_{x \to 4} \dfrac{\sqrt{x} - 2}{x - 4}$

7. $\lim_{x \to 1} \dfrac{|x - 1|}{x - 1}$

8. $\lim_{x \to 0} \dfrac{1}{x}$

9. $\lim_{x \to 0} \dfrac{2^x - 1}{x}$

10. $\lim_{x \to 0} \dfrac{3^x - 1}{x}$

Compute each of the following limits by first simplifying the expression for $f(x)$.

11. $\lim_{x \to 1} \dfrac{x^2 - 1}{x - 1}$

12. $\lim_{x \to -2} \dfrac{x^2 - 4}{x + 2}$

13. $\lim_{x \to 0} \dfrac{x^2 + x}{x}$

14. $\lim_{x \to 1} \dfrac{x^4 - 1}{x^2 - 1}$

15. $\lim_{x \to 1} \dfrac{x^{-1} - 1}{x - 1}$

16. $\lim_{h \to 0} \dfrac{(3 + h)^2 - 9}{h}$

★17. $\lim_{x \to 4} \dfrac{\sqrt{x} - 2}{x - 4}$ [*Hint:*

Rationalize the numerator.]

18. $\lim_{x \to -1/2} \dfrac{2x^2 - 3x - 2}{2x^2 + 3x + 1}$

19. $\lim_{x \to 0} \dfrac{x^2 - 3x}{x^2 + 2x}$

20. $\lim_{x \to 2} \dfrac{x^4 - 16}{x - 2}$ [*Hint:*

$(x^4 - 16) = (x^2)^2 - 4^2.$]

It is easy to verify that $\lim_{x \to 2} (2x - 1) = 3$. Likewise, $\lim_{x \to 2} (4x + 1) = 9$. Using the limit principles, compute each of the following.

21. $\lim_{x \to 2} \dfrac{4x + 1}{2x - 1}$

22. $\lim_{x \to 2} \dfrac{2x - 1}{4x + 1}$

23. $\lim_{x \to 2} \sqrt{4x + 1}$

24. $\lim_{x \to 2} \sqrt[3]{(2x - 1)(4x + 1)}$

25. $\lim_{x \to 2} \dfrac{(2x - 1)^2}{4x + 1}$

26. $\lim_{x \to 2} [(4x + 1)^2 - (2x - 1)]$

For $f(x) = \dfrac{x^2 - 9}{x - 3}$, compute:

27. $\lim_{x \to 1} f(x)$

28. $\lim_{x \to -3} f(x)$

For $f(x) = \dfrac{x-1}{x^2-1}$, compute:

29. $\text{limit}_{x \to 2}\, f(x)$ **30.** $\text{limit}_{x \to 1}\, f(x)$

★**31.** In example 6.2, we demonstrated that $\text{limit}_{x \to 1} \dfrac{x^6-1}{x-1} = 6$. It is not hard to show that $\text{limit}_{x \to 1}(x^3+1) = 2$. Using these two facts and the limit principles, compute:

(a) $\displaystyle\lim_{x \to 1} \frac{x-1}{x^6-1}$ (b) $\displaystyle\lim_{x \to 1} \frac{x^3-1}{x-1}$

[*Hint*: Write (x^6-1) as the difference of two squares to solve **(b)**.]

32. Given that $\displaystyle\lim_{x \to 1} \frac{x^3-1}{x-1} = 3$, find $\displaystyle\lim_{x \to 1} \frac{x^3-1}{x^2-1}$ using the limit principles.

Using the limit principles, compute each of the following limits.

33. $\displaystyle\lim_{x \to +\infty} \frac{3x-2}{x+1}$ **34.** $\displaystyle\lim_{x \to -\infty} \frac{3-4x}{2x-8}$

35. $\displaystyle\lim_{x \to +\infty} \frac{x^2-1}{x^2+1}$ **36.** $\displaystyle\lim_{x \to -\infty} \frac{x^3+2}{x^3-2}$

37. $\displaystyle\lim_{x \to -\infty} \frac{3x}{x^2-2}$ **38.** $\displaystyle\lim_{x \to +\infty} \frac{3x^2}{x^3-4x+2}$

39. $\displaystyle\lim_{x \to -\infty} \frac{3x}{\sqrt{x^2+8}}$ **40.** $\displaystyle\lim_{x \to +\infty} \frac{3x}{\sqrt{x^2+8}}$

★**41.** $\displaystyle\lim_{x \to -\infty} \frac{2x}{\sqrt[3]{x^3-4}}$ ★**42.** $\displaystyle\lim_{x \to +\infty} (\sqrt{x^2+x} - x)$

continuous functions and limits

We say that a function $y = f(x)$ is *continuous at $x = a$* if the graph of f does not break at the point $(a, f(a))$. Thus, in figure 7.1, $f(x)$ is continuous at all points *except a_1 and a_2*.

figure 7.1

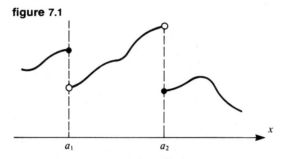

If we look at the branches of the graph to the right and to the left of $(a, f(a))$, several possibilities arise, as shown in figures 7.2, 7.3, 7.4, and 7.5.

figure 7.2 The two branches come together, but not at $(a, f(a))$. This type of discontinuity is called *removable*. If we redefine $f(a)$ to equal h, the function becomes continuous at $x = a$.

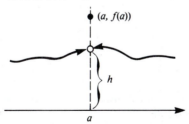

figure 7.3 The two branches do not come together. This represents a *jump discontinuity*.

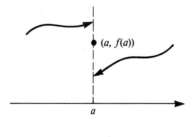

figure 7.4 A *vertical asymptote* exists at $x = a$.

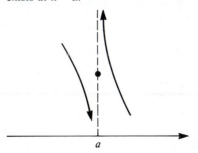

figure 7.5 The branches come together at $(a, f(a))$. The function f is continuous at a.

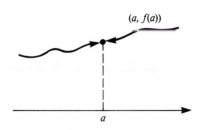

Shown in figures 7.6, 7.7, 7.8, and 7.9 are the graphs of the so-called elementary functions. *All are continuous at each point a in their domains of definition A.* We have included trigonometric functions, exponential functions, and logarithm functions, which will be studied in detail in sections 15 and 29.

figure 7.6(a)   Power functions
$f(x) = x^\kappa$. $f(x) = x^\kappa$, for $x \geq 0$.

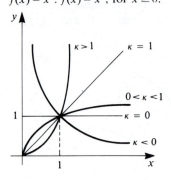

figure 7.6(b)   Power functions
$f(x) = x^\kappa$. $f(x) = x^\kappa$, κ even.

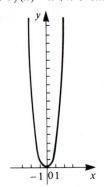

figure 7.6(c)   Power functions
$f(x) = x^\kappa$. $f(x) = x^\kappa$, κ odd.

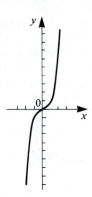

figure 7.7(a)   Exponential
functions.

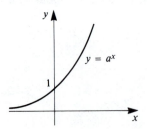

figure 7.7(b)   Logarithm
functions.

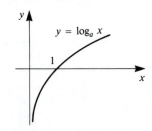

figure 7.8(a) Trigonometric functions $f(x) = \sin x$ and $f(x) = \cos x$.

figure 7.8(b) Trigonometric function $f(x) = \tan x$.

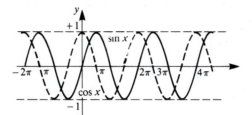

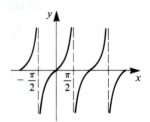

figure 7.9(a) Inverse trigonometric function $\sin^{-1} x$.

figure 7.9(b) Inverse trigonometric function $\cos^{-1} x$.

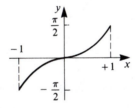

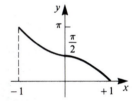

figure 7.9(c) Inverse trigonometric function $\tan^{-1} x$.

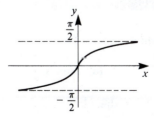

How does all of this relate to limits? Shown in figure 7.10 is a function f that is continuous at a. Note that as $x \to a$, the functional values approach the middle height, $f(a)$. Hence, if f is continuous at $x = a$, then $\text{limit}_{x \to a} f(x) = f(a)$.

figure 7.10

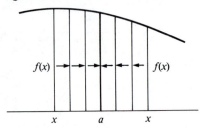

Conversely, suppose that $\text{limit}_{x \to a} f(x) = f(a)$. From figure 7.11, we see that both points P and Q are forced to approach the point $(a, f(a))$ since $(x, f(x)) \to$

figure 7.11

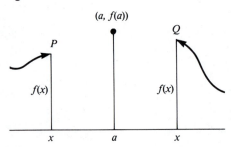

$(a, f(a))$. Hence, the two branches of the graph come together at $(a, f(a))$ and so f is continuous at $x = a$. In summary,

---

f is continuous at $x = a$ if and only if

$$\text{limit}_{x \to a} f(x) = f(a)$$

This key result provides a versatile tool for either computing limits or determining where a function is continuous, as demonstrated in the following examples.

example 7.1   Compute each of the following limits.

    (a) $\text{limit}_{x \to 3} \sqrt{x}$ **(b)** $\text{limit}_{x \to -1} 2^x$ ▲ **(c)** $\text{limit}_{x \to \pi/2} \sin x$

solution 7.1   Each of the elementary functions is continuous at all points in its domain.

Hence, since $\text{limit}_{x \to a} f(x) = f(a)$, we conclude that

$$\lim_{x \to 3} \sqrt{x} = \sqrt{3}, \qquad \lim_{x \to -1} 2^x = 2^{-1} = \frac{1}{2}, \qquad \text{and} \qquad \lim_{x \to \pi/2} \sin x = \sin \frac{\pi}{2} = 1$$

example 7.2 Determine where the function

$$f(x) = \begin{cases} x + 1, \ x < 0 \\ 1 - x, \ 0 \le x < 1 \\ x + 2, \ x \ge 1 \end{cases}$$

is continuous by sketching its graph.

solution 7.2 The graph consists of the three separate line segments shown in figure 7.12. We observe that $f(x)$ is continuous everywhere except at $x = 1$, where a jump discontinuity exists. Note that $\text{limit}_{x \to 0} f(x) = f(0) = 1$, since f is continuous at 0.

figure 7.12

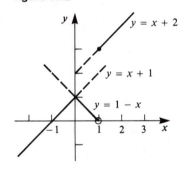

example 7.3 Calculate $\text{limit}_{x \to a} (x^3 - 3x + 1)$ to determine where $f(x) = x^3 - 3x + 1$ is continuous.

solution 7.3 Using limit principle 1, we can say

$$\lim_{x \to a} (x^3 - 3x + 1) = \lim_{x \to a} x^3 + \lim_{x \to a}(-3x + 1)$$
$$= a^3 - 3a + 1$$

In the last step, we used the fact that the elementary functions x^3 and $(-3x + 1)$ are continuous. Note that the limit is just $f(a)$. Hence, we can conclude that f is continuous for all values of x.

The last example shows how we can determine where a function $f(x)$ is continuous, even though we may have no idea what the graph looks like. We merely compute $\text{limit}_{x \to a} f(x)$ using the limit principles and see if the answer is $f(a)$. The generalization of example 7.3 is provided by theorem 7.1.

theorem 7.1   If $f(x)$ and $g(x)$ are continuous at $x = a$, then so are the functions:

$$f(x) + g(x), \qquad f(x) - g(x), \qquad f(x) \cdot g(x), \qquad \frac{f(x)}{g(x)} \quad \text{if } g(a) \neq 0,$$

and

$$\sqrt[n]{f(x)} \quad \text{provided} \quad \sqrt[n]{f(a)} \text{ is defined}$$

proof 7.1   Since f and g are continuous at $x = a$, we know

$$\lim_{x \to a} f(x) = f(a) \qquad \text{and} \qquad \lim_{x \to a} g(x) = g(a)$$

Hence, using the limit principles, we conclude $\lim[f(x) \cdot g(x)] = [\lim f(x)] \cdot [\lim g(x)] = f(a)g(a)$, which is just $f \cdot g$ evaluated at a. Hence, $f(x) \cdot g(x)$ is continuous at $x = a$. Proofs of the other parts are similar. ∎

Theorem 7.1 has the important consequence that any function made up of elementary functions connected together by $+$, $-$, $\cdot$, $\div$, and $\sqrt{}$ is continuous everywhere it is defined. Thus, the function

$$f(x) = \frac{x^3 + 3x^2 - 2^x}{x^2 - 1}$$

is continuous except at $x = \pm 1$. In particular, the following results hold:

1. A polynomial $p(x)$ is continuous for all real numbers x.
2. A rational function $f(x) = p(x)/q(x)$ is continuous at all real numbers except where $q(x) = 0$.

example 7.4   Compute the following limits.

$$\textbf{(a)} \ \lim_{x \to 2} \frac{x^3 - 1}{x - 1} \qquad \textbf{(b)} \ \lim_{x \to 1} \frac{x^3 - 1}{x - 1}$$

solution 7.4   **(a)** The rational function $f(x) = (x^3 - 1)/(x - 1)$ is continuous everywhere except at $x = 1$. Hence, $\lim_{x \to 2} f(x) = f(2) = 7$. To find the limit in **(b)**, we first simplify the fraction for $x \neq 1$:

$$\frac{x^3 - 1}{x - 1} = \frac{(x - 1)(x^2 + x + 1)}{(x - 1)} = x^2 + x + 1$$

Hence, $\lim_{x \to 1} f(x) = \lim_{x \to 1} (x^2 + x + 1) = 1 + 1 + 1 = 3$. ▤

We now state, in the box on page 71, a simple set of rules that will enable us to compute the limit of any function commonly encountered in calculus and its applications. *The function $f(x)$ can be any function made up of elementary functions as building blocks.*

---

procedure to compute limit $f(x)$
$$x \to a$$

Step 1. Determine whether f is defined at $x = a$. If $f(a)$ exists, this is the limit.

Step 2. If $f(a)$ does not exist (usually because a denominator equals 0), use algebra to rewrite the expression for $f(x)$ in a simpler form. Then *return to Step 1.*

Step 3. If you are unable to complete Step 2, estimate the limit numerically (as in section 6).

Let's illustrate the procedure with a few examples.

example 7.5   For a fixed x, compute $\lim_{h \to 0} (x^2 + xh + h^2)$.

solution 7.5   The given function is a function of h, not x. Hence, the limit is $x^2 + x \cdot 0 + 0^2 = x^2$.   ▤

example 7.6   Compute $\lim_{x \to 0} \dfrac{1}{x} \left(\dfrac{1}{2+x} - \dfrac{1}{2} \right)$.

solution 7.6   The expression is not defined at $x = 0$. Hence, we must simplify as follows:

$$\frac{1}{x} \left(\frac{1}{2+x} - \frac{1}{2} \right) = \frac{1}{x} \left[\frac{2 - (2+x)}{2(2+x)} \right] = \frac{-1}{2(2+x)}$$

Hence, the limit as $x \to 0$ is $-1/[2(2+0)] = -1/4$.   ▤

example 7.7   Compute $\lim_{x \to 0} (1+x)^{1/x}$.

solution 7.7   The function $f(x) = (1+x)^{1/x}$ is not defined at $x = 0$, and there does not appear to be any way of simplifying the original expression. We will construct the following input-output table:

table 7.1

x	.1	.01	.001	.0001	.00001
$(1+x)^{1/x}$	2.59374	2.70481	2.71692	2.71814	2.71825
x	$-.1$	$-.01$	$-.001$	$-.0001$	$-.00001$
$(1+x)^{1/x}$	2.86797	2.73199	2.719642	2.718417	2.71828

From table 7.1 it appears that, to three decimal places, $\text{limit}_{x \to 0}(1+x)^{1/x} = 2.718$. Shown in figure 7.13 is a computer-generated graph showing the values of $^{1/x}$

figure 7.13 $\text{Limit}(1+x)^{1/x}$

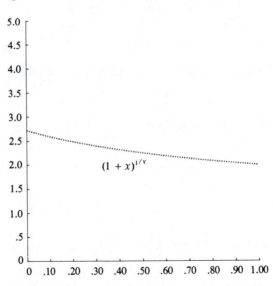

$(1+x)^{1/x}$ for $x = 1.00, .99, .98, \ldots, .02, .01$. Let us define the special number e to be $\text{limit}_{x \to 0}(1+x)^{1/x}$. You may find e on your calculator by keystroking:

$$1 \;\boxed{e^x} \qquad \text{or} \qquad 1 \;\boxed{\text{inv}} \;\boxed{\text{ln}}$$

You should obtain $2.718281828459\ldots.$

We are now in a position to justify the special technique for solving inequalities that we introduced in section 2. By first solving $f(x) = 0$ and then finding those points where f was *not* continuous, we formed open intervals I on which (1) f was continuous and (2) f was nonzero. The justification of the technique then follows from theorem 7.2.

theorem 7.2   Suppose $f(x)$ is continuous and nonzero on an open interval $I = (a, b)$. Then, either $f(x) > 0$ throughout I or $f(x) < 0$ throughout I.

proof 7.2   Suppose to the contrary that there are real numbers t_1 and t_2 in I for which $f(t_1) > 0$ and $f(t_2) < 0$. Let $P = (t_1, f(t_1))$ and $Q = (t_2, f(t_2))$. As can be seen in figure 7.14, the graph joining P to Q *does not break*, and hence must cross the x-axis in at least one place z. But then we would have $f(z) = 0$. This contradicts our hypothesis that f is nonzero throughout I. Hence, either $f(x) > 0$ throughout I or $f(x) < 0$ throughout I.

figure 7.14

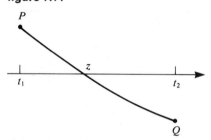

■

The Sign Chart Method breaks the real number line into intervals upon which $f(x)$ is nonzero and continuous. This is illustrated in the next example.

example 7.8   Solve the inequality $\dfrac{x^3 - 1}{x^2 - 3x + 1} > 0$.

solution 7.8   The function $f(x) = (x^3 - 1)/(x^2 - 3x + 1) = 0$ when $x^3 = 1$, i.e., when $x = \sqrt[3]{1} = 1$. The rational function $f(x)$ is not continuous at those places where $x^2 - 3x + 1 = 0$. Using the quadratic formula, we conclude that the discontinuities occur at

$$x = \frac{3 \pm \sqrt{9 - 4}}{2} = \frac{3 \pm \sqrt{5}}{2}$$

Hence, we can form the intervals upon which f is nonzero and continuous (figure 7.15):

figure 7.15

We have placed u's over $(3-\sqrt{5})/2 \approx .38$ and $(3+\sqrt{5})/2 \approx 2.62$ as reminders that f is *undefined* there. We next choose test points in the intervals (figure 7.16):

figure 7.16

Now, $f(0) = -1 < 0$, $f(.5) = 3.5 > 0$, $f(2) = -7 < 0$, and $f(3) = 26 > 0$. Hence, we can complete the sign chart for f as follows (figure 7.17):

figure 7.17

We can now conclude that $f(x) = \dfrac{x^3 - 1}{x^2 - 3x + 1} > 0$ for $\dfrac{3-\sqrt{5}}{2} < x < 1$ or $x > \dfrac{3+\sqrt{5}}{2}$.   ▤

Theorem 7.2 is a consequence of a more general theorem known as the **Intermediate Value Theorem**.

theorem 7.3   **intermediate value theorem**   Suppose $f(x)$ is a continuous function on a closed interval $[a, b]$. Then if c is a real number between $f(a)$ and $f(b)$, there exists at least one real number z in $[a, b]$ with $f(z) = c$.

It is not difficult to see from figure 7.18 why this theorem must be true. The graph, in traveling from $P = (a, f(a))$ to $Q = (b, f(b))$, must cut the line $y = c$ at least once. If $f(x)$ were not continuous, the graph could jump over the barrier at $y = c$. However, since $f(x)$ is continuous *at all points* of $[a, b]$, this possibility is eliminated.

figure 7.18

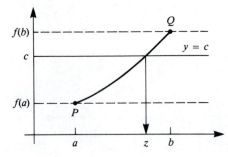

exercises for section 7

part A

Use the continuity of the elementary functions to evaluate each of the following limits. Use your calculator to evaluate the limit to three decimal places.

1. $\text{limit}_{x\to 4}\ \sqrt{x}$

2. $\text{limit}_{x\to 2}\ x^3$

▲**3.** $\text{limit}_{x\to\pi}\ (\sin x)$

▲**4.** $\text{limit}_{x\to\pi}\ (\cos x)$

5. $\text{limit}_{x\to 1/2}\ 3^x$

6. $\text{limit}_{x\to 2}\ (1/x^3)$

7. $\text{limit}_{x\to 2}\ (\log_{10} x)$

8. $\text{limit}_{x\to 3}\ \sqrt[3]{x}$

9. $\text{limit}_{x\to -2}\ 2^x$

▲★**10.** $\text{limit}_{x\to\pi/2}\ (\cos^{-1} x)$

Determine where each of the following functions in exercises 11–18 is continuous by first sketching the graph. If a function is *not* continuous at a given point, state whether a removable discontinuity, a jump discontinuity, or a vertical asymptote exists.

11. $f(x) = \begin{cases} x, & x<0 \\ 1-x, & 0\le x\le 1 \\ 1-x^2, & x>1 \end{cases}$

12. $f(x) = \begin{cases} x^3, & x<0 \\ x^2, & 0\le x\le 2 \\ 2x, & x>2 \end{cases}$

13. $f(x) = \dfrac{x^2-4}{x-2}$ for $x\neq 2$

14. $f(x) = \dfrac{x+1}{x}$ for $x\neq 0$

15. $f(x) = \dfrac{|x-2|}{x-2}$ for $x\neq 2$

16. $f(x) = 2+\dfrac{1}{x^2}$ for $x\neq 0$

17. $f(x) = \begin{cases} 1/x, & x<0 \\ 1/x^2, & 0<x<1 \\ x^2, & x\ge 1 \end{cases}$

18. $f(x) = \begin{cases} 3x-4, & x<-1 \\ 2x-5, & -1\le x\le 2 \\ -x+1, & x>2 \end{cases}$

Use the limit principles and the fact that the elementary functions are continuous to compute

$$\text{limit}_{x\to a}\ f(x)$$

Then determine where $f(x)$ is continuous. (See example 7.3.)

19. $f(x) = x^4 - 2x^3 + 1$

20. $f(x) = \dfrac{x}{x^2-1}$

21. $f(x) = x \cdot 2^x$

▲**22.** $f(x) = x^2 \sin x$

23. $f(x) = 3^x/x$

▲**24.** $f(x) = \dfrac{\sin x}{x^2+1}$

Use the procedure for computing limits on page 71 to find:

25. $\text{limit}_{h\to 0}\ (2xh + 3x^2)$

26. $\text{limit}_{x\to 0}\ (2xh + 3x^2)$

27. $\displaystyle\lim_{h\to 0}\frac{(1+h)^2-1}{h}$

28. $\displaystyle\lim_{h\to 0}\frac{(x+h)^2-x^2}{h}$

29. $\displaystyle\lim_{x\to 1}\frac{x^2+x-6}{x^2-6x+8}$

30. $\displaystyle\lim_{x\to 2}\frac{x^2+x-6}{x^2-6x+8}$

31. $\displaystyle\lim_{x\to 0}\frac{x^4-1}{x-1}$

32. $\displaystyle\lim_{x\to 1}\frac{x^4-1}{x-1}$

33. $\displaystyle\lim_{x\to 2}\frac{2x-1}{4x^2-1}$

34. $\displaystyle\lim_{x\to 1/2}\frac{2x-1}{4x^2-1}$

35. $\displaystyle\lim_{x\to 0}\frac{x^2+3x}{x}$

36. $\displaystyle\lim_{x\to 0}\frac{x^2+3}{x}$

37. $\displaystyle\lim_{x\to 0}\frac{1/(2+x)^2-1/4}{x}$

38. $\displaystyle\lim_{x\to 3}\frac{1/x-1/3}{x-3}$

39. $\displaystyle\lim_{x\to 2}\frac{\sqrt{x}-\sqrt{2}}{x-2}$

▲ **40.** $\displaystyle\lim_{x\to 0}(x\sin x)$

▲ **41.** $\displaystyle\lim_{x\to 0}\frac{\sin x}{x}$

▲ **42.** $\displaystyle\lim_{x\to 0}\frac{1-\cos x}{x^2}$

43. $\displaystyle\lim_{x\to 0}\frac{1}{x}\left(\frac{1}{x-1}+1\right)$

★ **44.** $\displaystyle\lim_{x\to 1}\frac{\log_{10} x}{x-1}$

★ **45.** $\displaystyle\lim_{x\to 0}(1-x)^{1/x}$

Using the sign chart method, solve each of the following inequalities.

46. $\displaystyle\frac{x-3}{(x^2+1)^2}>0$

47. $\displaystyle\frac{x-3}{(x^2-1)^2}<0$

48. $2^x(3x-2)<0$

49. $\displaystyle\frac{2^x}{x^2-1}<0$

50. $\displaystyle\frac{x-3}{x^2+2x-2}>0$

51. $\displaystyle\frac{x^2+2x-2}{x^3-1}>0$

★ **52.** $\displaystyle\frac{\log_{10} x}{x-4}<0$ [*Note*: $\log_{10} x=0$ only for $x=1$.]

▲★ **53.** $\displaystyle\frac{\sin x}{\cos x+1}<0$ [*Note*: $\sin x=0$ if an only if $x=n\pi$, $n=0,\pm 1,\pm 2,\dots$ and $\cos x=-1$ if and only if $x=(2n+1)\pi$, $n=0,\pm 1,\pm 2,\dots.$]

part B

54. Let $f(x)$ be a continuous function with domain of definition $[0, 1]$ and range $[0, 1]$. If $0 \leq z \leq 1$ and $f(z) = z$, we call z a *fixed point* of f.

figure 7.19

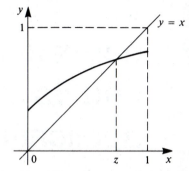

Apply either theorem 7.2 or theorem 7.3 to the continuous function $g(x) = f(x) - x$ and conclude that $f(x)$ must have a fixed point.

the derivative and its applications

prologue   We are now ready to begin our study of the derivative—one of the two main concepts in calculus. The derivative answers the question "How quickly is *y* changing with respect to *x*?" In terms of the graph, this is equivalent to determining how quickly the graph of $y = f(x)$ is rising at a given point P on the curve. For example, how steep is the graph at the point P, shown in figure 1? To answer this question we must first develop methods for finding the slope of the tangent line at P. Thus, one may interpret the derivative as the **slope of the tangent line.**

figure 1

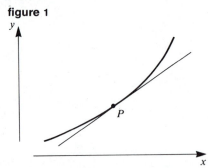

In many cases the variable *x* is interpreted as *time*. In these cases, the derivative gives a speedometer reading of how quickly *y* is changing at a given instant of time. The following three examples provide alternate interpretations of the derivative.

example 1 velocity   If y is the total distance traveled in miles and x is the time in hours, then the derivative is the **instantaneous velocity** or standard speedometer reading in mph (see figure 2).

figure 2(a)

Odometer

figure 2(b)

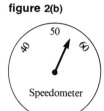

Speedometer

example 2 flow rate   If y is the present volume in a reservoir (in cubic feet) and x is time in minutes, then the derivative measures how quickly the volume y is increasing at a given moment (see figure 3). This is just the rate at which the liquid is entering—i.e., the **instantaneous flow rate**. This interpretation of the derivative is especially important in studying the respiratory and circulatory systems.

figure 3

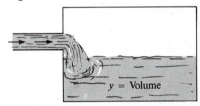

$y = \text{Volume}$

example 3 rate of population growth   Let y by the number of individuals in a population at time x. If we wish to determine how quickly the population is increasing (or possibly decreasing), we must compute a derivative (see figure 4).

figure 4

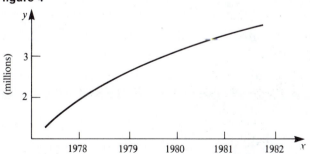

To begin our study of the derivative, we will emphasize the tangent line interpretation to motivate the definitions and theorems.

the derivative concept

Shown in figure 8.1 is a graph of $y = x^2$, for $-2 \le x \le 2$. Let $m(x)$ stand for the slope of the tangent line to the graph at the point $P = (x, x^2)$. Let's try to determine an expression for $m(x)$ by estimating various slopes from the graph.

figure 8.1

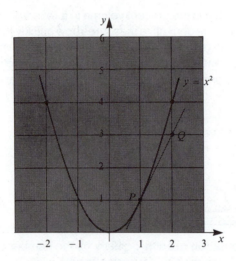

We will obtain these estimates by performing the following steps:

1. Place your ruler at point P in the tangent position.
2. Produce a second point Q on the tangent line using the rectangular grid.
3. Compute the slope of the tangent line using the formula $m = (y_2 - y_1)/(x_2 - x_1)$.

In figure 8.1 we have shown the case for estimating the slope of the tangent line at $x = 1$. A second point on the line is determined to be $Q = (2, 3)$. Hence, $m(1) = (3 - 1)/(2 - 1) = 2$. *Complete the missing entries in table* 8.1.

Table 8.1 produces strong evidence that $m(x) = 2x$ is the formula for the slope of the tangent line when $f(x) = x^2$. The function $m(x)$ is called the **derivative** of the function x^2. A more common notation for $m(x)$ is $f'(x)$. Therefore, if

$$f(x) = x^2, \qquad f'(x) = 2x$$

table 8.1

x	$P = (x, x^2)$	Q (on the tangent line)	$m(x)$
0	$(0, 0)$	$(2, 0)$	0
$\frac{1}{2}$	$(\frac{1}{2}, \frac{1}{4})$		
1	$(1, 1)$	$(2, 3)$	$2 = \dfrac{3-1}{2-1}$
$\frac{3}{2}$	$(\frac{3}{2}, \frac{9}{4})$		
2	$(2, 4)$		
$-\frac{1}{2}$	$(-\frac{1}{2}, \frac{1}{4})$	$(-\frac{1}{4}, 0)$	$-1 = \dfrac{0-1/4}{-1/4+1/2}$
-1	$(-1, 1)$		
$-\frac{3}{2}$	$(-\frac{3}{2}, \frac{9}{4})$	$(-\frac{3}{4}, 0)$	-3
-2	$(-2, 4)$		

Note then that *the derivative of a function is itself a function*. It may be thought of as a *formula that gives the slope of the tangent line.* For example, $f'(\sqrt{2}) = 2\sqrt{2} \approx 2.828$ is the slope of the tangent line at the point $(\sqrt{2}, 2)$ on the graph of $y = x^2$.

Formulas for $m(x) = f'(x)$ are in general not so easy to guess as in the previous example. We will now develop a method for finding $f'(x)$ based on the **limit concept**.

Instead of guessing a second point on the tangent line, we will choose a point Q *on the graph* (see figure 8.2), calculate the slope of PQ, and then slide Q down the curve toward P to get a better and better approximation to the slope of the tangent line. The procedure may be summarized, using mathematical notation, as follows. Let $P = (a, f(a))$ and $Q = (x, f(x))$. The line PQ has slope $[f(x) - f(a)]/(x - a)$. Now, as $x \to a$, the point Q will slide down toward P. Hence, the limit definition of a derivative is

definition of a derivative $\qquad f'(a) = m(a) = \lim_{x \to a} \dfrac{f(x) - f(a)}{x - a}$

figure 8.2

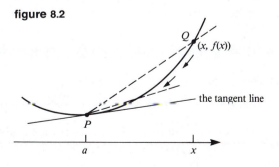

Note that in order to apply this definition, we must first rewrite $\dfrac{f(x) - f(a)}{x - a}$ using algebra, and then take the limit.

example 8.1   For $f(x) = x^2$, first show that $f'(2) = 4$, and then prove that $f'(x) = 2x$, using the limit definition of a derivative.

solution 8.1   For $a = 2$,

$$\frac{f(x) - f(2)}{x - 2} = \frac{x^2 - 4}{x - 2} = \frac{(x - 2)(x + 2)}{x - 2} = x + 2, \qquad \text{for } x \neq 2$$

Hence, $f'(2) = \lim_{x \to 2} (x + 2) = 4$. The general case is entirely similar:

$$\frac{f(x) - f(a)}{x - a} = \frac{x^2 - a^2}{x - a} = \frac{(x + a)(x - a)}{x - a} = x + a, \qquad \text{for } x \neq a$$

Hence, $f'(a) = \lim_{x \to a} (x + a) = 2a$. Replacing a by x yields $f'(x) = 2x$. ▤

example 8.2   For $f(x) = 1/x = x^{-1}$, prove that $f'(x) = -1/x^2 = -x^{-2}$, using the limit definition of a derivative.

solution 8.2   We will show that $f'(a) = -a^{-2}$. We first must simplify:

$$\frac{f(x) - f(a)}{x - a} = \frac{(1/x) - (1/a)}{x - a} = \frac{1}{x - a} \left(\frac{a - x}{xa} \right) = \frac{-1}{xa}$$

where we have used the relationship $a - x = -1(x - a)$. Hence, $f'(a) = \lim_{x \to a} -1/xa = -1/a^2 = -1a^{-2}$. ▤

An alternate definition of derivative can be given, based on relabeling the points P and Q in figure 8.2.

If we relabel $P = (x, f(x))$ and $Q = (x + h, f(x + h))$, then the slope of the line PQ becomes $[f(x + h) - f(x)]/[(x + h) - x] = [f(x + h) - f(x)]/h$. As $h \to 0$, the point Q will slide down toward P and the line PQ will approach the tangent line (see figure 8.3). Thus, we have the following alternate definition of a derivative:

---

alternate definition of a derivative $f'(x) = \lim\limits_{h \to 0} \dfrac{f(x + h) - f(x)}{h}$

As the next examples illustrate it is sometimes easier to use this alternate formulation in working out a derivative.

figure 8.3

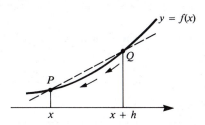

example 8.3 For $f(x) = x^3$, show that $f'(x) = 3x^2$. Then find the equation of the tangent line when $x = \frac{1}{2}$.

solution 8.3 First note that

$$(x + h)^3 = (x + h)^2 (x + h) = (x^2 + 2xh + h^2)(x + h)$$
$$= x^3 + 3xh^2 + 3x^2h + h^3$$

It follows that:

$$\frac{f(x + h) - f(x)}{h} = \frac{(x + h)^3 - x^3}{h} = \frac{x^3 + 3xh^2 + 3x^2h + h^3 - x^3}{h}$$
$$= \frac{h(3xh + 3x^2 + h^2)}{h} = 3x^2 + 3xh + h^2 \text{ for } h \neq 0$$

Hence, $f'(x) = \text{limit}_{h \to 0} (3x^2 + 3xh + h^2) = 3x^2$. When $x = \frac{1}{2}$, the slope of the tangent line is $f'(\frac{1}{2}) = \frac{3}{4}$ and the point P on the graph is $(\frac{1}{2}, \frac{1}{8})$. Then, using the point-slope formula we determine the equation of the tangent line:

$$\frac{y - 1/8}{x - 1/2} = \frac{3}{4}$$

which simplifies to $3x - 4y - 1 = 0$.

example 8.4 If $f(x) = mx + b$, show that $f'(x) = m$. Conclude that the derivative of x is 1 and that the derivative of any constant is 0.

solution 8.4 We first observe that $f(x + h) = m(x + h) + b = mx + mh + b$. Then it follows that

$$\frac{f(x + h) - f(x)}{h} = \frac{mx + mh + b - mx - b}{h} = m$$

Hence, $f'(x) = \text{limit}_{h \to 0} m = m$. This result is not surprising! The tangent line coincides with the original linear graph and hence has slope m. If $m = 1$ and $b = 0$, then $f(x) = x$ and $f'(x) = 1$. If $m = 0$, then $f(x) = b$ and $f'(x) = 0$.   ▤

Examples 8.1–8.4 suggest the following general derivative formula, which applies to *power functions*:

> **power rule**   The derivative of x^k is kx^{k-1}.

The rule is sometimes called "the pull-down rule":

$$k \longleftarrow x^{k-1}$$

The exponent k is pulled down in front. The original exponent is then decreased by 1.

warning!   The reader is warned that the derivative of 2^x is *not* $x2^{x-1}$ because 2^x *is not* a power function. A completely different rule applies to *exponential functions* such as 2^x, 3^x, and $(\frac{1}{2})^x$. The new rule will be developed in section 16, page 169.

example 8.5   Use the power rule to compute the derivatives of

   (a) $\sqrt{x}$ **(b)** $1/x^3$ **(c)** $x^{3/4}$

solution 8.5   In part **(a)**, we rewrite $\sqrt{x} = x^{1/2}$. Hence, the derivative $(1/2)x^{1/2-1} = (1/2)x^{-1/2} = 1/(2\sqrt{x})$.
In part **(b)**, note that $1/x^3 = x^{-3}$. Hence, $f'(x) = -3x^{-3-1} = -3x^{-4}$.
Finally, direct application of the power rule to part **(c)** yields $f'(x) = (3/4)x^{-1/4}$.
   ▤

The next two examples show that a continuous function need not have a derivative at every point.

example 8.6   For $f(x) = |x|$, show that $f'(0)$ does not exist.

solution 8.6   Shown in figure 8.4 is the graph of $y = |x|$. There appears to be more than one

figure 8.4

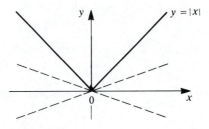

candidate for the tangent line at the point $x = 0$. Something is suspicious! The definition of derivative bears this out.

$$\frac{f(0 + h) - f(0)}{h} = \frac{|h| - |0|}{h} = \frac{|h|}{h} = \begin{cases} 1, & \text{for } h > 0 \\ -1, & \text{for } h < 0 \end{cases}$$

Hence, $f'(0) = \lim_{h \to 0} |h|/h$ does not exist. (See example 6.3 for details.) We call the point $(0, 0)$ a *sharp point* or *corner* of the graph.   ▤

example 8.7   For $f(x) = (x - 1)^{2/3}$, show that $f'(1)$ does not exist.

solution 8.7   Shown in figure 8.5 is the graph of $y = (x - 1)^{2/3} = \sqrt[3]{(x - 1)^2}$. Again, there are many lines that intersect the graph only at $(1, 0)$.

figure 8.5

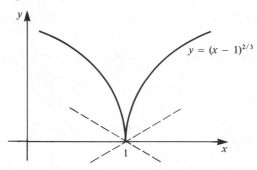

$$y = (x - 1)^{2/3}$$

Using the limit definition of a derivative, we have

$$\frac{f(1 + h) - f(1)}{h} = \frac{(1 + h - 1)^{2/3} - 0}{h} = \frac{h^{2/3}}{h} = \frac{1}{\sqrt[3]{h}}$$

Hence, $f'(1) = \lim_{h \to 0} 1/\sqrt[3]{h}$. But this limit does not exist, as the input-output table 8.2 shows.

table 8.2

h	.1	.001	.000001	−.1	−.001	−.000001
$1/\sqrt[3]{h}$	2.15	10	100	−2.15	−10	−100

The point $(1, 0)$ on the graph is called a *cusp point*.   ▤

For the function whose graph is shown in figure 8.6, $f(x)$ has a derivative for all values of x except a_1, a_2, and a_3. Although $f(x)$ is continuous at all three points, it is necessary that the graph of f be *smooth* at $(a, f(a))$ in order for a

figure 8.6

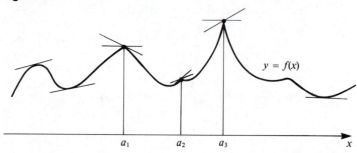

unique tangent line to exist. "Smooth" is a much stronger requirement than "continuous," as theorem 8.1 shows.

theorem 8.1   If $f'(a)$ exists, then $\text{limit}_{x \to a} f(x) = f(a)$, i.e., f is continuous at $x = a$.

proof 8.1   We are given that $\underset{x \to a}{\text{limit}} \dfrac{f(x) - f(a)}{x - a} = f'(a)$ exists. Note that we may write

$$f(x) = f(a) + (x - a)\frac{f(x) - f(a)}{x - a} \qquad \text{for } x \neq a$$

[The $(x - a)$ terms cancel, and then $f(a)$ cancels $-f(a)$.] Applying the limit principles yields

$$\underset{x \to a}{\text{limit}} f(x) = f(a) + \underset{x \to a}{\text{limit}}\, (x - a) \underset{x \to a}{\text{limit}} \frac{f(x) - f(a)}{x - a}$$

$$= f(a) + 0 \cdot f'(a) = f(a). \qquad \blacksquare$$

Developing techniques for finding $f'(x)$ will prove extremely useful in graphing functions.

1. Determining where the function $f'(x) = m(x) = 0$ will aid in locating vertices such as $V_1 - V_4$ in figure 8.7.

figure 8.7

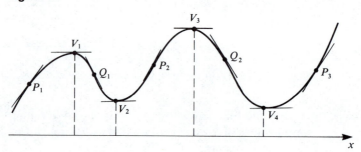

2. Determining where $f'(x) = m(x)$ is greater than 0 will tell us where the function $f(x)$ is increasing. See points P_1–P_3.

3. Determining where $f'(x) = m(x)$ is less than 0 will indicate where the function $f(x)$ is decreasing. See points Q_1 and Q_2.

We will return to sketching curves after we have mastered the art of finding derivatives.

exercises for section 8

part A

For each of the following functions, use the method outlined on page 80 to complete the table. Then try to guess the formula for $m(x) = f'(x)$.

1.

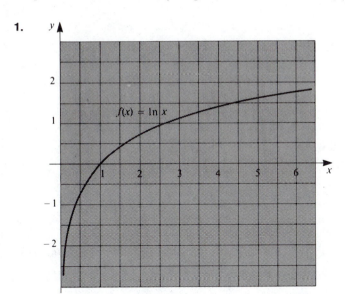

x	$P = (x, \ln x)$	Q (on tangent line)	$m(x)$
.5	$(.5, -.693)$		
1	$(1, 0)$		
2	$(2, .693)$		
4	$(4, 1.386)$		
.2	$(.2, -1.609)$		

2.

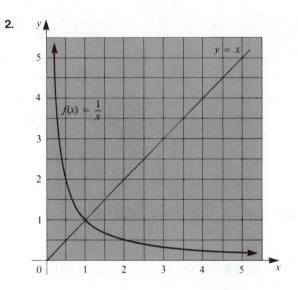

x	$P = (x, 1/x)$	Q (on tangent line)	$m(x)$
1	(1, 1)		
2	(2, .5)		
4	(4, .25)		
.5	(.5, 2)		
.25	(.25, 4)		

For each of the following functions in exercises 3–10, find $f'(1)$ by computing

$$\lim_{x \to 1} \frac{f(x) - f(1)}{x - 1}$$

3. $f(x) = x^3$

5. $f(x) = x^2 + x$

7. $f(x) = 3x^2$

★**9.** $f(x) = \log_{10} x$ (Use your calculator)

4. $f(x) = 1/x$

6. $f(x) = 3x - 2$

8. $f(x) = \sqrt{x}$

★**10.** $f(x) = 2^x$ (Use your calculator)

For each of the following functions in exercises 11–17, find $f'(x)$ by computing

$$\lim_{x \to a} \frac{f(x) - f(a)}{x - a}$$

11. $f(x) = 3x^2$

12. $f(x) = x^3$ [*Note* that $x^3 - a^3$ $= (x - a)(x^2 + ax + a^2)$.]

13. $f(x) = x^{-2}$ **14.** $f(x) = 3x - 4$

15. $f(x) = \frac{1}{2}x^2 + x$ **16.** $f(x) = x^2 + 3x + 2$

★**17.** $f(x) = \sqrt{x}$

For each of the following functions in exercises 18–23, find $f'(x)$ by computing

$$f'(x) = \lim_{h \to 0} \frac{f(x+h) - f(x)}{h}$$

18. $f(x) = 2x + 2$ **19.** $f(x) = x^3 - x$

20. $f(x) = 1/(x + 1)$ ★**21.** $f(x) = \sqrt{x+1}$ [*Hint*: rationalize the numerator.]

22. $f(x) = x^2 + x$ ★**23.** $f(x) = x/(x + 1)$

24.–30. For the functions in exercises 11–17, find the slope of the tangent line at $x = 2$.

31.–36. For the functions in exercises 18–23, determine where $f'(x) = 0$.

In exercises 37–48 use the power rule to compute the derivatives of:

37. x^4 **38.** x^9 **39.** $1/x$ **40.** $1/x^2$

41. $\sqrt[3]{x}$ **42.** $x\sqrt{x}$ **43.** $x^2\sqrt[3]{x}$ **44.** x^{100}

45. $x^{-4/5}$ **46.** $1/\sqrt{x}$ **47.** $(\sqrt[5]{x})^2$ **48.** x^{-4}

49. For $f(x) = |x - 1|$, show that $f'(1)$ does not exist by examining $[f(1+h) - f(1)]/h$.

50. For $f(x) = x^{2/3}$, show that $f'(0)$ does not exist by examining $[f(h) - f(0)]/h$.

★**51.** For $f(x) = x|x|$, show that $f'(0)$ exists.

52. Find those points x where $f'(x)$ does not exist in each of the following graphs.

(a)

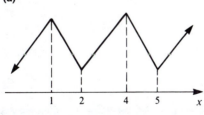

(b)

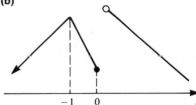

(c)

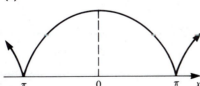

(d)

▲ **53.** If $f(x) = \sin x$, show that $f'(0) = 1$ by computing

$$\underset{h \to 0}{\text{limit}} \frac{\sin h - \sin 0}{h - 0}$$

with the aid of your calculator.

▲ **54.** If $f(x) = \cos x$, show that $f'(0) = 0$ by computing

$$\underset{h \to 0}{\text{limit}} \frac{\cos h - \cos 0}{h - 0}$$

with the aid of your calculator.

▲ **55.** One of the standard trigonometric identities is

$$\sin (x + y) = \sin x \cos y + \sin y \cos x.$$

(a) If $f(x) = \sin x$, show that

$$\frac{f(x + h) - f(x)}{h} = \cos x \left(\frac{\sin h}{h}\right) + \sin x \left(\frac{\cos h - 1}{h}\right).$$

(b) Use exercises 53 and 54 to conclude that $f'(x) = \cos x$.

56. If $f(x) = 3^x$, show that $f'(0) \approx 1.099$ by computing

$$\underset{h \to 0}{\text{limit}} \frac{3^h - 3^0}{h - 0}$$

with the aid of your calculator.

57. If $f(x) = 3^x$, note that $f(x + h) = 3^{x+h} = 3^x 3^h$.

(a) Show that $[f(x + h) - f(x)]/h = 3^x [(3^h - 1)/h]$.

(b) Use the result of exercise 56 to conclude that $f'(x) \approx 1.099 \ 3^x$.

9 formulas for obtaining derivatives

The building blocks of any function we will deal with are the elementary functions, and so it is appropriate that we find the derivatives of these functions first. We will derive many of the formulas shown in table 9.1 later in the text.

table 9.1

Elementary Function $f(x)$	Derivative $f'(x)$
c (a constant)	0
mx	m
x^k (k fixed)	kx^{k-1}
e^x ($e = 2.71828\ldots$)	e^x
a^x (a fixed)	$(\ln a)a^x$
$\ln x = \log_e x$	$1/x$
$\log_a x$	$\dfrac{1}{\ln a} \cdot \dfrac{1}{x}$
$\sin x$	$\cos x$
$\cos x$	$-\sin x$
$\sin^{-1} x$	$\dfrac{1}{\sqrt{1-x^2}}$
$\cos^{-1} x$	$\dfrac{-1}{\sqrt{1-x^2}}$
$\tan^{-1} x$	$\dfrac{1}{1+x^2}$

These formulas have been known for almost 300 years. Each may be derived using the limit definition of a derivative. We will use several alternate notations for $f'(x)$:

$$D_x f(x) \qquad D_x y \qquad \frac{dy}{dx} \qquad y' \qquad \dot{y}$$

Isaac Newton, one of the co-inventors of calculus, used the notation $\dot{y}$; the other co-inventor, Gottfried Wilhelm Leibniz, used the notation $\dfrac{dy}{dx}$. These alternate notations are used in examples 9.1–9.4.

example 9.1   Find $\dfrac{dy}{dx}$ if $y = 2$.

solution 9.1   Since the derivative of any constant is 0, we have $\dfrac{dy}{dx} = 0$.   ▤

example 9.2   Compute $D_x\, 6x$.

solution 9.2   Use of the second entry in table 9.1 yields $D_x\, 6x = 6$.   ▤

example 9.3   Find $\dot{y}$ if $y = \sqrt[4]{x}$. Find the slope of the tangent line when $x = 1$.

solution 9.3   Display $\sqrt[4]{x}$ as the power function $x^{1/4}$; then $\dot{y} = \frac{1}{4}x^{1/4-1} = \frac{1}{4}x^{-3/4}$. This is the general formula for the slope of the tangent line. When $x = 1$, $\dot{y} = \frac{1}{4}$.   ▤

example 9.4   Find y' if $y = \cos x$. Find the slope of the tangent lines at $x = 0$ and $x = 1$.

solution 9.4   From table 9.1, we conclude that $y' = -\sin x$. When $x = 0$, $y' = -\sin 0 = 0$. When $x = 1$, $y' = -\sin 1 = -.8415\ldots$ ▤

How do we compute the derivative of a simple polynomial like $2x - 3x^3 + 1$? The **Sum Rule**, presented in the accompanying box, allows us to concentrate on the elementary functions $2x$, x^3, and 1, take their derivatives and form:

$$[\,2\,] - 3[\,3x^2\,] + [\,0\,] = 2 - 9x^2$$

> **Sum Rule**   For constants a and b,
>
> $$D_x[a \cdot f(x) + b \cdot g(x)] = a \cdot D_x f(x) + b \cdot D_x g(x)$$

The Sum Rule is used in the examples 9.5–9.8.

example 9.5   Find the derivative of the polynomial $p(x) = 4x^3 - 2x^2 + x + 3$.

solution 9.5   In attacking the derivative of $4[\,x^3\,] - 2[\,x^2\,] + [\,x\,] + [\,3\,]$, we leave the coefficients alone and concentrate on the derivatives of the elementary functions to obtain:

$$p'(x) = 4[\,3x^2\,] - 2[\,2x\,] + 1 + 0$$
$$= 12x^2 - 4x + 1$$

▤

example 9.6   If $f(x) = x^3 - 3x + 1$, determine where $f'(x) = 0$, $f'(x) > 0$, and $f'(x) < 0$.

solution 9.6   The derivative is $f'(x) = 3x^2 - 3 = 3(x + 1)(x - 1)$. We may complete the solution by making a sign chart for f'. When $x = \pm 1$, $f'(x) = 0$. Using test points -2, 0, and 2, and noting that $f'(-2) = 3(-1)(-3) > 0$, $f'(0) = 3(1)(-1) < 0$, and $f'(2) = 3(3)(1) > 0$, we obtain figure 9.1. Hence, $f'(x) = 0$ for $x = \pm 1$, $f'(x) > 0$ for $x \in (-\infty, -1) \cup (1, +\infty)$, and $f'(x) < 0$ for $x \in (-1, 1)$.

figure 9.1

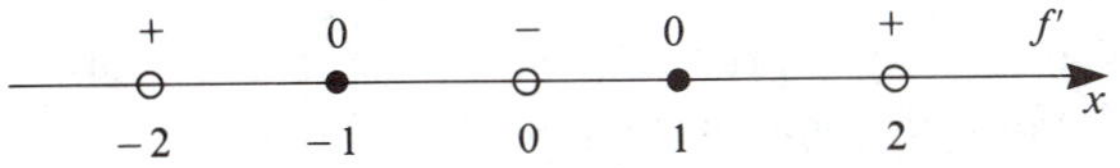

example 9.7   If $y = 3x^2 - 2 \ln x - \sin^{-1} x + \pi^3$, find $\dfrac{dy}{dx}$.

solution 9.7   The function of interest is $y = 3[\ x^2\ ] - 2[\ \ln x\ ] - [\ \sin^{-1} x\ ] + [\ \pi^3\ ]$. Making use of table 9.1, we obtain $\dfrac{dy}{dx} = 3[\ 2x\ ] - 2\left[\ \dfrac{1}{x}\ \right] - \left[\ \dfrac{1}{\sqrt{1 - x^2}}\ \right] + [\ 0\ ]$. Note that π^3 is a *constant* and therefore has derivative 0, although there may be a strong temptation to write $3\pi^2$.

example 9.8   If $y = a_2 x^2 + a_1 x + a_0$, determine where $y' = 0$.

solution 9.8   The derivative is $y' = 2a_2 x + a_1 + 0 = 2a_2 x + a_1$. Then $y' = 0$ when $2a_2 x = -a_1$ or $x = -a_1/(2a_2)$. This is just the formula for the x-coordinate of the vertex of the parabola. Notice that at the vertex, the slope of the tangent line, y', is 0.

In order to prove the Sum Rule we must return to the limit definition of a derivative. You may want to skip over the following proof at your first reading.

proof   **Sum Rule**   Let $F(x) = af(x) + bg(x)$. We must show that $F'(x) = af'(x) + bg'(x)$. We first will simplify:

$$\frac{F(x + h) - F(x)}{h} = \frac{[af(x + h) + bg(x + h)] - [af(x) + bg(x)]}{h}$$

$$= \frac{af(x + h) - af(x) + bg(x + h) - bg(x)}{h}$$

$$= a\left[\frac{f(x + h) - f(x)}{h}\right] + b\left[\frac{g(x + h) - g(x)}{h}\right]$$

We now take the limit as $h \to 0$ of both sides to obtain

$$F'(x) = af'(x) + bg'(x)$$

We can also form new functions such as $x^2/(x-1)$, $x^3 \sin x$, or $e^x/\sin x$, by taking products and quotients of elementary functions.

warning!   The derivatives of the above functions are not $\dfrac{2x}{1}$, $3x^2 \cos x$, or $\dfrac{e^x}{\cos x}$. The new rules (that are not easily guessed!) are needed to compute:

$$D_x f(x) \cdot g(x) \qquad \text{and} \qquad D_x \frac{f(x)}{g(x)}$$

The rule appropriate for obtaining the derivative of a product of functions is summarized in the accompanying box.

Product Rule $D_x f(x) \cdot g(x) = f(x) \cdot D_x g(x) + g(x) \cdot D_x f(x)$

Note that the derivative of $f(x)g(x)$ is *not* $f'(x)g'(x)$. The derivative of a product of functions is the first function times the derivative of the second plus the second times the derivative of the first. The functions "take turns being differentiated." "One watches as the other has its derivative taken."

$$\left(\boxed{1} \cdot \boxed{2}\right)' = \boxed{1} \cdot \boxed{2}' + \boxed{2} \cdot \boxed{1}'$$

Hence, the derivative of $x^3 \sin x$ is $x^3 \cdot [\sin x]' + \sin x \cdot [x^3]' = x^3 \cos x + (\sin x)(3x^2)$.

example 9.9   Find the derivative of:

$$\textbf{(a)} \ (x^2 + x + 1)(x^3 + 3x^2 - 2) \qquad \textbf{(b)} \ x^2 e^x$$

solution 9.9   For **(a)** using the Product Rule, $D_x[(x^2 + x + 1)(x^3 + 3x^2 - 2)]$ is $(x^2 + x + 1)(3x^2 + 6x) + (x^3 + 3x^2 - 2)(2x + 1)$.

For part **(b)**, according to the Product Rule, $D_x x^2 e^x = x^2 \cdot [e^x]' + e^x \cdot [x^2]' = x^2 e^x + 2x e^x$.   ▤

example 9.10   Find $\dfrac{dy}{dx}$ if $y = (x^3 + 1)(x^2 - 3x + 5)(x^4 - 4x + 3)$.

solution 9.10   Here we have the product of *three* functions. Rather than multiplying out the first two factors, we group them and temporarily think of them as a single function:

$$D_x[(x^3 + 1)(x^2 - 3x + 5)](x^4 - 4x + 3)$$
$$= [(x^3 + 1)(x^2 - 3x + 5)](4x^3 - 4) + (x^4 - 4x + 3)[(x^3 + 1)(x^2 - 3x + 5)]'$$

But $[(x^3+1)(x^2-3x+5)]' = (x^3+1)(2x-3)+(x^2-3x+5)(3x^2)$, using the Product Rule again. Hence, the final derivative is:

$$\frac{dy}{dx} = (x^3+1)(x^2-3x+5)(4x^3-4)$$
$$+ (x^4-4x+3)[(x^3+1)(2x-3)+3x^2(x^2-3x+5)]$$

The next rule is used to determine the derivative of the quotient of two functions.

Quotient Rule $\quad D_x \dfrac{f(x)}{g(x)} = \dfrac{g(x)\cdot D_x f(x) - f(x)\cdot D_x g(x)}{[g(x)]^2}$

There are two handy ways of remembering the Quotient Rule:

1. $D_x \dfrac{\text{top}}{\text{bottom}} = \dfrac{\text{bottom}\cdot D_x \text{ top} - \text{top}\cdot D_x \text{ bottom}}{(\text{bottom})^2}$ or

2. $D_x \dfrac{\text{hi}}{\text{ho}} = \dfrac{\text{ho } dhi - \text{hi } dho}{\text{ho ho}}$

Thus, the derivative of $\dfrac{x^2}{x-1}$ is $\dfrac{(x-1)(2x)-x^2(1)}{(x-1)^2} = \dfrac{x^2-2x}{(x-1)^2}$. There is no easy way, however, to see why this rule is valid except to derive the formula using the limit definition of the derivative. We will do this for the Product and Quotient Rules at the end of this section. Let's first practice the Quotient Rule:

example 9.11   Find the derivative of

(a) $\dfrac{x^2-1}{x^2+1}$ ▲ (b) $\tan x = \dfrac{\sin x}{\cos x}$

solution 9.11   For (a), application of the Quotient Rule yields

$$f'(x) = \frac{(x^2+1)(2x)-(x^2-1)(2x)}{(x^2+1)^2} = \frac{2x(x^2+1-x^2+1)}{(x^2+1)^2}$$
$$= \frac{4x}{(x^2+1)^2}$$

For (b), again using the Quotient Rule,

$$D_x \tan x = \frac{\cos x(\sin x)' - \sin x(\cos x)'}{(\cos x)^2}$$
$$= \frac{\cos x(\cos x) - \sin x(-\sin x)}{(\cos x)^2}$$
$$= \frac{1}{(\cos x)^2},$$

using the trigonometric identity $(\sin x)^2 + (\cos x)^2 = 1$. Hence, $D_x \tan x = 1/(\cos x)^2 = \sec^2 x$. 📄

Often one needs to use both the Product and Quotient Rules in the course of working out a single derivative, as illustrated in the next example.

example 9.12   Find $\dfrac{dy}{dx}$ if $y = \dfrac{(x^2 - x - 3)(x^2 - 4)}{4x^3 - 5x + 7}$

solution 9.12   Applying the Quotient Rule first, we have

$$\frac{dy}{dx} = \frac{(4x^3 - 5x + 7)\, D_x \text{ top} - \text{top} \cdot (12x^2 - 5)}{(4x^3 - 5x + 7)^2}$$

But, by the Product Rule, D_x top $= (x^2 - x - 3)(2x) + (x^2 - 4)(2x - 1)$. Hence, the final derivative is:

$$\frac{dy}{dx} = \frac{(4x^3 - 5x + 7)[(x^2 - x - 3)2x + (x^2 - 4)(2x - 1)] - (x^2 - x - 3)(x^2 - 4)(12x^2 - 5)}{(4x^3 - 5x + 7)^2}$$
📄

A special case of the Quotient Rule occurs frequently and is shown in the box.

> **Reciprocal Rule** $D_x \dfrac{1}{f(x)} = -\dfrac{D_x f(x)}{[f(x)]^2}$

example 9.13   Calculate the derivative of

$$\textbf{(a)}\ \frac{3}{(x-1)^2} \qquad \textbf{(b)}\ \frac{1}{x \cdot e^x}$$

solution 9.13   For **(a)** we rewrite the function $3/(x-1)^2$ as $3[1/(x^2 - 2x + 1)]$. Then, by application of the Reciprocal Rule,

$$y' = -3\left[\frac{2x - 2}{(x-1)^4}\right] = -\frac{6}{(x-1)^3}$$

For part **(b)**,

$$y' = -\frac{D_x x \cdot e^x}{(x \cdot e^x)^2} \qquad \text{by the Reciprocal Rule,}$$

$$= -\frac{x \cdot e^x + e^x \cdot 1}{(x \cdot e^x)^2} \qquad \text{by the Product Rule,}$$

$$= \frac{-e^x(x + 1)}{x^2 \cdot e^{2x}}$$
📄

Finally, we prove the Product, Reciprocal, and Quotient Rules.

proof   **the Product Rule**   Let $F(x) = f(x)g(x)$ and fix a at which both $f'(a)$ and $g'(a)$ exist. We will prove that $F'(a) = f(a)g'(a) + g(a)f'(a)$. To compute $\underset{x \to a}{\text{limit}} \dfrac{F(x) - F(a)}{x - a}$, we first simplify:

$$\frac{F(x) - F(a)}{x - a} = \frac{f(x)g(x) - f(a)g(a)}{x - a} = \frac{f(x)g(x) - f(x)g(a) + f(x)g(a) - f(a)g(a)}{x - a}$$

having added and subtracted $f(x)g(a)$ in the numerator. Hence,

$$\frac{F(x) - F(a)}{x - a} = f(x) \cdot \frac{g(x) - g(a)}{x - a} + g(a) \cdot \frac{f(x) - f(a)}{x - a}$$

Taking the limit of both sides as $x \to a$, we have:

$$F'(a) = \left(\underset{x \to a}{\text{limit}}\, f(x) \right) \cdot g'(a) + g(a) \cdot f'(a)$$

Since $f'(a)$ exists by hypothesis, we can use theorem 8.1 to conclude that $f(x)$ is continuous at $x = a$. Hence, $\text{limit}_{x \to a}\, f(x) = f(a)$ and we have $F'(a) = f(a) \cdot g'(a) + g(a) \cdot f'(a)$ as promised.   ∎

proof   **the Reciprocal Rule**   Let $F(x) = 1/f(x)$. As before, fix an a at which $f'(a)$ exists. We will show that $F'(a) = -f'(a)/[f(a)]^2$.

$$\frac{F(x) - F(a)}{x - a} = \frac{1/[f(x)] - 1/[f(a)]}{x - a} = \frac{1}{x - a}\left[\frac{f(a) - f(x)}{f(x)f(a)} \right]$$

$$= \frac{-1}{f(x)f(a)}\left[\frac{f(x) - f(a)}{x - a} \right]$$

As in the last proof, $\underset{x \to a}{\text{limit}}\, f(x) = f(a)$ and $\underset{x \to a}{\text{limit}} \dfrac{f(x) - f(a)}{x - a} = f'(a)$. Taking the limit of both sides as $x \to a$ yields

$$F'(a) = \frac{-1}{f(a) \cdot f(a)} \cdot f'(a) = \frac{-f'(a)}{[f(a)]^2}$$ ∎

proof   **the Quotient Rule**   To prove the Quotient Rule, we will make use of the Product and Reciprocal Rules:

$$D_x \frac{f(x)}{g(x)} = D_x f(x)\left[\frac{1}{g(x)} \right] = f(x)D_x\left[\frac{1}{g(x)} \right] + \frac{1}{g(x)}f'(x)$$

by the Product Rule

$$= f(x) \cdot \frac{-g'(x)}{[g(x)]^2} + \frac{f'(x) \cdot g(x)}{g(x) \cdot g(x)}$$

by the Reciprocal Rule

$$= \frac{f'(x)g(x) - f(x)g'(x)}{[g(x)]^2}$$

and the Quotient Rule has been demonstrated.   ■

exercises for section 9

part A

Find $\dfrac{dy}{dx}$ at $x = 2$ for:

1. $y = 3$ **2.** $y = 4x$ **3.** $y = x^3$ **4.** $y = \sqrt[3]{x}$

▲**5.** $y = \sin x$ **6.** $y = \ln x$ **7.** $y = 1/x$ ▲**8.** $y = \tan^{-1} x$

Compute:

9. $D_x x^6$ ▲**10.** $D_x \cos x$

★**11.** If $y = a^x$, show that y and y' are directly proportional.

▲**12.** If $y = \cos^{-1} x$, find $\dot{y}$ when $x = 0$.

▲**13.** If $y = \sin^{-1} x$, find y' when $x = 1/2$.

Using the Sum Rule, find the derivatives of:

14. $3x^4 - x^2 + 1$ **15.** $x^2 - 8x + 3$

16. $2x^6 - 3x + 4$ **17.** $3x^2 + 1/x$

18. $2\sqrt{x} + 1/x^2 + 1$ **19.** $3x^3 + 3^3$

20. $2x^3 + \pi^3$ **21.** $x^{3/2} + 3x^{1/4} - 2x^{-1/3}$

22. $(x^2 - 1)^2$ [*Hint*: Expand.]   **23.** $x^4 - x^2 - x - 1$

24. $3e^x$ **25.** $3 + \ln x$

26. $e^2 (e = 2.71828 \ldots)$ ▲**27.** $\tan^{-1} x - x^2$

▲**28.** $\sin x - 2 \cos x$ ▲**29.** $3 \sin x - 4 \sin^{-1} x$

30. $2^x - 3^x$ **31.** $(x - 1)^3$ [*Hint*: Expand.]

★**32.** $\sqrt{4x + 1}$ ★**33.** 3^{x+1}

34. $x^3 - 3 \ln x$ **35.** $3e^x - x^4 + 4$

For each of the following functions $f(x)$ determine where (a) $f'(x) = 0$, (b) $f'(x) > 0$, and (c) $f'(x) < 0$ by constructing the sign chart for $f'(x)$.

36. $f(x) = 3x - x^3$

37. $f(x) = x^4 - 4x + 3$

38. $f(x) = 1 - x^2 + x^6$

39. $f(x) = x^2 + 4/x$

40. $f(x) = 4x^3 - 27x^2 + 24x + 4$

Use the Product and Quotient Rules to find $\dfrac{dy}{dx}$ for:

41. $y = (x^2 + 1)(x^3 - 1)$

42. $y = (3x + 2)^2$

43. $y = (x^3 + 3x + 2)(x^2 - 4x + 3)$

44. $y = x^3 e^x$

45. $y = (x + 1/x)(x^2 - x)$

46. $y = x^2 \ln x$

▲47. $y = x \sin x$

▲48. $y = \sin x \cos x$

49. $y = x/(x^2 + 1)$

50. $y = e^x/(e^x + 1)$

51. $y = \dfrac{1}{(x^3 - 1)}$

52. $y = \dfrac{\sqrt{x} + 1}{\sqrt{x} - 1}$

53. $y = 5x - \dfrac{x}{x + 4}$

▲54. $y = \text{ctn } x = \dfrac{\cos x}{\sin x}$

55. $y = 10x - x \ln x$

56. $y = \dfrac{x^2 - 4}{x^2 + 4}$

57. $y = \dfrac{x^3 - x^2 - x}{x^5 - 1}$

58. $y = e^x/(x + 1)$

59. $y = (x^2 + 1)\dfrac{x + 1}{x - 1}$

★60. $y = \dfrac{(x^2 - 1)(x^2 + 3)}{(x - 2)^2}$

▲61. Use the Product Rule to compute $D_x\, xe^x \sin x$.

62. Find y' if $y = (x^3 + 2x + 4)(x^4 - 3x + 2)(x^2 - x + 8)$.

63. Find the slope of the tangent line at $x = 0$ for $y = \dfrac{(x^2 - 4)(x + 4)}{x^2 + x + 1}$

Use the Reciprocal Rule to find:

64. $D_x\, 1/(x - 1)^2$

65. $D_x\, e^{-x}$ [*Hint:* $e^{-x} = 1/e^x$.]

66. Shown here is the graph of $y = -x^3 + 9x^2 - 18x + 6$, for $0 \le x \le 4$.

   (a) Find the slopes of the tangent lines at the indicated points.

   (b) What is the x-coordinate of the point P shown, where the tangent line is horizontal?

   ★(c) Are there any other points on the complete graph where $f'(x) = 0$?

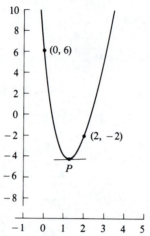

67. Shown here is the graph of $y = 8x^4 - 8x^2 + 1$. Find the coordinates of the peak and two valleys on the graph by determining where the tangent line is horizontal.

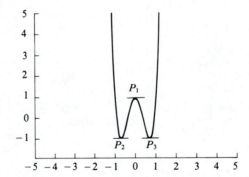

10 composite functions and the chain rule

There is another way of constructing a new function from two given functions f and g that we have not considered before now. As illustrated in figure 10.1, we form an "assembly line," let f act on x first, and then let g perform on the output $y = f(x)$. We will denote the resulting function by $g \circ f$. Thus, in computing $(g \circ f)(x)$, f performs first: $(g \circ f)(x) = g(f(x))$. Of course, we could let g perform first to form a different new function, $f \circ g$.

figure 10.1

As a simple example, consider the two function buttons $\boxed{1/x}$ and $\boxed{\sqrt{\ }}$ on your calculator. If we enter 4 and then hit $\boxed{\sqrt{\ }}$ followed by $\boxed{1/x}$, we are forming $1/2 = g(f(4))$, where $f(x) = \sqrt{x}$ and $g(y) = 1/y$.

If $g(x) = x + 1$ and $f(x) = \sqrt{x}$, then $(g \circ f)(x) = g(\sqrt{x}) = \sqrt{x} + 1$, while $(f \circ g)(x) = f(x + 1) = \sqrt{x + 1}$. Thus, $f \circ g$ is not necessarily the same as $g \circ f$. In addition, the domain of definition of $g \circ f$ is not necessarily the same as that of g. We must make sure that the first output $y = f(x)$ is acceptable as input to the second function g. We illustrate this point with an example.

example 10.1

Write $\sqrt{2x - 1}$ as the composite of two functions $g \circ f$, and determine the domain of definition. Then find $f \circ g$ and determine its domain.

solution 10.1

We can write $\sqrt{2x - 1} = g(f(x))$, where $f(x) = 2x - 1$ and $g(x) = \sqrt{x}$. In order for g to accept $(2x - 1)$ as input, we require $2x - 1 \geq 0$, i.e., $x \geq 1/2$. On the other hand, the composite function $(f \circ g)(x) = f(g(x)) = f(\sqrt{x}) = 2\sqrt{x} - 1$ has domain $x \geq 0$. ▤

It is easy to break up a given function f into the composite of two or more functions $f_1, f_2, \ldots, f_n$. Just list the steps you would take in carrying out the computation required by f. For example, $\sqrt{x^2 - 1} = \sqrt{\boxed{x^2 - 1}}$ is the composite of $x^2 - 1 = f_1(x)$ and $\sqrt{\boxed{}} = f_2(\boxed{})$, i.e., $f_2(x) = \sqrt{x}$. The function $\sqrt{\sin x^3}$ is the composite of $f_1(x) = x^3$, $f_2(x) = \sin x$, and $f_3(x) = \sqrt{x}$.

What, then, is the rule for finding the derivative of composites? Let's first illustrate with an example.

Let $y = (x^2 - 1)^9 = \boxed{x^2 - 1}^9$. We will temporarily think of y as $\boxed{}^9$. If $\boxed{}$ were the variable, the derivative would be $9\boxed{}^8$. This is almost the correct answer. To complete the computation, it is necessary to *multiply* by the derivative of what has been boxed—i.e., multiply by $2x$. Thus,

$$\frac{dy}{dx} = 9\boxed{}^8 \cdot 2x = 9(x^2 - 1)^8 \cdot 2x$$

In general, this technique can be summarized by the formula known as the **Chain Rule**, which is presented in the box.

Chain Rule

$$\frac{dy}{dx} = \frac{dy}{d\boxed{}} \cdot \frac{d\boxed{}}{dx}$$

The Chain Rule may be summarized in the following steps:

Step 1. Box up the inside function.

Step 2. With $\boxed{}$ as the variable, compute the derivative in the usual manner.

Step 3. Multiply the expression in Step 2 by the derivative of $\boxed{}$ with respect to x, that is, treating x as the variable.

example 10.2   Compute the derivative of $(x^3 - 9x + 2)^4$

solution 10.2   Application of the Chain Rule yields

$$D_x\boxed{x^3 - 9x + 2}^4 = 4\boxed{x^3 - 9x + 2}^3 (3x^2 - 9)$$
$$= 4(x^3 - 9x + 2)^3 (3x^2 - 9)$$

example 10.3   Compute the derivative of $\sqrt{x^3 - 9x + 2}$

solution 10.3   We may write the function as $\boxed{x^3 - 9x + 2}^{1/2}$. By treating $\boxed{}$ as the variable, we obtain the derivative:

$$\frac{1}{2}\boxed{}^{-1/2} (3x^2 - 9) = \frac{3x^2 - 9}{2}\frac{1}{\sqrt{x^3 - 9x + 2}}$$

▲example 10.4   Compute the derivative of $\sin(x^3 - 9x + 2)$

solution 10.4   The derivative is:

$$D_x \sin\boxed{x^3 - 9x + 2} = \cos(\boxed{})(3x^2 - 9)$$
$$= (3x^2 - 9)\cos(x^3 - 9x + 2)$$

example 10.5   Compute the derivative of $\ln (x^3 - 9x + 2)$

solution 10.5   Recall that the derivative of $\ln x$ is $1/x$. Hence, by the Chain Rule,

$$D_x \ln(\boxed{x^3 - 9x + 2}) = \frac{1}{\boxed{}} (3x^2 - 9)$$

$$= \frac{3x^2 - 9}{x^3 - 9x + 2}$$

example 10.6   Find $\dfrac{dy}{dx}$ if $y = \left(\dfrac{x-1}{x+1}\right)^3$.

solution 10.6   Here, we must use the Quotient Rule in conjunction with the Chain Rule:

$$D_x \boxed{\frac{x-1}{x+1}}^3 = 3 \boxed{\frac{x-1}{x+1}}^2 \cdot \frac{(x+1) \cdot 1 - (x-1) \cdot 1}{(x+1)^2}$$

$$= \frac{6(x-1)^2}{(x+1)^4}$$

example 10.7   If $y = (x-4)^4 (2x-3)^5$, determine where $y' = 0$.

solution 10.7   We begin by using the Product Rule:

$$y' = (x-4)^4 D_x(2x-3)^5 + (2x-3)^5 D_x(x-4)^4$$

But by the Chain Rule,

$$D_x \boxed{2x-3}^5 = 5 \boxed{2x-3}^4 \cdot 2 \quad \text{and} \quad D_x \boxed{x-4}^4 = 4 \boxed{x-4}^3 \cdot 1$$

Hence, the final derivative is

$$y' = (x-4)^4 (10)(2x-3)^4 + (2x-3)^5 4(x-4)^3$$
$$= (x-4)^3 (2x-3)^4 (10x - 40 + 8x - 12)$$
$$= (x-4)^3 (2x-3)^4 (18x - 52)$$

Setting $y' = 0$, we obtain the solutions $x = 4$, $3/2$, and $26/9$.

The Chain Rule can also be used to handle functions that are the composite of three or more elementary functions, as shown in the following example.

example 10.8   Find $f'(x)$ if $f(x) = (x + \sqrt{x^2 - 1})^3$.

solution 10.8   First write $f(x)$ as $\boxed{x + \sqrt{x^2 - 1}}^3$. Using the Chain Rule, we obtain

$$f'(x) = 3 \cdot \boxed{x + \sqrt{x^2 - 1}}^2 \cdot D_x [x + \sqrt{x^2 - 1}]$$

But

$$D_x\left(x + \boxed{x^2-1}^{1/2}\right) = 1 + \frac{1}{2}\,\boxed{x^2-1}^{-1/2} \cdot 2x = 1 + \frac{x}{\sqrt{x^2-1}}$$

Hence, the final derivative is

$$f'(x) = 3(x + \sqrt{x^2-1})^2\left(1 + \frac{x}{\sqrt{x^2-1}}\right)$$

It is a good idea to actually draw in the boxes until you feel comfortable with the Chain Rule. After a while your "mind's eye" will put in the boxes and you'll be able to quickly write the answer.

If we substitute the more conventional "u" in place of [⬛], then we may write the Chain Rule as shown in the box.

---

the Chain Rule

$$\frac{dy}{dx} = \frac{dy}{du} \cdot \frac{du}{dx}$$

The Chain Rule may be written in the usual function notation, as indicated in the box.

---

the Chain Rule

$$D_x g(\boxed{f(x)}) = g'(\boxed{f(x)})\, f'(x)$$

There are three special cases of the Chain Rule that occur frequently enough to be singled out.

1. $D_x[f(x)]^k = k[f(x)]^{k-1}\, f'(x)$

2. $D_x e^{f(x)} = e^{f(x)}\, f'(x)$

3. $D_x \ln[f(x)] = \dfrac{f'(x)}{f(x)}$

The special cases will be illustrated with three examples.

example 10.9   Find $\dfrac{dy}{dx}$ if $y = e^{x^2+x+1}$.

solution 10.9   According to the second special case of the Chain Rule, $\dfrac{dy}{dx} = e^{x^2+x+1}(2x+1)$.

example 10.10   Find $\frac{dy}{dx}$ if $y = u^9$ and $u = x^2 + x + 1$.

solution 10.10   Writing y as a function of x, we have $y = (x^2 + x + 1)^9$. Hence, $\frac{dy}{dx} = 9(x^2 + x + 1)^8 (2x + 1)$, using the first special case of the Chain Rule.

An alternative method of solving example 10.10 is by applying the Chain Rule to obtain $\frac{dy}{dx} = \frac{dy}{du} \cdot \frac{du}{dx} = 9u^8(2x + 1) = 9(x^2 + x + 1)^8 (2x + 1)$. 📃

example 10.11   Find $\frac{dy}{dx}$ if $y = x \ln (x^2 + 1)$.

solution 10.11   We apply the Product Rule to obtain

$$\frac{dy}{dx} = x D_x \ln (x^2 + 1) + \ln (x^2 + 1) \cdot 1$$

But, $D_x \ln (x^2 + 1) = 2x/(1 + x^2)$, according to the third special case of the Chain Rule. Thus, the derivative is

$$\frac{dy}{dx} = \frac{2x^2}{1 + x^2} + \ln (x^2 + 1)$$ 📃

Why is the Chain Rule valid? The following demonstration, which falls short of being a proof, gives some insight as to its validity.

Let $F(x) = f(g(x))$, and fix a at which $g'(a)$ and $f'(g(a))$ exist. Then,

$$\frac{F(x) - F(a)}{x - a} = \frac{f(g(x)) - f(g(a))}{x - a} = \frac{f(g(x)) - f(g(a))}{g(x) - g(a)} \cdot \frac{g(x) - g(a)}{x - a}$$

If we let $y = g(x)$ and $b = g(a)$, then the quotient can be rewritten as

$$\frac{F(x) - F(a)}{x - a} = \frac{f(y) - f(b)}{y - b} \cdot \frac{g(x) - g(a)}{x - a}$$

Now as x approaches a, $y = g(x)$ approaches $g(a) = b$, since $g(x)$ is continuous at a. Hence, taking the limit of both sides, we obtain

$$F'(a) = f'(b) g'(a) = f'(g(a)) g'(a)$$

which is the statement of the Chain Rule. [The above argument is a valid proof, provided $g(x)$ is not equal to $g(a)$ for x near a. Although this cannot always be guaranteed, it is true in most cases.]

exercises for section 10

part A

Write each of the following functions as $g(f(x))$, where f and g are elementary functions. Then compute $f(g(x))$.

	Function	$f(x)$	$g(x)$	$f(g(x))$
1.	$(2x + 3)^5$			
2.	$\sqrt{x - 1}$			
3.	$\left(\dfrac{x + 2}{x - 2}\right)^3$			
▲4.	$\sin(x^2 + 1)$			
5.	$\dfrac{1}{(3 - 4x)^4}$			

What functions are formed if the following calculator buttons are hit in succession? [Stroke sequences are in algebraic logic.]

▲**6.**  $\boxed{x^2}\ \boxed{1/x}\ \boxed{\sin}$

7. $\boxed{x^2}\ \boxed{x^2}\ \boxed{x^2}\ \boxed{+}\ \boxed{3}\ \boxed{=}$

8. $\boxed{x^2}\ \boxed{x^y}\ \boxed{8}\ \boxed{=}$

9. $\boxed{-}\ \boxed{3}\ \boxed{=}\ \boxed{1/x}$

10. $\boxed{+}\ \boxed{1}\ \boxed{=}\ \boxed{\sqrt{\ }}$

Use the Chain Rule to find the derivatives in **(a)** and **(b)**. Use these results to compute the derivative in **(c)**

	(a)	(b)	(c)
11.	$(x - 1)^5$	$(2x + 1)^3$	$(x - 1)^5 (2x + 1)^3$
12.	$(x^4 + 1)^3$	$(x - 8)^4$	$(x - 8)^4/(x^4 + 1)^3$
13.	$\sqrt{2x + 3}$	$(4 - x)^3$	$(4 - x)^3\sqrt{2x + 3}$
14.	$(2x - 1)^7$	$(3 - 2x)^3$	$(2x - 1)^7/(3 - 2x)^3$
15.	$\sqrt{x + 1}$	$(2 - 5x)^2$	$(2 - 5x)^2/\sqrt{x + 1}$
16.	$1/(x - 2)^3$	e^{2x}	$e^{2x}/(x - 2)^3$
▲**17.**	e^{-2x}	$\sin 2x$	$e^{-2x} \sin 2x$
18.	$\ln(x + 2)$	$(x + 2)^3$	$\ln(x + 2)/(x + 2)^3$
▲**19.**	$\cos 4x$	e^{4x}	$e^{4x} \cos 4x$
20.	$\ln(x^2 + x)$	$(x^2 + x)^2$	$(x^2 + x)^2 \ln(x^2 + x)$

Use the Chain Rule, in conjunction with the Sum, Product, and Quotient Rules, to compute y' for:

21. $y = (x + 2)^3/(x - 2)^3$

22. $y = 4(x + 2)^3 - 3(2x - 5)^4$

23. $y = (2x - 3)^{5/4}$

24. $y = \sqrt{x}/(x + 1)$

25. $y = x^4\sqrt{x+1}$

26. $y = \sqrt{x^3 - x + 3}$

27. $y = (2x^2 + 1)^2 (x + 1)^5$

28. $y = 3(2x^2 + 1)^2 - 4(x + 1)^5$

29. $y = (x^3 + 1)^5$

30. $y = [(x - 1)^2 (3x + 2)^3]^4$

31. $y = (x - 1)^3/(3x + 1)^2$

32. $y = x^7/(1 - 2x)^7$

33. $y = 1/(x^2 + 1)^3$

34. $y = \sqrt{(x - 1)^3 + 6}$

35. $y = [x - (x + 1)^5]^2$

36. $y = \sqrt{\sqrt{x} + x}$

37. $y = e^{-3x}$

38. $y = x^2 e^{-3x}$

39. $y = \ln(x^2 + 2)$

40. $y = x \ln(x^2 + 2)$

41. $y = (\ln x)^{10}$

42. $y = e^{x^2 + 1}$

43. $y = e^{-4x}/(x + 2)^5$

▲44. $y = \sin(x^3 + 1)$

▲45. $y = [\sin(x + 1)]^3$

46. $y = \ln(x^9 + 4x^5 + 7)^3$

47. $y = (e^{3x} + 1)^3$

48. $y = x^3 e^{-2x} + e^{5x}$

★49. $y = \ln[\ln(x)]$

★50. $y = e^{e^x}$

51. If $y = (x - 2)^3 (2x + 1)^2$, determine where $y' = 0$.

★52. If $y = x^2 e^{-2x}$, determine where $y' = 0$. [*Hint*: Use the fact that e^{-2x} is never 0.]

★53. If $y = \sqrt{(30 - x)^2 + 100}$, determine where $y' = 0$.

★54. If $y = x^n e^{-\alpha x}$, show that $y' = 0$ when $x = 0$ or n/α.

55. If $f(x) = (x^2 - 1)^5$, find the slope of the tangent line when $x = 0$.

★56. If $y = \sqrt{x + \sqrt{x + \sqrt{x}}}$, find $y'(1)$.

part B

An airplane is executing the maneuver $y = f(x)$ as shown here. The plane flies from left to right and fires its ammunition in the tangent direction. (Old World War 1 fighter planes shot their bullets through the propeller.)

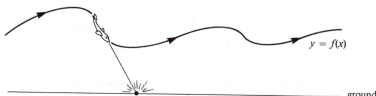

57. If $y = x^2 + 1$, where will the ammunition hit if the guns are fired at $(-1, 2)$? at $(0, 1)$?

★58. If $y = x^2 + 1$, where on the curve should the guns be fired to hit a target placed at $(3, 0)$?

59. If $y = 1/(x + 1)^3$ for $x > -1$, where will the ammunition hit if the guns are fired at $(0, 1)$?

★60. If $y = 1/(x + 1)^3$, where on the curve should the guns be fired to hit a target placed at $(10, 0)$?

 **the derivative as a rate of change**

We mentioned in the prologue that the derivative $\frac{dy}{dx}$ measures *how quickly* a given quantity y is changing with respect to a second quantity x. The derivative is then an "instantaneous rate of change." Let's examine why. Suppose x is changed from x_1 to x_2 as shown in figure 11.1. This change in x,

figure 11.1

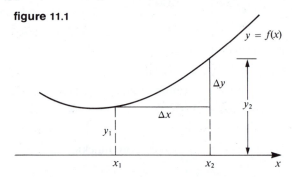

$x_2 - x_1$, is denoted by Δx. The corresponding change in y, $y_2 - y_1$, is denoted by Δy. Note that

$$\lim_{\Delta x \to 0} \frac{\Delta y}{\Delta x} = \lim_{\Delta x \to 0} \frac{f(x_1 + \Delta x) - f(x_1)}{\Delta x} = \frac{dy}{dx}$$

We will use this fact in the examples that follow.

example 11.1

the 100-meter dash   A participant in a track meet is about to run the 100-meter dash. Let t = time (in seconds) and let y = distance run (in meters) after t seconds. If he runs the race in 10.5 seconds, then $\Delta t = 10.5$, $\Delta y = 100$, and hence, his average velocity over the whole race is $\Delta y / \Delta t = 100/10.5 \approx$ 9.44 m/sec. However, this average velocity hardly tells the whole stroy of the race. Some runners are quick starters, while others are especially strong finishers. To measure the runner's quickness at the start, we might use high speed film and examine the frames. In figure 11.2, suppose that the film speed is 24 frames per second. Then, in this case, $\Delta y = .5$, $\Delta t = 1/24$ and hence $\Delta y / \Delta t = 12$ m/sec. Now, in figure 11.3, suppose that the film speed is 48 frames per second. Then $\Delta y = .2$, $\Delta t = 1/48$ and hence $\Delta y / \Delta t = 9.6$ m/sec. By choosing Δt smaller and smaller, we obtain (at least in theory) progressively better

figure 11.2

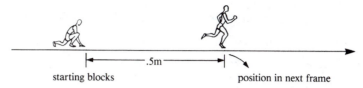

starting blocks   position in next frame

figure 11.3

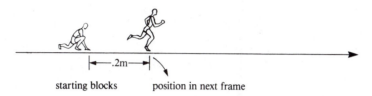

starting blocks   position in next frame

estimates of how quickly the runner starts. The perfect measure is

$$\lim_{\Delta t \to 0} \frac{\Delta y}{\Delta t} = \frac{dy}{dt}$$

the derivative of y with respect to t at time $t = 0$, in units of m/sec. Similarly, $\frac{dy}{dt}$ at time $t = 10.5$ measures how quickly he finishes—i.e., his *instantaneous velocity* as he crosses the finish line.

If we suppose that $y = .05t^3 + 5t$, then $y(0) = 0$, $y(10) = 100$, and $\frac{dy}{dt} = .15t^2 + 5$. Hence, the runner's velocity out of the blocks would be 5 m/sec, while his velocity at the finish line would be $(.15)(100) + 5 = 20$ m/sec (which would be very fast indeed).

If we let $v = \frac{dy}{dt}$, then the derivative $\frac{dv}{dt}$ measures how quickly the velocity changes and is called the *acceleration* at time t. The units of $a = \frac{dv}{dt}$ are (m/sec) per sec or m/(sec)^2.

example 11.2

The formula $y = -16t^2 + v_0 t + s_0$ gives the distance above the ground of an object moving under the influence of gravity alone. Note that $y(0) = s_0$. Hence, s_0 is the initial distance off the ground. The instantaneous velocity of the object at time t is given by $v = \frac{dy}{dt} = -32t + v_0$. Hence, $v_0 = v(0)$ is the initial velocity in the vertical direction, and $a = \frac{dv}{dt} = -32$ is the acceleration due to gravity.

A high diver jumps off a 60-foot tower into a pool below. In springing from the tower, she rises up an additional 4 feet off the platform.

(a) How long does it take to reach the pool (after reaching the peak of her jump)?

(b) What is her velocity upon entering the water?

figure 11.4

solution 11.2

We will let time $t = 0$ correspond to the time she is at the peak of her jump. Hence, $s_0 = 64$ and $v_0 = 0$ since she momentarily stops at the peak. It follows that $y = -16t^2 + 64$.

(a) When she hits the pool, $y = 0$. Hence, $0 = -16t^2 + 64$, which implies that $t^2 = 4$, or $t = 2$.

(b) The instantaneous velocity is $v = \dfrac{dy}{dt} = -32t$. Hence, $v(2) = -32(2) = -64$ feet/sec is the diver's velocity upon entering the pool. The minus sign indicates that the direction is downward.

example 11.3

how quickly does an organism grow?   Let W denote the weight in pounds of an organism, and let t denote its age in years. If, for example, an 18-year-old girl weighed 8 pounds at birth and now weighs 152 pounds, then $\Delta W = 152 - 8 = 144$ and $\Delta t = 18$. Hence, her average rate of growth over the first 18 years is $\Delta W / \Delta t = 144/18 = 8$ pounds/year. This figure does not convey much information. More insight may be gained by computing average rates of growth at specific times, as shown in table 11.1. To determine precisely how fast the child was growing at time $t = 1$ year, we would compute (if possible)

$$\frac{\Delta W}{\Delta t} = \frac{W(1 + \Delta t) - W(1)}{\Delta t}$$

for smaller and smaller values of Δt. Thus, the limit $\dfrac{dW}{dt}$ is the instantaneous rate of growth in weight. Likewise, if h is height in inches, then $\dfrac{dh}{dt}$ is the instantaneous rate of growth in height in units of inches/year.

table 11.1

Age	Medical Record Height (inches)	Weight (pounds)
0	20	8
6 weeks	22	13
.	.	.
.	.	.
.	.	.
1 year	31	22
13 months	31.5	26
.	.	.
.	.	.
16 years	67	124
16.5 years	67	134
.	.	.
.	.	.
18 years	67	152

$\left.\begin{array}{c} 22 \\ 26 \end{array}\right\} \dfrac{\Delta W}{\Delta t} = \dfrac{26-22}{13-12} = 4$ pounds/month

$\left.\begin{array}{c} 124 \\ 134 \end{array}\right\} \dfrac{\Delta W}{\Delta t} = \dfrac{134-124}{16.5-16} = 20$ pounds/year

example 11.4   If the expression $W = 152(t+1)/(t+20)$ describes the weight in pounds of an individual after t years, find the rate of growth when $t = 1$ and when $t = 16$.

solution 11.4   Using the Quotient Rule, we obtain the instantaneous rate of growth

$$\frac{dW}{dt} = 152\left[\frac{(t+20)\cdot 1 - (t+1)\cdot 1}{(t+20)^2}\right] = \frac{152\cdot 19}{(t+20)^2}$$

Hence, when $t = 1$, $\dfrac{dW}{dt} = \dfrac{(152)(19)}{(21)^2} = 6.55$ pounds/year. When $t = 16$, $\dfrac{dW}{dt} = \dfrac{(152)(19)}{(36)^2} = 2.23$ pounds/year.

example 11.5   Shown in figure 11.5 are the growth curves for boys between the ages of two and eighteen in the United States. Although we do not have explicit formulas for the graphs, we can still estimate $\dfrac{dW}{dt}$ and $\dfrac{dh}{dt}$. These growth rates are the slopes of the respective tangent lines, and we can estimate the slopes by selecting *two nearby* points on the graphs.

To estimate $\dfrac{dW}{dt}$ for a typical thirteen-year-old boy, we find the slope of the line segment AB. Now, $A = (13, 99)$ and $B = (14, 112)$. Hence, $\dfrac{dW}{dt} \approx 13$ pounds/year.

figure 11.5 (Adapted from: Hamill PVV, Drizd TA, Johnson CL, Reed RB, Roche AF, Moore WM: Physical growth: National Center for Health Statistics percentiles. AM J CLIN NUTR 32:607, 629, 1979. Data from the National Center for Health Statistics (NCHS) Hyattsville, Maryland. Reproduced by permission of Ross Laboratories, Columbus, Ohio, 1980.)

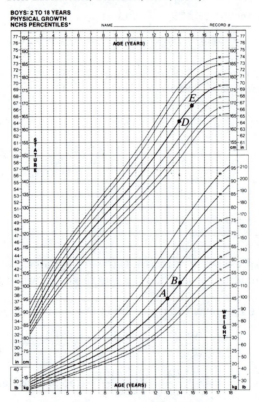

To estimate $\dfrac{dh}{dt}$ for a typical fourteen-year-old boy, we find the slope of the line segment DE. Now, $D = (14, 64.2)$ and $E = (15, 66.5)$. Hence, $\dfrac{dh}{dt} \approx$ 2.3 inches/year.

example 11.6   Let R denote the reaction of the body to a stimulus of strength x. The following are three of the many possibilities for R and x.

1. x could be the amount of drug (in cubic centimeters, for example) placed in the body, and R could be the corresponding change in pulse rate.

2. x could be the temperature of a hot object in contact with your finger, and R could be the corresponding frequency (or firing rate) of the sensory nerve impulses.

3. x could be the brightness of a light source, and R could be the resulting area of the pupil.

Suppose that the stimulus is increased from level x to level $x + \Delta x$, as shown in figure 11.6. The corresponding change in reaction is $\Delta R = R(x + \Delta x) - R(x)$.

figure 11.6(a) brightness level x.

iris

pupil

figure 11.6(b) brightness level $(x + \Delta x)$.

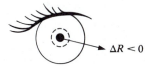

$\Delta R < 0$

The average change in reaction per change in stimulus is then $\Delta R / \Delta x$. We define the *sensitivity at stimulus level x* to be

$$\frac{dR}{dx} = \lim_{\Delta x \to 0} \frac{\Delta R}{\Delta x}$$

The empirical formula

$$R = \frac{40 + 23.7x^{.4}}{1 + 3.95x^{.4}}$$

describes the relationship between pupil area R in mm^2 and the brightness x of a light source. As x increases, R decreases from $R(0) = 40$ to $6 = \lim_{x \to +\infty} R(x)$. The sensitivity is

$$\frac{dR}{dx} = \frac{(1 + 3.95x^{.4})(9.48x^{-.6}) - (40 + 23.7x^{.4})(1.58x^{-.6})}{(1 + 3.95x^{.4})^2}$$

which simplifies to $\dfrac{-53.72}{x^{.6}} \cdot \dfrac{1}{(1 + 3.95x^{.4})^2}$ mm^2 per unit of brightness. As would

be expected, the sensitivity $S = \dfrac{dR}{dx}$ is much larger at lower levels of brightness.

In general, $\dfrac{dy}{dx}$ may be thought of as the **instantaneous rate of change** of y with respect to x. If we use $[x]$ to denote units in which a variable x is measured, then

$$\left[\frac{dy}{dx}\right] = [y] \text{ per } [x] = \frac{[y]}{[x]}$$

Listed in table 11.2 are some other interpretations of the derivative; typical units are given. Other interpretations will be presented in the exercises.

table 11.2

x	y	$\dfrac{dy}{dx}$
time (weeks)	volume (ft^3)	flow rate (ft^3/week)
time (hours)	velocity (mph)	acceleration (mph/h or miles/hr^2)
time (hours)	population size (number of bacteria)	population growth rate (number/hour)
time (years)	amount of radioactive material (grams)	rate of disintegration (grams/year)
distance from ocean's surface (feet)	amount of light reaching depth x (foot-candles)	rate of disintegration (foot-candles/ft)

exercises for section 11

[*Note to the student*: Unless otherwise noted, all the functions appearing in the problems are *purely hypothetical*. The sole purpose of the exercises is to give you practice with derivatives as rates of change. *In all the problems, specify the units for the derivative.*]

part A

A particle is traveling along the line shown here:

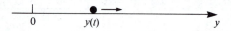

Its distance from 0 is given by $y = f(t)$, where t is in seconds and y is in inches. Find the formulas for the instantaneous velocity and acceleration if:

1. $f(t) = 4t + 5$

2. $f(t) = (t^2 - 1)^2$

3. $f(t) = \dfrac{t}{t+4}$

4. $f(t) = \sqrt{t+1}$

5. $f(t) = 5e^{-t}$

▲6. $f(t) = 3 + \cos 2t$

Two runners are set to run the 100-meter dash. The distances they will have run after t seconds are given by:

$$y_1(t) = \frac{1}{5}t^2 + 8t \qquad y_2(t) = \frac{1{,}100t}{t+100}, \qquad \text{for } t \ge 0$$

7. By setting $y_1(t) = 100$ and $y_2(t) = 100$, show that the race ends in a dead heat.

8. How fast does each runner start? Who is the faster starter?

9. How fast does each runner finish? Who is the faster finisher?

★10. In order to propel an object 1 mile (5,280 feet) straight up, what must the initial velocity be? [*Hint*: We ignore air resistance and use the formula in example 11.2. At the peak of the trajectory, $v = 0$ and $s = 5,280$.]

The great Zambini is shot out of a cannon onto a bed of hay, as depicted in the figure. His distance off the ground at time t is given by $y = -16t^2 + 15.5t + 10$, and he lands in the hay 1.3 seconds after the cannon is fired.

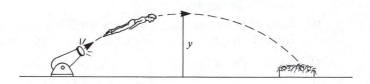

11. What was Zambini's initial *vertical* velocity?

12. What is the downward velocity at the time of impact?

★13. At the peak of his jump, the vertical velocity is 0. Find the maximum height attained.

▲ A weight is bobbing up and down on a spring, as illustrated in the figure. The distance (in inches) from the bottom of the weight to the support is given by

$$y = 24 + 12 \cos \pi t \qquad (t \geq 0)$$

where t is measured in seconds.

14. What is the initial velocity?

15. What is the velocity at time $t = 1$?

★16. When is the velocity equal to 0? [*Hint*: Recall that $\sin x = 0$ for $x = 0, \pm\pi, \pm 2\pi, \ldots$.]

17. At present, the world high dive record is 130.5 feet. Assuming the diver's initial vertical velocity is 0, use the formula in example 11.2 to estimate the speed in mph at the time of entry into the water.

★18. The record distance for a human cannonball is 175 feet. This feat was performed regularly by Emanuel Zacchini in 1940 for Ringling Brothers and Barnum and Bailey Circus. The initial vertical velocity imparted by the cannon was 56 feet/second and the flight lasted 3.75 seconds.

(a) How high off the ground was the cannon?

(b) What was the maximum height attained by Zacchini in flight?

If we measure weight W in pounds and time t in years, find the rates of growth if

19. $W = 25t^2/(t^2 + 1)$ at times 1, 5, and 10 years.

20. $W = (10 - 9/t)^3$ at times 1, 2, and 5 years.

★21. $W = 100(1 - .5e^{-t})$ for $t = 1$ and 10.

▲22. **the perennial dieter:** $W = 200 + 20 \sin 2\pi t$ for $t = 0, \frac{1}{4}, \frac{3}{4}$, and 1.

Use the height-weight curves for U.S. boys in figure 11.5 to answer the following:

23. For a boy in the 95th percentile of height, estimate the rate of growth at age 2, age 13, and age 18.

24. Rates of growth in weight are very small in young children aged two to six. Estimate the rate of growth in weight for an average three-year-old boy.

The actual formula for the growth in length of haddock is given by

$$L = 53(1 - .808e^{-.2t})$$

where L is in centimeters and t is in years.

★25. Compute the rate of growth in length at times 0, 1, 2, and 5 years.

★26. Show that $\dfrac{dL}{dt} = .2(53 - L)$. Conclude that the rate of growth is directly proportional to the length yet to be achieved.

★27. Use the result in exercise 26 to compute $\dfrac{dL}{dt}$ where $L = 30$ cm.

★28. The *blob* is spherical in shape and grows at the rate of 100 pounds/day. If each cubic foot of blob weighs 50 pounds and the blob is presently 10 feet in diameter, how fast is the diameter growing after 7 days, 14 days, and 30 days? [*Hint:* $V = (4/3)\pi r^3$ and $W = 50V$. Also, $W = 25{,}000\pi/3 + 100t$ where $V =$ volume, $r =$ radius, $W =$ weight, and $t =$ time (in days).]

Suppose that, if x cubic centimeters of drug are injected into the body, the corresponding change in pulse is given by

$$R(x) = x^2(10 - x)$$

29. Compute the sensitivity S when $x = 1$ and $x = 5$.

30. Show that $S(x)$ is largest when $x = 10/3$. [*Hint:* The graph of S is a parabola.]

31. For the formula relating pupil area R to brightness x in example 11.6, compare the sensitivities at levels $x = 1$ and $x = 2$.

32. If x is the amount of weight lifted and R is our own subjective estimate of the weight, Stevens* has shown experimentally that

$$R(x) = kx^{3/2}$$

---

* Stevens, S.S. *Science* (170), pp. 1043–1050, 1970.

Show that when the actual weight is multiplied by 4, our subjective estimates of weight are increased by a factor of 8, while our sensitivity to small weight changes is doubled.

33. If a subject is brought from a dark room into light, the relationship between the subjective estimate of brightness R and the actual brightness x is given by

$$R(x) = kx^{1/3}$$

Show that the sensitivity S is inversely proportional to R^2.

part B

miscellaneous interpretations of the derivative

The population P (in thousands) after t years in a community is given below. compute the rates of population growth at times $t = 0$, 1, 5, and 10 years.

34. $P = 1 + 3t$ **35.** $P = 3\sqrt{t+1}$

36. $P = 2/(t^2 + 1)$ **37.** $P = 5t/(1 + 2t)$

38. The total volume V of water (in cubic feet) in a small reservoir during spring runoff is given by

$$V = 3{,}000(t + 2)^2 \qquad (0 \le t \le 3)$$

where t is in months and $t = 0$ corresponds to March 21. Compute the rates of flow at $t = 0$, 1, 2, and 3.

★**39.** The total amount R of rainfall (in inches) in a city during the rainy season is given by

$$R = t^2(3 - t) \qquad \text{for } 0 \le t \le 2 \text{ (months)}$$

When did it rain the hardest?

40. The cost (in dollars) of producing x tons of sugar is given by $C = 1{,}600x^2 + .01x + 5{,}000$. The derivative $C'(x)$ is called the *marginal cost at level of production x.*

(a) Compute C' when $x = 1$, 10, and 25. Specify the units for C'.

(b) Explain what each of these marginal costs represents.

▲**41.** The temperature T during a hot summer day is given by

$$T = 75 + 20 \sin\left(\frac{\pi}{12}t\right)$$

where $t = 0$ corresponds to 6:00 A.M. Compute $T'(t)$ at 12 noon, 6:00 P.M., and 12 midnight. What is the interpretation of T'?

★**42.** Let I be the amount of light (in candles/m²) reaching an ocean depth of x meters. Compute the rate of disintegration at $x = 1$, 2, and 5 meters if

$$I = 10\left(\frac{1}{2}\right)^x$$

[*Hint*: See section 9 for $D_x\, a^x$.]

★**43.** Circle City is a city of 20,000 people spread out over a circular area with 100 people per square mile. The population is expected to grow at a steady rate of 10,000 per year, and city fathers want the population density to remain the same. How quickly is the distance from the outskirts of town to the center of the city growing after two years? [*Hint*: $A = \pi r^2$ and $A = N/100 = 200 + 100t$ where A is area (in square miles), N is number in the population, and t is time (in years).]

the first derivative test and curve sketching

To graph a function accurately, one must not only plot points but also know how to connect them. For example, in sketching the graph of $y = mx + b$, we need to plot only two points, since the general shape of the graph is a straight line. In this section, we will show how analyzing where $f'(x) = 0$, $f'(x) > 0$, and $f'(x) < 0$ (i.e., finding the *sign chart* for f') determines the *general shape* of the graph. *We will assume throughout that $f(x)$ has a continuous derivative in its domain.*

To begin, let's make four basic observations about the slopes of tangent lines.

observation 1   If f is increasing on (a, b), then $f'(x) \geq 0$, as shown in figure 12.1.

figure 12.1

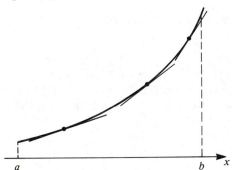

observation 2   If f is decreasing on (a, b), then $f'(x) \leq 0$, as illustrated in figure 12.2. It follows from observations 1 and 2 that if $f'(x) > 0$ on (a, b), then f must be an increasing function on (a, b). Likewise, if $f'(x) < 0$ on (a, b), f is a decreasing function.

observation 3   If a hill or **relative maximum** occurs at $x = c$, as in figure 12.3, then $f'(c) = 0$. Furthermore, $f'(x) > 0$ just to the left of c and $f'(x) < 0$ just to the right of c. Thus, in the near vicinity of c, the sign chart for f' appears as shown in figure 12.4.

figure 12.2

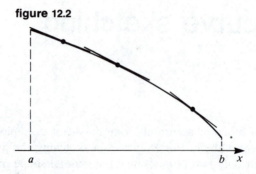

figure 12.3

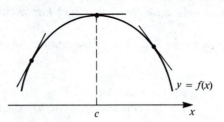

$y = f(x)$

c x

figure 12.4

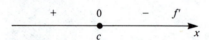

$+$ 0 $-$ f'

c x

observation 4   If a valley or **relative minimum** occurs at $x = c$, as in Figure 12.5, then $f'(c) = 0$. Furthermore, $f'(x) < 0$ just to the left of c and $f'(x) > 0$ just to the right of c. Thus, in the near vicinity of c, the sign chart for f' appears as shown in figure 12.6.

figure 12.5

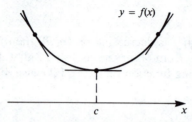

$y = f(x)$

c x

figure 12.6

$-$ 0 $+$ f'

c x

Hence, if the complete sign chart for f' were as shown in figure 12.7, then $f(x)$ would increase in $(-\infty, 0)$, then decrease in $(0, 2)$, and finally increase in $(2, \infty)$. This is a consequence of observations 1 and 2. Thus, f would have the *general shape* shown in figure 12.8.

figure 12.7

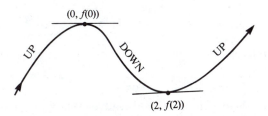

figure 12.8

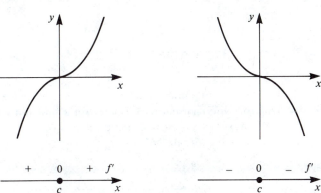

Thus, the first step in sketching a graph will be to solve the equation $f'(x) = 0$. Solutions to this equation are called **critical points**. Observations 3 and 4 indicate that *relative maxima and relative minima must be critical points*. There are, however, two other types of critical points that correspond to two other possibilities that can occur in the sign chart for f'. The functions $y = x^3$ and $y = -x^3$, graphed in figure 12.9, are two such examples. Note that in each case, the tangent line at $x = 0$ has slope 0.

figure 12.9(a)   **figure 12.9(b)**

To determine the nature of a critical point, we construct the sign chart for f' and examine the sign of the derivative to the left and right of c. The procedure, known as the **First Derivative Test**, is summarized in the box on page 122.

First Derivative Test

Suppose that $f'(x)$ exists in the vicinity of $x = c$ and that $f'(c) = 0$. Then:

1. 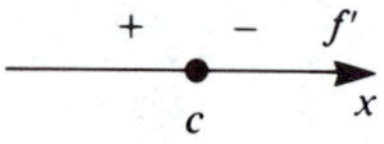 $\xrightarrow{\quad + \quad\bullet\quad - \quad} f' \atop x$ Implies a relative maximum occurs at $x = c$.

2. 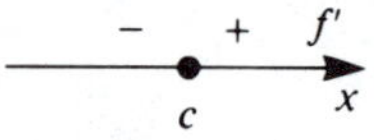 $\xrightarrow{\quad - \quad\bullet\quad + \quad} f' \atop x$ Implies a relative minimum occurs at $x = c$.

3a. 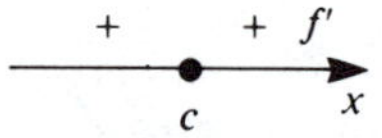 $\xrightarrow{\quad + \quad\bullet\quad + \quad} f' \atop x$ or 3b. 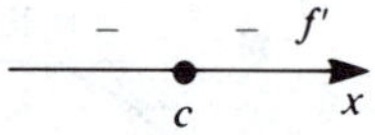 $\xrightarrow{\quad - \quad\bullet\quad - \quad} f' \atop x$ implies neither a

relative maximum nor a relative minimum occurs at $x = c$.*

We now use the First Derivative Test to sketch the graphs of selected polynomial and rational functions:

example 12.1   Classify the critical points of the function $f(x) = x^3 - 3x + 1$ and sketch the graph.

solution 12.1   This curve sketching problem was introduced in example 1.9 and we constructed the sign chart for $f'(x)$ in example 9.6. To find the critical points, we set $f'(x) = 3x^2 - 3 = 3(x - 1)(x + 1) = 0$. Thus, the critical points are $x = \pm 1$ and the sign chart for f' is as shown in figure 12.10. Hence, by the First Derivative

figure 12.10

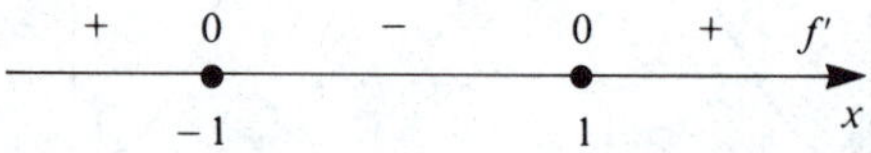

Test, a relative maximum occurs at $x = -1$, while a relative minimum occurs at $x = 1$. From the sign chart, we determine that the graph has shape "up-down-up." We next determine specific points, shown in table 12.1, *making sure that the critical points are included in the table.*

table 12.1

x	-3	-2	-1	$-.5$	0	$.5$	1	2	3
$f(x)$	-17	-1	3	2.375	1	$-.375$	-1	3	19

* We will name this third type of critical point in section 13.

figure 12.11

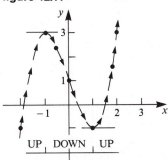

In the resulting graph, shown in figure 12.11, we place horizontal tangents at $(-1, 3)$ and $(1, -1)$ as a reminder to level off the curve at these points. Finally, we connect the plotted points using the sign chart.　📧

example 12.2　　Classify the critical points of the function $f(x) = x^4 - 4x^2 + 3$ and sketch the graph.

solution 12.2　　Solving the equation $f'(x) = 4x^3 - 8x = 4x(x^2 - 2) = 0$ gives critical points $x = 0$, $\pm\sqrt{2}$. Now, $f'(-2) = -16 < 0$, $f'(-1) = 4 > 0$, $f'(1) = -4 < 0$, and $f'(2) = 16 > 0$. The sign chart for f' therefore takes the form shown in figure 12.12. Hence,

figure 12.12

$$
\begin{array}{ccccccc}
- & 0 & + & 0 & - & 0 & + \quad f' \\
\circ\!\!-\!\!\bullet\!\!-\!\!\circ & & \bullet & & \circ\!\!-\!\!\bullet\!\!-\!\!\circ & & \\
-2 \quad -\sqrt{2} \quad -1 & & 0 & & 1 \quad \sqrt{2} \quad 2 & & x
\end{array}
$$

the general shape of the graph is "down-up-down-up." By the First Derivative Test, relative minima occur at $x = \pm\sqrt{2}$, while a relative maximum occurs at $x = 0$. Finally, we determine specific points, shown in table 12.2, being sure to include the critical points $\pm\sqrt{2}$. The final graph is illustrated in figure 12.13.

table 12.2

x	$x^4 - 4x^2 + 3$
0	3
± 1	0
± 2	3
± 3	48
$\pm\sqrt{2}$	-1

figure 12.13

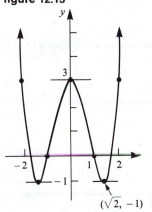

$(\sqrt{2}, -1)$　📧

example 12.3   Suppose $f'(x)$ has the sign chart shown in figure 12.14. What is the general shape of $f(x)$?

figure 12.14

solution 12.3   The general shape of f is "up-down-down-up-down", with the curve leveling off (i.e., with horizontal tangents) at $x = -2, -1, 0,$ and 1. Thus, we have the graph shown in figure 12.15. A relative minimum occurs at $x = 0$, while relative maxima occur at $x = -2$ and 1.

figure 12.15

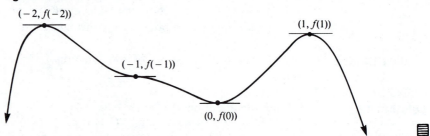

example 12.4   Sketch the graph of $f(x) = x/2 + 2/x^2$.

solution 12.4   We may write $f(x) = \frac{1}{2}x + 2x^{-2}$. Hence, $f'(x) = \frac{1}{2} - 4x^{-3}$. Setting $f'(x) = 0$ to find the critical points yields $1/2 = 4/x^3$, or $x^3 = 8$. Hence the only critical point occurs at $x = 2$. In constructing the sign chart for f' (see figure 12.16), we must be sure to include $x = 0$, the point where $f'(x)$ is not defined. Since $f'(-1) = \frac{9}{2} >$

figure 12.16

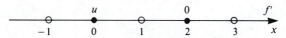

$0, f'(1) = -\frac{7}{2} < 0,$ and $f'(3) = \frac{1}{2} - \frac{4}{27} > 0,$ the complete sign chart for f' is as shown in figure 12.17. The graph has a vertical asymptote at $x = 0$, and it has the

figure 12.17

general shape "up-down-up." In addition, for x^2 large, $2/x^2 \approx 0$. Hence, $y \approx x/2$ for large values of x^2. We call the line $y = x/2$ a **linear asymptote**. Plotting the points shown in table 12.3 and using the sign chart, we obtain the final graph of figure 12.18. The important point in this example is that, in constructing the

table 12.3

x	-5	-4	-3	-2	-1	-.5	.5	1	2	3	4
$2/x + 2/x^2$	-2.4	-1.9	-1.3	-.5		1.5	7.9	8.1	2.5	1.5	1.7	2.1

figure 12.18

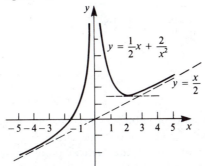

sign chart for f', *include not only the critical points but also all points where* f' *is not defined.* ▤

example 12.5 Sketch the graph of $f(x) = (x^2 + 1)/(x^2 - 1)$.

solution 12.5 The derivative of the function is

$$f'(x) = \frac{(x^2 - 1)2x - (x^2 + 1)2x}{(x^2 - 1)^2} = -\frac{4x}{(x^2 - 1)^2}$$

Thus, $f'(x) = 0$ when $x = 0$. Also, note that $f'(x)$ is not defined at $x = \pm 1$. Since $f'(-2) = \frac{8}{9} > 0$, $f'(-\frac{1}{2}) = 3.55 > 0$, $f'(\frac{1}{2}) = -3.55 < 0$, and $f'(2) = -\frac{8}{9} < 0$, the sign chart for f' takes the form shown in figure 12.19. Thus, a relative

figure 12.19

```
        +     u     +     0     -     u     -    f'
    ─────────●───────────●───────────●──────────►
            -1           0           1         x
```

maximum exists at $x = 0$. If we rewrite $f(x)$ as $(1 + 1/x^2)/(1 - 1/x^2)$, then it follows that $\lim_{x \to +\infty} f(x) = 1$ and $\lim_{x \to -\infty} f(x) = 1$. Thus, the line $y = 1$ is a horizontal asymptote. Plotting the points shown in table 12.4 and connecting them using the sign chart, we obtain the graph of figure 12.20.

table 12.4

x	± 4	± 3	± 2	± 1.5	± 1.1	$\pm .9$	$\pm .5$	0
$\dfrac{x^2+1}{x^2-1}$	$\dfrac{17}{15}$	$\dfrac{5}{4}$	$\dfrac{5}{3}$	2.6	10.52	-9.52	$\dfrac{-5}{3}$	-1

figure 12.20

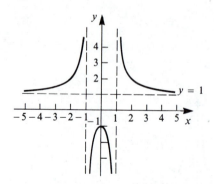

An important concept in biology (and science in general) is that of **stability**. Suppose that a ball is rolling along the curve $y = f(x)$ subject only to gravity. The ball may rest at any of the critical points c_1–c_5 shown in figure 12.21. These points, where $f'(x) = 0$, are the **stationary** or **equilibrium points**. The

figure 12.21

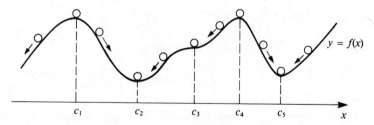

stationary points c_1, c_3, and c_4 are **unstable**; the slightest perturbation will cause the ball to roll away from the point and not return. The stationary points at c_2 and c_5 are called **locally stable**; if the ball is moved slightly from either of these points, it will return to the original position after a short time. From the figure we see that

A stationary point c with $f'(c) = 0$ is locally stable if and only if a relative minimum occurs at c.

example 12.6

Find the stationary points if a ball is rolling on the graph of $f(x) = 3x^5 - 5x^3 + 1$. Which of the stationary points are locally stable?

solution 12.6

We will use the First Derivative Test to categorize the stationary points. The derivative of the function is $f'(x) = 15x^4 - 15x^2 = 15x^2(x^2 - 1)$. Hence, the solutions to $f'(x) = 0$ are $x = 0, \pm 1$. The sign chart is then constructed, as illustrated in figure 12.22. A relative maximum occurs at $x = -1$, and hence this sta-

figure 12.22

tionary point is unstable. Similarly, $x = 0$ is unstable. The only locally stable point is $x = 1$, where a relative minimum occurs. Shown in figure 12.23 is the graph of f.

figure 12.23

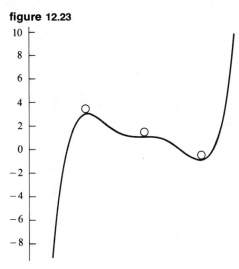

Higher-dimensional analogues of stationary points and stability occur in ecology (where an ecosystem may be "in balance"), in population dynamics (where several species of animals coexist), and in chemistry (where compounds may be stable or unstable).

The four observations made at the beginning of section 12 form the basis for our curve sketching techniques. These basic facts can be deduced from the **Mean Value Theorem**, an important theoretical tool in calculus. Shown in figure 12.24 is the graph of a continuous function $f(x)$ that has derivative $f'(x)$

figure 12.24

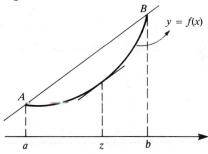

defined on (a, b). The slope of the line joining $A = (a, f(a))$ and $B = (b, f(b))$ is $m = [f(b) - f(a)]/(b - a)$. The Mean Value Theorem asserts that *there is a tangent line at a point on the graph between A and B with the same slope.*

theorem   **mean value**   If $f(x)$ is continuous on $[a, b]$ and differentiable on (a, b), then there is at least one z between a and b with

$$f'(z) = \frac{f(b) - f(a)}{b - a}$$

This theorem may also be interpreted in terms of rates of growth. It is always a source of amazement when parents examine the marks they have made on a wall to show their child's increase in height over a period of six months, as shown in figure 12.25.

figure 12.25

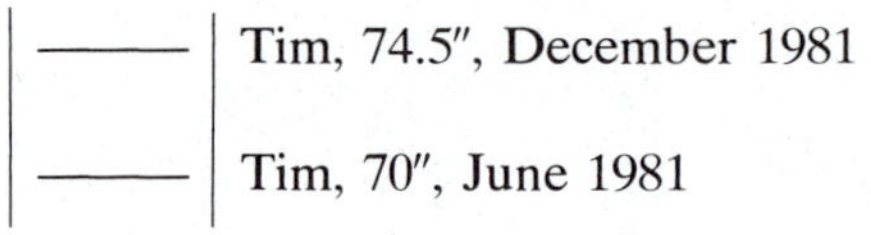

——— | Tim, 74.5″, December 1981

——— | Tim, 70″, June 1981

In this case,

$$\frac{f(b) - f(a)}{b - a} = \frac{\text{change in height}}{\text{change in time}} = \frac{4.5 \text{ inches}}{6 \text{ months}}$$

$$= \frac{3}{4} \text{inch/month}$$

The Mean Value Theorem asserts that at some time z between June and December, the instantaneous rate of growth $f'(z)$ matched the average rate of growth of $\frac{3}{4}$ inch/month.

To illustrate how this theorem can be used to justify the methods of this section, suppose that $f'(x) > 0$ on (c, d). We will show that f is increasing on (c, d). Let a and b be real numbers in (c, d) with $a < b$. Then, according to the Mean Value Theorem,

$$\frac{f(b) - f(a)}{b - a} = f'(z)$$

for some z between a and b. Since $f'(z) > 0$, it follows that $f(b) - f(a) > 0$. Hence, $f(a) < f(b)$ and so f is increasing on (c, d).

exercises for section 12

part A

Given the graphs in Exercises 1–4, construct the sign chart for f'.

1.
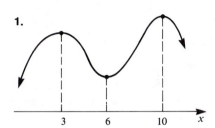

2.
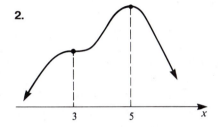

3.
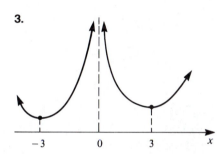

4.
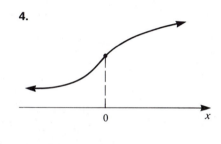

Find all critical points for the function f given in exercises 5–18. Classify the critical points using the First Derivative Test. Finally, use the sign chart for f' to sketch the graph of f.

polynomial functions

5. $f(x) = 4x - x^2$	**6.** $f(x) = 3x^2 - 12x + 5$
7. $f(x) = x^2 - 6x + 4$	**8.** $f(x) = 1 + 4x + x^2$
9. $f(x) = x^3 - 6x$	**10.** $f(x) = 4x - 4x^3$
11. $f(x) = 3x^4 - 4x^3 + 1$	**12.** $f(x) = 4x^3 + 12x^2$
13. $f(x) = (x^2 - 4)^2$	**14.** $f(x) = x^4 - 5x^2 + 4$
15. $f(x) = 4x - x^4$	**16.** $f(x) = x^3 + 3x + 1$
17. $f(x) = 2x^6 - 3x^4$	**★18.** $f(x) = x^2(x - 1)^3$
19. $f(x) = (x - 1)(x - 2)(x - 3)$	**★20.** $f(x) = 4x^5 + 5x^4 - 20x^2 + 8$

rational functions

21. $f(x) = x + 1/x$	**22.** $f(x) = 4x^2 + 1/x$
23. $f(x) = x/(x^2 - 1)$	**24.** $f(x) = (x^2 - 1)/(x^2 + 1)$

25. $f(x) = x/(x^2 - 4)$ **26.** $f(x) = (x + 2)^2/x$
   [*Hint*: See example 4.8.]

27. $f(x) = 9/x + x + 1$ **28.** $f(x) = (x^3 + 1)/(x^3 - 1)$

★**29.** $f(x) = x^2 - 27/x$ ★**30.** $f(x) = x^2[(6 - x)/(x + 2)]$

★**31.** $f(x) = (x^2 + 2)/(x - 3)$ ★**32.** $f(x) = \sqrt{x - 1} - x/2,$ for $x \geq 1$

★**33.** $f(x) = 2x/\sqrt{x^2 - 1}$ ★**34.** $f(x) = ax/(1 + bx),$ for $a, b > 0$

Given the sign charts for f' shown in exercises 35–38, determine the general shape of the graph of f. [*Hint*: See example 12.3.] Assume that a vertical asymptote occurs at any points where f and f' are undefined.

35.   **36.**

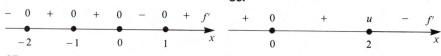

37.   **38.**

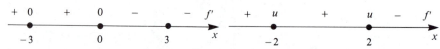

A ball is rolling along the curve $y = f(x)$ subject only to gravity. Find all stationary points, in exercises 39–48. Determine which of these points are locally stable, using the First Derivative Test.

39. $f(x) = x^6 - 16x^3 + 1$ **40.** $f(x) = (x^2 - 2)^4$

41. $f(x) = (x - 1)(x - 2)^2$ **42.** $f(x) = x^2\sqrt{x - 1},$ for $x \geq 1$

43. $f(x) = 12x^3 - 6x^2 - 3x + 1$ **44.** $f(x) = 2x + \sqrt{x^2 + 1}$

45. $f(x) = 2x^3 + 3x^2 - 72x + 20$ ★**46.** $f(x) = x^2e^{-x}$

★**47.** $f(x) = 5x - 4 \ln x,$ for $x > 0$ ▲**48.** $f(x) = 3 \sin 4x$

part B

★**49.** If $y = ax^3 + bx^2 + cx + d$, determine a, b, c, and d so that both
   (a) a relative minimum with value -3 occurs at $x = -1$ and
   (b) a relative maximum with value 2 occurs at $x = 1$.

★**50.** Prove that the polynomial $p(x) = ax^3 + bx^2 + cx + d$ has at most two critical points, and exactly two critical points if $b^2 - 3ac > 0$.

★**51.** If $y = ax/(1 + bx)^2$, find a and b so that a maximum value of 4 occurs at $x = 3$.

★**52.** If $y = ax^{2n}/(1 + bx^{2n})$, where $n = 1, 2, 3, \ldots$, then determine the general shape of the graph of f if $a > 0$ and $b > 0$.

▲**53.** Show that the critical points of $f(x) = A \cos (Bx + C)$ occur at $x = n\pi/B - C/B$ for $n = 0, \pm1, \pm2, \ldots$.

13 the second derivative and concavity

Suppose we know that the function $f(x)$ increases between the two points P_1 and P_2. In sketching a graph of the function, how do we connect the points? There are at least two ways, as illustrated in figure 13.1. How do we decide

figure 13.1

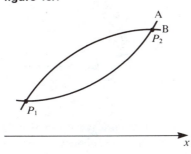

which case is applicable? Once again, the clue comes from examining tangent lines. In case A (see accompanying figure), the graph of f is "cupped upward" or concave upward between P_1 and P_2. As we move from left to right, the steepness increases, i.e., $f'(x)$ increases. In case B (see figure) the graph is concave downward between P_1 and P_2. As we move from left to right, the curve levels off, i.e., $f'(x)$ decreases.

case A: Concave upward.   **case B:** Concave downward.

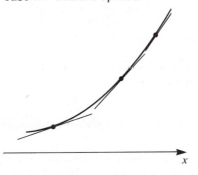 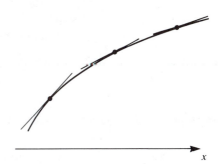

The figures for cases C and D show that a similar situation holds when $f(x)$ decreases between P_1 and P_2. In case C, $f'(x_1) < f'(x_2)$. Thus, f' is increasing, and the resulting graph is concave upward. In contrast, $f'(x_1) > f'(x_2)$ in case D.

Thus, f' is decreasing and the graph is seen to be concave downward. How can we determine when the function $f'(x)$ is increasing? The answer follows from the discussion in section 12: $f'(x)$ *is increasing when its derivative is positive,* i.e., when $f''(x) > 0$. Likewise, if $f''(x) < 0$ on an interval, then $f'(x)$ is decreasing on that interval. *Throughout this section we will assume that $f''(x)$ is continuous.*

case C: Concave upward   **case D:** Concave downward

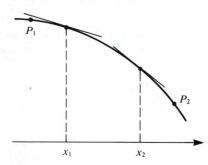

definition 13.1   We say that the graph of f is **concave upward** on the open interval (a, b) if $f''(x) > 0$ on (a, b). The graph of f is **concave downward** on (a, b) if $f''(x) < 0$ on (a, b).

example 13.1   Determine where the function $f(x) = x^4 - 6x^2 + x + 1$ is concave upward and concave downward.

solution 13.1   We must examine the sign of f''. The first derivative is $f'(x) = 4x^3 - 12x + 1$; thus, $f''(x) = 12x^2 - 12$. To solve the inequalities $12x^2 - 12 > 0$ and $12x^2 - 12 < 0$, we construct the sign chart for f''. Setting $f''(x) = 0$ implies $x = \pm 1$. Since $f''(-2) = 36 > 0$, $f''(0) = -12 < 0$, and $f''(2) = 36 > 0$, we have the sign chart shown in figure 13.2. Thus, f is concave upward on the intervals $(-\infty, -1)$ and $(1, +\infty)$ and concave downward on $(-1, 1)$.

figure 13.2

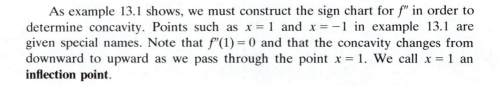

As example 13.1 shows, we must construct the sign chart for f'' in order to determine concavity. Points such as $x = 1$ and $x = -1$ in example 13.1 are given special names. Note that $f''(1) = 0$ and that the concavity changes from downward to upward as we pass through the point $x = 1$. We call $x = 1$ an **inflection point**.

definition 13.2   We say that an *inflection point* occurs at $x = d$ if $f''(d) = 0$ and the concavity of the graph changes as x passes through d.

The two possible cases are illustrated in the accompanying figures.

case I: Concave upward to concave downward.

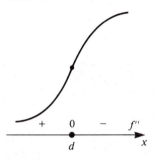

case II: Concave downward to concave upward.

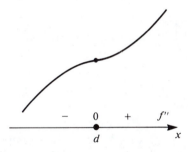

We will next refine the curve sketching techniques of section 12 by building concavity features into our graphs.

example 13.2   Sketch the graph of $f(x) = x^3 + x^2 + x + 1$ by first finding all relative extrema and then all inflection points.

solution 13.2   The first derivative is $f'(x) = 3x^2 + 2x + 1$. In applying the quadratic formula, we compute $\sqrt{b^2 - 4ac} = \sqrt{4 - 12} = \sqrt{-8}$. Thus, there are no solutions to $f'(x) = 0$, and hence no critical points. Since $f'(0) = 1$, the sign chart for f' is as shown in figure 13.3. We conclude that $f(x)$ is always increasing. The second derivative is $f''(x) = 6x + 2$. Setting $f''(x) = 0$ yields $x = -\frac{1}{3}$. Since $f''(-1) < 0$ and

figuro 13.3

$$\xrightarrow{\hspace{2cm} + \hspace{1.5cm} f'}$$

figure 13.4

$$\xrightarrow{\hspace{1cm} - \; 0 \; + \quad f''}$$
$$-1/3$$

$f''(0) > 0$, f'' has the sign chart shown in figure 13.4. We conclude that concavity changes from downward to upward as we move through $x = -\frac{1}{3}$. The complete sketch of the function is shown in figure 13.5.

figure 13.5

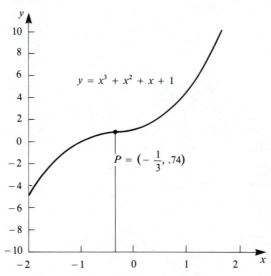

$y = x^3 + x^2 + x + 1$

$P = \left(-\frac{1}{3}, .74\right)$

example 13.3   Sketch the graph of $y = 3x^5 - 5x^3 + 1$ by first finding all relative extrema and then all inflection points.

solution 13.3   The first derivative is $f'(x) = 15x^4 - 15x^2$. In example 12.6 we showed that $x = -1$, 0, and 1 were the critical points and that f' had the sign chart shown in figure 13.6. The second derivative is $f''(x) = 60x^3 - 30x = 30x(2x^2 - 1)$. Setting $f''(x) = 0$, we have $x = 0$ and $\pm\sqrt{.5} \approx \pm.707$. Now, $f''(-1) = -30 < 0$, $f''(-.5) =$

figure 13.6

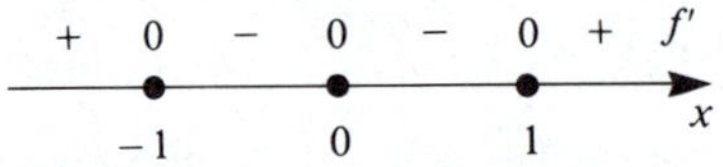

$$+ \quad 0 \quad - \quad 0 \quad - \quad 0 \quad + \quad f'$$
$$\overline{\qquad \bullet \qquad \bullet \qquad \bullet \qquad} \; x$$
$$-1 \qquad 0 \qquad 1$$

$7.5 > 0$, $f''(.5) = -7.5 < 0$, and $f''(1) = 30 > 0$. Hence, the sign chart for f'' is as shown in figure 13.7. We conclude that the general shape of the graph is "up-down-down-up," with horizontal tangents at 0, -1, and 1. Now, we plot the graph as in section 12, pencil it in light, and temporarily ignore the

figure 13.7

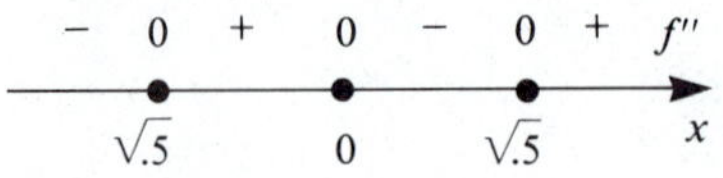

$$- \quad 0 \quad + \quad 0 \quad - \quad 0 \quad + \quad f''$$
$$\overline{\qquad \bullet \qquad \bullet \qquad \bullet \qquad} \; x$$
$$\sqrt{.5} \qquad 0 \qquad \sqrt{.5}$$

concavity features. From the sign chart for f'', we observe that there are inflection points at $\pm\sqrt{.5}$ and 0. Finally, we plot the inflection points as shown in figure 13.8 and draw the final graph, taking concavity into account.

figure 13.8

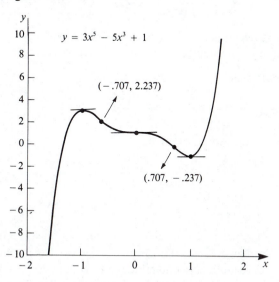

$y = 3x^5 - 5x^3 + 1$

$(-.707, 2.237)$

$(.707, -.237)$

The step-by-step procedure described in the accompanying box may prove helpful in sketching the graph of $f(x)$.

---

procedure for graphing $y = f(x)$

Step 1. Determine the domain of definition of $f(x)$. Vertical asymptotes may occur at points not in the domain.

Step 2. Set $f'(x) = 0$ to determine the critical points.

Step 3. Construct the sign chart for $f'(x)$.

Step 4. Determine whether horizontal asymptotes exist by examining $\lim_{x \to +\infty} f(x)$ and $\lim_{x \to -\infty} f(x)$.

Step 5. Plot points and pencil in the graph, temporarily ignoring concavity features.

Step 6. Compute $f''(x)$ and construct the sign chart for f''.

Step 7. Plot any inflection points and redraw the graph, incorporating concavity features.

Let's use this procedure to graph the rational function on page 136.

example 13.4   Sketch the graph of $y = \dfrac{x}{x^2 + 1}$. Where is the graph the steepest?

solution 13.4   Using the Quotient Rule, we obtain

$$y' = \frac{(x^2 + 1) \cdot 1 - x(2x)}{(x^2 + 1)^2} = \frac{1 - x^2}{(x^2 + 1)^2} = 0 \qquad \text{when } x = \pm 1$$

The first derivative y' is defined everywhere, since $x^2 + 1$ is never 0. It follows that y' has the sign chart shown in figure 13.9. To compute and simplify y''

figure 13.9

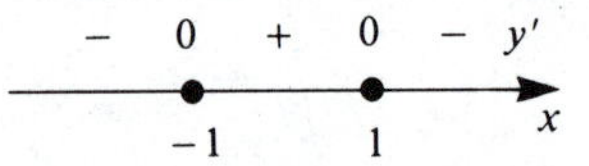

$$
\begin{array}{ccccc}
- & 0 & + & 0 & - \quad y'\\
\end{array}
$$

requires a bit more patience:

$$y'' = \frac{(x^2 + 1)^2 (-2x) - (1 - x^2)[2(x^2 + 1) \cdot 2x]}{(x^2 + 1)^4} = \frac{2x}{(x^2 + 1)^3}(-x^2 - 1 - 2 + 2x^2)$$

$$= \frac{2x(x^2 - 3)}{(x^2 + 1)^3} = 0 \qquad \text{when } x = 0 \quad \text{or } \pm\sqrt{3}$$

Since $y''(-2) < 0$, $y''(-1) > 0$, $y''(1) < 0$, and $y''(2) > 0$, y'' has the sign chart shown in figure 13.10. To summarize, $x = -1$ is a relative minimum, $x = 1$ is a relative maximum, and inflection points occur at $(-\sqrt{3}, -\sqrt{\tfrac{3}{4}})$, $(0, 0)$ and $(\sqrt{3}, \sqrt{\tfrac{3}{4}})$. Note that the graph, sketched in figure 13.11, appears to be steepest at $x = 0$.

figure 13.10

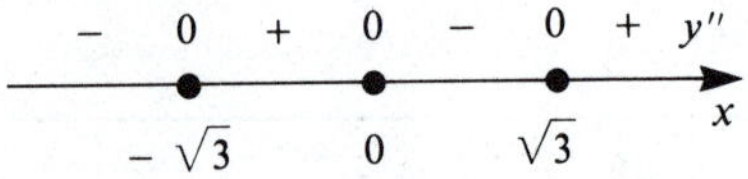

$$
\begin{array}{ccccccc}
- & 0 & + & 0 & - & 0 & + \quad y''\\
\end{array}
$$

figure 13.11

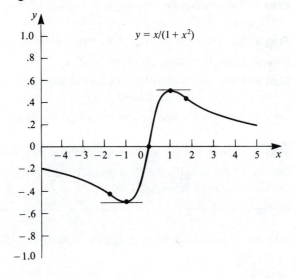

$$y = x/(1 + x^2)$$

To verify this, we must find the largest value of y'. If the largest value occurs at c, then a relative maximum occurs at $x = c$. Hence $(y')'(c) = y''(c) = 0$. From the sign chart for $(y')' = y''$ it can be seen that only $x = 0$ gives a relative maximum. Note also that $x = 0$ is an inflection point. [*Note*: An astute reader will note that we have assumed that a largest value of y' exists.]   ▤

The second derivative may be used to classify the critical points of a function $f(x)$, as described in the accompanying box.

> **Second Derivative Test**
>
> Assume that $f''(x)$ is continuous in the vicinity of c and that c is a critical point of $f(x)$.
> **1.** If $f''(c) > 0$, then a relative minimum occurs at $x = c$.
> **2.** If $f''(c) < 0$, then a relative maximum occurs at $x = c$.

It is easy to understand why the Second Derivative Test works. When $f''(c) > 0$, the graph is *concave upward* in the vicinity of c, and hence we must have a relative minimum. When $f''(c) < 0$, the graph is *concave downward* around c, and hence a relative maximum occurs.

example 13.5   Classify the critical points of $f(x) = 1 + x^2 - x^6$ using the Second Derivative Test.

solution 13.5   The first derivative is $f'(x) = 2x - 6x^5 = 2x(1 - 3x^4)$. Hence, the critical points are $x = 0$ and $x = \pm\sqrt[4]{\frac{1}{3}} \approx \pm.7598$. The second derivative is $f''(x) = 2 - 30x^4$, and $f''(0) = 2 > 0$. Hence, using the Second Derivative Test, we conclude that a relative minimum occurs at 0. For $x = \pm\sqrt[4]{\frac{1}{3}}$, $x^4 = \frac{1}{3}$ and hence $f''(\sqrt[4]{\frac{1}{3}}) = 2 - 30(\frac{1}{3}) = -8 < 0$. We conclude that relative maxima occur at $x = \pm\sqrt[4]{\frac{1}{3}}$. Shown in figure 13.12 is the graph of f.

figure 13.12

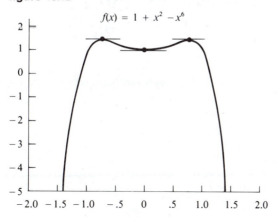

$f(x) = 1 + x^2 - x^6$

▤

If $f''(c) = 0$, the Second Derivative Test gives *no information*. The nature of the critical point remains undisclosed, and one must apply the First Derivative Test, as demonstrated in the next example.

example 13.6   Find and categorize the critical points of $f(x) = 3x^4 - 4x^3 + 1$.

solution 13.6   Setting the first derivative $f'(x) = 12x^3 - 12x^2 = 12x^2(x - 1) = 0$ gives $x = 0$ and 1 as critical points. The second derivative is $f''(x) = 36x^2 - 24x$. Now, $f''(1) = 12 > 0$. Hence, a relative minimum occurs at $x = 1$. Also, $f''(0) = 0$, and so the Second Derivative Test has failed. The sign chart for f' is as shown in figure 13.13. Hence, by the First Derivative Test, neither a relative maximum nor minimum occurs at $x = 0$.

figure 13.13

We have not yet given a special name to the type of critical point that arises from sign charts of the form shown in figures 13.14**(a)** and **(b)**. For such critical points, the Second Derivative Test tells us that $f''(c) = 0$. [If $f''(c) \neq 0$, we would have a relative maximum or minimum.] Furthermore, it can be shown that for $f''(x)$ continuous, a change in concavity takes place at c. Hence, we call this third type of critical point a **point of horizontal inflection**.

figure 13.14(a)   **figure 13.14(b)**

exercises for section 13

part A   For the functions whose graphs are shown in exercises 1–4, construct the sign chart for f''.

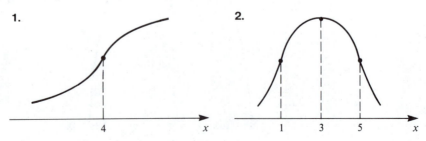

3.

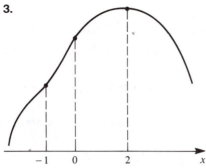

4.

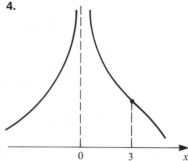

For each of the following functions, determine where the graph is concave upward, concave downward, and locate all inflection points on the graph. However, do not sketch the graphs.

5. $f(x) = x^3 - 6x - 4$

6. $f(x) = x^4 - x - 1$

7. $f(x) = x^6 - 16x^3 + 1$

8. $f(x) = x^2 e^{-x}$

9. $f(x) = (x + 2)^2/x$

10. $f(x) = 5x^2 + 5 \ln x,$ for $x > 0$

11. $f(x) = 12x^4 + 6x^2 - 9$

12. $f(x) = 12x^4 - 2x^3 + 12x + 3$

13. $f(x) = x^3(x - 2)^2$

14. $f(x) = 8x^2 + 1/x$

The following functions were graphed in the exercises of section 12. [The corresponding exercise in section 12 is given in parentheses.] Refine those sketches by determining inflection points and incorporating the concavity features.

15. (5) $f(x) = 4x - x^2$

16. (6) $f(x) = 3x^2 - 12x + 5$

17. (7) $f(x) = x^2 - 6x + 4$

18. (8) $f(x) = 1 + 4x + x^2$

19. (9) $f(x) = x^3 - 6x$

20. (10) $f(x) = 4x - 4x^3$

21. (11) $f(x) = 3x^4 - 4x^3 + 1$

22. (12) $f(x) = 4x^3 + 12x^2$

23. (13) $f(x) = (x^2 - 4)^2$

24. (14) $f(x) = x^4 - 5x^2 + 4$

25. (15) $f(x) = 4x - x^4$

26. (16) $f(x) = x^3 + 3x + 1$

27. (17) $f(x) = 2x^6 - 3x^4$

★28. (18) $f(x) = x^2(x - 1)^3$

29. (19) $f(x) = (x - 1)(x - 2)(x - 3)$

30. (21) $f(x) = x + 1/x$

31. (22) $f(x) = 4x^2 + 1/x$

32. (24) $f(x) = (x^2 - 1)/(x^2 + 1)$

33. (25) $f(x) = x/(x^2 - 4)$

★34. (20) $f(x) = x^2 - 27/x$

For each of the following functions, classify the critical points of f using the Second Derivative Test. If the test fails [$f''(c) = 0$], then use the First Derivative Test.

35. $f(x) = 3x^2 - 8x + 4$

36. $f(x) = x + 9/x$

37. $f(x) = x^3 - 9x^2 - 21x - 4$

38. $f(x) = 2x^3 + 3x^2 - 12x$

39. $f(x) = x^4 + 8x^3 + 18x^2 + 5$

40. $f(x) = 3x^5 - 20x^3 + 1$

41. $f(x) = x/(x^2 + 1)$

42. $f(x) = x^2(x - 1)^2$

43. $f(x) = 3x^8 - 8x^3 - 2$

44. $f(x) = 3x - 6 \ln x$ for $x > 0$

★**45.** $f(x) = x^3 e^{-x}$

46. $f(x) = x^2 + 128/x$

47. Show that the quadratic function $q(x) = ax^2 + bx + c$ is concave upward when $a > 0$.

48. Show that a third-degree polynomial always has a unique inflection point.

14 applied problems in optimization

What is the *largest value M* that a function $f(x)$ attains on a prescribed domain A? What is the *smallest value m* that $f(x)$ achieves? If $A = [a, b]$, a closed interval, and *if $f'(x)$ exists at each point of A*, then the method for finding M and m is straightforward, as described in the accompanying box.

---

determining the maximum and minimum values of a function over a given closed interval

Step 1. Calculate $f(a)$ and $f(b)$, the functional values at the endpoints of the interval.

Step 2. Find $f'(x)$ and then set $f'(x) = 0$ to determine all critical points $c_1, c_2, \ldots$, of f.

Step 3. Evaluate $f(c_1), f(c_2), \ldots$, for those critical points *inside* (a, b).

Step 4. Select the largest value M and the smallest value m of the numbers $f(a), f(b), f(c_1), f(c_2), \ldots$. These are the maximum and minimum values, respectively, achieved by the function over the interval.

---

example 14.1   Find the largest and smallest values of $f(x) = x - x^3$ for $0 \le x \le 1$.

solution 14.1   The functional values at the endpoints of the interval are $f(0) = 0$ and $f(1) = 0$. To find the critical points, we set $f'(x) = 1 - 3x^2 = 0$. Hence, $x^2 = \frac{1}{3}$ or $x = \pm\sqrt{\frac{1}{3}}$. Only $+\sqrt{\frac{1}{3}}$ is inside the domain $(0, 1)$, and $f(+\sqrt{\frac{1}{3}}) = \sqrt{\frac{1}{3}} - (\frac{1}{3})^{3/2} \approx .43$. Hence, $m = 0$ and $M \approx .43$.

It is easy to see why the technique works. It is possible that M and/or m occur at an endpoint, as in figure 14.1. We take this possibility into account by performing Step 1. If the minimum value occurs inside (a, b) (as in figure 14.2). then a relative minimum occurs at c and hence $f'(c) = 0$. Thus $m = f(c)$. We take this possibility into account by performing Steps 2–3.

Thus, in conducting our search for the absolute maximum M and the absolute minimum m, we may confine our efforts to endpoints and critical points inside (a, b). This procedure is illustrated in our next two examples.

figure 14.1   **figure 14.2**

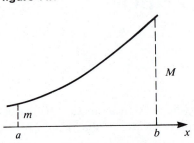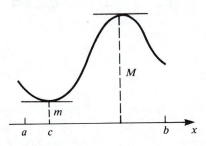

example 14.2   Find the maximum value of $f(x) = x^3 - 3x + 1$ over the domain $[-2, 2]$.

solution 14.2   The functional values at the endpoints of the domain are $f(-2) = -8 + 6 + 1 = -1$ and $f(2) = 8 - 6 + 1 = 3$. Since $f'(x) = 3x^2 - 3$, the critical points occur at $x = 1, -1$. The functional values at the critical points are $f(-1) = -1 + 3 + 1 = 3$ and $f(1) = 1 - 3 + 1 = -1$. Thus, $M = 3$, the value assumed by the function both at $x = 2$ and $x = -1$.   ▤

example 14.3   Find the minimum value of

$$f(x) = \frac{x}{30} + \frac{\sqrt{(x - 20)^2 + 100}}{15} \qquad \text{for } 0 \le x \le 20$$

solution 14.3   The functional values at the endpoints of the interval are $f(0) = \sqrt{500}/15 = 1.4907$ and $f(20) = \frac{2}{3} + \frac{10}{15} = \frac{4}{3} = 1.3\overline{3}$. To determine the critical points of this function, we compute the first derivative

$$f'(x) = \frac{1}{30} + \frac{1}{15}[(x - 20)^2 + 10^2]^{-1/2}(x - 20)$$

Hence, $f'(x) = 0$ implies

$$\frac{1}{30} = \frac{1}{15} \frac{(20 - x)}{\sqrt{(x - 20)^2 + 100}}$$

Squaring both sides and simplifying, we obtain

$$1 = \frac{4(x - 20)^2}{(x - 20)^2 + 100}$$

or $(x - 20)^2 + 100 = 4(x - 20)^2$

or $3(x - 20)^2 = 100$

Thus, the critical points are $20 \pm 10/\sqrt{3}$, or 14.226 and 25.773. Only 14.226 is inside the domain $[0, 20]$, and $f(14.226) = 1.244$. Of the three values 1.4907,

1.3$\overline{3}$, and 1.244, clearly 1.244 is the smallest. Hence, the minimum value of f is 1.244 and it occurs when $x = 14.226$. ▤

The technique for finding m and M is not so simple if the domain of definition is an infinite interval, such as $(0, +\infty)$. There is one commonly occurring special case that is easy to handle, as described in the accompanying box.

---

the sole critical point test

Suppose $f(x)$ has as its domain of definition an interval A, $f'(x)$ exists and is continuous at each point of A, and $x = c$ is the only critical point inside A.

1. If a relative minimum occurs at c, then the absolute minimum m occurs at c.

2. If a relative maximum occurs at c, then the absolute maximum M occurs at c.

Use either the First or Second Derivative Tests to categorize the sole critical point.

This technique is illustrated in the following example.

example 14.4   Find the minimum value of the function

$$f(x) = 20x + \frac{5}{x}$$

for $x > 0$.

solution 14.4   The first derivative of the function is $f'(x) = 20 - 5x^{-2}$. Setting $f'(x) = 0$, we obtain

$$20 = \frac{5}{x^2} \quad \text{or} \quad x^2 = \frac{1}{4}$$

The only critical point in the domain $(0, +\infty)$ is $x = \frac{1}{2}$. Since $f''(x) = 10/x^3$, $f''(1/2) > 0$. Hence, by the Second Derivative Test, a relative minimum occurs at $x = \frac{1}{2}$. Since $\frac{1}{2}$ is the sole critical point in the domain, the absolute minimum is $m = f(\frac{1}{2}) = 20$. ▤

What is the "best" way to perform a given task? Before we can begin to answer such questions, we must set up a *criterion* for judging whether one method is superior to another. Consider, for example, the case of a farmer who wishes to enclose his livestock in a grazing area that lies along a river, as illustrated in figure 14.3. Here, the farmer's objective is to maximize the

figure 14.3

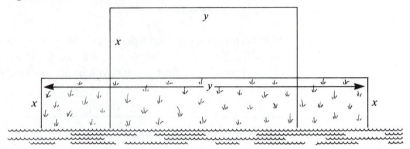

grazing area. Hence, he will judge the value of a given fencing arrangement by computing

the objective function $A = xy.$

Undoubtedly, the amount of fence he has is limited. If the farmer has 3,000 feet of fence available, and uses all the fence, then

the constraint $2x + y = 3,000$

Hence, $A = x(3,000 - 2x) = 3,000x - 2x^2$. Then $A'(x) = 3,000 - 4x$, and the sole critical point occurs at $x = 750$. Since $A''(750) = -4 < 0$, the maximum area occurs when $x = 750$ (and $y = 1,500$). A field $750' \times 1500'$ is the optimal design.

The problem described above is a simple case of a general class of problems called **problems of optimal design**. They occur naturally in all areas of science and technology.

The procedure described in the accompanying box may prove helpful in setting up the function to be maximized or minimized.

---

suggestions for constructing the objective function

Step 1. Draw a picture and label the various line segments with variables $x, y, z, \ldots$.

Step 2. Determine the *objective function* by answering the question, "what is the quantity Q to be maximized or minimized?"

Step 3. Write Q in terms of the variables $x, y, z, \ldots$.

Step 4. Look for *constraints*—i.e., determine relationships among the variables.

Step 5. Write the objective function Q as a function of *one variable only*, using the relationships found in Step 4.

Step 6. Apply the calculus procedures developed at the beginning the this section.

Let's use these suggested steps to set up and solve the following optimization problems.

example 14.5   A box is to contain 16 cubic feet and have a reinforced square base. If the base is twice as thick as the sides or top, find the optimal design—i.e., find the dimensions of the box of minimum surface area (and therefore minimum cost).

solution 14.5   Referring to figure 14.4, we see that the surface area of the top is xy and that of the bottom is $2xy$, since twice as much material is being used for the

figure 14.4

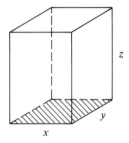

bottom. The sides have total area $(xz + yz + xz + yz)$. Hence, the quantity to be minimized is

$$S = 3xy + 2xz + 2yz$$

But the box is to contain 16 cubic feet and have a *square* base. Hence, we impose the constraints $xyz = 16$ and $x = y$. Thus, $z = 16/x^2$ and so

$$S = 3x^2 + 2x\,\frac{16}{x^2} + 2x\,\frac{16}{x^2} = 3x^2 + \frac{64}{x}, \qquad \text{for } x > 0$$

The first derivative of the objective function is $S'(x) = 6x - 64/x^2$. Critical points satisfy the equation

$$6x = \frac{64}{x^2} \qquad \text{or} \qquad x^3 = \frac{64}{6}$$

Thus, the only critical point is $x = \sqrt[3]{64/6} = 2.20128\ldots$. Now, the sign chart for S' may be constructed as shown in figure 14.5. The First Derivative Test will show that a relative minimum occurs at $x = 2.20$. Since 2.20 is the sole critical

figure 14.5

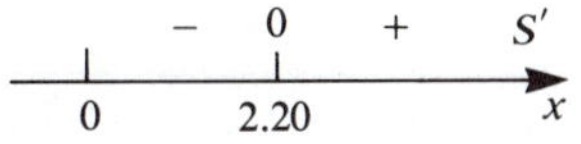

point in $(0, +\infty)$, the absolute minimum of S occurs at $x = 2.20$. The optimal design is then $x = 2.20128$, $y = 2.20128$, and $z = 16/x^2 = 3.30192$.

example 14.6

A plane crashes at a point 10 miles from the nearest highway. The nearest town is 20 miles down the road. Luckily for the survivors, Lassie is on board. She can run 30 mph along the highway, but only 15 mph over the rough terrain. This noble beast instinctively solves this time minimization problem, runs the optimal route, jumps up on the ranger's table and spells out "HELP" in the mashed potatoes. Find the optimal route.

solution 14.6

The objective function is

Total time = (time over the rough terrain) + (time along the highway)

$$T = \frac{y}{15} \qquad\qquad + \frac{x}{30}$$

We have used the formula

time = distance/velocity.

Examining the right triangle in the figure 14.6, we observe that

$$10^2 + (20 - x)^2 = y^2$$

figure 14.6

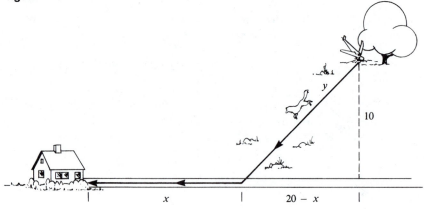

Hence, $y = \sqrt{(20 - x)^2 + 10^2}$ and so we must minimize

$$T = \frac{x}{30} + \frac{\sqrt{(20 - x)^2 + 100}}{15}$$

for $0 \leq x \leq 20$. (Lassie could run directly to the highway, or directly to the town.) In example 14.3 we showed that this function has minimum value 1.244, occurring when $x = 14.226 = 20 - 10/\sqrt{3}$. ▤

On a more serious note, essentially the same equations as those in example 14.6 have been applied to the biological setting in example 14.7.

▲ **example 14.7 cardiovascular branching**   The vascular system consists of the collection of arteries, veins, and capillaries that carry blood to every cell in the body. This incredibly complex system is over 60,000 miles in length, but some aspects of the system can be explained in terms of optimal design.

Refer to figure 14.7 and suppose it is necessary to supply point C with a certain quantity of blood per unit time. Let $R_1 =$ radius of the main vessel and $R_2 =$ radius of the smaller vessel. At what point D should the branching take place in order for the system to operate *optimally*? We must determine exactly what is to be optimized. The greater the resistance to flow, the harder the heart must pump. Hence, a reasonable *hypothesis* is that the optimal design would be one that minimizes resistance to flow (which is due mainly to friction along the wall of the vessel). Poiseuille's Law asserts that this resistance is directly proportional to the length of the vessel and inversely proportional to R_i^4.

figure 14.7

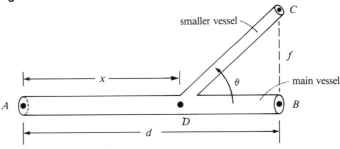

Hence, the objective function is

$$f(x) = k\,\frac{AD}{R_1^4} + k\,\frac{CD}{R_2^4}$$

$$= \frac{k}{R_1^4}\,x + \frac{k}{R_2^4}\,\sqrt{(d-x)^2 + f^2}$$

for $0 \le x \le d$. If we proceed as in example 14.3, we find that the unique critical point satisfies

$$\left(\frac{R_2}{R_1}\right)^4 = \frac{d-x}{\sqrt{(d-x)^2 + f^2}}$$

Note that in terms of the angle θ, $\cos\theta = DB/CD$, and $DB = d - x$. It follows from the Pythagorean Theorem that $CD = \sqrt{(d-x)^2 + f^2}$. Hence, the *optimal branching angle* satisfies

$$\cos \theta = (R_2/R_1)^4$$

or

$$\theta = \cos^{-1}(R_2/R_1)^4$$

Using the last formula, we can construct table 14.1, which gives the optimal branching angle in terms of R_1/R_2.

table 14.1

$\dfrac{R_2}{R_1}$	θ
.75	71.55°
.50	86.42°
.25	89.78°
.10	89.99°

Thus, our model predicts that small branches will come off the main vessel at an angle of about 90°. This is in agreement with the actual measurements on the cardiovascular system.

▲ **example 14.8**　A cylinder is to be made of an elastic material. To support the structure, a wire of fixed length L wraps around the cylinder as shown in figure 14.8. (A section of vacuum cleaner hose is a good example.) For a fixed length L, find the design with largest volume.

figure 14.8(a)

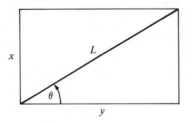

figure 14.8(b)

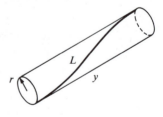

solution 14.8　The function to be maximized is $V = \pi r^2 y$. But note that if we cut and flatten the cylinder, then we have the constraints $x = \text{circumference} = 2\pi r$ and $x^2 + y^2 = L^2$. Hence,

$$V = \frac{1}{4\pi} y(L^2 - y^2) \qquad \text{for } 0 < y < L$$

The first derivative is $V'(y) = (1/4\pi)L^2 - (1/4\pi)(3y^2)$. Critical points satisfy the equation $V' = 0$, and hence $3y^2 = L^2$. Thus, $y = L/\sqrt{3}$ is the only critical point in the domain $(0, L)$. Since $V''(y) < 0$, a relative maximum, and hence the *absolute* maximum, occurs at $y = L/\sqrt{3}$. It follows that $x = \sqrt{2/3}L$ and hence $y/x = 1/\sqrt{2}$. Thus, $\tan \theta = \sqrt{2}$, which implies $\theta = 54.7°$.   ▤

The simple model of example 14.8 has been used to explain shape changes in worms. The sea worm *Amphiporus* can extend its body so that it is long and thin or contract it so that is short and fat. The maximum length is about five times the minimum length. A series of fine inflexible fibres under the epithelium encircle the body as shown in figure 14.9 and limit shape changes. When completely relaxed, this worm attains maximum volume. Actual measurements of θ are in the vicinity of 55°, in close agreement with the angle predicted by example 14.8.*

figure 14.9(a)   **figure 14.9(b)**

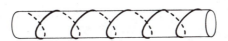
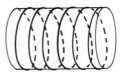

The **Principle of Optimal Design** asserts that an organism will evolve in such a manner that its parts become optimally designed with respect to the functions they must perform. It has been proposed by some biologists as a unifying principle that at least in theory can be used to explain the shape of things. Thus, if an organism remains unchanged over a long period of time, its present design is optimal—i.e., many maximization/minimization problems have been solved in the evolution to its present state. More examples will be presented in the exercises. As you may have surmised, the difficult task is to determine exactly what is being optimized in a given physical situation, and to recognize all of the constraints.

The final example shows that if there is a point on the graph of a function where the curve rises most rapidly, then the point must be an inflection point.

example 14.9   Determine where the function $f(x) = -x^3 + x^2 + x + 1$ is increasing the most rapidly—i.e. find the maximum value of $f'(x)$.

solution 14.9   We are asked to determine where the graph is the steepest. Hence, we must maximize $m(x) = f'(x) = -3x^2 + 2x + 1$. The first derivative of $m(x)$ is $m'(x) = f''(x) = -6x + 2$, and hence, the sole critical point is at $x = \frac{1}{3}$. To show that a relative maximum occurs at $\frac{1}{3}$, we use the First Derivative Test. The sign chart for $m' = f''$ is illustrated in figure 14.10. Since $x = \frac{1}{3}$ is the only critical point, the

figure 14.10

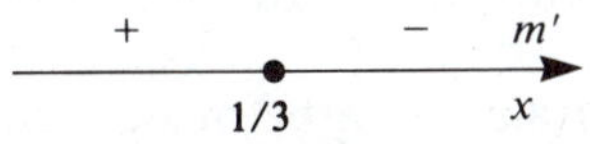

$$+ \qquad\qquad - \quad m'$$
$$\xrightarrow{\hspace{5cm}}$$
$$\quad 1/3 \qquad\qquad x$$

* For more information, see, R.M. Alexander, *Animal mechanics*. Seattle: University of Washington Press (1968), pp. 101–105.

absolute maximum occurs there. Notice also that an inflection point for the original function f occurs at $x = \frac{1}{3}$. ▤

If $f(x)$ has a derivative for all x in $[a, b]$, then the graph will be steepest at
e   from the procedure outlined in
the   as applied to $m(x) = f'(x)$. If $f(x)$ is differentiable on
the open interval (a, b) (either finite or infinite), and *if there is a place where the graph is steepest, then it will occur at an* inflection *point on the graph*.

exercises for section 14

part A

Find the largest value M and the smallest value m that $f(x)$ assumes over the prescribed interval $[a, b]$.

1. $f(x) = 2x - x^2$ on $[0, 2]$

2. $f(x) = 4x^3 - 3$ on $[-1, 1]$

3. $f(x) = x^5 - 5x + 2$ on $[-2, 2]$

4. $f(x) = x^4 - 8x^2$ on $[-1, 1]$

5. $f(x) = 2x^3 - 9x^2 + 12x$
on $[0, 3]$

6. $f(x) = x^3 + 6x - 9$ on $[0, 2]$

7. $f(x) = x^4 - 8x^2$
on $[-2.1, 2.1]$

8. $f(x) = x/(x^2 + 1)$ on $[-2, 2]$

9. $f(x) = x + 4/x$ on $[1, 3]$

10. $f(x) = x^3 e^{-x}$ on $[-1, 4]$

11. $f(x) = 3x^5 - 25x^3 + 60x$
on $[-2, 2]$

12. $f(x) = 2x - \sqrt{x^2 + 4}$ on $[0, 2]$

13. $f(x) = x/10 + \sqrt{[(10 - x)^2 + 1]}/5$
on $[0, 10]$

14. $f(x) = \dfrac{x}{5} + \dfrac{\sqrt{(10 - x)^2 + 1}}{10}$
on $[0, 10]$

15. $f(x) = x + \sqrt{x^2 + 1}$ on $[0, 2]$

Each of the following functions f has a sole critical point in the domain specified. Find the critical point and determine whether an absolute maximum or minimum occurs there.

16. $f(x) = x^2 + \dfrac{2}{x}$ for $x > 0$

17. $f(x) = 2/x - x^2$ for $x < 0$

18. $f(x) = x^3 - 6x$ for $x > 0$

19. $f(x) = x^3 + 3/x$ for $0 < x < 2$

20. $f(x) = x^2/(x^3 + 4)$ for $x > 0$

21. $f(x) = x - \sqrt{x - 1}$ for $x \geq 1$

22. $f(x) = x^2/(x^2 - 4)$ on $(-2, 2)$

23. $f(x) = \dfrac{x^2 - 1}{x^2 + 1}$ for $-\infty < x < +\infty$

24. $f(x) = x^2 e^{-x}$ for $x > 0$

25. $f(x) = x - 5 \ln x$ for $x > 0$

26. A rectangular picture frame is to be made from a piece of wood 16 feet long. How should the frame be cut in order to maximize the area enclosed by the frame?

27. A rancher wishes to make use of an existing 20-foot-long stone wall and 400 feet of fence to enclose the largest possible field, as illustrated in the figure. Find the optimal design.

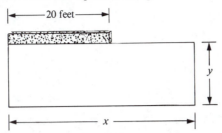

28. Refer to exercise 27. Suppose the rancher wishes to make use of the 20-foot stone wall, but has not yet purchased the fence. He needs to construct an enclosed field of area 10,000 square feet. Find the design that will minimize the total amount of fence needed.

29. A large rectangular section of pasture is to be enclosed by a fence that costs $2 per foot. Two short fences are to run down the middle to form corrals, as shown in the figure. This type of fence is much stronger and costs $5 per foot. Find the largest such field that can be enclosed for $5,000.

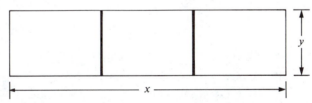

30. A rectangular field is to contain 3,000 square yards. Find the design that minimizes the diagonal distance across the field. [*Hint*: Minimize $z = (\text{diagonal})^2 = x^2 + y^2$.]

31. Find the area of the largest rectangle that can be placed inside the region shown in the figure.

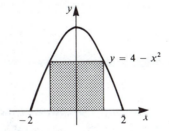

$y = 4 - x^2$

32. Shown in the figure on page 152 is the corner of a bluff overlooking the ocean.

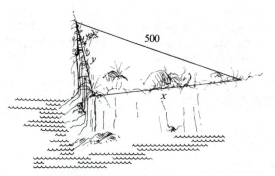

(a) Find the area of the largest triangular field that can be enclosed with 500 feet of fence running from cliff to cliff. [*Hint*: Instead of maximizing the area: A, since A is largest when A^2 is a maximum, we can maximize $A^2 = \frac{1}{4}x^2y^2$.]

(b) Could a larger area be enclosed if a rectangular design were used?

★**33.** A garden is to be built in the shape indicated in the figure. Determine the area of the largest garden that can be enclosed with 400 feet of fence. [*Hint*: Recall that circumference $C = \pi(\text{diameter})$.]

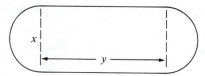

34. Rework example 14.5, assuming the box is to be open on the top.

35. A box is to be made from a $12'' \times 12''$ sheet of metal by cutting squares from the corners and turning up the sides as shown in the figure. Find the dimensions of the box of largest volume.

(a)   (b)

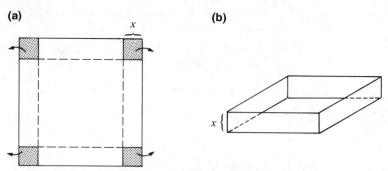

36. Rework exercise 35, assuming the sheet of metal is $16'' \times 21''$.

37. Refer to the figure on page 153, and find the volume of the largest paper cone that can be made from a circular disc of radius $s = 10''$. [*Hint*: $V = \frac{1}{3}\pi r^2 h$ and $r^2 + h^2 = s^2$.]

(a)   **(b)**

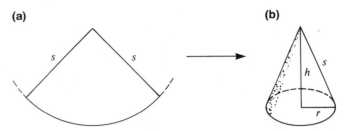

★**38.** Refer to the figure and find the dimensions of the cylindrical can with volume V_0 cubic inches that has a minimum total surface area. Show that $h = 2r$ in the optimal design. [*Hint*: $V_0 = \pi r^2 h$ and the total surface area is $S = 2\pi r^2 + (2\pi r)h$.]

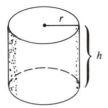

39. Rework exercise 38, assuming the top of the can is to be open.

40. A telephone cable must be laid across a river one mile in width to a town ten miles downstream, as illustrated in the figure. It costs $20,000 per mile to install the cable over water and $10,000 per mile over land. Determine the position of the point P that will result in minimum installation cost.

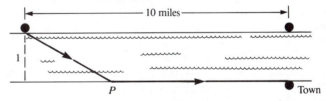

41. Imagine that you have just been given a vacation home situated on an island ten miles off the coast. The nearest town is sixty miles up the coast,

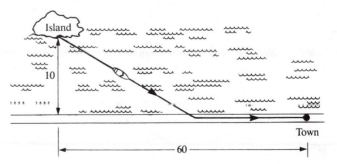

as shown in the figure. Every month you must travel into town to restock your supplies. Naturally, you have an amphibious auto. It travels 30 mph in water and 55 mph along the coastal highway. Find the route that is the quickest. [*Note*: In this problem we ignore heir resistance.]

Determine where each of the following functions is increasing most rapidly:

42. $f(x) = 15x^2 - x^3$ **43.** $f(x) = 80x - x^5$

44. $f(x) = x^2/(x^2 + 1)$ **45.** $f(x) = x^4 - 3x^2 + x + 1$
                                                for $-1 \le x \le 1$

part B

Express 36 as the sum of two positive numbers x and y such that each of the following quantities is maximized

46. xy **47.** xy^2

48. $(x^2 + y^2)$ ★ **49.** xe^y

★ **50.** As we have pointed out in example 4.8, brightness is inversely proportional to the square of the distance from the source. Three light sources of equal strength are placed as shown in the figure. At what point between the lights is brightness minimized?

51. For infants less than nine months old, the relationship between the rate of growth R (in pounds/month) and the present weight W is approximated by

$$R = cW(21 - W)$$

for some constant c. At what weight is the growth rate the largest?

52. Orville is trying to decide when to sell the turkey farm that his great uncle left him. All of that gobbling is driving him nuts. Right now, he has 1,000 turkeys ready for market which he can sell at $5 per turkey. If he waits t years, the number of marketable turkeys will grow to $N = 1,000 + 300t$, but the actual buying power derived from the sale of a single turkey diminishes according to $B = 5 - 1.2t$. When should Orville sell the turkeys?

53. At present forty apple trees are planted per acre. The final average yield is twenty bushels per tree. The experienced grower estimates that for every five new trees added per acre, the yield will decrease by two bushels per tree. How many trees should be planted per acre in order to maximize the harvest?

54. A cable television firm serves a community of 15,000 households. At present, the number of customers is 5,000 and the firm charges $20 per month. A market survey indicates that for each reduction in monthly price of $1, an increase of 500 customers will result. What price will result in maximum revenue for the company?

55. An area of five thousand square feet has been planned for the new rectangular seal exhibit at the local zoo. Decking 10 feet wide must be put along two opposite sides. In addition, decking 20 feet wide must be put along the other two sides. The enclosed area is for the pool, as shown in the figure. What design will maximize the area of the pool?

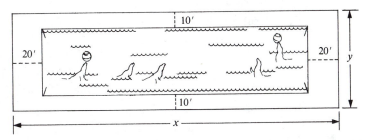

★56. A ladder is to reach over a ten-foot wall to a point three feet behind the wall, as shown in the illustration. Find the length of the shortest ladder that will work.

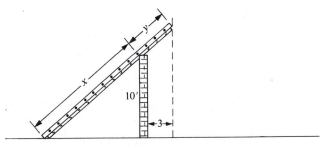

part C

57. In a simple model of territoriality, a single animal defends a circular territory of radius x (miles). The following assumptions are made:

   (1) The energy L spent per day in looking for food and defending the territory is directly proportional to the area of the region.

   (2) The energy G gained per day is directly proportional to the radius x.

   Suppose further that $L = 3,000$ calories and $G = 3,500$ calories when $x = 1$. Find the territorial size that will result in maximum benefit to the animal, i.e., maximize the net energy $G - L$.

58. When salmon struggle upstream to their spawning grounds, it is essential that they conserve energy, for they no longer feed once they have left the ocean. Let $v_0 =$ speed (in mph) of the current, $d =$ distance the salmon must travel, and $v =$ speed of the salmon. Hence, $(v - v_0)$ is the net velocity of the salmon upstream. (Refer to the figure on page 156.)

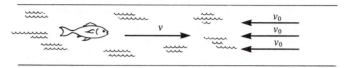

(a) Show that if the journey takes t hours, then $t = d/(v - v_0)$.

We next will assume that the amount of energy expended per hour when the speed of the salmon is v is directly proportional to v^α for some $\alpha > 1$. (Empirical data suggests this.)

(b) Show that the total energy T expended over the journey is given by

$$T = k \frac{v^\alpha}{v - v_0} \qquad \text{for } v > v_0$$

(c) Show that T is minimized by selecting velocity $v = \alpha v_0/(\alpha - 1)$.

(d) If $v_0 = 2$ mph and the salmon make the 20-mile journey by swimming for 40 hours, estimate α. What must you assume in order to do the computations?

59. For most living things, reproduction is seasonal—i.e., it can take place only at selected times of the year. Large whales, for example, reproduce every two years during a relatively short time span of about two months. Shown on the time axis in the figure are the reproductive periods. Let S = number of adults present during the reproductive period and let R = number of adults that return the next season to reproduce.

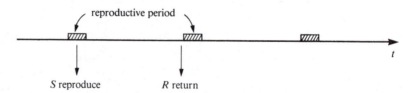

If we find a relationship between R and S, $R = f(S)$, then we have formed a *spawner-recruit* function or *parent-progeny* function. These functions are notoriously hard to develop because of the difficulty of obtaining accurate counts and because of the many hypotheses that can be made about the life stages. The actual curves devised by fishery scientists will be derived later in the text. We will simply suppose that the function f takes various forms.

If $R > S$, we can presumably harvest $H = R - S$ individuals, leaving S to reproduce. Next season, $R = f(S)$ will return and the harvesting process can again be repeated, as shown in the figure. Find the number of spawners S that will maximize the harvest

$$H = R - S = f(S) - S$$

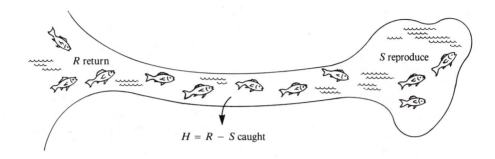

R return

S reproduce

$H = R - S$ caught

for:

(a) $f(S) = -.1S^2 + 11S$

(b) $f(S) = -S^2 + 2.2S$

(c) $f(S) = 15\sqrt{S}$

(d) $f(S) = 12S^{25}$

(e) $f(S) = 25S/(S + 2)$

(f) $f(S) = .999S$

Both R and S are measured in thousands.

60. a model of the airways during coughing It is well known that during coughing, the diameters of the trachea and bronchi decrease. Let r_0 be the normal radius of an airway at atmospheric pressure P_0. For the trachea, $r_0 \approx .5$ inch. The glottis is the opening at the entrance of the trachea through which air must pass as it either enters or leaves the lungs, as depicted in the illustration. Assume that, after a deep inspiration of air, the glottis is closed. Then pressure develops in the airways and the radius of the airway decreases. We will assume a simple linear relation between r

(Adapted from Barbara R. Landau, *Essential Anatomy and Physiology*, 2nd ed. Glenview, Ill.: Scott, Foresman, and Co., Copyright 1980.)

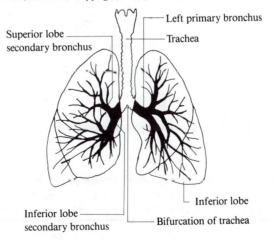

Left primary bronchus

Superior lobe
secondary bronchus

Trachea

Inferior lobe

Inferior lobe
secondary bronchus

Bifurcation of trachea

and P:

$$r - r_0 = a(P - P_0)$$

$$\text{or} \qquad \Delta r = a\Delta P$$

where $a < 0$ and $r_0/2 \leq r \leq r_0$.

When the glottis is opened, how does the air flow through these passages? We will assume that the flow is governed by Poiseuille's Laws:

(i) $v = \dfrac{P - P_0}{k}(r^2 - x^2), \qquad 0 \leq x \leq r$

(ii) $F = \dfrac{dV}{dt} = \dfrac{\pi(P - P_0)}{2k} r^4$

Here, v is the velocity at a distance x from the center of the airway (in cm/sec, e.g.) and F is the flow rate (in cm^3/sec, e.g.). The average velocity $\bar{v}$ over the circular cross section is given by

(iii) $\bar{v} = F/(\pi r^2) = \dfrac{P - P_0}{2k} r^2$

[Laws (ii) and (iii) can be derived from (i) using integrals. This will be done in section 21.]

(a) Write the flow rate as a function of r only. Find the value of r that maximizes the rate of flow.

(b) Write both the average velocity $\bar{v}$ and $v(0)$, the velocity in the center of the airway, as functions of r alone. Find the value of r that maximizes these two velocities.

(c) Use the results of parts **(a)** and **(b)** to discuss how a given quantity of air can be expelled from the lungs as efficiently as possible.*

---

* *References*: (1) J.H. Comroe, *Physiology of respiration*, 2nd ed., Chicago: Yearbook Medical Publishers, (1974), p. 230–1.

(2) B.B. Ross et al, "Physical dynamics of the cough mechanism", *J. Appl. Physiol.* (1955) **8**: 264–268.

15
exponential and logarithm functions

In this section, we will introduce two more important classes of functions that occur frequently in applications—namely, exponential and logarithm functions. An exponential function is a function of the form

$$f(x) = a^x$$

where $a > 0$ is a fixed real number called the *base*. Contrast this to the power functions x^κ, where the base is variable and the exponent κ is fixed. A logarithm function is, as we shall soon see, the *inverse* of an exponential function. In the examples that follow, we will use two key facts from section 5:

1. The graph of f^{-1} is the reflection of the graph of f in the line $y = x$.

2. $f^{-1}(f(x)) = x$ and $f(f^{-1}(x)) = x$, as illustrated in the figure.

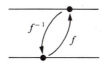

example 15.1 Sketch the graphs of $y = 2^x$ and $y = (\frac{1}{2})^x$ on the same axes.

solution 15.1 We use the law of exponents (see section 5) to construct table 15.1. The larger the value of x, the larger will be 2^x. Thus, $y = 2^x$ is an increasing function. Also, note that $2^x > 0$ for all x.

table 15.1

x	0	.5	1	2	3	-1	-2	-3
2^x	1	$\sqrt{2}$	2	4	8	.5	.25	.125

If $f(x) = (\frac{1}{2})^x$, then $f(-x) = (.5)^{-x} = 1/(.5)^x = 2^x$. Thus, the graph of $y = (\frac{1}{2})^x$ is the reflection of the graph of $y = 2^x$ in the y-axis, as shown in figure 15.1. Note that as $x \to +\infty$, $(\frac{1}{2})^x \to 0$. We call $y = 2^x$ an **exponential growth function**, and $y = (\frac{1}{2})^x$ an **exponential decay function**.

figure 15.1

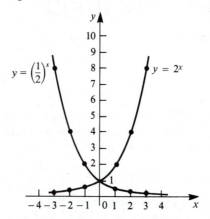

In general, the graph of $y = a^x$ takes one of the two forms illustrated in figure 15.2. Note that for an exponential decay function, $\lim_{x \to +\infty} a^x = 0$. Furthermore, in both cases, the function $f(x) = a^x$ is one-to-one. Hence, the inverse function $f^{-1}(x)$ exists.

figure 15.2(a)   **figure 15.2(b)**

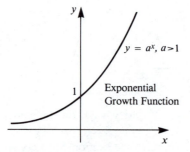
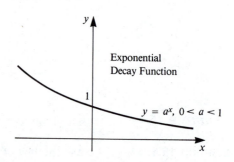

What is the nature of f^{-1}? If $f(x) = 2^x$, then from the formula $f(f^{-1}(x)) = x$, we obtain

$$2^{f^{-1}(x)} = x$$

Thus, $f^{-1}(x)$ is the power to which 2 must be raised in order to obtain x. For example, $f^{-1}(4) = 2$ since $2^2 = 4$, $f^{-1}(1) = 0$ since $2^0 = 1$, and $f^{-1}(\frac{1}{4}) = -2$ since $2^{-2} = \frac{1}{4}$. The more common name for $f^{-1}(x)$ is $\log_2 x$, *the logarithm of x to the base 2*. With this notation, $\log_2 4 = 2$, $\log_2 1 = 0$, and $\log_2 \frac{1}{4} = -2$. We will show later how to use a calculator to find, for example, $\log_2 6$, where $\log_2 6$ is the power x that satisfies:

$$2^x = 6$$

For now, we say only that x is somewhere between 2 and 3.

The inverse function for $f(x) = a^x$ is $\log_a x$. In most applications, we work with $\log_a x$ for $a > 1$. Its graph, the reflection of the graph of $y = a^x$ in the line $y = x$, is shown in figure 15.3.

figure 15.3

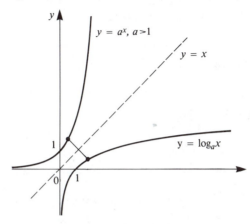

If $f(x) = a^x$ and $f^{-1}(x) = \log_a x$, then from the relationships $f(f^{-1}(x)) = x$ and $f^{-1}(f(x)) = x$. we obtain

$$a^{\log_a x} = x \qquad \text{for } x > 0$$

and

$$\log_a a^x = x \qquad \text{for all real } x$$

Thus, $\log_a x$ is the power to which the base a must be raised in order to obtain x. In the equation $a^? = x$, $? = \log_a x$.

The standard scientific calculator is equipped with two logarithm functions:

1. $\log_{10} x$, which is usually the button $\boxed{\log}$

2. $\ln x = \log_e x$ $(e = 2.7182818\ldots)$, which is denoted by $\boxed{\ln}$

You will notice from the graph of $\log_a x$ or from trying to compute logarithms on your calculator that the domain of definition of $\log_a x$ is $(0, +\infty)$. Hence, logarithms of nonpositive numbers *do not exist.*

Computations with $\log_{10} x$ are illustrated in the next two examples.

example 15.2 Compute (without the aid of a calculator) $\log_{10} (.0001)$ and $\log_{10} (1,000,000)$.

solution 15.2
$$\log_{10} (.0001) = \log_{10} (10^{-4}) = -4$$

and

$$\log_{10} (1,000,000) = \log_{10} 10^6 = 6$$

example 15.3   With the aid of a calculator, solve the equation $10^x = 27$ for x.

solution 15.3   The required power x is $\log_{10} 27 = 1.43136376 \ldots.$

Log$_a$ x is a "true logarithm," and the usual laws of logarithms, (see box), that you have encountered in an algebra course are still valid.

Laws of Logarithms

1. $\log_a (xy) = \log_a x + \log_a y$

2. $\log_a (x/y) = \log_a x - \log_a y$

3. $\log_a x^\kappa = \kappa \log_a x$

4. $\log_a a = 1$ and $\log_a 1 = 0$

All of these laws are quite easy to prove. To prove (1), first write $x = a^{\log_a x}$ and $y = a^{\log_a y}$. By the law of exponents,

$$xy = a^{(\log_a x + \log_a y)}$$

Thus, in order to obtain xy, we raise a to the power $(\log_a x + \log_a y)$. Hence, $\log_a (xy) = \log_a x + \log_a y$.

example 15.4   If $\log_a x = 2.1$ and $\log_a y = .45$, compute

(a) $\log_a x^3$ **(b)** $\log_a (x^3 y)$ **(c)** $\log_a \sqrt[3]{y}$

solution 15.4   Using Law (3), we obtain $\log_a x^3 = 3 \log_a x = 3(2.1) = 6.3$. Law (1) yields $\log_a (x^3 y) = \log_a x^3 + \log_a y = 6.3 + .45 = 6.75$. Finally, $\log_a \sqrt[3]{y} = \log_a y^{1/3} = \frac{1}{3} \log_a y = .15$.

The final two logarithm laws allow us to compute logarithms *to any base* using either $\boxed{\log}$ or $\boxed{\ln}$ on the calculator.

Laws of Logarithms

5. $\log_a x = (\log_{10} x)/(\log_{10} a)$

6. $\log_a x = (\ln x)/(\ln a)$

Laws 5 and 6 are not difficult to prove. To prove Law 5, we first write $x = a^{\log_a x}$ and $a = 10^{\log_{10} a}$. Hence,

$$x = a^{\log_a x} = (10^{\log_{10} a})^{\log_a x} = 10^{(\log_{10} a)(\log_a x)}$$

Thus, in order to obtain x, we raise 10 to the power $(\log_{10} a)(\log_a x)$. We conclude that $\log_{10} x = (\log_{10} a)(\log_a x)$, and the proof is complete.

example 15.5   Compute $\log_2 6$.

solution 15.5 $\log_2 6 = (\log_{10} 6)/(\log_{10} 2) = .77815125/.30102999 = 2.584962 \ldots$

example 15.6   Solve the equation $5^x = .23$.

solution 15.6   We are asked to find the power to which we must raise 5 in order to obtain .23. This is just the definition of $\log_5 .23$. As in example 15.5, $\log_5 .23 = (\log_{10} .23)/(\log_{10} 5) = -.63827216/.69897000 = -.91316102 \ldots$

At one time, logarithms were used as an aid in carrying out computations such as $(3.1)^{.43}$ or $(2.43)(.691)/(1.302)$. The modern scientific calculator has replaced this logarithmic method of computation. We will use laws (1)–(6) to solve exponential and logarithmic equations, as illustrated in examples 15.7 and 15.8.

example 15.7   Solve the equations

(a) $2^{3x} = 1.7$ **(b)** $\log_4 (3x) = 1.4$

solution 15.7   **(a)** We will take $\log_{10}$ of both sides and use Law (3): $\log_{10} 2^3 = \log_{10} 1.7$ implies $3x(\log_{10} 2) = \log_{10} 1.7$. Solving this linear equation for x, we have $x = (\log_{10} 1.7)/(3 \log_{10} 2) = .255178 \ldots$

For **(b)**, $\log_4 (3x) = 1.4$ implies that in order to obtain $3x$, we must raise 4 to the power 1.4. Thus, $3x = 4^{1.4}$ or $x = (\frac{1}{3})4^{1.4}$. Using the power button on your calculator, you should obtain $x = 2.321468 \ldots$

example 15.8   A radioactive material decays according to the law $N = 5e^{-.4t}$, where N is the number of grams present after t months.

(a) When will only 1 gram of the substance remain?
(b) What is the **half-life** of the substance—i.e., how long does it take for N grams to decay to $(\frac{1}{2})N$ grams?

solution 15.8

(a) We are being asked when $N = 5e^{-.4t}$ equals 1. Now, $5e^{-.4t} = 1$ implies $e^{-.4t} = .2$. Although we could proceed as in example 15.7 and take $\log_{10}$ of both sides, it is easier to use $\ln = \log_e$. Hence, $\ln (e^{-.4t}) = \ln (.2)$ or $-.4t \ln e = \ln (.2)$. But $\ln e = \log_e e = 1$. Hence, $t = -[\ln (.2)]/.4 = 4.0235 \ldots$ months.

For **(b)**, let τ denote the half-life. If $N = 5e^{-.4t}$, then $(\frac{1}{2})N = 2.5e^{-.4t}$. But at time $(t + \tau)$, $(\frac{1}{2})N$ grams remain, by definition of half-life. Hence, a second expression for $(\frac{1}{2})N$ is $5e^{-.4(t+\tau)}$. Hence,

$$2.5e^{-.4t} = 5e^{-.4(t+\tau)} = 5e^{-.4t}e^{-.4\tau}$$

It follows that $e^{-.4\tau} = .5$. Taking $\ln$ of both sides yields $\ln (e^{-.4\tau}) = \ln (.5)$ or $-.4\tau \ln e = -.4\tau = \ln (.5)$. Solving for τ, we obtain $\tau = [\ln (.5)]/(-.4) = 1.7328 \ldots$ months. Note that the value for τ *does not depend on N*. It takes 1.7328 months for 5 grams to decay to 2.5 grams, 1.7328 months for 2.5 grams to decay to 1.25 grams, and so forth.   ▤

Note that if $f(x) = ca^x$, then $f(x + 1) = ca^{x+1} = c \cdot a \cdot a^x = af(x)$, regardless of the value of x. This property can be used to explain why exponential functions occur so frequently in science. Here is one such example.

example 15.9

Beer-Lambert Law   Suppose that a light ray of intensity I_0 (as measured by a lightmeter, for example) strikes the surface of a medium as shown in figure 15.4. The medium could be glass or seawater, for example. How much light

figure 15.4

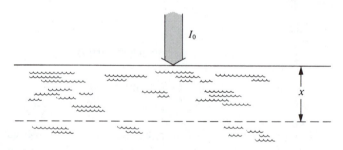

reaches depth x? For purposes of illustration, assume that $\frac{1}{4}$ of the original amount reaches a depth of 1 meter (as would be the case for very clear seawater). Thus, $I(1) = \frac{1}{4}I(0) = \frac{1}{4}I_0$.

Next let's compare $I(x + 1)$ to $I(x)$, as depicted in figure 15.5. The light reaching depth x has no memory of how it got there. In attempting to reach depth $x + 1$, it is as if a new free surface were at x. In other words, the situation should be the same as at the original surface. Hence, $I(x + 1) = \frac{1}{4}I(x)$. Thus, $I(x) = I_0(\frac{1}{4})^x$ describes the situation. In general, $I(x) = I_0 a^x$ where $0 < a < 1$ and a depends on the medium. This is known as the **Beer-Lambert Law**.

figure 15.5(a)

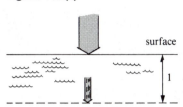

figure 15.5(b)

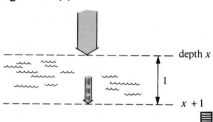

Logarithm functions are often used to scale down measurements that otherwise would stretch over a very large range. A good example of this is the scale used to measure the intensity of earthquakes, discussed in the next example.

example 15.10   **the Richter Scale**   A *seismograph* is a device that measures the actual amplitude of ground movement during an earthquake. These physical measurements vary so greatly that a "log scale" is used to report the results. More specifically, we define

$$\text{(Richter number)} \quad R = \log_{10}(A/A_0)$$

where A_0 is the smallest ground movement that the seismograph can detect. (An earthquake of amplitude A_0 might rattle a few dishes.) As an example, the famous 1971 Los Angeles earthquake measured 6.7. Thus, $6.7 = \log_{10}(A/A_0)$, which implies $A/A_0 = 10^{6.7}$. If an earthquake measured 7.7, then $A/A_0 = 10^{7.7} = 10(10^{6.7})$, equivalent to ten times the earth movement of the Los Angeles earthquake. The largest recorded earthquake (1933, in Japan) measured 8.9. Thus, $A/A_0 = 10^{8.9} = (10^{2.2})(10^{6.7}) \approx (160)10^{6.7}$. This earthquake then was about 160 times as strong as the Los Angeles earthquake, and $10^{8.9} \approx 3.1$ billion times as strong as an earthquake of threshold intensity A_0.

exercises for section 15

part A   Graph the following pairs of exponential functions on the same axes.

1. $y = (\tfrac{3}{2})^x$ and $y = (\tfrac{2}{3})^x$

2. $y = (2.7)^x$ and $y = (1/2.7)^x$

3. $y = 3(.5)^x$ and $y = 3(2^x)$

4. $y = (\tfrac{3}{4})^x$ and $y = 2(\tfrac{3}{4})^x$

5. $y = 1 - (.5)^x$ and $y = 1 - (.75)^x$
for $x > 0$

Graph each of the following logarithmic functions.

6. $y = \log_3 x$ **7** $y = \log_{10} x$

8. $y = \ln x = \log_e x$ (Use $\boxed{\log}$ on your calculator.)

    (Use $\boxed{\ln}$ on your calculator.)   **9.** $y = \log_4 x$

★**10.** $y = \log_2 (x - 1)$

If $y = a^x$, then $x = \log_a y$. Use this fact to convert each of the following exponential equations to logarithmic form.

11. $4^3 = 64$ **12.** $\frac{1}{4} = 2^{-2}$ **13.** $10,000 = 10^4$

14. $8^{-2/3} = \frac{1}{4}$ **15.** $3^x = 5$ **16.** $2(5^x) = 4$

17. $2^{3x} = 5$ **18.** $3^x = \frac{1}{27}$ **19.** $x = 2^8$

★**20.** $x^2 = 2^x$

If $y = \log_a x$, then $a^y = x$. Use this fact to compute each of the following logarithms.

21. $\log_2 128$ **22.** $\log_2 \sqrt[4]{2}$ **23.** $\log_3 27$

24. $\log_{\sqrt{10}} 100$ **25.** $\log_{\sqrt{2}} 8$ **26.** $\log_b (1/b)$

27. $\log_e 1$ **28.** $\log_x x^3$ **29.** $\log_a a$

30. $\log_{1.4} \sqrt{1.4}$

Use your calculator to compute (if possible) $\ln x$ and e^x for:

31. $x = 1.31$ **32.** $x = 18.4$ **33.** $x = -3.2$

34. $x = \pi$ **35.** $x = e$

Use your calculator and the formula $\log_a x = (\log_{10} x)/(\log_{10} a)$ to compute:

36. $\log_2 7$ **37.** $\log_7 2$ **38.** $\log_8 9$

39. $\log_9 8$ ★**40.** $\log_2 (\log_3 4)$

Solve each of the following exponential equations for x:

41. $3(2^x) = 12$ **42.** $4(3^{2x}) = 8$ **43.** $3(e^{-x}) = 4$

44. $6(.5)^x = 8$ **45.** $5 - 2^{3x} = 3$ **46.** $3 - 2^{3x} = 5$

47. $e^{.05x} = 6$ **48.** $4e^{-.2x} = 28$ **49.** $10(2^{-x}) = 6$

50. $3^{x^2-x} = 9$

Solve each of the following logarithmic equations for x:

51. $\log_{10} 2x = 50$ **52.** $\ln (2x + 3) = 2$ **53.** $\log_2 (x + 4) = 8$

54. $\log_8 (x + 4) = 2$ ★**55.** $\log_2 (\log_3 2x) = 4$

If $\log_a x = .5$ and $\log_a y = 3.2$, compute:

56. $\log_a (xy^{-2})$ **57.** $\log_a \sqrt[3]{xy}$ **58.** $\log_a (x^5 y^2)$

59. x if $a = 8$ **60.** a if $y = 4$

61. Radioactive iodine I^{131} is frequently used in tracer studies in the human body (and notably in tests of the thyroid gland). The substance decays according to the equation

$$N = N_0(\tfrac{1}{2})^{t/8}$$

Show that the half-life of I^{131} is about 8 days.

62. The radioactive substance Strontium 90 has a half-life of about 29 years. Use this information to find the constant k in the radioactive decay equation

$$A(t) = A_0 e^{-kt}$$

where $A_0 = A(0)$ is the initial amount present.

63. One thousand young trout are put in a fishing pond. Three months later, the owner estimates that there are about 600 left. Find an exponential function

$$N = N_0 a^t$$

fitting this information, and use it to estimate the number of trout left after one year.

64. The common bacteria *E. Coli* undergo cell division approximately every 20 minutes. If we start with 10,000 cells, and assume that an exponential function is appropriate, how many cells are present after 1.5 hours?

65. Ten thousand dollars are placed into a high interest account where interest is *compounded continuously* at the rate of 11% per year. The amount A in the account t years later is given by

$$A = 10,000 e^{.11t}$$

(a) When will the account contain \$35,000 (for your silver Maserati)?

(b) How long does it take your money to double in such an account?

The most important zone in the sea from the viewpoint of marine biology is the *photic zone*, the zone in which photosynthesis can take place. For marine phytoplankton, the photic zone must end at the depth where about 1 percent of the surface light penetrates.

66. Near Cape Cod, Massachusetts, the depth of the photic zone is about 16 meters. Find the constant a in the Beer-Lambert Law

$$I = I_0 a^x$$

67. In very clear waters in the Caribbean, 50 percent of the light at the surface reaches a depth of about 13 meters. Estimate the depth of the photic zone.

68. The 1964 earthquake in Alaska measured 8.4 on the Richter scale. Compare the actual amplitude A of the ground movement to that of

(a) the 1933 earthquake in Japan that measured 8.9.

(b) the 1971 Los Angeles earthquake that measured 6.7.

69. Solve the Richter scale equation $R = \log_{10}(A/A_0)$ for A.

70. The hydrogen ion concentration H^+ of a solution (measured in moles of hydrogen ion per liter of solution) may vary greatly over negative powers of ten. To scale down these measurements, chemists have invented the formula for pH:

$$pH = -\log_{10}[H^+]$$

For water at 24°C, pH = 7. Hence, $\log_{10}[H^+] = -7$, which gives $[H^+] = 10^{-7}$.

(a) Compute $[H^+]$ for a solution in which pH = 2.1.

(b) If $[H^+] = 1.5 \times 10^{-6}$, find the pH.

71. Normal or "clean" rain has a pH of 5.6. The lower the pH value, the higher the acidic content.

(a) A typical pH value for rain in the eastern United States is 3.8. How many times more acidic than clean rain is rain of pH 3.8?

(b) If the pH of lake water falls below 5, adult fish may fail to reproduce. Compare the acidic level in such a lake to that in a clean lake.

 # derivatives of exponential and logarithm functions

In this section we will derive the following formulas that were stated in section 9:

$$D_x a^x = (\ln a)a^x$$

$$D_x \log_a x = \frac{1}{\ln a} \cdot \frac{1}{x}$$

Here, $\ln a$ is $\log_e a$, the logarithm of a to the base $e = 2.7182818\ldots$. We first encountered the special number e in section 7, example 7.7, where it was defined as

$$e = \lim_{h \to 0} (1 + h)^{1/h}$$

Note that $\ln e = \log_e e = 1$. Hence, of all the exponential and logarithm functions, e^x and $\ln x$ have the simplest derivatives:

$$D_x e^x = e^x$$

$$D_x \ln x = \frac{1}{x}$$

It is this fact that explains why this strange number e has found a place in science. *From the point of view of calculus, e^x and $\ln x$ are the easiest to work with.*

In order to derive these formulas, we must return to the limit definition of a derivative. Let $f(x) = a^x$. Then

$$f'(x) = \lim_{h \to 0} \frac{f(x + h) - f(x)}{h} = \lim_{h \to 0} \frac{a^{x+h} - a^x}{h}$$

For $x = 0$, the last equation becomes

$$f'(0) = \lim_{h \to 0} \frac{a^h - 1}{h}$$

If we note that

$$\frac{a^{x+h} - a^x}{h} = \frac{a^x a^h - a^x}{h} = a^x \cdot \frac{a^h - 1}{h}$$

then it follows that $f'(x) = f'(0)a^x$. This is very close to the formula $f'(x) = (\ln a)a^x$ that we are trying to derive. We will first, however, derive the formula for the derivative of e^x, and use the resulting formula to complete the derivation for $D_x a^x$.

Next, we wish to find the special number a with $f'(0) = 1$. Recall that $f'(0)$ is just the slope of the tangent line at $x = 0$. As figure 16.1 shows, if $a = 2$, then the function $y = 2^x$ has tangent line with slope less than 1. In fact, the slope is about .7. Figure 16.2 shows $y = 3^x$ and the line $y = x + 1$. It looks like this line could be the tangent line at $(0, 1)$.

figure 16.1

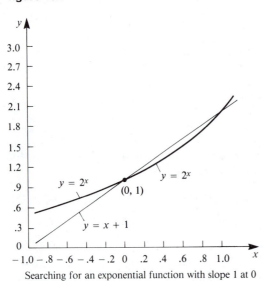

Searching for an exponential function with slope 1 at 0

figure 16.2

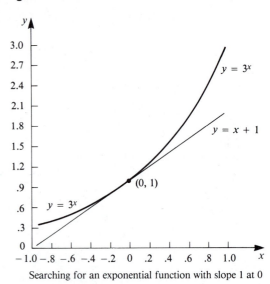

Searching for an exponential function with slope 1 at 0

Figure 16.3, however, shows that the tangent line to $y = 3^x$ has slope slightly greater than 1. Hence, the special number a with $f'(0) = 1$ is slightly less than 3. We will next show that if $a = e$, then $f'(0) = 1$ and consequently $f'(x) = D_x e^x = e^x$.

For h very small, $e \approx (1 + h)^{1/h}$. Hence, $e^h \approx 1 + h$ and so

$$\frac{e^h - 1}{h} \approx \frac{(1 + h) - 1}{h} = 1$$

Since the approximation of e by $(1 + h)^{1/h}$ improves as $h \to 0$, we have

$$f'(0) = \lim_{h \to 0} \frac{e^h - 1}{h} = 1$$

figure 16.3 A closer look around (0, .1)

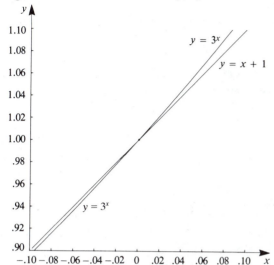

figure 16.4 The graphs of exp(x) and ln(x)

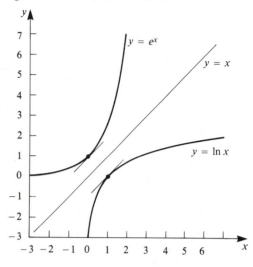

All of the remaining formulas can be derived from the following special case of the Chain Rule:

$$D_x e^{f(x)} = e^{f(x)} f'(x)$$

To derive the formula $D_x a^x = (\ln a)a^x$, write a as $e^{\ln a}$. Then

$$D_x a^x = D_x e^{(\ln a)x} = e^{(\ln a)x} \cdot (\ln a)$$
$$= (\ln a)a^x$$

To derive $D_x \ln x$, note that $x = e^{\ln x}$. Differentiating both sides with respect to x yields

$$1 = e^{\ln x} \cdot D_x \ln x = x \cdot D_x \ln x$$

Hence, $D_x \ln x = 1/x$. The graphs of e^x and $\ln x$ are shown in figure 16.4.

Finally, we may write $\log_a x = (\ln x)/(\ln a)$ using Law (6) of logarithms (see section 15). Hence, $D_x (\log_a x) = \dfrac{1}{x} \cdot \dfrac{1}{\ln a}$.

The following formulas are special cases of the formulas we have just derived:

$$D_x 2^x = (.693147\ldots)2^x$$
$$D_x e^{kx} = ke^{kx}$$
$$D_x \log_{10} x = (.434294\ldots)\frac{1}{x}$$

In addition, the Chain Rule gives:

$$D_x \ln f(x) = \frac{f'(x)}{f(x)}$$

Let's put these formulas to work in order to compute rates of change, sketch graphs involving exponentials and logarithms, and solve optimization problems.

example 16.1   A bacteria culture is doubling its size every two hours. Initially, the number of bacteria per mL is 10^6.

(a) Find an exponential function $N = ca^t$ that fits the data.
(b) Find the rate of growth of the population at times $t = 1$ and $t = 3$.
(c) Finally, show that $\frac{dN}{dt} = kN$ for some constant k. Conclude that the rate of growth is directly proportional to the population count.

solution 16.1   **(a)** We are given that when $t = 0$, $N = 10^6$ and when $t = 2$, $N = 2(10^6)$. If $N = ca^t$, then $10^6 = N(0) = ca^0 = c$. Hence, $N = 10^6 a^t$. Furthermore, $2(10^6) = N(2) = 10^6 a^2$. Therefore, $a^2 = 2$ and hence $a = 2^{1/2}$. The final function is then $N = 10^6 (\sqrt{2})^t$.

(b) To find rates of growth, we must compute the derivative of N

$$\frac{dN}{dt} = [10^6 \ln(\sqrt{2})](\sqrt{2})^t$$

When $t = 1$,

$$\frac{dN}{dt} = [\sqrt{2}\ln(\sqrt{2})]10^6 = 490,129 \text{ bacteria/hour}$$

When $t = 3$,

$$\frac{dN}{dt} = [(\sqrt{2})^3 \ln(\sqrt{2})]10^6 = 980,258 \text{ bacteria/hour}$$

(c) In **(b)** we showed $\frac{dN}{dt} = \ln(\sqrt{2})[10^6(\sqrt{2})^t] = kN$, with $k = \ln(\sqrt{2})$. Hence, $\frac{dN}{dt}$ and N are directly proportional.

example 16.2   **Weber-Fechner model**   The Weber-Fechner "Law" asserts that the response R of the body to an outside stimulus of strength x is given by:

$$R = \alpha \ln \frac{x}{x_0}, \qquad x \geq x_0$$

where α is a fixed constant and x_0 is called the **threshold intensity**—i.e., the intensity at which a response is first possible. Recall from example 11.6 that the sensitivity at level x, a measure of the ability to detect small stimulus changes, is defined by $S = \dfrac{dR}{dx}$. Writing $R = \alpha[\ln x - \ln x_0] = \alpha \ln x - \alpha \ln x_0$, we obtain

$$S = \frac{dR}{dx} = \frac{\alpha}{x} - 0 = \frac{\alpha}{x}$$

Hence, *sensitivity S is inversely proportional to stimulus strength x.* According to this model, doubling the stimulus level x would halve the ability S to detect small differences in stimuli.   ▤

example 16.3   The **unit normal curve** is one of the most widely used curves in statistics. Its exact equation is:

$$y = \frac{1}{\sqrt{2\pi}} e^{-x^2/2}$$

Sketch the graph of $y = e^{-x^2/2}$ and show that it has a "bell shape."

solution 16.3   We will proceed exactly as in sections 12 and 13. By the Chain Rule, $y' = e^{-x^2/2}(-x)$. Hence, since $e^{-x^2/2} > 0$, the only critical point is $x = 0$. Computing the second derivative by the Product Rule yields

$$y'' = e^{-x^2/2}(-1) + (-x)e^{-x^2/2}(-x)$$
$$= e^{-x^2/2}(-1 + x^2)$$

Hence, $y'' = 0$ implies $x = \pm 1$. Now $y''(0) = -1 < 0$. Hence, by the Second Derivative Test, a relative maximum occurs at 0. Since $y''(-2) = y''(2) > 0$, the sign chart for y'' is as shown in figure 16.5. Thus, inflection points occur at $x = \pm 1$. Now, $y'(1) < 0$ and $y'(-1) > 0$. Hence, the sign chart for y' is as illustrated in figure 16.6, and we conclude that the graph has general shape

figure 16.5

figure 16.6

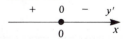

"up-down," with a relative maximum at 0. Finally, we plot the points in table 16.1 and obtain the graph shown in figure 16.7.

table 16.1

x	0	±1	±2	±3	±.5	±1.5
$e^{-x^2/2}$	1	.6065	.1353	.01	.8825	.3246

figure 16.7 P and R are inflection points.

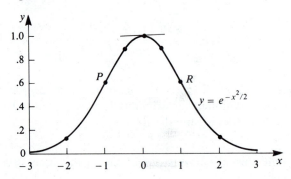

example 16.4   Shown in figure 16.8 is a computer-generated graph of $y = -x^2 \ln x$, for $0 < x < 1$. There appears to be a relative maximum somewhere near .6 and an inflection point between .2 and .3.

figure 16.8

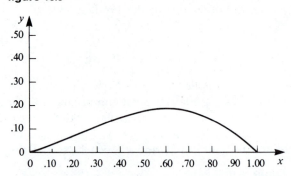

(a) Determine the exact location of the relative maximum.

(b) Determine the location of all inflection points.

solution 16.4

To compute y', we use the Product Rule:

$$y' = -x^2 \frac{1}{x} - 2x(\ln x) = -x(1 + 2 \ln x)$$

Hence, the nonzero critical point satisfies $1 + 2 \ln x = 0$ or $\ln x = -.5$. Thus, $x = e^{-.5} = .60653065 \ldots$ The second derivative is

$$y'' = -x \frac{2}{x} + (1 + 2 \ln x)(-1) = -3 - 2 \ln x$$

The only possible inflection point satisfies $\ln x = -1.5$. Hence, $x = e^{-1.5} = .22331016 \ldots$ To verify that $x = e^{-.5}$ is a relative maximum, note that $y''(e^{-.5}) = -3 - 2 \ln (e^{-.5}) = -3 + 1 = -2 < 0$. To verify that $x = e^{-1.5}$ is an inflection point, construct the sign chart for y'', as shown in figure 16.9.

figure 16.9

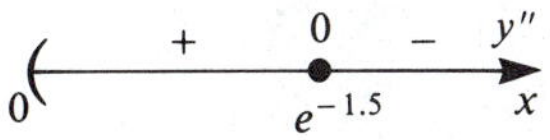

example 16.5

The owner of a trout farm purchases 5,000 fingerlings from the hatchery and places them in a large pond to grow. From tagging studies, he knows that the number $N(t)$ of fish alive t years later is well described by $N(t) = 5,000e^{-.6t}$. The weight (in pounds) of an individual trout is given by $W(t) = .01 + .5t$. When should the pond be harvested?

solution 16.5

The owner clearly wants to maximize the total number of pounds of trout, and must balance their natural growth with their dwindling numbers. The function to be maximized is

$$B(t) = N(t)W(t) = 5,000e^{-.6t}(.01 + .5t), \qquad \text{for } t > 0$$

The function $B(t)$ is called the **biomass**. Computing B', we obtain

$$B'(t) = 5,000e^{-.6t}(.5) + (.01 + .5t) \cdot 5,000e^{-.6t}(-.6)$$

$$= 5,000e^{-.6t}(.5 - .006 - .3t)$$

Since $e^{-.6t} > 0$, the unique critical point satisfies $.3t = .494$. Hence, $t = .494/.3 = 1.6\overline{46} \approx 1.65$ years. The reader can verify that a relative maximum occurs at $t = 1.6\overline{46}$ by using either the First or Second Derivative Test. Since $1.6\overline{46}$ is the only critical point, the absolute maximum occurs there. The maximum biomass is $B(1.6\overline{46}) = 1,551.3$ pounds.

exercises for section 16

part A

1. Estimate $f'(0)$ for $f(x) = 2^x$ by computing $\lim_{h \to 0} (2^h - 1)/h$ correct to three decimal places.

2. Estimate $f'(0)$ for $f(x) = 3^x$ by computing $\lim_{h \to 0} (3^h - 1)/h$ correct to three decimal places.

3. Compute $(e^h - 1)/h$ for $h = .01, .001, .0001$ to estimate $f'(0)$ for $f(x) = e^x$.

Compute the derivative of each of the following exponential and logarithmic functions.

4. 2^x

5. 3^{-x}

6. e^{4x}

7. e^{-3x+1}

8. e^{-2x}

9. e^{-ex}

10. $\frac{1}{2}(e^x - e^{-x})$

11. $x \, 2^x$

12. $(e^{2x} + 1)^5$

▲13. $\sin(e^{-x})$

14. $\ln(x + 1)$

15. $\ln(x^2 + x - 4)$

16. $\ln(x + e^x)$

17. $e^{x + \ln x}$

18. $20(1 - e^{-2x})$

19. $15(1 - e^{-5x})$

20. xe^{-x}

21. $x^2 e^{-2x}$

22. $(x + 4 \ln x)^3$

23. $e^{2x}/(x^2 + 1)$

24. $(\ln x)/x$

25. $(1/x) \, e^{7x-11}$

26. $e^{-x}/(1 - 2e^{-x})$

27. $x^5 \ln x$

▲28. $e^{-x} \cos x$

▲29. $e^{-2x} \sin x$

★30. x^x [*Hint*: Write $x = e^{\ln x}$.]

31. A population of animals is growing exponentially according to the formula

$$N = 100e^{.1t}$$

where t is measured in years. How fast is the population growing after 1 year? 2 years? Initially $(t = 0)$?

32. A cup of coffee of temperature $100°$ is placed in a large room of temperature $70°$ and begins to cool, as depicted in the figure. Its temperature t hours later is given by $T = 70 + 30e^{-2t}$. Compute the rate at which it is cooling after 30 minutes, 1 hour, and 2 hours.

33. The statistician D.G. Chapman,* in a study of a population of male fur seals, found that the population was well described by the *logistic* function

$$N = \frac{12,000}{1 + 4e^{-.314t}}$$

Find the rate of population growth after 0, 1 and 2 years.

34. For the Weber-Fechner Law (see example 16.2) $R = \alpha \ln(x/x_0)$, compare the sensitivities at stimulus levels x_0 and $2x_0$.

▲★**35.** The function $x(t) = e^{-t} \cos t$ gives the displacement from equilibrium of a weight bobbing on a spring after t seconds. Compute the velocity of the weight at times 0, $\pi/2$, and π.

36. The thriving community of Circle City is growing exponentially according the formula
$$N = 10,000e^{.1t}$$

To accommodate the incoming residents, the city will expand outward in a circle, but the population density is to remain at 100 residents per square mile.

(a) How far from the center of town will the outskirts be after 5 years?

(b) How fast is this distance increasing after 5 years?

37. Shown in the figure is a computer-generated graph of $y = xe^{-x}$.

(a) Determine where the maximum occurs and find the maximum value M.

(b) Is there an inflection point on the graph? If so, exactly where does it occur?

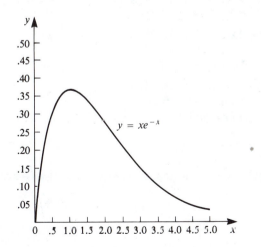

* D.G. Chapman, "Population Dynamics of the Alaska Fur Seal Herd", *Wildlife Conference*, Wash., (1961) *26*: 356–69.

38. Shown below is a computer-generated graph of $y = 10 \ln x - 1.5x$.

(a) Show that a relative maximum occurs at $x = 20/3$.

(b) Show that the graph is concave downward for $x > 0$.

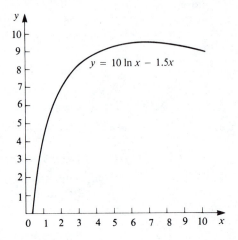

39. Sketch the graph of $f(x) = \ln(x^2 + 1)$. Where are the inflection points on the graph?

40. Sketch the graph of $f(x) = x \ln x$, for $0 < x < 1$.

★**41.** Functions of the form

$$f(x) = \frac{\alpha^{n+1}}{n!} x^n e^{-\alpha x}, \qquad x > 0$$

are called *gamma distributions* and play an important role in probability theory. Shown here are the graphs for $\alpha = 1$ and $n = 2, 3$ and 4. [Recall that $2! = 2$, $3! = 6$, and $4! = 24$.]

(a) Determine where the absolute maximum occurs in each case.

(b) Where are the functions increasing the most rapidly?

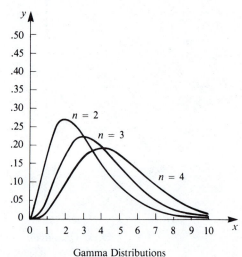

Gamma Distributions

42. Find the maximum value of $y = e^{-x} - e^{-2x}$ for $x \geq 0$.

★**43.** Determine the general shape of the graph of $f(x) = e^{-x} - \frac{5}{2}e^{-2x} + 2e^{-3x}$ for $x \geq 0$.

44. In the trout farm problem in example 16.5, suppose that the owner feeds the trout a special diet that results in a new weight function $W = .01 + .8t$. When should the pond be harvested?

45. Rework the trout farm problem in example 16.5, assuming $W = 4(1 - e^{-.25t})$.

★**46.** Rework the trout farm problem in example 16.5, assuming $W = 5(1 - e^{-.8t})^3$. This growth function is a *von Bertalanffy growth function*, commonly used in fishery science.

part B

Logarithmic Differentiation The derivative of

$$f(x) = \frac{4(x-1)^3}{(x-3)^4 (x-2)^3}$$

can be computed using the Quotient and Chain Rules. An easier way of finding $f'(x)$ is to first simplify $\ln f(x)$, using the logarithm laws, and then differentiate:

$$\ln f(x) = \ln 4 + 3 \ln(x-1) - 4 \ln(x-3) - 3 \ln(x-2)$$

and hence,

$$\frac{f'(x)}{f(x)} = D_x \ln f(x) = \frac{3}{x-1} - \frac{4}{x-3} - \frac{3}{x-2}$$

It follows that

$$f'(x) = f(x)\left(\frac{3}{x-1} - \frac{4}{x-3} - \frac{3}{x-2}\right)$$

This is the method known as **logarithmic differentiation**.
Use this method to compute $f'(x)$ for:

47. $f(x) = \dfrac{\sqrt{x-1}}{x^2+1}$

48. $f(x) = \dfrac{x(2x-3)^3}{(x+1)^4}$

49. $f(x) = \dfrac{x^2 e^x}{(x+4)^4}$

50. $f(x) = \dfrac{x^5(2x+3)^7 (x-1)^3}{(x+4)^6}$

51. The **logistic population curve** takes the form

$$N = \frac{K}{1 + ce^{-rt}} \qquad \text{where } r, K, \text{ and } c > 0$$

(a) Compute N' and show that $N' = rN(K-N)/(K)$. Conclude that the rate of population growth is 0 for $N = 0$ and $N = K$.)

(b) What is the limit of $N(t)$ as $t \to +\infty$?

(c) Show that the rate of growth $R = N' = rN(K - N)/K$ is largest when $N = K/2$.

optimal design in nerve fibers

52. Shown in the figure are two sketches of myelinated nerve fibers. These fibers vary in diameter from .001 mm to .02 mm. The *axon* is the actual transport line for the nerve impulses. A layer of the lipid *myelin* covers the axon. This layer (the *myelin sheath*) is an insulator that protects the axon from the outside environment and retards the flow of ions in the radial direction. As a result, the velocity of an impulse along the axon is increased. The *nodes of Ranvier* serve as relay stations along the fiber. The portion of fiber between two nodes resembles a cylindrical cable with insulation.

Let $x = r/R$. For cables, it is known from physics that the velocity v of an impulse is given by

$$v = -\alpha x^2 \ln x$$

where $0 < x \le 1$ and α is a constant. *Remark*: This formula is a consequence of Laplace's equation for an annular region. The derivation from first principles is lengthy.*

(Reprinted by permission from, Barbara R. Landau, *Essential Anatomy and Physiology.* 2nd ed. Glenview, Ill.: Scott, Foresman, and Co., copyright 1980.)

(a)   **(b)**   **(c)**

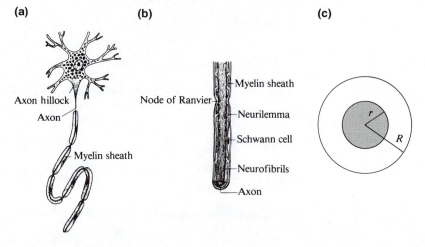

Axon hillock
Axon
Myelin sheath

Node of Ranvier
Myelin sheath
Neurilemma
Schwann cell
Neurofibrils
Axon

* A complete discussion of the material in this project from a physiological point of view is given in W.A.H. Rushton, "A Theory of the Effects of Fibre Size in Medullated Nerve" *J. Physiol.* (1951) *115*: 101–122. In particular, see pages 103–106.

(a) Show that $\lim_{x\to 0^+} v(x) = 0$ by letting $x = .01, .001, \ldots$.*

(b) Find the value of $x = r/R$ that maximizes the velocity v. The observed ratio between r and R for the majority of myelinated nerve fibers is around .6.

   A nerve cell does not fire until the charge on it exceeds a certain threshold value. Once this value is reached, a spike of fixed size is sent along the axon.

★(c) Suppose the charge is increased by one unit every τ seconds and that individual charges decay exponentially according to e^{-kt}.

(i) Find the the largest possible charge on the cell.

(ii) Sketch the graph of total charge versus time.

---

* The notation "$x \to 0^+$" means "x approaches 0 from the right."

17 applications of exponential functions

As we saw in section 16, the derivatives of exponential functions $y = a^x$ have a very special property: $y' = ky$, where $k = \ln a$. Thus, the original function y and its derivative y' are **directly proportional**. This special property can be used to explain the occurrence of exponential functions in a variety of situations.

In what follows, we will write $y = a^x$ in the form e^{kx}, where $k = \ln a$. To see that this is valid, note that $a = e^{\ln a}$. (This is just $f(f^{-1}(a)) = a$ again!) Hence, $a^x = e^{(\ln a)x}$.

If $f(x) = ce^{kx}$, then $f'(x) = ce^{kx} \cdot k = k(ce^{kx}) = kf(x)$. We will show in section 18 that ce^{kx} is the *only* type of function that satisfies $y' = ky$. Hence

$$y' = ky \text{ if and only if } y = ce^{kx} \text{ for some constant } c.$$

It is this key fact of calculus, together with the ability to solve exponential equations, that we will call upon in this section.

example 17.1   Find the function y that satisfies $y' = 2y$ with $y(0) = 4$.

solution 17.1   Since $y' = 2y$, we have $k = 2$, and so $y = ce^{2x}$. We are given that $y(0) = 4$. On the other hand, $y(0) = ce^0 = c$. Hence, $c = 4$ and so $y = 4e^{2x}$.   ▤

example 17.2   Find the function y that satisfies $y' = ky$ with $y(0) = 3$ and $y(1.5) = 6$.

solution 17.2   We know that y is of the form ce^{kx}, but we must determine the constants c and k. Now, $y(0) = ce^0 = c$ and we are given that $y(0) = 3$. Hence, $c = 3$ and $y = 3e^{kx}$. The value of $y(1.5) = 3e^{1.5k}$ is given as 6, and so $e^{1.5k} = 2$. Taking ln of both sides yields $1.5k = \ln 2$. Thus, $k = (\frac{2}{3}) \ln 2$ and we may write $y = 3e^{(\ln 4)(x/3)} = 3(4)^{x/3}$.   ▤

Equations such as $y' = ky$ are called **differential equations**. Conditions such as $y(0) = 3$ or $y(1.5) = 6$ in example 17.2 are known as **boundary conditions**.

the exponential growth model   The first model to be presented is a model for *growth in an unlimited environment*. If $N(t)$ is the number in the population at time t, then $N'(t)$ may be interpreted as the instantaneous rate of growth. The exponential growth model begins with the assumption given in the accompanying box.

---

exponential growth hypothesis

The rate of growth is directly proportional to the number present.

In the language of derivatives, the hypothesis states that $N'(t) = kN(t)$ for some constant k. Hence, $N = ce^{kt}$. If we are given N_0, the number present at time 0, then $N(0) = ce^0 = c$ will be N_0. The conclusion is stated in the box.

---

conclusion $N(t) = N_0 e^{kt}$ for some constant k.

Let's examine the exponential growth hypothesis more carefully. If the population size is increased from N to $2N$, we are assuming that the rate of growth will double, as illustrated in figure 17.1. If we can consider the growing

figure 17.1

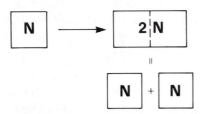

population of size $2N$ as the sum of two populations, each of size N, growing as before, then the rate of growth should be $N' + N' = 2N'$. There are several conditions implicit in the last statement, including the following:

1. There are no effects of crowding. Space and food are in ample supply. Death rates are not changed.

2. Doubling the population also doubles the number of individuals able to reproduce. (It is conceivable that the extra N individuals are all immature and so the rate of growth for the $2N$ individuals is still N'.) We are assuming that the *age structure* of the population is unchanged.

Undoubtedly you can think of other implicit conditions. Suffice it to say that the environment is *unlimited*.

example 17.3   A population of 100 pheasants is introduced on an island. Lacking any natural predators, the population flourished to 560 three years later. Assume that the exponential growth model is appropriate.

   (a) Predict $N(5)$.

   (b) Find the **doubling time**—i.e., determine how long it takes for N_0 pheasants to increase to $2N_0$.

solution 17.3   Since $N_0 = 100$, we have $N(t) = 100e^{kt}$. We will determine k using the fact that $N(3) = 560$:

$$560 = N(3) = 100e^{3k}$$

Hence,

$$e^{3k} = 5.6 \quad \text{or} \quad 3k = \ln 5.6 \quad \text{or} \quad k = (\tfrac{1}{3}) \ln 5.6 = .57425 \ldots$$

   (a) $N(5) = 100e^{5k} = 100e^{2.87127 \cdots} = 1{,}766$ pheasants

   (b) If $N = N_0 e^{kt}$, and if T is the doubling time, then $N(T) = 2N_0 = N_0 e^{kT}$. Hence, $e^{kT} = 2$ or $kT = \ln 2$. It follows that $T = (\ln 2)/k = (\ln 2)/.57425 \ldots \approx 1.21$ years.   ▤

   In general, the constant k in the exponential growth model is called the **intrinsic rate of growth**. The units of k are time $^{-1}$. In the case of example 17.3, we would say that the population is growing continuously at the rate of 57.4% per year. The general formula for T, the **doubling time**, is

$$T = \frac{\ln 2}{k}$$

   The exponential growth model works well for populations in their initial stages of growth—i.e., for the period of time in which the environment plays no role in limiting growth. Some examples are:

   1. Bacterial growth

   2. World population growth prior to 1900

   3. Growth of young timber in a forest [Here we are measuring the height $h(t)$ rather than $N(t)$.]

A situation similar to that described in example 17.3 was actually observed between 1939 and 1942 on Protection Island, Washington, in an experiment carried out by the Washington State Department of Game.*
   The differential equation $y' = ky$ can also be used to model radioactive decay.

---

* For details see, A. Einaren, *Murrelet, 26*: (1945) pp. 2–44.

model for radioactive decay   Radioactive substances decay due to the emission of α-particles. Let $A(t)$ be the number of grams of the substance at time t. Since $A(t)$ is decreasing, $A'(t)$ is negative and represents the rate of decay. The model begins with the assumption stated in the box.

radioactive decay hypothesis   The rate of decay is directly proportional to the amount present.

Hence, $A'(t) = -kA(t)$ for some constant $k > 0$. It follows that $A(t) = ce^{-kt}$ for some constant c. If A_0 is the initial number of grams, then $A_0 = A(0) = ce^0 = c$. The resulting conclusion is shown in the box.

conclusion $A(t) = A_0 e^{-kt}$ for some constant k.

The radioactive decay hypothesis is reasonable. Halving the number of grams present should halve the number of emissions about to take place, and should therefore halve the rate of decay.

The *half-life* $t_{1/2}$ is the amount of time it takes for A_0 grams to decay to $\frac{1}{2}A_0$ grams. Hence, $\frac{1}{2}A_0 = A_0 e^{-kt_{1/2}}$ which implies that $kt_{1/2} = \ln 2$. Hence, the half-life is

$$t_{1/2} = \frac{\ln 2}{k}$$

The constant k is called the **specific rate of disintegration**. The units for k are time^{-1}. Thus, if $k = .05$, we say that the substance is decaying continuously at the rate of 5 per unit time. Listed in table 17.1 are some common radioactive substances with their corresponding half-lives and specific rates of disintegration.

table 17.1

Substance	Half-life $t_{1/2}$	k
Barium 140	13 days	.0533
Carbon 14	5,760 years	.000120
Radium	1,620 years	.000428
Strontium 90	29 years	.0239

example 17.4   Five grams of a particular radioactive substance decay to 4.1 grams after 4 days. Find the half-life and the specific rate of decay.

solution 17.4   Since $A_0 = 5$, $A(t) = 5e^{-kt}$. We will use the condition $A(4) = 4.1$ to find k:

$$4.1 = A(4) = 5e^{-4k}$$

Hence, $e^{-4k} = .82$, which implies that $k = (\ln .82)/(-4) = .0496 \ldots$. From the formula for the half-life, we conclude that $t_{1/2} = (\ln 2)/k = 13.97$ days.   ◼

example 17.5   **carbon dating**   All living things contain small amounts of radioactive C^{14}. The ratio of C^{14} to the stable C^{12} is constant in the atmosphere. There are about $6(10^{10})$ atoms of C^{14} per gram of carbon. The C^{14} in the atmosphere is in the form of carbon dioxide, $C^{14}O_2$. Through the process of photosynthesis, this radioactive carbon becomes fixed in plant tissues and eventually makes its way through the food chain. Once death occurs, C^{14} is no longer taken in, and hence the amount in the organism begins its natural decay.

Let A_0 be the amount of C^{14} in a living organism and A_1 the present amount in the article to be dated. Then to determine the age of the object, we must solve the equation $A_1 = A_0 e^{-kt}$ for t. Hence, $e^{-kt} = A_1/A_0$ and so $-kt = \ln (A_1/A_0)$, or $t = (1/k) \ln (A_0/A_1)$. If we solve the half-life equation for k, and use the half-life value $t_{1/2} = 5{,}760$ from table 17.1, we obtain

carbon dating equation $t \approx 8{,}310 \ln \dfrac{A_0}{A_1}$

example 17.6   A human bone has been unearthed at an archeological site. The present C^{14} content is between 79% and 81% of the original C^{14} content. Estimate the age of the bone.

solution 17.6   We are given that $A_1/A_0 = .80 \pm .01$. If $A_1/A_0 = .79$, then the estimate of age is $8{,}310 \ln (1/.79) = 1{,}959$ years. If $A_1/A_0 = .81$, then $t = 8{,}310 \ln (1/.81) = 1{,}751$ years. Hence, we can say that the bone is between 1,750–1,960 years old, approximately.   ◼

Both the exponential growth and radioactive decay models used some form of the differential equation $y' = ky$ in their development. In the following model the closely related differential equation $y' = -k(y - y_0)$ will be solved to obtain the equation for cooling.

Newton's Law of cooling   An object of temperature T_1 is placed in a room of constant temperature T_0, as depicted in figure 17.2. If $T_1 > T_0$ (as would be the case for a cup of coffee), then the object begins to cool and will eventually reach the room temperature T_0. Let $T(t)$ denote the temperature of the object at time t. Then $T'(t)$ is the rate at which the temperature is changing,

figure 17.2

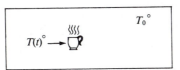

which is equal to the rate of cooling when $T_1 > T_0$. We begin with the assumption given in the box.

> **cooling hypothesis**   The rate of cooling is directly proportional to the difference between object temperature and room temperature

In mathematical language, we say $T'(t) = -k(T - T_0)$ for some constant $k > 0$. To solve this differential equation, set $y = T - T_0$. Then $y' = T' = -k(T - T_0) = -ky$. We have seen that the solution to $y' = -ky$ is $y = ce^{-kt}$. Hence, $T - T_0 = ce^{-kt}$ or $T = T_0 + ce^{-kt}$. To find c, we use the condition that $T(0) = T_1$. Hence, $T_1 = T_0 + ce^0$, and so $c = (T_1 - T_0)$. The result is summarized in the box.

> **conclusion** $T(t) = T_0 + (T_1 - T_0)e^{-kt}$

This cooling function has a graph of the form shown in figure 17.3. Hence, as t increases, the temperature $T(t) \rightarrow T_0$, the room temperature.

figure 17.3

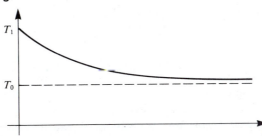

example 17.7   If the room temperature is 70° and the temperature of the coffee is 110°, find the temperature function if the coffee has cooled to 85° after 30 minutes.

solution 17.7   We are given $T_0 = 70$, $T_1 = 110$, and $T(\frac{1}{2}) = 85$. Hence, $T(t) = 70 + 40e^{-kt}$. Since

$85 = T(\frac{1}{2}) = 70 + 40e^{-.5k}$, we have $e^{-.5k} = \frac{15}{40} = \frac{3}{8}$, and so $k = [\ln(\frac{3}{8})]/(-.5) = 1.961658 \ldots$. Finally, $T(t) = 70 + 40e^{-1.961658t}$.

Under certain special conditions, Newton's Law of Cooling can be used to establish the time of death, as described in the following example.

example 17.8

A body is discovered in the basement of a building. The room temperature is 70° and the body temperature is 85°. One hour later, the body temperature is 80°. Determine the time of death. What assumptions do you have to make in order to apply the model?

solution 17.8

We will assume that the temperature of the victim at the time of death was a normal 98.6°. In addition, we must assume that the room temperature has been a constant 70° since the time of death. Let $t = 0$ correspond to the time of death, and let t be the time of discovery. We are therefore given the information summarized in figure 17.4. Hence, $T = 70 + 28.6e^{-kt}$.

figure 17.4

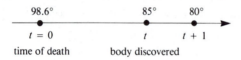

Since $T(t) = 85$ and $T(t + 1) = 80$, we have the system of equations:

$$85 = 70 + 28.6e^{-kt}$$
$$80 = 70 + 28.6e^{-k(t+1)}$$

Thus, $e^{-kt} = 15/28.6$ and, from the second equation, $10 = 28.6e^{-kt}e^{-k} = 28.6(15/28.6)e^{-k} = 15e^{-k}$. It follows that $e^{-k} = 2/3$. From the first equation, we have $15/28.6 = (e^{-k})^t = (2/3)^t$. The solution to this exponential equation is $t = [\log(15/28.6)]/\log(2/3) = 1.5916 \approx 1.6$ hours. Hence, death occurred 1.6 hours before the body was discovered.

Suppose that we have performed an experiment and gathered data. How can we determine whether the data can be described by an exponential function $y = ce^{kx}$? First, note that $\ln y = \ln(ce^{kx}) = \ln c + \ln(e^{kx}) = \ln c + kx$. Hence, if we graph $Y = \ln y$ versus $X = x$, the graph should be a straight line, with slope $m = k$, and Y-intercept $\ln c$. Let's illustrate this technique with an example from the medical literature.

example 17.9

To assess how quickly a wound was healing, its area was estimated every four days. The results are shown in table 17.2. Is the area decreasing exponentially? Fit an exponential function $y = ce^{kx}$ to the data.

table 7.2

time t (days)	0	4	8	12	16	20	24	28
area A (cm²)	107	88	75	62	51	42	34	27

solution 17.9

We first let $Y = \ln A$ and $T = t$ to form table 17.3. We must next fit a line through the data points. Although there are standard statistical procedures for doing so, we have fitted a line by sight. The Y-intercept is $4.67 = \ln(107)$. To

table 17.3

T	0	4	8	12	16	20	24	28
Y	4.67	4.48	4.32	4.13	3.93	3.74	3.53	3.295

figure 17.5

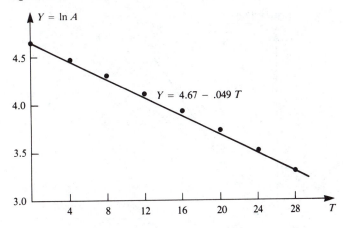

compute the slope, we will use the points $(0, 4.67)$ and $(28, 3.30)$. Hence, $m = (4.67 - 3.30)/(0 - 28) = -.049$. Thus, $k = -.049$ and $c = 107$, and so $A = 107e^{-.049t}$. The value of $k = -.049$ measures how quickly the wound is healing. Since the units for k are (day)$^{-1}$, we say that the wound is healing continuously at the rate of 4.9 percent per day. Presumably the effects of different methods of treatment could be studied by comparing their respective values of k. ▤

Essentially the same logarithmic method can be used to fit power functions to data. If $y = \alpha x^{\beta}$, then $\ln y = \ln(\alpha x^{\beta}) = \ln \alpha + \beta \ln x$. Hence, if we graph $Y = \ln y$ versus $X = \ln x$, the graph should be linear, with slope $m = \beta$, and Y-intercept equal to $\ln \alpha$. This method is illustrated in our final example.

example 17.10   Data collected on the length-weight relationship for yellow fin tuna taken in the Central Pacific are shown in table 17.4. Fit a power function $W = \alpha L^{\beta}$ to the data.

table 17.4

length L (cm)	70	80	90	100	110	120	130	140	160	180
weight W (lbs)	14.3	21.5	30.8	42.5	56.8	74.1	94.7	119	179	256

solution 17.10   We first let $Y = \ln W$ and $X = \ln L$ and construct table 17.5. The graph of Y versus X is given in figure 17.6. After fitting a line through these points by

table 17.5

X	4.25	4.38	4.50	4.61	4.70	4.79	4.87	4.94	5.08	5.19
Y	2.66	3.07	3.43	3.75	4.04	4.31	4.55	4.78	5.19	5.55

figure 17.6

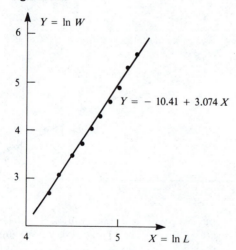

sight, we use the points (4.25, 2.66) and (5.19, 5.55) to compute the slope:

$$\beta = m = \frac{5.55 - 2.66}{5.19 - 4.25} = 3.074$$

The Y-intercept is $Y - mX = 2.66 - 3.074(4.25) = -10.41$. Hence, $\ln \alpha = -10.41$, which implies that $\alpha = e^{-10.41} = .0000301$. Hence, the power function $W = .0000301L^{3.074}$ fits the data.

exercises for section 17

part A

Solve each of the following differential equations subject to the given boundary conditions.

1. $y' = -2y$ with $y(0) = 3$
2. $y' = 2y$ with $y(1) = 3$
3. $y' = \frac{3}{2} y$ with $y(0) = 10$
4. $y' = \frac{3}{2} y$ with $y(1) = 15$
5. $y' = y$ with $y'(0) = 2$
6. $y' = ky$ with $y(0) = 2$ and $y(2) = 9$
7. $y' = ky$ with $y(2) = 10$ and $y(3) = 15$
8. $y' = ky$ with $y(0) = 10$ and $y(10) = 5$
9. $y' = ky$ with $y(0) = 5$ and $y'(0) = 2$
★10. $y' = 2 + y$ with $y(0) = 5$ [*Hint*: Let $z = 2 + y$. Show that $z' = z$.]

11. Under ideal conditions the bacteria *E. Coli* undergo cell division every 20 minutes. Estimate the intrinsic rate of growth k.

12. At present, the population of South America is growing continuously at the rate of $k = .023$ per year. Assuming exponential growth, how long does it take for the population to double?

13. Based on 1967 figures, the population of the United States is growing continuously at the rate of $k = .007$ per year. Assuming exponential growth, how long does it take for the population to double?

14. A population of deer is introduced into a large game preserve. In 1979, the population consisted of fifty deer. One year later, the herd had increased to sixty-five.

    (a) How long will it take for the herd to double in size?

    (b) Estimate the size of the herd in 1985.

15. In the early 1900s the world population of blue whales numbered between 150,000 and 200,000. Due to heavy exploitation by the whaling industry, numbers had been reduced to dangerously low levels in the early 1960s. In 1966, the International Whaling Commission protected the species. The best estimate for the intrinsic rate of growth is $k = .047$ per year, and in 1978, the population in the southern hemisphere was thought to number 5,000. Assuming exponential growth, how long will it take for the population to recover to a level of 50,000? 150,000?

16. Shown in the tables are the population counts of two populations. Are the data consistent with the hypothesis of exponential growth? [*Hint*: Use two of the data points to specify ce^{kt}. Then compare the third data point with the predicted value.]

(a)

t (years)	0	1	2
N	1,000	1,350	1,820

(b)

t (years)	0	1	2
N	1,000	1,456	1,910

17. Under ideal conditions, a population of Norwegian brown rats will double its numbers about every 1.5 months. What differential equation does the population satisfy?

18. Two mg of radioactive material decays to 1.3 mg after 10 days. Find the half-life of the substance.

19. Radioactive strontium 90 is potentially very dangerous to man. If deposited into pasture land by "acid rain," it will eventually make its way into plants and into cow's milk. About 3% of the strontium 90 entering the human body becomes a permanent part of bone tissue, where its radioactivity could cause cancer. If it is has been determined that the strontium 90 level in the pasture is three times the safe level, how many years will it take before the pasture can again be used for grazing?

Use the C^{14} dating equation to estimate the age of the following artifacts.

20. Charcoal from the occupied level of a cave dwelling shows a 79%–81% loss in C^{14} content.

21. A human skull uncovered at an archeological site shows a 97.5%–98.5% loss in C^{14} content.

22. The temperature in a freezer is $-10°$. Water of temperature $60°$ is placed in the freezer and after 15 minutes reaches $50°$. When will the water reach the freezing temperature of $32°$?

23. In example 17.8, we assumed that the temperature of the victim was normal at the time of death. How would the conclusion of example 17.8 be altered if it were known that the victim had been running a temperature of $101°$ at the time of death?

★24. The rate at which an object will cool is directly proportional not only to the difference between object temperature and room temperature, but also to the *surface area S* of the object. Thus, a cup of coffee with twice the exposed surface area will cool at double the rate, as depicted in the figure. Thus, $T'(t) = cS(T - T_0)$ for some constant c.

(a) Show that $T = T_0 + (T_1 - T_0)e^{-cSt}$ where $T_1 = T(0)$.

(b) The room temperature is $70°$ and the initial temperature of the coffee is $110°$. Thirty minutes later, the temperature in the smaller cup is $85°$. What is the temperature in the larger cup, which has twice the exposed surface area?

25. Shown in the table are data on U.S. population growth prior to 1900. Fit an

exponential function $N = ce^{kt}$ to the data, where $t = 0$ corresponds to the year 1750.

Year	1750	1800	1850	1900
Population (millions)	1.6	5.3	40.0	75.9

26. A study of land and freshwater birds in an island group yields the data shown in the table. Fit a power function $S = \alpha A^\beta$ to the data.

A = area of the island (square miles)	50	125	350	700	1,025
S = number of species	13	17	24	29	32

27. The tabled data on the fiddler crab are taken from Thompson (1942).* Fit a power function $C = \alpha B^\beta$ to the data.

B = Weight of body minus claw (g)	58	300	536	1,080	1,449	2,233
C = Weight of claw (g)	5	78	196	537	773	1,380

28. A standard test in cardiovascular physiology is the dye-dilution technique. A bolus of dye (such as indocyanine green) is suddenly injected near the right atrium of the heart and samples of blood are taken from an artery (such as the aorta). Shown in the table are the results of such a test. Fit an exponential decay function $C = ce^{-kt}$ to the data.

Time t (sec)	0	2	4	6	8	20
Dye concentration C (mg/liter)	12	8.5	5.9	4.1	2.8	2.0

29. Shown in the table are data on the length-weight relationship for Pacific halibut. Fit a power function $W = \alpha L^\beta$ to the data.

Length L (m)	.5	1.0	1.5	2.0	2.5
Weight W (kg)	1.3	10.4	35.0	82.0	163.0

30. On page 184, we mentioned a study of a pheasant population on Protection Island in the San Juan de Fuca Straits of Washington. The actual data are shown in the table on page 194. Fit an exponential function $N = ce^{kt}$ to the data. Let $t = 0$ correspond to 1939.

* d'Arcy Thompson, *On growth and form*, Cambridge Univ. Press, (1961).

Year	1939	1940	1941	1942
Population	81	282	641	1,194

31. In the exponential growth model, the population parameter k is called the *intrinsic rate of increase* and depends not only on the particular species, but also on the complete ecosystem in which the population develops. To illustrate this, consider the following hypothetical problem.

In locale 1, a population of rabbits is increasing exponentially at the continuous rate of $k = .1$ per year. Deaths that occur are due to predation by foxes and natural causes. Locale 2 is entirely similar in nature, but no foxes are present. Here, the population increases very rapidly and doubles every two years.

(a) In locale 1, what percentage of the year's initial population N_0 dies during the year due to predation by foxes?

(b) If the population of foxes in locale 1 now numbers about 1,000, how many more foxes need to be present to completely stabilize the population of rabbits? What assumptions are you making in order to carry out the computations?

32. This exercise is based on a 1968–69 study of nuclear fallout in a Puerto Rican rain forest. Radioactive strontium 90 makes its way into a forest ecosystem by rainfall ("acid rain"). Once in the ecosystem, it is eliminated by natural radioactive decay (strontium 90 has a half-life of about 29 years) and runoff from the soil into streams that lead out of the forest.

If $A(t) =$ total strontium 90 per hectare in the forest, then

$$A'(t) = (\text{rate of natural decay}) + (\text{rate of removal by runoff})$$
$$= -k_1 A(t) - k_2 A(t) \qquad \text{by assumption}$$

(a) Solve the differential equation for $A(t)$.

(b) If the half-life of strontium 90 in the ecosystem is to be 20 years, what percent of strontium 90 must be removed each year by runoff?

33. Let A_1 be the true C^{14} content of an artifact, t_1 the true age, and $x_1 = A_1/A_0$. Then $t_1 = -8,310 \ln x_1$. This exercise is designed to show that if x_1 is small, then a small error in measuring x_1, Δx, can produce a large error in estimating t_1.

(a) Show that $\dfrac{dt}{dx} = -\dfrac{8,310}{x}$.

(b) For Δx small, $\dfrac{dt}{dx} \approx \dfrac{\Delta t}{\Delta x}$. Conclude that $\Delta t \approx -8,310 \dfrac{\Delta x}{x}$.

(c) Complete the table on page 195.

x_1	error Δx	Δt, error in dating t_1
.100	±.01	
.020	±.01	
.001	±.0005	
.001		±500
.001		±100

★ **34.** A couple has been found dead. The woman is 5′6″ and weighs about 130 pounds. The man is 6′0″ and weighs about 170 pounds. The room temperature is 70°, while the body temperatures of the woman and man are 90° and 85°, respectively. Were the two killed at the same time? State all assumptions you are making in your calculations. [*Hint*: Use exercise 24 and the Du Bois surface area formula of exercise 62, section 5.]

part C

Ricker's spawner-recruit model   Mature salmon return from the open sea to reproduce in the freshwater streams where their life began. For the sockeye salmon, this final journey begins when they are about four years old. There is a particular timetable that must be maintained in order to ensure successful spawning. The fish cease to feed when they leave the ocean and rely on the energy that has been stored within their bodies to struggle upstream.

Those fortunate enough to reach their spawning grounds deposit their eggs (3,000–4,000 for the sockeye) in the gravel, and then, dramatically, they die. The eggs will hatch in the spring and young salmon may spend a year in freshwater lakes and streams before returning to the sea. Those escaping predation and surviving to maturity (the *recruits*) will return to the mouths of the rivers to start the life cycle once again.

A *spawner-recruit* model attempts to find a relationship $R = f(S)$ between the number of spawners S and the number of future recruits R. At least in theory, if $R > S$, then $(R - S)$ fish may be taken by fishermen, leaving the same number S to spawn as before.

Many different functions f are possible, depending on what hypotheses are made about the various stages in the life cycle. Little is known about the early life history of many fish so many hypotheses are conceivable. The model to be developed is known as Ricker's Model.* We will make the assumptions stated in the box on page 196.

* W.E. Ricker, *J. Fish Research Board Can.*, (1954), *11*: 559–623.

Spawner – Recruit Model Hypotheses

1. The number of eggs E is directly proportional to the number of spawners S.

2. The length of time T spent in an early "vulnerable" stage is directly proportional to the number of eggs E. Presumably, the food resources of a small stream are limited and hence, the increased competition for these resources lengthens the time of development and extends exposure to predation by other species.

3. The mortality rate $\dfrac{1}{N} \cdot \dfrac{dN}{dt}$ is constant for $0 \le t \le T$.

4. A fixed percentage of those passing through this "vulnerable" stage survive to the recruitment stage.

---

35. Show that assumptions 1–4 imply that $R = \alpha S e^{-\beta S}$ for some α and $\beta > 0$.

36. Sketch the graph of $R = \alpha S e^{-\beta S}$ and find the maximum value of R. Find S when $R = S$. (See accompanying figure.)

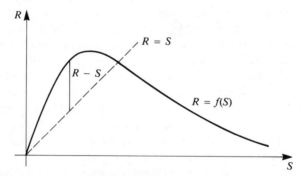

37. Show that the value of S that maximizes $(R - S)$ satisfies the equation $\alpha e^{-\beta S}(1 - \beta S) = 1$.

38. If $\alpha = 3$ and $\beta = .1$, find the value of S that maximizes $(R - S)$ by trial and error solution of the equation in exercise 37. (The units of R and S are thousands.)

Note that the Ricker model predicts low recruitment from very high stocks. This has been observed for some species, the arctic cod being a prime example. This function is just one of many with this property that have been proposed.

the integral
and its
applications

prologue   The second main concept in calculus is the *integral*. The word "integral" is derived from the Latin verb "integro," which means "to make whole." The mathematical integral is a magnificent adder. It takes the very small parts making up a physical quantity, measures them, and finally adds them, hence measuring the whole. The standard symbol for the integral is

$$\int_a^b f(x)\, dx$$

The symbol $\int$ is an elongated S, standing for "sum." It adds terms of the form $f(x)\, dx$ from $x = a$ to $x = b$. Let's illustrate these principles with two simple examples.

example 1 the area under a curve   What is the area under the curve shown in figure 1? We might consider the area to be made up of an extremely large number of slender reeds, one next to the other. If dx denotes the very small width of a typical reed (see figure 2), then $f(x)\, dx$ is the area of a typical reed. The integral *adds* all terms of the form $f(x)\, dx$ from $x = 0$ to $x = 3$ to give the whole area under the curve. Hence, the area is $A = \int_0^3 f(x)\, dx$.

figure 1

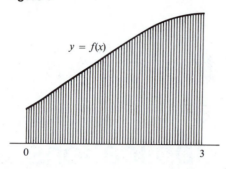

$y = f(x)$

0   3

figure 2

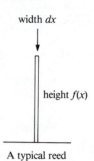

width dx

height $f(x)$

A typical reed

example 2 measuring abundance in the sea Shown in figure 3 is the *water column*, a typical one-by-one-meter square column extending from the ocean surface to the ocean floor. It is impossible to measure directly the total

figure 3

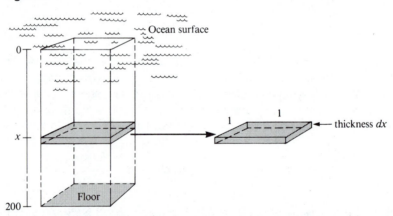

Ocean surface

0

x

1   1

thickness dx

200

Floor

amount of, say, phytoplankton in the column. By taking water samples, however, we can measure the *concentration* or *density* at various depths. Let $f(x) =$ density (in number/m³) at depth x. We may imagine that the water column consists of an extremely large number of thin square layers, each of height dx, piled on one another. The number of organisms in a typical layer is

$$f(x)\left(\frac{\text{number}}{\text{m}^3}\right) \cdot 1^2 \, dx \, (\text{m}^3) = f(x) \, dx$$

The integral adds all these terms from $x = 0$ to $x = 200$ to obtain $\int_0^{200} f(x) \, dx$, the total number of organisms in the water column.

Remarkably, the actual method of computing an integral involves first finding an *antiderivative*. We begin Part III with this topic.

 finding antiderivatives

In our prior work, we have concentrated on the rules for computing $f'(x)$ for a given function $f(x)$. Now we will reverse the process. Given only the expression for $f'(x)$, can we determine the original function?

What is a function whose derivative is $2x$? Clearly, x^2 is one such function. We say x^2 is an **antiderivative** for $2x$ and write either

$$x^2 = D_x^{-1} 2x$$

or
$$x^2 = \int 2x \, dx$$

The second notation will not make much sense until we have discussed the integral in section 19. For now, think of it as just another way of denoting that x^2 is an antiderivative for $2x$. Ignore the symbol "dx" temporarily and interpret "$\int$" as "is an antiderivative for."

definition 18.1   If $F'(x) = f(x)$, we say $F(x)$ is an *antiderivative* for $f(x)$ and write:

$$F(x) = D_x^{-1} f(x)$$

or
$$F(x) = \int f(x) \, dx$$

Antiderivatives can often be discovered by trial-and-error.

example 18.1   Discover antiderivatives for

(a) x^1 (b) e^{3x}

solution 18.1   For (a), observe that $D_x x^5 = 5x^4$. Hence, $D_x(\frac{1}{5}x^5) = \frac{1}{5}(5x^4) = x^4$. Thus, an antiderivative for x^4 is $x^5/5$ and so we write $x^5/5 = \int x^4 \, dx$. For (b), we will determine k and c so that $D_x ce^{kx} = cke^{kx} = e^{3x}$. Hence, set $k = 3$ and $ck = 1$. Thus, $c = 1/k = 1/3$ and so $D_x^{-1} e^{3x} = (1/3) e^{3x}$.   ▤

The generalizations of the formulas in example 18.1 are:

$$D_x^{-1} x^k = \frac{x^{k+1}}{k+1}, \qquad k = -1 \qquad \text{and } D_x^{-1} e^{\alpha x} = \frac{1}{\alpha} e^{\alpha x}$$

Every formula for a derivative mentioned in sections 9 and 16 gives rise to a corresponding formula for an antiderivative. For example, we know that $D_x \ln x = 1/x$. Thus, $\ln x$ is an antiderivative for $1/x$:

$$\ln x = \int \frac{1}{x}\, dx \qquad \text{for } x > 0$$

Listed in table 18.1 are the antiderivatives of some of the elementary functions. Each of these entries can be verified by differentiating $F(x)$ to see that

table 18.1

Elementary Function $f(x)$	$F(x) = \int f(x)\, dx$
0	c (any constant)
m	mx
x^κ	$x^{\kappa+1}/(\kappa + 1)$ $\qquad$ for $\kappa \neq -1$
$\dfrac{1}{x}$	$\ln x$ $\qquad$ for $x > 0$
$\sin x$	$-\cos x$
$\cos x$	$\sin x$
$e^{\alpha x}$	$\dfrac{1}{\alpha} e^{\alpha x}$ $\qquad$ for $\alpha \neq 0$

the derivative is indeed $f(x)$. For example, is it true that $D_x^{-1} x(1 + x^2)^2 = \frac{1}{6}(1 + x^2)^3$? We can check by computing $D_x \frac{1}{6}(1 + x^2)^3 = (\frac{1}{6})3(1 + x^2)^2(2x) = x(1 + x^2)^2$. Hence the statement is true. In general, we can say

$$\frac{d}{dx}\left(\int f(x)\, dx \right) = f(x) \qquad \text{or} \qquad D_x(D_x^{-1}f(x)) = f(x)$$

In the next example we will check our antiderivatives after we have applied the formulas in table 18.1.

example 18.2

Find antiderivatives for each of the following functions and check your answers:

$$\textbf{(a) } 1/x^2 \qquad \textbf{(b) } \sqrt{x} \qquad \textbf{(c) } e^{.1x}$$

solution 18.2

(a) Write $1/x^2$ as x^{-2}. We add one to the exponent and divide by that new exponent to obtain $x^{-2+1}/(-2+1) = x^{-1}/(-1) = -1/x$.
For (b), write $\sqrt{x} = x^{1/2}$. Hence, $D_x^{-1} x^{1/2} = x^{1/2+1}/(\frac{1}{2}+1) = (\frac{2}{3})x^{3/2}$.
Finally, (c) follows directly from the last entry in table 18.1, with $\alpha = .1$:

$$\int e^{.1x}\, dx = e^{.1x}/.1 = 10 e^{.1x}$$

We can check each answer by differentiating. For **(a)**, $D_x(-1x^{-1}) = (-1)(-1)x^{-2} = x^{-2}$. For **(b)**, $D_x(\frac{2}{3})x^{3/2} = \frac{2}{3} \cdot \frac{3}{2}x^{1/2} = \sqrt{x}$. Finally, $D_x 10e^{.1x} = 10e^{.1x}(.1) = e^{.1x}$. ◼

To find the antiderivative for a polynomial such as $p(x) = x^3 - 4x + 3 = [\ x^3\ ] - 4[\ x\ ] + [\ 3\ ]$, we concentrate on x^3, x, and 3, and take their antiderivatives to obtain $[x^4/4] - 4[x^2/2] + [3x]$. The next two rules allow us to bypass the constants and to compute antiderivatives by concentrating on the simpler expressions in the formula.

$$\int [f(x) + g(x)]\, dx = \int f(x)\, dx + \int g(x)\, dx$$

$$\int cf(x)\, dx = c\left(\int f(x)\, dx \right)$$

These rules are illustrated in our next example.

example 18.3   Compute

(a) $\int (x^2 - 4x + 1)\, dx$ **(b)** $\int \left(x - \frac{1}{x} \right) dx$ **(c)** $\int (x^3 - 1)^2\, dx$

solution 18.3   The answers to **(a)** and **(b)** are $x^3/3 - 4(x^2/2) + x = x^3/3 - 2x^2 + x$ and $x^2/2 - \ln x$, respectively.

Before we attack **(c)**, we must expand $(x^3 - 1)^2 = (x^3)^2 - 2x^3 + 1 = x^6 - 2x^3 + 1$. Hence, $\int (x^3 - 1)^2\, dx = x^7/7 - 2x^4/4 + x = x^7/7 - x^4/2 + x$. ◼

warning!   At this point, we warn the student that $\int f(x)g(x)\, dx$ is not $(\int f(x)\, dx)(\int g(x)\, dx)$. Just as the special Product Rule is used to compute $D_x f(x)g(x)$, so special techniques are needed to find $\int f(x)g(x)\, dx$.

Each of the rules for derivatives presented in sections 9 and 16 gives rise to a formula for an antiderivative. These are summarized in table 18.2.

example 18.4   Using table 18.2 compute

(a) $\int xe^{x^2}\, dx$ **(b)** $\int 2x/(1 + x^2)\, dx$ **(c)** $\int (1 + x)^5\, dx$

solution 18.4   **(a)** The expression $\int xe^{x^2}\, dx$ is almost of the form $\int f'(x)e^{f(x)}\, dx$ with $f(x) = x^2$. We can write:

$$\int xe^{x^2}\, dx = \frac{1}{2} \int (2x)e^{x^2}\, dx = \frac{1}{2} e^{x^2}$$

table 18.2

Derivative Rule	Antiderivative Rule
$D_x e^{f(x)} = f'(x) e^{f(x)}$	$\int f'(x) e^{f(x)} \, dx = e^{f(x)}$
$D_x [f(x)]^n = n f'(x) [f(x)]^{n-1}$	$\int f'(x) [f(x)]^{n-1} \, dx = \dfrac{[f(x)]^n}{n}$
$D_x \ln [f(x)] = f'(x)/f(x)$	$\int f'(x)/f(x) \, dx = \ln [f(x)]$
Chain Rule: $D_x f(g(x)) = f'(g(x)) g'(x)$	$\int f'(g(x)) \cdot g'(x) \, dx = f(g(x))$
Product Rule: $D_x f(x) g(x)$ $= f(x) g'(x) + g(x) f'(x)$	$\int [f'(x) g(x) + g'(x) f(x)] \, dx$ $= f(x) g(x)$

using the first antiderivative rule in table 18.2. Note that $D_x(\frac{1}{2} e^{x^2}) = \frac{1}{2} e^{x^2} \cdot (2x) = x e^{x^2}$, and so the antiderivative checks.

For **(b)**, note that $2x/(1 + x^2) = f'(x)/f(x)$ with $f(x) = 1 + x^2$. Hence, by the third antiderivative rule, $\int 2x/(1 + x^2) \, dx = \ln (1 + x^2)$.

Finally, for **(c)** $(1 + x)^5 = f'(x)[f(x)]^{n-1}$ with $f(x) = 1 + x$ and $n = 6$. Thus, $f'(x) = 1$, and by the second antiderivative formula, $\int (1 + x)^5 \, dx = (1 + x)^6/6$.

The formula arising from the Chain Rule $\int f'(g(x)) g(x) \, dx = f(g(x))$ gives rise to a special technique called the **substitution method**, which we will study in section 22. The very last antiderivative formula in table 18.2 gives rise to the technique known as **integration by parts**, which we will study in section 24.

We now return to the first illustration in this section. Are there functions other than x^2 whose derivative is $2x$? Clearly, $x^2 + 1$, $x^2 + 1.5, \ldots$ are additional such functions since D_x (constant) $= 0$. Are there any functions other than $x^2 + c$ whose derivative is $2x$? The next theorem shows that the answer is no. The proof depends on the following fact:

If a function has derivative *always* 0, then that function must be a constant.

If $f'(x)$ is constant 0, then $f(x)$ cannot increase (for then $f'(x) > 0$ for some x) and $f(x)$ cannot decrease (for then $f'(x) < 0$ for some x). Hence, $f(x) \equiv c$. We will now show that in order to find the *most general antiderivative* for $f(x)$, all we need to do is find one antiderivative $F_1(x)$, and then form $F_1(x) + c$.

theorem 18.1 If $F_1(x)$ is an antiderivative for $f(x)$, and if $F_2(x)$ is a second antiderivative for $f(x)$, then

$$F_2(x) = F_1(x) + c$$

for some constant c.

proof

We know that $F_1'(x) = f(x)$ and $F_2'(x) = f(x)$ by definition of antiderivative. Hence, $D_x [F_2(x) - F_1(x)] = f(x) - f(x) = 0$. Since the derivative of $F_2 - F_1$ is always 0, $F_2(x) - F_1(x) \equiv c$ for some constant c. Hence, $F_2(x) = F_1(x) + c$.   ∎

In some books, you will see $\int 2x\, dx = x^2 + c$. This is the most general antiderivative for $2x$. In this context, the interpretation of "$\int$" is "the most general antiderivative for."

We are now in a position to justify the key fact used in section 17— namely, that the solution to $y' = ky$ is $y = ce^{kx}$.

theorem 18.2

If $y = f(x)$ satisfies the differential equation $y' = ky$, then $f(x) = ce^{kx}$ for some constant c.

proof

Showing $f(x) = ce^{kx}$ is equivalent to showing $f(x)e^{-kx} = c$ for some constant c. We will do this by showing $D_x [f(x)e^{-kx}] \equiv 0$. We have $D_x [f(x)e^{-kx}] = f(x)e^{-kx}(-k) + e^{-kx}f'(x) = e^{-kx}[f'(x) - kf(x)] = e^{-kx} \cdot 0$ since $f'(x) = kf(x)$. Hence, $f(x)e^{-kx} \equiv c$ for some constant c.   ∎

We will next use theorem 18.1 to solve simple differential equations.

example 18.5

Solve the equation $y' = x - 1$ given that $y = 4$ when $x = 2$.

solution 18.5

The equation $y' = x - 1$ implies that y is an antiderivative for $x - 1$. Hence, $y = x^2/2 - x + c$ for some constant c. The condition $y(2) = 4$ gives $4 = 4/2 - 2 + c$, or $c = 4$. The solution is then $y = x^2/2 - x + 4$.   ▣

example 18.6

Solve the equation $y'' = 4$ given the boundary conditions $y(0) = 1$ and $y'(0) = 4$.

solution 18.6

The equation $y'' = (y')' = 4$ implies that y' is an antiderivative for 4. Hence, $y' = 4x + c$ for some constant c. The condition $y'(0) = 4$ gives $4 = 4(0) + c$. Thus, $y' = 4x + 4$. Then y is an antiderivative for $4x + 4$ and so $y = 2x^2 + 4x + c$. Finally, $y(0) = 1$ implies $1 = 2(0)^2 + 4(0) + c$. Hence, $c = 1$ and so the required solution is $y = 2x^2 + 4x + 1$.   ▣

example 18.7

Let $R(x)$ be the reaction of the body to a stimulus of intensity x. The function $S(x) = R'(x)$ is the sensitivity at stimulus level x (see example 11.6). We expect

$S(x)$ to decrease as x increases. Suppose we assume that $S(x)$ is inversely proportional to x and that $R(x_0) = 0$. Find the expression for $R(x)$.

solution 18.7

We are given that $R'(x) = S(x) = \alpha/x$ for some constant of proportionality α. Hence, $R(x)$ is an antiderivative for α/x. It follows that $R(x) = \alpha \ln x + c$. The condition $R(x_0) = 0$ implies $0 = \alpha \ln x_0 + c$. Hence, $c = -\alpha \ln x_0$ and so

$$R(x) = \alpha \ln x - \alpha \ln x_0 = \alpha(\ln x - \ln x_0) = \alpha \ln (x/x_0)$$

This is just the Weber-Fechner Law we mentioned in example 16.2.

example 18.8

Suppose that an object is tossed into the air with initial velocity v_0 ft/sec in the vertical direction, and is released s_0 feet off the ground, as illustrated in figure 18.1. Let $s(t) =$ height off the ground at time t. Then $v(t) = s'(t)$ is the velocity in

figure 18.1

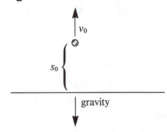

the vertical direction, and $a(t) = v'(t) = s''(t)$ is the acceleration. *Assume that gravity is the only force acting on the object.* What is the correct expression for $s(t)$? Since the acceleration due to gravity is -32 ft/sec^2, we have $s''(t) = -32$. From here on, the problem is identical to example 18.6. Now, $s''(t) = -32$ implies $s'(t) = -32t + c$. But $v(0) = s'(0)$ is given as v_0. Hence, $v_0 = -32(0) + c$. Thus, $c = v_0$ and so $s'(t) = -32t + v_0$. Thus, $s(t)$ is an antiderivative for $-32t + v_0$ and so $s(t) = -16t^2 + v_0 t + c$. We are also given that $s(0) = s_0$. Hence, $s_0 = -16(0) + v_0(0) + c = c$. It follows that

> **Falling Object Law** $s(t) = -16t^2 + v_0 t + s_0$

example 18.9

A baseball is tossed into the air and returns to the ground 4 seconds later. Assume that the baseball was released 6 feet off the ground.

   (a) What was the initial velocity of the ball?

   (b) What was the maximum height reached?

solution 18.9

Since s_0 is given as 6, we apply the Falling Object Law to obtain $s(t) =$

$-16t^2 + v_0t + 6$. Since the object returns to earth after 4 seconds, $s(4) = 0$. Hence, $0 = -16(4)^2 + v_0(4) + 6$, and so $4v_0 - 250 = 0$. It follows that $v_0 = 62.5$ ft/sec. The complete expression for $s(t)$ is $-16t^2 + 62.5t + 6$. To find the maximum height, set $s'(t) = v(t) = 0$: $s'(t) = -32t + 62.5 = 0$ implies $t = 62.5/32 = 1.953125$ seconds. After 1.953125 seconds, the height is $s(1.953125) = 67.035$ feet.

exercises for section 18

part A

Find each of the following antiderivatives in exercises 1–24. Check your answers by differentiating.

1. $\int x^6 \, dx$

2. $\int e^{2x} \, dx$

3. $\int 1/x \, dx$

4. $\int \frac{1}{x^3} \, dx$

5. $\int 3x^5 \, dx$

6. $\int (1 + 1/x) \, dx$

7. $\int e^{-3x} \, dx$

8. $\int 2e^{-.1x} \, dx$

9. $\int 3 \, dx$

10. $\int \sqrt{x} \, dx$

11. $\int 1/\sqrt[3]{x} \, dx$

12. $\int (3x + 2) \, dx$

13. $\int (x^2 + 2x) \, dx$

14. $\int (x^2 - 1)^2 \, dx$

15. $\int (x^3 - 4x + 3) \, dx$

16. $\int (x - 2/x) \, dx$

17. $\int (x + 1/x)^2 \, dx$

18. $\int \frac{e^x + e^{-x}}{2} \, dx$

19. $\int x(3x - 2) \, dx$

20. $\int \frac{x^5 + 3x^3}{x} \, dx$

21. $\int (3x^2 - 3/x^2) \, dx$

22. $\int (3 - x) \, dx$

23. $\int (3 - x)^2 \, dx$

24. $\int (\pi^2 - x^2) \, dx$

25. $\int (e^{2x} - e^2) \, dx$

26. If $D_x(x \ln x - x) - x \cdot \frac{1}{x} + \ln x$ $1 = 1 \mid \ln x$ $1 = \ln x$, what is $\int \ln x \, dx$?

27. If $D_x(x^2 - 1)^5 = 10x(x^2 - 1)^4$, what is $\int x(x^2 - 1)^4 \, dx$?

28. If $D_x(xe^x - e^x) = xe^x + e^x \cdot 1 - e^x = xe^x$, what is $\int xe^x \, dx$?

29. Verify that $\int \frac{x}{\sqrt{9 - x^2}} \, dx = -\sqrt{9 - x^2}$.

30. Verify that $\int \frac{1}{x^2 - 4} \, dx = \frac{1}{4} \ln \left(\frac{x - 2}{x + 2} \right)$.

Use the antiderivative formula $\int f'(x)e^{f(x)}\,dx = e^{f(x)}$ to find:

31. $\displaystyle\int x^2 e^{x^3}\,dx$

32. $\displaystyle\int e^{x+1}\,dx$

33. $\displaystyle\int 2xe^{x^2-1}\,dx$

34. $\displaystyle\int \frac{1}{x^2}e^{1/x}\,dx$

▲**35.** $\displaystyle\int (\cos x)e^{\sin x}\,dx$

Use the antiderivative formula $\int f'(x)[f(x)]^n\,dx = \dfrac{[f(x)]^{n+1}}{n+1}$, $n \neq -1$, to find:

36. $\displaystyle\int x(1+x^2)^3\,dx$

37. $\displaystyle\int 2x^2(1+x^3)^4\,dx$

38. $\displaystyle\int (2x+3)^4\,dx$

39. $\displaystyle\int x\sqrt{1+x^2}\,dx$

40. $\displaystyle\int 1/(x-4)^3\,dx$

Use the antiderivative formula $\int \dfrac{f'(x)}{f(x)}\,dx = \ln[f(x)]$ to find:

41. $\displaystyle\int \frac{1}{x+1}\,dx$

42. $\displaystyle\int \frac{2}{3-2x}\,dx$

43. $\displaystyle\int \frac{x}{x^2-4}\,dx$

44. $\displaystyle\int \frac{x^2}{x^3+1}\,dx$

45. $\displaystyle\int \frac{x}{2-x^2}\,dx$

Find *all* functions $y = f(x)$ that satisfy:

46. $y' = x - 2$

47. $y' = e^{-2x}$

48. $y'' = 2$

49. $y'' = 1/x^2$

50. $y' = x^2 - x + 1$

Solve each of the following differential equations subject to the given boundary conditions.

51. $y' = x - 2$ with $y(1) = 3$

52. $y' = e^{-2x}$ with $y(0) = 4$

53. $y'' = 2$ with $y(0) = 1$ and $y'(0) = 2$

★**54.** $y'' = 1/x^2$ with $y(1) = 1$ and $y(4) = 3$

55. $y' = x^2 - x + 1$ with $y(1) = 0$

In example 18.7, we assumed that the sensitivity S at stimulus level x was inversely proportional to x. The Weber-Fechner formula resulted. Many other hypotheses can be formulated. In each of the following cases, determine $R(x)$ given that $R(0) = 0$.

56. $S(x)$ is inversely proportional to x^k, where $0 < k < 1$. [*Note*: The resulting $R(x)$ gives a *Steven's Power Law*. For loudness, $k = \frac{1}{3}$. For hardness, $k = \frac{1}{4}$.]

57. $S(x)$ is directly proportional to e^{-kx} for some $k > 0$.

58. $S(x)$ is directly proportional to x^k for some $k > 0$. [*Note*: The resulting $R(x)$ again gives a Steven's Power Law. For taste (either sucrose or salt), $k = \frac{1}{2}$. For temperature (heat on the arm), $k = \frac{1}{2}$.]

59. $S(x) = \alpha x(b - x)$, for $0 \le x \le b$. Thus, sensitivity rises to a maximum at $x = b/2$ and then decreases to 0 at $x = b$. [*Note*: The resulting $R(x)$ is a common drug dose-response curve.]

The force of gravity on the moon is about $\frac{1}{6}$ that on the earth. The resulting acceleration due to gravity is then $a = -\frac{32}{6}$ feet/second^2.

60. Starting with the acceleration equation $s''(t) = -\frac{32}{6}$, derive the Falling Object Law for the moon:

$$s(t) = -\frac{8}{3} t^2 + v_0 t + s_0$$

On the earth, a young baseball player can impart an initial vertical velocity of 50 feet/second to the baseball. He releases the ball 5 feet off the ground.

61. How high does the ball travel?

62. On the moon, how high would the ball travel?

63. The tallest office building in the world is the Sears Tower in Chicago. It rises 1,454 feet. If a ball is dropped from the building, what would the impact velocity be?

★**64.** (See exercise 18, section 11) Assume that Emmanuel Zacchini could perform his human cannonball stunt on the moon.

(a) How high off the moon's surface would he rise?

(b) How long would his flight last?

[*Note*: The cannon rises 15 feet off the ground and the initial vertical velocity is 56 feet/second.]

65. Suppose that applying the brakes in a car results in a constant deceleration of -12 feet/second^2. Find the *braking distance d* (see figure) if the car is traveling at 60 mph (88 feet/second).

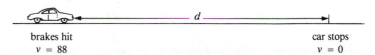

brakes hit   car stops
$v = 88$ $v = 0$

★**66.** In exercise 65, what initial velocities will result in a braking distance of 100 feet or less?

part B   Find the function passing through $(3, 4)$ whose slope y' is given by:

67. $y' = 3 - x$ **68.** $y' = 1/x$

69. $y' = (x - 1)^2$ **70.** $y' = e^{-x}$

Which of the following hypotheses about the rate of growth of an organism imply an upper limit on the size of the organism? Let $W(t) =$ weight in pounds at time t.

71. The rate of growth $W'(t)$ is inversely proportional to t for $t \geq 1$.

72. The rate of growth $W'(t)$ is inversely proportional to t^2 for $t \geq 1$.

73. $W'(t)$ is directly proportional to $W(t)$.

74. $W'(t) = ce^{-kt}$ for some c and $k > 0$.

75. The rate at which water is flowing into a reservoir is continually monitored, and is given by $V'(t) = 5{,}000t(2 - t)$ cubic feet/week, where $0 \leq t \leq 2$. How much water flows in during the two-week period?

 summation and the integral

We are now ready to study the second main concept of calculus, the **integral**. As mentioned in the prologue, its main job is to add an extremely large (in fact, infinite) number of quantities, and, like the derivative, it too is a limit. We begin this section with a discussion of sums and the **summation notation**. You will use this notation not only in calculus but also in statistics.

If $x_1, x_2, x_3, \ldots, x_n$ are n real numbers, we may form their sum $S = x_1 + x_2 + x_3 + \cdots + x_n$. The shorthand notation for this sum is

$$\sum_{i=1}^{n} x_i \qquad \text{or, more simply,} \qquad \sum x$$

The Greek symbol Σ stands for "sum." the "$i = 1$" reminds us that we start the addition with x_1. Finally, the "n" indicates that the final term in the sum is x_n.

The **mean** or **average value** of the numbers $x_1, x_2, \ldots, x_n$ is defined as $\bar{x} = (\sum x)/n$.

example 19.1   The heights (in inches) of the 1979–80 Los Angeles Lakers basketball team were

$$74, 81, 86, 78, 81, 83, 78, 75, 75, 79, 81$$

Compute $\sum x$ and $\bar{x}$.

solution 19.1   The sum of the heights, $\sum x$, is easily seen to be 871. Hence $\bar{x}$, the average height, is $871/11 = 79.2$ inches.

If $f(x)$ is a function and $x_1, x_2, \ldots, x_n$ are n real numbers in the domain of f, we may compute $f(x_1), f(x_2), \ldots, f(x_n)$ and then form their sum $\sum_{i=1}^{n} f(x_i) = f(x_1) + f(x_2) + \cdots + f(x_n)$. When it is understood from the context what the numbers $x_1, x_2, \ldots, x_n$ are, we simply write $\sum f(x)$. For the heights in example 19.1, we could form $\sum x^2$, $\sum x^3$, $\sum e^x$, and so forth. The special number

$$\sigma^2 = \frac{\sum x^2}{n} - (\bar{x})^2$$

is known as the **variance** of $x_1, x_2, \ldots, x_n$.

example 19.2   For the heights in example 19.1, compute $\Sigma\, x^2$ and then the variance σ^2.

solution 19.2   The squares of the heights are

$$5{,}476,\ 6{,}561,\ 7{,}396,\ 6{,}084,\ 6{,}561,\ 6{,}889,$$
$$6{,}084,\ 5{,}625,\ 5{,}625,\ 6{,}241,\ 6{,}561$$

Hence, $\Sigma\, x^2$, the sum of the squared heights, is 69,103. It follows that $\sigma^2 = 69{,}103/11 - (79.2)^2 = 9.45$.

Given two sets of n real numbers

$$x_1,\ x_2,\ x_3,\ \ldots,\ x_n$$
$$y_1,\ y_2,\ y_3,\ \ldots,\ y_n$$

we can calculate the sum of *all* the numbers in two ways. We could add the numbers within each row and then add the two results: $\Sigma_{i=1}^{n}\, x_i + \Sigma_{i=1}^{n}\, y_i$. Or, we could add the numbers within columns and then add the results: $\Sigma_{i=1}^{n}\, (x_i + y_i)$. It follows that $\Sigma\, x + \Sigma\, y = \Sigma\, (x + y)$.

In addition, $c(\Sigma_{i=1}^{n}\, x_i) = c(x_1 + x_2 + \cdots + x_n) = cx_1 + cx_2 + \cdots + cx_n = \Sigma_{i=1}^{n}\, cx_i$. If $x_i = 1$, for all i, we obtain $\Sigma_{i=1}^{n}\, c = cn$. These results are summarized in the accompanying box, and illustrated in example 19.3.

Laws of Summation

$$\Sigma\, (x + y) = \Sigma\, x + \Sigma\, y$$

$$\Sigma\, cx = c\left(\Sigma\, x\right)$$

$$\Sigma\, c = nc,\ \text{where } n \text{ is the number of terms}$$

---

example 19.3   Given the numbers 1, 2, 3, 4, $\ldots$, 9, 10, compute $\Sigma\, x$ and then $\Sigma\, (2x - 1) = 1 + 3 + 5 + \cdots + 19$ using the Laws of Summation.

solution 19.3   With the aid of a calculator, we obtain $\Sigma\, x = 55$. Hence, $\Sigma\, (2x - 1) = \Sigma\, 2x - \Sigma\, 1 = 2(\Sigma\, x) - 10 = 2(55) - 10 = 100$.

To motivate the definition of an integral, consider the following challenging problem. Shown in figure 19.1 is the graph of $f(x) = x^2$, for $0 \le x \le 1$. What is the average of all the heights that stretch from the x-axis up to the curve, over the interval from $x = 0$ to $x = 1$? We must somehow average the infinitely many heights. A first step might be to divide the interval $[0, 1]$ into ten pieces,

figure 19.1

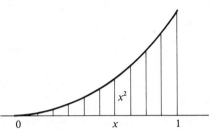

where the change in x value is $\Delta x = .1$, and form table 19.1. The average of the functional values in the table $\Sigma x^2/10 = 3.85/10 = .385$. An even better approach would be to divide $[0, 1]$ into 100 pieces, where the change in x value is $\Delta x = .01$, and form table 19.2. The average of these 100 functional values is found (with much effort) to be $\Sigma x^2/100 = 33.835/100 = .33835$.

table 19.1

x	.1	.2	.3	.4	.5	.6	.7	.8	.9	1
x^2	.01	.04	.09	.16	.25	.36	.49	.64	.81	1

table 19.2

x	.01	.02	.03		.97	.98	.99	1
x^2	.0001	.0004	.0009		.9409	.9604	.9801	1

At least in theory, we can continue this process using more and more values of x. If we divide $[0, 1]$ into n pieces, then $\Delta x = 1/n$ and the average of the functional values is $\Sigma x^2/n = (1/n) \Sigma x^2 = \Delta x \Sigma x^2 = \Sigma x^2 \Delta x$. The *true average value* shoud be

$$\lim_{\Delta x \to 0} \left(\Sigma x^2 \Delta x \right)$$

How does one compute such a limit? The answer is "by computing an *integral*." Like the derivative, the integral has many interpretations—e.g., average value, area, or volume. Initially, we will concentrate on the interpretation as the area under a curve. We will return to average values in section 21 of the text.

Shown in figure 19.2 is the area bounded by the graph of $y = f(x)$ and the interval $[a, b]$. In the following steps we first approximate the area by rectangles and then take a limit:

Step 1. Divide $[a, b]$ into n pieces, each of length $\Delta x = (b - a)/n$, as shown in figure 19.3.

Step 2. Let $x_1 = a + \Delta x$, $x_2 = a + 2\Delta x$, ..., $x_n = a + n\Delta x = a + (b - a) = b$. Compute $f(x_1)$, $f(x_2)$, ..., $f(x_n)$.

Step 3. Form the sum $\sum_{i=1}^{n} f(x_i)\Delta x = \sum f(x)\Delta x$. Note that $f(x)\Delta x$ is just the area of a typical rectangle.

Step 4. Compute $\lim_{\Delta x \to 0} (\sum f(x)\Delta x)$.

Note that as $\Delta x \to 0$, two things happen: the rectangles become more and more slender, and they become more and more numerous. The net result is that we obtain a *progressively better fit* to the area under the curve, as shown in figure 19.4.

figure 19.2

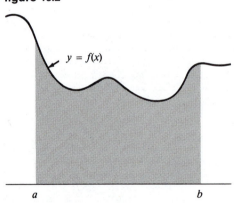

figure 19.3

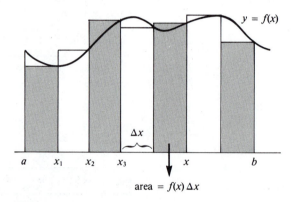

area $= f(x)\,\Delta x$

figure 19.4(a) $\Delta x = .5$

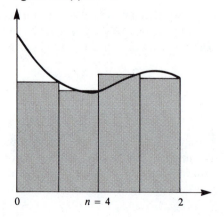

$n = 4$

figure 19.4(b) $\Delta x = .25$

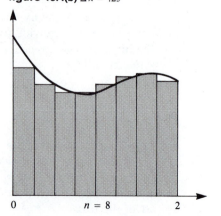

$n = 8$

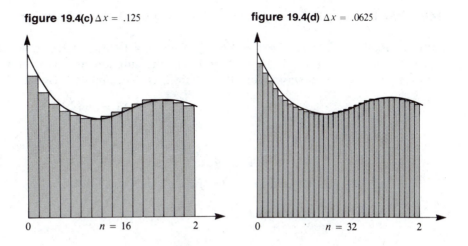

figure 19.4(c) $\Delta x = .125$ **figure 19.4(d)** $\Delta x = .0625$

definition 19.1   For $f(x)$ defined on $[a, b]$, we define $\int_a^b f(x)\,dx$, the *integral of f from a to b*, to be

$$\int_a^b f(x)\,dx = \lim_{\Delta x \to 0} \sum f(x)\Delta x$$

provided the limit exists.

In the definition, the symbol $\int$ is an elongated "S," standing for "sum." This infinite summation begins at $x = a$ and ends at $x = b$. The symbol "dx" reminds us that Δx is getting smaller and smaller. Originally, Newton called the symbol "dx" an "infinitesimal," and thought of the term "$f(x)dx$" as the area of a rectangle of height $f(x)$ and width dx.

For $f(x)$ continuous on $[a, b]$, the integral $\int_a^b f(x)\,dx$ always exists. The key question is, How is its value found? At this stage of the development, all we can do is to compute (possibly with the aid of a calculator or computer) the summation $\sum f(x)\Delta x$ for smaller and smaller Δx, to estimate the limit. By the summation laws, we have

$$\sum f(x)\Delta x = \Delta x \left(\sum f(x) \right)$$

We will use this fact in the examples that follow.

example 19.4   Estimate $\int_1^2 (1/x)\, dx$ by computing $\Sigma\, (1/x)\Delta x$ with $\Delta x = .1$.

solution 19.4   Here, $a = 1$ and $b = 2$. Then $x_1 = 1 + \Delta x = 1.1$, and we obtain 1.2, 1.3, 1.4, 1.5, 1.6, 1.7, 1.8, 1.9, and 2 as the subsequent values of x. These ten values are shown in table 19.3 where $\Sigma\, 1/x$ is computed. Hence,

$$\Sigma \frac{1}{x} \Delta x = \Delta x \left(\Sigma \frac{1}{x} \right)$$
$$= .1(6.68771)$$
$$= .668771,$$

and our estimate for $\int_1^2 (1/x)\, dx$ is .668771.

table 19.3

x	$1/x$
1.1	.90909
1.2	.83333
1.3	.76923
1.4	.71428
1.5	.66666
1.6	.62500
1.7	.58823
1.8	.55555
1.9	.52631
2.0	.50000
$\Sigma = 6.68771$	

Shown in table 19.4 are the results of computations done on a computer for estimating $\int_1^2 (1/x)\, dx$. Thus, it appears that $\int_1^2 (1/x)\, dx = .69$ to two decimal places. The table also shows how very slowly the limit is reached. Fortunately, there is a simple, direct way of computing $\int_a^b f(x)\, dx$ based on antiderivatives.

table 19.4

Δx	$\Sigma\, (1/x)\, \Delta x$
.1 ($n = 10$)	.668771
.05 ($n = 20$)	.680803
.01 ($n = 100$)	.690377
.005 ($n = 200$)	.691899
.001 ($n = 1{,}000$)	.692896
.0001 ($n = 10{,}000$)	.693120

This method, called the **Fundamental Theorem of Calculus**, is presented in section 20. There we will show that $\int_1^2 (1/x)\, dx = \ln 2 = .6931471805 \ldots$.

example 19.5

Estimate $\int_0^1 \sqrt{1+x^4}\, dx$ by computing $\Sigma \sqrt{1+x^4}\, \Delta x$ with $\Delta x = .2$.

solution 19.5

Here, $a = 0$, $b = 1$, and $x_1 = 0 + \Delta x = .2$. The other values of x are .4, .6, .8, and 1. In table 19.5 we compute the sum $\Sigma \sqrt{1+x^4}$. Hence,

$$\Sigma \sqrt{1+x^4}\, \Delta x = \Delta x \left(\Sigma \sqrt{1+x^4} \right)$$

$$= .2(5.67782)$$

$$= 1.13556$$

and we estimate $\int_0^1 \sqrt{1+x^4}\, dx$ to be 1.13556.

table 19.5

x	$\sqrt{1+x^4}$
.2	1.00080
.4	1.01272
.6	1.06283
.8	1.18727
1.0	1.41421
$\Sigma = 5.67782$	

Shown in table 19.6 are the results of computations done on a computer for estimating $\int_0^1 \sqrt{1+x^4}\, dx$. [The required program in BASIC is shown in the accompanying box.] Thus, it appears that $\int_0^1 \sqrt{1+x^4}\, dx \approx 1.09$ to two decimal places. The table once again shows how slowly the limit is being reached.

table 19.6

Δx	$\Sigma \sqrt{1+x^4}\, \Delta x$
.1	1.11132
.05	1.10008
.01	1.09151
.005	1.08985
.001	1.08964

Unfortunately, there is no easy way to compute the integral for this example. We will, however, study methods of estimating integrals that *converge faster* to the limit in Section 27.

```
program in BASIC for computing Σ f(x)Δx

10    A=0
20    B=1
30    DEF FNA(X)=SQR(1+X↑4)     The function f(x), the
40    N=100                      endpoints a and b, and
50    D=(B-A)/N                   n are entered.


60    S=0
70    X=A
80    FOR I=1 TO N
90       LET X=X+D                The sum Σ f(x)Δx is
100      LET S=S+FNA(X)*D         computed.
110   NEXT I

120   PRINT "THE APPROX FOR DELTA X=";D; "IS";S
```

exercises for section 19

part A

1. The glory years of the Boston Celtics were the early 1960s. The heights of the starting players (in inches) were

$$73, 74, 81, 77, 79$$

Compute $\sum x$, $\bar{x}$, $\sum x^2$, and σ. Can you name the starting five?

2. Test scores on a calculus exam are shown below:

$$91, 54, 82, 97, 80, 99, 72, 75, 77, 53$$

Compute $\sum x$, $\bar{x}$, $\sum x^2$, and σ^2.

3. To obtain a rough estimate of life expectancy (in years) in a certain locale, the ages at death of ten adult males were found:

$$73, 61, 47, 29, 71, 46, 58, 83, 49, 59$$

Compute $\sum x$, $\bar{x}$, $\sum x^2$ and σ^2.

Given one hundred numbers with $\sum x = 231.6$, $\sum x^2 = 621.4$, and $c = .2$, use the Laws of Summation to compute:

4. $\sum (x - 1)$ 5. $\sum (x^2 + x)$ 6. $\sum cx^2$ 7. $\sum c(x - c)$

For $x = .1, .2, .3, \ldots, .9, 1.0$, compute $\sum f(x)$ for:

8. $f(x) = 2x$ **9.** $f(x) = \sqrt{x}$ **10.** $f(x) = \sqrt{x^2 + 1}$

Estimate $\int_0^1 f(x)\,dx$ by computing $\sum f(x)\,\Delta x$ with $\Delta x = .1$ for:

11. $f(x) = 2x$ **12.** $f(x) = \sqrt{x}$ **13.** $f(x) = \sqrt{x^2 + 1}$

14. Estimate $\int_1^3 \ln x\,dx$ by computing $\sum \ln x\,\Delta x$ with $\Delta x = .2$.

15. Estimate $\int_{-1}^0 \dfrac{1}{x+2}\,dx$ by computing $\sum \dfrac{1}{x+2}\Delta x$ with $\Delta x = .1$.

16. Estimate $\int_1^2 e^x\,dx$ by computing $\sum e^x\,\Delta x$ with $\Delta x = .2$.

17. Estimate $\int_1^4 \sqrt{2x+1}\,dx$ by computing $\sum \sqrt{2x+1}\,\Delta x$ with $\Delta x = .2$.

part B   Implement the program shown in page 216 or write a program (on a programmable calculator, for example) to compute

$$\sum_{i=1}^n f(x_i)\Delta x = \sum f(x)\Delta x$$

Use your program to compute $\sum f(x)\Delta x$ for $\Delta x = .1, .01,$ and $.001$ and thereby estimate:

18. $\displaystyle\int_0^2 10e^{-x}\,dx$ **19.** $\displaystyle\int_{-1}^1 \sqrt{1+x^3}\,dx$

20. $\displaystyle\int_1^4 \dfrac{1}{x+1}\,dx$ ▲**21.** $\displaystyle\int_0^1 \sin x\,dx$

22. $\displaystyle\int_0^1 \dfrac{x(x-1)}{x^4+1}\,dx$

 computing integrals:
fundamental theorem
of calculus

In section 19, we defined $\int_a^b f(x)\, dx = \text{limit}_{\Delta x \to 0} \Sigma f(x)\Delta x$. If you are willing to expend a great amount of energy (or if you have access to a computer), you could approximate $\int_a^b f(x)\, dx$ by computing a sum $\Sigma f(x)\Delta x$ corresponding to a small Δx. The smaller the value of Δx, the better the approximation, but the more terms that must be added.

Fortunately, there is a beautiful and simple theorem courtesy of Isaac Newton that allows us to calculate the *exact value* of most integrals fairly easily.

---

the Fundamental Theorem of Calculus

If $f(x)$ is continuous on $[a, b]$, and if $F(x) = \int f(x)\, dx$ is an antiderivative for $f(x)$, then

$$\int_a^b f(x)\, dx = F(b) - F(a)$$

Find an antiderivative and you can compute the integral! Thus, for example, $\int_0^1 x^2\, dx = 1^3/3 - 0^3/3 = 1/3$, since $x^3/3 = \int x^2\, dx$. Why should this result be true? We will give a demonstration of this famous and important theorem at the end of this section.

special notation   In place of $F(b) - F(a)$, it is customary to write $F(x)\big|_a^b$.

Hence, we will carry out the computation of $\int_0^1 x^2\, dx$ as follows:

$$\int_0^1 x^2\, dx = \frac{x^3}{3}\bigg|_0^1 = \frac{1}{3} - 0 = \frac{1}{3}$$

example 20.1   Compute each of the following integrals using the Fundamental Theorem.

$$\textbf{(a)}\ \int_{-1}^{1} x^4\, dx \qquad \textbf{(b)}\ \int_{-1}^{4} (x^2 + x)\, dx \qquad \textbf{(c)}\ \int_0^1 e^{2x}\, dx$$

solution 20.1   Antiderivatives are $x^5/5$, $x^3/3 + x^2/2$, and $e^{2x}/2$, respectively. Hence,

$$\textbf{(a)}\ \int_{-1}^{1} x^4\, dx = \frac{x^5}{5}\bigg|_{-1}^{1} = \frac{1}{5} - \frac{(-1)^5}{5} = \frac{1}{5} + \frac{1}{5} = \frac{2}{5}$$

(b) $\int_{-1}^{4} (x^2 + x)\, dx = \left(\frac{x^3}{3} + \frac{x^2}{2}\right)\Big|_{-1}^{4} = \left(\frac{64}{3} + \frac{16}{2}\right) - \left(\frac{-1}{3} + \frac{1}{2}\right)$

$$= \frac{65}{3} + \frac{15}{2} = \frac{175}{6}$$

warning!   A common error in solving this problem is to evaluate $(x^3/3 + x^2/2)|_{-1}^{4} = 64/3 + 16/2 - (-1)/3 + 1/2$. *Be generous in your use of parentheses!*

(c) $\int_{0}^{1} e^{2x}\, dx = \frac{e^{2x}}{2}\Big|_{0}^{1} = \frac{e^2}{2} - \frac{e^0}{2} = \frac{1}{2}(e^2 - 1) \approx 3.1945$

example 20.2   Explain what is wrong with the following application of the Fundamental Theorem:

$$\int_{-1}^{1} (1/x^2)\, dx = -1/x|_{-1}^{1} = (-1) - (1) = -2$$

solution 20.2   Mechanically, all steps in the calculation are correct: $\int x^{-2}\, dx = x^{-2+1}/-1 = -1/x$, for example. We have failed, however, to verify the hypotheses of the theorem. The function $f(x) = 1/x^2$ is *not continuous* on the interval $[-1, 1]$. A vertical asymptote exists at 0, and so *the fundamental theorem does not apply.*

example 20.3   Find the area under the curve $y = 1/x^2$ from $x = 1$ to $x = b$. What is the area from 1 to $+\infty$?

solution 20.3

figure 20.1

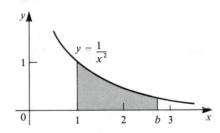

$$\int_{1}^{b} x^{-2}\, dx = -x^{-1}|_{1}^{b} = -b^{-1} - (-1) = 1 - \frac{1}{b}$$

Note that as b increases, $1/b \to 0$. Hence, the total area from 1 to $+\infty$ is one square unit. We write $\int_{1}^{+\infty} x^{-2}\, dx = 1$.

▲ **example 20.4**   Find the area under one of the "humps" of the sine curve $y = \sin x$, shown in figure 20.2, (page 220).

figure 20.2

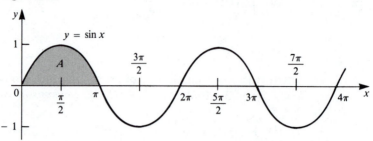

solution 20.4

There are many ways we could compute the area. Perhaps the easiest is to compute

$$A = \int_0^\pi \sin x \, dx = -\cos x \big|_0^\pi = -\cos \pi - (-\cos 0)$$

Now, $\cos 0 = 1$ and $\cos \pi = -1$. Hence, the area $A = -(-1) - (-1) = 2$ square units.

example 20.5

Find the area of the region bounded by the lines $y = 0$, $x = 0$, $x = 1$, and the graph of $y = 1/\sqrt{x}$.

solution 20.5

Since $y = 1/\sqrt{x}$ is not continuous on $[0, 1]$, we cannot apply the Fundamental Theorem directly. Instead, we will compute $\int_a^1 1/\sqrt{x} \, dx$, for a near 0, as illustrated in figure 20.3:

$$\int_a^1 x^{-1/2} \, dx = 2x^{1/2} \big|_a^1 = 2 - 2\sqrt{a}$$

Hence, as $a \to 0$, the desired area approaches 2. The area under the graph of $y = 1/\sqrt{x}$ from 0 to 1 is therefore 2 square units.

figure 20.3

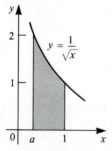

If $f(x) < 0$ on $[a, b]$, then $\int_a^b f(x) \, dx$ is negative. This is because each term $f(x) \Delta x$ in the sum $\Sigma f(x) \Delta x$ is negative. The area A shown in figure 20.4 is

figure 20.4

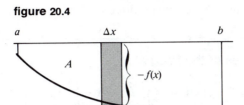

$-\int_a^b f(x)\,dx$. In other words,

> When $f(x) < 0$ for $a \le x \le b$, the integral
>
> $$\int_a^b f(x)\,dx = -A$$
>
> where A is the area bounded by the graph of $f(x)$ and $[a, b]$.

The following example illustrates the computation of an area using this new formula.

example 20.6   Find the area bounded by the y-axis and the graph of $f(x) = x(x-2)$.

solution 20.6   The graph of the parabola $y = x(x-2) = x^2 - 2x$ is shown in figure 20.5.

figure 20.5

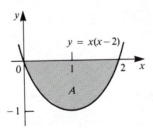

The area A is $-\int_0^2 (x^2 - 2x)\,dx$, and

$$\int_0^2 (x^2 - 2x)\,dx = \frac{x^3}{3} - x^2\Big|_0^2$$

$$= \left(\frac{8}{3} - 4\right) - 0$$

$$= -\frac{4}{3}$$

Hence, $A = 4/3$ square units.

The integral has the following three useful properties:

---

properties of the integral

1. $\displaystyle\int_a^b [f(x) + g(x)]\, dx = \int_a^b f(x)\, dx + \int_a^b g(x)\, dx$

2. $\displaystyle\int_a^b cf(x)\, dx = c \int_a^b f(x)\, dx$

3. For $a < c < b$, $\displaystyle\int_a^b f(x)\, dx = \int_a^c f(x)\, dx + \int_c^b f(x)\, dx$

The first two properties are the analogues of the first two basic summation laws given in section 19. Note that

$$\sum [f(x) + g(x)]\,\Delta x = \sum f(x)\,\Delta x + \sum g(x)\,\Delta x$$

Thus, if we let $\Delta x \to 0$ and recall the definition of an integral, then property 1 follows. Property 3 makes sense if we remember that $\int_a^b$ is a sum. To sum from a to b, we may first sum from a to c, and then sum from c to b.

figure 20.6

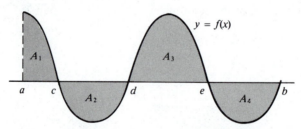

If $y = f(x)$ has the graph shown in figure 20.6, what is a geometric interpretation of $\int_a^b f(x)\, dx$? The answer follows as a consequence of property 3:

$$\int_a^b f(x)\, dx = \int_a^c f(x)\, dx + \int_c^d f(x)\, dx + \int_d^e f(x)\, dx + \int_e^b f(x)\, dx$$

$$= A_1 \qquad\quad - A_2 \qquad\quad + A_3 \qquad\quad - A_4$$

The next two examples show how these properties can be used to compute the areas of more complicated figures.

example 20.7   Find the area bounded by $y = x^2 - 1$, $x = -1$, $x = 2$, and $y = 0$.

solution 20.7   A sketch of the region is shown in figure 20.7. We compute the areas A_1 and

A_2 as follows:

figure 20.7

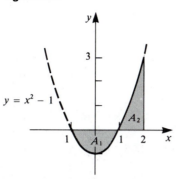

$y = x^2 - 1$

$$A_1 = -\int_{-1}^{1} (x^2 - 1)\, dx = -\left(\frac{x^3}{3} - x\right)\Big|_{-1}^{1}$$

$$= -\left[\left(\frac{1}{3} - 1\right) - \left(\frac{-1}{3} + 1\right)\right] = -\left(\frac{-2}{3} - \frac{2}{3}\right)$$

$$= \frac{4}{3} \text{ square units}$$

$$A_2 = \int_{1}^{2} (x^2 - 1)\, dx = \left(\frac{x^3}{3} - x\right)\Big|_{1}^{2}$$

$$= \left(\frac{8}{3} - 2\right) - \left(\frac{1}{3} - 1\right)$$

$$= \frac{4}{3} \text{ square units}$$

The total area is therefore 8/3 square units.

example 20.8   Compute $\int_0^2 f(x)\, dx$ if

$$f(x) = \begin{cases} x^2, & 0 \le x \le 1 \\ 2 - x, & 1 \le x \le 2 \end{cases}$$

solution 20.8

$$\int_0^2 f(x)\, dx = \int_0^1 f(x)\, dx + \int_1^2 f(x)\, dx, \qquad \text{by property (3)}$$

$$= \int_0^1 x^2\, dx + \int_1^2 (2 - x)\, dx$$

figure 20.8

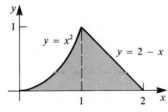

$y = x^2$

$y = 2 - x$

But $\int_0^1 x^2\, dx = \frac{1}{3}$ and $\int_1^2 (2 - x)\, dx = \frac{1}{2}$, using the Fundamental Theorem. Hence, $\int_0^2 f(x)\, dx = \frac{1}{3} + \frac{1}{2} = \frac{5}{6}$.

To end this section, we will present a demonstration of the Fundamental Theorem of Calculus based on the relationship between a car's *odometer* and *speedometer* (as shown in figure 20.9). Let $s(t)$ be the odometer reading at time t, and let $v(t)$ be the speedometer reading at time t, as indicated in figure

figure 20.9(a)   Odometer.   **figure 20.9(b)**   Speedometer.

20.9. We pointed out in section 11 that $s'(t) = v(t)$. Suppose that we begin a trip at time $t = a$ and end it at time $t = b$. How can we compute the total distance we have traveled?

One simple method is to take the *difference in odometer readings*:

$$s(b) - s(a) = \text{distance traveled between times}$$
$$t = a \quad \text{and} \quad t = b$$

There is another method of calculating this distance, based on speedometer readings alone. Divide the time interval $[a, b]$ into smaller time intervals, each of length Δt, as depicted in figure 20.10. Between times t and $t + \Delta t$, the speed

figure 20.10

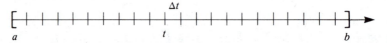

of the car may vary a bit, but for Δt very small (e.g., a millisecond), it is for all practical purposes constant. Hence, the distance traveled between times t and $(t + \Delta t)$ is approximately $v(t)\Delta t$, and the total distance traveled between times a and b is approximately $\Sigma \, v(t)\Delta t$. If we let $\Delta t \to 0$, then

$$\int_a^b v(t) \, dt = \text{distance traveled between times}$$
$$t = a \quad \text{and} \quad t = b$$

Hence, equating the two expressions for this distance yields $s(b) - s(a) = \int_a^b v(t) \, dt$. If we set $f(t) = v(t)$ and $F(t) = s(t)$, then $F(t)$ is an antiderivative for $f(t)$ and

$$\int_a^b f(t) \, dt = F(b) - F(a)$$

We have thus established the Fundamental Theorem for the case $f(t) = v(t) \geq 0$. We have, in the process, discovered a second interpretation of the integral: For the velocity $v(t) \geq 0$, $\int_a^b v(t) \, dt$ is the total distance traveled between times $t = a$ and $t = b$. This new interpretation is illustrated in our final example.

example 20.9   If the speed of a moving particle (in ft/sec) is given by $v(t) = 50/t$, for $t \geq 1$, find the total distance the particle travels between times $t = 1$ and $t = 6$.

solution 20.9   The distance traveled is $\int_1^6 (50/t)\, dt = (50 \ln t)|_1^6 = 50 \ln 6 - 50 \ln 1 = 89.587$ ... feet.

The Fundamental Theorem reduces the problem of calculating $\int_a^b f(x)\, dx$ to that of finding an antiderivative $\int f(x)\, dx$. In sections 22 and 24 we will study special techniques for finding antiderivatives. In subsequent sections, we will show that the integral may be interpreted in many ways, and is a valuable tool in many areas of science, including biology.

exercises for section 20

part A   Using the Fundamental Theorem of Calculus, compute each of the following integrals.

1. $\int_1^2 x^2\, dx$

2. $\int_1^4 \sqrt{x}\, dx$

3. $\int_2^4 \frac{1}{x^2}\, dx$

4. $\int_2^4 1/\sqrt{x}\, dx$

5. $\int_{-1}^1 2\, dx$

6. $\int_{-1}^1 3x\, dx$

7. $\int_0^1 e^{-x}\, dx$

8. $\int_{-2}^0 2e^{-2x}\, dx$

9. $\int_3^5 (4x - x^2)\, dx$

10. $\int_1^2 (x^2 - 1)^2\, dx$

11. $\int_1^3 \left(3x^2 - \frac{1}{x}\right) dx$

12. $\int_0^1 10e^{.5t}\, dt$

13. $\int_{-1}^1 (x^3 - 4x + 3)\, dx$

14. $\int_0^1 \pi\, dx$

15. $\int_0^1 (3 - x)\, dx$

16. $\int_0^1 (c - x)\, dx$

17. $\int_3^5 (3 - x)^2\, dx$

▲ **18.** $\int_0^{\pi/2} \cos x\, dx$

★ **19.** $\int_{-R}^R (R^2 - x^2)\, dx$

★ **20.** $\int_0^2 \frac{x}{1 + x^2}\, dx$

21. Compute $\int_0^1 xe^{-x}\, dx$ given that $\int xe^{-x}\, dx = -(x + 1)e^{-x}$.

22. Compute $\int_1^3 \ln x\, dx$ given that $\int \ln x\, dx = x \ln x - x$.

Determine whether the Fundamental Theorem has been applied correctly in each of the following situations.

23. $\displaystyle\int_{-1}^{1} \frac{1}{x^3}\,dx = \frac{x^{-2}}{-2}\Big|_{-1}^{1} = -\frac{1}{2} - \left(-\frac{1}{2}\right) = 0$

24. $\displaystyle\int_{-1}^{1} (2x - 1)\,dx = (x^2 - x)\big|_{-1}^{1} = 1 - 1 - 1 + 1 = 0$

25. What is the area under the curve $y = 1/x^3$ from $x = 1$ to $x = b$? What is the area from $x = 1$ to $+\infty$?

26. What is the area under the curve $y = 1/x$ from $x = 1$ to $x = b$? What is the area from $x = 1$ to $+\infty$?

★**27.** Determine under what conditions the area beneath the curve $y = x^{-\kappa}$ from $x = 1$ to $+\infty$ is finite.

28. Find the area of the region bounded by $y = 0$, $x = 0$, $x = 1$, and the graph of $y = 1/\sqrt[3]{x}$.

29. Find the area of the region bounded by $y = 0$, $x = 0$, $x = 1$, and the graph of $y = 1/x$.

★**30.** Determine under what conditions the area of the region bounded by $y = 0$, $x = 0$, $x = 1$, and $y = x^{-\kappa}$ is finite.

Find the area of the shaded region depicted in each of the figures.

31.

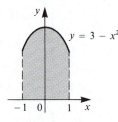

$y = 3 - x^2$

32.

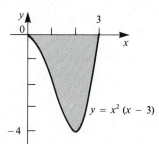

$y = x^2(x - 3)$

33.

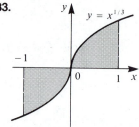

$y = x^{1/3}$

34.

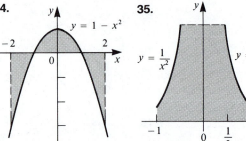

$y = 1 - x^2$

35.

$y = \dfrac{1}{x^2}$ $y = \dfrac{1}{x^2}$

36. Compute $\displaystyle\int_{0}^{2} f(x)\,dx$ where

$$f(x) = \begin{cases} 2 - x, & 0 \le x \le 1 \\ x^2, & 1 < x \le 2 \end{cases}$$

37. If $f(x) = \begin{cases} x^2, & x < 0 \\ x^3, & x \geq 0 \end{cases}$, compute $\int_{-1}^{1} f(x)\, dx$.

38. Compute $\int_{0}^{3} f(x)\, dx$ if

$$f(x) = \begin{cases} 3 - x^2, & 0 \leq x \leq 1 \\ 4 - x, & 1 < x \leq 2 \\ 4/x, & x > 2 \end{cases}$$

39. The speed of a moving particle is given by $v(t) = 4 - 1/t^2$ for $t \geq 1$. Find the total distance traveled between times $t = 1$ and $t = 5$.

part B

Use the Fundamental Theorem of Calculus to discover the meaning of each of the following integrals.

40. $W'(t)$ is the rate of growth in pounds/year. What does $\int_{a}^{b} W'(t)\, dt$ represent?

41. The acceleration of a particle at time t is given by $a(t)$. What does $\int_{c}^{d} a(t)\, dt$ represent?

42. If $F(t)$ is the flow rate in ft^3/min, what does $\int_{a}^{b} F(t)\, dt$ represent?

21 average values, areas, and volumes

In this section, we will use the fact that

$$\int_a^b f(x)\,dx = \lim_{\Delta x \to 0} \sum f(x)\,\Delta x$$

to give alternate interpretations of the integral. Although we interpreted $f(x)\,\Delta x$ as the area of a slender rectangle, we are not bound by this interpretation. Our line of attack is to approximate a small portion of a physical quantity Q by an expression of the form $f(x)\,\Delta x$, add the approximations to form $\sum f(x)\,\Delta x$, and then let $\Delta x \to 0$ to obtain an integral.

the average value of a function   In section 19, we introduced the idea of the average value for a function $f(x)$ on $[a, b]$. The formula for the average value $\bar{Y}$ is given by:

$$\bar{Y} = \frac{1}{b-a}\int_a^b f(x)\,dx$$

Let's see how the formula is derived. We divide $[a, b]$ into n pieces, each of length $\Delta x = (b-a)/n$, as illustrated in figure 21.1, and first average the n heights $f(x_1), f(x_2), \ldots, f(x_n)$. Hence, an approximation to $\bar{Y}$ is given by

figure 21.1

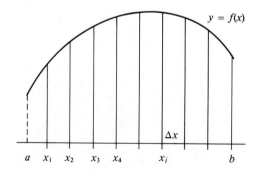

$y = f(x)$

Δx

$a \quad x_1 \quad x_2 \quad x_3 \quad x_4 \qquad x_i \qquad\qquad b$

$$\frac{f(x_1) + f(x_2) + \cdots + f(x_n)}{n} = \frac{1}{n} \sum f(x) = \frac{\Delta x}{b-a} \sum f(x)$$

since $1/n = \Delta x/(b-a)$. Rewriting the last expression, we obtain

$$\bar{Y} \approx \frac{1}{b-a} \sum f(x)\,\Delta x$$

Now, as $\Delta x \to 0$, more and more values of $f(x)$ are averaged and $\sum f(x)\,\Delta x \to \int_a^b f(x)\,dx$. The average value formula is illustrated in the following two examples.

example 21.1   Find the average value of $1{,}000e^{2t}$ over the interval $[0, 2]$. Interpret the result in terms of population growth.

solution 21.1   Let $N(t) = 1{,}000e^{2t}$. We may interpret $N(t)$ as the number of individuals in a population at time t, and we are assuming that the population is described by an exponential growth model. Hence, $\bar{N}$, the average number in the population from $t = 0$ to $t = 2$, is given by:

$$\bar{N} = \frac{1}{2-0} \int_0^2 1{,}000e^{2t}\,dt = 500\left(\frac{e^{2t}}{2}\right)\Big|_0^2$$
$$= 250e^4 - 250e^0 = 250(e^4 - 1) \approx 13{,}400 \text{ individuals} \qquad \blacksquare$$

example 21.2   For the high diver in example 11.2, we showed that $v(t) = -32t$ is her velocity t seconds after reaching the peak of her jump. If she hits the water 4 seconds later, find the average velocity during the dive.

solution 21.2   The average velocity is $\bar{V} = \dfrac{1}{4-0}\left(\displaystyle\int_0^4 -32t\,dt\right) = -8\left(\dfrac{t^2}{2}\right)\Big|_0^4 = -4t^2\,\Big|_0^4 = -64$ ft/sec. $\blacksquare$

the area between two curves   Suppose that $f(x) \ge g(x)$ for $a \le x \le b$, as shown in figure 21.2. The length of a cross section is $[f(x) - g(x)]$. We can

figure 21.2

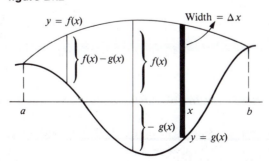

imagine that the *total area* is composed of a large number of slender reeds, where the area of a typical reed is $[f(x) - g(x)]\Delta x$. Hence, the total area is approximated by

$$\sum [f(x) - g(x)]\Delta x \to \int_a^b [f(x) - g(x)] \, dx$$

as $\Delta x \to 0$. Thus,

If $f(x) \geq g(x)$ for $a \leq x \leq b$, then the area bounded by the two graphs and the lines $x = a$ and $x = b$ is given by

$$A = \int_a^b [f(x) - g(x)] \, dx$$

---

example 21.3   Find the area bounded by the graphs of $y = x^2$ and $y = \sqrt{x}$.

solution 21.3   We must sketch the graphs to determine which curve is on top, and to determine the values of a and b. From the graphs shown in figure 21.3, we see

figure 21.3

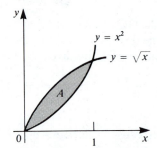

that $a = 0$ and $b = 1$. In addition $\sqrt{x} \geq x^2$ for $0 \leq x \leq 1$. Hence

$$A = \int_0^1 (\sqrt{x} - x^2) \, dx = \left(\frac{2}{3} x^{3/2} - \frac{x^3}{3} \right) \Big|_0^1$$

$$= \frac{2}{3} - \frac{1}{3} = \frac{1}{3} \text{ square unit}$$

example 21.4   Find the area bounded by the curves $y = 1 - x^2$ and $y = -1 + x^2$.

solution 21.4   The two curves, shown in figure 21.4, intersect at points $P = (x, y)$ such that $1 - x^2 = -1 + x^2$. Thus, $2x^2 = 2$ or $x = \pm 1$.

figure 21.4

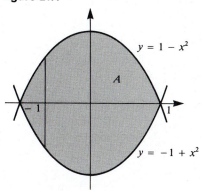

Hence

$$A = \int_{-1}^{1} [(1 - x^2) - (-1 + x^2)] \, dx = \int_{-1}^{1} (2 - 2x^2) \, dx = \left(2x - \frac{2}{3}x^3\right)\Big|_{-1}^{1}$$

$$= \left(2 - \frac{2}{3}\right) - \left(-2 + \frac{2}{3}\right) = 4 - \frac{4}{3} = \frac{8}{3} \text{ square units}$$

Since most common regions can be broken up into regions of the type shown in figure 21.2, the formula $A = \int_a^b [f(x) - g(x)] \, dx$ represents a powerful tool for computing areas. Our next formula can be used to compute the volume of many solids.

the volume of a solid of revolution   A solid may be formed from the graph of a function $y = f(x)$, $a \le x \le b$, by rotating the bounded area about the x-axis, as illustrated in figure 21.5. Note that the cross section of any such solid is a circle, and therefore has the area of $\pi r^2 = \pi [f(x)]^2$. Common solids such as

figure 21.5(a)

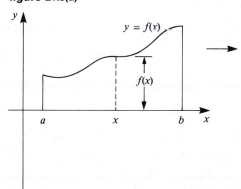

figure 21.5(b)

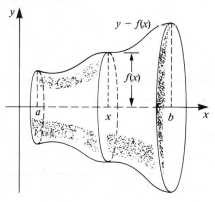

figure 21.6(a)   **figure 21.6(b)**   **figure 21.6(c)**

spheres, cones, and bottles can be formed in the manner depicted in figure 21.6. The formulas for the volumes of the solids shown in the figures can all be derived from one general integral:

$$V = \int_a^b \pi [f(x)]^2 \, dx$$

Let's examine why this formula is valid. We may consider the solid of figure 21.5**(b)** as the sum of many extremely thin discs or wafers, one piled on another, as depicted in figure 21.7. The typical wafer has volume = (area of the

figure 21.7(a)   **figure 21.7(b)**

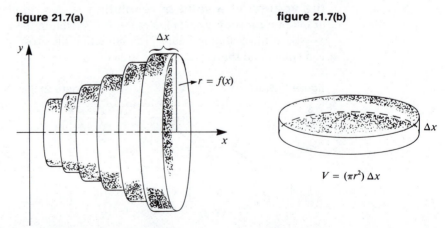

$$V = (\pi r^2) \, \Delta x$$

base) · (height) = $\pi r^2 \Delta x$. The total volume of the solid is approximated by $\Sigma \, \pi [f(x)]^2 \Delta x$. As $\Delta x \to 0$, the wafers become extremely thin and the fit becomes better and better. In addition, by the definition of an integral,

$$\sum \pi[f(x)]^2 \Delta x \to \int_a^b \pi[f(x)]^2 \, dx$$

This formula is used in our next three examples.

example 21.5   Rotate $y = \sqrt{x}$, $0 \le x \le 4$, about the x-axis and find the volume of the resulting solid.

solution 21.5   We have $f(x) = \sqrt{x}$, and so $\pi[f(x)]^2 = \pi x$. Hence, the volume of the solid shown in figure 21.8 is given by

$$V = \int_0^4 \pi x \, dx = \frac{\pi x^2}{2} \bigg|_0^4$$

$$= 8\pi \text{ cubic units}$$

figure 21.8

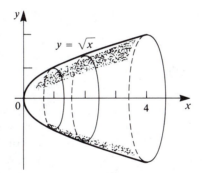

example 21.6   Rotate the circle $x^2 + y^2 = R^2$ about the x-axis and derive the formula for the volume of a sphere of radius R.

solution 21.6   If we set $y = 0$ in the equation $x^2 + y^2 = R^2$, we obtain $x^2 = R^2$ or $x = \pm R$. Thus, the x-intercepts are $x = -R$ and $x = R$. Since $[f(x)]^2 = y^2 = R^2 - x^2$, the volume of the solid shown in figure 21.9 is given by:

figure 21.9

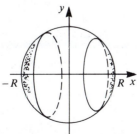

$$V = \int_{-R}^{R} \pi(R^2 - x^2) \, dx = \pi\left(R^2 x - \frac{x^3}{3}\right)\bigg|_{-R}^{R}$$

$$= \left(\pi R^3 - \frac{\pi R^3}{3}\right) - \left(-\pi R^3 + \frac{R^3}{3}\right)$$

$$= \frac{2}{3} \pi R^3 - \left(-\frac{2}{3} \pi R^3\right) = \frac{4}{3} \pi R^3$$

We have derived the well-known formula $V = \frac{4}{3}\pi R^3$.

example 21.7   Show that the volume of the paraboloid of revolution of height a and base radius b is given by $V = (\pi/2)ab^2$.

figure 21.10(a)

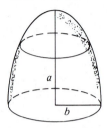

figure 21.10(b)

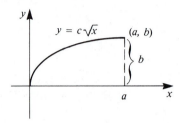

solution 21.7   To generate the solid, we rotate the graph of $y = c\sqrt{x}$ for $0 \le x \le a$ about the x-axis. Since $f(a) = b$, $b = c\sqrt{a}$. It follows that $c = b/\sqrt{a}$. Hence, $[f(x)]^2 = y^2 = b^2 x/a$. Hence,

$$V = \int_0^a \frac{\pi b^2}{a} x \, dx = \frac{\pi b^2}{a} \int_0^a x \, dx = \frac{\pi b^2}{a} \cdot \frac{a^2}{2} = \frac{\pi}{2} ab^2$$

As a final application, we will use the volume formula of example 21.7 to derive an important law in biophysics.

Poiseuille's Law   In the 1840s, the French physician Poiseuille discovered experimental formulas to describe the flow of a liquid through cylindrical tubing. As long as the velocities reached are not too large, the flow is **laminar**—i.e., the paths or streamlines of flow are parallel to the walls of the tube, as indicated in figure 21.11. Naturally, the flow is freest—i.e., the velocity

figure 21.11

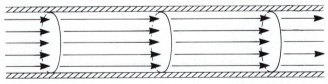

is greatest—along the center of the tube and decreases as the liquid approaches the wall of the tube. Let R be the radius of the tube and let r be the distance from a point to the center axis of the tube. Poiseuille discovered that the relationship between the velocity v and r is **parabolic**, as illustrated in figure 21.12. If we assume that $v(R) = 0$ (the velocity along the wall is 0), then it follows that:

Poiseuille's Law $v = k(R^2 - r^2)$ for $0 \le r \le R$

---

figure 21.12

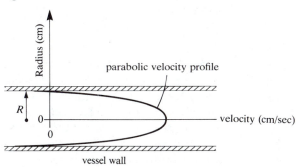

The constant k depends on the length of the tube, the pressure of the liquid, and the nature of the fluid itself. This law can also be derived from basic physics.

Let us now compute the amount of liquid that passes through a circular cross section. For purposes of illustration, suppose that v is measured in cm/sec. After one second, a particle has traveled $x = k(R^2 - r^2)$ cm. When $r = 0$, $x = kR^2$. Figure 21.13 shows that we must compute the volume of a

figure 21.13

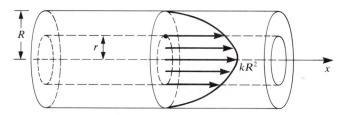

paraboloid of revolution. If we use the formula derived in example 21.7, with $a = kR^2$ and $b = R$, we obtain $V = \dfrac{\pi}{2} ab^2 = \dfrac{k\pi}{2} R^4$ cm³. If we let F denote **flow rate** (in cm³/sec, e.g.), we have derived:

Poiseuille's Law for rate of flow $F = \dfrac{k\pi}{2} R^4$

---

example 21.8   Compare the flow rates of water through a $\frac{1}{2}''$ diameter pipe versus a more standard $\frac{5}{8}''$ diameter pipe.

solution 21.8   If $F_1 = (k\pi/2)R_1^4$ and $F_2 = (k\pi/2)R_2^4$, then $F_1/F_2 = (R_1/R_2)^4$. Hence, if $R_1 = 5/16$ and $R_2 = 1/4$, then $F_1/F_2 = (5/4)^4 = 2.44$. Hence, $F_1 = 2.44F_2$. Remarkably, increasing the diameter by only $\frac{1}{8}''$ will result in a rate of delivery that is more than doubled!

The Poiseuille equations are used by cardiovascular physiologists to demonstrate how vascular smooth muscle can regulate circulation in arteries and arterioles, as discussed in example 21.9.

example 21.9   **Poiseuille's model for arterial blood flow**   Blood is, strictly speaking, a suspension of cells (on the order of 10μ in diameter) in the plasma, the fluid portion. Thus, we expect that Poiseuille's Laws are only approximately true and will not work well in very small vessels such as capillaries. Arteries, the vessels that carry blood away from the heart, are surrounded by circular smooth muscle and elastic tissue. Thus, the radius R of the vessel may be increased or decreased through muscular action.

If we assume that $F = \dfrac{k\pi}{2} R^4$, then it follows that $\dfrac{F_1}{F_2} = \left(\dfrac{R_1}{R_2}\right)^4$. This relationship can be used to show that *a small change in radius can have a large effect in regulating the rate of flow*, as the calculations in table 21.1 demonstrate. The velocity profile of aortic blood has been determined experimentally and is not strictly parabolic. Nevertheless, the model does give some insight into how vascular muscle can regulate blood flow.

table 21.1

R_1/R_2	F_1/F_2
.95	.8145
.90	.6561
.80	.4096
.70	.2401
1.05	1.2155
1.10	1.4641
1.20	2.0736

exercises for section 21

part A   Find the average value of each of the following functions over the prescribed interval.

1. $f(x) = x^2$ on $[1, 3]$ **2.** $f(x) = e^{-2x}$ on $[0, .5]$

3. $f(x) = 2\sqrt{x}$ on $[0, 4]$ ▲**4.** $f(x) = \sin x$ on $[0, \pi]$

5. $f(x) = \ln x$ on $[1, 10]$ given that $\int \ln x \, dx = x \ln x - x$

6. The population of Circle City (see section 17, exercise 36) is increasing according to the formula $N(t) = 10,000e^{.1t}$. Find the average population over the next 5 years.

7. One thousand dollars is deposited in a high-interest account that pays 12.5% interest compounded continuously. The amount $A(t)$ in the account t years later is given by $A(t) = 1{,}000e^{.125t}$. Estimate the average daily balance over the first year.

8. If the buying power of the dollar t years from now is given by $B(t) = (.9)^t$, find the average buying power over the next two years.

★9. If a ball is dropped from a height of s_0 feet, show that the average velocity over the ball's journey to the ground is $-4\sqrt{s_0}$ feet/sec.

★10. The Beer-Lambert Law can be written in the form $I = I_0 e^{-kx}$, where I_0 is the light intensity at the surface and x is the depth (see example 15). Find the average light intensity in the photic zone, that zone from the ocean's surface to the depth where $I = .01I_0$.

Find the area bounded by each of the following graphs.

11. $y = x^3$ and $y = \sqrt[3]{x}$ **12.** $y = 2x^2$ and $y = x + 1$

13. $y = x^2$, $y = \frac{1}{2}x^2$, and $x = 1$ **14.** $y = |x|$ and $y = 2 - x^2$

★15. $y = 1/x$ and $y = x(2 - x)$

Find the area of the region shown in each of the figures.

★16.

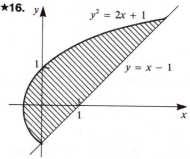

★17.

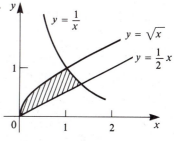

Find the volume of the solid generated by rotating each of the following graphs about the x-axis. Sketch the solids formed.

18. $y = 2x^2$ on $[0, 1]$ **19.** $y = 4$ on $[0, 2]$

20. $y = 1/x$ on $[1, 3]$ **21.** $y = e^{-.5x}$ on $[0, 2]$

22. $y = 3x + 2$ on $[0, 1]$ **23.** $y = \sqrt[3]{x}$ on $[-1, 1]$

24. $y = 1 - x^2$ on $[-1, 1]$ **25.** $y = |x|$ on $[-2, 2]$

26. $y = x^n$ on $[0, 1]$ with $n > 0$ **27.** $y = \sqrt{4 - x^2}$ on $[-2, 2]$

28. Derive the formula for the volume of a cone of radius r and height h by rotating the line shown in the figure about the x-axis (on page 238).

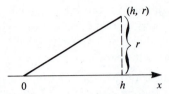

★**29.** Show that the volume of the spherical cap of radius r and height h shown in the figure is given by the formula $V = \frac{1}{3}\pi h^2(3r - h)$.

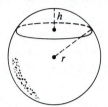

30. Water is flowing through a $\frac{5}{8}''$ diameter pipe at the rate of 1 cubic foot per minute. Determine the new flow rates for pipes of the following diameters.

    (a) $\frac{1}{2}''$ **(b)** $\frac{3}{4}''$ **(c)** $1''$

31. Hypertension (high blood pressure) is caused by constriction of the arteries. To increase the flow rate, the heart is forced to pump much harder, increasing the pressure in the cardiovascular system. The formula

$$F = \frac{k\pi}{2} R^4$$

was developed on page 235. The constant k is directly proportional to the pressure P on the liquid. Hence,

$$F = \frac{c\pi P}{2} R^4$$

Thus, for a fixed radius, twice the pressure will result in twice the rate of flow.

    (a) If F_1 and P_1 are normal flow rates and pressure, show that

$$\frac{F_2}{F_1} = \frac{P_2}{P_1} \left(\frac{R_2}{R_1}\right)^4$$

    (b) If the heart pumps harder in order to increase the new flow rate F_2 back to F_1, show that

$$\frac{P_2}{P_1} = \left(\frac{R_1}{R_2}\right)^4$$

and fill in the accompanying table. [*Note:* The heart can increase its pumping rate (and hence the blood pressure) by a factor of at most 4.]

R_2/R_1	P_2/P_1
.9	
.8	
.7	
.6	
.5	

part B

32. Poiseuille discovered that the velocity function for laminar fluid flow is parabolic, with maximum velocity in the center of the tube. Thus $v = ar^2 + br + c$ and $v'(0) = 0$. Show that if we assume $v(R) = 0$, then v can be written in the form

$$v = k(R^2 - r^2), \qquad 0 \le r \le R$$

33. Show that the volume obtained by rotating the graph of the power function shown in the figure is given by $V = \pi \dfrac{n}{n+2} ab^2$

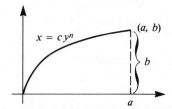

$$x = c y^n$$ (a, b) b a

34. As mentioned in example 21.9, the velocity profile $v(r)$ is not strictly parabolic, but has a much flatter shape around the vertex, as illustrated in the figures. This suggests using a higher-order power function $v = k(R^n - r^n)$ for the velocity function. Use exercise 33 to derive the generalized flow rate formula

$$F = \pi \dfrac{n}{n+2} kR^{n+2}$$

(a) A typical velocity profile

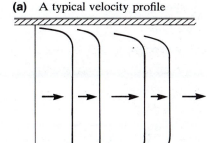

(b) $v = k(R^n - r^n)$

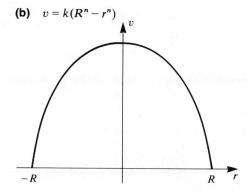

basic integration methods

Working out an integral $\int_a^b f(x)\,dx$ becomes difficult if it is hard to find an antiderivative $F(x) = \int f(x)\,dx$ for the given function $f(x)$. A large number of special techniques called **integration methods** have been developed for finding $\int f(x)\,dx$. One of the most widely used is the **substitution method** or the **change of variable method**. It is based on the Chain Rule, which states that if $F'(x) = f(x)$, then $D_x F(g(x)) = F'(g(x))g'(x) = f(g(x))g'(x)$. This gives the antiderivative formula:

$$\int f(g(x))g'(x)\,dx = F(g(x)) \qquad \text{where } F(x) = \int f(x)\,dx$$

Thus, for example, if we were asked to find $\int 2x(1+x^2)^9\,dx$, we would write $2x(1+x^2)^9 = f(g(x))g'(x)$, where $g(x) = 1 + x^2$ and $f(x) = x^9$. Then according to the formula,

$$\int 2x(1+x^2)^9\,dx = F(g(x)) = \frac{(1+x^2)^{10}}{10}$$

since $F(x) = \int x^9\,dx = x^{10}/10$.

Fortunately, there is a more straightforward, mechanical way of implementing the antiderivative formula, as described in the box.

---

the Substitution Method

Step 1. Let $u = g(x)$. Usually, $g(x)$ is the *most bothersome* term in the integrand, the function to be integrated.

Step 2. Take the derivative $\dfrac{du}{dx} = g'(x)$. Imagine $\dfrac{du}{dx}$ to be a fraction and write $du = g'(x)\,dx$.

Step 3. Replace every expression in x by the corresponding expression in u:

$$\int f(g(x))g'(x)\,dx = \int f(u)\,du$$

Step 4. Find $\int f(u)\,du = F(u)$ and replace u by $g(x)$ to obtain the final answer $F(g(x))$.

The accompanying box serves as a dictionary to be used to translate the expression in x into one in u which is hopefully easier to handle.

$u - x$ dictionary

$$u = g(x)$$
$$du = g'(x)\, dx$$

The substitution method is illustrated in the next six examples.

example 22.1   Find the antiderivative $\int x^2(1 + x^2)^3\, dx$, using the substitution method.

solution 22.1   Let $u = (1 + x^3)$. Then $\dfrac{du}{dx} = 3x^2$ or $du = 3x^2\, dx$. Thus,

$$\int x^2(1 + x^3)^3\, dx = \int u^3 x^2\, dx = \int u^3 \left(\frac{1}{3}\, du\right) = \frac{u^4}{12}$$

In the third step, we have substituted $x^2\, dx = \frac{1}{3}\, du$, which follows from the fact that $du = 3x^2\, dx$. Finally, we let $u = (1 + x^3)$ to obtain an antiderivative $(1 + x^3)^4/12$.

example 22.2   Find the antiderivative $\displaystyle\int \frac{1}{(4 - 3x)^2}\, dx$, using the substitution method.

solution 22.2   Let $u = 4 - 3x$. Then $\dfrac{du}{dx} = -3$ or $du = -3\, dx$. Solving for dx yields $dx = -\frac{1}{3}\, du$. Thus,

$$\int \frac{1}{(4 - 3x)^2}\, dx = \int \frac{1}{u^2}\left(-\frac{1}{3}\, du\right) = \int -\frac{u^{-2}}{3}\, du = \frac{u^{-1}}{3}$$

Hence, letting $u = 4 - 3x$, we obtain an antiderivative $\dfrac{1}{3}\left(\dfrac{1}{4 - 3x}\right)$.

example 22.3   Find the antiderivative $\int x\sqrt{1 + x^2}\, dx$, using the substitution method.

solution 22.3   Let $u = 1 + x^2$. Then $\dfrac{du}{dx} = 2x$ or $du = 2x\, dx$. Hence,

$$\int x\sqrt{1 + x^2}\, dx = \int \sqrt{u}(x\, dx) = \int \sqrt{u}\left(\frac{1}{2}\, du\right) = \frac{1}{3}\, u^{3/2}$$

Letting $u = 1 + x^2$, we obtain $\frac{1}{3}(1 + x^2)^{3/2}$ as an antiderivative.

The same method works just as well for trigonometric, exponential, and logaritham functions, as the following examples demonstrate.

example 22.4   Compute the exact value of the integral $\int_0^1 xe^{-x^2/2} \, dx$.

solution 22.4   We will first find $\int xe^{-x^2/2} \, dx$. Let $u = -\dfrac{x^2}{2}$. Then $\dfrac{du}{dx} = -\dfrac{2x}{2} = -x$. Thus, $du = -x \, dx$ or $x \, dx = -du$. Hence,

$$\int xe^{-x^2/2} \, dx = \int e^u(x \, dx) = \int -e^u \, du = -e^u$$

Hence, an antiderivative is $-e^{-x^2/2}$. Using the Fundamental Theorem, we obtain

$$\int_0^1 xe^{-x^2/2} \, dx = -e^{-x^2/2}\Big|_0^1 = -e^{-.5} - (-1)$$

The exact value is $1 - e^{-.5}$, or approximately $.393469 \ldots$.

▲ example 22.5   Compute the exact value of the integral $\int_0^\pi x \sin x^2 \, dx$.

solution 22.5   To find $\int x \sin x^2 \, dx$, let $u = x^2$. Then $\dfrac{du}{dx} = 2x$ or $x \, dx = \dfrac{1}{2} du$. Hence,

$$\int x \sin x^2 \, dx = \int \sin u(x \, dx) = \int \frac{1}{2} \sin u \, du = -\frac{1}{2} \cos u$$

Letting $u = x^2$, we see that $(-1/2)\cos x^2$ is an antiderivative. Applying the Fundamental Theorem yields

$$\int_0^\pi x \sin x^2 \, dx = -\frac{1}{2} \cos x^2\Big|_0^\pi = -\frac{1}{2} \cos \pi^2 - \left(-\frac{1}{2} \cos 0\right) \approx .9513 \ldots$$

example 22.6   Compute the exact value of the integral $\displaystyle\int_1^2 \frac{x \ln(1 + x^2)}{1 + x^2} \, dx$.

solution 22.6   The natural substitution to make is $u = 1 + x^2$. Since $x \, dx = \frac{1}{2} du$, we have

$$\int \frac{x \ln(1 + x^2)}{1 + x^2} \, dx = \int \frac{1}{2} \frac{\ln u}{u} \, du$$

Although the second antiderivative is somewhat simpler, it is still not manageable. Since our substitution has not made the antiderivative in u elementary, we will try another substitution.

Let $u = \ln(1 + x^2)$. Then $\dfrac{du}{dx} = \dfrac{1}{1 + x^2} 2x$ or $\dfrac{x}{1 + x^2} dx = \dfrac{1}{2} du$. Hence,

$$\int \frac{x \ln(1 + x^2)}{1 + x^2} dx = \frac{1}{2} \int u \, du = \frac{u^2}{4} = \frac{1}{4} [\ln(1 + x^2)]^2$$

Finally, we apply the Fundamental Theorem to obtain

$$\frac{1}{4} [\ln(1 + x^2)]^2 \Big|_1^2 = \frac{1}{4} (\ln 5)^2 - \frac{1}{4} (\ln 2)^2 \approx .527459 \ldots$$

As the next examples show, it is often very difficult to recognize directly f and g in the expression $\int f(g(x)) g'(x) \, dx$, but the step-by-step method outlined in the box on page 240, makes this unnecessary.

example 22.7   Compute $\displaystyle\int \frac{x}{\sqrt{1 + x}} \, dx$.

solution 22.7   If we let $u = 1 + x$, then $\dfrac{du}{dx} = 1$ and so $du = dx$. Hence,

$$\int \frac{x}{\sqrt{1 + x}} dx = \int \frac{1}{\sqrt{u}} x \, du = \int u^{-1/2} (u - 1) \, du$$

Note that in the last step we solved $u = 1 + x$ for x in order to complete the translation into u's. Finally, the last antiderivative is

$$\int (u^{1/2} - u^{-1/2}) \, du = \frac{2}{3} u^{3/2} - 2u^{1/2} = \frac{2}{3} (1 + x)^{3/2} - 2\sqrt{1 + x}$$

example 22.8   Compute $\displaystyle\int_0^1 \frac{x^2}{\sqrt{3x + 1}} \, dx$.

solution 22.8   Let $u = 3x + 1$. Then $\dfrac{du}{dx} = 3$ or $dx = \dfrac{1}{3} du$. Also, note that $x = \dfrac{u - 1}{3}$. Hence,

$$\int \frac{x^2}{\sqrt{3x - 1}} dx = \int \frac{1}{\sqrt{u}} x^2 \left(\frac{1}{3} du \right) = \int \frac{1}{3} u^{-1/2} \left(\frac{u - 1}{3} \right)^2 du$$

$$= \int \frac{1}{27} u^{-1/2} (u^2 - 2u + 1) \, du$$

$$= \int \frac{1}{27} (u^{3/2} - 2u^{1/2} + u^{-1/2}) \, du$$

$$= \frac{1}{27} \left(\frac{2}{5} u^{5/2} - \frac{4}{3} u^{3/2} + 2u^{1/2} \right)$$

$$= \frac{1}{27} \left[\frac{2}{5} (3x + 1)^{5/2} - \frac{4}{3} (3x + 1)^{3/2} + 2\sqrt{3x + 1} \right]$$

Hence,

$$\int_0^1 \frac{x^2}{\sqrt{3x+1}}\, dx = F(1) - F(0)$$

$$= \frac{1}{27}\left[\frac{2}{5}(2^5) - \frac{4}{3}(2^3) + 2\sqrt{4}\right] - \frac{1}{27}\left(\frac{2}{5} - \frac{4}{3} + 2\right)$$

$$= \frac{1}{27} \cdot \frac{76}{15} = .1876543\ldots$$

Not all antiderivatives can be successfully attacked using the substitution method. For example, the expression $\int x^2\sqrt{1+x^2}\, dx$ does not simplify if $u = x^2$ or $u = 1 + x^2$. We will study more methods of integration in section 24.

The following antiderivative formulas can be demonstrated by using the substitution method with $u = x - a$:

$$\int \frac{1}{(x-a)^n}\, dx = \frac{(x-a)^{-n+1}}{1-n} \qquad \text{for } n \neq 1$$

$$\int \frac{1}{x-a}\, dx = \ln|x-a|$$

These formulas, together with a special algebraic technique known as **partial fractions decomposition**, can be used to find antiderivatives for many rational functions, as indicated in the next two examples.

example 22.9   Find $\int \dfrac{1}{(x+3)(x-1)}\, dx$.

solution 22.9   We first will find real numbers A and B so that

$$\frac{1}{(x+3)(x-1)} = \frac{A}{x+3} + \frac{B}{x-1}$$

Then

$$\int \frac{1}{(x+3)(x-1)}\, dx = A \ln|x+3| + B \ln|x-1|$$

Now,

$$\frac{A}{x+3} + \frac{B}{x-1} = \frac{A(x-1) + B(x+3)}{(x+3)(x-1)}$$

The numbers A and B must be chosen so that $A(x-1) + B(x+3) = 1$. To find A, set $x = -3$. Then $A(-4) + B(0) = -4A = 1$, and thus $A = -\frac{1}{4}$. Setting $x = 1$ yields $A(0) + B(4) = 4B = 1$, and hence $B = \frac{1}{4}$. The antiderivative is then

$$\frac{1}{4}\ln|x-1| - \frac{1}{4}\ln|x+3|$$

example 22.10   Use the partial fractions method to find $\int \dfrac{3x-2}{x^2-3x+2}\,dx$.

solution 22.10   We can rewrite $\dfrac{3x-2}{x^2-3x+2}$ in the form

$$\frac{A}{x-1}+\frac{B}{x-2}=\frac{A(x-2)+B(x-1)}{x^2-3x+2}$$

We must find A and B so that $A(x-2)+B(x-1)=3x-2$. Setting $x=1$ implies that $A(-1)+B(0)=-A=3(1)-2=1$, and hence $A=-1$. Letting $x=2$ yields $B=6-2=4$. It follows that

$$\int \frac{3x-2}{x^2-3x+2}\,dx = \int \left(\frac{-1}{x-1}+\frac{4}{x-2}\right)dx$$
$$= -\ln|x-1|+4\ln|x-2| \qquad \blacksquare$$

If $f(x)$ is the rational function $p(x)/q(x)$, where degree $p \geq$ degree q, we must first divide $q(x)$ into $p(x)$ before applying the partial fractions method, as the next example demonstrates.

example 22.11   Find the antiderivative for $\dfrac{x^2}{(x-1)(x-2)}$.

solution 22.11   We must first divide x^2-3x+2 into x^2:

$$
\begin{array}{r}
1 \\
x^2-3x+2\overline{)x^2} \\
x^2-3x+2 \\
\hline
3x-2
\end{array}
$$

Hence,

$$\frac{x^2}{(x-1)(x-2)}=1+\frac{3x-2}{(x-1)(x-2)}$$

As we saw in example 22.6,

$$\frac{3x-2}{(x-1)(x-2)}=\frac{-1}{x-1}+\frac{4}{x-2}$$

Hence,

$$\int \left(1+\frac{-1}{x-1}+\frac{4}{x-2}\right)dx = x-\ln|x-1|+4\ln|x-2| \qquad \blacksquare$$

When a factor of $(x - a)^n$ occurs in the denominator of the rational function, we must include in the decomposition terms of the form

$$\frac{A_1}{x - a} + \frac{A_2}{(x - a)^2} + \cdots + \frac{A_n}{(x - a)^n}$$

This technique is illustrated in the final example.

example 22.12   Find $\int \dfrac{x^2}{(x + 2)^2 (x + 3)}\, dx$.

solution 22.12   The partial fractions decomposition now takes the form

$$\frac{x^2}{(x + 2)^2 (x + 3)} = \frac{A}{x + 2} + \frac{B}{(x + 2)^2} + \frac{C}{x + 3}$$

$$= \frac{A(x + 2)(x + 3) + B(x + 3) + C(x + 2)^2}{(x + 2)^2 (x + 3)}$$

Thus, A, B, and C must be chosen so that

$$x^2 = A(x + 2)(x + 3) + B(x + 3) + C(x + 2)^2$$

Setting $x = -2$, we obtain $4 = B(-2 + 3) = B$. Letting $x = -3$ yields $9 = C(-3 + 2)^2 = C$. Finally, to find A, let $x = 0$. $0 = 6A + 3B + 4C = 6A + 3(4) + 4(9)$, from which it follows that $A = -8$. Therefore,

$$\frac{x^2}{(x + 2)^2 (x + 3)} = \frac{-8}{x + 2} + \frac{4}{(x + 2)^2} + \frac{9}{x + 3}$$

and the antiderivative is

$$-8 \ln |x + 2| - 4(x + 2)^{-1} + 9 \ln |x + 3|$$

exercises for section 22

part A

Using the substitution method, find the following antiderivatives.

1. $\displaystyle\int \frac{1}{(2x - 5)^3}\, dx$ 2. $\displaystyle\int \frac{1}{(2x - 5)}\, dx$ 3. $\displaystyle\int \sqrt{3x - 2}\, dx$

4. $\displaystyle\int x(4 - x^2)^4\, dx$ 5. $\displaystyle\int \frac{x^4}{(1 + x^5)^2}\, dx$ 6. $\displaystyle\int \frac{x^4}{1 + x^5}\, dx$

7. $\displaystyle\int \sqrt[3]{3 - 2x}\, dx$ 8. $\displaystyle\int \frac{x}{x^2 - 9}\, dx$ 9. $\displaystyle\int \frac{e^x}{1 + e^x}\, dx$

10. $\displaystyle\int x e^{-x^2}\, dx$ 11. $\displaystyle\int (3x - 2)^8\, dx$ 12. $\displaystyle\int x(3x - 2)^8\, dx$

13. $\displaystyle\int \frac{\ln x}{x}\, dx$ 14. $\displaystyle\int \frac{x^3}{(3-x^4)^3}\, dx$ 15. $\displaystyle\int x^3\sqrt{1+x^4}\, dx$

▲16. $\displaystyle\int \cos x\,(\sin x)^4\, dx$ ▲17. $\displaystyle\int x \cos x^2\, dx$ 18. $\displaystyle\int x\sqrt{1+x}\, dx$

19. $\displaystyle\int \frac{x}{3-x}\, dx$ ★20. $\displaystyle\int \frac{x^2}{(1+x)^3}\, dx$

Compute each of the following integrals by first finding an antiderivative.

21. $\displaystyle\int_0^1 \frac{x}{(3+x^2)^3}\, dx$ 22. $\displaystyle\int_2^3 \frac{(\ln x)^3}{x}\, dx$ 23. $\displaystyle\int_0^2 x\sqrt{4-x^2}\, dx$

24. $\displaystyle\int_0^3 e^{-x+2}\, dx$ 25. $\displaystyle\int_1^4 \frac{(1+\sqrt{x})}{\sqrt{x}}\, dx$ ▲26. $\displaystyle\int_{\pi/2}^{\pi} \sin 2x\, dx$

Discover the substitution that simplifies each of the following antiderivatives.

27. $\displaystyle\int \frac{x-1}{\sqrt{x-2}}\, dx$ 28. $\displaystyle\int \frac{\ln(\sqrt{x+1})}{x+1}\, dx$ 29. $\displaystyle\int \frac{e^{4x}}{1+e^{2x}}\, dx$

30. $\displaystyle\int x^2\sqrt{2-x}\, dx$ 31. $\displaystyle\int \frac{1}{x \ln x}\, dx$ 32. $\displaystyle\int \frac{x}{(x+3)^5}\, dx$

33. Find the area under the curve $y = x/(1+x^2)$ from $x = 0$ to $x = 5$.

34. Find the volume of the solid of revolution formed by rotating the graph of
$y = \sqrt[3]{3x+2},\ 0 \le x \le 2$ around the x-axis.

Use the partial fractions decomposition method to find an antiderivative for each of the following rational functions.

35. $\displaystyle\frac{1}{x(x+2)}$ 36. $\displaystyle\frac{1}{x^2-5x+4}$ 37. $\displaystyle\frac{x}{1-x^2}$

38. $\displaystyle\frac{x}{x^2-4}$ 39. $\displaystyle\frac{x}{(x+2)^2}$ 40. $\displaystyle\frac{3x}{(x-1)^2}$

41. $\displaystyle\frac{3x}{(x+1)^2(x+2)}$ 42. $\displaystyle\frac{x-2}{x^2+x}$ 43. $\displaystyle\frac{x^2}{x^2-1}$

44. $\displaystyle\frac{x^3}{x\ \ 1}$ 45. $\displaystyle\frac{x^3}{x^2-3x+2}$ 46. $\displaystyle\frac{1}{x^3-4x}$

★47. Use the partial fractions decomposition method to establish

$$\int \frac{1}{x(ax+b)}\, dx = \frac{1}{b}\ln\left|\frac{x}{ax+b}\right|$$

★48. Use the partial fractions decomposition method to establish

$$\int \frac{1}{x^2(ax+b)}\, dx = -\frac{1}{bx} + \frac{a}{b^2}\ln\left|\frac{ax+b}{x}\right|$$

23 work

Perhaps the strongest man who has ever lived is Paul Anderson of the United States, the 1956 Olympic heavyweight champion; he has bench pressed 627 pounds. This weight must be lifted a distance of about 2 feet, as shown in

figure 23.1

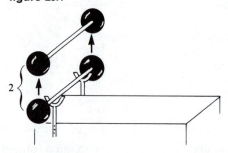

figure 23.1. To measure the enormity of such tasks, physicists have developed the formula for **work**:

$$W = F \cdot d \qquad \text{for a constant force } F \text{ applied over a distance } d.$$

For example, if a constant force of $F = 627$ pounds has been applied over a distance of $d = 2$ feet, then the work done is $627(2) = 1,254$ foot-pounds, where the unit **foot-pound** is the amount of work done in applying a force of 1 pound over a distance of 1 foot. To use the formula, it is necessary that the force applied be *constant* over the journey, as depicted in figure 23.2.

figure 23.2

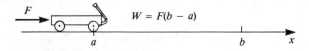

Following are two simple situations in which the force applied *varies* as the distance is traversed:

1. Gas is sealed in a cylinder by a movable piston, as shown in figure 23.3. The force $F(x)$ needed to move the piston increases greatly as $x \to 0$.

figure 23.3

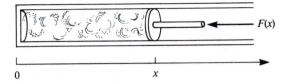

2. Water is to be pumped up and over the top of the cylinder shown in figure 23.4. The job becomes easier and easier—i.e., $F(x)$ decreases—as x approaches the top of the cylinder.

figure 23.4

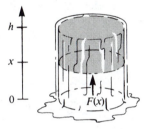

How is the work computed when the force applied is variable? The answer is given by the simple integral:

$$W = \int_a^b F(x)\,dx$$

It is easy to see why this formula is valid by referring to figure 23.5. The work done in moving from x to $x + \Delta x$ is approximately $F(x)\Delta x$. This is true because over an extremely small distance Δx, the force applied $F(x)$ is almost constant. The total work done is $W \approx \Sigma F(x)\Delta x$. As $\Delta x \to 0$, the approximation improves and the sum approaches $\int_a^b F(x)\,dx$.

figure 23.5

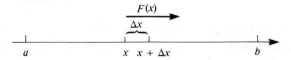

As the following three examples show, the major task is to determine the force function $F(x)$.

example 23.1   A force of 40 pounds is necessary to stretch a large spring 1 foot. Assuming that the spring obeys Hooke's Law (see example 3.6), how much work is done in stretching the spring from its natural length of 3 feet to a length of 6 feet?

solution 23.1   If x is the amount the spring is stretched, then Hooke's Law asserts that the needed force $F(x)$ is given by $F(x) = kx$ for some constant k. Since $F(1) = 40$, $k = 40$. Hence, $F(x) = 40x$. Since x is the amount the spring is stretched, x varies from 0 to 3 in order to attain a length of 6 feet. Hence,

$$W = \int_0^3 40x\,dx = 20x^2\big|_0^3 = 180 \text{ ft-lbs}$$

example 23.2   A gas is sealed in a cylinder by a movable piston, as shown in figure 23.6. When the length of the cylinder is 1 foot, the pressure within the cylinder is 15 lbs/ft². Assuming that the gas obeys Boyle's Law (see example 4.7), how much work is done in moving the piston from $x = 1$ to $x = \frac{1}{2}$?

figure 23.6

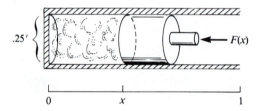

solution 23.2   Boyle's Law asserts that the pressure $P = c/V$, where V is the volume of the cylinder. The circular cross section has area $A = \pi r^2 = \pi(.125)^2$. The force exerted by the gas on the far right base of the cylinder is therefore

$$F(x) = PA = c\frac{A}{V} = c\frac{\pi(.125)^2}{\pi(.125)^2 x} = \frac{c}{x}$$

We are given that when $x = 1$, $F(1) = PA = 15\pi(.125)^2 = 15\pi/64$ lbs. Hence, $c = 15\pi/64$. It follows that

$$W = \int_{.5}^1 \frac{c}{x}\,dx = c\ln x\big|_{.5}^1 = \frac{15\pi}{64}(-\ln .5)$$

The work done is approximately .5104 ft-lb.

example 23.3   Water is placed in a cylindrical tube of length 3 feet and diameter 1 foot. If water weighs 62.4 pounds per cubic foot, how much work is done in pumping the water up and over the top?

solution 23.3   Refer to figure 23.4 on page 249. If $h = 3$ and $r = \frac{1}{2}$, then when the base has been lifted up x feet, the weight of the water remaining is 62.4

(lbs/ft^3) $\cdot$ $\pi(1/2)^2(3-x)$(ft^3) $= 62.4(\pi/4)(3-x)$ lbs. Hence, an upward force of $F(x) = 62.4(\pi/4)(3-x)$ pounds must be exerted. It follows that the total work performed is

$$W = \int_0^3 \frac{62.4\pi}{4}(3-x)\,dx = \frac{62.4\pi}{4}\left(3x - \frac{x^2}{2}\right)\Big|_0^3$$

$$= \frac{62.4\pi}{4}\cdot\frac{9}{2} = 220.54 \text{ ft-lbs}$$

The formula derived in the last example can easily be extended to a more general solid, as indicated in the accompanying box. Let us give a demonstration of the formula as Isaac Newton might have argued. The thin layer of water at height x has volume $A(x)\,dx$ (see figure 23.7), where dx is the "infinitesimal thickness." We might consider dx as the diameter of a water molecule. This thin layer has weight $\rho A(x)\,dx$ and must be lifted a distance of $(h-x)$. The work required is $\rho A(x)(h-x)\,dx$. Adding these "small works" from $x = 0$ (the bottom layer) to $x = a$ (the top layer), we obtain $\int_0^a \rho A(x)(h-x)\,dx$.

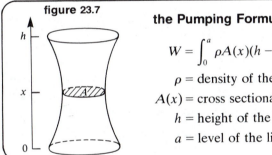

figure 23.7

the Pumping Formula

$$W = \int_0^a \rho A(x)(h-x)\,dx \text{ where}$$

ρ = density of the liquid (in lbs/ft^3, e.g.)

$A(x)$ = cross sectional area at height x

h = height of the solid

a = level of the liquid in the solid

Several of the exercises can be done easily using the Pumping Formula. Formulas for work can be used to study the heart as a pump, the breathing process, and the work of the kidney.*

exercises for section 23

part A

Compute the work done (in ft-lbs) given the force function $F(x)$ and the interval $[a, b]$ in each of the following situations.

1. $F(x) = c$, for $x \le x \le b$

2. $F(x) = \dfrac{k}{x}$, for $a \le x \le b$

3. $F(x) = k/x^2$, for $1 \le x \le b$ (What happens if $b \to +\infty$?)

---

* For more information, see G.B. Benedek, and M.H. Villars. *Physics with illustrative examples with medicine and biology.* Vol. 1, Reading, Mass.: Addison-Wesley, (1973), pp. 5–136 to 5–146.

4. $F(x) = kx^\alpha$, for $\alpha > 1$ and $0 \le x \le b$

5. Find the work done in stretching a spring obeying Hooke's Law 3 feet beyond its natural spring length, given that a 20-pound weight stretches the spring 1 foot.

6. For any spring obeying Hooke's Law, show that the work done in stretching the spring d units beyond its natural spring length is given by $W = \frac{1}{2}kd^2$.

7. Fifty foot-pounds of work are required to stretch a spring 2 feet beyond the natural spring length.

    (a) Find the spring constant k if the spring obeys Hooke's Law.

    (b) Find the spring constant k if the spring obeys a law of the form $F(x) = kx^{3/2}$.

★8. Two identical springs, each of natural spring length 4 feet, are attached as shown in the figure. The springs obey Hooke's Law with $k = .3$. Find the work done in moving the point P a distance of 3 feet to the right.

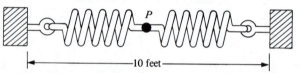

9. When the length of the cylinder shown in the figure is 3 feet, the pressure within the cylinder is 20 lbs/ft^2. Assuming the gas obeys Boyle's Law, find the work done in moving the piston from $x = 3$ to $x = 1$.

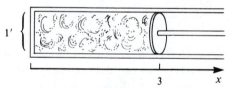

10. If the energy given off during compression of the cylinder raises the temperature of the system, then Boyle's Law will not hold. Rework exercise 9, assuming that the gas obeys the adiabatic expansion law $PV^{1.4} = c$.

11. Assuming Boyle's Law, show that the work done in compressing a gas from volume V_1 to volume V_2, as indicated in the figure, is given by

$$W = a \ln (V_1/V_2)$$

for some constant a. [*Hint*: See example 23.2 and generalize.]

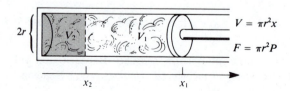

12. Rework exercise 11, assuming that the gas satisfies an adiabatic expansion law of the form $PV^\alpha = c$, where $\alpha \neq 1$.

In the following two problems, use the Pumping Formula (on page 251) to compute the amount of work done.

(a)

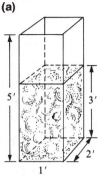

(b)

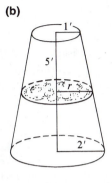

(c)

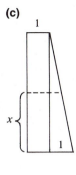

13. Find the amount of work done in pumping the water over the top of the rectangular tank shown in figure **(a)**, if the tank is filled to the 3-foot level.

★ **14.** The tank shown in figure **(b)** is a truncated cone with circular cross sections, and is full of water. Find the work done in emptying the tank by pumping up and over the top. [*Hint*: To find $A(x)$, which is the cross sectional area at height x, first show that $r = 2 - x/5$, using similar triangles].

15. A large cable hoists cargo from the dock into awaiting ships. The cable itself weighs 5 pounds per foot. How much work is done in lifting 1,000 pounds of cargo up 50 feet if the top of the pulley is 75 feet above the dock?

part B

16. The rate at which work is performed, $\dfrac{dW}{dt}$, is known as *power*. A spring is stretched from its natural length at the rate of 3 inches per second. Find power as a function of t, if the spring obeys Hooke's Law.

17. A body of mass m is to be accelerated from rest ($v_0 = 0$) to velocity $v_1 > 0$, as depicted in the figure. The total work done is $\int_0^{x_1} F(x)\, dx$, where $F(x)$ is the force acting at position x.

(a) Let $x = s(t)$ and use the substitution method to show that $W = \int_0^{t_1} F(s(t))s'(t)\, dt$, where $s(t_1) = x_1$.

(b) Newton's Law asserts that $F = ma = ms''(t)$. Thus, $F(s(t)) = ms''(t)$. Use the substitution $u = s'(t)$ to conclude that $W = \frac{1}{2}mv_1^2$. [*Note*: The *kinetic energy* of a body with velocity v is defined to be $\frac{1}{2}mv_1^2$.]

 **additional integration techniques and integral tables**

The substitution method introduced in section 22 was based on the Chain rule. A second commonly employed integration method is **integration by parts**, a method whose justification comes from the Product Rule:

$$D_x f(x)g(x) = f(x)g'(x) + f'(x)g(x)$$

Hence,

$$\int [f(x)g'(x) + f'(x)g(x)]\, dx = f(x)g(x)$$

This equation can be rearranged as follows:

$$\int f(x)g'(x)\, dx = f(x)g(x) - \int f'(x)g(x)\, dx$$

To demonstrate how this formula for integration by parts is used, let's find $\int xe^x\, dx$. We will set $f(x) = x$ and $g'(x) = e^x$—i.e., $g(x) = e^x$. The formula then becomes

$$\int xe^x\, dx = xe^x - \int 1 \cdot e^x\, dx = xe^x - e^x$$

The success of the technique depends upon obtaining an antiderivative $\int f'(x)g(x)\, dx$ on the right side that is easier to compute than the one on the left side.

Just as with the substitution method, there is a mechanical way of implementing the formula for integration by parts. We will manipulate $\dfrac{du}{dx}$ and $\dfrac{dv}{dx}$ as if they were fractions. If $u = f(x)$ and $v = g(x)$, then $\dfrac{du}{dx} = f'(x)$ or $du = f'(x)\, dx$, and $\dfrac{dv}{dx} = g'(x)$ or $dv = g'(x)\, dx$. Hence, the integration by parts formula may be written:

$$\int u\, dv = uv - \int v\, du$$

We will practice this new version of the formula in our next three examples.

example 24.1   Find $\int xe^{ax}\, dx$ using integration by parts.

solution 24.1   We could set $u = e^{ax}$ or $u = x$. The substitution $u = x$ leads to a simple integrand on the right-hand side. Let $u = x$ and $dv = e^{ax}\, dx$. Hence, $\dfrac{du}{dx} = 1$ or

$du = dx$ and $\dfrac{dv}{dx} = e^{ax}$, or $v = \int e^{ax}\, dx = e^{ax}/a$. It follows that

$$\int xe^{ax}\, dx = \overset{u}{x} \cdot \overset{v}{\frac{e^{ax}}{a}} - \int \overset{v}{\frac{e^{ax}}{a}}\,\overset{du}{dx} = \frac{xe^{ax}}{a} - \frac{e^{ax}}{a^2}$$

Hence, we have established the commonly used integration formula:

$$\int xe^{ax}\, dx = \frac{e^{ax}}{a}\left(x - \frac{1}{a}\right)$$

example 24.2   Find $\int x^2 e^{ax}\, dx$ using integration by parts.

solution 24.2   Let $u = x^2$ and $dv = e^{ax}\, dx$. It follows that $du = 2x\, dx$ and, as in example 24.1, $v = e^{ax}/a$. Hence,

$$\int x^2 e^{ax}\, dx = \frac{x^2 e^{ax}}{a} - \frac{2}{a}\int xe^{ax}\, dx = \frac{x^2 e^{ax}}{a} - \frac{2}{a}\cdot\frac{e^{ax}}{a}\left(x - \frac{1}{a}\right)$$

In the last step, we have used the formula derived in example 24.1.

example 24.3   Find $\int x \ln x\, dx$ using integration by parts.

solution 24.3   Let's first try the substitution $u = x$ and $dv = \ln x\, dx$. Then $du = dx$ and since $\dfrac{dv}{dx} = \ln x$, $v = \int \ln x\, dx$. But what is this last antiderivative? We have quickly reached a dead end. Next, let's try the substitution $u = \ln x$ and $dv = x\, dx$. Hence, $\dfrac{du}{dx} = \dfrac{1}{x}$, or $du = \dfrac{1}{x}\, dx$. Then $\dfrac{dv}{dx} = x$ gives $v = \dfrac{x^2}{2}$. It follows that

$$\int x \ln x\, dx = \overset{u}{(\ln x)}\,\overset{v}{\frac{x^2}{2}} - \int \overset{v}{\frac{x^2}{2}}\cdot\overset{du}{\frac{1}{x}}\, dx = \frac{x^2}{2}\ln x - \frac{x^2}{4}$$

The next example shows how the substitution method, in conjunction with integration by parts, can be used to simplify an antiderivative.

example 24.4   Compute $\int e^{\sqrt{x}}\, dx$.

solution 24.4   First, let $u = \sqrt{x}$. Then $x = u^2$ and so $\dfrac{dx}{du} = 2u$ or $dx = 2u\, du$. Hence,

$$\int e^{\sqrt{x}}\, dx = \int 2ue^u\, du = 2\int ue^u\, du = 2(ue^u - e^u)$$

using the formula of example 24.1. Finally, letting $u = \sqrt{x}$, we obtain

$$\int e^{\sqrt{x}}\, dx = 2(\sqrt{x}\,e^{\sqrt{x}} - e^{\sqrt{x}})$$

Over the years, mathematicians have devised many special tricks to compute antiderivatives. The results of their labor are conveniently summarized in special tables called *integration tables*. A short table of integrals is included at the end of this section. To use this table to compute $\int f(x)\, dx$, we must first classify the integrand $f(x)$ into the appropriate category. For example, four of the categories in our integral tables are:

1. Forms involving e^{ax} (such as $\int x^3 e^{2x}\, dx$)
2. Forms involving $\ln x$ (such as $\int x \ln x\, dx$)
3. Forms involving $\sin ax$ or $\cos ax$ (such as $\int x \cos 2x\, dx$)
4. Forms involving $\sqrt{a^2 - x^2}$ (such as $\int \sqrt{9 - x^2}\, dx$)

We then find an entry similar to the integrand of interest and set the parameters $a, b, \ldots,$ equal to the appropriate constants. To find $\int \sqrt{9 - x^2}\, dx$, for example, we would use formula 44 with $a = 3$.

The following four examples provide practice in using an integral table.

example 24.5   Find $\int \sqrt{x^2 + 4}\, dx$ using the integral tables.

solution 24.5   Using formula 89 with $a = 2$ gives

$$\int \sqrt{x^2 + 4}\, dx = \frac{x\sqrt{x^2 + 4}}{2} + \frac{4}{2}\ln(x + \sqrt{x^2 + 4})$$

example 24.6   Compute $\displaystyle\int_2^3 \frac{1}{x^2\sqrt{4x^2 - 9}}\, dx$.

solution 24.6   Formula 56 in the table comes closest to fitting the situation, but before we can apply this formula, we must write

$$\sqrt{4x^2 - 9} = \sqrt{4(x^2 - 9/4)} = 2\sqrt{x^2 - (3/2)^2}$$

Then setting $a = 3/2$ in formula 56 yields

$$\int \frac{1}{x^2\sqrt{4x^2-9}}\,dx = \frac{1}{2}\frac{\sqrt{x^2-(3/2)^2}}{9x/4} = \frac{2\sqrt{x^2-(3/2)^2}}{9x} = \frac{\sqrt{4x^2-9}}{9x}$$

Hence,

$$\int_2^3 = \frac{\sqrt{4x^2-9}}{9x}\Big|_2^3 = \frac{\sqrt{27}}{27} - \frac{\sqrt{7}}{18} = .04546.\ldots$$

example 24.7 Find $\int \frac{1}{x(3x-2)}\,dx$ using the integral tables.

solution 24.7 The integrand $1/x(3x-2)$ is of the form $1/(ax+b)(px+q)$ with $a = 1$, $b = 0$, $p = 3$, and $q = -2$. Hence, using formula 78 in the table, we obtain

$$\int \frac{1}{x(3x-2)}\,dx = \frac{1}{0(3)-1(-2)}\ln\left|\frac{3x-2}{x}\right| = \frac{1}{2}\ln\left|\frac{3x-2}{x}\right|$$

It is also possible to use formula 69.

example 24.8 Compute $\int_0^2 \frac{1}{x^2-9}\,dx$.

solution 24.8 Using formula 115 with $a = 3$ gives

$$\int_0^2 \frac{1}{x^2-9}\,dx = \frac{1}{6}\ln\left|\frac{x-3}{x+3}\right|\Big|_0^2 = \frac{1}{6}\ln\left|-\frac{1}{5}\right| - \frac{1}{6}\ln|-1| = -\frac{1}{6}\ln 5$$

Often, an integral that does not appear directly in the table can be converted to a tabled entry by first using the substitution method or integration by parts, as demonstrated in the following examples.

example 24.9 Find $\int \frac{1}{x(x^4+4)}\,dx$.

solution 24.9 Since there are many tabled entries of the form $x^2 + a^2$, let $u = x^2$. Then $du = 2x\,dx$ and so

$$\int \frac{1}{x(x^4+4)}\,dx = \int \frac{x}{x^2(x^4+4)}\,dx = \frac{1}{2}\int \frac{1}{u(u^2+4)}\,du$$

Using formula 100 with $a = 2$, we obtain

$$\frac{1}{2}\left[\frac{1}{2(4)}\ln\left(\frac{u^2}{u^2+4}\right)\right] = \frac{1}{16}\ln\left(\frac{x^4}{x^4+4}\right)$$

example 24.10   Find $\int \dfrac{\ln{(x^2+1)}}{x^3}\,dx$.

solution 24.10   Using the integration by parts formula, let $u = \ln{(x^2+1)}$ and $dv = x^{-3}\,dx$. Hence, $du = 2x/(x^2+1)\,dx$ and $v = -x^{-2}/2 = -1/(2x^2)$. Integrating by parts, we obtain

$$\int \frac{\ln{(x^2+1)}}{x^3}\,dx = \frac{-1}{2x^2}\ln{(x^2+1)} - \int \frac{-1}{x(x^2+1)}\,dx$$

Again, using formula 100, we obtain the antiderivative

$$\frac{-1}{2x^2}\ln{(x^2+1)} + \frac{1}{2}\ln{\left(\frac{x^2}{x^2+1}\right)}$$

◼

▲ **example 24.11**   Compute $\int x \sin^2 x\,dx$.

solution 24.11   Let $u = x$ and $dv = \sin^2 x\,dx$. Hence, $du = dx$ and from $\dfrac{dv}{dx} = \sin^2 x$, we obtain

$v = \displaystyle\int \sin^2 x\,dx = \dfrac{x}{2} - \dfrac{\sin 2x}{4}$ by using formula 25 in the table. Hence, integration by parts yields

$$\int x \sin^2 x\,dx = x\left(\frac{x}{2} - \frac{\sin 2x}{4}\right) - \int \left(\frac{x}{2} - \frac{\sin 2x}{4}\right)dx$$

$$= \frac{x^2}{2} - \frac{x \sin 2x}{4} - \frac{x^2}{4} - \frac{\cos 2x}{8}$$

In the last line we used formula 21 to obtain $\displaystyle\int \frac{1}{4}\sin 2x\,dx = -\frac{\cos 2x}{4(2)}$.   ◼

exercises for section 24

part A   Use integration by parts to compute each of the following antiderivatives.

▲ **1.** $\displaystyle\int x \sin x\,dx$ and then $\displaystyle\int x^2 \cos x\,dx$ [*Hint*: Recall that $D_x \sin x = \cos x$ and $D_x \cos x = -\sin x$.]

▲ **2.** $\displaystyle\int x \cos x\,dx$ and then $\displaystyle\int x^2 \sin x\,dx$

3. Use the formula derived in example 24.2 to find $\displaystyle\int x^3 e^{-2x}\,dx$.

4. $\displaystyle\int x^5 \ln x\,dx$ **5.** $\displaystyle\int x^{-1/2}\ln x\,dx$

6. Use the result of example 24.3 and integration by parts to find $\int x(\ln x)^2\, dx$.

7. $\int (2x+3)\ln x\, dx$ **8.** $\int x \ln (2x)\, dx$

★**9.** Let $u = x^2$ in the integration by parts formula to find

(a) $\int x^3\sqrt{x^2+1}\, dx$ (b) $\int \dfrac{x^3}{\sqrt{1+x^2}}\, dx$

10. Integrate by parts repeatedly to find $\int x^4 e^x\, dx$.

Use the substitution method and then integration by parts to compute each of the following.

▲**11.** $\int \sin\sqrt{x}\, dx$ ▲**12.** $\int \cos\sqrt{x}\, dx$

13. $\int x^3 e^{-x^2}\, dx$ **14.** $\int e^{-3\sqrt{x}}\, dx$

15. $\int \sqrt{x}\,e^{\sqrt{x}}\, dx$ **16.** $\int \dfrac{1}{x^3}\,e^{x^{-1}}\, dx$

17. $\int x^5 e^{x^3}\, dx$ **18.** $\int (x^3+x)\ln (x^2+1)\, dx$

Use the table of integrals to compute each of the following.

19. $\int \sqrt{x^2+16}\, dx$ **20.** $\int \dfrac{1}{x(x^2+1)}\, dx$ **21.** $\int x^3 \ln x\, dx$

22. $\int_0^2 \sqrt{x^2+9}\, dx$ **23.** $\int_2^3 x^2\sqrt{x^2-1}\, dx$ **24.** $\int_0^2 \sqrt{3x+2}\, dx$

25. $\int \dfrac{x}{4x+3}\, dx$ **26.** $\int \dfrac{1}{x^2\sqrt{4-x^2}}\, dx$ **27.** $\int x^2 e^{3x}\, dx$

28. $\int \dfrac{1}{(x+2)(x+3)}\, dx$ **29.** $\int x\sqrt{4x-3}\, dx$ **30.** $\int \dfrac{x^2}{36-25x^2}\, dx$

31. $\int_1^2 \dfrac{1}{x(x+2)^2}\, dx$ **32.** $\int_{-2}^1 \dfrac{1}{x^2\sqrt{4x^2+9}}\, dx$ **33.** $\int_0^1 \dfrac{1}{4-x^2}\, dx$

34. $\int x^3\sqrt{4-25x^2}\, dx$ **35.** $\int \dfrac{\sqrt{3-2x^2}}{x}\, dx$ **36.** $\int \dfrac{1}{\sqrt{x^2+2}}\, dx$

▲**37.** $\int x^2 \cos 3x\, dx$ ▲**38.** $\int \sin^2 4x\, dx$

Use the substitution method or integration by parts to reduce the given antiderivative to an entry in the integral tables.

39. $\int x \ln (3x+1)\, dx$ ▲**40.** $\int x \cos^2 3x\, dx$ **41.** $\int \sqrt{e^{2x}+1}\, dx$

part B

42. $\int x\sqrt{x^4+1}\; dx$ **43.** $\int \dfrac{1}{\sqrt{x}+4}\, dx$ **44.** $\int \dfrac{x^5}{\sqrt{x^4+1}}\, dx$

45. Use the substitution method with $u = ax + b$ to verify formula 66 in the integral tables.

46. Use the substitution $u = ax + b$ to establish formula 137 in the integral tables.

47. Use the substitution method with $u = x^2 + a^2$ to verify formula 99 in the integral tables.

48. Use integration by parts to establish the formula for $\int x^k \ln (ax)\, dx$.

49. Use integration by parts with $u = x^2$ to find $\int x^3\sqrt{x^2 - a^2}\; dx$.

table of integrals

Elementary forms

1. $\displaystyle\int u^n\,du = \frac{u^{n+1}}{n+1}, \qquad n \neq -1$

2. $\displaystyle\int \frac{du}{u} = \ln|u|$

3. $\displaystyle\int e^u\,du = e^u$

4. $\displaystyle\int a^u\,du = \frac{a^u}{\ln a}, \qquad \text{for } a > 0, \qquad a \neq 1$

5. $\displaystyle\int \sin u\,du = -\cos u$

6. $\displaystyle\int \cos u\,du = \sin u$

Integrals involving e^{ax}

7. $\displaystyle\int e^{ax}\,dx = \frac{e^{ax}}{a}$

8. $\displaystyle\int xe^{ax}\,dx = \frac{e^{ax}}{a}\left(x - \frac{1}{a}\right)$

9. $\displaystyle\int x^2 e^{ax}\,dx = \frac{e^{ax}}{a}\left(x^2 - \frac{2x}{a} + \frac{2}{a^2}\right)$

10. $\displaystyle\int x^n e^{ax}\,dx = \frac{x^n e^{ax}}{a} - \frac{n}{a}\int x^{n-1}e^{ax}\,dx$
$$= \frac{e^{ax}}{a}\left(x^n - \frac{nx^{n-1}}{a} + \frac{n(n-1)x^{n-2}}{a^2} - \ldots \frac{(-1)^n n!}{a^n}\right)$$

11. $\displaystyle\int e^{ax}\sin bx\,dx = \frac{e^{ax}(a\sin bx - b\cos bx)}{a^2 + b^2}$

12. $\displaystyle\int e^{ax}\cos bx\,dx = \frac{e^{ax}(a\cos bx + b\sin bx)}{a^2 + b^2}$

Integrals involving $\ln x$

13. $\displaystyle\int \ln x\,dx = x\ln x - x$

14. $\displaystyle\int x\ln x\,dx = \frac{x^2}{2}\left(\ln x - \frac{1}{2}\right)$

15. $\displaystyle\int x^m \ln x\,dx = \frac{x^{m+1}}{m+1}\left(\ln x - \frac{1}{m+1}\right)$

16. $\displaystyle\int \frac{\ln x}{x}\,dx = \frac{1}{2}\ln^2 x$

17. $\displaystyle\int \frac{\ln x}{x^2}\,dx = -\frac{\ln x}{x} - \frac{1}{x}$

18. $\displaystyle\int \ln^2 x\,dx = x\ln^2 x - 2x\ln x + 2x$

19. $\displaystyle\int \frac{\ln^n x}{x}\,dx = \frac{\ln^{n+1} x}{n+1}$

20. $\displaystyle\int \frac{dx}{x\ln x} = \ln(\ln x)$

Integrals involving $\sin ax$

21. $\displaystyle\int \sin ax\,dx = -\frac{\cos ax}{a}$

22. $\displaystyle\int x\sin ax\,dx = \frac{\sin ax}{a^2} - \frac{x\cos ax}{a}$

23. $\displaystyle\int x^2 \sin ax\,dx = \frac{2x}{a^2}\sin ax + \left(\frac{2}{a^3} - \frac{x^2}{a}\right)\cos ax$

24. $\displaystyle\int x^3 \sin ax\,dx = \left(\frac{3x^2}{a^2} - \frac{6}{a^4}\right)\sin ax + \left(\frac{6x}{a^3} - \frac{x^3}{a}\right)\cos ax$

25. $\displaystyle\int \sin^2 ax\,dx = \frac{x}{2} - \frac{\sin 2ax}{4a}$

table integrals (continued)

Integrals involving cos ax

26. $\displaystyle\int \cos ax\, dx = \frac{\sin ax}{a}$

27. $\displaystyle\int x \cos ax\, dx = \frac{\cos ax}{a^2} + \frac{x \sin ax}{a}$

28. $\displaystyle\int x^2 \cos ax\, dx = \frac{2x}{a^2} \cos ax + \left(\frac{x^2}{a} - \frac{2}{a^3}\right) \sin ax$

29. $\displaystyle\int x^3 \cos ax\, dx = \left(\frac{3x^2}{a^2} - \frac{6}{a^4}\right) \cos ax + \left(\frac{x^3}{a} - \frac{6x}{a^3}\right) \sin ax$

30. $\displaystyle\int \cos^2 ax\, dx = \frac{x}{2} + \frac{\sin 2ax}{4a}$

Integrals involving inverse trigonometric functions

31. $\displaystyle\int \sin^{-1}\frac{x}{a}\, dx = x \sin^{-1}\frac{x}{a} + \sqrt{a^2 - x^2}$

32. $\displaystyle\int x \sin^{-1}\frac{x}{a}\, dx = \left(\frac{x^2}{2} - \frac{a^2}{4}\right) \sin^{-1}\frac{x}{a} + \frac{x\sqrt{a^2 - x^2}}{4}$

33. $\displaystyle\int x^2 \sin^{-1}\frac{x}{a}\, dx = \frac{x^3}{3} \sin^{-1}\frac{x}{a} + \frac{(x^2 + 2a^2)\sqrt{a^2 - x^2}}{9}$

34. $\displaystyle\int \cos^{-1}\frac{x}{a}\, dx = x \cos^{-1}\frac{x}{a} - \sqrt{a^2 - x^2}$

35. $\displaystyle\int x \cos^{-1}\frac{x}{a}\, dx = \left(\frac{x^2}{2} - \frac{a^2}{4}\right) \cos^{-1}\frac{x}{a} - \frac{x\sqrt{a^2 - x^2}}{4}$

36. $\displaystyle\int x^2 \cos^{-1}\frac{x}{a}\, dx = \frac{x^3}{3} \cos^{-1}\frac{x}{a} - \frac{(x^2 + 2a^2)\sqrt{a^2 - x^2}}{9}$

Integrals involving $\sqrt{a^2 - x^2}$

37. $\displaystyle\int \frac{dx}{\sqrt{a^2 - x^2}} = \sin^{-1}\frac{x}{a}$

38. $\displaystyle\int \frac{x\, dx}{\sqrt{a^2 - x^2}} = -\sqrt{a^2 - x^2}$

39. $\displaystyle\int \frac{x^2\, dx}{\sqrt{a^2 - x^2}} = -\frac{x\sqrt{a^2 - x^2}}{2} + \frac{a^2}{2} \sin^{-1}\frac{x}{a}$

40. $\displaystyle\int \frac{x^3\, dx}{\sqrt{a^2 - x^2}} = \frac{(a^2 - x^2)^{3/2}}{3} - a^2\sqrt{a^2 - x^2}$

41. $\displaystyle\int \frac{dx}{x\sqrt{a^2 - x^2}} = -\frac{1}{a} \ln\left(\frac{a + \sqrt{a^2 - x^2}}{x}\right)$

42. $\displaystyle\int \frac{dx}{x^2\sqrt{a^2 - x^2}} = -\frac{\sqrt{a^2 - x^2}}{a^2 x}$

43. $\displaystyle\int \frac{dx}{x^3\sqrt{a^2 - x^2}} = -\frac{\sqrt{a^2 - x^2}}{2a^2 x^2} - \frac{1}{2a^3} \ln\left(\frac{a + \sqrt{a^2 - x^2}}{x}\right)$

44. $\displaystyle\int \sqrt{a^2 - x^2}\, dx = \frac{x\sqrt{a^2 - x^2}}{2} + \frac{a^2}{2} \sin^{-1}\frac{x}{a}$

45. $\displaystyle\int x\sqrt{a^2 - x^2}\, dx = -\frac{(a^2 - x^2)^{3/2}}{3}$

46. $\displaystyle\int x^2\sqrt{a^2 - x^2}\, dx = -\frac{x(a^2 - x^2)^{3/2}}{4} + \frac{a^2 x\sqrt{a^2 - x^2}}{8} + \frac{a^4}{8} \sin^{-1}\frac{x}{a}$

47. $\displaystyle\int x^3\sqrt{a^2 - x^2}\, dx = \frac{(a^2 - x^2)^{5/2}}{5} - \frac{a^2(a^2 - x^2)^{3/2}}{3}$

48. $\displaystyle\int \frac{\sqrt{a^2 - x^2}}{x}\, dx = \sqrt{a^2 - x^2} - a \ln\left(\frac{a + \sqrt{a^2 - x^2}}{x}\right)$

49. $\displaystyle\int \frac{\sqrt{a^2 - x^2}}{x^2}\, dx = -\frac{\sqrt{a^2 - x^2}}{x} - \sin^{-1}\frac{x}{a}$

50. $\displaystyle\int \frac{\sqrt{a^2 - x^2}}{x^3}\, dx = -\frac{\sqrt{a^2 - x^2}}{2x^2} + \frac{1}{2a} \ln\left(\frac{a + \sqrt{a^2 - x^2}}{x}\right)$

Integrals involving $\sqrt{x^2-a^2}$

51. $\displaystyle\int \frac{dx}{\sqrt{x^2-a^2}} = \ln\left(x+\sqrt{x^2-a^2}\right)$

52. $\displaystyle\int \frac{x\,dx}{\sqrt{x^2-a^2}} = \sqrt{x^2-a^2}$

53. $\displaystyle\int \frac{x^2\,dx}{\sqrt{x^2-a^2}} = \frac{x\sqrt{x^2-a^2}}{2} + \frac{a^2}{2}\ln\left(x+\sqrt{x^2-a^2}\right)$

54. $\displaystyle\int \frac{x^3\,dx}{\sqrt{x^2-a^2}} = \frac{(x^2-a^2)^{3/2}}{3} + a^2\sqrt{x^2-a^2}$

55. $\displaystyle\int \frac{dx}{x\sqrt{x^2-a^2}} = \frac{1}{a}\sec^{-1}\left|\frac{x}{a}\right|$

56. $\displaystyle\int \frac{dx}{x^2\sqrt{x^2-a^2}} = \frac{\sqrt{x^2-a^2}}{a^2x}$

57. $\displaystyle\int \frac{dx}{x^3\sqrt{x^2-a^2}} = \frac{\sqrt{x^2-a^2}}{2a^2x^2} + \frac{1}{2a^3}\sec^{-1}\left|\frac{x}{a}\right|$

58. $\displaystyle\int \sqrt{x^2-a^2}\,dx = \frac{x\sqrt{x^2-a^2}}{2} - \frac{a^2}{2}\ln\left(x+\sqrt{x^2-a^2}\right)$

59. $\displaystyle\int x\sqrt{x^2-a^2}\,dx = \frac{(x^2-a^2)^{3/2}}{3}$

60. $\displaystyle\int x^2\sqrt{x^2-a^2}\,dx = \frac{x(x^2-a^2)^{3/2}}{4} + \frac{a^2x\sqrt{x^2-a^2}}{8} - \frac{a^4}{8}\ln\left(x+\sqrt{x^2-a^2}\right)$

61. $\displaystyle\int x^3\sqrt{x^2-a^2}\,dx = \frac{(x^2-a^2)^{5/2}}{5} + \frac{a^2(x^2-a^2)^{3/2}}{3}$

62. $\displaystyle\int \frac{\sqrt{x^2-a^2}}{x}\,dx = \sqrt{x^2-a^2} - a\sec^{-1}\left|\frac{x}{a}\right|$

63. $\displaystyle\int \frac{\sqrt{x^2-a^2}}{x^2}\,dx = -\frac{\sqrt{x^2-a^2}}{x} + \ln\left(x+\sqrt{x^2-a^2}\right)$

64. $\displaystyle\int \frac{\sqrt{x^2-a^2}}{x^3}\,dx = -\frac{\sqrt{x^2-a^2}}{2x^2} + \frac{1}{2a}\sec^{-1}\left|\frac{x}{a}\right|$

Integrals involving $ax+b$

65. $\displaystyle\int \frac{dx}{ax+b} = \frac{1}{a}\ln|ax+b|$

66. $\displaystyle\int \frac{x\,dx}{ax+b} = \frac{x}{a} - \frac{b}{a^2}\ln|ax+b|$

67. $\displaystyle\int \frac{x^2\,dx}{ax+b} = \frac{(ax+b)^2}{2a^3} - \frac{2b(ax+b)}{a^3} + \frac{b^2}{a^3}\ln|ax+b|$

68. $\displaystyle\int \frac{x^3\,dx}{ax+b} = \frac{(ax+b)^3}{3a^4} - \frac{3b(ax+b)^2}{2a^4} + \frac{3b^2(ax+b)}{a^4} - \frac{b^3}{a^4}\ln|ax+b|$

69. $\displaystyle\int \frac{dx}{x(ax+b)} = \frac{1}{b}\ln\left|\frac{x}{ax+b}\right|$

70. $\displaystyle\int \frac{dx}{x^2(ax+b)} = -\frac{1}{bx} + \frac{a}{b^2}\ln\left|\frac{ax+b}{x}\right|$

71. $\displaystyle\int \frac{dx}{x^3(ax+b)} = \frac{2ax-b}{2b^2x^2} + \frac{a^2}{b^3}\ln\left|\frac{x}{ax+b}\right|$

72. $\displaystyle\int \frac{dx}{(ax+b)^2} = \frac{-1}{a(ax+b)}$

73. $\displaystyle\int \frac{x\,dx}{(ax+b)^2} = \frac{b}{a^2(ax+b)} + \frac{1}{a^2}\ln|ax+b|$

74. $\displaystyle\int \frac{x^2\,dx}{(ax+b)^2} = \frac{ax+b}{a^3} - \frac{b^2}{a^3(ax+b)} - \frac{2b}{a^3}\ln|ax+b|$

75. $\displaystyle\int \frac{x^3\,dx}{(ax+b)^2} = \frac{(ax+b)^2}{2a^4} - \frac{3b(ax+b)}{a^4} + \frac{b^3}{a^4(ax+b)} + \frac{3b^2}{a^4}\ln(ax+b)$

76. $\displaystyle\int \frac{dx}{x(ax+b)^2} = \frac{1}{b(ax+b)} + \frac{1}{b^2}\ln\left|\frac{x}{ax+b}\right|$

77. $\displaystyle\int \frac{dx}{x^2(ax+b)^2} = \frac{-a}{b^2(ax+b)} - \frac{1}{b^2x} + \frac{2a}{b^3}\ln\left|\frac{ax+b}{x}\right|$

Integrals involving $ax+b$ and $px+q$

78. $\displaystyle\int \frac{dx}{(ax+b)(px+q)} = \frac{1}{bp-aq}\ln\left|\frac{px+q}{ax+b}\right|$

79. $\displaystyle\int \frac{x\,dx}{(ax+b)(px+q)} = \frac{1}{bp-aq}\left\{\frac{b}{a}\ln|ax+b| - \frac{q}{p}\ln|px+q|\right\}$

80. $\displaystyle\int \frac{dx}{(ax+b)^2(px+q)} = \frac{1}{bp-aq}\left\{\frac{1}{ax+b} + \frac{p}{bp-aq}\ln\left|\frac{px+q}{ax+b}\right|\right\}$

81. $\displaystyle\int \frac{x\,dx}{(ax+b)^2(px+q)} = \frac{1}{bp-aq}\left\{\frac{q}{bp-aq}\ln\left|\frac{ax+b}{px+q}\right| - \frac{b}{a(ax+b)}\right\}$

table of integrals (continued)

Integrals involving $\sqrt{x^2+a^2}$

82. $\displaystyle\int \frac{dx}{\sqrt{x^2+a^2}} = \ln(x+\sqrt{x^2+a^2})$

83. $\displaystyle\int \frac{x\,dx}{\sqrt{x^2+a^2}} = \sqrt{x^2+a^2}$

84. $\displaystyle\int \frac{x^2\,dx}{\sqrt{x^2+a^2}} = \frac{x\sqrt{x^2+a^2}}{2} - \frac{a^2}{2}\ln(x+\sqrt{x^2+a^2})$

85. $\displaystyle\int \frac{x^3\,dx}{\sqrt{x^2+a^2}} = \frac{(x^2+a^2)^{3/2}}{3} - a^2\sqrt{x^2+a^2}$

86. $\displaystyle\int \frac{dx}{x\sqrt{x^2+a^2}} = -\frac{1}{a}\ln\left(\frac{a+\sqrt{x^2+a^2}}{x}\right)$

87. $\displaystyle\int \frac{dx}{x^2\sqrt{x^2+a^2}} = -\frac{\sqrt{x^2+a^2}}{a^2x}$

88. $\displaystyle\int \frac{dx}{x^3\sqrt{x^2+a^2}} = -\frac{\sqrt{x^2+a^2}}{2a^2x^2} + \frac{1}{2a^3}\ln\left(\frac{a+\sqrt{x^2+a^2}}{x}\right)$

89. $\displaystyle\int \sqrt{x^2+a^2}\,dx = \frac{x\sqrt{x^2+a^2}}{2} + \frac{a^2}{2}\ln(x+\sqrt{x^2+a^2})$

90. $\displaystyle\int x\sqrt{x^2+a^2}\,dx = \frac{(x^2+a^2)^{3/2}}{3}$

91. $\displaystyle\int x^2\sqrt{x^2+a^2}\,dx = \frac{x(x^2+a^2)^{3/2}}{4} - \frac{a^2x\sqrt{x^2+a^2}}{8} - \frac{a^4}{8}\ln(x+\sqrt{x^2+a^2})$

92. $\displaystyle\int x^3\sqrt{x^2+a^2}\,dx = \frac{(x^2+a^2)^{5/2}}{5} - \frac{a^2(x^2+a^2)^{3/2}}{3}$

93. $\displaystyle\int \frac{\sqrt{x^2+a^2}}{x}\,dx = \sqrt{x^2+a^2} - a\ln\left(\frac{a+\sqrt{x^2+a^2}}{x}\right)$

94. $\displaystyle\int \frac{\sqrt{x^2+a^2}}{x^2}\,dx = -\frac{\sqrt{x^2+a^2}}{x} + \ln(x+\sqrt{x^2+a^2})$

95. $\displaystyle\int \frac{\sqrt{x^2+a^2}}{x^3}\,dx = -\frac{\sqrt{x^2+a^2}}{2x^2} - \frac{1}{2a}\ln\left(\frac{a+\sqrt{x^2+a^2}}{x}\right)$

Integrals involving x^2+a^2

96. $\displaystyle\int \frac{dx}{x^2+a^2} = \frac{1}{a}\tan^{-1}\frac{x}{a}$

97. $\displaystyle\int \frac{x\,dx}{x^2+a^2} = \frac{1}{2}\ln(x^2+a^2)$

Integrals involving x^2+a^2 (continued)

98. $\displaystyle\int \frac{x^2\,dx}{x^2+a^2} = x - a\tan^{-1}\frac{x}{a}$

99. $\displaystyle\int \frac{x^3\,dx}{x^2+a^2} = \frac{x^2}{2} - \frac{a^2}{2}\ln(x^2+a^2)$

100. $\displaystyle\int \frac{dx}{x(x^2+a^2)} = \frac{1}{2a^2}\ln\left(\frac{x^2}{x^2+a^2}\right)$

101. $\displaystyle\int \frac{dx}{x^2(x^2+a^2)} = -\frac{1}{a^2x} - \frac{1}{a^3}\tan^{-1}\frac{x}{a}$

102. $\displaystyle\int \frac{dx}{x^3(x^2+a^2)} = -\frac{1}{2a^2x^2} - \frac{1}{2a^4}\ln\left(\frac{x^2}{x^2+a^2}\right)$

103. $\displaystyle\int \frac{dx}{(x^2+a^2)^2} = \frac{x}{2a^2(x^2+a^2)} + \frac{1}{2a^3}\tan^{-1}\frac{x}{a}$

104. $\displaystyle\int \frac{x\,dx}{(x^2+a^2)^2} = \frac{-1}{2(x^2+a^2)}$

105. $\displaystyle\int \frac{x^2\,dx}{(x^2+a^2)^2} = \frac{-x}{2(x^2+a^2)} + \frac{1}{2a}\tan^{-1}\frac{x}{a}$

106. $\displaystyle\int \frac{x^3\,dx}{(x^2+a^2)^2} = \frac{a^2}{2(x^2+a^2)} + \frac{1}{2}\ln(x^2+a^2)$

107. $\displaystyle\int \frac{dx}{x(x^2+a^2)^2} = \frac{1}{2a^2(x^2+a^2)} + \frac{1}{2a^4}\ln\left(\frac{x^2}{x^2+a^2}\right)$

108. $\displaystyle\int \frac{dx}{x^2(x^2+a^2)^2} = -\frac{1}{a^4x} - \frac{x}{2a^4(x^2+a^2)} - \frac{3}{2a^5}\tan^{-1}\frac{x}{a}$

109. $\displaystyle\int \frac{dx}{x^3(x^2+a^2)^2} = -\frac{1}{2a^4x^2} - \frac{1}{2a^4(x^2+a^2)} - \frac{1}{a^6}\ln\left(\frac{x^2}{x^2+a^2}\right)$

110. $\displaystyle\int \frac{dx}{(x^2+a^2)^n} = \frac{x}{2(n-1)a^2(x^2+a^2)^{n-1}} + \frac{2n-3}{(2n-2)a^2}\int \frac{dx}{(x^2+a^2)^{n-1}}$

111. $\displaystyle\int \frac{x\,dx}{(x^2+a^2)^n} = \frac{-1}{2(n-1)(x^2+a^2)^{n-1}}$

112. $\displaystyle\int \frac{dx}{x(x^2+a^2)^n} = \frac{1}{2(n-1)a^2(x^2+a^2)^{n-1}} + \frac{1}{a^2}\int \frac{dx}{x(x^2+a^2)^{n-1}}$

113. $\displaystyle\int \frac{x^m\,dx}{(x^2+a^2)^n} = \int \frac{x^{m-2}\,dx}{(x^2+a^2)^{n-1}} - a^2\int \frac{x^{m-2}\,dx}{(x^2+a^2)^n}$

114. $\displaystyle\int \frac{dx}{x^m(x^2+a^2)^n} = \frac{1}{a^2}\int \frac{dx}{x^m(x^2+a^2)^{n-1}} - \frac{1}{a^2}\int \frac{dx}{x^{m-2}(x^2+a^2)^n}$

Integrals involving $x^2 - a^2$

115. $\displaystyle\int \frac{dx}{x^2-a^2} = \frac{1}{2a}\ln\left|\frac{x-a}{x+a}\right|$

116. $\displaystyle\int \frac{x\,dx}{x^2-a^2} = \frac{1}{2}\ln|x^2-a^2|$

117. $\displaystyle\int \frac{x^2\,dx}{x^2-a^2} = x + \frac{a}{2}\ln\left|\frac{x-a}{x+a}\right|$

118. $\displaystyle\int \frac{x^3\,dx}{x^2-a^2} = \frac{x^2}{2} + \frac{a^2}{2}\ln|x^2-a^2|$

119. $\displaystyle\int \frac{dx}{x(x^2-a^2)} = \frac{1}{2a^2}\ln\left|\frac{x^2-a^2}{x^2}\right|$

120. $\displaystyle\int \frac{dx}{x^2(x^2-a^2)} = \frac{1}{a^2x} + \frac{1}{2a^3}\ln\left|\frac{x-a}{x+a}\right|$

121. $\displaystyle\int \frac{dx}{x^3(x^2-a^2)} = \frac{1}{2a^2x^2} - \frac{1}{2a^4}\ln\left|\frac{x^2}{x^2-a^2}\right|$

122. $\displaystyle\int \frac{dx}{(x^2-a^2)^2} = \frac{-x}{2a^2(x^2-a^2)} - \frac{1}{4a^3}\ln\left|\frac{x-a}{x+a}\right|$

123. $\displaystyle\int \frac{x\,dx}{(x^2-a^2)^2} = \frac{-1}{2(x^2-a^2)}$

124. $\displaystyle\int \frac{x^2\,dx}{(x^2-a^2)^2} = \frac{-x}{2(x^2-a^2)} + \frac{1}{4a}\ln\left|\frac{x-a}{x+a}\right|$

125. $\displaystyle\int \frac{x^3\,dx}{(x^2-a^2)^2} = \frac{-a^2}{2(x^2-a^2)} + \frac{1}{2}\ln|x^2-a^2|$

126. $\displaystyle\int \frac{dx}{x(x^2-a^2)^2} = \frac{-1}{2a^2(x^2-a^2)} + \frac{1}{2a^4}\ln\left|\frac{x^2}{x^2-a^2}\right|$

127. $\displaystyle\int \frac{dx}{x^2(x^2-a^2)^2} = -\frac{1}{a^4x} - \frac{x}{2a^2(x^2-a^2)} - \frac{3}{4a^5}\ln\left|\frac{x-a}{x+a}\right|$

128. $\displaystyle\int \frac{dx}{x^3(x^2-a^2)^2} = -\frac{1}{2a^4x^2} - \frac{1}{2a^4(x^2-a^2)} + \frac{1}{a^6}\ln\left|\frac{x^2}{x^2-a^2}\right|$

129. $\displaystyle\int \frac{dx}{(x^2-a^2)^n} = \frac{-x}{2(n-1)a^2(x^2-a^2)^{n-1}} - \frac{2n-3}{(2n-2)a^2}\int \frac{dx}{(x^2-a^2)^{n-1}}$

130. $\displaystyle\int \frac{x\,dx}{(x^2-a^2)^n} = \frac{-1}{2(n-1)(x^2-a^2)^{n-1}}$

Integrals involving $\sqrt{ax+b}$

131. $\displaystyle\int \frac{dx}{\sqrt{ax+b}} = \frac{2\sqrt{ax+b}}{a}$

132. $\displaystyle\int \frac{x\,dx}{\sqrt{ax+b}} = \frac{2(ax-2b)}{3a^2}\sqrt{ax+b}$

133. $\displaystyle\int \frac{x^2\,dx}{\sqrt{ax+b}} = \frac{2(3a^2x^2-4abx+8b^2)}{15a^3}\sqrt{ax+b}$

134. $\displaystyle\int \frac{dx}{x\sqrt{ax+b}} = \begin{cases} \dfrac{1}{\sqrt{b}}\ln\left|\dfrac{\sqrt{ax+b}-\sqrt{b}}{\sqrt{ax+b}+\sqrt{b}}\right| & \text{for } b>0 \\[2ex] \dfrac{2}{\sqrt{-b}}\tan^{-1}\sqrt{\dfrac{ax+b}{-b}} & \text{for } b<0 \end{cases}$

135. $\displaystyle\int \frac{dx}{x^2\sqrt{ax+b}} = -\frac{\sqrt{ax+b}}{bx} - \frac{a}{2b}\int \frac{dx}{x\sqrt{ax+b}}$

136. $\displaystyle\int \sqrt{ax+b}\,dx = \frac{2\sqrt{(ax+b)^3}}{3a}$

137. $\displaystyle\int x\sqrt{ax+b}\,dx = \frac{2(3ax-2b)}{15a^2}\sqrt{(ax+b)^3}$

138. $\displaystyle\int x^2\sqrt{ax+b}\,dx = \frac{2(15a^2x^2-12abx+8b^2)}{105a^3}\sqrt{(ax+b)^3}$

 **density functions with illustrations from oceanography**

A common problem in trying to describe quantitatively various features of an ocean is to measure how its various occupants are distributed. These "occupants" might include salts, nitrates, plankton, and fish, for example. Suppose that we examine 1 square meter of ocean surface all the way down to the ocean floor—the so-called *water column*, illustrated in figure 25.1.

figure 25.1

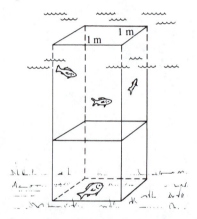

Since the ocean is about 96% H_2O, we may consider this column as a container with various types of solutes in the solvent H_2O. If it were possible to mix thoroughly the contents of this water column, then it would be quite easy to determine the total quantity T of some solute (e.g., salt):

1. Take a small sample (e.g., 1 liter) of sea water and determine the amount of solute ρ in the solution. The sample can be taken from any region in the water column.

2. Calculate T by computing $\rho V = \rho h$, where h is the height of the column and V is the volume of the water column. If ρ is measured in grams per liter and h is in meters, then $T = 1,000\rho h$, since 1 cubic meter is 1,000 liters.

Unfortunately, such thorough mixing is seldom the case. The concentration of a given object will usually vary with depth. Very few plankton will be found in lower depths. The saltiness (or *salinity*) of the water will vary somewhat with depth. A particular species of fish, such as halibut, may be predominantly bottom feeders. Shown in figure 25.2 is the vertical distribution of a species of *calanus,* the so-called "insects of the sea." As you can see, the concentration is greatest around a depth of 35 meters.

figure 25.2 Depth profile of *Calanus cristatus* showing the actual echogram obtained with a 200 K Hz recorder (top) and depth samples (○) using Miller nets (bottom). (Redrawn from Barraclough et al., 1969.) (Adapted from Barraclough et al., (1969). Reproduced in T.R. Parsons and M. Takahashi, *Biological oceanographic processes.* New York: Pergamon Press, 1973).

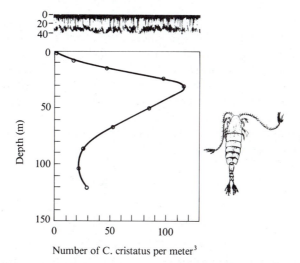

Number of C. cristatus per meter³

Let $A(x)$ be the total amount of a given solute (say, salt in grams) from the ocean surface to the depth x. If, as shown in figure 25.3, we increase the depth by a small amount Δx, then $A(x + \Delta x) - A(x)$ is the amount of salt between depths x and $x + \Delta x$. Since the total volume of solution in this region is

figure 25.3

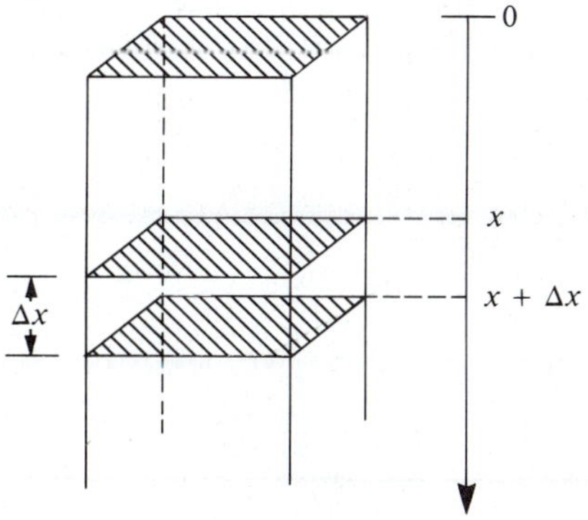

$1^2 \Delta x = \Delta x$ cubic meters, the average concentration or density is

$$\rho(x, \Delta x) = \frac{A(x + \Delta x) - A(x)}{\Delta x} \text{ g/m}^3$$

Admittedly for Δx small, this average density will not vary much. In fact, experimentally, we may not be able to detect any difference at all between $\rho(x, \Delta x)$ and $\rho(x, \Delta x/2)$. Nevertheless, to be mathematically correct, we define the *density of the solute at depth x* as

$$\rho(x) = \lim_{\Delta x \to 0} \frac{A(x + \Delta x) - A(x)}{\Delta x}$$

Hence, the *density function* $\rho(x)$ is a derivative. The function $A(x)$, which gives the total amount of the solute down to depth x, is called the **distribution function**. The relationship between $\rho(x)$ and $A(x)$ is $\rho(x) = A'(x)$. Now, which of these two quantities are we able to measure? By taking samples at various depths, we can estimate various values of $\rho(x)$. There is *no practical way* of directly measuring $A(x)$. However, we will suppose that, on the basis of our samples, we have determined a functional form for $\rho(x)$. Is there any way of capturing $A(x)$ from $\rho(x)$? Yes! The Fundamental Theorem of Calculus asserts that we can then determine $A(x)$ as follows:

$$A(d) - A(c) = \int_c^d \rho(x)\, dx$$

and

$$A(x) = \int_0^x \rho(y)\, dy$$

The following alternate way of obtaining this result also provides insight into the Fundamental Theorem. Divide $[c, d]$ into subintervals of length Δx, as shown in figure 25.4. The amount of solute between depths x and $x + \Delta x$ is

figure 25.4

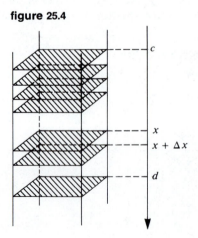

approximately $\rho(x)\,\Delta x$. This approximation will improve as $\Delta x \to 0$. The total amount of solute is given by $\Sigma\,\rho(x)\,\Delta x$. As $\Delta x \to 0$, the sum approaches $\int_c^d \rho(x)\,dx$. But $[A(d) - A(c)]$ also gives the total amount of solute between depths c and d. Hence, $A(d) - A(c) = \int_c^d \rho(x)\,dx$.

The following examples from oceanography illustrate the formulas developed in this section.

example 25.1

Suppose that the density of sardines (number of fish per cubic meter) is given by $\rho(x) = .005x(75 - x)$, where $0 \le x \le 75$. We can determine the total number of sardines in the water column by computing

$$\int_0^{75} .005x(75 - x)\,dx = .005\left(37.5x^2 - \frac{x^3}{3}\right)\Bigg|_0^{75} = 351.56$$

Hence, the total number of sardines in the water column is about 352.

The density of the sardines is largest at depth 37.5 meters. This can be seen by noting that $\rho(0) = \rho(75) = 0$ and $\rho'(x) = .005(75 - 2x)$. Thus, $\rho'(x) = 0$ when $x = 37.5$ meters. Shown in figure 25.5 is a fisherman trawling for sardines. The net has an opening that is 10 meters wide and 10 meters deep. It is lowered down between depths 32.5 and 42.5 meters in an attempt to capture the most fish. If the normal trawling speed is 20 meters per minute, how many sardines could be caught in 15 minutes?

figure 25.5

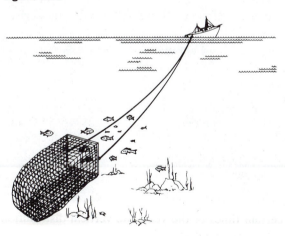

solution 25.1

Between depths 32.5 and 42.5 meters in a 1 square meter water column, the number of sardines is

$$\int_{32.5}^{42.5} .005x(75 - x)\,dx = .005\left(37.5x^2 - \frac{x^3}{3}\right)\Bigg|_{32.5}^{42.5} \approx 69.9$$

Hence, if the net moves through 1 meter, it can capture $10(69.9) = 699$ sardines. Over a 15-minute period, the net is moved through $20(15) = 300$ meters. Hence, in theory, it would contact $300(699) \approx 209,700$ sardines. However, since the escape rate is probably quite high, we would expect to capture only a small percentage of this number.

example 25.2

The density of salt in the ocean (*salinity*) can sometimes be represented by a function of the form

$$s(x) = s_\infty[1 - ae^{-bx}(1 + cx)]$$

where a, b, c, and s_∞ are positive constants. The function $s(x)$ is usually measured in units of "grams of salt per kilogram of seawater." By computing $s'(x)$, it is easy to see that $s(x)$ is increasing when $c < b$. For the Central Pacific Ocean (in October), the values $s_\infty = 34.7$, $a = .0176$, $b = .05$, and $c = .04$ provide a good fit with experimental data. Using these values, compute the total number of kilograms of salt in a 1 square meter water column from $x = 0$ to $x = 100$ meters. Use the approximation 1 m³ seawater $= 1,000$ kg.

solution 25.2

The density of the salt in kg/m³ is given by

$$\rho(x) = .001s(x) \text{ (kg/kg seawater)} \cdot 1,000 \text{ (kg seawater/m}^3)$$
$$= 34.7 - .611e^{-.05x} - .0244xe^{-.05x} \text{ (kg/m}^3)$$

In computing $\int_0^{100} \rho(x) \, dx$, we will use the antiderivative formulas

$$\int e^{ax} \, dx = \frac{e^{ax}}{a} \quad \text{and} \quad \int xe^{ax} \, dx = \frac{e^{ax}}{a}\left(x - \frac{1}{a}\right)$$

Hence,

$$\int_0^{100} \rho(x) \, dx = 34.7x + 12.2144e^{-.05x} + .488e^{-.05}(x + 20)\big|_0^{100}$$

where we have substituted $a = -.05$ into the antiderivative formulas. With the aid of a calculator, we obtain the result $3,470.477 - 21.974 = 3,448.50$ kg.

solution 25.3

We are given that $\rho(x)$ is largest when $x = 40$. Hence, $\rho'(40) = 0$ and $\rho(40) = 10^8$. This information will enable us to find a and k:

$$\rho'(x) = ax\left[e^{-x^2/(2k^2)}\frac{-x}{k^2}\right] + e^{-x^2/(2k^2)} \cdot a$$

$$= ae^{-x^2/(2k^2)}\left(-\frac{x^2}{k^2} + 1\right)$$

Hence, $\rho'(x) = 0$ when $x = \pm k$. Since $\rho'(40) = 0$, we must have $k = 40$. Furthermore, $\rho(40) = 10^8 = a(40)e^{-.5}$ which implies that $a = 10^8 \, e^{.5}/40$.

Before we compute $\int_0^{200} \rho(x)\,dx$, we must find $\int xe^{-x^2/(2k^2)}\,dx$. Let $u = -\dfrac{x^2}{2k^2}$. Then $\dfrac{du}{dx} = -\dfrac{2x}{2k^2} = -\dfrac{x}{k^2}$. Hence, $du = -\dfrac{x^2}{k^2}\,dx$ or $x\,dx = -k^2\,du$. It follows that

$$\int xe^{-x^2(2k^2)}\,dx = \int -k^2 e^u\,du = -k^2 e^u = -k^2 e^{-x^2/(2k^2)}$$

Hence,

$$\int_0^{200} \rho(x)\,dx = -ak^2 e^{-x^2/(2k^2)}\big|_0^{200} = ak^2(1 - e^{-12.5})$$

Using $a = 10^8 e^{.5}/40$, we obtain $40e^{.5} \cdot 10^8 (1 - .00000372) = 6.595 \cdot 10^9$ cells. 📖

example 25.4

The blue whale feeds on krill, a small shrimplike animal of length 2–5 centimeters. Suppose that the density of krill (number per cubic meter is given by $\rho(x) = .70xe^{-.001x}$, where x is the distance (in meters) from the Antarctic coast. The distribution in this problem is *horizontal* rather than vertical, but this does not change the fact that $\int_c^d \rho(x)\,dx$ gives the total number of krill from distance c to distance d. If the whale acts as a strainer with a cross sectional area of one square meter (see figure 25.6), then the number of krill

figure 25.6

the whale can catch in a run from 1,500 meters off the coast is

$$\int_0^{1,500} .70xe^{-.001x}\,dx = .70 \cdot \frac{e^{-.001x}}{-.001}(x + 1,000)\big|_0^{1,500}$$

$$= -700(2,500)e^{-1.5} + 700(1,000)$$

$$= 309,522 \text{ krill}$$

The density $\rho(x)$ of krill is largest at distance $x = 1,000$ from the coast. This may be seen from the computation:

$$\rho'(x) = .7[xe^{-.001x}(-.001) + e^{-.001x}] = .7e^{.001x}(1 - .001x)$$

Hence, $\rho'(x) = 0$ when $x = 1,000$. Since $\rho(1,000) = 257.5$ krill/m³, the whale could catch about $1,500(257.5) = 386,273$ krill in a 1,500-meter run through the zone of highest density (as shown in figure 25.7).

figure 25.7

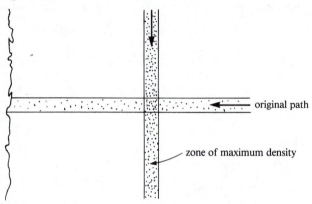

original path

zone of maximum density

exercises for section 25

The total number $A(x)$ of organisms from the ocean's surface to a depth x meters into the water column is given below. Find the corresponding density function $\rho(x)$ in each situation.

1. $A(x) = 5x$

2. $A(x) = x^2 + x$

3. $A(x) = 75x^2 - x^3$, $0 \le x \le 50$

4. $A(x) = 50(1 - e^{-2x})$

5. $A(x) = -10xe^{-.1x} - 100e^{-.1x} + 100$

The water column measures 50 m in a certain part of the sea. Given $\rho(x)$, the hypothetical density function (number of cod per cubic meter) below, answer each of the following.

(a) Find the total number of cod in the water column.

(b) Find the depth where the density is largest.

(c) Find the number of cod between 20 and 30 meters.

(d) Find the percentage of the total number of cod in the water column that are between 20 and 30 meters.

6. $\rho(x) = 10$

7. $\rho(x) = \frac{1}{10} x(50 - x)$

8. $\rho(x) = 25 - \frac{1}{2}x$

9. $\rho(x) = 150x/(x^2 + 900)$

10. $\rho(x) = xe^{-.025x}$

▲**11.** $\rho(x) = 5 \sin(\pi x/50)$

▲**12.** Rework the sardine problem of example 25.1, assuming that $\rho(x) = 8.9 \sin(\pi x/75)$.

13. For $s(x)$ in example 25.2, show that $s(x)$ is increasing when $c < b$.

14. Rework example 25.3, assuming that the maximum density is 10^7 cells per cubic meter at depth 25 meters.

15. Let $\rho(x) = \alpha x e^{-\beta x}$, where $\alpha > 0$ and $\beta > 0$.

(a) Show that the maximum density occurs at $x = 1/\beta$.

(b) Find α and β so that a maximum density of 115 organisms per cubic meter occurs at a depth of 35 meters.

(c) Find the total number of organisms between $x = 0$ and $x = 50$ meters.

part B

16. Let $\rho(x) = c(e^{-\alpha x} - e^{-\beta x})$ where α, β, and c are positive and $\alpha < \beta$.

(a) Show that $\rho(x)$ takes on its maximum value at

$$x = \frac{1}{\beta - \alpha} \ln\left(\frac{\beta}{\alpha}\right)$$

(b) Show that the graph has an inflection point at

$$x = \frac{2}{\beta - \alpha} \ln\left(\frac{\beta}{\alpha}\right)$$

(c) Find $A(x)$, the total number in the water column down to depth x.

17. Let $\rho(x) = ax/(x^2 + b^2)$, where $a > 0$ and $b > 0$.

(a) Show that $\rho(x)$ takes on its maximum value at $x = b$.

(b) Find the total number in the water column down to depth b.

26 probability density functions

In section 25, we computed the total number of organisms or the total quantity of a given substance between depths $x = a$ and $x = b$ by computing $\int_a^b \rho(x)\, dx$, where $\rho(x)$ is the density function, as illustrated in figure 26.1. If the water

figure 26.1

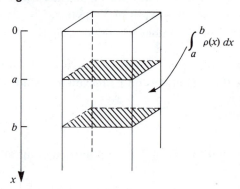

$$\int_a^b \rho(x)\, dx$$

column is of height h, then the total amount of the substance in the water column is given by $T = \int_0^h \rho(x)\, dx$. If we are interested in the *proportion* or *percentage* P of the total that lies between depths $x = a$ and $x = b$, then we can compute:

$$P = \frac{\int_a^b \rho(x)\, dx}{T} = \frac{1}{T}\int_a^b \rho(x)\, dx = \int_a^b \frac{\rho(x)}{T}\, dx$$

If we let $f(x) = \rho(x)/T$, then $P = \int_a^b f(x)\, dx$. We call $f(x)$ a **probability density function**. If we let $F(x) = A(x)/T$, then $F(x)$ represents the percentage of the inhabitants of the water column down to depth x. The function $F(x)$ is called a **probability distribution function**. Using the fact that $A'(x) = \rho(x)$, we have

$$F'(x) = \frac{A'(x)}{T} = \frac{\rho(x)}{T} = f(x)$$

Thus, we have the following results:

1. The derivative of the probability distribution function $F(x)$ is the probability density function $f(x)$.

2. The proportion between $x = a$ and $x = b$ is

$$P = \int_a^b f(x) \, dx = F(b) - F(a)$$

example 26.1   The density of a species of zooplankton (organisms per cubic meter) is assumed to be given by $\rho(x) = 25x(50 - x)$, where $0 \le x \le 50$.

  (a) Find the probability density function $f(x)$.
  (b) Find the distribution function $F(x)$.
  (c) Determine the proportion between depths 0 and 20 meters.
  (d) Determine the proportion between depths 20 and 40 meters.

solution 26.1   The total number T of zooplankton in the water column is given by

$$T = \int_0^{50} 25x(50 - x) \, dx = \int_0^{50} 25(50x - x^2) \, dx$$

$$= 25\left(25x^2 - \frac{x^3}{3}\right)\Big|_0^{50} = \frac{1{,}562{,}500}{3} = 520{,}833.3\bar{3}$$

Hence, $f(x) = \rho(x)/T = (75/1{,}562{,}500) \; x(50 - x) = .000048x(50 - x)$. The distribution function has derivative $f(x)$. Hence, $F(x) = \int f(x) \, dx = .000048 \, (25x^2 - x^3/3) + c$. Since $F(0) = A(0)/T = 0$, c must be 0. It follows that

$$F(x) = .000048\left(25x^2 - \frac{x^3}{3}\right)$$

The proportion of zooplankton between depths 0 and 20 meters is therefore $F(20) - F(0) = .352 - 0 = .352$, while the proportion of zooplankton between depths 20 and 40 meters is $F(40) - F(20) = .896 - .352 = .544$. ∎

By a **population** we mean the totality of objects under consideration. This population may be an actual population of animals, or a purely hypothetical population, such as the collection of all light bulbs that could be produced by a company using a particular process. In the case of a human population, we may be interested in various numerical characteristics of its members, such as H = height, W = weight, A = age, or N = number of children, for example. A numerical measurement X on the members of a population is given the special name **random variable**. If the random variable can take on any value in some *interval* such as (a, b) or $[a, b]$, then we call X a **continuous** random variable. The random variables H, W, and A are continuous random variables, but the random variable N defined previously can assume only the values $0, 1, 2, \ldots$, and is called *discrete. In this section, we will confine our attention to continuous random variables.*

In the case of a human population, we may be interested in the *proportion* of individuals who have, for example, height H between certain limits. For example, we may wish to determine the percentage of females who are between 5'2" and 5'8" tall. By "$P(a < X < b)$" we mean that fraction of the members of a population with numerical characteristic X between a and b. The expression $P(a < X < b)$ may also be interpreted as the *chance* or *probability* that a given individual will have a value of X between a and b. How is $P(a < X < b)$ computed? The answer comes from our earlier discussion of proportions in the water column.

definition 26.1   Given a continuous random variable X, we define the **distribution function** $F(x)$ to be $P(X \leq x)$. If $F'(x) = f(x)$ exists, then we call $f(x)$ the **probability density function** of X. Finally, we compute

$$P(a < X < b) = \int_a^b f(x)\, dx = F(b) - F(a)$$

If X can take on values only between x_1 and x_2, then

$$1 = P(x_1 < X < x_2) = \int_{x_1}^{x_2} f(x)\, dx$$

figure 26.2

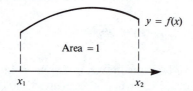

As shown in figure 26.2 $F(x_1) = 0$, since X cannot take on values less than x_1. Also, $F(x_2) = 1$ since X is always less than or equal to x_2. These are the key facts to be used in the examples that follow.

example 26.2   Let $f(x) = c(1 - x)$ for $0 \leq x \leq 1$.

   (a) Find c so that $f(x)$ is a probability density function.
   (b) Find the distribution function $F(x)$.
   (c) Compute $P(0 < X < \frac{1}{2})$.

solution 26.2   It is easily seen that $\int_0^1 c(1 - x)\, dx = c/2$. Hence, for this integral to equal 1, we must have $c = 2$. Therefore, $f(x) = 2(1 - x)$, and so $F'(x) = 2(1 - x)$. Hence, $F(x) = 2x - x^2 + c_1$. Since $F(0) = 0$, we must have $c_1 = 0$. Finally, $P(0 < X < \frac{1}{2})$ $= F(\frac{1}{2}) - F(0) = .75 - 0 = .75$.

example 26.3 **the uniform distribution** Let $f(x) = 1/(b-a)$ for $a \le x \le b$. Then $\int_a^b f(x)\, dx = 1$. A random variable X with this probability density function is said to have a *uniform distribution*. If $a \le c < d \le b$, then

$$P(c < X < d) = \int_c^d \frac{1}{b-a}\, dx = \frac{d-c}{b-a}$$

Thus, probability is proportional to the length of the interval (c, d) in question. Thus, if intervals I_1, I_2, and I_3 have the same length, as shown in figure 26.3,

figure 26.3

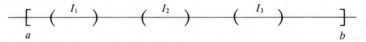

then $P(X \text{ lies in } I_1) = P(X \text{ lies in } I_2) = P(X \text{ lies in } I_3)$. For this reason, we say that "X is chosen *at random* from the interval $[a, b]$."

example 26.4 **the exponential distribution** If X represents the *life length* of an item or an organism, a common distribution for X is the *exponential distribution*. The probability density function takes the form $f(x) = \alpha e^{-\alpha x}$ for $x > 0$, where $\alpha > 0$ is a fixed constant. To show that $f(x)$ is indeed a density function, we must show that the area under the curve [see figure 26.4(a)] from $x = 0$ to $+\infty$ is 1.

figure 26.4(a)   **figure 26.4(b)**

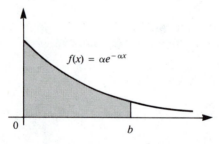 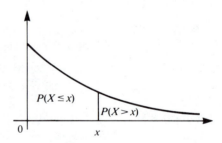

To do this, we first compute

$$\int_0^b \alpha e^{-\alpha x}\, dx = -e^{-\alpha x}\big|_0^b = 1 - e^{-\alpha b}$$

As $b \to +\infty$, $1 - e^{-\alpha b} \to 1$ since the assumption that $\alpha > 0$ makes $e^{-\alpha b}$ an exponential decay function. Now, $F'(x) = f(x) = \alpha e^{-\alpha x}$. Hence, $F(x) = -e^{-\alpha x} + c$. Since $0 = F(0) = -e^{-\alpha \cdot 0} + c = -1 + c$, we have $c = 1$ and so the distribution function is $F(x) = 1 - e^{-\alpha x}$. We may interpret $P(X > x)$ as the probability that an individual lives more than x units of time. Figure 26.4**(b)**

shows   that $P(X>x)+P(X\leq x)=1$.   Hence, $P(X>x)=1-P(X\leq x)=1-F(x)=e^{-\alpha x}$.   ▤

The exponential distribution describes well the life length of many s̶p̶ such as songbirds. Part B exercise 25 will demonstrate that it is approp̶ ate when the forces of mortality are constant over time. As the next examp̶le illustrates, this distribution is also used to describe the life length of large mammals.

example 26.5

The average life length of a whale is estimated to be 50 years. If life length X is assumed to be exponentially distributed, we will show in example 26.10 that the proper value of α is .02.

   (a) Find the proportion of whales that die between 40 and 60 years of age.
   (b) Find the probability that a whale will live for more than 80 years.

solution 26.5

In (a) we are asked to find $P(40<X<60)$, which is $F(60)-F(40)$. Since $F(x)=1-e^{-.02x}$, we obtain $.6988-.5507=.1481$.
For (b), we use the formula $P(X>x)=e^{-\alpha x}$ developed in example 26.4. Thus, $P(X>80)=e^{-.02(80)}=.2019$.   ▤

The general beta distribution has probability density function of the form $f(x)=cx^{r-1}(1-x)^{s-1}$, for $0\leq x\leq 1$, where r, s, and c are positive. One special case is given in example 26.6.

example 26.6

The probability density function $f(x)=12x(1-x)^2$ for $0\leq x\leq 1$ is a special case of a class of distributions known as **beta distributions**.

   (a) Verify that $\int_0^1 f(x)\,dx = 1$.
   (b) Calculate $P(\frac{1}{4}<X<\frac{3}{4})$.

solution 26.6

Expanding $f(x)$, we obtain $f(x)=12(x-2x^2+x^3)$. Hence,

$$\int_0^1 f(x)\,dx - 12\left(\frac{x^2}{2}-\frac{2}{3}x^3 + \frac{x^4}{4}\right)\Big|_0^1 = 12\left(\frac{1}{2}-\frac{2}{3}+\frac{1}{4}-0\right) - 1$$

Finally,

$$P\left(\frac{1}{4}<X<\frac{3}{4}\right)=\int_{1/4}^{3/4} f(x)\,dx = 12\left(\frac{x^2}{2}-\frac{2}{3}x^3+\frac{x^4}{4}\right)\Big|_{1/4}^{3/4}$$
$$= 12(.07910\ldots-.021809\ldots)$$
$$= 12(.05729\ldots)=.6875$$ ▤

We will now use the probability density function to compute two key parameters associated with a distribution.

expectation and variance   For a random variable X, what is the average value that X assumes over the entire population? We will show that when X has probability density function $f(x)$ for $a \le x \le b$, this average or *mean* value is given by the formula:

$$\mu = \int_a^b xf(x)\, dx$$

μ is also referred to as the *expected value* of X. To motivate this formula, we first consider the following simple example.

example 26.7   A survey of 100 families in a small community determined the number of children X in each family. The results are shown in table 26.1. The average number of children per family is given by

$$\frac{\text{total number of children}}{\text{number of families}} = \frac{0(20) + 1(31) + 2(29) + 3(10) + 4(5) + 5(3) + 6(2) + 7(0)}{100}$$

$$= \frac{166}{100} = 1.66 \text{ children per family}$$

table 26.1

X (number of children)	0	1	2	3	4	5	6	7
f (frequency)	20	31	29	10	5	3	2	0

The general formula for the mean of a *discrete random variable* X that takes on values $x_1, x_2, \ldots, x_n$ with frequencies $F_1, F_2, \ldots, F_n$ is

$$\mu = \frac{\sum_{i=1}^{n} F_i x_i}{N} = \frac{\sum Fx}{N}$$

where $N = \sum_{i=1}^{n} F_i$ is the total number in the population. If X is a continuous random variable with probability density function $f(x)$ defined on $[a, b]$, divide $[a, b]$ into n pieces, each of length Δx, as indicated in figure 26.5. Then, $P(x < X < x + \Delta x) = \int_x^{x+\Delta x} f(x)\, dx \approx f(x)\Delta x$ since, for Δx very small, the area of interest is almost a rectangle. Thus, if there are N individuals in the population, then $F = N[f(x)\Delta x]$ of them have an X value between x and $(x + \Delta x)$. By the previous formula, the average value μ of X is approximately

$$\frac{\sum Fx}{N} = \frac{\sum N[f(x)\Delta x]x}{N} = \sum xf(x)\Delta x$$

figure 26.5

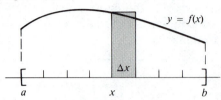

Now, as $\Delta x \to 0$, all of the approximations improve and $\Sigma\, xf(x)\,\Delta x \to \int_a^b xf(x)\,dx$ by definition of the integral. Use of this new formula is illustrated in the next three examples.

example 26.8   Compute μ for the beta distribution in example 26.6.

solution 26.8   For $0 \le x \le 1$, $xf(x) = 12x^2(1-x)^2 = 12(x^2 - 2x^3 + x^4)$. Hence,

$$\mu = \int_0^1 xf(x)\,dx = \int_0^1 12(x^2 - 2x^3 + x^4)\,dx$$

$$= 12\left(\frac{x^3}{3} - \frac{x^4}{2} + \frac{x^5}{5}\right)\Big|_0^1$$

$$= 12\left(\frac{1}{3} - \frac{1}{2} + \frac{1}{5}\right) - 0 = .4$$

example 26.9   The probability density function of X is of the form $f(x) = \alpha x^{\alpha-1}$ for $0 \le x \le 1$. If $\mu = 1/3$, find α.

solution 26.9   We have

$$\frac{1}{3} = \mu = \int_0^1 xf(x)\,dx$$

$$= \int_0^1 \alpha x^\alpha\,dx = \frac{\alpha}{\alpha+1}\, x^{\alpha+1}\Big|_0^1$$

$$= \frac{\alpha}{\alpha+1} \qquad \text{for } \alpha+1 > 0$$

Hence, $1/3 = \alpha/(\alpha+1)$ or $\alpha+1 = 3\alpha$. Solving for α, we obtain $\alpha = 1/2$.

example 26.10   Verify that the mean of an exponentially distributed random variable is given by $\mu = 1/\alpha$.

solution 26.10   The probability density function is $f(x) = \alpha e^{-\alpha x}$ for $x > 0$. To find μ, we must first compute $\int_0^b xf(x)\,dx = \int_0^b x\alpha e^{-\alpha x}\,dx$. Let $u = x$ and $dv = \alpha e^{-\alpha x}\,dx$. Then $du = dx$ and $v = -e^{-\alpha x}$. Using the integration by parts formula, we obtain

$$\int xe^{-\alpha x}\,dx = -xe^{-\alpha x} + \int e^{-\alpha x}\,dx = -xe^{-\alpha x} - \frac{1}{\alpha}e^{-\alpha x}$$

Hence,

$$\int_0^b xe^{-\alpha x}\,dx = -be^{-\alpha b} - \frac{1}{\alpha}e^{-\alpha b} + \frac{1}{\alpha}$$

Now, as $b \to +\infty$, $e^{-\alpha b}$ decays to 0. With the aid of your calculator, you can convince yourself that $\lim_{x \to +\infty} xe^{-x} = 0$. We may rewrite $-be^{-\alpha b}$ as $(-1/\alpha)(\alpha b)e^{-\alpha b}$. If we set $x = \alpha b$, then as $b \to +\infty$, $x \to +\infty$, and $-be^{-\alpha b} = (-1/\alpha)xe^{-x} \to 0$. Thus, if $b \to +\infty$ in the last display, it follows that $\mu = 1/\alpha$. ∎

Shown in figure 26.6 are two probability density functions f_1 and f_2 with a common mean μ. However, distribution f_1 is much more tightly packed around the mean—i.e., the characteristic being measured does not *vary* nearly as much

figure 26.6

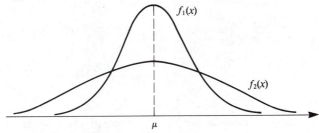

as for distribution f_2. The precise way of measuring *variance* about the mean is defined below:

If the random X has probability density function $f(x)$ for $a \le x \le b$, then the *variance* is computed as

$$\sigma^2 = \int_a^b (x - \mu)^2 f(x)\,dx$$

The quantity σ^2 may be interpreted as the *average squared distance* from the mean μ. Since $(x - \mu)^2 f(x) = x^2 f(x) - 2\mu x f(x) + \mu^2 f(x)$, we have

$$\sigma^2 = \int_a^b x^2 f(x)\,dx - \int_a^b 2\mu x f(x)\,dx + \int_a^b \mu^2 f(x)\,dx$$

But $\mu = \int_a^b x f(x)\,dx$ and $\int_a^b f(x)\,dx = 1$. It follows that

$$\sigma^2 = \int_a^b x^2 f(x)\, dx - \mu^2$$

Usually, it is easier to work with this second formula for variance. This is demonstrated in our next two examples.

example 26.11   Find the variance of the uniform distribution of example 26.3.

solution 26.11   The probability density function for the uniform distribution is $f(x) = 1/(b-a)$, for $a \le x \le b$. Hence,

$$\mu = \int_a^b x\, \frac{1}{b-a}\, dx = \frac{1}{b-a} \left(\frac{x^2}{2}\right)\Bigg|_a^b = \frac{1}{b-a}\left(\frac{b^2 - a^2}{2}\right) = \frac{a+b}{2}$$

and
$$\int_a^b x^2 f(x)\, dx = \int_a^b \frac{x^2}{b-a}\, dx = \frac{1}{b-a}\left(\frac{x^3}{3}\right)\Bigg|_a^b = \frac{b^3 - a^3}{3(b-a)}$$

From the above formula it follows that

$$\sigma^2 = \frac{(b-a)(b^2 + ab + a^2)}{3(b-a)} - \frac{(a+b)^2}{4} = \frac{4(b^2 + ab + a^2) - 3(a+b)^2}{12}$$

$$= \frac{4b^2 + 4ab + 4a^2 - 3a^2 - 6ab - 3b^2}{12} = \frac{(b-a)^2}{12}$$

example 26.12   Find the formula for the variance of the probability density function $f(x) = \alpha x^{\alpha-1}$ for $0 \le x \le 1$.

solution 26.12   In example 26.9 we showed that $\mu = \alpha/(\alpha + 1)$. We next compute

$$\int_0^1 x^2 (\alpha x^{\alpha-1})\, dx = \int_0^1 \alpha x^{\alpha+1}\, dx = \frac{\alpha x^{\alpha+2}}{(\alpha + 2)}\Bigg|_0^1 = \frac{\alpha}{\alpha + 2}$$

Hence, using the second formula for the variance, we obtain

$$\sigma^2 = \frac{\alpha}{\alpha + 2} - \frac{\alpha^2}{(\alpha + 1)^2}$$

we are faced with a dilemma. The term $e^{-x^2/2}$ resists attack by all the anti-derivative techniques we have introduced. In fact, it is known that there is no elementary antiderivative for $e^{-x^2/2}$. Since this distribution (and others like it) is of such importance in probability and statistics, we need to develop effective ways of estimating integrals such as the one above that *do not depend* on the Fundamental Theorem of Calculus. In section 27, we will discuss **numerical integration techniques**.

exercises for section 26

part A

For each of the following hypothetical density functions $\rho(x)$ for the vertical distribution of shrimp (per cubic meter), find **(a)** the probability density function $f(x)$ and the distribution function $F(x)$. **(b)** the proportion between depths 10 and 20 meters.

1. $\rho(x) = 5$, for $0 \le x \le 50$ **2.** $\rho(x) = 2x + 1$, for $0 \le x \le 20$

3. $\rho(x) = 3x(50 - x)$, for $0 \le x \le 50$ **4.** $\rho(x) = 100e^{-2x}$, for $x > 0$

For each nonnegative function $f(x)$ in exercises 5–10 **(a)** find c so that $f(x)$ is a probability density function, **(b)** find the distribution $F(x)$, **(c)** compute $P(X > 1)$. $P(\tfrac{1}{2} < X < 1)$

5. $f(x) = c$, for $0 \le x \le 2$ **6.** $f(x) = cx$, for $0 \le x \le 2$

7. $f(x) = c/(x + 1)$, for $0 \le x \le 2$ **8.** $f(x) = cx(2 - x)^2$, for $0 \le x \le 2$

9. $f(x) = ce^{-2x}$, for $x > 0$ **★10.** $f(x) = cxe^{-x}$, for $x > 0$

11. Most computer systems have a command which selects a number X "at random" between 0 and 1.

  (a) What is the probability density function of X?

  (b) What is the probability that the first digit of X is 8?

12. Assuming that human births are uniformly distributed through the year, find the proportion of births that occur:

  (a) in January,   *Between Sept. 1 and Sept 20 (inclusive)*

  (b) in February (with 28 days), and

  (c) in the Spring.

In one of the standard models of fishery science (see section 36), it is assumed that the natural life length T of many common commercial fish is *exponentially distributed*—i.e., T has probability density function

$$f(t) = \alpha e^{-\alpha t} \qquad \text{for some } \alpha > 0$$

(b)

13. For Pacific halibut, $\alpha \approx .20$. What is the probability that a halibut will live more than 10 years?

(a) Find the proportion halibut that die between 2 and 7 years. of age   Also problem 21

14. For the North Sea plaice, $\alpha \approx .10$. If our fishing nets can capture only fish aged 4 years or older, what percentage of the plaice population is subject to fishing?

15. Compute μ and σ^2 for the probability density function in exercise 5.

16. Compute μ and σ^2 for the probability density function in exercise 6.

★17. Compute μ and σ^2 for the probability density function in exercise 7.

18. Compute μ and σ^2 for the probability density function in exercise 8.

19. Compute μ and σ^2 for the probability density function in exercise 9.

★20. Compute μ and σ^2 for the probability density function in exercise 10.

21. What is the average life length of the Pacific halibut in exercise 13?

22. A species of robin has average life length 6 years. Assuming that the life length T is exponentially distributed, compute $P(T > 10)$.

23. If X has probability density function of the form $f(x) = 1/\alpha$, for $0 \le x \le \alpha$, and $\mu = 2$, find the parameter α.

24. Suppose that X has probability density function of the form $f(x) = ax^2 + c$, where $c = 1/2 - a/3$, for $-1 \le x \le 1$. (See accompanying figure.)

  (a) Show that $\int_{-1}^{1} f(x)\,dx = 1$ and $\mu = 0$.

  ★ (b) If $\sigma^2 = \frac{1}{4}$, find the parameter a.

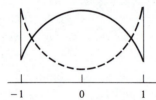

part B   **25.** This exercise is designed to show that the exponential distribution is appropriate when the forces of mortality are constant over time. To demonstrate this, we will follow a large cohort of N_0 newborns through their lifetimes.

  (a) If the life length T is exponentially distributed, show that the number still alive at time t is predicted to be

$$N(t) = N_0 e^{-\alpha t}$$

  (b) Show that, *of those still alive at time t, the proportion who die within the next Δt units of time is given by*

$$d(\Delta t) = 1 - e^{-\alpha \Delta t}$$

Since this proportion is independent of t, a fixed percentage of the population is removed every Δt units of time. We will next show that the exponential distribution is the *only continuous distribution* with this property.

(c) If $d(\Delta t) = \dfrac{N(t) - N(t + \Delta t)}{N(t)}$ is independent of t, then its derivative with respect to t is 0. Show that this implies

$$\frac{N'(t + \Delta t)}{N(t + \Delta t)} = \frac{N'(t)}{N(t)}$$

(d) Conclude that $N'(t)/N(t)$ is constant so $N(t) = N_0 e^{-\alpha t}$ for some $\alpha > 0$. Then show that the probability density function for T is exponential.

In the following two exercises, use the fact that $\text{limit}_{x \to +\infty} x^n e^{-\alpha x} = 0$ for $\alpha > 0$ and n a positive integer.

26. Show that the variance of an exponentially distributed random variable is given by $\sigma^2 = 1/\alpha^2$.

27. Suppose the random variable X has a *gamma distribution* with probability density function of the form $f(x) = c x^n e^{-\alpha x}$, for $x > 0$. Find c so that $f(x)$ is a valid probability density function. [*Hint*: Use integral table formula 10 and the above limit.]

numerical integration techniques

The successful application of the Fundamental Theorem of Calculus depends upon being able to find an antiderivative $F(x)$ for the given function $f(x)$. When this is not possible, **numerical integration techniques** may be used. Recall that

$$\int_a^b f(x)\,dx = \lim_{\Delta x \to 0} \sum_{i=1}^n f(x_i)\,\Delta x$$

where $\Delta x = (b-a)/n$ and $x_i = a + i\,\Delta x$. As shown in figure 27.1, the problem with computing $\sum f(x)\,\Delta x$ to approximate $\int_a^b f(x)\,dx$ is that, in general, $\sum f(x)\,\Delta x \to \int_a^b f(x)\,dx$ fairly slowly. If, for example, we estimate $\int_1^2 1/x\,dx$ by

figure 27.1(a)

figure 27.1(b)

$$a \quad x_1 \ x_2 \ x_3 \ x_4 \ x_5 \ x_6 \ x_7 \ x_8 \ x_9 \ x_{10} \ x_n = b \qquad\qquad x_1 \qquad x_2$$

computing $\sum (1/x)\,\Delta x$ (see example 19.4), we obtain the results shown in table 27.1. The exact value of this integral is $\ln 2 = .6931471805\ldots$; thus, a tremendous effort is required to obtain three correct digits!

table 27.1

Δx	$\sum (1/x)\,\Delta x$
.1 $(n = 10)$	.668771
.01 $(n = 100)$	.690377
.001 $(n = 1,000)$	.692896
.0001 $(n = 10,000)$	.693120

To remedy the problem of slow convergence, we next fit trapezoids rather than rectangles, as illustrated in figure 27.2(**a**). The area of the trapezoid shown in figure 27.2(**b**) is $\frac{1}{2}$ (height) (base$_1$ + base$_2$) = $\frac{1}{2}\Delta x(y_1 + y_2)$. This can be seen by

figure 27.2(a)   **figure 27.2(b)**

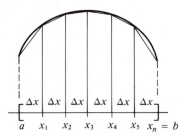   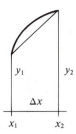

integration, but we have used the familiar formula from plane geometry. Applying this formula to each of the trapezoids in figure 27.2(**a**) yields

$$A \approx \frac{1}{2}\Delta x[f(a) + f(x_1)] + \frac{1}{2}\Delta x[f(x_1) + f(x_2)] + \frac{1}{2}\Delta x[f(x_2) + f(x_3)]$$

$$+ \cdots + \frac{1}{2}\Delta x[f(x_{n-2}) + f(x_{n-1})] + \frac{1}{2}\Delta x[f(x_{n-1}) + f(b)]$$

$$= \frac{1}{2}\Delta x[f(a) + 2f(x_1) + 2f(x_2) + \cdots + 2f(x_{n-1}) + f(b)]$$

To summarize, we have shown:

Trapezoid Rule

$$\int_a^b f(x)\, dx \approx \frac{\Delta x}{2}[f(a) + 2f(x_1) + 2f(x_2) + \cdots + 2f(x_{n-1}) + f(x_n)]$$

Our next example shows how the Trapezoid Rule speeds up convergence.

example 27.1   Estimate $\int_1^2 1/x\, dx$ by using the Trapezoid Rule with $\Delta x = .1$.

solution 27.1   We must compute values of $f(x) = 1/x$ for $x = 1, 1.1, 1.2, \ldots, 1.9, 2.0$. These values are shown in table 27.2. These functional values must be multiplied by either $m = 1$ or $m = 2$. The correct multipliers are also given in table 27.2. Hence, using the Trapezoid Rule, we have

$$\int_1^2 \frac{1}{x}\, dx \approx \frac{.1}{2}(13.875428) = .693771 \ldots$$

Note that the first three digits are already correct.

table 27.2

x	$f(x) = 1/x$	m (multiplier)	$mf(x)$
1.0	1.000000	1	1.000000
1.1	.909090	2	1.818181
1.2	.833333	2	1.666666
1.3	.769230	2	1.538461
1.4	.714285	2	1.428571
1.5	.666666	2	1.333333
1.6	.625000	2	1.250000
1.7	.588235	2	1.176470
1.8	.555555	2	1.111111
1.9	.526315	2	1.052631
2.0	.500000	1	.500000

$$\Sigma = 13.875428$$

Shown in table 27.3 are Trapezoid Rule approximations done with the aid of a computer.

table 27.3

Δx	Trapezoid Rule approximation
.1	.693771403
.01	.69315343
.001	.69314721

A slightly more clever approach to numerical integration is to use small stretches of parabolic arc instead of line segments, as shown in figure 27.3. Two

figure 27.3(a)

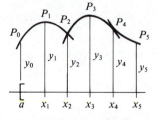

figure 27.3(b)

points determine a line, and, as we shall see at the end of this section, three points are needed to determine a parabola. By working with points P_0, P_1, and P_2, and then with P_2, P_3, and P_4, and so forth, we can fit parabolic arcs along the original curve $y = f(x)$. If we add the areas under the parabolic arcs, we obtain:

Simpson's Rule If n, the number of subintervals, is even, then

$$\int_a^b f(x)\, dx \approx \frac{\Delta x}{3}\, [f(a) + 4f(x_1) + 2f(x_2) + 4f(x_3) + 2f(x_4)$$
$$+ \cdots + 2f(x_{n-2}) + 4f(x_{n-1}) + f(b)]$$

We will give a complete derivation of Simpson's Rule at the end of this section.

Note that the multipliers in Simpson's Rule follow the pattern:

$$1 \quad 4 \quad 2 \quad 4 \quad 2 \quad 4 \quad \ldots\ldots \quad 2 \quad 4 \quad 1$$

example 27.2 Estimate $\int_1^2 1/x\, dx$ by using Simpson's Rule with $n = 10$, and hence $\Delta x = .1$.

solution 27.2 As shown in table 27.4, this problem is set up as in example 27.1, except that the sequence of multipliers changes and $\Delta x/3$ is used in place of $\Delta x/2$. By Simpson's Rule, we estimate:

$$\int_1^2 \frac{1}{x}\, dx \approx \frac{.1}{3}\, (20.794507) = .693150231 \ldots$$

table 27.4

x	$f(x) = 1/x$	m (multiplier)	$mf(x)$
1.0	1.000000	1	1.000000
1.1	.909090	4	3.636363
1.2	.833333	2	1.666666
1.3	.769230	4	3.076923
1.4	.714285	2	1.428571
1.5	.666666	4	2.666666
1.6	.625000	2	1.250000
1.7	.588235	4	2.352941
1.8	.555555	2	1.111111
1.9	.526315	4	2.105263
2.0	.500000	1	.500000
			$\Sigma = 20.794507$

Note that $|.693150231 \ldots - \ln 2| = .0000030505$, remarkable accuracy for so little work!

Shown in table 27.5 are Simpson's Rule approximations for $\int_1^2 1/x\, dx$ done with the aid of a computer. In most cases, Simpson's Rule converges more quickly than the Trapezoid Rule formula. We will use it in the next two examples.

table 27.5

Δx	Simpson's Rule approximation
.1 ($n = 10$)	.693150231
.05 ($n = 20$)	.693147375
.02 ($n = 50$)	.693147185

example 27.3

Estimate $\int_0^1 \sqrt{1 + x^4}\, dx$ correct to three decimal places by using Simpson's Rule with $n = 4$ and then $n = 10$.

solution 27.3

For $n = 4$, $\Delta x = (b - a)/n = \frac{1}{4}$. Hence, the values of x in table 27.6 are 0, .25, .50, .75, and 1.0. The Simpson's Rule approximation with $n = 4$ is $\frac{.25}{3}$ (13.072959) = 1.089413 . . . .

table 27.6

x	$f(x) = \sqrt{1 + x^4}$	m	$mf(x)$
0	1.000000	1	1.000000
.25	1.0019512	4	4.007804
.50	1.0307764	2	2.0615528
.75	1.1473474	4	4.589389
1.0	1.4142135	1	1.4142135
			$\Sigma = 13.072959$

For $n = 10$, we have $\Delta x = .1$, and the resulting calculations are shown in table 27.7. The Simpson's Rule approximation with $n = 10$ is (.1/3)(32.682881) = 1.089429 . . . . Thus, it appears that $\int_0^1 \sqrt{1 + x^4}\, dx = 1.089$ to three decimal places.

table 27.7

x	$f(x)$	m	$mf(x)$
0	1.000000	1	1.000000
.1	1.000050	4	4.000200
.2	1.000799	2	2.001599
.3	1.0040418	4	4.0161673
.4	1.012719	2	2.025438
.5	1.030776	4	4.123105
.6	1.062826	2	2.125652
.7	1.113597	4	4.454391
.8	1.187265	2	2.3745315
.9	1.286895	4	5.1475819
1.0	1.414213	1	1.4142135
			$\Sigma = 32.682881$

In section 26, we encountered normal probabilities such as

$$P(0 < X < .5) = \frac{1}{\sqrt{2\pi}} \int_0^{.5} e^{-x^2/2} \, dx$$

If we use Simpson's Rule with $n = 10$, we obtain $.1917\ldots$ as the estimate of the probability. To aid in the computation of such normal probabilities, mathematicians have used numerical integration techniques to evaluate

$$N(z) = \frac{1}{\sqrt{2\pi}} \int_0^{z} e^{-x^2/2} \, dx$$

for various values of z. Shown in table 27.8 is a short version of such a table.

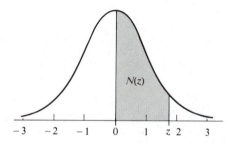

table 27.8

z	.0	.1	.2	.3	.4	.5	.6
$N(z)$	.0000	.0398	.0793	.1179	.1554	.1915	.2257
z	.7	.8	.9	1.0	1.1	1.2	1.3
$N(z)$	.2580	.2881	.3159	.3413	.3643	.3849	.4032
z	1.4	1.5	1.6	1.7	1.8	1.9	2.0
$N(z)$	.4192	.4332	.4452	.4554	.4641	.4713	.4772
z	2.1	2.2	2.3	2.4	2.5	2.6	2.7
$N(z)$	.4821	.4861	.4893	.4918	.4938	.4953	.4965

We will use table 27.8 to compute the normal probabilities given in the following example.

example 27.4   Suppose that X is a random variable with probability density function $f(x) = (1/\sqrt{2\pi})e^{-x^2/2}$. Compute each of the following probabilities with the aid of table 27.8.

(a) $P(0 < X < 2)$ (b) $P(X > 1.5)$ (c) $P(-2 < X < 2)$
(d) $P(1.2 < X < 2.4)$

solution 27.4   We first depict each probability as an area. This will give some insight as to how the table can be used.

figure 27.4(a) $P(0<X<2)$ is just $N(2) = .4772$ from table 27.8.

figure 27.4(b) By the symmetry of the bell, an area of .5000 lies to the right of 0. Hence, $P(X>1.5) = .5000 - N(1.5) = .5000 - .4332 = .0668.$

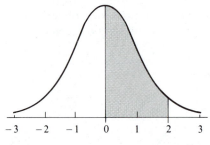

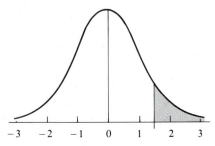

figure 27.4(c) $P(-2<X<2)$ is just $2N(2)$, since $\int_{-2}^{0} f(x)\,dx = \int_{0}^{2} f(x)\,dx$. Hence, $P(-2<X<2) = 2(.4772) = .9544.$

figure 27.4(d) $P(1.2<X<2.4) = \int_{1.2}^{2.4} f(x)\,dx = \int_{0}^{2.4} f(x)\,dx - \int_{0}^{1.2} f(x)\,dx = N(2.4) - N(1.2) = .4918 - .3849 = .1069.$

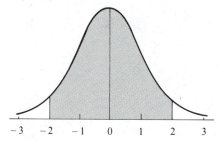

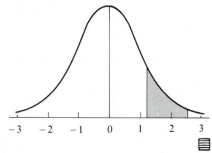

Finally, we will give a complete derivation of the Simpson's Rule formula. The general equation of a parabolic function is

$$y = ax^2 + bx + c$$

Since this equation contains three unknowns, three points are needed to determine the parabola. We wish to compute $\int_{x_0}^{x_2} (ax^2 + bx + c)\,dx$. By translating the parabola so that x_1 is moved to 0, we will instead compute the equal area

$$\int_{-\Delta x}^{\Delta x} (ax^2 + bx + c)\,dx$$

where a, b, and c are to be chosen so that the parabola passes through $(-\Delta x, y_0)$, $(0, y_1)$, and $(\Delta x, y_2)$, as shown in figure 27.5. Hence,

$$y_0 = a(\Delta x)^2 - b(\Delta x) + c$$
$$y_1 = c$$
$$y_2 = a(\Delta x)^2 + b(\Delta x) + c$$

figure 27.5(a)   **figure 27.5(b)**

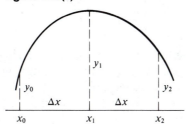

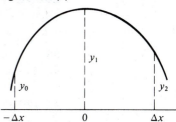

Hence, $y_0 + y_2 = 2a(\Delta x)^2 + 2c = 2a(\Delta x)^2 + 2y_1$ and $y_2 - y_0 = 2b(\Delta x)$. It follows that

$$a = \frac{y_0 + y_2 - 2y_1}{2(\Delta x)^2} \quad \text{and} \quad b = \frac{y_2 - y_0}{2\Delta x}$$

Hence.

$$\int_{-\Delta x}^{\Delta x} (ax^2 + bx + c)\, dx = a\frac{x^3}{3} + b\frac{x^2}{2} + cx\Big|_{-\Delta x}^{\Delta x}$$

$$= 2a\frac{(\Delta x)^3}{3} + 2c\,\Delta x$$

Substituting for a and c, we obtain

$$2\left(\frac{y_0 + y_2 - 2y_1}{2(\Delta x)^2}\right)\frac{(\Delta x)^3}{3} + 2y_1\,\Delta x = \frac{\Delta x}{3}\,(y_0 + y_2 - 2y_1 + 6y_1)$$

$$= \frac{\Delta x}{3}\,(y_0 + 4y_1 + y_2)$$

If we apply this formula to each of the parabolas in figure 27.6, we obtain

$$\int_a^b f(x)\, dx \approx \frac{\Delta x}{3}\,(y_0 + 4y_1 + y_2) + \frac{\Delta x}{3}\,(y_2 + 4y_3 + y_4) + \frac{\Delta x}{3}\,(y_4 + 4y_5 + y_6)$$

$$+ \cdots + \frac{\Delta x}{3}\,(y_{n-2} + 4y_{n-1} + y_n)$$

$$= \frac{\Delta x}{3}\,(y_0 + 4y_1 + 2y_2 + 4y_3 + 2y_4 + \cdots + 4y_{n-1} + y_n)$$

figure 27.6

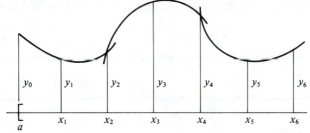

exercises for section 27

part A

For each of the following integrals:

(a) Compute the integral using the Fundamental Theorem.
(b) Estimate the integral using the Trapezoid Rule with $n = 10$.
(c) Estimate the integral using Simpson's Rule with $n = 10$. Which approximation is closer to the exact value?

1. $\displaystyle\int_0^1 x^2 \, dx$ **2.** $\displaystyle\int_1^2 (4x - 3) \, dx$ **3.** $\displaystyle\int_1^3 \frac{1}{x^2} \, dx$

4. $\displaystyle\int_2^3 \ln x \, dx$ **5.** $\displaystyle\int_{-1}^1 (3 - 2x) \, dx$ **6.** $\displaystyle\int_0^2 e^x \, dx$

7. $\displaystyle\int_1^4 \sqrt{x} \, dx$ ★ **8.** $\displaystyle\int_{-1}^2 |x| \, dx$

Estimate each of the following integrals correct to two decimal places by using Simpson's Rule with $n = 4$ and then $n = 10$.

9. $\displaystyle\int_2^3 \sqrt{1 + x^3} \, dx$ **10.** $\displaystyle\int_1^2 e^{-x^2/2} \, dx$

11. $\displaystyle\int_0^2 \sqrt{e^x + 1} \, dx$ **12.** $\displaystyle\int_{-1}^3 \sqrt[4]{x^2 + 1} \, dx$

13. $\displaystyle\int_2^3 \frac{1}{\ln x} \, dx$ ▲★ **14.** $\displaystyle\int_0^1 \sin (x^2) \, dx$

Frequently in the biosciences, the actual function $f(x)$ is not known, but experimental data have been collected to reveal functional values at a few values of x. Numerical integration can then be used to estimate

$$\int_a^b f(x) \, dx$$

without establishing a specific functional form for $f(x)$.

15. Estimate $\int_0^3 f(x) \, dx$ using Simpson's Rule and the data shown in the table.

x	0	.5	1.0	1.5	2.0	2.5	3.0
$f(x)$	.3	.8	1.4	1.9	2.1	1.4	.7

16. Densities of a species of zooplankton were estimated by taking water samples at the depths shown in the table. Use Simpson's Rule to estimate the total number of zooplankton in a square meter of the water column.

x (meters)	0	5	10	15	20	25	30	35	≥ 40
$\rho(x)$ (number/m³)	0	200	310	425	275	100	20	5	0

Use Simpson's Rule to estimate the total number of zooplankton in a square meter of the water column.

17. Temperatures at the beach on a summer afternoon were recorded every half hour, as shown in the table. Estimate the average temperature between noon and 4:00 by

(a) computing a simple average.

(b) using Simpson's Rule and the average value formula.

time	12:00	12:30	1:00	1:30	2:00	2:30	3:00	3:30	4:00
T (°F)	75	78	80	82	85	84	80	77	75

★**18.** Use a ruler to estimate the lengths (in centimeters) of the cross sections shown in the figure. Then estimate the area of the region using the formula $A = \int_a^b [f(x) - g(x)]\, dx$ and Simpson's Rule.

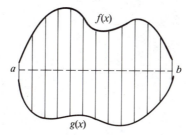

Use the values of $N(z)$ given in table 27.8 to compute each of the following.

19. $P(X > 1.2)$ **20.** $P(-.2 < X < .3)$ **21.** $P(X > -.4)$

22. $P(2.3 < X < 2.7)$ **23.** Find z so that $P(X > z) \approx .05$.

★**24.** Find z so that $P(-z < X < z) \approx .90$.

part B

25. Write a program for either a programmable calculator or a computer to implement Simpson's Rule. Compute the approximations for $n = 10, 20, 30, \ldots, 100$. Using these approximations, estimate the integral correct to six decimal places.

(a) $\int_0^1 x^2(1 - x)^7\, dx$ **(b)** $\int_1^2 \frac{\ln x}{x^2 + 2}\, dx$

The general form of the normal probability density function is

$$f(x) = \frac{1}{\sqrt{2\pi}\sigma} \exp\left[\frac{-(x - \mu)^2}{2\sigma^2}\right]$$

where $\exp[\ ] = e^{[\ ]}$, μ is the mean and σ^2 is the variance.

26. By letting $z = \dfrac{x - \mu}{\sigma}$, show that $P(a < X < b) = N\left(\dfrac{b - \mu}{\sigma}\right) - N\left(\dfrac{u - \mu}{\sigma}\right)$,

where $N(z) = \displaystyle\int_0^z \frac{1}{\sqrt{2\pi}} e^{-x^2/2}\, dx$.

27. The height X (in inches) for adult males in the United States is normally distributed with $\mu = 69.7$ and $\sigma = 2.6$. Use either table 27.8 or a Simpson's Rule calculation to estimate:

    (a) the proportion of adult males in the population who are at least 6 feet tall.

    (b) the proportion of adult males who are between 5′5″ and 6′3″ tall. [*Note*: This is the height range that most department stores attempt to accommodate.]

 a model for measuring
cardiac output

In this section, we present a simple model that is used to compute cardiac output—i.e., the rate of flow of blood along the aorta. The technique falls under a general category of physiological measurement techniques called **dye-dilution** methods.

the model   A fluid is moving at a rate of F (liters per minute, e.g.) into and out of the tank shown in figure 28.1. The flow rate F and the volume V of the tank are assumed to be unknown but constant. At time $t = 0$, a known quantity A_0 (grams, e.g.) of dye is placed either into the tank or into the vessel entering the tank. The contents of the tank are thoroughly mixed at all times t. (We might imagine a blender placed in the tank.)

figure 28.1

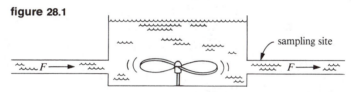

Let $c(t)$ be the concentration of dye at time t (in grams per liter, e.g.) and let T be the time when all the dye has been cleared from the tank. If we divide $[0, T]$ into equal intervals of length Δt (see figure 28.2), then, between times t

figure 28.2

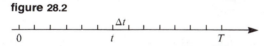

and $(t + \Delta t)$, $F \Delta t$ liters leave the tank, and for Δt very small, the concentration of dye in the fluid leaving the tank is approximately $c(t)$ grams per liter. Hence, the amount of dye leaving the tank between times t and $(t + \Delta t)$ is approximately $c(t)F \Delta t = Fc(t) \Delta t$. The total quantity of dye A_0 is well approximated by $\Sigma Fc(t) \Delta t = F \Sigma c(t) \Delta t$. As $\Delta t \to 0$, all of the approximations improve and $\Sigma c(t) \Delta t \to \int_0^T (c(t) \, dt$. Thus, $A_0 = F \int_0^T c(t) \, dt$ or

flow rate formula $$F = \frac{A_0}{\displaystyle\int_0^T c(t) \, dt}$$

Although A_0 is known, the mathematical expression for $c(t)$ will not generally be known. Suppose that at times $t_0 = 0$, $t_1 = \Delta t$, $t_2 = 2\Delta t, \ldots, t_n = n\,\Delta t = T$, samples of the fluid are taken and, from these samples, concentrations $c(t_0)$, $c(t_1)$, $c(t_2), \ldots, c(t_n)$ are determined. Then $\int_0^T c(t)\,dt$ can be approximated using Simpson's Rule:

$$\int_0^T c(t)\,dt \approx \frac{\Delta t}{3}\left[c(t_0) + 4c(t_1) + 2c(t_2) + \cdots + c(t_n)\right]$$

The above formula can then be used to estimate F. This is illustrated in example 28.1.

example 28.1   Ten grams of dye are mixed in the tank shown in figure 28.1 and concentrations are measured at a sampling site every ten seconds, with the results shown in table 28.1. Estimate the flow rate F.

table 28.1

t (sec)	0	10	20	30	40	50	60	70	80
$c(t)$ (g/l)	1	1.5	2.5	3.0	2.5	1.0	.5	.2	0

solution 28.1   Here, $\Delta t = 10$, $T = 80$, and $A_0 = 10$. Using Simpson's Rule, we have

$$\int_0^{80} c(t)\,dt \approx \frac{10}{3}[1 + 4(1.5) + 2(2.5) + 4(3.0) + 2(2.5) + 4(1.0) + 2(.5) + 4(.2) + 0]$$
$$= 116$$

Hence, using the formula for F, we obtain $F \approx 10/116$ liters per second, or 5.17 liters per minute.   ▤

We now show how this basic model can be used to approximate the flow rate of blood through the heart.

application to measuring cardiac output   Shown in figure 28.3(a) is a representation of the human cardiovascular system as a system of pumps and tubes. Blood enters the right heart from the veins and is pumped into the lungs where it is oxygenated. From there it is pumped into the left heart which pushes the oxygenated blood into the aorta. Organs are represented in the diagram by square boxes. A more anatomically correct diagram of the cardiovascular system is shown in figure 28.3(b).

To estimate the slow rate F into the aorta, a bolus of dye is injected into a vein as close as possible to the right side of the heart. If possible, the dye is injected into the right atrium, which serves as the mixing chamber. The sampling site is selected near the aorta. In practice, blood is drawn rapidly by a

figure 28.3(a) Adapted from: T.C. Ruch, and H. D. Patton, *Physiology and Biophysics*, Vol. II (*Circulation, respiration, and fluid balance*). 20th ed., Philadelphia: W.B. Saunders Co. (1974).

figure 28.3(b) Adapted from J.M. Ford, and J.E. Monroe, *Living systems*, 2nd ed. by New York: Harper & Row, Publishers, Inc., (1974). Reprinted by permission of Harper & Row, Publishers, Inc.

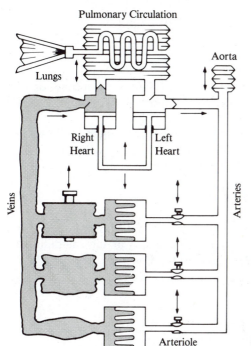

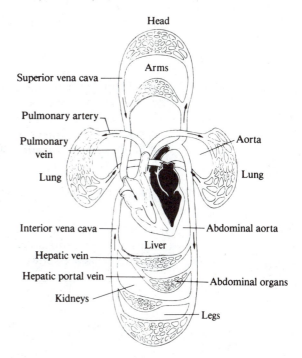

motor-driven syringe and concentrations are measured by a special device known as a *densitometer*. Most tests will be completed in less than thirty seconds.

Our model assumes that once the dye has left the tank it cannot return. Because of recirculation in the cardiovascular system, it is important to use a dye that will be removed by an organ. It is also important to select a dye that is not lost in the lungs. A commonly used dye is *indocyanine green*, which is rapidly metabolized by the liver. The computation of the flow rate F is illustrated in example 28.2.

example 28.2   A 5-milligram bolus of indocyanine green dye is injected into the right atrium. Concentration samples are then measured each second, as reported in table 28.2. Estimate cardiac output F.

table 28.2

t (sec)	0	1	2	3	4	5	6	7	8	9	10	11	12
$c(t)$ (mg/l)	0	0	1.7	5.6	9.2	8.4	5.2	3.8	2.1	1.0	.5	.2	0

solution 28.2   Here, $\Delta t = 1$, $T = 12$, and $A_0 = 5$. Using Simpson's Rule, we have

$$\int_0^{12} c(t)\, dt \approx \frac{1}{3}[0 + 4(0) + 2(1.7) + 4(5.6) + 2(9.2) + 4(8.4) + 2(5.2)$$
$$+ 4(3.8) + 2(2.1) + 4(1.0) + 2(.5) + 4(.2) + 0]$$
$$= \frac{113.4}{3} = 37.8$$

We then apply the formula for F to obtain $F \approx 5/37.8$ liters per second, or 7.94 liters per minute. Such a cardiac output is considered healthy.   ▤

Additional examples of dye-dilution measurement techniques in physiology will be presented in section 39.

exercises for section 28

part A

1. One hundred fifty grams of dye are placed in the tank shown in figure 28.1 and concentrations are measured every ten seconds at the sampling site. The results are shown in the table. Estimate the flow rate F.

t (sec)	0	10	20	30	40	50	60	70	80
$c(t)$ (g/l)	0	15	22	42	21	10	1	0	0

2. An unknown quantity A_0 of waste has been dumped into a river. The river flows at a rate of 1,500 cubic feet per second, or approximately $F = 3.67 \times 10^9$ liters per day. In an effort to estimate A_0, biologists measure concentrations of waste each day, as reported in the table. Using the model of this section, estimate A_0 in kilograms.

t (day)	0	1	2	3	4	5	6	7	8	9	10
$c(t)$ (g/l)	0	.032	.030	.026	.022	.018	.012	.009	.006	.002	0

A 5-milligram bolus of the dye indocyanine green is injected into the right atrium. In the following exercises, use the concentration measurements shown in the table (in mg/l) to estimate the cardiac output F. Give your answers in liters per minute.

3.

t (sec)	0	1	2	3	4	5	6	7	8	9	10	11	12
$c(t)$	0	1.3	3.8	6.4	8.9	7.0	6.1	4.1	2.9	1.3	.8	.6	0

4.

t (sec)	0	2	4	6	8	10	12	14	16	18	20
$c(t)$	0	1.6	3.7	6.8	5.2	3.1	2.4	1.8	.6	.2	0

5.

t (sec)	0	2	4	6	8	10	12	14	16	18	20	22	24
$c(t)$	0	3.2	6.4	7.0	7.2	6.8	5.4	3.1	2.6	2.0	1.2	.4	0

part B

Establish a formula for the flow rate F in terms of A_0, and positive constants a and k, given that:

6. $c(t) = ae^{-kt}$. Let $T \to +\infty$.

7. $c(t) = ate^{-kt}$. Let $T \to +\infty$ and use the fact that

$$\lim_{x \to +\infty} x\, e^{-kx} = 0$$

where $k > 0$.

 **the trigonometric functions**

The trigonometric functions (including sine, cosine, and tangent) are those special functions used in the biological and physical sciences to model *repetitive* or *periodic* phenomena. Simple examples of such phenomena include the oscillations of a pendulum, ocean tides, heart beat, and breathing. In most cases, one of the variables in question is time, t, and we will write $\sin t$ and $\cos t$ in our development.

The trigonometric functions are also called the **circular functions**, since their definitions can be based on movement around a circle. Shown in figure 29.1 is a circle of radius $r = 1$ with center at the origin $(0, 0)$. We might imagine

figure 29.1

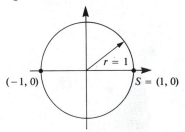

this circle to be a *racetrack* along which races of various lengths will be run. All the races will *start* at $S = (1, 0)$. The distance around the track is just the circumference $2\pi r = 2\pi$. Thus, for example, if we were to run a "π unit dash," then the race would end at $F = (-1, 0)$. We are now ready to define $\sin t$ and $\cos t$.

For $t > 0$, we race t units along the circle in the *counterclockwise* direction. Let F_t be the final destination point, as illustrated in figure 29.2. If F_t has

figure 29.2(a) The case $t > 0$.   **figure 29.2(b)** The case $t < 0$.

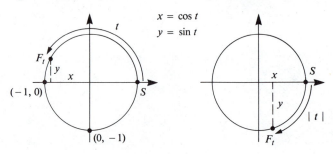

coordinates (x, y), we define $\cos t$ to be x, and $\sin t$ to be y. If $t < 0$ we race $|t|$ units in the *clockwise* direction and let F_t be the finishing point. As before, we define $\cos t$ as the x-coordinate of F_t and $\sin t$ as the y-coordinate of F_t. These definitions are illustrated in our next two examples.

example 29.1   Use the above definitions of the functions sine and cosine to find:

    (a) $\cos \pi$ and $\sin \pi$ **(b)** $\cos (-\pi/2)$ and $\sin (-\pi/2)$

    (c) $\cos 8\pi$ and $\sin 8\pi$

solution 29.1   **(a)** Marching counterclockwise through π units, we arrive at $F_\pi = (-1, 0)$. Hence, $\cos \pi = -1$ and $\sin \pi = 0$.
(b) For $-\pi/2$, we race clockwise through $\pi/2$ units, arriving at $F_{-\pi/2} = (0, -1)$. Hence, $\cos (-\pi/2) = 0$ and $\sin (-\pi/2) = -1$.
(c) The destination $F_{8\pi}$ is just $S = (1, 0)$, the starting point. We have made four complete trips around the circle. Hence, $\cos 8\pi = 1$ and $\sin 8\pi = 0$.   ▤

example 29.2   Use the definitions of sine and cosine to locate on the unit circle the approximate locations of

    (a) $\cos 1$ and $\sin 1$ **(b)** $\cos (-7.5)$ and $\sin (-7.5)$

Then compute an exact value for each of the above with the aid of your calculator. ·

solution 29.2   **(a)** Since $\pi/2 = 1.57 \ldots$ and $\pi/4 = .785 \ldots$, the point F_1 is approximately in the position shown in figure 29.3. To determine the exact values of the coordinates,

figure 29.3

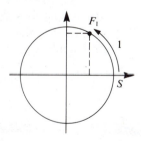

figure 29.4

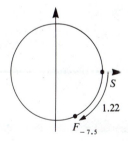

make sure your calculator is in *radian* mode. You should obtain $\cos 1 = .5403023 \ldots$ and $\sin 1 = .8414709 \ldots$.
For **(b)**, since $2\pi = 6.28 \ldots$, we march 7.5 units clockwise by making one complete trip around and stepping through about 1.22 additional units. Thus, $F_{-7.5}$ is approximately in the position shown in figure 29.4. Using the calculator, we obtain $\cos (-7.5) = .3466353 \ldots$ and $\sin (-7.5) = -.93799 \ldots$.   ▤

Most of the standard trigonometric identities are fairly simple consequences of our "racetrack" definitions of $\cos t$ and $\sin t$. First, note that F_t is the same as $F_{t+2\pi}$, since traveling an additional 2π units will take us once around the circle and back to F_t, as illustrated in figure 29.5. Hence, $F_{t+2\pi} = F_t$,

figure 29.5

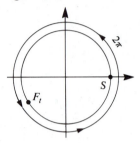

or $(\cos (t + 2\pi), \sin (t + 2\pi)) = (\cos t, \sin t)$. We then have the following identities:

$$\cos (t + 2\pi) = \cos t$$
$$\sin (t + 2\pi) + \sin t$$

We will now use these identities to sketch the graphs of $f(t) = \sin t$ and $g(t) = \cos t$. Refer to figure 29.6 and observe that, as t increases from 0 to $\pi/2$, the y-coordinate of F_t (namely, $\sin t$) increases steadily from 0 to 1. For $\pi/2 \le t \le \pi$, the y-coordinate decreases from 1 to 0. Once t increases beyond π, the y-coordinate becomes negative, decreasing to -1 and then increasing to 0 as t approaches 2π. Since $\sin (t + 2\pi)$ is just $\sin t$, the graph simply repeats itself every 2π additional units. The familiar complete graph is shown in figure 29.7.

figure 29.6

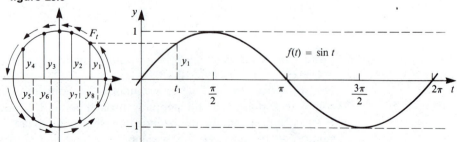

A similar analysis can be performed with the x-coordinates of F_t (namely, $\cos t$) giving the graph shown in figure 29.8. Since $\cos(t + 2\pi) = \cos t$, the graph repeats itself on subsequent intervals of length 2π. The complete graph is shown in figure 29.9. The quantity $A = 1$ is called the **amplitude** of the wave, while $p = 2\pi$ is called the **period**. One complete cycle takes place on the

figure 29.7

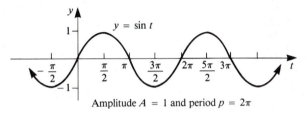

Amplitude $A = 1$ and period $p = 2\pi$

figure 29.8

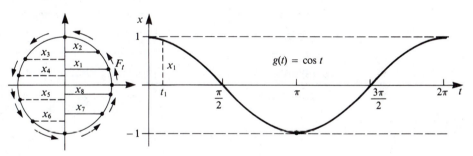

figure 29.9

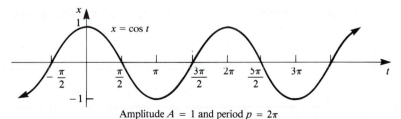

Amplitude $A = 1$ and period $p = 2\pi$

interval $[0, 2\pi]$. Note the similarity between the graphs of $\sin t$ and $\cos t$. In fact, the graph of $\cos t$ is just a "sine wave" starting at $t = -\pi/2$ rather than at $t = 0$. This observation translates into the identity:

$$\cos t = \sin\left(t + \frac{\pi}{2}\right)$$

Since the point F_t is always on the circle of radius 1, as shown in figure 29.10, we have $(|x|)^2 + (|y|)^2 = x^2 + y^2 = 1$. Since $x = \cos t$ and $y = \sin t$, we have

figure 29.10

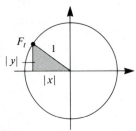

the identity:

$$\sin^2 t + \cos^2 t = 1$$

Figures 29.11 and 29.12 demonstrate that F_t and F_{-t} have the same x-

figure 29.11

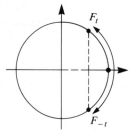

figure 29.12

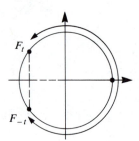

coordinate, but their y-coordinates are opposite in sign. This produces the following identities:

$$\cos(-t) = \cos t$$
$$\sin(-t) = -\sin t$$

From the graph of sin t, it is easy to see that:

1. sin $t = 0$ if and only if $t = \pm n\pi$ for $n = 0, 1, 2, \ldots$

2. sin $t = 1$ if and only if $t = \pi/2 \pm 2n\pi$

3. sin $t = -1$ if and only if $t = -\pi/2 \pm 2n\pi$

To give an alternate demonstration of **(3)**, for example, note that we are trying to determine which "t unit runs" will finish at $F = (0, -1)$. This is the only point on the unit circle with y-coordinate equal to -1, as shown in figure 29.13.

figure 29.13

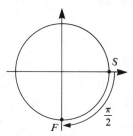

One such race corresponds to $t = -\pi/2$, but we could also make complete trips from $(0, -1)$ in either the clockwise or counterclockwise direction and return to F. Thus, $t = -\pi/2 \pm 2n\pi$.

Likewise, from the graph of cos t, it is easy to see that:

4. cos $t = 0$ if and only if $t = \pi/2 \pm n\pi$

5. cos $t = 1$ if and only if $t = \pm 2n\pi$

6. cos $t = -1$ if and only if $t = \pm(2n + 1)\pi$

Facts **1–6** can be used to solve trigonometric equations:

example 29.3   Solve the trigonometric equations

$$\textbf{(a)} \ \sin(3t) = 0 \qquad \textbf{(b)} \ \cos(2t - \pi) = -1$$

solution 29.3   **(a)** The equation $\sin(3t) = 0$ implies $3t = \pm n\pi$. Thus, $t = \pm n(\pi/3)$, for $n = 0, 1, 2, \ldots$

(b) The equation $\cos(2t - \pi) = -1$ implies $2t - \pi = \pm(2n + 1)\pi$. Hence, $2t = \pi \pm (2n + 1)\pi = \pi[\pm(2n - 1) + 1]$. Hence, $t = (\pi/2)[\pm(2n + 1) + 1]$. But $(2n + 1) + 1 = 2n + 2$ and $-(2n + 1) + 1 = -2n$. Hence, $[\pm(2n + 1) + 1]$ gives multiples of ± 2, and thus $t = (\pi/2)(\pm 2k) = \pm \pi k$.   ■

The following two identities, which date back to Ptolemy (150 A.D.), are known as the **addition formulas**. They will prove useful in section 30 when we prove $D_t \sin t = \cos t$.

$$\sin (t + h) = \sin t \cos h + \sin h \cos t$$
$$\cos (t + h) = \cos t \cos h - \sin t \sin h$$

The proof, which is outlined in part B exercise 52, is based on the unit circle definition of cos t and sin t.

We will now demonstrate that the graph of $f(t) = A \sin \omega t$, for $A > 0$ and $\omega > 0$, takes the form shown in figure 29.14. This is easy to see. As $\tau = \omega t$

figure 29.14   The graph of $A \sin \omega t$.

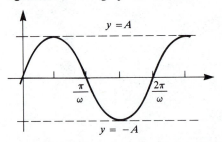

Amplitude $= A$ and period $p = 2\pi/\omega$

increases to $\pi/2$ [i.e., as t increases to $\pi/(2\omega)$], sin $\omega t = \sin \tau$ increases to 1. As $\tau = \omega t$ increases from $\pi/2$ to π (i.e., as t increases to π/ω), sin $\omega t = \sin \tau$ decreases from 1 to 0, and so forth. The constant A is called the *amplitude* of the wave, while the constant $p = 2\pi/\omega$ is called the *period* of the wave. One complete cycle takes place on $[0, p]$ and curve repeats its behavior every additional p units. Hence, $f(t + p) = f(t)$.

Similarly, the graph of $f(t) = A \cos \omega t$ takes the form shown in figure 29.15.

figure 29.15   The graph of $A \cos \omega t$.

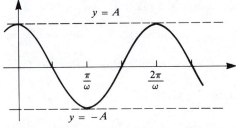

Amplitude $= A$ and period $p = 2\pi/\omega$

example 29.4   Sketch the graphs of $f(t) = 2 \sin t$ and $g(t) = \sin 2t$ on the same axes.

solution 29.4   The function $2 \sin t$ has period 2π and amplitude $A = 2$. The function $\sin 2t$ has

figure 29.16

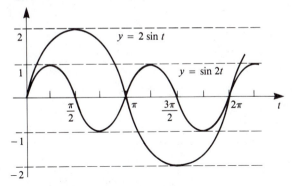

period $p = 2\pi/2 = \pi$ and amplitude $A = 1$; one full cycle takes place on $[0, \pi]$. The two graphs are shown in figure 29.16. Note that we divided $[0, \pi]$ into four equal subintervals to find the key points.

example 29.5   Find a function of the form $f(t) = A \cos \omega t$ with amplitude 3 and period 4 seconds.

solution 29.5   Since the period is $p = 2\pi/\omega$, we have $4 = 2\pi/\omega$ or $\omega = 2\pi/4 = \pi/2$. Hence, $f(t) = 3 \cos (\pi/2) t$.

If $f(t) = A \cos \omega(t - t_0)$, then the graph is similar to the graph of $A \cos \omega t$, but note that $f(t_0) = A \cos 0 = A$. Thus, t_0 is a new starting point for the cosine wave.* The wave still repeats itself every $p = 2\pi/\omega$ units and a complete cycle takes place on $[t_0, t_0 + p]$, as shown in figure 29.17.

figure 29.17 The graph of $A \cos \omega(t - t_0)$.

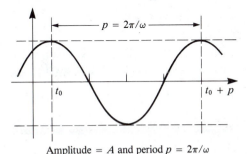

Amplitude $= A$ and period $p = 2\pi/\omega$

Likewise, the graph of $f(t) = A \sin \omega(t - t_0)$ takes the form shown in figure 29.18. One such function is graphed in example 29.6.

* Note that t_0 is not unique. Although we could have started the wave at $t_0 + p$, for example, t_0 is usually selected to be the smallest positive starting point.

figure 29.18   The graph of $A \sin \omega(t - t_0)$.

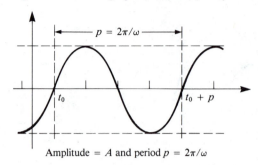

Amplitude $= A$ and period $p = 2\pi/\omega$

example 29.6

Sketch the graph of the sine curve $f(t) = 3/2 \sin (2t - \pi/2)$ for $0 \leq t \leq 2\pi$.

solution 29.6

Here, $A = 3/2$ and $\omega = 2$. To find t_0, we must write $2t - \pi/2 = 2(t - \pi/4)$. Hence, $t_0 = \pi/4$ and the period is $p = 2\pi/2 = \pi$. Hence, we must draw a sine wave starting at $\pi/4$, with amplitude $3/2$, and with one complete cycle taking place on $[\pi/4, 5\pi/4]$. The graph is shown in figure 29.19.

figure 29.19

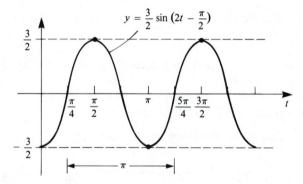

Finally, we can shift the graph of $A \sin \omega(t - t_0)$ up (or down) B units by forming $f(t) = B + A \sin \omega(t - t_0)$, as illustrated in figure 29.20.

figure 29.20   The graph of $B + A \sin \omega(t - t_0)$.

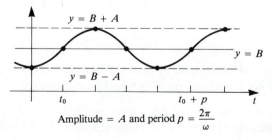

Amplitude $= A$ and period $p = \dfrac{2\pi}{\omega}$

We continue this section with applications of sine curves from physiology and ecology.

example 29.7

The rhythmic process of breathing consists of alternating periods of inhaling and exhaling. The duration of one complete cycle is about 5 seconds. The air flow F into and out of the lungs has a maximum of about .5 liter per second. Find a function $F = A \sin \omega t$ that fits this information.

solution 29.7

The amplitude A is the maximum of $A \sin \omega t$. Thus, $A = .5$. Also, $p = 5 = 2\pi/\omega$. Hence, $\omega = 2\pi/5$ and so $F = .5 \sin (2\pi/5) t$.

example 29.8

If D is the length of the day in hours and t is time in days, then D is well approximated by the formula

$$D = 12 + \frac{K}{2} \sin \frac{2\pi}{365} (t - 79)$$

where $t = 0$ corresponds to January 1, and K depends on the latitude. (For Los Angeles, $K \approx 5$, while for Seattle, $K \approx 7.5$, for example.)

(a) What are the period and amplitude?
(b) What is the interpretation of the quantity 79 in the formula?
(c) Find the times when the longest and shortest days occur.

solution 29.8

Here we have $B = 12$, $A = K/2$, $\omega = 2\pi/365$, and $t_0 = 79$. The amplitude is therefore $K/2$. Hence, we may interpret $K = 2A$ as the total variation in day length. The period $p = 2\pi/\omega = 365$, as expected. When $t = 79$, $D = 12 + (K/2) \sin 0 = 12$ hours. The time $t = 79$ corresponds to approximately March 20, which is the beginning of spring. As figure 29.21 shows, the maximum day

figure 29.21

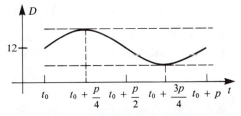

length occurs at $t_0 + p/4 = 79 + 365/4 = 170.25$. This corresponds to June 19, the beginning of summer. The minimum day length occurs at $t_0 + \frac{3}{4}p = 352.75$, which corresponds to December 19, the onset of winter.

example 29.9

In constructing models of an ecosystem, ecologists often use sine functions to simulate temperature variations during the day or between the seasons. Construct a function of the form

$$T = B + A \sin \omega(t - t_0)$$

that fits the following information:

(a) The temperature reaches a maximum of $80°$ at 2 p.m.

(b) The minimum temperature is $40°$.

(c) The time $t = 0$ corresponds to 12 midnight.

solution 29.9   We are given that $p = 24$, and therefore $\omega = 2\pi/p = \pi/12$. In addition, $B + A = 80$ and $B - A = 40$. Hence, $A = 20$ and $B = 60$. To find t_0, recall that the maximum value of $B + A \sin \omega(t - t_0)$ occurs first at $t = t_0 + p/4 = t_0 + 6$. Since 2 P.M. corresponds to $t = 14$, we have $t_0 + 6 = 14$ or $t_0 = 8$. Thus, $T = 60 + 20 \sin(\pi/12)(t - 8)$. The graph of T is shown in figure 29.22.

figure 29.22

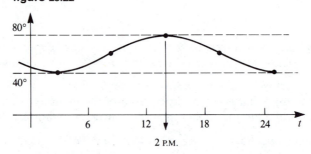

2 P.M.

Most rhythmic processes in biology are not governed by the simple sine waves presented in this section. It is known, however, that any *continuous periodic function* (see figure 29.23) can be approximated by a sum of sine functions of the form

$$a_0 + a_1 \sin \omega_1(t - t_1) + a_2 \sin \omega_2(t - t_2) + \cdots + a_n \sin \omega_n(t - t_n)$$

This is the famous theorem of Fourier. It is important, therefore, to first become familiar with the more elementary sine waves.

figure 29.23

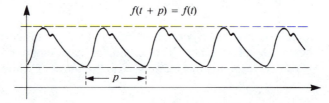

$$f(t + p) = f(t)$$

The functions $\sin t$ and $\cos t$ are not one-to-one functions. If, however, we restrict their domains to $[-\pi/2, \pi/2]$ and $[0, \pi]$, respectively, as shown in figures 29.24 and 29.25, then each function is one-to-one and therefore possesses an

figure 29.24

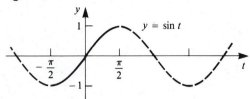

figure 29.25

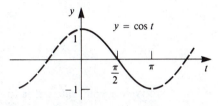

inverse function. For $f(t) = \sin t$, $-\pi/2 \le t \le \pi/2$, and if $k = \sin t$, then $-1 \le k \le 1$ and $f^{-1}(k) = f^{-1}(f(t)) = t$. We denote the inverse function by $\sin^{-1}$. Thus, $\sin^{-1} k = t$. To summarize:

> For $-1 \le k \le 1$, $t = \sin^{-1} k$ is that unique real number in $[-\pi/2, \pi/2]$ whose sine is k.

Use ⌊sin⁻¹⌋ or ⌊inv⌋ ⌊sin⌋ button on your calculator to fill in the missing entries in table 29.1.

table 29.1

k	-1.2	-1	$-.7$	$-.3$	0	$.4$	$.8$	1
$\sin^{-1} k$		$-\pi/2$		$-.3047$	0		$.9273$	

If we let $f^{-1}(x) = \sin^{-1} x$, then the graph of f^{-1} is the reflection of the graph of $y = \sin x$, $-\pi/2 \le x \le \pi/2$, in the line $y = x$. Alternately, note that if $y = \sin^{-1} x$, then $\sin y = \sin(\sin^{-1} x) = x$ and $-\pi/2 \le y \le \pi/2$. Hence, the graph of $\sin^{-1} x$ is that part of the sine wave (along the y-axis) $x = \sin y$ for $-\pi/2 \le y \le \pi/2$. The graphs are shown in figure 29.26.

In a similar way, one can show:

> For $-1 \le k \le 1$, $t = \cos^{-1} k$ is that unique real number in $[0, \pi]$ whose cosine is k.

figure 29.26(a)

figure 29.26(b)

figure 29.26(c)

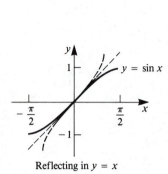

Reflecting in $y = x$

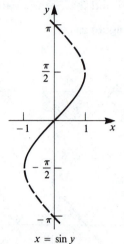

$x = \sin y$

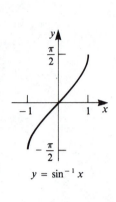

$y = \sin^{-1} x$

Use $\boxed{\cos^{-1}}$ or $\boxed{\text{inv}}$ $\boxed{\cos}$ on your calculator to fill in the missing entries in table 29.2.

table 29.2

k	-1.1	-1	$-.8$	$-.4$	0	$.3$	$.7$	1
$\cos^{-1} k$		$-\pi$		1.9823		1.2661		

The graph of $y = \cos^{-1} x$, which has domain $[-1, 1]$ and range $[0, \pi]$, is that part of the cosine wave $x = \cos y$ between $y = 0$ and $y = \pi$, as shown in figures 29.27 and 29.28.

figure 29.27

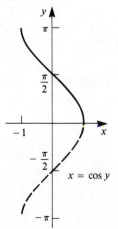

$x = \cos y$

figure 29.28

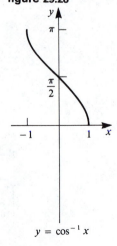

$y = \cos^{-1} x$

The inverse trigonometric functions $\sin^{-1}$ and $\cos^{-1}$ are especially useful in solving trigonometric equations. For $-1 \le k \le 1$, the solutions to $\sin x = k$ are shown graphically in figure 29.29. The value x_1 is just $\sin^{-1} k$, while $x_2 =$

figure 29.29

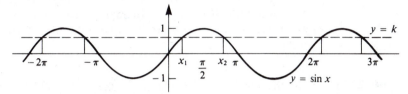

$\pi - \sin^{-1} k$. All other solutions differ from these two by multiples of 2π. To summarize:

> To solve $\sin x = k$, let $x_1 = \sin^{-1} k$. Then the solutions are of the form:
>
> $$x_1 \pm 2n\pi \qquad \text{and} \qquad (\pi - x_1) \pm 2n\pi$$

figure 29.30

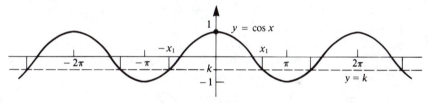

From the graph of $y = \cos x$ shown in figure 29.30, it is easy to establish the following result.

> To solve $\cos x = k$, let $x_1 = \cos^{-1} k$. Then the solutions are of the form:
>
> $$x_1 \pm 2n\pi \qquad \text{and} \qquad -x_1 \pm 2n\pi$$

These formulas are used in our final two examples.

example 29.10   Solve the trigonometric equations:

$$\textbf{(a)} \cos t = -.5 \qquad \textbf{(b)} \sin 3t = -.2 \qquad \textbf{(c)} \sin t + \sin 2t = 0$$

solution 29.10   For **(a)**, we have $\cos^{-1}(-.5) = 2.09439\ldots$. Hence, $t = 2.09439 \pm 2n\pi$ or $t = -2.09439 \pm 2n\pi$.

For **(b)**, let $x = 3t$. Then $\sin x = -.2$. Since $\sin^{-1}(-.2) = -.20135\ldots$, $x = 3t$ is either of the form $-.20135 \pm 2n\pi$ or $\pi + .20135 \pm 2n\pi$. Solving for t, we obtain $t = -.06712 \pm 2n\pi/3$ or $t = 1.11432 \pm 2n\pi/3$.

For **(c)**, $\sin 2t = \sin (t + t) = \sin t \cos t + \cos t \sin t = 2 \sin t \cos t$. Hence, $\sin t + \sin 2t = \sin t + 2 \sin t \cos t = \sin t (1 + 2 \cos t)$. The solution therefore consists of the solution to $\sin t = 0$ (namely $\pm n\pi$) and the solution to $\cos t = -1/2$, which was found in **(a)**. Hence, $t = \pm n\pi$, $t = -2.09493 \pm 2n\pi$, or $t = 2.09439 \pm 2n\pi$.

example 29.11   For the temperature function in example 29.9, determine those times of the day when $T = 50°$.

solution 29.11   We must find those solutions to $T = 60 + 20 \sin (\pi/12)(t - 8) = 50$ for $0 \le t \le 24$. Hence, $\sin (\pi/12)(t - 8) = -1/2$. Let $x = (\pi/12)(t - 8)$. Then $\sin x = -1/2$ has solutions of the form

$$x = \sin^{-1}(-1/2) \pm 2n\pi \qquad \text{or} \qquad x = \pi - \sin^{-1}(-1/2) \pm 2n\pi$$

Hence, solving for t, we obtain

$$t = 8 + \frac{12 \sin^{-1}(1/2)}{\pi} \pm 24n \qquad \text{or} \qquad t = 8 + \frac{12}{\pi}\left[\pi - \sin^{-1}\left(-\frac{1}{2}\right)\right] \pm 24n$$

With the aid of a calculator, we obtain $t = 8 - 2 \pm 24n$ or $t = 8 + (12 + 2) \pm 24n$. The only solutions in $[0, 24]$ are $t = 6$ and $t = 22$, which correspond to 6 A.M. and 10 P.M., respectively.   ▤

exercises for section 29

part A   Use the racetrack definitions of sine and cosine to compute $\cos t$ and $\sin t$ for each of the following values of t.

1. $\pi/2$ **2.** $3\pi/2$ **3.** -2π **4.** 7π **5.** $-5\pi/2$

The lines $y = x$ and $y = -x$ intersect the unit circle $x^2 + y^2 = 1$ at the points P_1, P_2, P_3, and P_4 shown in the figure.

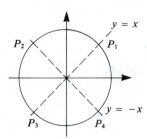

6. Show that $P_1 = (1/\sqrt{2}, 1/\sqrt{2})$ and find $\cos(\pi/4)$.

7. Show that $P_4 = (1/\sqrt{2}, -1/\sqrt{2})$ and find $\sin(7\pi/4)$.

8. Show that $P_2 = (-1/\sqrt{2}, 1/\sqrt{2})$ and find $\cos(3\pi/4)$.

9. Find $\sin(5\pi/4)$ and $\cos(5\pi/4)$.

10. Find $\sin(-13\pi/4)$ and $\cos(-13\pi/4)$.

Locate the approximate position of $P_t = (\cos t, \sin t)$ on the unit circle. Then use your calculator to evaluate $\cos t$ and $\sin t$ correct to four decimal places for each of the following values of t.

11. .5   **12.** 4.2   **13.** −2.1   **14.** −5.1   **15.** 2.7   **16.** −1.5

17. Use the racetrack definition of $\cos t$ to show that $\cos t = -1$ if and only if $t = \pm(2n+1)\pi$.

18. Use the racetrack definition of $\sin t$ to show that $\sin t = 0$ if and only if $t = \pm n\pi$.

Find *all solutions* to each of the following trigonometric equations.

19. $\sin 2t = 0$ **20.** $\cos 2t = 1$ **21.** $\cos t/2 = -1$

22. $\sin t/2 = 1$ **23.** $\sin(t + \pi) = 1$ **24.** $\cos(t - \pi) = 0$

25. Find a function of the form $f(t) = A \sin \omega t$ with period 2 and maximum value 4.

26. Find a function of the form $f(t) = A \cos \omega t$ with period 4π and $f(2\pi) = -3$.

27. Find a function of the form $f(t) = A \sin \omega(t - t_0)$ with maximum value 4, period 2, and $f(1) = 0$.

★**28.** If $f(t) = A \cos \omega(t - t_0)$, show that $f(t) = A \sin \omega(t - t_1)$ for some constant t_1. [*Hint*: $\cos t = \sin(t + \pi/2)$]

Sketch the graphs of each of the following trigonometric functions.

29. $2 \cos t$ and $\cos 2t$, on the same axes

30. $\sin t/2$ and $2 \sin t/2$, on the same axes

31. $2 \sin t$ and $2 \sin(t - \pi/4)$, on the same axes

32. $\cos 2t$ and $\cos(2t - \pi)$, on the same axes

33. $\sin 2\pi t$ **34.** $2 \cos(\pi/4) t$ **35.** $2 + \sin 2\pi t$

36. $4 + 2 \cos(\pi/4) t$ **37.** $2 \cos(t/2 - \pi/2)$ **38.** $\frac{1}{2} \sin(2\pi t - \pi) + 1$

39. The number in a particular animal population varies yearly between 200 and 350 animals. If the present ($t = 0$) population is 200, find a function of the form

$$N = A + B \sin \omega(t - t_0)$$

that fits this information.

40. Using the formula for day length given in example 29.8, estimate the day length in Los Angeles ($K \approx 5$) on:

(a) December 25   (b) July 4   (c) the beginning of fall

41. Construct a temperature function of the form

$$T = A + B \sin \omega(t - t_0)$$

that fits the following information:

(a) The temperature reaches a maximum of $100°$ at 3:00 P.M..

(b) The total variation in temperature during the day is $40°$.

(c) The time $t = 0$ corresponds to midnight.

42. The light intensity I at the ocean's surface has a maximum value of I_0 when the altitude of the sun is highest. If $t = 0$ corresponds to sunrise and $t = D$ (day length) corresponds to sunset, find a function of the form $I = A \sin \omega t$ that fits this information.

43. A flying insect makes twenty wing-beats per second. The angle θ that the wings make with the line of flight (see the figure) varies between $-60°$ and $60°$. Find a function of the form $\theta = A \sin \omega t$ that fits this information.

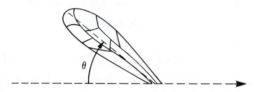

Use the inverse sine and inverse cosine buttons on your calculator to aid in solving the following trigonometric equations.

44. $\sin t = .5$ **45.** $\cos 2t = -1$.   **46.** $\sin (t + 2) = .3$

47. $\sin^2 t = 1/4$ **48.** $2 \sin^2 t + \sin t - 1 = 0$

49. $\cos t + \cos 2t = 0$ [*Hint*: $\cos 2t = 2 \cos^2 t - 1$.]

50. For the temperature function in example 29.9, determine those times of the day when $T = 70°$.

51. For the temperature function in exercise 41, determine the period of the day when the temperature is $90°$ or higher.

part B

52. This exercise demonstrates the formula $\cos (t + h) = \cos t \cos h + \sin t \sin h$ in the case where h is small. Recall that $P_t = (\cos t, \sin t)$, $P_{t+h} = (\cos (t + h), \sin (t + h))$, and $P_h = (\cos h, \sin h)$.

(a) Compute the length of the chord $\overline{P_t P_{t+h}}$ (see the figure) using the distance formula

$$d(P, Q) = \sqrt{(x_1 - x_2)^2 + (y_1 - y_2)^2}$$

where $P = (x_1, y_1)$ and $Q = (x_2, y_2)$.

(a)   **(b)**

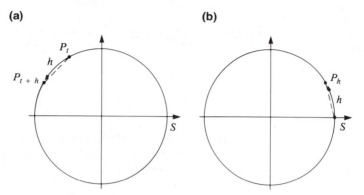

(b) Compute the length of the chord $\overline{SP_h}$.

(c) Since each of the arcs has length h, $\overline{P_tP_{t+h}} = \overline{SP_h}$. Show that this relationship, together with the trigonometric identity $\sin^2 t + \cos^2 t = 1$ gives the addition formula for $\cos(t + h)$.

53. Shown in the figure is a graph* of the rise and fall of the tide at Boston Harbor over a 24-hour period. Find a function of the form $y = B + A \sin \omega(t - t_0)$ that approximates the graph.

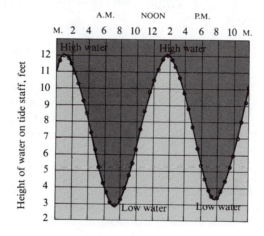

* Graph adapted from A.N. Strahler, *Physical Geography*, 2nd ed. New York: John Wiley and Sons, (1960).

30 calculus of the trigonometric functions

To begin this section, we will derive the formulas for the derivatives of the trigonometric functions. As in section 16, the demonstration that $D_t \sin t = \cos t$ depends on two special limits:

1. $\displaystyle \lim_{h \to 0} \frac{\sin h}{h} = 1$

2. $\displaystyle \lim_{h \to 0} \frac{\cos h - 1}{h} = 0$

To demonstrate (1) and (2), let $f(t) = \sin t$ and $g(t) = \cos t$. Then, from the definition of derivative,

$$f'(0) = \lim_{h \to 0} \frac{\sin h - \sin 0}{h} = \lim_{h \to 0} \frac{\sin h}{h}$$

and

$$g'(0) = \lim_{h \to 0} \frac{\cos h - \cos 0}{h} = \lim_{h \to 0} \frac{\cos h - 1}{h}$$

Hence, (1) asserts that $f'(0) = 1$ while (2) states that $g'(0) = 0$. That $g'(0) = 0$ follows from the fact that a relative maximum for $\cos t$ exists at $t = 0$. There is no quick demonstration that $\lim_{h \to 0} (\sin h)/h = 1$. Perhaps the easiest way to convince yourself is to use your calculator to compute $(\sin h)/h$ for small values of h. A formal proof is outlined in the exercises. Shown in figure 30.1 is a computer-generated graph of $(\sin h)/h$ for $h = 1.0, .99, \ldots, .02, .01$.

figure 30.1

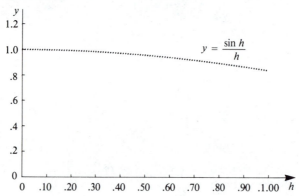

With these two limits, the proof that $D_t \sin t = \cos t$ is fairly straightforward:

$$\frac{f(t+h)-f(t)}{h} = \frac{\sin(t+h)-\sin t}{h}$$

$$= \frac{\sin t \cos h + \sin h \cos t - \sin t}{h}$$

$$= \frac{\sin t (\cos h - 1)}{h} + \frac{\sin h \cos t}{h}$$

$$= \sin t \frac{\cos h - 1}{h} + \cos t \frac{\sin h}{h}$$

Letting $h \to 0$, we obtain $f'(t) = \sin t \cdot 0 + \cos t \cdot 1 = \cos t.$

To show $D_t \cos t = -\sin t$, recall that $\cos t = \sin (t + \pi/2)$. Hence, by the Chain Rule,

$$D_t \cos t = \cos (t + \pi/2) \cdot 1$$

$$= \cos t \cos \pi/2 - \sin t \sin \pi/2 \quad \text{using the addition formula}$$

$$= \cos t \cdot 0 - \sin t \cdot 1$$

$$= -\sin t$$

Again, from the Chain Rule, we have the useful formulas:

$$D_t \sin f(t) = [\cos f(t)]f'(t)$$
$$D_t \cos f(t) = [-\sin f(t)]f'(t)$$

For example, $D_t \sin (t^2 + 1) = [\cos (t^2 + 1)]2t = 2t \cos (t^2 + 1)$. These two formulas are illustrated in the next three examples.

example 30.1   Find the derivatives of

(a) $e^{3 \sin t}$ (b) $t^2 \cos^2 t$ (c) $\dfrac{\cos 2t}{t^2 + 1}$

solution 30.1   (a) By the Chain Rule, $D_t e^{3 \sin t} = e^{3 \sin t} D_t (3 \sin t) = e^{3 \sin t} (3 \cos t)$.
For (b), first recall that $t^2 \cos^2 t = t^2 (\cos t)^2$. Hence, by the Product Rule,
$D_t (t^2 \cos^2 t) = t^2 \cdot 2(\cos t)(-\sin t) + 2t(\cos t)^2 = -2t^2 \sin t \cos t + 2t \cos^2 t$.
For (c), the Quotient Rule gives

$$D_t \frac{\cos 2t}{t^2 + 1} = \frac{(t^2 + 1)(-\sin 2t) \cdot 2 - (\cos 2t) \cdot 2t}{(t^2 + 1)^2}$$

example 30.2   Shown in figure 30.2 is a computer-generated graph of $f(t) = e^{\sin (t^2 + 1)}$. Locate the relative maxima and minima.

solution 30.2   The first derivative of $f(t)$ is $f'(t) = e^{\sin(t^2+1)} D_t \sin (t^2 + 1) = e^{\sin(t^2+1)} \cdot 2t \cos (t^2 + 1)$.

figure 30.2

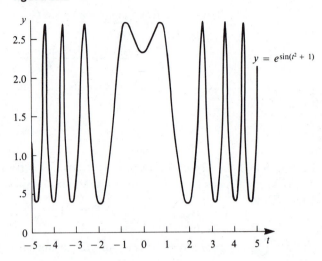

$$y = e^{\sin(t^2 + 1)}$$

Hence, $f'(t) = 0$ implies $t = 0$ or $\cos(t^2 + 1) = 0$. Hence, $t = 0$ or $t^2 + 1 = \dfrac{\pi}{2} \pm n\pi$. Since $t^2 + 1 > 0$, we disregard the negative solutions. Thus, $t = 0$ or $t = \pm\sqrt{(\pi/2) - 1 + n\pi}$. With the aid of a calculator, we obtain

$$t = 0, \quad \pm.755, \quad \pm 1.927, \quad \pm 2.618, \quad \pm 3.162, \quad \pm 3.624, \quad \pm 4.0347, \ldots$$

Note that the graph of $f(t)$ oscillates between e^{-1} and e.

example 30.3   Show that $y = A \sin \omega(t - t_0)$ satisfies the differential equation $y'' = -\omega^2 y$.

solution 30.3   The first derivative of y is $y' = A \cos \omega(t - t_0)D_t\,\omega(t - t_0) = A\omega \cos \omega(t - t_0)$, and $y'' = -A\omega^2 \sin \omega(t - t_0)$. Rewriting y'' as $-\omega^2[A \sin \omega(t - t_0)]$, we see that $y'' = -\omega^2 y$.

It is known that $y = A \sin \omega(t - t_0)$ is the *complete* solution to $y'' = -\omega^2 y$. Thus, $y'' = -\omega^2 y$ if and only if $y = A \sin \omega(t - t_0)$ for some constants A and t_0. This differential equation can be used to derive the equations for the movement of a weight on a spring.

example 30.4   **simple harmonic motion**   A weight of W pounds is attached to a spring as shown in figure 30.3. Hooke's Law states that when the spring is stretched y feet from equilibrium, as shown in figure 30.4, the restoring force F_s of the spring is given by $F_s = -ky$ for some constant $k > 0$. If we ignore all other possible forces that could affect the system (including frictional forces), then

figure 30.3

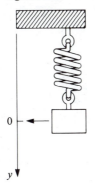

figure 30.4

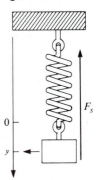

Newton's Second Law asserts that

$$F_s = (\text{mass})\,(\text{acceleration}) = my''(t)$$

The mass m is related to the weight W by $W = mg$, where $g = 32\text{ ft/sec}^2$. Hence, $my'' = -ky$ or $y'' = (-k/m)\,y$. Thus, y satisfies $y'' = -\omega^2 y$ with $\omega^2 = k/m$. It follows that

$$y = A \sin\left[\sqrt{k/m}\;(t - t_0)\right]$$

for some constants A and t_0. The amplitude of the motion is $|A|$ and the period is $p = 2\pi/\omega = 2\pi\sqrt{m/k}$. The familiar graph is shown in figure 30.5.

figure 30.5

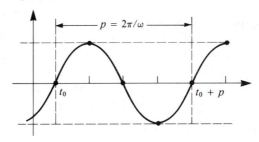

example 30.5   A 16-pound weight stretches a spring 1 foot to the equilibrium position. The spring is then stretched an additional 2 feet and released. Assuming the model of example 30.4 applies, find the function y governing the simple harmonic motion and determine the amplitude and period.

solution 30.5   Since $W = mg$, $m = \frac{16}{32} = \frac{1}{2}$. When the spring is stretched 1 foot, $F_s = -16$ since the weight and F_s balance one another. This gives $k = 16$ as the spring constant. Hence, $y = A \sin\left[\sqrt{32}(t - t_0)\right]$. We are given that the weight is lowered an additional 2 feet and *released*. Hence, $y(0) = -2$ and $y'(0) = 0$. Now, $y' = \sqrt{32}A \cos [\sqrt{32}(t - t_0)]$. Hence, $y'(0) = \sqrt{32} \cos [\sqrt{32}(-t_0)] = \sqrt{32} \cos \sqrt{32}t_0$ $= 0$. Hence, $\sqrt{32}t_0 = \pi/2$. Since $y(0) = -2$, $-2 = A \sin [\sqrt{32}(-t_0)] = -A \sin (\sqrt{32}t_0)$ $= -A \sin \pi/2 = -A$. Thus $A = 2$, and

$$y = 2 \sin [\sqrt{32}(t - t_0)]$$

where $t_0 = \pi/(2\sqrt{32}) \approx .277 \dots$. The amplitude is then 2 and the period is $p = 2\pi/\sqrt{32} \approx 1.1107 \dots$ seconds.   ▤

Since $D_t \sin \omega t = \omega \cos \omega t$ and $D_t \cos \omega t = -\omega \sin \omega t$, we have the following antiderivative formulas:

$$\int \sin \omega t \, dt = -\frac{1}{\omega} \cos \omega t$$

$$\int \cos \omega t \, dt = \frac{1}{\omega} \sin \omega t$$

These formulas are used in the next two examples.

example 30.6   If $F(t)$ denotes the rate of flow (in cubic feet per day) of a stream into a reservoir, then $\int_a^b F(t) \, dt$ denotes the new amount of water introduced between times $t = a$ and $t = b$. If $F(t) = 1,000 \sin (\pi/60)t$ for $0 \le t \le 60$, how much additional water is introduced between times $t = 0$ and $t = 60$?

solution 30.6   The change in volume is given by $\displaystyle\int_0^{60} 1,000 \sin \frac{\pi}{60} t \, dt = -1,000 \frac{60}{\pi} \cos \frac{\pi}{60} t \Big|_0^{60} =$

$-1,000 \dfrac{60}{\pi} (\cos \pi - \cos 0) = -1,000 \dfrac{60}{\pi} (-2) = 38,197$ cubic feet.   ▤

example 30.7   Using the formula for day length in example 29.8, estimate the *average* day length during spring ($t = 79$ to $t = 170.25$) in Los Angeles.

solution 30.7   For Los Angeles, $D = 12 + \dfrac{K}{2} \sin \dfrac{2\pi}{365} (t - 79)$, with $K \approx 5$. Using the formula for *average value* found in section 21, we obtain

$$\text{Average day length in spring} = \frac{1}{170.25 - 79} \int_{79}^{170.25} D(t) \, dt$$

To find $\displaystyle\int \left[12 + \frac{K}{2} \sin \frac{2\pi}{365} (t - 79) \right] dt = 12t + \frac{K}{2} \int \sin \frac{2\pi}{365} (t - 79) \, dt$, let $u = (t - 79)$ in the last antiderivative. Then $du = dt$ and

$$\int \sin \frac{2\pi}{365} (t - 79) \, dt = \int \sin \frac{2\pi}{365} u \, du = -\frac{365}{2\pi} \cos \frac{2\pi}{365} u$$

Substituting $u = (t - 79)$, we have:

$$\text{Average day length} = \frac{1}{91.25}\left[12t - \frac{K}{2} \cdot \frac{365}{2\pi} \cos \frac{2\pi}{365}(t - 79)\right]\Big|_{79}^{170.25}$$

$$= \frac{1}{91.25}\left\{\left[2{,}043 - \frac{365K}{4\pi}\cos\frac{\pi}{2}\right] - \left(948 + \frac{365K}{4\pi}\right)\right\}$$

$$= \frac{1}{91.25}\left(1{,}095 + \frac{365K}{4\pi}\right) = 12 + \frac{K}{\pi}$$

Using the value $K = 5$, we find that the average day length during spring in Los Angeles is approximately 13.6 hours.

The remaining trigonometric functions are defined in terms of the basic functions $\sin t$ and $\cos t$:

$$\tan t = \frac{\sin t}{\cos t} \qquad \textbf{tangent}$$

$$\text{ctn } t = \frac{\cos t}{\sin t} \qquad \textbf{cotangent}$$

$$\csc t = \frac{1}{\sin t} \qquad \textbf{cosecant}$$

$$\sec t = \frac{1}{\cos t} \qquad \textbf{secant}$$

Using the Quotient Rule, it is easy to establish the following derivative formulas:

$$D_t \tan t = \sec^2 t$$

$$D_t \text{ ctn } t = -\csc^2 t$$

$$D_t \csc t = -\text{ctn } t \csc t$$

$$D_t \sec t = \tan t \sec t$$

As an example, $D_t \csc t = D_t \dfrac{1}{\sin t} = \dfrac{-\cos t}{\sin^2 t} = -\dfrac{\cos t}{\sin t} \cdot \dfrac{1}{\sin t} = -\text{ctn } t \csc t$. These new derivative formulas are used in our next example.

example 30.8   Find the derivatives of

        (a) $\tan^3 t$ **(b)** $\ln (\sin t)$ **(c)** $\sec t \tan t$

solution 30.8   **(a)** Since $\tan^3 t = (\tan t)^3$, we have, by the Chain Rule,

$$D_t(\tan t)^3 = 3(\tan t)^2 \sec^2 t$$

For **(b)**, $D_t \ln (\sin t) = (1/\sin t) D_t \sin t = \cos t/\sin t = \tan t$.
Finally, applying the Product Rule to **(c)** yields

$$D_t \sec t \tan t = \sec t(\sec^2 t) + \tan t(\tan t \sec t)$$
$$= \sec^2 t(\sec^2 t + \tan^2 t)$$

the tangent and inverse tangent functions   Although inverse functions can be constructed for ctn t, csc t, and sec t, the most useful inverse function is $\tan^{-1} t$. We begin by sketching the graph of $y = \tan t$, using the methods of sections 12 and 13.

example 30.9   Sketch the graph of $y = \tan t$ for $-\pi/2 < t < \pi/2$, and discuss concavity features of the graph.

solution 30.9   The first derivative of y is $y' = \sec^2 t = 1/\cos^2 t$. Hence, $y' > 0$ for $-\pi/2 < t < \pi/2$, and $y = \tan t$ is therefore *increasing* on this interval. The second derivative is, $y'' = 2 \sec t(\tan t \sec t) = 2 \sec^2 t \tan t = 2 \sin t/\cos^3 t$. Hence, $y'' = 0$ for $\sin t = 0$, that is, for $t = 0$. (All other solutions are outside the domain.) The sign chart for y'' is therefore as shown in figure 30.6. It is easy to see, with the aid of your

figure 30.6

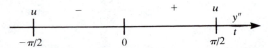

calculator, that $x = \dfrac{\pi}{2}$ and $x = \dfrac{-\pi}{2}$ are vertical asymptotes for $\tan t$. The graph is shown in figure 30.7. Thus $t = 0$ is an inflection point for $\tan t$.

figure 30.7

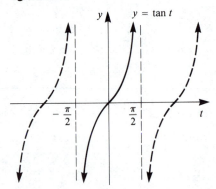

The period of the tangent function is π. This can be seen by using the addition formulas:

$$\tan (t + \pi) = \frac{\sin (t + \pi)}{\cos (t + \pi)} = \frac{\sin t \cos \pi + \cos t \sin \pi}{\cos t \cos \pi - \sin t \sin \pi}$$

$$= \frac{-\sin t}{-\cos t} = \tan t$$

Thus, the graph of $y = \tan t$ repeats itself on subsequent intervals of length π: $(\pi/2, 3\pi/2)$, $(-3\pi/2, -\pi/2)$, ...

If we confine our attention to the interval $(-\pi/2, \pi/2)$, then $f(t) = \tan t$ is increasing and therefore one-to-one. Denoting the inverse function by $f^{-1}(k) = \tan^{-1} k$, we have $k = f(f^{-1}(k)) = \tan (\tan^{-1} k)$. Hence,

> For a real number k, $t = \tan^{-1} k$ is that unique real number between $-\pi/2$ and $\pi/2$ whose tangent is k.

To find $\tan^{-1} k$, use the $\boxed{\tan^{-1}}$ or $\boxed{\text{inv}}$ $\boxed{\tan}$ button on your calculator. Fill in the missing entries in table 30.1.

table 30.1

k	-10	-5	-1	0	$.5$	4	100
$\tan^{-1} k$		-1.373		0		1.3258	

If we let $f^{-1}(x) = \tan^{-1} x$, the graph of f^{-1} can be obtained by reflecting the graph of $y = \tan x$ in the line $y = x$. An easier way to discover the shape of the graph is to let $y = \tan^{-1} x$ and note that $\tan y = x$, and $-\pi/2 < y < \pi/2$. The graph is shown in figure 30.8.

figure 30.8 $y = \tan^{-1} x$

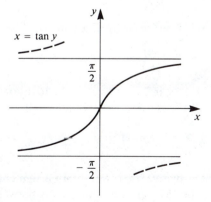

Figure 30.9 shows that to solve the trigonometric equation $\tan x = k$, we first let $x_1 = \tan^{-1} k$. The solutions are then of the form $x_1 \pm n\pi$.

figure 30.9 Solving $\tan x = k$

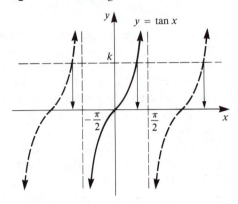

example 30.10   Solve the trigonometric equation $\tan 2t = 4$ for t.

solution 30.10   If $x = 2t$, then $\tan x = 4$. Hence, $x = \tan^{-1} 4 \pm n\pi$. Since $t = x/2$, we have $t = (\tan^{-1} 4)/2 \pm n(\pi/2) = .6629088 \pm n(\pi/2)$.

example 30.11   When frictional forces are considered in the simple harmonic motion model of example 30.4, the amplitude of the oscillations will decrease as t increases. The displacement function $y(t)$ is often of the form

$$y(t) = Ae^{-\alpha t} \sin \omega(t - t_0)$$

Sketch the graph of the *damped oscillation* $y = e^{-t/2} \sin 2t$ for $t \geq 0$.

solution 30.11   By the Product Rule,

$$y' = e^{-t/2}(\cos 2t) \cdot 2 + (\sin 2t)e^{-t/2}(-1/2)$$

$$= e^{-t/2}\left(2 \cos 2t - \frac{1}{2} \sin 2t\right)$$

Hence, $y' = 0$ if and only if $2 \cos 2t = \frac{1}{2} \sin 2t$. It follows that $\tan 2t = 4$. As we saw in example 30.10, $t = .6629 \pm n\pi/2$. Since $t \geq 0$, the critical points are

$$t_1 = .6629, \qquad t_1 + \frac{\pi}{2} = 2.2337, \qquad t_1 + \pi = 3.8045, \qquad t_1 + \frac{3\pi}{2} = 5.3753, \ldots$$

If we use the First Derivative Test, we obtain a sign chart of the form shown in figure 30.10. Thus, the critical points alternate between relative maxima and

figure 30.10

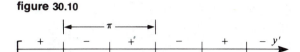

relative minima. Note that $|y| \le e^{-t/2}$ and the y-values at the critical points are given by .6964, $-.3175$, .1447, $-.0660$, .0301, ... The complete graph is shown in figure 30.11.

figure 30.11

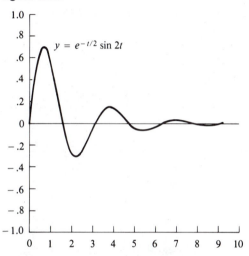

$y = e^{-t/2} \sin 2t$

It is interesting to note that the time elapsed between two successive relative maxima is π. This is just the period of the simple harmonic motion $y = \sin 2t$.

The derivatives of the inverse trigonometric functions are given below:

> **1.** $D_x \sin^{-1} x = 1/\sqrt{1 - x^2}$
> **2.** $D_x \cos^{-1} x = -1\sqrt{1 - x^2}$
> **3.** $D_x \tan^{-1} x = 1/(1 + x^2)$

To prove **(3)**, for example, let $f(x) = \tan^{-1} x$. Then $\tan f(x) = \tan (\tan^{-1} x) = x$. Taking the derivative of both sides yields $[\sec^2 f(x)]f'(x) = 1$. If we let $t = f(x)$ in the trigonometric identity $\sec^2 t = \tan^2 t + 1$, then we obtain

$$\sec^2 f(x) = \tan^2 f(x) + 1 = [\tan (\tan^{-1} x)]^2 + 1$$

Hence, $(1 + x^2)f'(x) = 1$ or $f'(x) = 1/(1 + x^2)$. These new derivative formulas are used in our next example.

example 30.12   Compute the derivatives of

$$\textbf{(a)} \ \tan^{-1} x^2 \qquad \textbf{(b)} \ (\sin^{-1} x)^3$$

solution 30.12   For **(a)**, $D_x \tan^{-1} x^2 = \dfrac{1}{1+(x^2)^2} \cdot 2x$, using the Chain Rule .

For **(b)**,

$$D_x (\sin^{-1} x)^3 = 3(\sin^{-1} x)^2 D_x \sin^{-1} x$$
$$= \frac{3(\sin^{-1} x)^2}{\sqrt{1-x^2}}$$

Formulas **(1)** and **(3)** give rise to the following important anti-derivative formulas:

$$\int \frac{1}{\sqrt{1-x^2}} \, dx = \sin^{-1} x$$
$$\int \frac{1}{1+x^2} \, dx = \tan^{-1} x$$

These two formulas have been incorporated into the integral tables found on page 261, and explain why $\sin^{-1}$ and $\tan^{-1}$ appear in many of the table entries of the form $\sqrt{a^2 - x^2}$ and $x^2 + a^2$ respectively.

example 30.13   With the aid of the integral tables, find:

$$\textbf{(a)} \ \int x^2 \sqrt{9-x^2} \, dx \qquad \textbf{(b)} \ \int \frac{\sqrt{x}}{x+1} \, dx$$

solution 30.13   **(a)** Using formula 46 with $a = 3$, we obtain

$$\int x^2 \sqrt{9-x^2} \, dx = -\frac{x(9-x^2)^{3/2}}{4} + \frac{9}{8} x \sqrt{9-x^2} + \frac{81}{8} \sin^{-1} (x/3)$$

For **(b)**, we let $u = \sqrt{x}$ and so $x = u^2$. Hence, $\dfrac{dx}{du} = 2u$ or $dx = 2u \, du$. It follows that

$$\int \frac{\sqrt{x}}{x+1} \, dx = \int \frac{u}{u^2+1} 2u \, du = 2 \int \frac{u^2}{u^2+1} \, du$$

If we now use formula 98 with $a = 1$, we obtain

$$2(u - \tan^{-1} u) = 2(\sqrt{x} - \tan^{-1}\sqrt{x})$$

exercises for section 30

part A

Compute the derivatives of each of the following trigonometric functions. [*Hint*: Recall that $\sin^n t$ means $(\sin t)^n$]

1. $\cos(2t - 1)$ **2.** $\cos^2(2t - 1)$ **3.** $\sin^3 t$

4. $\sin t^3$ **5.** $e^{\sin t}$ **6.** $\sin e^t$

7. $t^2 \cos 2t$ **8.** $t \cos^2 t$ **9.** $\sin[t/(t+2)]$ **10.** $\dfrac{\sin t}{\cos t}$

Find and classify all critical points of each of the following functions in exercises 11–15.

11. $e^{\cos t}$ **12.** $\sin t^2$ **★13.** $\sin t + \frac{1}{2}\sin 2t, \ 0 < t < 2\pi$

14. $t + \cos t$ **15.** $\sin e^t, \ 0 < t < 1$

16. Find *all* solutions to $y'' = -4y$ with $y(\pi) = 0$.

17. Solve $y'' + 9y = 0$ subject to $y(0) = 0$ and $y'(0) = 1$.

18. For the oscillating spring in example 30.4, what effect does doubling the weight have on the period of the motion?

19. A 16-pound weight stretches a spring 2 feet. The weight is displaced an additional 3 feet and then released. Find the function $y = f(t)$ governing the simple harmonic motion, where y is the displacement from equilibrium.

20. For the spring in exercise 19, what weight should be attached in order for the period of the motion to be 2 seconds?

Use the $\boxed{\tan^{-1}}$ button on your calculator to aid in solving each of the following trigonometric equations

21. $\sin t = \cos t$ **22.** $\sin t = 2\cos t$ **23.** $\tan 2t = 4$

24. $\tan^2 2t = 4$ **25.** $\tan^2 t = \tan t$ **26.** $\csc^2 t = \sec^2 t$

Use the formula for day length given in example 29.8 to solve each of the following exercises.

27. Find the formula for average day length during spring.

28. Find the formula for the average day length during summer ($t = 170.25$ to $t = 261.5$). What value of K results in an average day length of 18 hours?

29. Find the formula for average day length during winter.

30. The flow rate (in cubic feet per day) from streams into a lake during the first sixty days of Spring is given by

$$F(t) = 5{,}000 \sin\frac{\pi}{60}t$$

for $0 \le t \le 60$. Here, $t = 0$ corresponds to the beginning of spring. Find the total volume introduced into the lake.

★**31.** The function shown in the figure is sometimes used to model systolic and diastolic flow from the left ventricle to the aorta. Let n be the number of

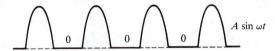

beats per minute. Find a formula for the volume per minute introduced into the aorta.

Compute the derivatives of each of the following trigonometric functions in exercises 32–40.

32. $\tan^2 t$

33. $e^{-t} \tan t$

34. $\tan 2t$

35. $\ln (\operatorname{ctn} t)$

36. $\sec^2 t$

37. $(\sec t)^2$

38. $\sin t \csc t$

39. $(\sec t)/(\cos t)$

40. $\sqrt{t} \tan \sqrt{t}$

41. Sketch the graph of $f(t) = \operatorname{ctn} t$ for $0 < t < \pi$, and discuss concavity features of the graph.

42. Sketch the graph of $f(t) = \sec t$ for $-\pi/2 < t < \pi/2$, and discuss concavity features of the graph.

Sketch the graphs of each of the damped oscillations given in exercises 43–44.

43. $y = e^{-t} \sin t$

44. $y = e^{-.1t} \sin 2t$

45. Show that the distance between two consecutive relative maxima for the damped oscillation $y = Ae^{-\alpha t} \sin \omega(t - t_0)$ is $p = 2\pi/\omega$.

46. Find the absolute minimum of $f(t) = 4/(\sin t) + 8/(\cos t)$ for $0 < t < \pi/2$.

Find the derivative of each of the following functions in exercises 47–52.

47. $t \sin^{-1} t$

48. $t \cos^{-1} t$

49. $(\tan^{-1} t)/(1 + t^2)$

50. $\sin^{-1} t^2$

51. $\tan^{-1} t^3$

52. $(\tan^{-1} t)^3$

53. Show that $D_t(\sin^{-1} t + \cos^{-1} t) = 0$. Conclude that $\sin^{-1} t + \cos^{-1} t = \pi/2$ for $0 \le t \le \pi/2$.

Use the table of integrals to find:

54. $\displaystyle\int \frac{1}{\sqrt{4 - x^2}} \, dx$

55. $\displaystyle\int \sqrt{9 - x^2} \, dx$

56. $\displaystyle\int \frac{1}{x^2 + 25} \, dx$

57. $\displaystyle\int \frac{1}{(x^2 + 4)^2} \, dx$

58. $\displaystyle\int \frac{x^2}{x^2 + 1} \, dx$

59. $\displaystyle\int \frac{x^2}{\sqrt{2 - x^2}} \, dx$

part B

60. This exercise demonstrates that limit$_{t\to 0}$ (sin t)/$t = 1$ in the special case where $t \to 0$ through positive values.

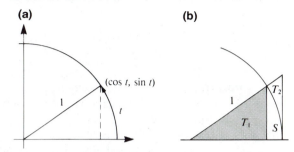

(a)   **(b)**

$(\cos t, \sin t)$

(a) Refer to the figures. Show that area $T_1 = \dfrac{1}{2}\sin t \cos t$ and area $T_2 = \dfrac{1}{2}\dfrac{\sin t}{\cos t}$.

(b) Show that area $S = t/2$.

(c) Conclude that $\sin t \cos t < t < \dfrac{\sin t}{\cos t}$ and so $\cos t < \dfrac{t}{\sin t} < \dfrac{1}{\cos t}$.

(d) Since limit$_{t\to 0}$ cos $t = 1$, conclude that limit$_{t\to 0}\dfrac{\sin t}{t} = 1$.

61. Find all relative maxima and minima of $f(t) = \sin (1/t)$. What does the graph look like?

triangle trigonometry with calculus applications

The word "trigonometry" is derived from the Greek for "triangle measurement". Many significant applications of the trigonometric functions depend on interpreting the variable θ (in "sin θ," for example) as an angle in a triangle.

An **angle** is by definition two rays with a common endpoint. One of the rays is designated as the **initial side** of the angle, while the other ray is called the **terminal side**. We might imagine that the initial side has been rotated to form the terminal side, as shown in figure 31.1. We will denote angles by Greek letters, such as α, β, and θ.

figure 31.1

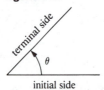

The standard way of measuring angles is in degrees. In calculus, however, it is much more convenient to use **radian measure**, as illustrated in figure 31.2.

figure 31.2(a)　　　　　　　　　　　　　**figure 31.2(b)**

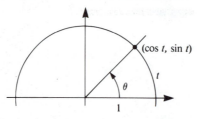　　　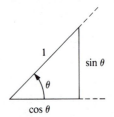

Using the unit circle, position the angle θ so that the initial side coincides with the positive x-axis. The *radian measure* of θ is defined to be the length t of the intercepted arc. In addition, we define

$$\sin\theta = \sin t \quad\text{and}\quad \cos\theta = \cos t$$

Shown in figure 31.3 are common angles with their corresponding radian measure. The relationship between radian and degree measure is given by the formulas:

$$x° = \frac{x}{180}\,\pi \text{ radians} \qquad x \text{ radians} = \left(\frac{x}{\pi}\,180\right)°$$

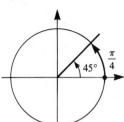

figure 31.3(a)

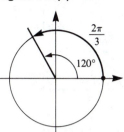

figure 31.3(b)

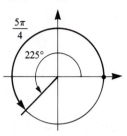

figure 31.3(c)

example 31.1

Convert 70° to radians, and 3 radians to degrees.

solution 31.1

We apply the conversion formulas to obtain $70° = \dfrac{70}{180}\pi = 1.22173\ldots$ radians and 3 radians $= \left(\dfrac{3}{\pi}\, 180\right)° = 171.887\ldots°$.

It is customary not to distinguish between an angle and its radian measure. Thus, we may refer to "the angle $\pi/3$" when technically we mean "an angle of radian measure $\pi/3$."

When α is an angle in a right triangle, the trigonometric functions can be computed by taking ratios of side lengths, as illustrated in figure 31.4.

figure 31.4

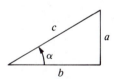

$$\sin\alpha = a/c = \frac{\text{opposite side}}{\text{hypotenuse}}$$

$$\cos\alpha = b/c = \frac{\text{side adjacent}}{\text{hypotenuse}}$$

$$\tan\alpha = a/b = \frac{\text{opposite side}}{\text{side adjacent}}$$

Referring to figure 31.5, we can verify these formulas by using the fact that $\triangle OAB$ is similar to $\triangle OCD$. Hence, the corresponding sides are proportional.

figure 31.5

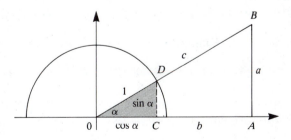

$$\sin\alpha = \frac{\sin\alpha}{1} = \frac{a}{c}$$

$$\cos\alpha = \frac{\cos\alpha}{1} = \frac{b}{c}$$

example 31.2

Determine all sides and angles in the right triangle shown in figure 31.6.

figure 31.6

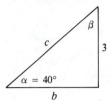

solution 31.2   Since $\alpha + \beta = 90°$, $\beta = 50°$. Furthermore, $\sin 40° = 3/c$. Hence, $c = 3/\sin 40° = 3/(.64278\ldots) = 4.667\ldots$. (Be sure your calculator is in *degree mode*.) Finally, $\sin 50° = b/c$, and hence $b = c \sin 50° = (4.667)(.76604\ldots) = 3.575\ldots$.

The remainder of this section is devoted to calculus applications involving triangle trigonometry.

example 31.3   **projectile motion**   A body s_0 feet off the ground is given an initial velocity *of V feet per second in the direction shown in figure 31.7(a). If we assume that gravity is the only force acting on the body*, then we can determine the position (x, y) of the body as a function of time t.

figure 31.7(a)   **figure 31.7(b)**

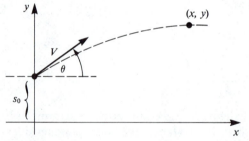

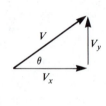

We must first resolve the velocity vector into its horizontal and vertical components: $\sin \theta = V_y/V$ and $\cos \theta = V_x/V$, as shown in figure 31.7**(b)**. Hence,

$$V_y = V \sin \theta \quad \text{and} \quad V_x = V \cos \theta$$

If we apply the Falling Object Law of section 18 with initial vertical velocity $v_0 = V \sin \theta$, we obtain

$$y = -16t^2 + (V \sin \theta)t + s_0$$

In addition, the body is traveling at a constant $V \cos \theta$ feet per second in the x-direction. Hence,

$$x = (V \cos \theta)t$$

Note that when y is expressed in terms of x [letting $t = x/(V \cos \theta)$], a quadratic function results. The actual path of the body is therefore a *parabola*.

example 31.4   Refer to the projectile motion situation described in example 31.3. Find the maximum height which the projectile attains as a function of the angle θ.

solution 31.4   The first derivative of y is $y' = -32t + V \sin \theta$. Setting $y' = 0$ yields $t = (V \sin \theta)/32$. Since $y'' = -32$, a maximum occurs at $t = (V \sin \theta)/32$. Finally, letting $t = (V \sin \theta)/32$ in the formula for y, we obtain the maximum height H:

$$H(\theta) = \frac{1}{64} V^2 \sin^2 \theta + s_0, \qquad 0 \le \theta \le \frac{\pi}{2}$$

example 31.5   Assuming $s_0 = 0$ in example 31.4, find the angle θ that maximizes the *horizontal distance* traveled.

solution 31.5   We have $y = -16t^2 + (V \sin \theta)t = t(-16t + V \sin \theta)$. Hence, the object strikes the ground when $t = (V \sin \theta)/16$. The total horizontal distance traveled is therefore given by

$$X(\theta) = (V \cos \theta) \frac{V \sin \theta}{16} = \frac{V^2}{16} \cos \theta \sin \theta$$

If we use the identity $\sin 2\theta = 2 \sin \theta \cos \theta$, then $X(\theta)$ can be rewritten as $(V^2/32) \sin 2\theta$. Hence, $X'(\theta) = (V^2/32)(2 \cos 2\theta) = 0$ when $2\theta = \pm \pi/2$, $\pm 3\pi/2, \ldots$. Thus, $\theta = \pm \pi/4, \pm 3\pi/4, \ldots$. Only $\theta = \pi/4$ is in the prescribed domain. Since $X''(\theta) = -V^2(\sin 2\theta)/8$, $X''(\pi/4) < 0$. Hence, the absolute maximum occurs at $\theta = \pi/4$. To summarize, the maximum horizontal distance of $V^2/32$ occurs when θ is $\pi/4$ or $45°$.

The equations we have developed can be used as first approximations to the motion of leaping animals, as indicated in examples 31.6 and 31.7.

example 31.6   A porpoise leaps from the water at angle of about $75°$. If the initial velocity is 35 feet per second, how high does the porpoise rise? How far does it travel in the horizontal direction during the leap?

figure 31.8

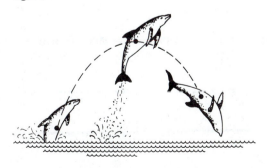

solution 31.6   From example 31.4, we know that the maximum height attained is $H(75°) = \frac{1}{64}(35)^2 \sin^2 75° = 17.9$ feet. From example 31.5, we conclude that the distance traveled is $X(75°) = \frac{1}{32}(35)^2 \sin [2(75)]° = 19.14$ feet.   ▤

example 31.7   The longest recorded frog jump is a single leap of 17.5 feet made in 1975 by "Ex Lax" at the famous Calaveras County Frog Jubilee. Assume that the frog leaped at the optimal angle of 45°, as shown in figure 31.9.

figure 31.9 Photographs of a jumping frog. Note that the take-off angle appears to be close to 45 degrees. (Adapted from, James Gray, *How animals move*. London: Cambridge University Press, 1960.)

(a) What was the initial velocity?
(b) How high off the ground did "Ex Lax" jump?

solution 31.7   From example 31.5, $V^2/32 = 17.5$ feet. Hence, $V = \sqrt{(17.5)(32)} = 23.66$ feet per second, or approximately 16 miles per hour. From example 31.4, the height off the ground is given by $H(45°) = \frac{1}{64}(23.66)^2 \sin^2 45° = 4.4$ feet.   ▤

The following two examples present optimization problems involving angles.

example 31.8   An animal photographer wishes to determine where she should position herself in order to obtain the "best view" of that portion of the tree shown in figure 31.10(a). By "best view," we mean the position x that maximizes the viewing angle θ. Let d be the length of the viewing area and let h be the vertical distance from the eye of the observer to the bottom of the viewing area.

Referring to the right triangles shown in figure 31.10(b), we see that $\tan \beta = h/x$ and $\tan \alpha = (d + h)/x$. Hence, $\theta = \alpha - \beta = \tan^{-1}(d + h)/x - \tan^{-1} h/x$.   This is the function we must maximize. The first derivative of θ is

$$\theta'(x) = \frac{-(d + h)/x^2}{1 + [(d + h)/x]^2} - \frac{-h/x^2}{1 + (h/x)^2} = \frac{-(d + h)}{x^2 + (d + h)^2} + \frac{h}{x^2 + h^2}$$

$$= \frac{-dx^2 + hd(d + h)}{[x^2 + (d + h)^2](x^2 + h^2)}$$

figure 31.10(a)

figure 31.10(b)

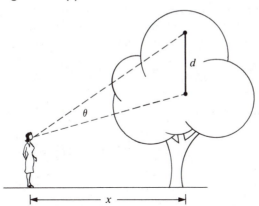

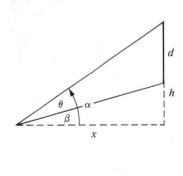

Hence, $\theta'(x) = 0$ implies $x^2 = h(d + h)$. Thus, $x = \sqrt{h(d + h)}$ is the sole critical point in the domain $x > 0$. By examining the numerator of $\theta'(x)$, it is easy to determine by the First Derivative Test that a relative maximum occurs at $x = \sqrt{h(d + h)}$.

As an illustration, if $h = 30$ feet and $d = 6$ feet, then $x = \sqrt{30(36)} = 32.9$ and the corresponding value of θ is 5.2°. Note that when d is small relative to h, x is approximately h.

example 31.9   **arterial bifurcation**   A common type of cardiovascular branching is *bifurcation*, where a large vessel of radius R splits into two smaller vessels of radius r is shown in figure 31.11. If, as in example 14.7, we assume that the optimal

figure 31.11

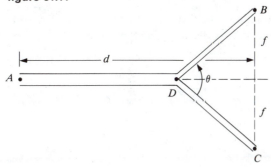

design is one that *minimizes resistance to flow*, then this total resistance is given by

$$F = \frac{kAD}{R^4} + \frac{kDB}{r^4} + \frac{kDC}{r^4}$$

(This is a form of Poiseuille's Law.)

We wish to determine the bifurcation angle θ that minimizes F. Note that $\sin \theta/2 = f/DB$. Hence, $DB = f \csc \theta/2$. Likewise, $DC = f \csc \theta/2$. In addition, $\tan \theta/2 = f/(d - AD)$. Therefore, $d - AD = f \operatorname{ctn} \theta/2$, and so $AD = d - f \operatorname{ctn} \theta/2$. The expression for total resistance can now be written:

$$F(\theta) = \frac{k}{R^4}\left[d - f \operatorname{ctn}\left(\frac{\theta}{2}\right)\right] + 2\frac{k}{r^4}f \csc \frac{\theta}{2}$$

The first derivative is

$$F'(\theta) = \frac{k}{R^4}\left[f\left(\csc^2 \frac{\theta}{2}\right)\frac{1}{2}\right] + 2\frac{k}{r^4}f\left(-\csc \frac{\theta}{2}\operatorname{ctn}\frac{\theta}{2}\right)\frac{1}{2}$$

Setting $F'(\theta) = 0$, we obtain

$$\frac{\csc^2 \theta/2}{2R^4} = \frac{\csc \theta/2 \operatorname{ctn} \theta/2}{r^4}$$

or

$$\frac{\operatorname{ctn} \theta/2}{\csc \theta/2} = \cos \frac{\theta}{2} = \frac{1}{2}\left(\frac{r}{R}\right)^4$$

Hence, $\theta/2 = \cos^{-1}\left[\frac{1}{2}\left(\frac{r}{R}\right)^4\right]$ or $\theta = 2\cos^{-1}\left[\frac{1}{2}\left(\frac{r}{R}\right)^4\right]$. Note that this optimal angle depends only on R and r, and *not* on the lengths of the arteries involved.

Other optimality considerations* lead to the conclusion that $r/R \approx .8$. Hence, $\theta \approx 2\cos^{-1}(.5(.8)^4) \approx 156°$. Thus, $\theta/2 \approx 78°$. These mathematical conclusions agree well with actual observations on small arteries.

example 31.10   A population of birds is driven from a small island one mile from the coastline. If θ is the angle shown in figure 31.12, then the hypothesis that the

figure 31.12(a)

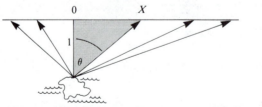

figure 31.12(b)

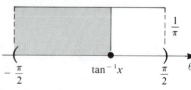

birds have dispersed at random is just the assumption that θ is uniformly distributed on $(-\pi/2, \pi/2)$. Let X be the position of a bird along the coast. How is X distributed? Compute $P(-1 < X < 1)$, the proportion of the population within one mile of the point 0.

* R. Rosen, *Optimality principles in biology*, London: Butterworth, (1967), pp. 48–52.

solution 31.10   From the figure, we see that $X = \tan \theta$. The distribution function is $F(x) = P(X \le x)$. But $X = \tan \theta \le x$ precisely when $\theta \le \tan^{-1} x$ and this occurs with probability $\dfrac{\tan^{-1} x - (\pi/2)}{\pi} = \dfrac{1}{\pi} \left(\tan^{-1} x + \dfrac{1}{2}\right)$. The density function is then $F'(x) = \dfrac{1}{\pi} \dfrac{1}{1+x^2}$. This the *Cauchy distribution*. Finally,

$$P(-1 < X < 1) = \int_{-1}^{1} \frac{1}{\pi} \cdot \frac{1}{1+x^2}\, dx = \frac{1}{\pi}[\tan^{-1} 1 - \tan^{-1}(-1)] = \frac{1}{2}$$

exercises for section 31

part A   The following angles have been measured in degrees. Find the radian measure for each.

1. $45°$ **2.** $60°$ **3.** $80°$ **4.** $135°$ **5.** $160°$

Convert each of the following radian measures to degrees.

6. $.5$ **7.** 2 **8.** $.25$ **9.** 1.5 **10.** 3.1

Determine all remaining sides and angles in each triangle.

11.
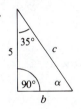

12.
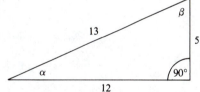

13.
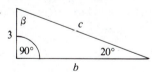

14.
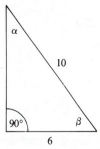

15.
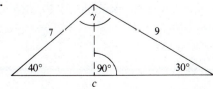

Use the formulas developed in examples 31.3–31.5 to solve each of the following in exercises 16–19.

16. A porpoise can leap 20 feet out of the water, leaving the water at an angle of 70°. Estimate the initial velocity out of the water in miles per hour.

17. The red kangaroo is able to make horizontal leaps of over 40 feet. Assume that the takeoff angle is 45°.

    (a) Estimate the maximum height attained during a 42-foot leap.

    (b) Estimate the takeoff velocity in miles per hour.

18. The German shepherd "Crumstone Danka" in 1942 scaled a wall 11 feet 3 inches high. From the pictures, it appears that the takeoff angle was about 90°. Estimate the vertical takeoff velocity.

★19. In 1977, Alain Jean Prieur recorded the longest distance yet achieved in motorcycle long jumping—212 feet. Assume that the ramp is similar to the one shown in the figure. Estimate the maximum height off the ground during the flight.

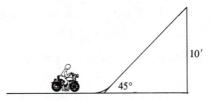

20. An animal photographer wishes to get the best view of a family of orangutans living in a portion of a tree 35–45 feet above eye level. Where should she position herself for the best view? [*Hint*: See example 31.8.]

21. A family of orangutans wishes to get the best view of a photographer living in the area shown in the figure. Where should the orangutans position themselves in the tree? [*Hint*: Rotate this page through 90°.]

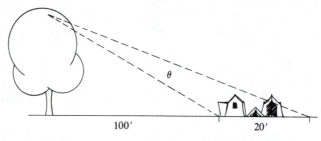

22. For the bird population in example 31.10, define X = position of a bird along the coast.

    (a) Compute $P(X > 1)$.

    (b) Compute $P(|X| < 2)$.

23. Refer to the figure. Show that the isosceles triangle with largest area for a given perimeter P_0 is an equilateral triangle.

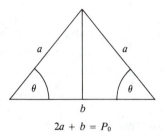

$$2a + b = P_0$$

24. A force F acts on a weight of W pounds in the direction θ shown in the figure. In order to move the weight, it is known that the force F must be at least

$$F = \frac{\mu W}{\cos \theta + \mu \sin \theta}$$

where μ is the coefficient of friction. Which direction θ results in minimizing the force F needed?

25. A herd of antelope is spotted by an expedition at a position 5 miles due south of their jeep. The antelope are running east at a speed of 25 mph, while the jeep, due to the rough terrain, can travel only 20 mph. The jeep will travel in a direction θ southeast until it is directly behind the herd, as shown in the figure.

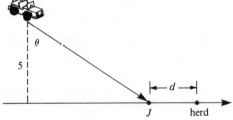

(a) Show that it takes $(\sec \theta)/4$ hours to travel to the point J.

(b) Show that $d = (25/4) \sec \theta - 5 \tan \theta$ and find θ so that d is minimized.

26. The initial velocity V obtained when jumping after a running start may depend on the actual takeoff angle θ. Assume that

$$V(\theta) = A \cos \omega\theta, \qquad 0 < \theta < \frac{\pi}{2}$$

where $0 < \omega \leq 1$.

(a) Find the maximum height attained as a function of θ.

(b) Assuming that $s_0 = 0$, show that the angle θ that maximizes the horizontal distance traveled satisfies

$$\tan \theta \tan \omega\theta = 1/\omega$$

(c) For $\omega = 1$, show that $\theta = 45°$.

27. A cylinder of fixed radius r is to contain a specified volume V_0, but will be sealed* off by adding a cone as shown in the figure.

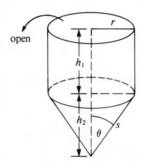

$h_1 =$ height of the cylinder
$h_2 =$ height of the cone
$S =$ total surface area excluding the open top
$s =$ slant height of the cone

(a) Show that $V_0 = \pi r^2(h_1 + h_2/3)$ and $h_2 = r \operatorname{ctn} \theta$.

(b) Show that the total surface area S is given by

$$S = \frac{2V_0}{r} + \pi r^2\left(\csc \theta - \frac{2}{3}\operatorname{ctn} \theta\right)$$

(c) Show that S is minimized when $\theta = \cos^{-1}\frac{2}{3} \approx 48.2°$.

---

* This *sealing problem* is simllar to the famous honeycomb-structure problem for bees. For more details (and a good project), see E. Batschelet, *Introduction to mathematics for life scientists.* Springer-Verlag New York: pages 233–235.

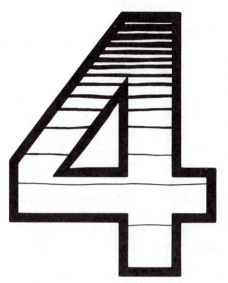

an introduction to differential equations and modeling

prologue   We have already encountered a few **differential equations** in our prior work. The differential equation $y' = ky$ arose in section 17 in discussing exponential growth and decay. There we showed that the only functions $y = f(t)$ that satisfy $f'(t) = kf(t)$ are functions of the form ce^{kt} with c an arbitrary constant. The function $y = ce^{kt}$ is called the *general solution* of the differential equation.

A **first order differential equation** is an equation involving y, y', and t. Each of the following is a first order differential equation:

$$y' = y^2 + t$$
$$t^2(y')^2 - ty = 4$$

A solution to the first equation is therefore a function $y = f(t)$ with $f'(t) = [f(t)]^2 + t$. A function $y = g(t)$ satisfies the second differential equation if $t^2[g'(t)]^2 - tg(t) = 4$.

A **second order differential equation** involves y, y', y'', and t. Thus,

$$y'' + 3y' + 2y = t$$

and
$$y'y'' + 3y + 4t = 0$$

are examples of second order differential equations. In section 30, we discovered that both $y = \cos \omega t$ and $y = \sin \omega t$ are solutions to $y'' = -\omega^2 y$.

In the next few sections, we will study techniques for finding *all functions* $y = f(t)$ that satisfy a given first or second order differential equation.

Mathematical models of physical or biological systems are constructed with particular goals in mind. For example, we may wish to understand the mechanisms responsible for cell growth, or we may wish to predict radioactivity levels in an ecosystem years in the future. We must first try to determine what variables x, y, t, . . . are responsible for changing the system. We may first decide to incorporate into the model only a few of the variables. In this step we are specifying the **level of resolution** of the model.

A model is a collection of hypotheses or reasonable assumptions about the mechanisms for change in the system. As we will show, these assumptions, when expressed mathematically, frequently lead to differential equations involving the original variables. Having solved the differential equations, we then see if the model's predictions are consistent with experimental data or known facts about the behavior of the system. If predictions are poor, we may either increase the level of resolution of the model or make alternate assumptions about the mechanisms for change. The steps in the modeling process are then repeated, as depicted in figure 1.

figure 1

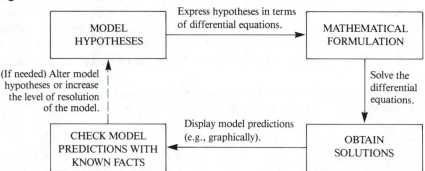

High resolution models typically lead to collections of differential equations that require the computer for their solution. However, we will study several of the **low resolution models** that have been found to be useful in the biosciences.

 implicit differentiation
and related rates

We have developed a large number of special methods for finding the derivative $\frac{dy}{dt}$ when two variables y and t are related by an *explicit formula* $y = f(t)$. Often, however, the relationship between the variables will be hidden in an equation such as $y^3 + y^2t - t^3y = 3$. It may be difficult or even impossible to solve the equation for y in terms of t. Nevertheless, there is a special method called **implicit differentiation** by which $\frac{dy}{dt}$ can be computed *without first solving* for y in terms of t.

As a first example, suppose $y^3 + t^2 = 1$. In this instance, we could first solve for y:

$$y = \sqrt[3]{1 - t^2} = (1 - t^2)^{1/3}$$

and then take the derivative:

$$\frac{dy}{dt} = \frac{1}{3}(1 - t^2)^{-2/3} \cdot (-2t)$$

The alternate implicit differentiation method is based on the Chain Rule formula:

$$D_t y^n = ny^{n-1} \cdot \frac{dy}{dt}$$

Differentiating both sides of $y^3 + t^2 = 1$ *with respect to t*, we obtain

$$3y^2 \frac{dy}{dt} + 2t = 0 \qquad \text{or} \qquad \frac{dy}{dt} = \frac{-2t}{3y^2}$$

Note that the derivative formula $\frac{dy}{dt}$ contains both variables y and t, rather than the single variable t. In most applied problems, however, this is not a serious drawback.

If $y = g(t)$, then $f(y) = f(g(t))$. The Chain Rule asserts that $D_t f(y) = f'(g(t))g'(t) = f(y)\frac{dy}{dt}$. To summarize:

$$D_t f(y) = f'(y)\frac{dy}{dt}$$

347

This is the cornerstone of the implicit differentiation method: *Take the usual derivative and then multiply by* $\dfrac{dy}{dt}$. *The method is illustrated in our next example.*

example 32.1   Find $\dfrac{dy}{dt}$ if y and t are related by:

$$\textbf{(a)} \ \sin y + 2t^2 = y \qquad \textbf{(b)} \ ty^2 - 3y = 4 \qquad \textbf{(c)} \ ye^{ty} = 1$$

solution 32.1   **(a)** Differentiating both sides with respect to t, we have $\cos y \cdot \dfrac{dy}{dt} + 4t = \dfrac{dy}{dt}$.

Solving for $\dfrac{dy}{dt}$, we have $\dfrac{dy}{dt} = \dfrac{4t}{(1 - \cos y)}$.

(b) To handle the term ty^2, we must use the Product Rule: $D_t ty^2 = t\left(2y\dfrac{dy}{dt}\right) + y^2(1)$. From the equation $ty^2 - 3y = 4$, we obtain

$$2ty\frac{dy}{dt} + y^2 - 3\frac{dy}{dt} = 0$$

Hence,

$$\frac{dy}{dt}(2ty - 3) = -y^2 \qquad \text{or} \qquad \frac{dy}{dt} = \frac{-y^2}{2ty - 3}$$

(c) First note that $D_t ye^{ty} = yD_t e^{ty} + e^{ty}\dfrac{dy}{dt}$ by the Product Rule. But the Chain Rule gives

$$D_t e^{ty} = e^{ty}D_t(ty) = e^{ty}\left(t\frac{dy}{dt} + y \cdot 1\right)$$

Differentiating both sides of $ye^{ty} = 1$ with respect to t yields

$$ye^{ty}\left(t\frac{dy}{dt} + y\right) + e^{ty}\frac{dy}{dt} = 0$$

Hence,

$$\frac{dy}{dt}(tye^{ty} + e^{ty}) = -y^2 e^{ty}$$

Solving for $\dfrac{dy}{dt}$ and simplifying, we have $\dfrac{dy}{dt} = \dfrac{-y^2}{(1 + ty)}$.

When the equations involve two or more variables that depend on t, the technique is essentially the same, as shown in the following example.

example 32.2   Find the relationship between $\dfrac{dx}{dt}$ and $\dfrac{dy}{dt}$ if x, y, and t are related by the equation:

$$\textbf{(a)}\ \ y = 4x^3 \qquad \textbf{(b)}\ \ x^2 + y^2 = 4 + t^2$$

solution 32.2   For **(a)**, we differentiate both sides of $y = 4x^3$ with respect to t to obtain $\dfrac{dy}{dt} = 12x^2 \dfrac{dx}{dt}$.

For **(b)**, we have $2x\dfrac{dx}{dt} + 2y\dfrac{dy}{dt} = 2t$. If we were to solve for $\dfrac{dy}{dt}$ we would have

$$\frac{dy}{dt} = \frac{\left(2t - 2x\dfrac{dx}{dt}\right)}{2y}$$

The remainder of this section is devoted to applied problems that use this special derivative technique. Recall that $\dfrac{dy}{dt}$ may be interpreted as the *instantaneous rate of change* of y with respect to t. The following rules, each reformulations or special cases of the Chain Rule, are very useful in working with problems that involve rates of change:

$$\frac{dy}{dt} = \frac{dy}{dx} \cdot \frac{dx}{dt} \qquad \text{or} \qquad \frac{dy}{dx} = \frac{dy/dt}{dx/dt}$$

$$\frac{dx}{dy} = 1 \Big/ \frac{dy}{dx}$$

Note that if we were to formally manipulate these expressions *as if they were true fractions*, then the above formulas would result. This is the beauty of the $\dfrac{dy}{dt}$ notation for derivative. As the next example shows, implicit differentiation is a useful tool in solving *related rates* problems.

example 32.3   A spherical balloon is being inflated at the rate of 20 cubic inches per second. At what rate is the radius increasing when the diameter of the balloon is 10 inches?

solution 32.3   Let V denote the volume of the balloon, and let r be the radius. We are given that $\dfrac{dV}{dt} = 20$ and are asked to find $\dfrac{dr}{dt}$ when the diameter is 10—i.e., when

$r = 5$. Since $V = \frac{4}{3}\pi r^3$, it follows that

$$20 = \frac{dV}{dt} = 4\pi r^2 \frac{dr}{dt}$$

Setting $r = 5$, we have $\dfrac{dr}{dt} = \dfrac{1}{(5\pi)} = .0636$ inch per second.   ▤

To solve a **related rates** problem such as example 32.3, the steps in the accompanying box may prove useful.

---

procedure for solving related rates problems

Step 1. Write down each of the given rates as derivatives.

Step 2. Find the relationship between the variables involved, but *do not substitute* the particular numerical values given.

Step 3. Use implicit differentiation and solve for the unknown rate.

Step 4. Use the numerical information in Step 1 and substitute into the rate equation from Step 3.

This procedure is illustrated in examples 32.4–32.6.

example 32.4   After takeoff, an airplane maintains a velocity of 200 feet per second in a direction 30° from ground level, as shown in figure 32.1. How quickly is the plane rising vertically?

figure 32.1

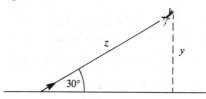

solution 32.4   We are given that $\dfrac{dz}{dt} = 200$ and wish to find $\dfrac{dy}{dt}$. Since $\sin 30° = \dfrac{y}{z}$, we have $y = z \sin 30°$. Hence, $\dfrac{dy}{dt} = (\sin 30°)\dfrac{dz}{dt} = 200 \sin 30° = 100$ feet per second.   ▤

example 32.5   A lighthouse is located 1 mile offshore. Its beacon light rotates once every 4 seconds. How fast is the light moving along the shore at a point 2 miles directly down the coastline?

figure 32.2

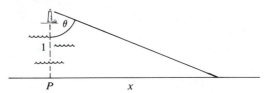

solution 32.5

We are given that the light rotates 90° per second. To apply calculus techniques however, we must measure θ in radians. Hence, we are given that $\dfrac{d\theta}{dt} = \dfrac{\pi}{2}$ radians per second, and we are asked to find $\dfrac{dx}{dt}$ when $x = 2$. From the triangle shown in figure 32.2, we see that $x = \tan \theta$. Hence, $\dfrac{dx}{dt} = \sec^2 \theta \left(\dfrac{d\theta}{dt}\right) = (\sec^2 \theta)\left(\dfrac{\pi}{2}\right)$ miles per second. When $x = 2$, the dimensions of the corresponding triangle are as shown in figure 32.3. Hence, $\sec \theta = \sqrt{5}$ and so when $x = 2$, $\dfrac{dx}{dt} = \dfrac{5\pi}{2} = 7.85$ miles per second.

figure 32.3

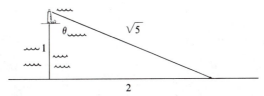

example 32.6

When fat deposits collect on the walls of an artery, the effective radius of the artery is reduced and the heart is forced to pump harder to maintain the needed flow rate. The administration of various drugs can increase the radius of the vessel. Use Poiseuille's Law (see section 21, page 234) to find a relationship between $\dfrac{dF}{dt}$ and $\dfrac{dR}{dt}$.

solution 32.6

Since $F = \dfrac{k\pi}{2} R^4$, we have $\dfrac{dF}{dt} = \dfrac{k\pi}{2} 4R^3 \dfrac{dR}{dt} = 2k\pi R^3 \dfrac{dR}{dt}$. If we divide both sides of the last equation by F, we obtain:

$$\frac{1}{F}\frac{dF}{dt} = \frac{1}{F} 2k\pi R^3 \frac{dR}{dt} = \left(\frac{2}{k\pi}\frac{1}{R^4} 2k\pi R^3\right)\frac{dR}{dt} = 4\frac{1}{R}\frac{dR}{dt}$$

Now, both $\dfrac{1}{R}\dfrac{dR}{dt}$ and $\dfrac{1}{F}\dfrac{dF}{dt}$ are percentage rates of increase. When we divide

$\dfrac{dR}{dt}$ by R at a particular time t, we are simply comparing the rate of increase to the present radius. Thus, if the percentage rate of increase of R is 10%, then the equation $\dfrac{1}{F}\dfrac{dF}{dt} = 4\left(\dfrac{1}{R}\dfrac{dR}{dt}\right)$ tells us that the percentage increase in flow rate would be 40%. ▤

example 32.7   **allometric growth**   A large amount of empirical data has been collected showing that if y and x are measurements of parts of an organism (e.g., length and weight, or limb length and organ weight), then y and x are related by a function of the form $y = \alpha x^k$ for some $\alpha > 0$ and $k > 0$. This is known as an *allometric law.*

The respective rates of growth are $\dfrac{dy}{dt}$ and $\dfrac{dx}{dt}$. The units used to measure $\dfrac{dy}{dt}$ and $\dfrac{dx}{dt}$ may not be comparable. If we form the percentage rates of growth $\dfrac{1}{x}\dfrac{dx}{dt}$ and $\dfrac{1}{y}\dfrac{dy}{dt}$, then the units for each are $(\text{time})^{-1}$. Implicit differentiation can be used to establish a relationship between these rates of change:

$$\frac{dy}{dt} = \alpha k x^{k-1}\frac{dx}{dt}$$

Dividing both sides by $y = \alpha x^k$ yields

$$\frac{1}{y}\frac{dy}{dt} = \frac{1}{\alpha x^k}\,\alpha k x^{k-1}\frac{dx}{dt} = k\left(\frac{1}{x}\frac{dx}{dt}\right)$$

The two percentage rates of growth are therefore directly proportional.

The allometric law $y = \alpha x^3$ is a common length-weight relationship for many varieties of fish. A 10% rate of increase for the length x would translate into a 30% rate of increase for the weight y.

In section 33, we will show that the steps in example 32.7 can be reversed—i.e., if the percentage rates of growth are directly proportional, then an allometric law will always result.

example 32.8   **measuring blood flow to a limb**   In this example, we will use implicit differentiation to derive a formula that is the basis of a physiological measurement technique known as **strain-gauge plethysmography**. This is a commonly employed method for measuring the rate of arterial blood flow to limbs such as fingers, arms, and legs. Our model of a limb is a right circular cylinder of radius r and length L, as illustrated in figure 32.4. Suppose that the cylinder is constrained in such a way that only its radius r can increase when the volume V

figure 32.4

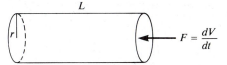

of the cylinder is increased. Hence, the length L of the cylinder is assumed to be constant.

Since $V = \pi r^2 L$, we have $\dfrac{dV}{dt} = 2\pi r L \dfrac{dr}{dt}$. We differentiate the expression $C = 2\pi r$ for circumference to obtain $\dfrac{dC}{dt} = 2\pi \dfrac{dr}{dt}$. Substituting for $\dfrac{dr}{dt}$ in the first equation yields

$$\frac{dV}{dt} = rL\left(2\pi \frac{dr}{dt}\right) = rL\frac{dC}{dt}$$

If we divide both sides of this equation by $V = \pi r^2 L$, we obtain

$$\frac{1}{V}\frac{dV}{dt} = \frac{1}{\pi r^2 L}\left(rL\frac{dC}{dt}\right) = \frac{1}{\pi r}\frac{dC}{dt} = \frac{2}{2\pi r}\frac{dC}{dt}$$

Solving for $\dfrac{dV}{dt}$, we have

$$\frac{dV}{dt} = V\left(\frac{2}{C}\frac{dC}{dt}\right)$$

To implement the formula above, we use the fact that there is a large difference in blood pressure between the arterial system and the venous system. The cuffs shown in figure 32.5 can therefore be adjusted so that arterial

figure 32.5

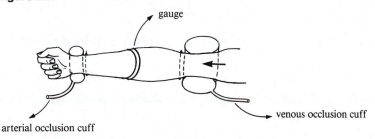

blood still flows to the limb, but so that no blood is returned from the limb to the heart. As a result, the limb will begin to swell. A mercury strain gauge is placed around the limb, as illustrated in figure 32.6. This sensitive device consists of thin mercury-filled latex or silicone tubing. Any change in the circumference of the

figures 32.6 **(a)** and **(b)** show a method of mounting a gauge on forearm and calf. **(c)** Illustrations showing two instruments with mercury-filled tubes applicable to the finger*.

(a)   **(b)**   **(c)**

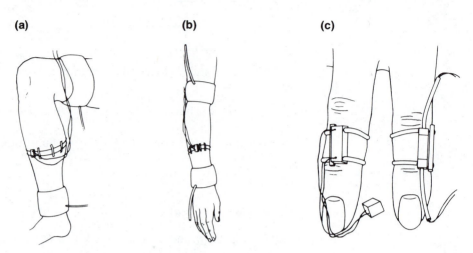

limb lengthens the tubing and produces a change in the length of the column of mercury inside. Thus, the device can be used to measure how quickly the circumference increases when the limb begins to swell.

If we use the formula $\dfrac{dV}{dt} = V\left(\dfrac{2}{C}\dfrac{dC}{dt}\right)$ to compute the flow rate at $t = 0$, we need measure only the normal limb circumference, the normal limb length, and $\dfrac{C(\Delta t) - C(0)}{\Delta t}$ $\left(\text{to estimate } \dfrac{dC}{dt}\right)$. Shown in figure 32.7 is typical output from the

figure 32.7

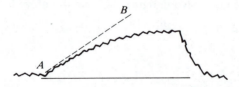

device. The slope of line AB estimates $\dfrac{dC}{dt}$ at $t = 0$. This technique is simple, painless, inexpensive, and remarkably exact.** A disadvantage is that it measures flow to *all tissues* in the limb.

Implicit differentiation can also be used to solve optimization problems that otherwise would be unwieldy. We first illustrate the method with a simple example.

* Adapted from *Journal of Physiology*, 121: 1–27 Cambridge University Press (1953).
** Whitney, *J. Physiology*, *121* (1953), pp. 1–27 and *125* (1954), pp. 1–24.

example 32.9   Find the minimum value of $z = x^2 + y^2$ subject to the constraints $xy = 4$ and $x > 0$.

solution 32.9   We will compute $\dfrac{dz}{dx}$ by implicit differentiation and then set $\dfrac{dz}{dx} = 0$:

$$\frac{dz}{dx} = 2x + 2y\frac{dy}{dx} = 0.$$

Hence, at a critical point, $\dfrac{dy}{dx} = -\dfrac{x}{y}$. Another formula for $\dfrac{dy}{dx}$ can be obtained from the constraints. Since $xy = 4$, we have $x\dfrac{dy}{dx} + y \cdot 1 = 0$, which implies $\dfrac{dy}{dx} = -\dfrac{y}{x}$. Equating the two expressions for $\dfrac{dy}{dx}$, we must have $-\dfrac{x}{y} = -\dfrac{y}{x}$ at a critical point. Hence, $x = y$ since $x > 0$ and $y > 0$. Since $xy = 4$, we conclude that $x = y = 2$, and $z = 8$.   ▤

Technically, in example 32.9, we should have verified that $x = 2$ actually gives a relative minimum. Often, however, we are convinced that a minimum or maximum value exists from the physical nature of the problem, as demonstrated in the following example.

example 32.10   **Snell's Law**   Snell's Law determines how a light ray will bend in passing from one medium (such as air) into a second medium (such as water), as shown in figure 32.8. If v_1 is the velocity of light in the first medium and v_2 is the

figure 32.8

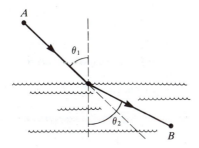

velocity in the second medium, then Snell's Law asserts that

$$\frac{v_2}{v_1} = \frac{\sin \theta_1}{\sin \theta_2}$$

We now give a derivation of this law based on the *minimum principle,* which states that light follows the path from A to B that minimizes the time of travel. From figure 32.9, it can be seen that $\cos \theta_1 = a/AO$. Hence, $AO =$

figure 32.9

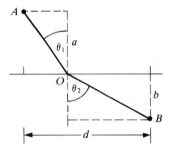

$a \sec \theta_1$. Likewise, $BO = b \sec \theta_2$. The total travel time T is therefore given by

$$T = \frac{AO}{v_1} + \frac{BO}{v_2} = \frac{a}{v_1} \sec \theta_1 + \frac{b}{v_2} \sec \theta_2$$

In addition, $d = a \tan \theta_1 + b \tan \theta_2$. This serves as the constraint. Differentiating both sides with respect to θ_1, we obtain

$$0 = a \sec^2 \theta_1 + b \sec^2 \theta_2 \frac{d\theta_2}{d\theta_1}$$

Hence, $\dfrac{d\theta_2}{d\theta_1} = -\dfrac{a \sec^2 \theta_1}{b \sec^2 \theta_2}$. This formula holds for *all* values of θ_1. At a critical value of θ_1,

$$0 = \frac{dT}{d\theta_1} = \frac{a}{v_1} \sec \theta_1 \tan \theta_1 + \frac{b}{v_2} \sec \theta_2 \tan \theta_2 \frac{d\theta_2}{d\theta_1}$$

Hence,

$$\frac{d\theta_2}{d\theta_1} = -\frac{a \, v_2 \sec \theta_1 \tan \theta_1}{b \, v_1 \sec \theta_2 \tan \theta_2}$$

If we equate the two expressions for $d\theta_2/d\theta_1$, then it follows that, at a critical point,

$$\frac{v_2 \tan \theta_1}{v_1 \tan \theta_2} = \frac{\sec \theta_1}{\sec \theta_2}$$

Since $\dfrac{\sec \alpha}{\tan \alpha} = \dfrac{1}{\sin \alpha}$, the above expression simplifies to

$$\frac{v_2}{v_1} = \frac{\sin \theta_1}{\sin \theta_2}$$

exercises for section 32

part A

In exercises 1–10, x and y are functions of t. Compute:

1. $D_t y^2$ **2.** $D_t x^3$ **3.** $D_t(ty)$ **4.** $D_t \dfrac{x}{t}$

5. $D_t e^x$ **6.** $D_t \sin y$ **7.** $D_t(xy)$ **8.** $D_t(t^2 y^2)$

9. $D_t \ln(tx)$ **10.** $D_t(x^2 + y^2)$

Find $\dfrac{dy}{dt}$ using implicit differentiation if y and t are related by:

11. $t^2 + y^2 = 4$ **12.** $t^3 - 3y^2 = 1$ **13.** $t^2 y + y = 1$

14. $ty^2 + y^3 = 2t$ **15.** $\sin ty = t$ **16.** $\cos ty = y$

17. $te^y = y$ **18.** $t^2 + ty = y$ **19.** $y^3 - ty + 4 = 0$

Find the relationship between $\dfrac{dx}{dt}$ and $\dfrac{dy}{dt}$ if x and y are related by:

20. $x^2 + y^3 = t^2$ **21.** $y = 5x^4$ **22.** $x = 2y^3$

23. $xy = 4$ **24.** $tx^2 = y$ **★25.** $\sin(xy) = t$

26. Given that $\dfrac{dx}{dt} = 3$ and $\dfrac{dy}{dx} = 2$ at some time t, compute $\dfrac{dy}{dt}$.

27. Given that $\dfrac{dx}{dt} = 4$ and $\dfrac{dy}{dt} = 8$ at some time t, compute $\dfrac{dx}{dy}$.

28. A spherical balloon is being inflated at a rate of 20 cubic inches per second. How fast is the diameter of this balloon increasing when the diameter is 10 inches?

29. Oil is leaking from a large tanker at the rate of 5 cubic feet per second. It then spreads out, forming a circular area of thickness 1 inch. Find an expression for dA/dt, the rate at which the area A is increasing, and find dr/dt, the rate at which the radius is increasing.

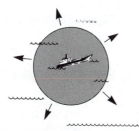

30. A small boy is flying a kite and letting out string at the rate of 2 feet per second. The kite rises at a constant angle of 70°, as shown in the figure. How quickly is the kite rising vertically?

31. A conical tank (see figure) has height 10 feet and radius 4 feet. Water is being pumped into the tank at the rate of 5 cubic feet per minute. How fast is the water level rising when the water level is 5 feet? [*Hint*: Show that $r/h = 2/5$.]

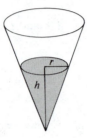

32. Circle City is increasing at the steady rate of 10,000 residents per year. City fathers have required that the population density remain 500 residents per square mile. Consequently, the city continues to expand outward, forming a circle.

(a) How fast is the area of the city growing?

(b) How fast is the distance from the center of town increasing when the population is 50,000?

33. An airplane flies 100 miles per hour at an altitude of 2 miles. Its position is tracked by a telescope as shown in the figure. Find the angular velocity $\dfrac{d\theta}{dt}$ in terms of θ.

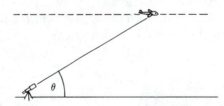

34. The relationship between the weight H of a rat's heart and its body weight W is given by a relationship of the form

$$H = cW^{.7}$$

Compare the two specific rates of growth $\dfrac{1}{H}\dfrac{dH}{dt}$ and $\dfrac{1}{W}\dfrac{dW}{dt}$.

35. The formula relating the length L (in feet) and the weight W (in metric tons) of sperm whales is

$$W = .000137L^{3.18}$$

Compare the two specific rates of growth $\dfrac{1}{L}\dfrac{dL}{dt}$ and $\dfrac{1}{W}\dfrac{dW}{dt}$.

36. The height h of a tree is related to its diameter d by a function of the form

$$h = cd^{2/3}$$

Compare the two specific rates of growth.*

Use the strain-gauge plethysmography formula in example 32.8 to estimate the arterial blood flow rates to the limbs described below:

37. The normal length is 10 cm, the normal circumference is 6.1 cm, and the swollen circumference after 2 seconds is 6.3 cm.

38. The normal limb is 51 cm, the normal circumference is 22 cm, and the swollen circumference after 2 seconds is 23.8 cm.

Use the implicit differentiation method presented in example 32.9 to solve each of the following optimization problems.

39. Maximize $z = x^2y$ subject to the constraint $x^2 + y^2 = 1$.

40. Minimize $z = x^3 + y^2$ subject to the constraints $xy = 4$ and $x > 0$.

41. Maximize $z = xy$ subject to the constraint $x + y = 1$.

42. Find the area of the largest rectangle that can be placed inside a circle of radius 1.

part B   **a model for the flight of homing pigeons**

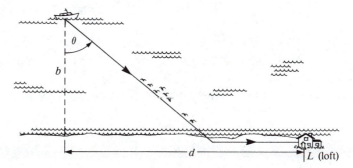

A homing pigeon is released from a boat at point B, which is b units (e.g., miles) from a fairly straight shoreline, as illustrated in the figure. The loft is d units downshore at point L. Due to the extra moisture in the air over the water, the pigeon must expend more energy to fly one unit over air than over land.

* T. McMahon, "Size and Shape in Biology", *Science 179* (1979), p. 1201.

Let e_1 be the number of calories used to fly one unit over water, and let e_2 be the similar quantity over land. Then $e_1 > e_2$.

43. Show that the total energy required to fly from B to L is given by $E = e_1 b \sec \theta + e_2(d - b \tan \theta)$. If v_1 is the pigeon's velocity over water and v_2 is the velocity over land, find the expression for the total time T it takes to fly from B to L.

44. Show that E is minimized by choosing $\theta = \sin^{-1}(e_2/e_1)$.

discussion: If the flight from B to L is long, it seems plausible that the pigeon may fly in a manner that minimizes its energy output. A nice feature of the minimal angle is that it does not depend on the distances involved, but only on the energy ratio e_2/e_1. One would seriously doubt the applicability of this model if the angle depended on the bird's ability to judge distances. It is much more reasonable that the pigeon can learn a single angle θ that applies regardless of the distances. The value of θ would be an invariant property of the species.

 separation of variables

One of the most widely used methods for solving a differential equation $y' = f(t, y)$ is the technique known as **separation of variables**. The technique is appropriate when the original differential equation can be rearranged into the form:

$$h(y)y' = g(t)$$

If H is an antiderivative for the function h, then $D_tH(y) = H'(y)y' = h(y)y'$ by implicit differentiation. The differential equation can then be rewritten $D_tH(y) = g(t)$. Hence, $H(y) = \int g(t)\, dt + c = G(t) + c$. We then (if possible) solve this new equation for y in terms of t.

There is, however, an easy way of implementing the above discussion based on treating $y' = \dfrac{dy}{dt}$ as if it were a fraction. The method is described in the accompanying box.

---

the separation of variables method

Step 1. Write the differential equation as

$$h(y)\frac{dy}{dt} = g(t)$$

Hence, $h(y)\, dy = g(t)\, dt$.

Step 2. Take antiderivatives of both sides to obtain

$$\int h(y)\, dy = \int g(t)\, dt$$

Step 3. Therefore, conclude $H(y) = G(t) + c$.

The separation of variables method is illustrated in the following three examples.

example 33.1   Solve the differential equation $y' = t/y^2$ for $y \neq 0$.

solution 33.1   Separating the variables yields $y^2 \dfrac{dy}{dt} = t$ or $y^2\, dy = t\, dt$. Hence, $\int y^2\, dy = \int t\, dt + c$, or $y^3/3 = t^2/2 + c$. Solving for y, we obtain

$$y = \sqrt[3]{3t^2/2 + 3c} = \sqrt[3]{3t^2/2 + c_1}$$

where $c_1 = 3c$ is an arbitrary constant.   ▤

warning!   Before proceeding to the next example, we warn the reader that the formula $\int \dfrac{1}{x}\, dx = \ln x$ holds only for $x > 0$. The more general formula is

$$\int \frac{1}{x}\, dx = \ln |x| \qquad \text{for } x \neq 0$$

Unless the variables are known to be positive, use this new formula.

example 33.2   Solve the differential equation $y' = ky$ of section 17 using the separation of variables method.

solution 33.2   We have $\dfrac{dy}{dt} = ky$ and so $\int \dfrac{1}{y}\, dy = k\, dt$. Hence, $\int \dfrac{1}{y}\, dy = \int k\, dt$, or $\ln |y| = kt + c$. Raising both sides to the power e, we obtain

$$|y| = e^{kt+c} = e^c e^{kt} = c_1 e^{kt}$$

where $c_1 = e^c > 0$. Removing absolute values yields $y = \pm c_1 e^{kt} = c_2 e^{kt}$. The constant c_2 can now take on any nonzero value. In the very first step, we implicitly assumed $y \neq 0$ in order to rearrange the equation. Note that $y \equiv 0$ is also a solution, because $y' = 0 = k(0)$. Hence, the general solution can be written as $y = c e^{kt}$, where c can take on *any real value*, including 0.   ▤

example 33.3   Solve the differential equation $y' = -x/y$ subject to the boundary condition $y(2) = 1$.

solution 33.3   We rewrite the differential equation as $y \dfrac{dy}{dx} = -x$ or $y\, dy = -x\, dx$. Hence, $\int y\, dy = \int -x\, dx$, or $y^2/2 = -x^2/2 + c$. Thus, $x^2 + y^2 = 2c$. Since $y \neq 0$ in the differential equation, $2c > 0$. Let $c_1^2 = 2c$. Then $x^2 + y^2 = c_1^2$ and thus the solutions are circles with centers at $(0, 0)$, as shown in figure 33.1. Since $y(2) = 1$, the solution passes through the point $(2, 1)$. Hence, $2^2 + 1^2 = c_1^2$ and so the solution is the circle $x^2 + y^2 = 5$.

figure 33.1

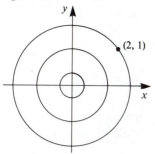

When rearranging the original differential equation into the form $h(y)y' = g(t)$, we can "lose" solutions. It is important to keep track of the additional assumptions that are needed in order to do the algebraic manipulations, as demonstrated in the next example.

example 33.4   Find *all* solutions to the differential equation $\dfrac{dy}{dt} = -y + 1$.

solution 33.4   We have $\dfrac{dy}{dt} = -(y - 1)$ and thus *for* $y \neq 1$, $\dfrac{1}{y-1}\, dy = -dt$. Taking antiderivatives of both sides yields

$$\ln|y - 1| = -t + c \qquad \text{or} \qquad |y - 1| = e^c e^{-t}$$

By removing absolute values, we obtain $y - 1 = \pm e^c e^{-t}$ or $y = 1 + c_1 e^{-t}$, where c_1 can take on any nonzero value. The solutions are shown in figure 33.2. In manipulating the original differential equation, we lost a solution. Note that $y \equiv 1$ is an additional solution: $y' = 0 = 1 - 1$. The general solution can then be written $y = 1 + c_1 e^{-t}$, where c_1 is an arbitrary constant.

figure 33.2

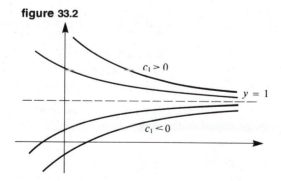

In the next example, we will use the integral tables to help us with the antiderivatives.

example 33.5   Find *all* solutions to the differential equation $y' = y(10 - y)$. Then find the particular solution that satisfies $y(0) = 5$.

solution 33.5   For $y \neq 0$ *and* $y \neq 10$, the equation can be rearranged as follows:

$$\frac{1}{y(10 - y)} \frac{dy}{dt} = 1 \quad \text{or} \quad \frac{1}{y(10 - y)} dy = dt$$

If we apply antiderivative formula 78 in the integral tables (pp. 261–265) with $a = 1$, $b = 0$, $p = -1$, and $q = 10$, we obtain

$$\frac{-1}{10} \ln \left| \frac{10 - y}{y} \right| = t + c$$

Hence, $\ln |(10 - y)/y| = -10t + 10c$ or $|(10/y) - 1| = e^{-10t} e^{10c}$. Removing absolute value signs yields

$$\frac{10}{y} - 1 = \pm e^{10c} e^{-10t} = c_1 e^{-10t}$$

where c_1 is any nonzero constant. Solving for y, we obtain $y = 10/(1 + c_1 e^{-10t})$. The two additional solutions lost in the manipulations are $y \equiv 0$ and $y \equiv 10$.

    If we impose the boundary condition $y(0) = 5$, then $5 = 10/(1 + c_1 e^0)$ or $1 + c_1 = 2$. Hence, $c_1 = 1$ and so $y = 10/(1 + e^{-10t})$. ▤

    The remainder of this section is devoted to mathematical models in biology and chemistry.

example 33.6   **allometric growth**   If x and y are measurements of parts of an organism, then $\frac{dx}{dt}$ and $\frac{dy}{dt}$ are their respective rates of growth. However, much more biological information is conveyed by the *specific rates of growth* $\frac{1}{x} \frac{dx}{dt}$ and $\frac{1}{y} \frac{dy}{dt}$. For example, the significance of a growth rate of $\frac{dx}{dt} = 5$ pounds per month would be much different for a human than for an elephant! By forming $\frac{dx}{dt} \Big/ x$ we can compare the rate of growth to the present status of the organism. We begin with the hypothesis stated in the box. ▤

---

growth hypothesis   The specific rates of growth $\frac{1}{x} \frac{dx}{dt}$ and $\frac{1}{y} \frac{dy}{dt}$ are directly proportional over the time inerval (a, b).

Hence, $\dfrac{1}{y}\dfrac{dy}{dt} = k\left(\dfrac{1}{x}\dfrac{dx}{dt}\right)$ for some positive constant k. It follows that

$$\frac{dy}{dx} = \left(\frac{dy}{dt}\bigg/\frac{dx}{dt}\right) = k\,\frac{y}{x}$$

where $x > 0$ and $y > 0$. Separating variables, we obtain $(1/y)\,dy = k\,(1/x)\,dx$. Hence, $\ln y = k \ln x + c$. Raising both sides to the power e yields

$$y = e^{k\ln x}e^{c} = \alpha x^{k}$$

where $\alpha = e^{c} > 0$. Hence, we have the conclusion shown in the box.

---

conclusion   The variables y and x are related by an allometric law $y = \alpha x^{k}$ where $\alpha > 0$ and $k > 0$.

Note that it was not necessary to have explicit formulas for x and y in terms of t in order to derive the above law. In fact, $x(t)$ and $y(t)$ could be quite complicated. It is reasonable to assume that, at least for a restricted period of time in an organism's development, percentage rates of growth should be directly proportional. This is borne out by the many occurrences of allometric laws found in the literature.

example 33.7   **chemical kinetics**   A gas consists of molecules of type A. When the gas is heated, a second substance B is formed as a result of molecular collision: two molecules of A collide and form two molecules of B. Let $A(t)$ be the number of molecules of A present at time t and let $B(t)$ be the number of molecules of B at time t. It is reasonable to assume that the rate at which the reaction proceeds depends directly on the likelihood of collision. This should depend on the number of possible pairs of molecules of A that could collide. The number of such pairs is given by $\{A(t)[A(t)-1]\}/2$. Since $A(t)$ is so large, we can use the approximation $A(t)-1 \approx A(t)$. Therefore, for simplicity, we will make the assumption stated in the accompanying box.

---

reaction rate hypothesis   The rate $\dfrac{dB}{dt}$ at which the substance B is formed is directly proportional to $[A(t)]^{2}$.

If A_0 molecules of A are present initially, then $A(t) + B(t) = A_0$. Hence, $\dfrac{dB}{dt} = k[A(t)]^{2} = k[A_0 - B(t)]^{2}$. Separating variables, we obtain

$$\frac{1}{(A_0 - B)^2} \, dB = k \, dt \qquad \text{or} \qquad \int \frac{1}{(A_0 - B)^2} \, dB = \int k \, dt$$

Hence, $[A_0 - B(t)]^{-1} = kt + c$, or $A_0 - B(t) = 1/(kt + c)$. But $B(0) = 0$. Hence, $A_0 - 0 = 1/c$. Therefore, $c = 1/A_0$ and so

$$B(t) = A_0 - \frac{1}{kt + (1/A_0)} = \frac{A_0^2 kt}{1 + A_0 kt}$$

We have arrived at the conclusion summarized in the box.

---

conclusion

$$B(t) = \frac{A_0^2 kt}{1 + A_0 kt}$$

where A_0 is the initial amount of gas A.

The graph of $B(t)$ is shown in figure 33.3. This model works well for some gases (see example 43.5), although many other hypotheses are possible.

figure 33.3

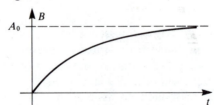

The graph of $B(t)$ is shown in figure 33.3. This model works well for some
gases (see example 43.5), although many other hypotheses are possible.

example 33.8   **a model for the growth of tumors—Gompertz Growth Curve**   If $V(t)$ is the volume of a tumor at time t, then $\dfrac{dV}{dt}$ is the rate of growth in volume. If we assume that $\dfrac{dV}{dt}$ is directly proportional to the size V of the tumor, then an exponential (and therefore unlimited) growth function results. To produce a function with $\lim_{t \to \infty} V(t)$ finite, we must ensure that the specific rate of growth

$$k(t) = \frac{1}{V} \frac{dV}{dt}$$

decreases rapidly as t increases. One possible assumption is that $k(t)$ is an *exponential decay function*, as stated in the box.

> **growth hypothesis** The specific rate of growth $\frac{1}{V}\frac{dV}{dt}$ decreases exponentially as t increases.

Hence, $\frac{1}{V}\frac{dV}{dt} = ke^{-\alpha t}$ for some $k > 0$ and $\alpha > 0$. Separating variables, we obtain $\frac{1}{V}\,dV = ke^{-\alpha t}\,dt$. Taking antiderivatives yields $\ln V = -\frac{k}{\alpha}e^{-\alpha t} + c$. Solving for V and letting $c_1 = e^c$ produces

$$V = c_1 \exp\left(-\frac{k}{\alpha}e^{-\alpha t}\right)$$

where $c_1 > 0$. Note that as t increases, $e^{-\alpha t} \to 0$. Hence, $\text{limit}_{t\to\infty} V(t) = c_1 \exp\left(-\frac{k}{\alpha}0\right) = c_1 e^0 = c_1$. If we denote the maximum volume by V_M, then we have the conclusion shown in the box.

> **conclusion** $V(t) = V_M \exp\left(-\frac{k}{\alpha}e^{-\alpha t}\right)$, where $k > 0$, $\alpha > 0$ and V_M is the maximum volume of the tumor.

This is an example of a *Gompertz growth curve.* To determine the shape of the graph, we will work with the differential equation $\frac{dV}{dt} = (ke^{-\alpha t})V$ rather than the explicit expression for V. Since $\frac{dV}{dt} > 0$, the function V is increasing. In addition,

$$\begin{aligned} V''(t) &= (ke^{-\alpha t})\frac{dV}{dt} + (-k\alpha e^{-\alpha t})V \\ &= (ke^{-\alpha t})^2 V - \alpha(ke^{-\alpha t})V \qquad \text{using } \frac{dV}{dt} = (ke^{-\alpha t})V \\ &= ke^{-\alpha t}V(ke^{-\alpha t} - \alpha) \end{aligned}$$

Setting $V''(t) = 0$ gives $ke^{-\alpha t} = \alpha$ or $t = (1/\alpha)\ln(k/\alpha)$. If $k > \alpha$, then $t > 0$ and an inflection point results. The sign chart for V'' takes the form shown in figure 33.4.

figure 33.4

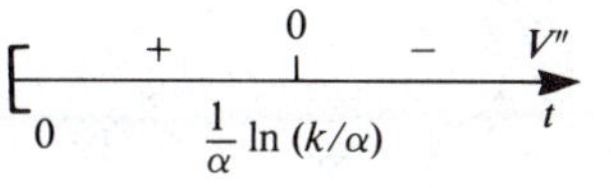

A sketch of $V(t)$ is shown in figure 33.5. When $t = (1/\alpha)\ln(k/\alpha)$—i.e., when $e^{-\alpha t} = \alpha/k$, then $V = V_M e^{-1} \approx .3678 V_M$. Thus, the tumor is growing at the greatest rate when its volume is about 37 percent of the maximum volume. This model actually does quite well in describing how a cancerous tumor increases in size.*

figure 33.5

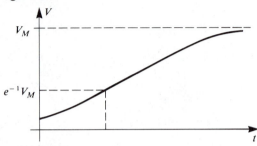

exercises for section 33

Using the separation of variables technique, find all solutions to each of the following differential equations in exercises 1–10.

1. $y' = \dfrac{t}{y}$, for $y \neq 0$

2. $y' = 2(1 - y)$

3. $y' = y^2$

4. $y' = t^2 y$

5. $y' = y/(1 + t)$, for $t \neq -1$

6. $y' = 5y/t$, for $t \neq 0$

7. $y' = e^{t-y}$

8. $y' = 2y^2/t$, for $t \neq 0$

9. $y' = t^3/(y + 1)^2$, for $y \neq -1$

10. $y' = y \sin t$

Find the solution to each of the following boundary value problems in exercises 11–15.

11. $y' = -t/y$, with $y(3) = 4$

12. $y' = 2ty^2$, with $y(0) = 2$

13. $y' = ty^{1/3}$, with $y(1) = 1$

★ **14.** $y' = k(50 - y)$, with $y(0) = 10$ and $y(1) = 20$

15. $y' = t^2/y^2$, with $y(2) = 4$

Use the table of integrals (on page 261), to assist in solving each of the following boundary value problems.

16. $y' = (y^2 - 1)/(t^2 - 1)$, with $y(2) = 9$

17. $y' = y(1 - y)$, with $y(0) = 2$

18. $y' = y^2 - 9$, with $y(0) = 2$

19. $y' = t\sqrt{1 - y^2}$, for $-1 < y < 1$ and $y(0) = 0$

20. $y' = (2t - 1)(y^2 + 4)$, with $y(0) = 2$

* For more information see, A.K. Laird, *British Journal of Cancer*, Vol. 19, (1965), p. 278 and M. Braun, *Mathematical Modeling*, Vol. 1, (1980), pp. 27–31.

21. A spherical cell has volume V and surface area S. A simple model* for cell growth before mitosis assumes that the rate dV/dt at which the cell grows is directly proportional to the surface area of the cell.

(a) Show that $\dfrac{dV}{dt} = kV^{2/3}$ for some constant $k > 0$.

(b) Solve the resulting differential equation and sketch the graph given that $V(0) = V_0$.

22. At high temperatures, nitrogen dioxide NO_2 decomposes into NO and O_2. If $y(t)$ is the concentration of NO_2 (in moles per liter), then, at 600°K, $y(t)$ changes according to the second order reaction law

$$\frac{dy}{dt} = -.05y^2,$$

where t is measured in seconds. Find y in terms of t and y_0, the initial concentration.

23. A model for seasonal population growth begins with the assumption that the percentage rate of growth can be expressed in the form

$$\frac{1}{N}\frac{dN}{dt} = k \sin 2\pi t$$

where t is measured in years and $t = 0$ corresponds to the beginning of spring. Thus, the population will increase during spring and summer, but decline during fall and winter.

(a) Solve the resulting differential equation given that $N(\tfrac{1}{2}) = N_0$.

(b) According to the model, when does the maximum population level occur?

24. As shown in the figure, a cell is in liquid containing a solute (e.g., potassium) of constant concentration c_0. Let $c(t)$ be the concentration in the cell at time t. Fick's Principle for passive diffusion across the cell membrane asserts that the rate at which the concentration increases is directly proportional to $c_0 - c(t)$, the difference in the two concentrations.

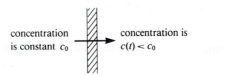

concentration is constant c_0 concentration is $c(t) < c_0$

(a) Show that $\dfrac{dc}{dt} = k\,(c_0 - c)$ for some $k > 0$.

---

* This model has also been applied to the fetal growth of mammals. See Payne and Wheeler, *Nature 215*, (1967) p. 849.

(b) Solve the differential equation and sketch the graph of a typical solution.

[Fick's Model in which concentrations on either side of the cell wall may vary will be presented in Section 39.]

25. The family of curves $y = ce^{-x}$ constitutes the solution to $y' = -y$. To construct a family of curves *orthogonal* (perpendicular) to the given family, solve the differential equation $y' = 1/y$. Sketch both families on the same axes.

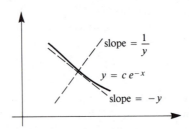

26. In a simple model for the spread of information through a population of y_0 individuals, it is reasonable to assume that the rate at which information spreads depends directly on the number of possible encounters between individuals who possess the information and those who do not. Let $y(t)$ be the number who possess the information at time t, and let $x(t)$ be the number who do not. Assume then that

$$\frac{dy}{dt} = k(xy), \quad \text{for some } k > 0$$

(a) Show that $y' = ky(y_0 - y)$.

(b) Show that $y = y_0(1 + ce^{-ky_0t})$, for some c.

27. A drug is infused into the body at a constant rate of I cubic centimeters per minute. On the other hand, the rate at which the body removes the drug (by metabolism and/or excretion) is assumed to be directly proportional to the amount $A(t)$ in the body at time t.

(a) Show that $A'(t) = I - kA(t)$, for some $k > 0$.

(b) Solve this differential equation for $A(t)$ given that $A(0) = 0$.

(c) What is $\lim_{t \to \infty} A(t)$?

28. In example 33.8, we modeled the growth in volume of a tumor by assuming that the specific rate of growth was an exponential decay function. Now assume $\frac{1}{V}\frac{dV}{dt} = \kappa(t)$, where $\kappa(t)$ is a positive decreasing function.

(a) Show that $V(t) = ce^{K(t)}$ where $K'(t) = \kappa(t)$.

★**(b)** Show that $\lim_{t \to \infty} V(t)$ is finite if and only if $\int_0^\infty \kappa(t)\, dt$ is finite.

(c) If $\kappa(t) = k/(t+1)^n$, determine when $\lim_{t \to \infty} V(t)$ is finite.

29. In a simple model for the growth of tissue, let $A(t)$ be the area of the tissue culture at time t (see the figure). The majority of cell divisions take place on the peripheral portion of the tissue. We will assume that the number of cells on the periphery is directly proportional to $\sqrt{A(t)}$. Further, we assume that the rate of growth in area is jointly proportional to both $\sqrt{A(t)}$ and $K - A(t)$, where K is the final area of the tissue when growth is completed.*

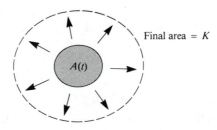

Final area $= K$

$A(t)$

(a) Show that the tissue is growing the fastest when $A = K/3$. (See example 33.8.)

(b) Find $\displaystyle\int \frac{1}{\sqrt{y}(K - y)}\, dy$ by letting $u = \sqrt{y}$ and using the integral tables.

(c) Solve the differential equation $\dfrac{dA}{dt} = c\sqrt{A}(K - A)$.

(d) Compute $\text{limit}_{t \to \infty}\, A(t)$.

part B   **30.** A simple generalization of the exponential growth model assumes that

$$y' = ky^\alpha$$

where $k > 0$, $\alpha > 0$, and $y(t)$ denotes the size of the population at time t.

(a) Solve the differential equation when $\alpha \neq 1$.

(b) For $\alpha > 1$, show that there is a time t_1 for which $\text{limit}_{t \to t_1}\, y(t) = +\infty$.

31. In a simple model for free fall with air resistance, we will assume that the object experiences an upward force due to the air that is directly proportional to its downward velocity v.

(a) Show that the net force on the body is

$$mg - kv$$

for some $k > 0$, where $m = $ mass of the body and $g = -32\,\text{ft/sec}^2$.

(b) Use Newton's Second Law (see example 30.4) to conclude that

$$m\frac{dv}{dt} = mg - kv$$

* Based on *V.A. Kostitzin, Mathematical Biology*, London: Harrap, (1939).

(c) Solve for v given that $v(0) = 0$.

(d) Solve for $y(t)$, the distance off the ground at time t, given that $s(0) = s_0$.

32. In a chemical reaction, one molecule of type A reacts with one molecule of type B to produce a single molecule of the product C. Let $x(t)$ be the amount of product C (in moles, e.g.) formed at time t, and let a and b be the initial amounts of the reactants A and B. The numbers of molecules of A and B left when x molecules of the product have been formed are $(a - x)$ and $(b - x)$, respectively. Assume that the rate of formation is

$$\frac{dx}{dt} = k(a - x)(b - x)$$

for some $k > 0$, and that $x(0) = 0$.

(a) Solve for $x(t)$ when $a = b$. What is $x_\infty = \lim_{t \to +\infty} x(t)$?

(b) Solve for $x(t)$ when $0 < a < b$. What is $x_\infty = \lim_{t \to +\infty} x(t)$?

33. Rework exercise 32, assuming that two molecules of A react with one molecule of B to produce a single molecule of the product C.

34 the von Bertalanffy growth model

In the next two sections, for we will develop commonly used models for **limited growth**, models for the growth of individual organisms, and models for the growth of populations. There are at least two ways of constructing such models. One method is to concoct mathematical expressions, without regard to biological considerations, that are extremely versatile in fitting field data. Another approach (and the one we will adopt) is to make reasonable assumptions about the mechanisms for growth, solve the corresponding differential equations, and then see how well the predictions of the model match reality.

The following two expressions for the length L and weight W of an organism as a function of time are due to von Bertalanffy (1938). They are widely used in the fishery sciences.

$$L(t) = L_\infty(1 - e^{-K(t-t_0)})$$
$$W(t) = W_\infty(1 - e^{-K(t-t_0)})^\nu$$

where L_∞, W_∞, K, and $\nu > 0$, and $t_0 \leq 0$

Of what significance are these constants, and what assumptions about the growth mechanisms lead to these mathematical expressions?

First note that since $e^{-K(t-t_0)} \to 0$ as t increases, $\lim_{t \to \infty} L(t) = L_\infty$ and $\lim_{t \to \infty} W(t) = W_\infty$. Hence, L_∞ and W_∞ are the largest possible values of length and weight, respectively. As L approaches its limit L_∞, we want the rate of growth $\frac{dL}{dt}$ to decrease. The model begins with the hypothesis stated in the box.

> **growth in length hypothesis**   The rate of growth in length is directly proportional to $L_\infty - L$, the length yet to be achieved (see figure 34.1).

figure 34.1

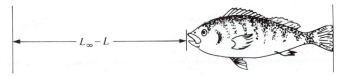

$$L_\infty - L$$

Hence, $\dfrac{dL}{dt} = K(L_\infty - L)$ for some $K > 0$. Separating variables, we obtain

$$\frac{1}{L_\infty - L}\, dL = K\, dt$$

Taking antiderivatives yields $-\ln (L_\infty - L) = Kt + c$ or $L_\infty - L = e^{-c}e^{-Kt}$. It follows that $L = L_\infty - c_1 e^{-Kt}$, where $c_1 = e^{-c} > 0$. The constant c_1 can be evaluated in terms of the initial length $L(0)$: $L(0) = L_\infty - c_1$. Hence,

$$L = L_\infty - [L_\infty - L(0)]e^{-Kt} = L_\infty\left[1 - \left(\frac{L_\infty - L(0)}{L_\infty}\right)e^{-Kt}\right]$$

Since $0 < [L_\infty - L(0)]/L_\infty \le 1$, we may write $[L_\infty - L(0)]/L_\infty$ as e^{Kt_0} where $t_0 \le 0$, as can be seen in figure 34.2. Hence we have the conclusion given in the box.

figure 34.2

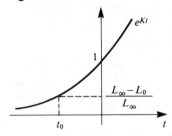

---

conclusion

$$L(t) = L_\infty[1 - e^{-K(t-t_0)}]$$

where $e^{Kt_0} = (L_\infty - L_0)/L_\infty$.

The constant K will depend upon the particular species and will determine how quickly the species matures. For the small North Sea haddock, $K \approx .2$, while for the large Pacific halibut, $K \approx .18$. No physical significance is attached to the constant t_0. The graph of $L(t)$ takes the form shown in figure 34.3.

figure 34.3

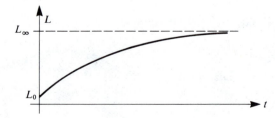

figure 34.4 Growth in length of Haddock. Average length of certain age-groups and the fitted von Bertalanffy curve with parameter values $L_\infty = 53$ cm., $K = 0.20$, $t_0 = -1.066$ yrs from Raitt (1933, 1939).*

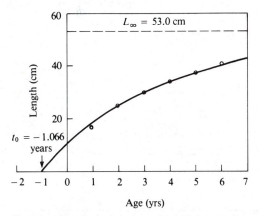

figure 34.5 Growth in length of Plaice. Average length of fish in each age-group from Lowestoft and Grimsby market samples, 1929–38. The curve is obtained by fitting the von Bertalanffy equation with parameter values $L_\infty = 68.5$ cm., $K = 0.095$, $t_0 = -0.815$ yrs.*

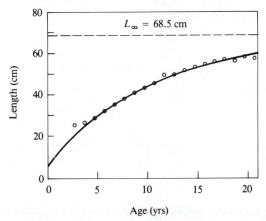

The von Bertalanffy model does quite well in predicting lengths of many commercial fish. Shown in figures 34.4 and 34.5, respectively, are the results of two case studies involving haddock and plaice.

* Adapted from, R.J.H. Beverton, and S.J. Holt, *On the dynamics of exploited fish populations.* Fishery Investigations, Series II, Vol. XIX. London: H.M.S.O., British Crown Copyright, 1957.

example 34.1   For the North Sea haddock, how fast is a fish growing:

(a) when its length is 30 cm?   (b) at birth?

solution 34.1   For (a) the easiest way to compute $\dfrac{dL}{dt}$ is through the original differential

equation $\dfrac{dL}{dt} = .20(53 - L)$. Hence, when $L = 30$, $\dfrac{dL}{dt} = .20(53 - 30) = 4.6 \text{ cm/yr.}$

For (b) at birth, $L(0) = 53(1 - e^{Kt_0}) = 53(1 - e^{-.2132}) = 10.18 \text{ cm.}$ Thus, $\dfrac{dL}{dt} =$

$.20(53 - 10.18) = 8.56 \text{ cm/yr.}$ ▤

It is reasonable to assume (see example 33.6) that the relationship between weight W and length L is given by an allometric law, as hypothesized in the box.

---

length-weight hypothesis W and L are related by an allometric law $W = \alpha L^\nu$, where $\alpha > 0$ and $\nu > 0$.

Thus, $W = \alpha(L_\infty)^\nu[1 - e^{-K(t-t_0)}]^\nu$. Since $\lim_{t \to \infty} W(t) = \alpha(L_\infty)^\nu$, we have the conclusion given in the box.

---

conclusion

$$W(t) = W_\infty[1 - e^{-K(t-t_0)}]^\nu$$

where $\nu > 0$ and W_∞ is the maximum attainable weight.

When $\nu = 3$, we have what is called **isometric growth**. This will occur when the organism grows in such a way that its body parts are always in the *same proportions*. To illustrate this, suppose that the rod-shaped organism shown in figure 34.6 grows in such a way that $\dfrac{r}{L} = c$. Hence, $V = \pi r^2 L = (\pi c^2)L^3$. If we assume that weight is directly proportional to volume, then $W = dV = (d\pi c^2)L^3 = \alpha L^3$. Shown in figure 34.7 is the length-weight relationship for plaice.

figure 34.6(a)   **figure 34.6(b)**

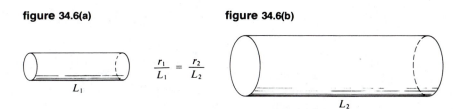

$$\frac{r_1}{L_1} = \frac{r_2}{L_2}$$

figure 34.7 (Adapted from, R.J.H. Beverton, and S.J. Holt, *On the dynamics of exploited fish populations.* Fishery Investigations, Series II, Vol. XIX. London: H.M.S.O., British Crown Copyright, 1957.)

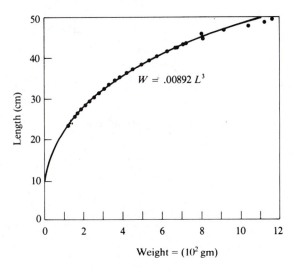

To determine the shape of the graph of $W(t)$ when $\nu = 3$, we will work with the formula $W = \alpha L^3$ and $\dfrac{dL}{dt} = K(L_\infty - L)$:

$$\frac{dW}{dt} - 3\alpha L^2 \frac{dL}{dt} = 3\alpha L^2 K(L_\infty - L) > 0$$

Hence, $W(t)$ is increasing. In addition,

$$W''(t) = (3\alpha L^2 K)\left(-\frac{dL}{dt}\right) + (L_\infty - L)\left(6\alpha LK \frac{dL}{dt}\right)$$

$$= 3\alpha LK \frac{dL}{dt}[-L + 2(L_\infty - L)] = 3\alpha LK \frac{dL}{dt}(2L_\infty - 3L)$$

Hence, $W''(t) = 0$ when $L = \frac{2}{3}L_\infty$. When $L = \frac{2}{3}L_\infty$, $W = \alpha L^3 = \alpha(\frac{2}{3})^3 L_\infty^3 = \frac{8}{27}W_\infty \approx .296 W_\infty$. To find the time when this occurs, we consider

$$\frac{2}{3} L_\infty = L = L_\infty[1 - e^{-K(t-t_0)}]$$

hence, $e^{-K(t-t_0)} = \frac{1}{3}$, which gives $t = t_0 + (\ln 3)/K$. When $t_0 + (\ln 3)/K > 0$, the graph has an inflection point, as in figure 34.8. Shown in figure 34.9 is the von Bertalanffy growth curve fitted to data on the North Sea sole. As you can see, the model works quite well in this case.

figure 34.8   The rate of growth in weight is greatest when $W = (8/27)W_\infty$, $L = (2/3)L_\infty$, and this occurs at time $t_0 + (\ln 3)/K$.

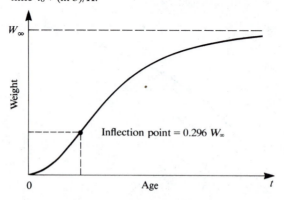

figure 34.9   (Adapted from, R.J.H. Beverton, and S.J. Holt, *On the dynamics of exploited fish populations.* Fishery Investigations, Series II, Vol. XIX. London: H.M.S.O., British Crown Copyright, 1957.)

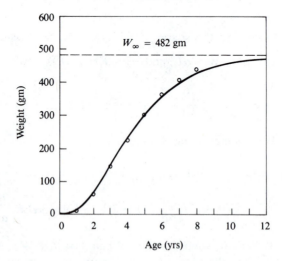

example 34.2 The Pacific Halibut Commission reports that the parameters in the von Bertalanffy growth model have current values $L_\infty = 200$ cm, $W_\infty = 83$ kg, $t_0 = -.25$, $K = .18$, and $\nu = 3$.

(a) How fast is a typical halibut growing in length and weight when $L = 100$ cm?

(b) When is the rate of growth in weight the largest? What is the length at this time?

(c) The commission recommends that only halibut 4 years or older be taken by commercial fishermen. Find the length and weight of a typical 4-year-old halibut.

solution 34.2 We know $\dfrac{dL}{dt} = K(L_\infty - L)$ and $\dfrac{dW}{dt} = 3\alpha L^2 \dfrac{dL}{dt}$. Since $W_\infty = \alpha(L_\infty)^3$, we have

$$\alpha = \frac{83}{200^3} = .000010375.$$

For **(a)**, when $L = 100$, $\dfrac{dL}{dt} = .18(200 - 100) - 18$ cm/yr. Also, $\dfrac{dW}{dt} = 3(.000010375)(100)^2(18) = 5.60$ kg/yr.

For **(b)**, $\dfrac{dW}{dt}$ is largest at $t = t_0 + \dfrac{\ln 3}{K} = -.25 + 6.1034 = 5.85$ yr. At this time,

$$L = \frac{2}{3} L_\infty = 133.3 \text{ cm}.$$

Finally, part **(c)** asks for $L(4)$ cm and $W(4)$. Substituting in the two formulas, you should find that $L(4) = 106.93$ cm and $W(4) = 12.69$ kg. ▤

exercises for section 34

part A For the Pacific herring, the parameters in the von Bertalanffy growth model have approximate values

$$L_\infty = 38.3 \text{ cm}, \qquad W_\infty = 440 \text{ g}, \qquad K = .21, \qquad t_0 = -2.4, \qquad \nu = 3$$

1. How fast is a typical herring growing in length at birth?

2. Show that the length-weight relationship is $W = .0078 L^3$.

3. When is the rate of growth in weight the largest? What is the weight at this time?

4. How fast is a typical herring growing in length and weight at age 6 years?

For the North Sea plaice, the parameters in the von Bertalanffy growth model have approximate values

$$L_\infty = 68.5 \text{ cm}, \qquad W_\infty = 2,870 \text{ g}, \qquad K = .095, \qquad t_0 = -.815, \qquad \nu = 3$$

5. What are the expected length and weight of a 15-year-old plaice? Find the rate of growth in length at age 15.

6. When is the rate of growth in weight the largest? What is the weight at this time?

7. Compute $L(0)$, $W(0)$, $L'(0)$, and $W'(0)$.

★8. When does the von Bertalanffy growth model predict that the rate of growth in length will be the greatest?

9. For $W(t) = W_\infty[1 - e^{-K(t-t_0)}]^\nu$, use implicit differentiation and the method on page 377 to show that $W(t)$ has an inflection point when

$$L = \left(1 - \frac{1}{\nu}\right)L_\infty \qquad \text{and} \qquad W = \left(1 - \frac{1}{\nu}\right)^\nu W_\infty$$

35 models for limited population growth

If $N(t)$ denotes the size of a population at time t, the model for exponential growth begins with the assumption that $N'(t) = rN(t)$ for some $r > 0$. Thus, the specific or relative rate of growth $\dfrac{1}{N}\dfrac{dN}{dt}$ is assumed to be a constant, r. True cases of exponential growth over long periods of time are hard to find, because the environment, which has limited resources, will at some point in time exert restrictions on growth. Thus, $\dfrac{1}{N}\dfrac{dN}{dt}$ can be expected to decrease as N increases in size.

The assumption that $\dfrac{1}{N}\dfrac{dN}{dt} = f(N)$ or $\dfrac{dN}{dt} = Nf(N)$ for some function f, is called the **density-dependent hypothesis**. The rate at which the population is growing (or declining) is assumed to *depend only upon the number present* and not on any time-dependent mechanisms such as seasonal phenomena. This hypothesis has generated much controversy among ecologists, but, nevertheless, is widely assumed in most models of animal populations.

Suppose that the environment is capable of sustaining no more than K individuals in the given population. The quantity K is called the **carrying capacity** of the environment. Hence, $f(N) = 0$ when $N = K$. In addition, we will let $f(0) = r$. Figure 35.1 shows three functions that satisfy these two conditions.

figure 35.1   Possible functions $f(N)$

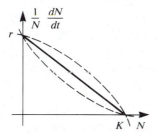

The simplest assumption we can make is that $f(N)$ is linear, that is, $f(N) = mN + b$. If we use the two conditions $f(0) = r$ and $f(K) = 0$, then f takes the form:

$$f(N) = r\,\frac{K - N}{K}$$

Thus, $\dfrac{dN}{dt} = rN \dfrac{K-N}{K}$. This equation is known as the **logistic differential equation** for population growth. The required hypotheses are summarized in the box.

---

logistic model hypotheses

1. Population growth is density-dependent, that is, $\dfrac{dN}{dt} = Nf(N)$ for some function f.

2. The relative rate of growth $f(N)$ is linear with $f(0) = r$ and $f(K) = 0$.

Note that for K very large, $\dfrac{(K-N)}{K} \approx 1$. Thus, $\dfrac{dN}{dt} \approx rN$ and we essentially have exponential growth. This is a nice feature of the model and justifies our interpretation of K.

Separating variables yields $1/[N(K-N)]\, dN = (r/K)\, dt$. In taking anti-derivatives, we use formula 78 in the integral tables with $a = 1$, $b = 0$, $p = -1$, and $q = K$ to obtain:

$$\frac{-1}{K} \ln \left| \frac{K-N}{N} \right| = \frac{r}{K} t + c_1$$

It follows that $|(K/N) - 1| = e^{-c_1} e^{-rt}$. Removing absolute values, we obtain $(K/N) - 1 = \pm e^{-c_1} e^{-rt} = ce^{-rt}$, where c is any nonzero real number. Solving for N, we obtain:

$$N = \frac{K}{1 + ce^{-rt}}$$

The additional solutions are $N \equiv 0$ and $N \equiv K$. If $N(0) = N_0$ is given, then $N_0 = K(1 + c)$, which gives $c = (K - N_0)/N_0$. Since the solution $N \equiv 0$ is of no interest, we have the conclusion shown in the box.

---

conclusion

$$N(t) = \frac{K}{1 + ce^{-rt}}$$

where $c = \dfrac{K - N_0}{N_0}$. If $N_0 = K$, the solution is $N(t) \equiv K$.

We may write $N(t) = N_0 e^{rt} \left[\dfrac{K}{N_0 e^{rt} + (K - N_0)} \right]$. For values of K much larger than $N_0 e^{rt}$, the second term is approximately equal to 1, and $N(t) \approx N_0 e^{rt}$.

To determine the shape of the graph of $N(t)$, we must divide our argument into three cases:

case I $0 < N_0 < K$

It follows that $c = (K - N_0)/N_0 > 0$. Hence, $1 + ce^{-rt} > 1$. As a result, $0 < N(t) < K$ for all $t \geq 0$, and, from the original logistic differential equation,

$$\frac{dN}{dt} = \frac{r}{K} N(K - N) > 0$$

Consequently, $N(t)$ increases on the interval $(0, +\infty)$.

Using implicit differentiation, we obtain $N''(t) = \dfrac{r}{K}(K - 2N)\dfrac{dN}{dt}$. Setting $N''(t) = 0$, we have $N = K/2$. This will occur when $1 + ce^{-rt} = 2$, that is, when $ce^{-rt} = 1$. Solving for t yields $t = (1/r)\ln c$, which will be positive provided $c = (K - N_0)/N_0 > 1$, or $N_0 < K/2$. When $N_0 < K/2$, there is no inflection point in the graph for $t > 0$.

case II $N_0 \equiv K$

The solution is $N(t) \equiv K$.

case III $N_0 > K$

An analysis similar to case I will show that $N(t)$ is decreasing and always concave upward. Note that in this case $c = (K - N_0)/N_0 < 0$ and so $N(t) > K$. The cases are summarized in figure 35.2.

figure 35.2

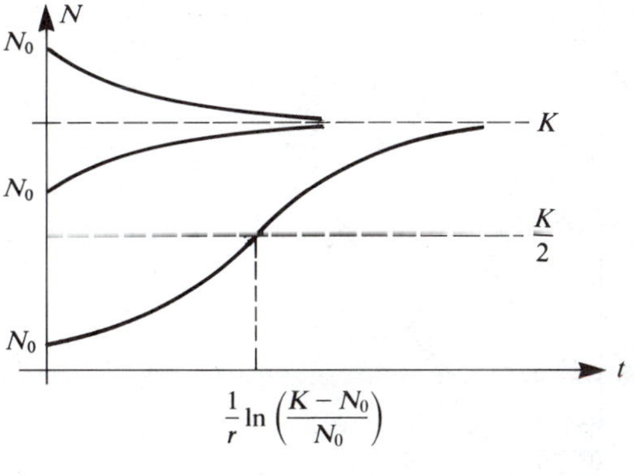

Shown in figure 35.3 are sketches of logistic curves with $K = 1$. Increasing r from .5 to 1 results in the population reaching K more rapidly. In each case, the population is increasing the most rapidly when $N = K/2 = .5$. For $r = 1$, this occurs when $t = \ln 4 = 1.386 \ldots$; for $r = .5$, this occurs at $t = 2 \ln 4 = 2.7725 \ldots$

figure 35.3

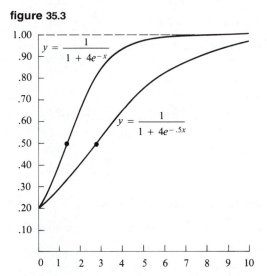

The following example presents one of the earliest applications of the logistic model to a natural population.

example 35.1   In a famous application of the logistic model to the growth of yeast cells, Pearl* (1925) took earlier data of Carlson† and fitted the logistic curve

figure 35.4   The logistic growth of a laboratory population of yeast cells. (From Pearl.)   (Adapted from, W.C., Allee et al., *Principles of Animal Ecology*. Philadelphia: W.B. Saunders Co., 1949.)

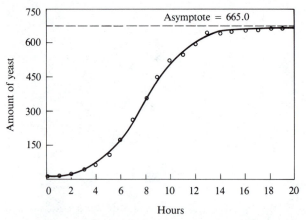

* R. Pearl, *Biology of Population Growth*, A.A. Knopf, New York, 1925.

† T. Carlson, "Über Geschwindigkeit und Grösse der Hefevermehrung in Würze", *Biochemische Zeitschrift, 57* (1913).

$$N = \frac{665}{1 + 66e^{-.5355t}}$$

The carrying capacity is $K = 665$, while the specific rate of growth is .5355. If K were increased greatly, we should experience exponential growth $N(t) = N(0)e^{rt} = 9.925e^{.5355t}$ for a much longer period of time. The population is increasing most rapidly when $N = K/2 = 332.5$ and this occurs when $t = (1/r) \ln c = (\ln 66)/.5355 \approx 7.8$ hours. ▤

The final form of the logistic equation $\frac{dN}{dt} = rN\left(\frac{K-N}{K}\right)$ was motivated mainly by mathematical considerations. Several attempts have been made to make this equation seem reasonable from a biological point of view. Three such attempts are discussed briefly below.

1. If the environment is capable of sustaining K individuals, then $(K - N)/K$ can be considered as that remaining portion of the environment yet to be utilized. The logistic equation therefore states that the rate of growth is jointly proportional to the number present and to the unused potential.

2. The rate of growth = (birth rate) − (death rate). We can assume that the birth rate is proportional to N. On the other hand, animals compete for common resources (*interspecific competition*). The encounter rate should be proportional to the *number of pairs* of individuals, which is $[N(N - 1)]/2$. We may assume that a fixed percentage of these encounters result in death. Thus,

$$\frac{dN}{dt} = bN - c\,\frac{N(N-1)}{2} = N\left[\left(b + \frac{c}{2}\right) - \frac{c}{2}N\right]$$

which is in the form of the logistic equation.

3. Schoener* has given an interesting derivation of the logistic equation based on simple assumptions about an animal's energy budget (energy lost in searching for food, energy lost in body maintenance, energy gained in feeding, and so forth).

The logistic model is but one of a multitude of population models. It is used mainly because of its simple mathematical properties and its ability to

* Schoener, *Theoretical Population Biology*, Vol. 4 (1973), pp. 56–84.

reasonably predict the growth of lower order organisms. It is the density-dependent growth model that is one step up in complexity from the simple exponential growth model.

We close this section with a completely different population growth model based on resource renewal.

example 35.2

As might be the case for laboratory populations, a resource is supplied at the steady rate of I calories per day. For an individual in the population, the maintenance cost is m calories per day. The animals divide the daily resources in an equitable way; the caloric ration per animal is I/N. The net energy of the entire population is therefore

$$N\left(\frac{I}{N} - m\right) \text{ calories per day}$$

Finally, we assume that this net energy is directly proportional to $\frac{dN}{dt}$, the rate of population growth. Presumably, if this net energy is positive, then it can be converted into offspring. Thus,

$$\frac{dN}{dt} = kN\left(\frac{I}{N} - m\right) = k(I - mN)$$

for some $k > 0$. If we separate variables and solve for N, we will obtain:

$$N = \frac{1}{m}(I - ce^{-mkt})$$

The carrying capacity (or limiting population) is therefore I/m and the graph has the form shown in figure 35.5. If $m = 100$ calories per day and we wish to eventually sustain a population of 50 animals, then, according to the model, the daily ration should be $50(100) = 5,000$ calories per day. This special functional form for $N(t)$ has actually been observed for some laboratory populations.

figure 35.5

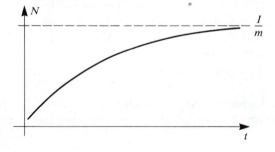

exercises for section 35

part A

In 1934, Gause carried out a series of famous competition experiments involving two species of protozoa: **(a)** *Paramecium caudatum*, and **(b)** *Paramecium aurelia* (see the figures). In one such experiment, two populations were cultivated separately and were found to grow according to the logistic equations

(a)　　　　　　　　　　**(b)**

$$N_1(t) = \frac{105}{1 + 34e^{-1.1244t}}$$

$$N_2(t) = \frac{64}{1 + 15e^{-.794t}}$$

where t is measured in days.

1. Compute $N_1(0)$, $N_2(0)$, K_1, and K_2. What are the interpretations of these numbers?

2. What differential equations do the populations satisfy?

3. Find the time when each population is growing the fastest.

Shown in the accompanying figure are the data of Bodenheimer (1937) on the growth of two populations of bees.

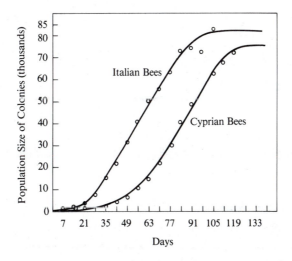

From the graph for the Italian bee population, it appears that $N(0) = 1$ (thousand), $K \approx 82$ (thousand), and $N = K/2$ when $t \approx 56$ days.

4. Using this information, estimate r and write the equation of the logistic function fitting the data.

From the graph for Cyprian bees, it appears that $N(0) = 1$, $K \approx 74$, and $N = K/2$ when $t \approx 84$ days.

5. Using this information, estimate r and write the equation of the logistic function fitting the data.

If we let the percentage rate of growth $f(N) = (dN/dt)/N$ be decreasing and linear, then the logistic equation results. Many other expressions for $f(N)$ have been proposed in the literature. A few of these are explored in the following exercises:

6. The Gompertz population model assumes $f(N) = -\beta \ln \left(\dfrac{N}{K} \right)$.

   (a) Rewrite this equation as $\dfrac{d}{dt} (\ln N) = -\beta (\ln N) + \beta \ln K$.

   (b) Let $u = \ln N$, and solve the differential equation for u.

   (c) Compute $\lim_{t \to \infty} N(t)$.

   (d) Find where the inflection point in the graph of $N(t)$ occurs.

7. If $f(N) = rN^{-k}$, where $r > 0$ and $k > 0$, solve the corresponding differential equation and show that population growth is *not* limited.

8. In a study of the water flea *Daphnia magna*, Smith[*] found that $f(N)$ took the form shown in the figure. Motivated by considerations of how the rate of food

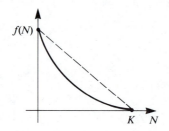

supply diminishes, Smith arrived at

$$\frac{1}{N} \frac{dN}{dt} = f(N) = r \frac{K - N}{K + aN}$$

where r, K, and $a > 0$.

   (a) Sketch the graph of $f(N)$ for $N \geq 0$.

[*] F.E. Smith, *Ecology* (1963) *44*, pp. 651–663.

(b) Show that the graph of $N(t)$ has an inflection point when

$$N = \left(\frac{\sqrt{1+a}-1}{a}\right)K$$

(c) In one of Smith's experiments, $r = .44$, $K = 228$, and $a = 3.46$. How large is the population when the rate of growth is the greatest?

★**(d)** Solve the original differential equation with the aid of the integral tables, (page 261–5).

9. Suppose $f(N)$ takes the special form shown in the figure.

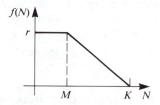

(a) By solving the corresponding differential equation, show that the population grows exponentially until it reaches size M.

(b) Show that, for $N \geq M$, the population follows a logistic equation and approaches K.

★**(c)** Determine conditions under which the graph has an inflection point.

36 models for wildlife management

In this section, we will present two models for wildlife management that have been used predominantly by fishery scientists. In each case, the model will contain parameters that are under the control of the resource manager. The goal of the manager is to select those parameters that maximize the yield from the resource and at the same time protect the population over the long run.

Let $N(t)$ be the size of the population at time t, and let $B(t)$ be the biomass (or total weight) of the resource at time t. There are two ways of keeping track of the rate at which the resource is being exploited:

1. The *rate of catching*, $\dfrac{dC}{dt}$, measures the numbers C being removed from the population per unit time.

2. The *rate of yield*, $\dfrac{dY}{dt}$, measures the weight or biomass Y being removed per unit time.

If we imagine fishing nets combing a given area of the sea, it is reasonable to assume that the catch per unit time depends directly upon the actual abundance of the resource at time t and the total effort $E(t)$ expended at time t. The effort may be measured in various ways. We might keep track of the number of nets, the number of hooks, or the number of hunters, for example.

In each model, we will make the assumption stated in the box.

---

catch-effort hypothesis

$$\frac{dC}{dt} = q_1 E(t) N(t) \qquad \text{and} \qquad \frac{dY}{dt} = q_2 E(t) B(t)$$

for some $q_1 > 0$ and $q_2 > 0$.

---

the Schaeffer model (1954)   In this model we keep track only of the total biomass of the population. We will assume that, when left unexploited, the resource grows according to the logistic equation

$$\frac{dB}{dt} = rB \frac{B_\infty - B}{B_\infty}$$

where $B_\infty = \text{limit}_{t\to\infty} B(t)$. In addition, we will assume that the effort $E(t)$ expended in exploiting the resource is constant over time. Letting $F = q_2E$, we have $\dfrac{dY}{dt} = FB(t)$. The value of F is called the **relative rate of exploitation** or the **exploitation mortality**. When the resource is being taken,

$$\frac{dB}{dt} = (\text{natural rate of growth}) - (\text{rate of yield})$$

$$= rB\,\frac{B_\infty - B}{B_\infty} - FB$$

For $0 < F < r$, this differential equation may be put in logistic form

$$\frac{dB}{dt} = (r - F)B\,\frac{(r - F)(B_\infty/r) - B}{(r - F)(B_\infty/r)}$$

$$= r'B\,\frac{K' - B}{K'}$$

where $r' = r - F > 0$ and $K' = (r - F)(B_\infty/r) = (1 - F/r)B_\infty$. If $F \geq r$, one can show that the population is driven to extinction. Thus, we may conclude $B(t) = K'/[1 + ce^{-r't})$, and so $\text{limit}_{t\to\infty} B(t) = (1 - F/r)B_\infty$.

For t very large, $\dfrac{dY}{dt} = FB \approx F\left(1 - \dfrac{F}{r}\right)B_\infty$. This is the **long-term rate of yield** and is usually measured in kilograms per year. We define the **sustainable yield SY** as:

$$SY = F\left(1 - \frac{F}{r}\right)B_\infty, \qquad \text{where } 0 \leq F \leq r$$

Presumably, the parameter $F = q_2E$ can be controlled by the manager. What choice of F maximizes the sustainable yield SY? Since the graph of SY(F) is parabolic, as shown in figure 36.1, the maximum will occur at $F = r/2$ with

figure 36.1

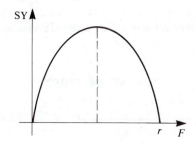

$SY = (r/4)B_\infty$. Note that for this choice of F, $\lim_{t\to\infty} B(t) = (1 - 1/2)B_\infty = B_\infty/2$. The conclusion is summarized in the box.

The maximum sustainable yield (MSY) is given by

$$\text{MSY} = \frac{r}{4} B_\infty$$

and this occurs when $F = \dfrac{r}{2}$ and $B = \dfrac{B_\infty}{2}$.

We showed in section 35 (see figure 35.2 on page 383), that the population biomass is growing naturally at the greatest rate when $B = B_\infty/2$. Thus, for maximum sustainable yield the Schaeffer model predicts that the population biomass should be kept at the level where its natural rate of growth is maximized.

example 36.1   Schaeffer (1967) applied his model to a population of Eastern Pacific yellowfin tuna. The best estimates for the population parameters are $r = 2.61$ and $B_\infty = 1.34 \times 10^8$ kilograms. The maximum sustainable yield is $(r/4)B_\infty = 8.74 \times 10^7$ kilograms per year. This occurs for fishing mortality $r/2 = 1.305$.

example 36.2   A resource with $r = .5$ and $B_\infty = 10,000$ kilograms is giving a sustainable yield SY of 1,000 kilograms per year. Estimate F.

solution 36.2   We are given that $SY = F(1 - F/r)B_\infty = 1,000$. Using the given parameters, we have $F(1 - 2F)10,000 = 1,000$. Hence, $2F^2 - F + .1 = 0$. The quadratic formula yields the solutions $F = \dfrac{1 \pm \sqrt{1 - 8(.1)}}{4} = .138$ and $.362$. If $F = .362$, then less than half the effort ($F = .138$) will result in the same long-term yearly yield.

The Schaeffer model assumes that the population will constantly strive to recover its numbers, but the model does not address the issue of how the reproductive mechanism takes place. It treats the population as an irrepressible "blob." It is clear that if exploitation upsets the *age structure* of the population, then any reproductive mechanism will be altered considerably. The model should therefore be used with extreme caution.

the Beverton-Holt year class model (1957)   In this model, we will take the animals resulting from one annual reproductive period (a *year class*) and follow them through their lifetimes. It might be helpful to think of a trout pond that is stocked yearly with small fish from the hatchery.

If $W(t)$ is the weight of an individual at time t, then $B(t) = W(t)N(t)$. As t increases, $N(t)$ will decrease but, on the other hand, $W(t)$ will increase. We will make the assumption given in the box.

---

population decay hypothesis

$$N(t) = N_0 e^{-Zt}$$

where $Z > 0$ is the relative mortality rate.

To determine when the population biomass is the largest, we compute the first derivative of $B(t)$:

$$B'(t) = N(t)W'(t) + N'(t)W(t) = N(t)W'(t) - ZN(t)W(t)$$

Setting $B'(t) = 0$ yields $N(t)[W'(t) - ZW(t)] = 0$ or $W'(t) = ZW(t)$. Thus, we have the conclusion shown in the box.

---

conclusion   If the biomass $B(t)$ has a maximum, it will occur at time t such that $W'(t)/W(t) = Z$.

This special formula is used in the next example to determine when the biomass is maximized.

example 36.3   A trout farm places $N_0 = 5{,}000$ young trout in a pond and lets them grow naturally. Tagging experiments indicate that 3,000 remain after one year. Their growth in weight (in pounds) is given by $W(t) = .1 + .6t$. Assuming the population decay hypothesis, when is the population biomass maximized?

solution 36.3   The function for population size is given by $N(t) = 5{,}000e^{-Zt}$. Since we are given $N(1) = 3{,}000$, it follows that $e^{-Z} = .6$. Hence, $Z = -\ln .6 = .5108$. Now, $W'(t) = .6$, and so we must solve the equation $.6/(.1 + .6t) = Z = .5108$. Thus, $.6 = .05108 + .30648t$ or $t = .54892/.30648 = 1.79$ years. It is not hard to show that the absolute maximum actually occurs at this sole critical point.   ▤

When the von Bertalanffy growth curve is used, the biomass is maximized at that time t satisfying:

$$\frac{W'(t)}{W(t)} = \frac{3Kx}{1 - x} = Z \quad \text{where} \quad x = e^{-k(t-t_0)}.$$

The result follows directly from computing $W'(t)/W(t)$ and simplifying. The special case is applied to the Pacific halibut in the next example.

example 36.4   The Pacific Halibut Commission gives $W_\infty = 83$ kilograms, $t_0 = -.25$, and $K = .18$ as the parameters for the von Bertalanffy growth curve. In addition the natural mortality rate (no fishing!) is estimated to be .2. Hence, $N(t) = N_0 e^{-.2t}$ when no fishing takes place. If fishing were suspended, when would a given year class reach its maximum biomass?

solution 36.4   The biomass will attain its maximum value at time t such that $\dfrac{W'(t)}{W(t)} = \dfrac{3(.18)x}{1-x}$, where $x = e^{-.18(t+.25)}$. Setting $\dfrac{W'(t)}{W(t)} = .2$, we obtain

$$\frac{.54x}{1-x} = .2 \quad \text{or} \quad .54x = .2 - .2x$$

Solving for x yields $x = .27027$. Hence, $-.18(t + .25) = \ln (.27027)$ and so $t = \ln (.27027)/(-.18) - .25 = 7.0185 \approx 7$ years. ▤

As in the Schaeffer model, we will assume that the effort $E(t)$ is constant E. Thus, if we let $F = q_1 E$, then $\dfrac{dC}{dt} = FN(t)$. Let t_c be the time when the year class is first susceptible to exploitation efforts. This gives an additional parameter that can be controlled by the resource manager. For example, the mesh size of the fishing nets or the hook size may determine the size, and hence the age, of the fish that may be caught. Or hunters may be instructed to capture only those animals that meet certain size requirements.

We are therefore assuming that

$$\frac{dN}{dt} = \begin{cases} -MN(t), & \text{for } 0 \le t < t_c \\ -(F + M)N(t), & \text{for } t \ge t_c \end{cases}$$

If we impose the initial conditions $N(0) = N_0$ and $N(t_c) = N_c$, then it follows that

$$N(t) = \begin{cases} N_0 e^{-Mt} & \text{for } 0 \le t < t_c \\ N_c e^{-(M+F)(t-t_c)}, & \text{for } t \ge t_c \end{cases}$$

Since $N_c = N_0 e^{-Mt_c}$, we may write $N(t) = N_0 e^{Ft_c} e^{-(M+F)t}$ for $t \ge t_c$.

We wish to determine expressions for $C(t_c, F)$ and $Y(t_c, F)$, the total catch and total yield over the lifetime of the year class, in terms of the two control parameters t_c and F.

For $t_1 \ge t_c$ and $t_2 \ge t_c$, the total catch between times t_1 and t_2 is given by

$$\int_{t_1}^{t_2} \frac{dC}{dt}\, dt = \int_{t_1}^{t_2} FN(t)\, dt = \int_{t_1}^{t_2} F[N_0 e^{Ft_c} e^{-(M+F)t}]\, dt$$

$$= -FN_0 e^{Ft_c} \frac{1}{M+F} e^{-(M+F)t} \Big|_{t_1}^{t_2} = \frac{F}{M+F} N_0 e^{Ft_c}[e^{-(M+F)t_1} - e^{-(M+F)t_2}]$$

If we let $t_1 = t_c$ and let $t_2 \to \infty$, then $e^{-(M+F)t_2} \to 0$, and we have the result given in the box.

---

catch equation   The total catch over the lifetime of the year class is given by

$$C(t_c, F) = \frac{F}{M+F} N_0 e^{-t_c M}$$

When $t_c = 0$, the catch equation takes the simple form $C(0, F) = \dfrac{F}{M+F} N_0$, which is used in the following example.

example 36.5   At Happy Harry's Trout Farm, a pond is stocked with 5,000 young trout. The value of M is estimated to be .3. By the end of the season, 1,500 trout have been caught and the pond is "fished out."
  (a) Estimate the fishing mortality F if $t_c = 0$.
  (b) If business doubles next year, how many trout will be caught under the same conditions?

solution 36.5   We are given $1,500 = C(0, F) = 5,000\, F/(.3 + F)$. It follows that $F = .09 + .3F$ or $.7F = .09$. Thus, $F = .128 \approx .13$. If business doubles, the effort expended in fishing at time t will double. Hence, the new F will double. If we use $F = .26$ in the catch equation, then $C(0, .26) = (.26)(5,000)/(.3 + .26) = 2,321$ trout. ▤

For $t_1 \geq t_c$ and $t_2 \geq t_c$, the *total yield* between times t_1 and t_2 is given by

$$\int_{t_1}^{t_2} \frac{dY}{dt}\, dt = \int_{t_1}^{t_2} FW(t)N(t)\, dt$$

If we use a von Bertalanffy weight function, then we obtain

$$FW(t)N(t) = FN_0 e^{Ft_c} e^{-(M+F)t} W_\infty[1 - e^{-K(t-t_0)}]^3$$
$$= FN_0 e^{Ft_c} e^{-(M+F)t} W_\infty(1 - 3ce^{-Kt} + 3c^2 e^{-2Kt} + c^3 e^{-3Kt})$$

where $c = e^{Kt_0}$. Multiplying out by $e^{-(M+F)t}$, we obtain

$$FW(t)N(t) = FN_0 W_\infty e^{Ft_c}[e^{-(M+F)t} - 3ce^{-(M+F+K)t} + 3c^2 e^{-(M+F+2K)t} + c^3 e^{-(M+F+3K)t}]$$

It follows that

$$\int_{t_1}^{t_2} \frac{dY}{dt} \, dt = \left(FN_0 W_\infty e^{Ft_c} \sum_{n=0}^{3} \frac{-a_n c^n}{M + F + nK} e^{-(M+F+nK)t} \right) \Bigg|_{t_1}^{t_2}$$

where $a_0 = 1$, $a_1 = -3$, $a_2 = 3$, and $a_3 = -1$.

If we let $t_2 \to \infty$, then the entire term goes to 0. Letting $t_1 = t_c$, we obtain the result shown in the box.

---

the yield equation   The total yield over the lifetime of the year class is given by

$$Y(t_c, F) = N_0 W_\infty e^{-Mt_c} F \sum_{n=0}^{3} \frac{a_n c^n e^{-nKt_c}}{M + F + nK}$$

where $c = e^{Kt_0}$, $a_0 = 1$, $a_1 = -3$, $a_2 = 3$, and $a_3 = -1$

The following three standard optimization problems are connected with this yield equation:

1. For a fixed F, determine t_c so that $Y(t_c, F)$ is maximized. This is referred to as Allen's problem.
2. With t_c fixed, determine F so that $Y(t_c, F)$ is maximized.
3. With $Y(t_c, F)$ designated, determine t_c so that F (and hence effort) is minimized.

Each of these problems can be solved, in theory, using the standard calculus methods. The calculations involved usually require a computer, however, as the following example of Allen's problem shows.

example 36.6   For the Pacific halibut, use the parameters given in example 36.4 and $F = .5$ to solve Allen's problem.

solution 36.6   The yield equation becomes

$$Y = N_0(83)e^{-.2t_c}(.5)\left[\frac{1}{.7} - 3\frac{1}{.88}(.955997)e^{-.18t_c} \right.$$

$$\left. + 3\frac{1}{1.06}(.91393118)e^{-.36t_c} - \frac{1}{1.24}(.8737145)e^{-.54t_c} \right]$$

$$= N_0[59.2857e^{-.2t_c} - 135.2518e^{-.38t_c} + 107.3438e^{-.56t_c} - 29.24125e^{-.74t_c}].$$

Computing the derivative with respect to t_c, we obtain

$$\frac{dY}{dt_c} = N_0(-11.85714e^{-.2t_c} + 51.3957e^{-.38t_c} - 60.1125e^{-.56t_c} + 21.6385e^{-.74t_c})$$

$$= N_0e^{-.2t_c}(-11.85714 + 51.3957x - 60.1125x^2 + 21.6385x^3)$$

where $x = e^{-.18t_c}$. Setting $dY/dt_c = 0$, we must find the root of the polynomial $p(x) = 21.6385x^3 - 60.1125x^2 + 51.3957x - 11.85714$ with $0 < x = e^{-.18t_c} < 1$. With the aid of a computer, (or using Newton's method in section 51) we find this root to be $x = .368365$, and so $t_c = \ln(.368365)/(-.18) = 5.548 \approx 5.55$ years. The corresponding yield is $Y = 7.44N_0$. ▤

In each of our two models, we have assumed that the effort $E(t)$ expended in tapping the resource is constant over time. This assumption is rarely satisfied. A more realistic model would attempt to find the *function* $E(t)$ that maximizes yield. This is a problem in *dynamic optimization*, a topic far beyond the scope of this text.*

exercises for section 36

part A

1. For the population of Pacific halibut south of Cape Spencer, Alaska, the best estimates for r and B_∞ in the Schaeffer model are $r = .71$ and $B_\infty = 80.5 \times 10^6$ kg. Find the maximum sustainable yield and the corresponding fishing mortality.

2. For the population of Antarctic fin whales, rough estimates are given by $r = .08$ and $B_\infty = 400,000$ whales. Find the smallest fishing mortality F that results in a sustainable yield of 10,000 whales per year. What is the maximum sustainable yield?

3. At present, the world population of blue whales is protected. Rough estimates for r and B_∞ in a logistic model are $r = .05$ and $B_\infty = 150,000$.

   (a) In 1978, the population was estimated to be about 5,000. Assuming $F = 0$, how long will it take to restore the population to a level of 50,000?

   (b) What value of F will result in a long-term population of 50,000?

4. A resource with $r = .4$ and $B_\infty = 20,000$ kg presently gives a sustainable yield of 1,500 kg/yr.

   (a) Estimate the possible values of F.

   (b) For each of the two cases arising in **(a)**, what sustainable yield will result from doubling the effort?

* Solutions involving the Schaeffer and Beverton-Holt models can be found in Colin Clark, *Mathematical bioeconomics: The optimal management of renewable resources.* New York: John Wiley and Sons, 1976.

5. Ten thousand young trout are placed into a pond and their numbers dwindle according to the formula

$$N(t) = 10,000e^{-.4t}$$

Their growth in weight (in pounds) is given by

$$W(t) = .05 + .7t$$

When is the population biomass maximized?

6. For the North Sea plaice the parameters in the von Bertalanffy growth model are $W_\infty = 2,870$ g, $t_0 = -.815$, and $K = .095$. In addition, the natural mortality rate is estimated to be $M = .1$. If fishing were suspended, when would a given year class reach its maximum biomass?

7. For the North Sea haddock, the parameters in the von Bertalanffy growth model are $W_\infty = 1,209$ g, $t_0 = -1.066$, and $K = .20$. The natural mortality is estimated to be $M = .2$. If fishing were suspended, when would a given year class reach its maximum biomass?

8. At Happy Harry's Trout Farm, a pond is stocked with 10,000 young trout and M is estimated to be .4. By the end of the season, 2,800 trout have been caught and the pond is "fished out."

   (a) If $t_c = 0$, estimate the fishing mortality F.

   (b) If business doubles the next year, how many trout will be caught under the same conditions?

9. Rework Exercise 8, assuming that $M = .3$ and $t_c = 1$.

★10. For the North Sea plaice population in exercise 6, the fishing mortality is assumed to be .7.

   (a) Show that the yield Y is given by

$$Y = 2,009N_0e^{-.1t_c}(1.25 - 3.62179x + 3.537826x^2 - 1.1626425x^3)$$

   where $x = e^{-.095t_c}$.

   (b) Show that

$$Y' = 2,009N_0e^{-.1t_c}(-.125 + .7062490x - 1.025969x^2 + .4476173x^3)$$

   where $x = e^{-.095t_c}$.

   (c) The only positive root of the polynomial in (b) is $x = .271185$. Find t_c so that the yield is maximized.

part B   11. In the Schaeffer model, show that if $F \geq r$, then $\lim_{t \to \infty} B(t) = 0$.

 **the first order linear differential equation**

The **first order linear differential equation** is an equation of the form

$$y' = \alpha(t) \cdot y + \beta(t)$$

where $\alpha(t)$ and $\beta(t)$ are continuous functions on an open interval I. Note that, for a fixed t, the right hand side is a linear function of y.

When $\beta(t) \equiv 0$, we have the **homogeneous equation** $y' = \alpha(t)y$. This special case can be solved by the separation of variables method. First note that $y \equiv 0$ is a solution. For $y \neq 0$,

$$\frac{1}{y}\, dy = \alpha(t)\, dt$$

If $A(t)$ is an antiderivative for $\alpha(t)$, then $\ln|y| = A(t) + c_1$. Hence, $y = \pm e^{c_1} e^{A(t)} = ce^{A(t)}$, where c is a nonzero constant. When $c = 0$, we obtain the zero solution. To summarize,

> The general solution to the homogeneous equation $y' = \alpha(t)y$ is
>
> $$y = ce^{A(t)}$$
>
> where $A'(t) = \alpha(t)$.

example 37.1 Solve each of the following differential equations subject to the given boundary condition.

(a) $y' = 2ty$, with $y(0) = 4$ **(b)** $ty' = 2y$, with $y(1) = 2$

solution 37.1 For **(a)**, we have $\alpha(t) = 2t$ and so $A(t) = t^2$. The general solution is therefore $y = ce^{t^2}$. Now, $4 = y(0) = ce^0 = c$. Hence, $y = 4e^{t^2}$.

For **(b)**, we must first rewrite $ty' = 2y$ in the form $y' = 2t^{-1}y$. Hence, $\alpha(t) = 2t^{-1}$ so that $A(t) = 2\ln|t|$. The general solution is therefore $y = ce^{2\ln|t|} = c(c^{\ln|t|})^2 = c|t|^2 = ct^2$. Since $y(1) = 2$, c must be 2. Hence $y = 2t^2$. ∎

To solve the general non-homogeneous case $y' = \alpha(t)y + \beta(t)$, first note that if $y_1(t)$ and $y_2(t)$ are solutions, then $y(t) = y_2(t) - y_1(t)$ satisfies the homogeneous equation $y' = \alpha(t)y$:

$$y'(t) = y_2'(t) - y_1'(t) = [\alpha(t)y_2(t) + \beta(t)] - [\alpha(t)y_1(t) + \beta(t)]$$
$$= \alpha(t)[y_2(t) - y_1(t)] = \alpha(t)y(t)$$

Hence, $y = ce^{A(t)}$ and so $y_2(t) = y_1(t) + ce^{A(t)}$. Consequently, if we can find *one* particular solution $y_1(t)$ to $y' = \alpha(t)y + \beta(t)$, *then the general solution is* $y_1(t) + ce^{A(t)}$.

To find a single solution, we will attempt to find a solution of the form $y_1(t) = u(t)e^{A(t)}$, where $u(t)$ is yet to be determined. This method is known as the **variation of parameters** technique. Computing the derivative of $y_1(t)$, we obtain:

$$y_1'(t) = u(t)[\alpha(t)e^{A(t)}] + u'(t)e^{A(t)}$$
$$= \alpha(t)y_1(t) + [u'(t)e^{A(t)}]$$

Then $y_1(t)$ will be a solution provided $u'(t)e^{A(t)} = \beta(t)$. It follows that $u'(t)$ must be $\beta(t)e^{-A(t)}$. To summarize,

> Let $u(t) = \int \beta(t)e^{-A(t)}\,dt$, where $A'(t) = \alpha(t)$. Then the general solution to $y' = \alpha(t)y + \beta(t)$ is
>
> $$y = u(t)e^{A(t)} + ce^{A(t)}$$

This technique is used in our next three examples.

example 37.2   Find the general solution to the differential equation $y' = -\dfrac{2}{t}y + t^3$ for $t \neq 0$.

solution 37.2   For this differential equation, $\alpha(t) = -2/t$ and so $A(t) = \int (-2/t)\,dt = -2 \ln |t|$. Then $e^{A(t)} = e^{-2 \ln |t|} = |t|^{-2} = 1/t^2$, and $e^{-A(t)} = t^2$. To find a particular solution, let $u(t) = \int t^3 t^2\,dt = t^6/6$, and form $u(t)e^{A(t)}$. Consequently, the general solution is $y = (t^6/6)(1/t^2) + c/t^2 = t^4/6 + c/t^2$. ■

example 37.3   Solve the linear differential equation $y' = -3t^2 y + t^2$ subject to the initial condition $y(0) = -4$.

solution 37.3   In this equation, we have $\alpha(t) = -3t^2$ so that $A(t) = -t^3$. To find a particular solution, let $u(t) = \int t^2 e^{t^3}\,dt$. We can use the substitution method to obtain $u(t) = \frac{1}{3}e^{t^3}$, and a particular solution is $u(t)e^{A(t)} = \frac{1}{3}$. The general solution is then given by $y = \frac{1}{3} + ce^{-t^3}$.

If we apply the initial condition $y(0) = 4$, then $-4 = \frac{1}{3} + c$. Hence, $c = -13/3$ and so $y = 1/3 - (13/3)e^{-t^3}$. ■

example 37.4　　Solve the linear differential equation

$$y' = -y + A \cos \omega t$$

subject to $y(0) = 0$.

solution 37.4　　We have $A(t) = -t$ and $u(t) = \int (A \cos \omega t) e^t \, dt$. Using formula 12 in the integral tables, we obtain

$$u(t) = A e^t \frac{(\cos \omega t + \omega \sin \omega t)}{1 + \omega^2}$$

The general solution is then $u(t) e^{-t} + c e^{-t} = [A/(1 + \omega^2)](\cos \omega t + \omega \sin \omega t) + c e^{-t}$. Finally, $0 = y(0) = A/(1 + \omega^2) + c$. Hence,

$$y = \frac{A}{1 + \omega^2} (\cos \omega t + \omega \sin \omega t - e^{-t})$$

▤

　　The remainder of this section is devoted to applications of first order linear differential equations to biology. In particular, we now introduce the general **one compartment model**.

　　In the compartment shown in figure 37.1, we have material of a fixed size V. This material is commonly a liquid, such as water or blood plasma, but could even be the gene pool in a population genetics model. The material flows into and out of the compartment at the rate $F(t)$. If V were measured in liters, then $F(t)$ might be measured in liters per minute. A second substance is introduced and mixed into the tank at the rate $I(t)$. In physiological applications, this substance (called a *tracer*) may be a dye or radioactive substance. Typical units for $I(t)$ are grams per minute.

figure 37.1

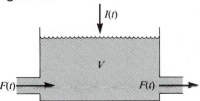

　　Let $y(t)$ be the amount of tracer in the compartment at time t, and let $c(t) = \dfrac{y(t)}{V}$ be the concentration at time t. We will now demonstrate that $y(t)$ satisfies the linear differential equation

$$y' = -\frac{F(t)}{V} y + I(t)$$

Between times t and $(t + \Delta t)$, the amount of tracer that enters the tank is approximately $I(t) \Delta t$. On the other hand, liquid in the amount of $F(t) \Delta t$ has left the tank. Of this amount, approximately $c(t)[F(t) \Delta t]$ is tracer. It follows that

$$y(t + \Delta t) - y(t) = \text{amount entering} - \text{amount leaving}$$
$$\approx I(t) \Delta t - c(t)F(t) \Delta t$$

Hence,

$$\frac{y(t + \Delta t) - y(t)}{\Delta t} \approx -c(t)F(t) + I(t)$$

As $\Delta t \to 0$, all of the approximations improve and the left hand side approaches $y'(t)$ by the definition of derivative. Since $c(t) = y(t)/V$, we obtain

$$y' = -\frac{F(t)}{V} y + I(t)$$

Note that when $I(t) \equiv 0$ and $F(t) \equiv F$, then $y' = -(F/V)y$ and so

$$y(t) = y_0 e^{-(F/V)t}$$

Thus, a **simple exponential decay** results. In order to keep tracer in the system $I(t)$ must be nonzero. The function $I(t)$ is commonly called a **driving function** for the system.

example 37.5

Water flows through a 500-gallon tank at the rate of 10 gallons per minute. Initially, 5 gallons of chlorine are placed into the the tank and additional chlorine is added to the tank at the constant rate of I gallons per minute.
(a) Solve the resulting differential equation for y, the amount of chlorine in the tank at time t.
(b) What rate of input I results in a long-range chlorine concentration of 1 part chlorine per 1,000 parts water?

solution 37.5

(a) The function $y(t)$ satisfies $y' = -.02y + I$. Thus, $\alpha(t) = -.02$ and so $A(t) = -.02t$. To find a particular solution, let

$$u(t) = \int e^{.02t} I \, dt = 50I e^{.02t}$$

A particular solution is then $u(t)e^{A(t)} = 50I$, and the general solution takes the form $y = ce^{-.02t} + 50I$. We are given the initial condition $y(0) = 5$. Hence, $c + 50I = 5$ and so $c = 5 - 50I$. The final solution is then

$$y = (5 - 50I)e^{-.02t} + 50I$$

(b) First note that $\lim_{t \to \infty} y(t) = 50I$. The long-range concentration is therefore $50I/500 = 0.1I$. In order to have $0.1I = .001$, we must let $I = .01$ gallon of chlorine per minute.

example 37.6   **the glucose tolerance test**   To measure carbohydrate metabolism in a subject, glucose is infused intravenously into the bloodstream at a known rate of I milligrams per minute. If we set $\lambda = F/V$ in our basic model, then glucose is metabolized at a rate of λy milligrams per minute. The constant λ is to be estimated from concentration measurements. The differential equation $y' = -\lambda y + I$ has solution

$$y(t) = ae^{-\lambda t} + \frac{I}{\lambda}$$

and so the glucose concentration is given by

conclusion $\qquad c(t) = \dfrac{a}{V} e^{-\lambda t} + \dfrac{I}{V\lambda}$

Hence, the long-term glucose concentration level is $\lim_{t \to \infty} c(t) = I/(V\lambda)$.

To be more specific assume $I = 300$ milligrams per minute, $c(0) = 850$ milligrams per liter and the long-range or *steady state* concentration has been measured at 1,400 milligrams per liter. Unfortunately, this is not enough information to solve for the key parameter λ. Although we can write $V\lambda = 300/1,400 = 3/14$, V itself is unknown.

Note, however, that $1,400 - c(t) = (a/V) e^{-\lambda t}$ and so $\ln[1,400 - c(t)] = \ln(a/V) - \lambda t$. Hence, if we graph $Y = \ln[1,400 - c(t)]$ versus t, then $-\lambda$ is the slope of the resulting linear function, and it can be estimated from concentration measurements at times $t_1, t_2, \ldots, t_n$.

If, for example, $c(30) = 1,250$, then $Y = \ln 150$. Since $c(0) = 850$, we have $Y(0) = 550$. Hence, an estimate for λ is $-\left[\dfrac{\ln 150 - \ln 550}{30 - 0}\right] = .043$.

In practice, λ is best estimated from many more than two concentration measurements. We will return to this curve fitting problem in section 43.

example 37.7   **a model for the distribution of a drug in the tissues**   If a fixed dose y_0 of a drug is injected into the bloodstream, the drug will eventually make its way into organ(s) where it is destroyed through metabolism and/or excretion. The compartment model for the bloodstream predicts a simple exponential decay $y_0 e^{-\lambda_1 t}$. Thus, the drug leaves the bloodstream at the rate $I(t) = y_0 \lambda_1 e^{-\lambda_1 t}$. This function serves as the driving function for the second compartment, as illustrated in figure 37.2. If $y(t)$ is the amount of drug in the organ, then

$$y' = -\lambda_2 y + y_0 \lambda_1 e^{-\lambda_1 t}$$

and we will assume that there is no drug in the compartment initially.

figure 37.2

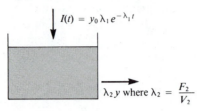

$$I(t) = y_0 \lambda_1 e^{-\lambda_1 t}$$

$$\lambda_2 y \text{ where } \lambda_2 = \frac{F_2}{V_2}$$

To find a particular solution, note that $\alpha(t) = -\lambda_2$ and so $A(t) = -\lambda_2 t$. Hence,

$$\int e^{-A(t)} B(t)\, dt = \int e^{\lambda_2 t} y_0 \lambda_1 e^{-\lambda_1 t}\, dt = \frac{y_0 \lambda_1}{\lambda_2 - \lambda_1}\, e^{(\lambda_2 - \lambda_1)t}$$

provided $\lambda_1 \neq \lambda_2$. Multiplying by $e^{A(t)} = e^{-\lambda_2 t}$, we obtain $[y_0 \lambda_1/(\lambda_2 - \lambda_1)]\, e^{-\lambda_1 t}$ as a particular solution. The general solution is then

$$y = ce^{-\lambda_2 t} + \frac{y_0 \lambda_1}{\lambda_2 - \lambda_1}\, e^{-\lambda_1 t}$$

Since $y(0) = 0$, $c = -y_0 \lambda_1/(\lambda_2 - \lambda_1)$. The final solution is then

---

conclusion $$y = \frac{-y_0 \lambda_1}{\lambda_2 - \lambda_1}[e^{-\lambda_2 t} - e^{-\lambda_1 t}]$$

It is not hard to show that $y(t)$ has the shape shown in figure 37.3, with a single maximum at $t_0 = \dfrac{1}{\lambda_2 - \lambda_1}\ln\left(\dfrac{\lambda_2}{\lambda_1}\right)$.

figure 37.3

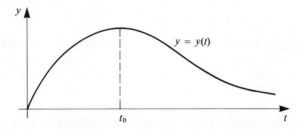

$$y = y(t)$$

example 37.8   **a model for blood flow through the aorta**   In the following simple model for blood flow through the aorta (see figure 37.4), the aorta is represented as an elastic container whose volume depends on blood pressure.

figure 37.4

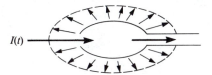

The rate of inflow from the left ventricle to the aorta will be represented by

$$I(t) = \begin{cases} A \sin \omega t, \ 0 \le t \le \dfrac{\pi}{\omega} \\[2mm] 0, \ \dfrac{\pi}{\omega} \le t \le T \end{cases}$$

The interval $[0, \pi/\omega]$ represents the *systolic phase*, in which blood rushes into the aorta and the pressure rises. The heart does not pump during the *diastolic phase* $[\pi/\omega, T]$. In general, the diastolic phase lasts more than π/ω seconds. Typical lengths for the systolic and diastolic phases for an individual at rest are .27 second and .53 second respectively.

The rate of outflow $F(t)$ is governed by Poiseuille's Law: $F(t) = kP$, where P is the blood pressure at time t. (As we have seen in Section 21, k is actually dependent on the radius R of the vessel.) The simplest assumption to make is that *the relationship between the pressure P in the aorta and its volume V is linear.* Thus $\dfrac{dP}{dV} = m$ for some constant $m > 0$. It follows that $\dfrac{dV}{dt} = I(t) - F(t) = I(t) - kP$. It is easier, however, to first solve the differential equation for P:

$$\frac{dP}{dt} = \frac{dP}{dV}\frac{dV}{dt} = m\frac{dV}{dt}$$
$$= -mkP + mI(t)$$

Let $\lambda = mk$. During the diastolic phase, $I(t) \equiv 0$ and so the solution takes the form

---

diastolic phase $P = c_2 e^{-\lambda t}$

To solve the differential equation during the systolic phase, we may proceed as in example 37.4 to find:

---

systolic phase $P = c_1 e^{-\lambda t} + Am \dfrac{\lambda \sin \omega t - \omega \cos \omega t}{\lambda^2 + \omega^2}$

If we let P_0 and P_1 be the aortic pressures (measured in millimeters of mercury) at the beginning and the end of the systolic phase, then c_1 and c_2 can be evaluated in terms of P_0 and P_1:

$$c_2 = P_1 e^{\lambda \pi / \omega}$$

and

$$c_1 = P_0 + Am \frac{\omega}{\lambda^2 + \omega^2}$$

The basic shape of the graph of P (and also of V) is shown in figure 37.5. This graph has many (but not all) of the geometric characteristics of the actual graph of $P(t)$ that has been determined experimentally. Typical values for P_0 and P_1 are 80 and 100, millimeters of mercury respectively, with a maximum of about 120.

figure 37.5

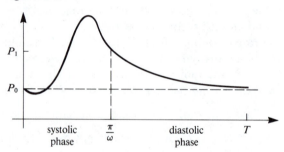

exercises for section 37

part A

Solve each of the following homogeneous linear differential equations subject to the given initial condition.

1. $y' = t^2 y$, with $y(0) = 4$ 2. $y' = (\sin t)y$, with $y(0) = 5$
3. $ty' = 5y$, for $t > 0$, and subject to $y(1) = -2$
4. $(t+1)y' = 2y$, for $t > -1$, and subject to $y(0) = 3$
5. $y' = (2 \ln t)y$, with $y(1) = \frac{3}{2}$

Find the general solution to each of the following nonhomogeneous linear differential equations.

6. $y' = 2y/t + t^2$ 7. $y' = -2ty + t$
8. $y' = t - 3y$ 9. $ty' = y - t^2$, for $t > 0$
10. $y' = -y + 2 \sin t$

Solve each of the following linear differential equations subject to the given boundary condition. Use the integral tables when needed.

11. $y' = y - e^t$, with $y(0) = 0$

12. $y' = -2y - e^{3t}$, with $y(0) = -1$

13. $y' = -3y + e^{-t}$, with $y(0) = -2$

14. $y' = y + \sin t$, with $y(0) = 4$

15. $ty' - 2y = 3t^2$, with $y(1) = 2$

16. $y' = 2t - y$, with $y(1) = 1$

17. $y' + (\tan t)y = 2\cos t$, for $-\pi/2 < t < \pi/2$, with $y(0) = 1$

18. $(1 + t^2)y' = 1 - 2ty$, with $y(0) = 5$

19. $y' = (-2/t)\,y + (\cos t)/t^2$, with $y(2\pi) = 0$

★20. $\dfrac{dy}{dt} = \dfrac{1}{y - t}$ with $y(2) = 0$

$$\left[Hint: \text{ Find } \dfrac{dt}{dy} \text{ and consider } y \text{ as the input variable.} \right]$$

21. Water is flowing through a tank containing 500 gallons of water at the rate of 10 gallons per minute. If 20 pounds of salt are mixed into the tank, how long will it take until just 5 pounds of salt remain?

22. Salt water of concentration .1 pound per gallon is pumped into a 100-gallon tank at the rate of 25 gallons per minute. Water leaves this tank at the same rate. Let $y(t)$ denote the amount of salt in the tank at time t.

(a) Find $y(t)$ if $y(0) = 0$.

(b) Compute $y_\infty = \text{limit}_{t \to \infty}\, y(t)$.

23. Solve the differential equation $y' = -\lambda y + I$ that arose in the glucose tolerance test in example 37.6.

24. In a simple model for radioactive buildup, suppose strontium 90 is being introduced into the environment at the steady rate of $I(t) = .5$ gram per day. The half-life of strontium 90 is 29 years.

(a) Set up the appropriate differential equation and solve for $y(t)$, the amount in the environment at time t.

(b) Compute $y_\infty = \text{limit}_{t \to \infty}\, y(t)$.

25. A population grows according to the differential equation

$$y' = .05(1{,}000 - y)$$

where $y(t)$ is the population count at time t.

(a) Solve for $y(t)$ and compute $\text{limit}_{t \to \infty}\, y(t)$.

(b) A fixed number h are harvested per unit time from the population.

Thus,

$$y' = .05(1{,}000 - y) - h$$

Solve for y and compute $\lim_{t \to \infty} y(t)$.

(c) What harvesting rate h will result in a long-term population of 300?

26. A bird population experiences seasonal growth following the equation

$$y' = [5 \sin (2\pi t)]y$$

where t is measured in years and $t = 0$ corresponds to the beginning of spring. On the other hand, migration to and from the locale is also seasonal at the rate $I(t) = 1{,}000 \sin 2\pi t$ birds per year.

(a) What differential equation does $y(t)$ satisfy? Solve for $y(t)$ if $y(0) = 200$.

(b) When is the population level the largest?

27. Show that the maximum value of the drug concentration curve in example 37.7 occurs at

$$t = \frac{1}{\lambda_2 - \lambda_1} \ln \left(\frac{\lambda_2}{\lambda_1} \right)$$

28. Carry out the details of solving the differential equation

$$P' = -\lambda P + mI(t)$$

of example 37.8 for the systolic phase.

29. R.H. Rainey* has presented a single compartment model to study pollution effects in the Great Lakes. The model assumes that pollutants are being added to a given lake at a constant rate of $I(t) = I$ and that pollutants are thoroughly mixed into the lake. If annual precipitation into the lake matches evaporation, we may assume that the flow rate $F(t)$ is constant. Let $c(t)$ denote the concentration of the pollutant at time t.

(a) Show that $c(t)$ satisfies the differential equation

$$c' = -\frac{F}{V}c + \frac{1}{V}$$

(b) Solve for c given that $c(0) = c_0$ and determine $c_\infty = \lim_{t \to \infty} c(t)$.

(c) For Lake Erie, $V \approx 458 \text{ km}^3$ and $F \approx 175 \text{ km}^3/\text{yr}$. If pollution were stopped completely, determine how many years it would take for $c(t)$ to drop from c_∞ to $.1c_\infty$.

(d) Rework **(c)** for Lake Superior, for which $V \approx 12{,}221 \text{ km}^3$ and $F \approx 65.2 \text{ km}^3/\text{yr}$.

* R. H. Rainey, *Science*, **155**, (1969), pp. 1,242–1,243.

30. The model presented in example 37.7 has been applied by C.S. Gist and F.W. Whicker* to study the effect of radioactive iodine on deer populations. The iodine 131, produced by nuclear fission, settles on vegetation and, after being eaten, accumulates primarily in the thyroid gland of the grazing animal. If the concentration of iodine 131 in the thyroid is sufficiently high, destruction of tissue and cancerous tumors may result. The model, then, takes the form illustrated in the figure. Let $y(t)$ be the

amount of iodine 131 in the thyroid after t days, and suppose I_0 is the initial amount deposited in the vegetation that will be eaten by a single animal.

(a) Show that $y' = -\lambda_2 y + I_0\lambda_1 e^{-\lambda_1 t}$ and solve.

(b) The effect on a Colorado deer population of the Chinese nuclear test of 1964 was studied. Estimates for the two parameters were $\lambda_1 = .126$ and $\lambda_2 = .107$. Determine when the iodine 131 in the thyroid is a maximum and find the maximum value in terms of I_0.

part B
Many bodily functions vary periodically over a 24-hour period. Urine output, for example, is low at night and reaches a maximum in midday. This suggests that the flow rate function should involve trigonometric functions. The single compartment model could then take the form

$$y' = -(B + A \sin \omega t)y + I(t)$$

31. If $I(t) \equiv 0$ and $y(0) = y_0$, solve for $y(t)$.

★32. If $I(t) \equiv I$, find a formula for $I(t)$.

A differential equation of the form $y' = \alpha(t)y + \beta(t)y^n$ for $n \neq 1$ is known as a *Bernoulli differential equation*. This equation has appeared in population growth models.

33. Let $w - y^{1-n}$ and show that w satisfies the first order linear differential equation

$$w' = (1 - n)\alpha(t)w + (1 - n)\beta(t)$$

34. Use part B exercise 33 to solve $y' = -y + t^2 y^2$.

35. In the model of example 37.8, the systolic pressure function satisfied the differential equation

$$P' = -mkP + m(A \sin \omega t)$$

* C.S. Gist and F.W. Whicker, *Journal of Wildlife Management*, **35** (1971), pp. 461–468.

The basic shape of the graph of P can be found by analyzing P' for $0 \le t \le \pi/\omega$.

(a) When $t = 0$, $P = P_0$. Show that $P'(0) < 0$.

(b) When $t = \pi/\omega$, $P = P_1$. Show that $P'(\pi/\omega) < 0$.

(c) As pointed out in example 37.8, $P_0 < P_1$. Conclude that $P(t)$ must have a relative maximum and minimum strictly between 0 and π/ω.

★**(d)** Conclude that P' has a sign chart of the form illustrated in the figure.

36. In a form of Fick's Principle that is used in the study of organs, dye or tracer of constant concentration c_0 flows into the organ. The organ removes the dye at the rate of R units per minute. Let $y(t)$ be the amount in the organ at time t, and let $c(t) = [y(t)]/V$.

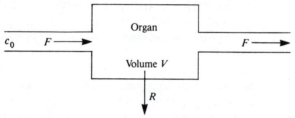

(a) Using pages 401 and 402 as a guide, show that $y(t)$ satisfies

$$y' = -\frac{F}{V} y - R + c_0 F$$

(b) Solve this differential equation and show that:

$$c_\infty = \lim_{t \to \infty} c(t) = \frac{c_0 F - R}{F}$$

(c) Conclude that for very large t, $F \approx \dfrac{R}{c_0 - c_\infty}$.

This is a general form of Fick's Principle. c_0 and c_∞ are measured using blood samples. If R is known, the equation estimates F. If F is known, $R = F(c_0 - c_\infty)$ estimates the rate of removal R.

37. The following simple model has been proposed as a first step in studying the process of breathing.* Air is forced down the airway into an elastic balloon. This balloon resides in a diaphragm where the pressure is a constant P_1 (see the figure). Define $P(t) = $ pressure in the balloon at time t,

* Based on John A. Jacquez, *Respiratory Physiology*, New York: Hemisphere Publishing Co., McGraw-Hill, (1979), pp. 56–59 and 70–72.

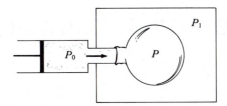

$V(t)$ = volume of the balloon at time t, and $P_0(t)$ = pressure at the airway opening at time t. We will make the following two assumptions:

1. The rate of flow $\dfrac{dV}{dt}$ into or out of the balloon is directly proportional to the pressure difference $(P_0 - P)$.

2. The volume V of the balloon is directly proportional to the pressure difference $(P - P_1)$.

(a) Conclude that $V' + (\alpha/\beta)\, V = \alpha[P_0(t) - P_1]$ for some constants $\alpha > 0$ and $\beta > 0$.

(b) Solve the differential equation in case $P_0(t) \equiv P_0$. What is $\lim_{t \to +\infty} V(t)$?

(c) Solve the differential equation in case $[P_0(t) - P_1]$ is of the form $A \cos \omega t$.

 the second order linear differential equation

The **second order linear differential equation** is an equation of the form

$$y'' + \alpha_1(t)y' + \alpha_2(t)y = \beta(t)$$

where $\alpha_1(t)$, $\alpha_2(t)$, and $\beta(t)$ are continuous on an open interval I. A *solution* is then a twice differentiable function $f(t)$ that satisfies

$$f''(t) + \alpha_1(t)f'(t) + \alpha_2(t)f(t) = \beta(t)$$

When $\beta(t) \equiv 0$, we call the equation **homogeneous**. Otherwise, the equation is called **nonhomogeneous**.

Unfortunately, there are no explicit formulas (as in Section 37) for solving this general equation. In this section, we will concentrate on solving various special cases that commonly occur in applications.

the homogeneous equation with constant coefficients   The first special case occurs when $\beta(t) \equiv 0$ and $\alpha_1(t)$ and $\alpha_2(t)$ are constants. By clearing out fractions if necessary, we can rewrite the equation in the form

$$ay'' + by' + cy = 0$$

where $a \neq 0$. As we have seen in section 37, a first order linear differential equation with constant coefficients (such as $y' = ay$) has a solution involving a simple exponential function ($y = ce^{at}$). This suggests that we try $y = e^{mt}$ as a solution in the present situation:

$$ay'' + by' + cy = a(m^2 e^{mt}) + b(me^{mt}) + ce^{mt}$$
$$= e^{mt}(am^2 + bm + c)$$

Hence, e^{mt} will be a solution provided $am^2 + bm + c = 0$. The polynomial $am^2 + bm + c$ is called the **characteristic polynomial**.

example 38.1   Find exponential solutions to each of the following differential equations.

$$\text{(a) } y'' - 5y' + 4y = 0 \qquad \text{(b) } 2y'' + 5y' + 2y = 0$$

solution 38.1   (a) The function $y = e^{mt}$ is a solution provided $m^2 - 5m + 4 = (m - 1)(m - 4) = 0$. Hence, $m = 1$ and $m = 4$, and so e^t and e^{4t} are solutions.

For **(b)**, the characteristic polynomial is $2m^2 + 5m + 2 = (2m+1)(m+2)$. Hence, $m = -\frac{1}{2}$ and $m = -2$ are roots. Solutions are then $e^{-t/2}$ and e^{-2t}.   ▤

definition 38.1

Two nonzero solutions y_1 and y_2 to the equation $ay'' + by' + cy = 0$ are called **linearly independent solutions** provided $y_2(t)$ is not a constant multiple of $y_1(t)$. We then call $y_1(t)$ and $y_2(t)$ a **fundamental set of solutions**.

Thus, in example 38.1**(a)**, the solutions e^t and e^{4t} are linearly independent and constitute a fundamental set of solutions for $y'' + 5y' + 2y = 0$.

Next note that if $y_1(t)$ and $y_2(t)$ are solutions, then $y(t) = c_1 y_1(t) + c_2 y_2(t)$ is a solution for any choice of constants c_1 and c_2:

$$
\begin{aligned}
ay''(t) + by'(t) + cy(t) &= a[c_1 y_1''(t) + c_2 y_2''(t)] + b[c_1 y_1'(t) + c_2 y_2'(t)] \\
&\quad + c[c_1 y_1(t) + c_2 y_2(t)] \\
&= c_1(ay_1'' + by_1' + cy_1) + c_2(ay_2'' + by_2' + cy_2) \\
&= c_1 \cdot 0 + c_2 \cdot 0 = 0
\end{aligned}
$$

Thus, in example 38.1**(a)**, for any choice of constants c_1 and c_2, $y = c_1 e^t + c_2 e^{4t}$ is again a solution. Are there any others? The key theorem of this section asserts that *there are no others*.

theorem 38.1

Let $y_1(t)$ and $y_2(t)$ be linearly independent solutions to $ay'' + by' + cy = 0$. Then the general solution is $y(t) = c_1 y_1(t) + c_2 y_2(t)$, where c_1 and c_2 are arbitrary constants.

The proof of the theorem is beyond the level of the text. As the following example shows, two boundary conditions are needed to determine the constants c_1 and c_2.

example 38.2

Find the general solution to $y'' - 5y' + 6y = 0$. Then determine the solution that satisfies the boundary conditions $y(0) = 1$ and $y'(0) = -2$.

solution 38.2

The characteristic polynomial $m^2 - 5m + 6 = (m-2)(m-3)$ has roots $m = 2$ and $m = 3$. Hence, the general solution takes the form $y = c_1 e^{2t} + c_2 e^{3t}$. Hence, $y' = 2c_1 e^{2t} + 3c_2 e^{3t}$. Since $y(0) = 1$ and $y'(0) = -2$, we obtain:

$$c_1 + c_2 = 1$$
$$2c_1 + 3c_2 = -2$$

This system has solution $c_1 = 5$ and $c_2 = -4$, and so $y = 5e^{2t} - 4e^{3t}$.   ▤

The characteristic polynomial of $y'' - 2y' + y = 0$ is $(m-1)^2$. Hence, $y_1(t) = e^t$ is a solution. According to Theorem 38.1, we need a second independent solution before we can write the general solution. The following argument will establish that te^t is the additional solution needed. You may want to skip over the derivation at your first reading.

Assume that the characteristic polynomial is $(m - k)^2 = m^2 - 2km + k^2$. Thus, in the equation $ay'' + by' + cy = 0$, we have $a = 1$, $b = -2k$, and $c = k^2$. We will try to find an additional solution of the form $y = u(t)e^{kt}$, where $u(t)$ is yet to be determined. This method is another case of the **variation of parameters** technique. Using the Product Rule, we obtain

$$y'' - 2ky' + k^2y = (k^2ue^{kt} + 2ku'e^{kt} + u''e^{kt}) - 2k(u'e^{kt} + kue^{kt}) + k^2(ue^{kt})$$
$$= e^{kt}(k^2u + 2ku' + u'' - 2ku' - 2k^2u + k^2u)$$
$$= e^{kt}u''$$

Thus, $y = u(t)e^{kt}$ will be a solution provided $u'' = 0$. Note that $u(t) = t$ satisfies this condition so that $y_2(t) = te^{kt}$ is a second linearly independent solution.

The results accumulated thus far in this section are summarized in table 38.1.

table 38.1   Solutions to $ay'' + by' + cy = 0$

Roots of $am^2 + bm + c$	Fundamental Solutions	General Solution
1. distinct real roots m_1 and m_2	e^{m_1t} and e^{m_2t}	$c_1e^{m_1t} + c_2e^{m_2t}$
2. a single real root k	e^{kt} and te^{kt}	$c_1e^{kt} + c_2te^{kt}$

example 38.3   Solve the differential equation $y'' + 4y' + 4y = 0$ subject to the boundary conditions $y(0) = 4$ and $y(1) = 1$. (This is known as a **two-point boundary value problem**.)

solution 38.3   The characteristic polynomial is $m^2 + 4m + 4 = (m + 2)^2$. Then $m = -2$ is the single root. Hence, the general solution is $y = c_1e^{-2t} + c_2te^{-2t}$. Since $y(0) = 4$, we have $c_1 + c_2 \cdot 0 = c_1$. Hence, $c_1 = 4$ and so $y = 4e^{-2t} + c_2te^{-2t}$. Since $y(1) = 1$, we have $1 = 4e^{-2} + c_2e^{-2} = e^{-2}(4 + c_2)$. Hence, $4 + c_2 = e^2$ or $c_2 = e^2 - 4$. The final solution is then

$$y = 4e^{-2t} + (e^2 - 4)te^{-2t} \qquad \blacksquare$$

The characteristic polynomial of $y'' + y = 0$ is $m^2 + 1$. There are no real solutions to $m^2 + 1 = 0$. Recall that the solutions are the *imaginary numbers* $\pm i$, where $i = \sqrt{-1}$. We have come up empty! In Section 30, however, we discovered that both $y_1(t) = \sin \omega t$ and $y_2(t) = \cos \omega t$ are solutions to $y'' = -\omega^2y$, that is, $y'' + \omega^2y = 0$. This suggests that the solutions corresponding to roots $\alpha + \beta i$ and $\alpha - \beta i$ might involve trigonometric functions.

When $m = \alpha + \beta i$ and $m = \alpha - \beta i$ are the two complex roots arising from the quadratic formula solution of $am^2 + bm + c = 0$, it can be shown that $y_1(t) = e^{\alpha t} \sin \beta t$ and $y_2(t) = e^{\alpha t} \cos \beta t$ are solutions (see Part B exercise 49). This final case is summarized in table 38.2 and illustrated in the next two examples.

table 38.2 Solutions to $ay'' + by' + cy = 0$

Roots of $am^2 + bm + c$	Fundamental Solutions	General Solution
3. complex roots $\alpha + \beta i$ and $\alpha - \beta i$	$e^{\alpha t} \sin \beta t$ and $e^{\alpha t} \cos \beta t$	$e^{\alpha t}(c_1 \sin \beta t + c_2 \cos \beta t)$

example 38.4   Find the general solution to the differential equation $y'' - 6y' + 13y = 0$.

solution 38.4   The characteristic polynomial is $m^2 - 6m + 13$. Using the quadratic formula, we obtain the roots

$$m = \frac{6 \pm \sqrt{36 - 4(13)}}{2} = \frac{6 \pm \sqrt{-16}}{2} = \frac{6 \pm 4i}{2} = 3 \pm 2i$$

Thus, $y = e^{3t}(c_1 \sin 2t + c_2 \cos 2t)$ is the general solution.   ▤

example 38.5   Find the solution to $y'' + 2y' + 2y = 0$ that satisfies the boundary conditions $y(0) = -1$ and $y'(0) = 3$.

solution 38.5   The characteristic polynomial $m^2 + 2m + 2 = 0$ has solutions $m = (-2 \pm \sqrt{4 - 8})/2 = (-2 \pm 2i)/2 = -1 \pm i$. The general solution is then $y = e^{-t}(c_1 \sin t + c_2 \cos t)$. Now, $-1 = y(0) = e^0(c_1 \sin 0 + c_2 \cos 0) = c_2$. Hence, $c_2 = -1$ and so $y = e^{-t}(c_1 \sin t - \cos t)$. Now, $y' = e^{-t}(c_1 \cos t + \sin t) - e^{-t}(c_1 \sin t - \cos t)$. Hence, $3 = y'(0) = c_1 - (-1) = c_1 + 1$, and so $c_1 = 2$. Therefore, the final solution is $y = e^{-t}(-\cos t + 2 \sin t)$.   ▤

the nonhomogeneous linear differential equation   If $y_1(t)$ and $y_2(t)$ are solutions to the nonhomogeneous differential equation

$$ay'' + by' + cy = \beta(t)$$

then $y(t) = y_2(t) - y_1(t)$ is a solution to the corresponding homogeneous equation, as the following proof demonstrates:

$$\begin{aligned} ay'' + by' + cy &= a(y_2 - y_1)'' + b(y_2 - y_1)' + c(y_2 - y_1) \\ &= ay_2'' + by_2' + cy_2 - (ay_1'' + by_1' + cy_1) \\ &= \beta(t) - \beta(t) \\ &= 0 \end{aligned}$$

Hence, $y_2(t) = y_1(t) + y(t)$, where $y(t)$ is a solution to $ay'' + by' + cy = 0$. Thus to find the general solution to the nonhomogeneous equation, we follow the steps described in the accompanying box.

---

procedure for obtaining solutions to $ay'' + by' + cy = \beta(t)$

Step 1. Solve the homogeneous differential equation $ay'' + by' + cy = 0$. Let $f_1(t)$ and $f_2(t)$ be fundamental solutions.

Step 2. Construct *one particular solution* $y_p(t)$ to $ay'' + by' + cy = \beta(t)$.

Step 3. Form $y = y_p(t) + c_1 f_1(t) + c_2 f_2(t)$. This is the *general solution*.

---

example 38.6

Verify that $y_p(t) = \frac{1}{2}t - \frac{3}{4}$ is a solution to the differential equation $y'' + 3y' + 2y = t$. Finally, construct the general solution.

solution 38.6

The first and second derivatives of y_p are $y_p'(t) = \frac{1}{2}$ and $y_p''(t) = 0$. Hence,

$$y_p'' + 3y_p' + 2y_p = 0 + 3\left(\frac{1}{2}\right) + 2\left(\frac{1}{2}t - \frac{3}{4}\right)$$

$$= \frac{3}{2} + t - \frac{3}{2} = t$$

The general solution to $y'' + 3y' + 2y = 0$ is $c_1 e^{-t} + c_2 e^{-2t}$. This follows from solving the characteristic polynomial equation $m^2 + 3m + 2 = 0$. We may then conclude that the general solution to the original equation is $y = (\frac{1}{2}t - \frac{3}{4}) + c_1 e^{-t} + c_2 e^{-2t}$ for arbitrary constants c_1 and c_2.   ▤

How then can a single solution be constructed? One of the most widely used techniques is the **method of undetermined coefficients**.

Given $y'' + 3y' + 2y = t$, it is likely that a solution may take the form $y_p(t) = A + Bt$, where the constants A and B are *to be determined*. Since $y_p'(t) = A$ and $y_p''(t) = 0$,

$$y_p'' + 3y_p' + 2y_p = 3B + 2(A + Bt) = (3B + 2A) + 2Bt$$

For y_p to be a solution, we must have $3B + 2A = 0$ and $2B = 1$. Hence, $B = \frac{1}{2}$ and $A = \frac{1}{2}(-3B) = -\frac{3}{4}$. Thus, $y_p(t) = -\frac{3}{4} + \frac{1}{2}t$. This is the solution used in example 38.6.

Table 38.3 gives the form of the particular solution for a given $\beta(t)$. The capital letters in the table indicate *coefficients to be determined*. Notice the similarity between a given $\beta(t)$ and the form of the trial solution. If we substitute the trial solution form into the original differential equation, we will *usually* obtain a system of linear equations that has a solution.

table 38.3 The Method of Undetermined Coefficients

For $\beta(t)$ of the form:	Try a particular solution of the form:
constant a	A
$at + b$	$At + B$
$at^2 + bt + c$	$At^2 + Bt + C$
$a_n t^n + a_{n-1} t^{n-1} + \cdots + a_0$	$A_n t^n + A_{n-1} t^{n-1} + \cdots + A_0$
ae^{kt}	Ae^{kt}
$a \sin bt \quad$ or $\quad a \cos bt$	$A \sin bt + B \cos bt$
$e^{at} \cos bt \quad$ or $\quad e^{at} \sin bt$	$e^{at}(A \sin bt + B \cos bt)$

This method is illustrated in the next two examples.

example 38.7

Using the method of undetermined coefficients, construct particular solutions to each of the following nonhomogeneous differential equations:

$$\textbf{(a)} \ y'' - 3y = 5, \qquad \textbf{(b)} \ y'' - 3y = 2e^{-t} \qquad \textbf{(c)} \ y'' - 3y = 3 \cos 2t$$

solution 38.7

(a) Letting $y_p = A$, we obtain $y_p'' = 0$ and so $y_p'' - 3y_p = -3A$. Hence, $-3A = 5$, and it follows that $y_p(t) = -5/3$.

(b) Letting $y_p = Ae^{-t}$, we obtain $y_p' = -Ae^{-t}$ and $y_p'' = Ae^{-t}$. Consequently, $y_p'' - 3y_p = Ae^{-t} - 3Ae^{-t} = -2Ae^{-t}$. To match up with $\beta(t) = 2e^{-t}$, set $A = -1$. The particular solution is then $y_p = -e^{-t}$.

(c) Let $y_p = A \sin 2t + B \cos 2t$. Then $y_p' = 2A \cos 2t - 2B \sin 2t$, and so $y_p'' = -4A \sin 2t - 4B \cos 2t$. Hence,

$$y_p'' - 3y_p = (-4A \sin 2t - 4B \cos 2t) - 3(A \sin 2t + B \cos 2t)$$
$$= -7A \sin 2t - 7B \cos 2t$$

Since $\beta(t) = 3 \cos 2t$, set $A = 0$ and $B = -\dfrac{3}{7}$. Then $y_p(t) = -\dfrac{3}{7} \cos 2t$.

example 38.8

Find the general solution to the differential equation $y'' - 2y' + y = 4e^{-3t}$.

solution 38.8

If we let $y_p = Ae^{-3t}$, then $y_p' = -3Ae^{-3t}$, and $y_p'' = 9Ae^{-3t}$. Hence, $y_p'' - 2y_p' + y_p = (9Ae^{-3t}) - 2(-3Ae^{-3t}) + Ae^{-3t} = 16Ae^{-3t}$. Since $\beta(t) = 4e^{-3t}$, we can find a particular solution by setting $A = 1/4$.

For the homogeneous equation, the characteristic polynomial is $m^2 - 2m + 1 = (m - 1)^2$. Hence, e^t and te^t are fundamental solutions. The general solution to $y'' - 2y' + y = 4e^{-3t}$ is therefore given by $y = \frac{1}{4}e^{-3t} + c_1 e^t + c_2 t e^t$.

It should be pointed out that there are a few instances in which our trial solutions will fail. *This will always occur when $\beta(t)$ is itself a solution to $ay'' + by' + cy = 0$. In these situations, the trick is to multiply the usual table entry by t,* as demonstrated in the final example.

example 38.9   Find the general solution to the nonhomogeneous differential equation $y'' - 3y' + 2y = 2e^t$.

solution 38.9   The solution to $y'' - 3y' + 2y$ is $c_1e^t + c_2e^{2t}$ and so $\beta(t) = 2e^t$ is itself a solution. Instead of trying $y_p = Ae^t$ (which fails), set $y_p = Ate^t$. Hence, $y_p' = A(te^t + e^t)$ and $y_p'' = A(te^t + 2e^t)$. Substituting into the differential equation, we obtain

$$y_p'' - 3y_p' + 2y_p = (Ate^t + 2Ae^t) - 3(Ate^t + Ae^t) + 2(Ate^t)$$
$$= e^t[(A - 3A + 2A)t + (2A - 3A)]$$
$$= -Ae^t$$

Setting $A = -2$, we obtain $-2te^t$ as a particular solution, and $y = c_1e^t + c_2e^{2t} - 2te^t$ as the general solution.   ▤

exercises for section 38

part A   Find two fundamental solutions to each of the following second order linear differential equations.

1. $y'' - y' - 2y = 0$ 2. $2y'' - 3y' + y = 0$

3. $y'' - 5y' = 0$ 4. $y'' + 6y' = 0$

5. $y'' + 4y = 0$ 6. $y'' - 2y' + y = 0$

7. $y'' + 6y' + 9y = 0$ 8. $y'' + 2y' + 2y = 0$

9. $y'' - 4y' + 13y = 0$ 10. $4y'' + 12y' + 9y = 0$

Find the general solution to each of the following second order linear differential equations. Then find the solution that satisfies $y(0) = 1$ and $y'(0) = 0$.

11. $y'' + 9y = 0$ 12. $y'' - 9y = 0$

13. $y'' - 2y' = 0$ 14. $y'' - 10y' + 25y = 0$

15. $2y'' - 7y' - 4y = 0$

Find the solution to each of the given differential equations that satisfies the specified boundary conditions.

16. $y'' - y = 0$, with $y(0) = 1$ and $y'(0) = -2$

17. $2y'' - 5y' + 2y = 0$, with $y(0) = 0$ and $y'(0) = 1$

18. $y'' + 4y = 0$, with $y(\pi) = 1$ and $y'(\pi) = 0$

19. $y'' - 2y' + y = 0$, with $y(0) = 3$ and $y'(0) = 1$

20. $y'' - 8y' + 16y = 0$, with $y(0) = 2$ and $y'(0) = -1$

21. $y'' + 2y' + 2y = 0$, with $y(0) = 0$ and $y'(0) = 2$

22. $y'' + 2y' + 5y = 0$, with $y(0) = 0$ and $y'(0) = -1$

23. $y'' - y' = 0$, with $y(0) = 0$ and $y(\ln 2) = 4$

24. $4y'' + 4y' + y = 0$, with $y(0) = 1$ and $y(1) = 0$

25. $y'' + 4y = 0$, with $y(0) = 0$ and $y(\pi/4) = 2$

Verify that the given function $y_p(t)$ is a particular solution to the given nonhomogeneous differential equation. Then find the general solution.

26. $y'' + 3y' + 2y = 2t^2 + 1$ and $y_p(t) = t^2 - 3t + 4$

27. $y'' - y' - 12y = e^{-t}$ and $y_p(t) = -\dfrac{1}{10}e^{-t}$

28. $4y'' - 12y' + 9y = e^{2t}$ and $y_p(t) = e^{2t}$

29. $y'' + 8y = 4\sin 2t$ and $y_p(t) = \sin 2t$

30. $y'' - 2y' = -6$ and $y_p(t) = 3t$

Use the method of undetermined coefficients to construct particular solutions to each of the following nonhomogeneous linear differential equations.

31. $y'' - 5y' + 4y = 7$ **32.** $y'' + 4y = 2e^{3t}$

33. $y'' - 2y' + y = 5 - 4t$ **34.** $y'' - y' + y = t^2 - 4$

35. $y'' + y = 3\cos 2t$ **36.** $y'' - 8y' + 6y = 3t$

37. $y'' - y' + 2y = 3t^2$ ★**38.** $2y'' - 7y' - 4y = te^t$

39. $y'' - 3y' = 6$ **40.** $y'' + 5y' + 6y = 3e^{-2t}$

Find the solution to the nonhomogeneous linear differential equation that satisfies the given initial conditions.

41. $y'' - 5y' + 4y = 7$, with $y(0) = 1$ and $y'(0) = 2$ [*Hint:* See exercise 31.]

42. $y'' + 4y = 2e^{3t}$, with $y(0) = 0$ and $y'(0) = -3$ [*Hint:* See exercise 32.]

43. $y'' + 3y' + 2y = 2t^2 + 1$, with $y(0) = 0$ and $y'(0) = 0$ [*Hint:* See exercise 26.]

44. $y'' - 3y' = 6$, with $y(0) = -1$ and $y'(0) = 2$ [*Hint:* See exercise 39.]

★**45.** $y'' + y = \sin t$, with $y(\pi) = 0$ and $y'(\pi) = 1$

part B

If m_1 and m_2 are real roots to $am^2 + bm + c = 0$, then, for any t_0,

$$e^{m_1(t-t_0)} \quad \text{and} \quad e^{m_2(t-t_0)}$$

are fundamental solutions to $ay'' + by' + cy = 0$. Use this fact to aid in solving the following initial value problems of the form $y(t_0) - a$ and $y'(t_0) - b$.

46. $y'' - 4y' - 5y = 0$, with $y(1) = 2$ and $y'(1) = 0$

47. $-y'' - 2y' + 35y = 0$, with $y(4) = 3$ and $y'(4) = -1$

48. If $ay_1'' + by_1' + cy = \beta_1(t)$ and $ay_2'' + by_2' + cy_2 = \beta_2(t)$, then $y = y_1 + y_2$ is a particular solution to:

$$ay'' + by' + cy = \beta_1(t) + \beta_2(t)$$

Thus, to find a particular solution to $y'' + y' - 6y = t + \cos t$, we first find particular solutions to $y'' + y' - 6y = t$ and $y'' + y' - 6y = \cos t$. Complete the solution of this problem.

49. Suppose that $m = \alpha \pm \beta i$ are complex roots of $y'' + by' + cy = 0$.

(a) Show that $b = -2\alpha$ and $c = \alpha^2 - \beta^2$.

(b) Let $w = e^{-\alpha t}y$ and show that w satisfies $w'' + \beta^2 w = 0$.

(c) Conclude that $w = c_1 \sin \beta t + c_2 \cos \beta t$ and so

$$y = e^{\alpha t}(c_1 \sin \beta t + c_2 \cos \beta t)$$

(d) Alternately, conclude (see example 30.3) that $w = A \sin \beta(t - t_0)$ for some constants A and t_0, and show

$$y = Ae^{\alpha t} \sin \beta(t - t_0)$$

50. Determine the conditions for which the two-point boundary problem

$$y'' + \omega^2 y = 0, \text{ with } y(0) = 0 \quad \text{and} \quad y(T) = b$$

has a solution.

 # an introduction to linear systems and compartmental models

This section gives a brief introduction to systems of linear differential equations. In particular, we will concentrate on 2×2 linear systems and on two-compartment models. Let x and y be two variables, each functions of time t. It might be helpful to think of two interacting populations of sizes $x(t)$ and $y(t)$, respectively. Alternately, when a radioactive tracer is injected into the bloodstream, we may be interested in keeping track of the amount $x(t)$ in the bloodstream and the amount $y(t)$ in a particular organ such as the liver.

A **2×2 linear system** is a set of equations of the form:

$$x' = ax + by$$
$$y' = cx + dy$$

Here $x' = \dfrac{dx}{dt}$, $y' = \dfrac{dy}{dt}$, and a, b, c, and d are constants. A *solution* consists of two differentiable functions $x = f(t)$ and $y = g(t)$ that satisfy

$$f'(t) = af(t) + bg(t)$$
$$g'(t) = cf(t) + dg(t)$$

Note that a solution $f(t)$ necessarily possesses a second derivative $f''(t) = af'(t) + bg'(t)$. *Our basic line of attack is to derive a second order linear differential equation that $x = f(t)$ satisfies.* We will then apply the techniques of Section 38.

Note that $x'' = ax' + by' = ax' + b(cx + dy)$, using the second equation of the system. From the first equation, we have $by = x' - ax$. Hence,

$$x'' = ax' + bcx + d(by) = ax' + bcx + d(x' - ax)$$
$$= (a + d)x' + (bc - ad)x$$

By rearranging, we obtain $x'' - (a + d)x' + (ad - bc)x = 0$. Once x is known, y can be found from the equation $by = x' - ax$. The procedure is summarized in the box.

---

procedure for solving the linear system

$$\begin{aligned} x' &= ax + by \\ y' &= cx + dy \end{aligned} \quad \text{when } b \neq 0$$

> **Step 1:** Solve the second order equation $x'' - (a + d)x' + (ad - bc)x = 0$ by applying the techniques of Section 38. If $f_1(t)$ and $f_2(t)$ are fundamental solutions, then
>
> $$x = c_1 f_1(t) + c_2 f_2(t)$$
>
> **Step 2:** Find y from the relationship $by = x' - ax$, where x is the known function $c_1 f_1(t) + c_2 f_2(t)$.

example 39.1　　Find the general solution to the linear system

$$x' = -2x + y$$
$$y' = x - 2y$$

solution 39.1　　We have $a + d = -4$, and $ad - bc = 3$. Hence, x must satisfy $x'' + 4x' + 3x = 0$. The characteristic polynomial $m^2 + 4m + 3$ has roots $m = -1$ and $m = -3$. Hence, $x = c_1 e^{-t} + c_2 e^{-3t}$.

From the first equation, we conclude $y = x' + 2x = (-c_1 e^{-t} - 3c_2 e^{-3t}) + 2(c_1 e^{-t} + c_2 e^{-3t}) = c_1 e^{-t} - c_2 e^{-3t}$. ▤

example 39.2　　Find the solution to the linear system

$$x' = 2x - 3y$$
$$y' = x - 2y$$

that satisfies the initial conditions $x(0) = 10$ and $y(0) = 6$.

solution 39.2　　Here, $a + d = 0$ and $ad - bc = -1$. Thus, x must satisfy $x'' - x = 0$. Hence, $x = c_1 e^t + c_2 e^{-t}$. From the first equation, we have

$$-3y = x' - 2x = (c_1 e^t - c_2 e^{-t}) - 2(c_1 e^t + c_2 e^{-t})$$
$$= -c_1 e^t - 3c_2 e^{-t}$$

Hence, $y = \frac{1}{3} c_1 e^t + c_2 e^{-t}$. Using the initial conditions $x(0) = 10$ and $y(0) = 6$, we obtain

$$10 = c_1 + c_2$$
$$6 = \frac{1}{3} c_1 + c_2$$

This system has solution $c_1 = 6$ and $c_2 = 4$. Hence, $x = 6e^t + 4e^{-t}$ and $y = 2e^t + 4e^{-t}$. ▤

A linear system of the form

$$x' = ax + by + \beta_1(t)$$
$$y' = cx + dy + \beta_2(t)$$

is known as a **nonhomogeneous linear system**. If $\beta_1(t)$ is differentiable, then computing x'' leads to a nonhomogeneous second order differential equation. This is illustrated in the following example.

example 39.3   Find the general solution to the nonhomogeneous linear system

$$x' = -4x - y + t$$
$$y' = 2x - y - 1$$

solution 39.3   We have $x'' = -4x' - y' + 1 = -4x' - (2x - y - 1) + 1$, using the second equation. From the first equation, $y = -x' - 4x + t$. Hence,

$$x'' = -4x' - 2x + y + 2 = -4x' - 2x + (-x' - 4x + t) + 2$$
$$= -5x' - 6x + (t + 2)$$

Hence, x must satisfy $x'' + 5x' + 6x = t + 2$. By applying the procedure outlined in the box on page 417, we see that $x_p = \frac{7}{36} + (\frac{1}{6})t$ is a particular solution, and hence

$$x = c_1 e^{-2t} + c_2 e^{-3t} + \left[\frac{7}{36} + \left(\frac{1}{6}\right)t \right]$$

To find y, recall that $y = -x' - 4x + t = -(-2c_1 e^{-2t} - 3c_2 e^{-3t} + 1/6) - 4[c_1 e^{-2t} + c_2 e^{-3t} + (7/36 + t/6)] + t$. This simplifies to

$$y = -2c_1 e^{-2t} - c_2 e^{-3t} - \frac{17}{18} + \left(\frac{1}{3}\right)t$$

modeling the interaction of two species   The linear differential equation $y' = ky$ with exponential solution $y = ce^{kt}$ was used in Section 17 to model a single species population. This model was the simplest (and also the most unrealistic) model of population growth. Likewise, the linear system

$$x' = ax + by$$
$$y' = cx + dy$$

is the starting point for modeling the interaction of two species. Here, $x(t)$ is the number in population 1 at time t and $y(t)$ is the number in population 2, which interacts with population 1.

Note that when species 2 is absent ($y \equiv 0$), the first equation reduces to $x' = ax$. If species 1 *competes* with species 2 for common food resources, then we will set $a > 0$. In the absence of species 2, species 1 thrives and grows exponentially. On the other hand, if species 1 *preys* exclusively upon species 2, then we will set $a < 0$. When $y \equiv 0$, species 1 will starve and decrease exponentially.

The term by in the first equation represents the attempt to model the type of interaction. If species 2 competes with or preys upon species 1, then we will set $b < 0$. Thus, in a **predator-prey** interaction, we are assuming that the number of prey eaten per unit time varies directly with the number of predators y. On the other hand, in a **symbiotic relationship**, the presence of species 2 may increase the survival rate of species 1. A member of species 2 could provide refuge for a member of species 1. We would then set $b > 0$. Two such interactions are presented in examples 39.4 and 39.5.

example 39.4   If x is the number of prey and y is the number of predators, assume

$$x' = x - 2y$$
$$y' = 2x - 4y$$

with $x(0) = 500$ and $y(0) = 100$. Solve the linear system and determine the fate of each of the two populations.

solution 39.4   Here $a + d = -3$ and $ad - bc = 0$. Hence, x must satisfy $x'' + 3x' = 0$. Since $m = 0$ and $m = -3$ are roots of the characteristic polynomial, we have $x(t) = c_1 + c_2 e^{-3t}$. Now $2y = x - x' = (c_1 + c_2 e^{-3t}) - (-3c_2 e^{-3t}) = c_1 + 4c_2 e^{-3t}$. Hence, we have shown that

$$x = c_1 + c_2 e^{-3t}$$
$$y = \frac{1}{2}c_1 + 2c_2 e^{-3t}$$

We now use the initial conditions to obtain $500 = x(0) = c_1 + c_2$ and $100 = y(0) = \frac{1}{2}c_1 + 2c_2$. It follows that $c_1 = 600$ and $c_2 = -100$.

If we let $t \to +\infty$, then $x(t) \to c_1 = 600$ and $y(t) \to \frac{1}{2}c_1 = 300$.   ▤

example 39.5   Two species that compete for a common resource are governed by the linear system

$$x' = x - 2.5y$$
$$y' = -1.6x + y$$

Initially, $x(0) = 1{,}500$ and $y(0) = 1{,}000$. Which species wins the competition?

solution 39.5 Since $a + d = 2$ and $ad - bc = 1 - (-1.6)(-2.5) = -3$, x satisfies $x'' - 2x' - 3x = 0$, whose solution is $x = c_1 e^{-t} + c_2 e^{3t}$. Now

$$2.5y = x - x' = (c_1 e^{-t} + c_2 e^{3t}) - (-c_1 e^{-t} + 3c_2 e^{3t})$$
$$= 2c_1 e^{-t} - 2c_2 e^{3t}$$

Hence, $y = .8c_1 e^{-t} - .8c_2 e^{3t}$. Using the initial conditions, we obtain

$$x(0) = 1,500 = c_1 + c_2$$
$$y(0) = 1,000 = .8c_1 - .8c_2$$

This system has solution $c_1 = 1,375$ and $c_2 = 125$. It follows that $x = 1,375e^{-t} + 125e^{3t}$ and $y = 1,100e^{-t} - 100e^{3t}$. Now $x(t)$ is always positive, but $y(t) = 0$ when $1,100e^{-t} = 100e^{3t}$. Thus, $e^{4t} = 11$, and so $t = \frac{1}{4}\ln 11 \approx .6$. At this time, $x = 1,510$. Thus for $t \geq .6$, species 2 becomes extinct and species 1 grows according to $x' = x$. The two graphs are shown in figure 39.1.

figure 39.1

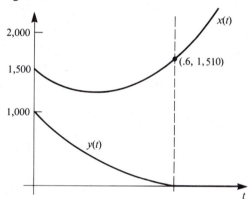

The remainder of this section is devoted to compartment models in physiology and systems ecology.

example 39.6 **Fick's model for passive diffusion across a membrane** Shown in figure 39.2 is a compartmental representation of the cell membrane, the extracellular fluid, and the fluid within the cell. Those materials necessary for cell growth

figure 39.2

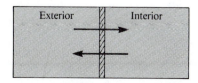

make their way past the membrane. One of the transport mechanisms, known as **passive diffusion**, is governed by Fick's Law: *the rate of transport is directly proportional to the difference in the concentrations in the two compartments.*

Let V_1 and V_2 be the volumes of the compartments, and let $A_1(t)$ and $A_2(t)$ be the amounts of the particular substance in question in the respective compartments. Let $x(t) = A_1(t)/V_1$ and $y(t) = A_2(t)/V_2$. Fick's Law then asserts:

Fick's Law

$$A_1' = \kappa(y - x)$$
$$A_2' = \kappa(x - y)$$

The constant κ is known as the **permeability factor**. Note that when $y > x$, A_1' will be positive and so A_1 will increase.

Since $A_1' = x'V_1$ and $A_2' = y'V_2$, we may rewrite the system in the form:

$$x' = \frac{\kappa}{V_1}(y - x) \qquad y' = \frac{\kappa}{V_2}(x - y)$$

Thus, in our linear system notation, $a = -\kappa/V_1$, $b = \kappa/V_1$, $c = \kappa/V_2$, and $d = -\kappa/V_2$. Hence, $ad - bc = 0$ and $a + d = -(\kappa/V_1 + \kappa/V_2)$. Let $\alpha = \kappa/V_1 + \kappa/V_2$. Then x satisfies $x'' + \alpha x' = 0$, which has characteristic polynomial $m^2 + \alpha m$. Thus, $x(t) = c_1 + c_2 e^{-\alpha t}$. From the first equation, we have

$$y = \frac{V_1}{\kappa} x' + x = \frac{V_1}{\kappa}(-\alpha c_2 e^{-\alpha t}) + (c_1 + c_2 e^{-\alpha t})$$

$$= c_1 + \left(-\alpha \frac{V_1}{\kappa} + 1\right) c_2 e^{-\alpha t}$$

Note that $-\alpha \dfrac{V_1}{\kappa} + 1 = -\left(\dfrac{\kappa}{V_1} + \dfrac{\kappa}{V_2}\right)\dfrac{V_1}{\kappa} + 1 = -\dfrac{V_1}{V_2}$. Thus, $y = c_1 - (V_1/V_2) c_2 e^{-\alpha t}$. Observe that $c_1 = \lim_{t \to \infty} x(t) = \lim_{t \to \infty} y(t)$. This constant may be evaluated in terms of $x_0 = x(0)$ and $y_0 = y(0)$. A bit of careful algebra will show that $c_1 = (x_0 V_1 + y_0 V_2)/(V_1 + V_2)$. The graphs of x and y are shown in figure 39.3. Note that the larger the value of κ, the more quickly $e^{-\alpha t}$ goes to 0, and so the more quickly the equilibrium concentrations are reached.

figure 39.3

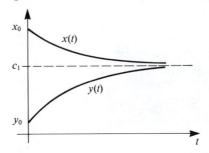

the general two-compartment model   Material flows between two tanks of sizes V_1 and V_2, respectively. In the diagram shown in figure 39.4, F_{01}, F_{12}, F_{21}, F_{10}, and F_{20} denote flow rates. Typical units might be liters per minute. Note that F_{ij} denotes the flow rate from tank i to tank j. These models are often called **bathtub models**.

figure 39.4

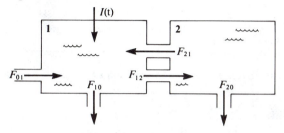

Next, suppose a second substance (the *tracer*) is infused into the first compartment at known rate $I(t)$. We will assume that the tracer is thoroughly mixed in both compartments at all times t. The rate $I(t)$ might be measured in grams per minute.

Let $x(t)$ be the amount of tracer in compartment 1 and let $y(t)$ denote the amount in compartment 2. The concentrations are given by $c_1(t) = x(t)/V_1$ and $c_2(t) = y(t)/V_2$. It follows that:

$$x' = -(F_{12} + F_{10})c_1(t) + F_{21}c_2(t) + I(t)$$
$$y' = F_{12}c_1(t) - (F_{21} + F_{20})c_2(t)$$

To see that the first differential equation is valid, note that in a very small time interval of length Δt, the following events occur.

1. The approximate volume that flows out is $(F_{12} + F_{10}) \Delta t$, of which about $(F_{12} + F_{10}) \Delta t \, c_1(t)$ is tracer.

2. The volume that flows in from compartment 2 is approximately $F_{21} \Delta t$, of which about $F_{21} \Delta t \, c_2(t)$ is tracer.

3. $I(t) \Delta t$ units of tracer are introduced into the first compartment.

Thus, the change in the amount of tracer in compartment 1 is given by

$$x(t + \Delta t) - x(t) \approx -(F_{12} + F_{10})c_1(t) \Delta t + F_{21}c_2(t) \Delta t + I(t) \Delta t$$

and so

$$\frac{x(t + \Delta t) - x(t)}{\Delta t} \approx -(F_{12} + F_{10})c_1(t) + F_{21}c_2(t) + I(t)$$

As $\Delta t \to 0$, all approximations improve and the left hand side approaches $x'(t)$ by definition of a derivative.

Let $a_{12} = F_{12}/V_1$, $a_{21} = F_{21}/V_2$, $a_{10} = F_{10}/V_1$, and $a_{20} = F_{20}/V_2$. These are known as **transfer coefficients**. The system then takes the form

$$x' = -(a_{12} + a_{10})x + a_{21}y + I(t)$$
$$y' = a_{12}x - (a_{21} + a_{20})y$$

and the original compartment diagram can be simplified as shown in figure 39.5.

figure 39.5

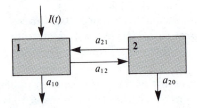

This basic model has many applications in physiology, some of which will be explored in the exercises. We close this section with an application from *systems ecology*.

example 39.7   Compartment models have been used in ecology to model how matter (such as a nutrient or pollutant) or energy is transported between the various components of an ecosystem. In modeling radioactive fallout, for example, we might divide the ecosystem into convenient categories (soil, trees, animals, plants, and waste, for example) and investigate how a harmful substance such as strontium 90 is passed from compartment to compartment.

Realistic models necessarily require more than two compartments. Nevertheless, as a simple introductory example, consider the model for the cycling of DDT residue that is illustrated in figure 39.6.

When the plants are sprayed, part of the DDT goes directly to the soil. The DDT makes its way from the plants to the soil through leaf fall and leaching. On the other hand, DDT in the soil makes its way back into plants as a part of

figure 39.6

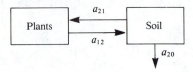

normal growth. We will assume that DDT is lost to the system through water runoff from the soil.

To implement the model, the ecologist must estimate the transfer coefficients $a_{ij} = F_{ij}/V_i$ by measuring F_{ij} and V_i directly from field data. We will assume this difficult task has been performed, giving values $a_{12} = .25$ per month, $a_{21} = .02$, and $a_{20} = .05$.

Of the total T sprayed on the crops, let $x(t)$ be the percentage of DDT in or on the plants, and let $y(t)$ be the percentage in the soil. Then

$$x' = -.25x + .02y$$
$$y' = .25x - .07y$$

Assume further that $x(0) = .60$ and $y(0) = .40$ after spraying. Solve this system.

solution 39.7

We first find the second order differential equation for $x(t)$. x satisfies $x'' + .32x' + .0125x = 0$. The characteristic polynomial $m^2 + .32m + .0125$ has roots $m = -.046$ and $m = -.274$. Hence,

$$x = c_1 e^{-.046t} + c_2 e^{-.274t}$$

Since $.02y = x' + .25x$, we have

$$y = 50(x' + .25x) = 50(-.046c_1 e^{-.046t} - .274c_2 e^{-.274t})$$
$$+ 12.5(c_1 e^{-.046t} + c_2 e^{-.274t})$$
$$= 10.2c_1 e^{-.046t} - 1.2c_2 e^{-.274t}$$

Since $x(0) = .6$ and $y(0) = .4$, it follows that $c_1 + c_2 = .6$ and $10.2c_1 - 1.2c_2 = .4$. This system has solution $c_1 = .098$ and $c_2 = .502$. Hence, we may conclude

$$x(t) = .098e^{-.046t} + .502e^{-.274t}$$
$$y(t) = 1.00e^{-.046t} - .60e^{-.274t}$$

The two graphs are shown in figure 39.7. The model predicts a maximum value in the soil of 65%, occurring at $t = 5.6$. The DDT residue steadily decreases in the plants.

figure 39.7

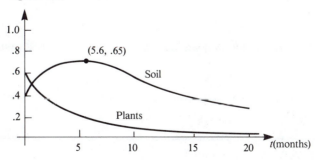

exercises for section 39

Find the general solution to each of the following linear systems.

1. $x' = 2x + 6y$
$y' = -2x - 5y$

2. $x' = x + 3y$
$y' = 5x + 3y$

3. $x' = -3x - 5y$
$y' = 5x + 3y$

4. $x' = x + 8y$
$y' = -2x - 7y$

5. $x' = y$
$y' = -4x$

6. $x' = 6x - y$
$y' = 5x + 2y$

Solve each of the following linear systems subject to the given boundary conditions.

7. $x' = 5x - y$
$y' = 3x + y$
$x(0) = 4$ and $y(0) = -2$

8. $x' = x - 3y$
$y' = -2x + 2y$
$x(0) = 5$ and $y(0) = 15$

★9. $x' = -3x - y$
$y' = 2x - y$
$x(0) = 4$ and $y(0) = 2$

10. $x' = -y$
$y' = x + 2y$
$x(0) = 2$ and $y(0) = 3$

11. $x' = x - 4y$
$y' = 4x - 7y$
$x(0) = 2$ and $y(0) = 1$

12. $x' = x - 5y$
$y' = x - 3y$
$x(0) = 1$ and $y(0) = 1$

Following the line of attack outlined in example 39.3, find the general solution to each of the following nonhomogeneous linear systems.

13. $x' = 6x - 3y + 8$
$y' = 2x + y$

14. $x' = x + y$
$y' = 4x + y + 7$

15. $x' = -2x + y - 3t$
$y' = x - 2y$

16. $x' = -4x - 6y + e^{-3t}$
$y' = x + y + 2e^{-3t}$

★17. $x' = -x + y + \sin t$
$y' = x - y$ [*Hint*: Try a particular solution of the form $x_p = A \sin t + B \cos t$.]

The following are linear models for the interaction of two species. Let x denote the number of "Hatfields" and y the number of "McCoys."
  (a) Solve the resulting systems for $x(t)$ and $y(t)$.
  (b) Determine the fates of the two populations. Who wins in the long run?

18. $x' = 4x - 2y$
$y' = 2x - y$
$x(0) = 400$ and $y(0) = 100$

19. $x' = x - 2.5y$
$y' = -1.6x + y$
$x(0) = 1,000$ and $y(0) = 1,600$

20. $x' = 2x - y$
$y' = -x + 2y$
$x(0) = 100$ and $y(0) = 300$

21. $x' = -x + 2y$
$y' = x - 2y$
$x(0) = 450$ and $y(0) = 150$

★22. $x' = -2y + 1,000$
$y' = x - 3y$
$x(0) = 3,000$ and $y(0) = 1,500$

23. A bass population migrates back and forth between two adjoining lakes. Let $x(t)$ and $y(t)$ denote the numbers in lakes 1 and 2, respectively. The migration rate between the two lakes is assumed to be governed by Fick's Law. Lake 1 has volume $V = 2\,\text{km}^3$, while lake 2 has volume $V = 1\,\text{km}^3$. At the beginning of the fishing season, 30,000 young bass are stocked in lake 1.

(a) Find the corresponding linear system and solve.

(b) Find the final number of fish in each lake, assuming no fishing and no natural deaths.

24. In the general two-compartment model, suppose $I(t) \equiv 0$ and $F_{01} = F_{10} = F_{20} = 0$. Show that the amount of tracer in the system satisfies Fick's Law.

25. A two-tank system is shown in the figure. If 50 grams of dye are deposited

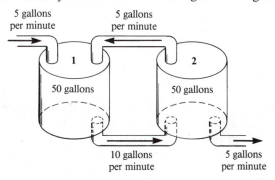

5 gallons per minute   5 gallons per minute

50 gallons   50 gallons

10 gallons per minute   5 gallons per minute

into tank 1, find expressions for $x(t)$ and $y(t)$, the amounts of dye in each tank at time t.

★26. In exercise 25, dye of concentration 1 gram per gallon flows into tank 1. Assume $x(0) = 0$ and $y(0) = 0$.

(a) Find $x(t)$ and $y(t)$.

(b) Compute $x_\infty = \text{limit}_{t \to \infty}\, x(t)$ and $y_\infty = \text{limit}_{t \to \infty}\, y(t)$.

27. The burning of fossil fuels by industry adds carbon dioxide to the earth's atmosphere at a rate of approximately 6 billion metric tons per year. This excess carbon must be absorbed by the oceans in order to avoid long-term buildups. Shown in the figure is a simple two-compartment model for carbon dioxide cycling. For this system $x(0) = 7$ billion metric tons and

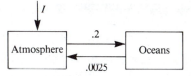

I

Atmosphere   .2   Oceans

.0025

$y(0) = 35{,}000$ billion metric tons are the rough estimates given by Bolin.*

* B. Bolin, "The Carbon Cycle", *Scientific American*, **223–3**, (1970), pp. 125–132.

The corresponding linear system is

$$x' = -.2x + .0025y + I$$
$$y' = .2x - .0025y$$

(a) If $I \equiv 0$, solve for $x(t)$ and $y(t)$. Determine $x_\infty = \lim_{t \to \infty} x(t)$ and $y_\infty = \lim_{t \to \infty} y(t)$.

★(b) Solve for $x(t)$ and $y(t)$ if $I = 6$ (billion metric tons per year). Discuss the solution for large t. [*Hint*: Try a particular solution of the form $x_p(t) = At$.]

28. Solve the linear system arising from the two-compartment model shown in the figure.

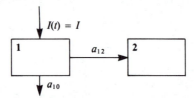

29. The drug kanamycin is used in the treatment of pneumonia and bacterial infections. The drug is injected into the hip, from which it makes its way to all parts of the body through the circulatory system. It is eliminated from the body through the urine. If high concentrations of the drug collect in the inner ear, however, hearing loss can result. Assuming an initial dose of x_0, formulate, but do not solve, an appropriate two-compartment model with driving function $I(t)$.

30. A herd of cattle is grazing on pasture land. Formulate, but do not solve, a two-compartment model that keeps track of the biomass of the grass and the cattle, and also takes into account photosynthesis and respiration.

part B   **compartment models in physiology**   In physiologicals application of compartment models, the transfer coefficients are rarely known and may be impossible to measure directly. In fact, the point of the model may be to estimate a_{ij}, V_i, or F_{ij} to judge the health of the system. This is the problem of *estimation of parameters*.

Samples of the tracer are usually taken from one of the compartments and a curve is fit using exponential functions. As we will show in the following exercises, this will produce the roots m_1 and m_2 of the characteristic polynomial from which the key parameters may be estimated.

31. In the bloodstream, potassium ions move back and forth from the plasma to the red blood cells. This suggests that the two-compartment model* shown in

* C.W. Sheppard and W.R. Martin, *Journal of General Physiology*, **33**, (1950), pp. 703–722.

the figure might be appropriate. A fixed quantity x_0 of radioactive potassium K^{42} is injected into the bloodstream and we will assume K^{42} is not lost to the system. Then:

$$x' = -a_{12}x + a_{21}y$$
$$y' = a_{12}x - a_{21}y$$

(a) Solve for $x(t)$ and $y(t)$ given that $x(0) = x_0$ and $y(0) = 0$.

(b) Find $x_\infty = \text{limit}_{t \to \infty}\, x(t)$.

(c) From plasma samples and radioactivity comparisons, a curve fit to the data produced

$$\frac{x(t)}{x_0} = .06 + .94e^{-.3t}$$

Find a_{21} and a_{12}.

32. *Creatinine* * is a component of urine originally formed from creatinine phosphate, a substance that supplies energy to the muscles. When a dose of creatinine is injected into the bloodstream, subsequent blood samples reveal that creatinine concentration in the bloodstream can be represented by a function of the form

$$c_1 e^{-m_1 t} + c_2 e^{-m_2 t}$$

This suggests that the two-compartment model shown in the figure might be appropriate. The corresponding system is

$$x' = -(a_{12} + a_{10})x + a_{21}y$$
$$y' = a_{12}x - a_{21}y$$

```
┌─────────┐   a₁₂    ┌─────────┐
│  Blood  │ ──────→  │ Tissues │
│         │ ←──────  │         │
└─────────┘   a₂₁    └─────────┘
     │
 a₁₀ │  Elimination
     │  through the
     ↓  kidneys
```

(a) Show that x satisfies $x'' + (a_{12} + a_{21} + a_{10})x' + a_{10}a_{21}x = 0$.

(b) Prove that both roots of the characteristic equation are negative.

Factor the characteristic polynomial as $(m + m_1)(m + m_2)$

---

* This problem is based on, L.A. Sapirstein, D.G. Vidt, M.J. Mandel, and G. Hanusek, "Volumes of Distribution and Clearances of Intravenously Injected Creatinine in the Dog", *American Journal of Physiology*, **181**, (1955) pp. 330–336.

(c) Show that $m_1 + m_2 = a_{12} + a_{21} + a_{10}$ and $m_1 m_2 = a_{10} a_{21}$.

(d) Write $x(t)$ as $c_1 e^{-m_1 t} + c_2 e^{-m_2 t}$ and then find $y(t)$.

(e) If $x(t) = x_0$ and $y(t) = 0$, show that

$$x(t) = \frac{x_0}{m_2 - m_1} [(a_{21} - m_1)e^{-m_1 t} + (m_2 - a_{21})e^{-m_2 t}]$$

$$y(t) = \frac{x_0 a_{12}}{m_2 - m_1} [e^{-m_1 t} - e^{-m_2 t}]$$

(f) In an experiment carried out by Sapirstein et al., $x_0 = 2$ grams, and from blood samples, the creatinine concentration (in grams per liter) in the first compartment was determined to be:

$$c(t) = \frac{x(t)}{V_1} = .188 e^{-.0161 t} + .321 e^{-.1105 t}$$

Show that $V_1 = 3.93$ liters and then use **(c)** and **(e)** to estimate a_{12}, a_{21}, a_{10}, and V_2.

5 an introduction to multivariate calculus

prologue   The two important calculus concepts of *derivative* and *integral* can be extended to functions of more than one variable. As we will show in Section 40, the graph of a function $z = f(x, y)$ of two variables is a **surface**.

figure 1

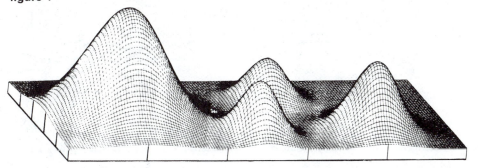

Surfaces are mathematically formed "mountain ranges." As in the single variable case, we will develop techniques for finding where the hills and valleys occur and for computing volumes formed by portions of the surface. As the computer-generated graph in figure 1 clearly shows, a two-dimensional graph is formed by piecing together many one-dimensional graphs called **sections**. These sections are defined by functions of one variable and so ordinary derivatives and integrals can be computed. The analysis of the derivative and integral of sections is the key step in solving calculus problems for functions of several variables.

40 functions of several variables

Most realistic models of biological or physical phenomena depend upon many variables. Plant growth z during the day is directly dependent upon the amount of sunlight x_1 and the proper nutrients $x_2, x_3, \ldots$ in the soil. Figure 40.1 is a world map that shows the average air temperature z as a function of longitude x and latitude y. When $x = -60$ (60° west) and $y = 0$, z is about 80°F. This method of representing a function of two variables is known as a *contour map*.

figure 40.1   (Adapted from, A.N. Strahler, *Physical Geography*, New York: John Wiley and Sons, 1960.)

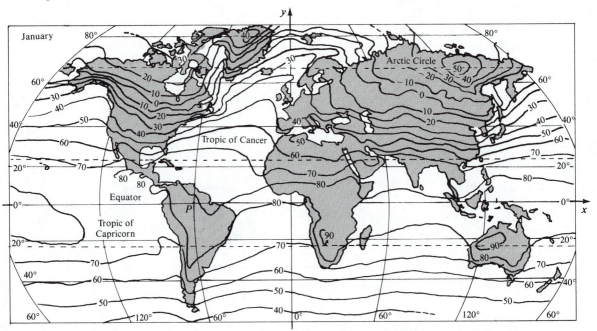

The following are simple examples of functions of two or more variables:

1. The function $V = V(x, y, z) = xyz$ gives the volume of a cube in terms of its length x, width y, and height z. For example, $V(2, 3, 5) = (2)(3)(5) = 30$ cubic units.

2. The function of two variables given on your calculator by the power button $\boxed{x^y}$ is $z = f(x, y) = x^y$. Thus, $f(2, 3) = 2^3 = 8$, while $f(3, 2) = 3^2 = 9$. Note that the input is not just two positive real numbers, but an **ordered pair** (x, y) of positive real numbers.

3. An animal's diet consists of four items, A, B, C, and D. The numbers of calories per gram for each of these items are 3, 5, 10, and 7, respectively. If an animal consumes x grams of A, y grams of B, z grams of C, and w grams of D, then the total number of calories T is given by

$$T = f(x, y, z, w) = 3x + 5y + 10z + 7w$$

T is a function of the four variables x, y, z, and w.

figure 40.2

4. Poiseuille's Law (see section 21) gives the flow rate F of a liquid through cylindrical tubing as a function of R, the radius, ΔP, the pressure difference between the ends of the tubing, and L, the length of the tubing, as illustrated in figure 40.2. In particular, $F = f(R, \Delta P, L) = (c\,\Delta P/L)\,R^4$ for some constant c.

Throughout the remainder of this section we will confine our attention to functions of two or three variables.

definition 40.1   Let $\mathscr{A}$ be a collection of ordered pairs (x, y) in the xy-plane. Then a function f from $\mathscr{A}$ into the real numbers $\mathscr{R}$ is a rule that assigns to each (x, y) in $\mathscr{A}$ a **single real number** $f(x, y)$. The set $\mathscr{A}$ is called the **domain of definition** of f. This definition is illustrated in figure 40.3.

figure 40.3

A similar definition holds for a function $w = f(x, y, z)$ of three real variables.

example 40.1   Find the domain of definition of the function $f(x, y) = \sqrt{4 - x^2 - y^2}$, and compute (if possible) $f(0, 1)$, $f(2, 3)$, and $f(-1, -1)$.

solution 40.1   The function $f(x, y)$ is defined provided $4 - x^2 - y^2 \geq 0$. Hence $x^2 + y^2 \leq 4$. Now, $x^2 + y^2 = 4$ is just a circle of radius 2 about the origin, and so $x^2 + y^2 \leq 4$ defines the *disk* shown in figure 40.4. The point $(2, 3)$ is not in the domain of definition

figure 40.4

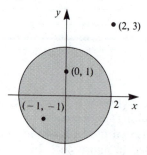

and so $f(2, 3)$ is not defined. We can compute $f(0, 1) = \sqrt{4 - 0^2 - 1^2} = \sqrt{3}$ and $f(-1, -1) = \sqrt{4 - (-1)^2 - (-1)^2} = \sqrt{2}$. ▤

example 40.2   Find the domain of definition of the function $f(x, y) = \ln(x^2 y)$ and determine where $f(x, y) = 0$.

solution 40.2   We must make sure that $x^2 y > 0$, since the natural logarithm is defined for positive real numbers only. Since $x^2 > 0$ for $x \neq 0$, y must also be positive. The domain consists of the upper half-plane shown in figure 40.5, with the line $x = 0$ *excluded*. Finally, $f(x, y) = \ln(x^2 y) = 0$ implies $x^2 y = 1$. Hence $y = 1/x^2$, and

figure 40.5

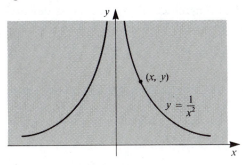

$f(x, y) = 0$ for all ordered pairs on the graph shown in the figure. The graph of $y = 1/x^2$ is called a **level curve** of f.

definition 40.2   The collection of all ordered triples (x, y, z) where $z = f(x, y)$ and (x, y) varies over the domain of definition $\mathcal{A}$ is called the *graph of the function f.*

To plot points on the graph of f, we construct the three-dimensional coordinate system shown in figure 40.6. Imagine the xy-plane to be tiled with squares as might be the floor of a room. To plot $A = (3, 4, 2)$, we first locate $(3, 4)$ in the xy-plane and proceed *directly upward* 2 units in the z direction. To locate $B = (4, -2, -1)$, find $(4, -2)$ and to *downward* 1 unit. The points $C = (0, 0, 4)$ and $D = (0, 5, 0)$ are also shown in the figure.

When all points on the graph of f are plotted, the points collectively will

figure 40.6

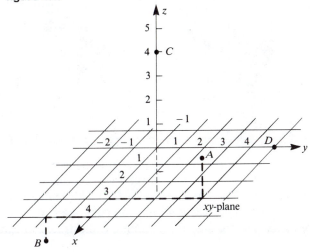

form a *surface*, as shown in figure 40.7. How can we determine what the surface looks like? The remainder of this section is devoted to methods for picturing the graph of $z = f(x, y)$.

figure 40.7

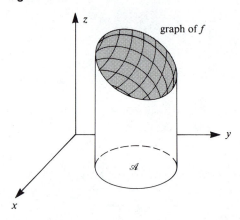

the method of sectioning   The equations $x = a$, $y = b$, and $z = c$ define planes in three-dimensional space that are parallel to the yz-, zx-, and xy-planes, respectively. These are pictured in figure 40.8.

figure 40.8(a)   **figure 40.8(b)**   **figure 40.8(c)**

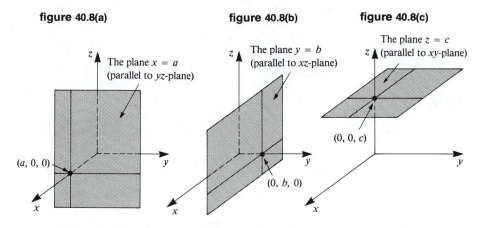

We will *slice* the surface defined by $z = f(x, y)$ using planes of the form $x = a$ and $y = b$ and look at the resulting one-dimensional graphs, as shown in figure 40.9. When we cut the surface with many planes of the form $x = a$ and $y = b$, we obtain figure 40.10. The curves $z = f(a, y)$ and $z = f(x, b)$ are called **sections**. The method of sectioning is illustrated in examples 40.3–40.5.

figure 40.9(a)

$z = f(a, y)$

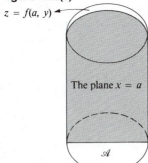

The plane $x = a$

$\mathscr{A}$

figure 40.9(b)

$z = f(x, b)$

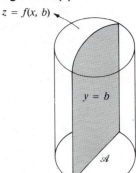

$y = b$

$\mathscr{A}$

figure 40.10

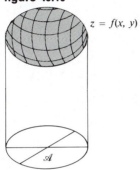

$z = f(x, y)$

$\mathscr{A}$

example 40.3   Sketch the graph of $f(x, y) = x^2 + y^2$ by finding sections.

solution 40.3   Setting $x = a$, we obtain $z = f(a, y) = a^2 + y^2$. The section is a parabola with vertex at $(a, 0, a^2)$. Setting $y = b$ yields $z = f(x, b) = x^2 + b^2$, a parabola with

figure 40.11

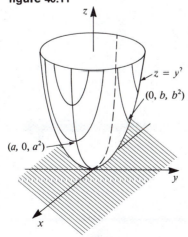

z

$z = y^2$

$(0, b, b^2)$

$(a, 0, a^2)$

y

x

vertex at $(0, b, b^2)$. If we set $a = 0$ and $b = 0$, then we obtain the sections $z = y^2$ and $z = x^2$. These should be sketched first. The final sketch is shown in figure 40.11. Note that a "valley" or **relative minimum** occurs when $x = 0$ and $y = 0$. This is one type of critical point for a function of two variables. ▤

example 40.4   Using sections, sketch the graph of the function $f(x, y) = 10e^{-(x^2+y^2)/2}$.

solution 40.4   Setting $x = a$ yields $z = f(a, y) = (10e^{-a^2/2})e^{-y^2/2}$. As we have seen in Section 16, this is a bell-shaped curve. The maximum value of $10e^{-a^2/2}$ occurs when $y = 0$, as illustrated in figure 40.12(a). Setting $y = b$ yields the bell-shaped curve

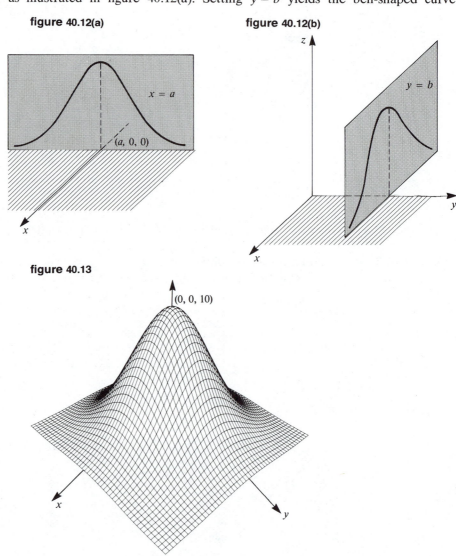

figure 40.12(a)

figure 40.12(b)

figure 40.13

$z = f(x, b) = (10e^{-b^2/2})e^{-y^2/2}$. Note from figures 40.12(a) and (b) that as a or b increases, the peak of the bell decreases. The maximum value of 10 occurs when $x = 0$ and $y = 0$. Shown in figure 40.13 is the final computer-generated sketch. Note that a "hill" or **relative maximum** occurs at $(0, 0)$. This is a second type of critical point for a function of two variables.

example 40.5   Sketch the graph of $z = f(x, y) = y^2 - x^2$ by finding sections.

solution 40.5   Setting $x = a$, we have $f(a, y) = y^2 - a^2$. This gives a parabola opening upward, with vertex at $(a, 0, -a^2)$. Setting $y = b$ produces $z = f(x, b) = b^2 - x^2$. Unlike the first collection of parabolas, these open downward, with vertex at $(0, b, b^2)$. Note that as b increases, the vertex rises in the z direction, as illustrated in figure 40.14. The final sketch is shown in figure 40.15. The point $(0, 0, 0)$ on the

figure 40.14

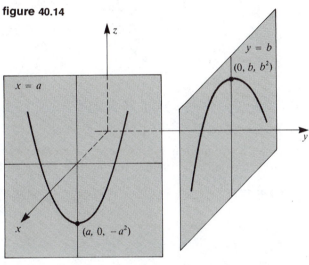

figure 40.15

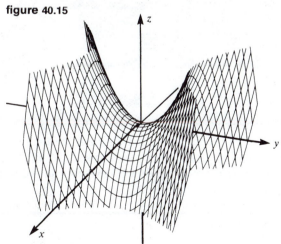

graph is called a **saddle point**. This is the third type of critical point in this setting.

the method of level curves   Sketching the final surface from the sections really does require some artistic ability. A far more common way to depict a function $z = f(x, y)$ is to sketch the **level curves** of f in the xy-plane. Here, we intersect planes $z = c$ with the graph of f and project the resulting curve down to the xy-plane, as illustrated in figure 40.16. The curve in the

figure 40.16(a)   **figure 40.16(b)**

$z = c$

$f(x, y) = c$

xy-plane has equation $f(x, y) = c$. This method is used in constructing *topographic maps* of mountainous regions (see figure 40.17).

figure 40.17(a)   **figure 40.17(b)**

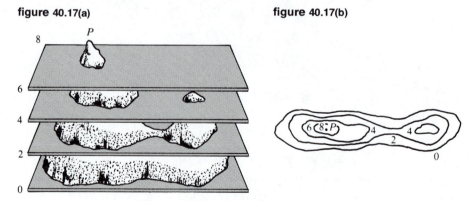

The method of level curves is also used in constructing weather maps, profiles of the ocean floor, rainfall maps, and maps indicating how the number of species of mammals varies with longitude and latitude.

We will use the method of level curves in the next two examples.

example 40.6   Find and sketch the level curves of the function $f(x, y) = x^2 + y^2$ that was graphed in example 40.3.

solution 40.6   Setting $z = f(x, y) = c$, we obtain $x^2 + y^2 = c$. For $c > 0$, these level curves are just circles of radius $\sqrt{c}$ about the origin, as shown in figure 40.18.

figure 40.18(a)   **figure 40.18(b)**

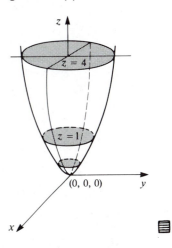

example 40.7   Sketch the level curves of the function $z = f(x, y) = .1091x^{.425}y^{.725}$. This is an approximate formula for the surface area z (in square feet) of the human body, given the weight x (in pounds) and the height y (in inches).

solution 40.7   The equation $.1091x^{.425}y^{.725} = c$ implies $y^{.725} = 9.1659cx^{-.425}$. Hence, the level curves are the graphs of

$$y = \frac{(9.1659c)^{1/.725}}{x^{.425/.725}}$$

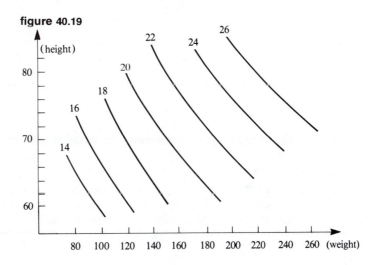

figure 40.19

These curves are of the form $y = a/x^k$ and were graphed in section 5. Shown in figure 40.19 are the level curves corresponding to $c = 14, 16, 18, 20, 22, 24,$ and 26. With this sketch can you roughly approximate your own body surface area?

computer graphics   The age of the computer has provided a third method for picturing graphs of functions. The availability of computer plotters has increased greatly over the past few years and gives an exciting tool to study functions of several variables. Computer graphics devices can plot many sections $z = f(a, y)$ and $z = f(x, b)$ and piece them together. Some of these graphs are strikingly beautiful. Shown in figure 40.20 is the graph of $f(x, y) = 10 \exp(-\sqrt{x^2 + y^2}) \cos(\sqrt{x^2 + y^2})$.

figure 40.20

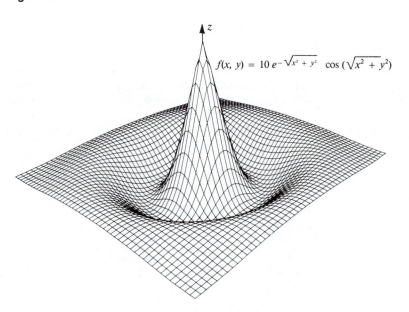

$$f(x, y) = 10\, e^{-\sqrt{x^2 + y^2}} \cos(\sqrt{x^2 + y^2})$$

Many functions of two variables (such as the function $Y = f(F, t_c)$ that appeared in section 36) are nearly impossible to sketch using either sections or level curves. The mathematical expressions may be too complex. A computer graphics picture can provide a revealing look at these more complex surfaces. Shown in figure 40.21 is a two-dimensional surface that arises from the model presented in section 36 applied to the Pacific halibut.

figure 40.21

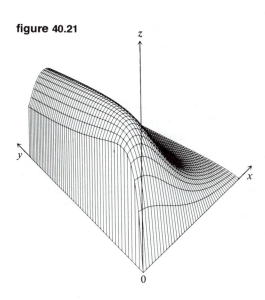

exercises for section 40

part A

Using the temperature contour map in figure 40.1, estimate the temperature z, given the following values of longitude x and latitude y.

1. $x = -120$, $y = 20$ 2. $x = 120$, $y = 40$

3. $x = 0$, $y = 0$ 4. $x = -120$, $y = -40$

5. Shown in the figure is a contour map giving the yield z (in cubic feet per acre) when x trees per acrea are planted and then harvested t years later. If we write $z = f(x, t)$, estimate $f(900, 20)$ and $f(950, 15)$.

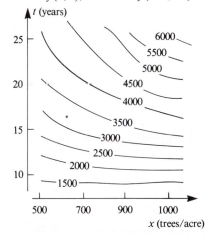

Compute (if possible) $f(0, 0)$, $f(1, -2)$, and $f(-2, 1)$ for each of the following functions.

6. $f(x, y) = 3x - 4y + 8$ 　　**7.** $f(x, y) = \ln(xy^2)$ 　　**8.** $f(x, y) = \sqrt{2 - x + y}$

9. $f(x, y) = e^{x-y}$ 　　　　**10.** $f(x, y) = (x^2 + 1)^y$

11. A rectangular cage has length x feet, width y feet, and height z feet. If wire is to be placed around the sides of the cage, express the amount A of wire needed as a function of x, y, and z.

12. An animal's diet consists of three items, A, B, and C. The caloric contents per gram are 10, 20, and 50 calories, respectively. Let x, y, and z denote the numbers of grams of A, B, and C, respectively, eaten per day. Write the total number T of calories consumed per day as a function of x, y, and z.

13. A model for the *bacillus* bacteria consists of a cylinder of radius x and height y with spherical caps attached to the two ends. Write the volume V as a function of x and y.

14. Two concentric circles have radii x and y inches, respectively. Write the area A between the two circles as a function of x and y.

15. Express the height h of the mountain in terms of the two angles of inclination shown in the figure.

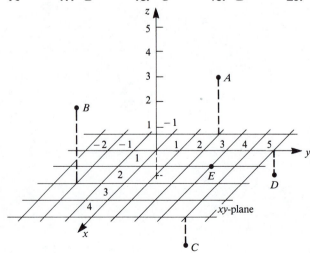

Give the (x, y, z) coordinates of the points shown in the accompanying figure.

16. A 　　**17.** B 　　**18.** C 　　**19.** D 　　**20.** E

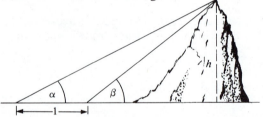

Find the domain of definition of each of the following functions $f(x, y)$. In addition, sketch the level curve $f(x, y) = 1$.

21. $\sqrt{2 - x^2 - y^2}$ **22.** $\ln (x + y)$ **23.** $\sqrt{x - y}$

24. xe^y **25.** $\tan^{-1} (y/x)$ **26.** $|x - y|$

27. $\dfrac{y}{x^2 - 1}$ **★ 28.** $\sin (xy)$

Using the method of sections, sketch the graphs of each of the following functions.

29. $z = 2x^2 + y^2$ **30.** $z = 2 - x^2 - y^2$ **31.** $z = x^2 - y^2$

32. $z = \dfrac{1}{x^2 + y^2 + 1}$ **33.** $z = ye^{-x}$, for $x \geq 0$ and $y \geq 0$

34. $z = \dfrac{1}{x + y}$, for $x \geq 0$ and $y \geq 0$ **35.** $z = e^{x^2 + y^2}$

★ 36. $z = x^y$, for $x > 0$ and $y > 0$

Find and sketch the level curves $f(x, y) = c$ for $c = 1, 2,$ and 4, for each of the following functions.

37. $f(x, y) = 2x^2 + y^2$ **38.** $f(x, y) = 2 - x^2 - y^2$ **39.** $f(x, y) = x^2 - y^2$

40. $f(x, y) = \dfrac{1}{x^2 + y^2 + 1}$ **41.** $f(x, y) = ye^{-x}$ **42.** $f(x, y) = 1/(x + y)$

43. The **ideal gas law** asserts that $PV = nRT$, where V is the volume in liters, T is the temperature in degrees Kelvin, P is the pressure in atmospheres, n is the number of moles of the gas, and R is the universal gas constant. If $nR = .32$, sketch the level curves $V = c$, for $c = 1, 2, 5,$ and 10. Let P range from 0 to 10 atmospheres, and let T vary from 0 to 350 degrees Kelvin.

44. **Graham's Law** gives the relationship between the average velocity U of an ideal gas and the pressure P, volume V, and molecular weight M:

$$U = \sqrt{\frac{3PV}{M}}$$

If $M = 3$, sketch the level curves $U = c$ for $c = 1, 2, 5,$ and 10.

45. If the temperature T at a point in the plane is given by

$$T(x, y) = 50e^{-(x^2 + y^2)}$$

sketch the level curves $T = 25$, $T = 10$, and $T = 5$. Curves where the temperature is constant are called **isotherms**.

41 partial derivatives

One of the most important applications of the derivative was to solve optimization (maximization or minimization) problems. As we have seen in examples 40.3 and 40.4, a function of several variables can have a maximum or minimum value occurring at a point. The first step in determining where such values can occur is to compute **partial derivatives**.

If $f(x, y) = x^2y + 3xy^3$, we can set $y = 2$, for example, and then compute the usual derivative:

$$z = 2x^2 + 24x, \qquad \text{when} \quad y = 2$$

Hence,

$$\frac{dz}{dx} = 4x + 24 \qquad \text{when} \quad y = 2.$$

In general, *for a fixed value of y*, we can write

$$z = y[\,x^2\,] + y^3[\,3x\,]$$

and take the usual derivative with respect to x to obtain $y[\,2x\,] + y^3[\,3\,] = 2xy + 3y^3$. This derivative is denoted by $\frac{\partial f}{\partial x}$ or f_x. Likewise, *for a fixed value of x*, we can view $f(x, y)$ as $x^2[\,y\,] + 3x[\,y^3\,]$ and compute the partial derivative with respect to y:

$$\frac{\partial f}{\partial y} = x^2 + 3x[\,3y^2\,] = x^2 + 9xy^2$$

Hence, in computing this partial derivative, we treat the variable x as a constant when we apply the usual derivative rules.

In general,

To compute $\dfrac{\partial f}{\partial x_1}$, or f_{x_1}, for $w = f(x_1, x_2, \ldots, x_n)$, treat the variables $x_2, \ldots, x_n$ as constants and compute the usual derivative with respect to x_1.

We call $\partial f/\partial x_1$ the *partial derivative of f with respect to x_1*. Methods for computing partial derivatives are practiced in the examples 41.1–41.5.

example 41.1   For $w = f(x, y, z) = x^2yz^3$, compute the three partial derivatives $\dfrac{\partial w}{\partial x}$, $\dfrac{\partial w}{\partial y}$, and $\dfrac{\partial w}{\partial z}$.

solution 41.1   To compute $\dfrac{\partial w}{\partial x}$, regard y and z as fixed and view w as $w = yz^3[\, x^2\,]$. Hence $\dfrac{\partial w}{\partial x} = yz^3[\, 2x\,] = 2xyz^3$. Likewise, $w = x^2z^3[\, y\,]$ gives $\dfrac{\partial w}{\partial y} = x^2z^3[\, 1\,] = x^2z^3$, and $\dfrac{\partial w}{\partial z} = x^2y[\, 3z^2\,] = 3x^2yz^2$.   📑

example 41.2   If $f(x, y) = (4x + 1)/y^2$, find f_x and f_y.

solution 41.2   To compute f_x, write $f(x, y) = (1/y^2)[4x + 1]$. Hence, $f_x = (1/y^2)[4] = 4/y^2$. Viewing $f(x, y)$ as $(4x + 1)[\, y^{-2}\,]$, we have $f_y = (4x + 1)[-2y^{-3}] = (-8x - 2)/y^3$.   📑

example 41.3   If $w = 3x^2 + 4y^3 + 2z^3$, find $\dfrac{\partial w}{\partial z}$.

solution 41.3   Since $3x^2$ and $4y^3$ are treated as constants, $\dfrac{\partial w}{\partial z} = 6z^2$.   📑

Often the Product, Quotient, and Chain Rules are used to compute partial derivatives, as demonstrated in examples 41.4 and 41.5.

example 41.4   If $f(x, y) = x \ln (x^2 + y^2)$, find f_x and f_y.

solution 41.4   The partial derivative with respect to y is

$$f_y = x[\text{derivative of } \ln (x^2 + y^2) \text{ with respect to } y]$$

$$= x\left[\frac{1}{(x^2 + y^2)}\right]\frac{\partial}{\partial y}[\, x^2 + y^2\,] \qquad \text{by the Chain Rule}$$

$$= x\left[\frac{1}{(x^2 + y^2)}\right][\, 2y\,] = \frac{2xy}{(x^2 + y^2)}$$

We must use the Product Rule to compute f_x:

$$f_x = x\left(\frac{2x}{x^2 + y^2}\right) + [\ln (x^2 + y^2)] \cdot 1$$

$$= \frac{2x^2}{(x^2 + y^2)} + \ln (x^2 + y^2)$$ 📑

example 41.5   If $w = f(x, y, z) = \dfrac{4xy}{yz + xz}$, find $\dfrac{\partial w}{\partial x}$.

solution 41.5   Using the Quotient Rule, we obtain

$$\frac{\partial w}{\partial x} = \frac{(yz + xz)\dfrac{\partial}{\partial x}(4xy) - 4xy\dfrac{\partial}{\partial x}(yz + xz)}{(yz + xz)^2}$$

$$= \frac{(yz + xz)(4y) - 4xy(z)}{(yz + xz)^2} = \frac{4y^2z}{(yz + xz)^2}$$

For a function $z = f(x, y)$ of two variables, the partial derivatives $\dfrac{\partial f}{\partial x}$ and $\dfrac{\partial f}{\partial y}$ have simple geometric interpretations. To emphasize further that x is fixed in the computation of $\dfrac{\partial f}{\partial y}$, set $x = a$. Then we must take the derivative of $f(a, y)$ with respect to y. Note, however, that $z = f(a, y)$ is the section of the graph of f shown in figure 41.1. Thus, $\dfrac{\partial f}{\partial y}$ at (a, y) is just the *slope of the tangent line of the*

figure 41.1

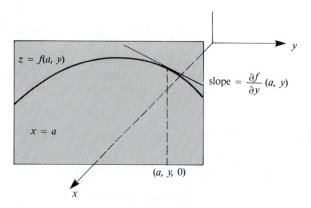

section $z = f(a, y)$ at a particular value of y. When $\dfrac{\partial f}{\partial y}(a, b)$ is computed, we obtain the slope of the tangent line in the y direction. Likewise, $\dfrac{\partial f}{\partial x}(a, b)$ gives the slope of the tangent line in the x direction, as depicted in figure 41.2.

It is not surprising that if a relative maximum or relative minimum occurs at (a, b), then both partial derivatives must vanish. As illustrated in the next two examples, setting $\dfrac{\partial f}{\partial x} = 0$ and $\dfrac{\partial f}{\partial y} = 0$ and solving the resulting system is the first step in finding out where these special points on the graph of f *might* occur.

figure 41.2

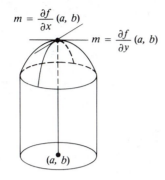

$$m = \frac{\partial f}{\partial x}(a, b)$$

$$m = \frac{\partial f}{\partial y}(a, b)$$

(a, b)

example 41.6   If $f(x, y) = 3x^2 - 6xy + y^2 + 12x - 16y + 2$, determine where $\dfrac{\partial f}{\partial x} = 0$ and $\dfrac{\partial f}{\partial y} = 0$.

solution 41.6   We first compute the partial derivatives $\dfrac{\partial f}{\partial x} = 6x - 6y + 12$ and $\dfrac{\partial f}{\partial y} = -6x + 2y$ $- 16$. Setting both derivatives equal to 0, we have the system

$$6x - 6y = -12$$
$$-6x + 2y = 16$$

Hence, $-4y = 4$ or $y = -1$. It follows that $x = -3$. However, thus far we have no way of knowing whether a hill or valley actually occurs at $(-3, -1)$. We will return to the problem of classifying critical points in Section 42.   ▤

example 41.7   If $f(x, y) = x^2 + y^3 - 12y$, determine where $\dfrac{\partial f}{\partial x}$ and $\dfrac{\partial f}{\partial y}$ are both 0.

solution 41.7   The partial derivatives are $\dfrac{\partial f}{\partial x} = 2x$ and $\dfrac{\partial f}{\partial y} = 3y^2 - 12$. Setting the partials equal to 0 yields $x = 0$ and $y^2 = 4$. Hence $y = \pm 2$ and so the partials both vanish at $(0, 2)$ and $(0, -2)$.   ▤

second order partial derivatives   Once we have computed $\dfrac{\partial f}{\partial x}$ and $\dfrac{\partial f}{\partial y}$, we can repeat the process to compute

$$\frac{\partial}{\partial x}\left(\frac{\partial f}{\partial x}\right), \quad \frac{\partial}{\partial y}\left(\frac{\partial f}{\partial x}\right), \quad \frac{\partial}{\partial x}\left(\frac{\partial f}{\partial y}\right), \quad \text{and} \quad \frac{\partial}{\partial y}\left(\frac{\partial f}{\partial y}\right)$$

The shorthand notation for these *second order partial derivatives* is

$$\frac{\partial^2 f}{\partial x^2}, \qquad \frac{\partial^2 f}{\partial y\,\partial x}, \qquad \frac{\partial^2 f}{\partial x\,\partial y}, \qquad \frac{\partial^2 f}{\partial y^2}, \qquad \text{respectively}$$

or

$$f_{xx}, \qquad f_{xy}, \qquad f_{yx}, \qquad \text{and} \qquad f_{yy}$$

The computation of second order partial derivatives is illustrated in examples 41.8 and 41.9.

example 41.8   Compute all first and second order partial derivatives for $f(x, y) = x^3 - 2x^4y^2 + y^2$.

solution 41.8   The first order partial derivatives are

$$f_x = \frac{\partial f}{\partial x} = 3x^2 - 8x^3y^2 \qquad \text{and} \qquad f_y = \frac{\partial f}{\partial y} = -4x^4y + 2y$$

Hence,

$$f_{xx} = \frac{\partial^2 f}{\partial x^2} = \frac{\partial}{\partial x}(3x^2 - 8x^3y^2) = 6x - 24x^2y^2$$

$$f_{yy} = \frac{\partial^2 f}{\partial y^2} = \frac{\partial}{\partial y}(-4x^4y + 2y) = -4x^4 + 2$$

$$f_{xy} = \frac{\partial^2 f}{\partial y\,\partial x} = \frac{\partial}{\partial y}(3x^2 - 8x^3y^2) = -16x^3y$$

$$f_{yx} = \frac{\partial^2 f}{\partial x\,\partial y} = \frac{\partial}{\partial x}(-4x^4y + 2y) = -16x^3y$$

Note that in example 41.8, $f_{xy} = f_{yx}$. This usually occurs! In fact, it is difficult to produce examples where $f_{xy} \neq f_{yx}$, although such special examples have been constructed by mathematicians.

example 41.9   Compute all first and second order partial derivatives of $f(x, y) = \ln(x^2 + y^2)$.

solution 41.9   The first order partial derivatives are

$$f_x = \frac{2x}{(x^2 + y^2)} \qquad \text{and} \qquad f_y = \frac{2y}{(x^2 + y^2)}$$

By the Quotient Rule, we have

$$f_{xx} = \frac{\partial}{\partial x}\left(\frac{2x}{x^2 + y^2}\right) = \frac{(x^2 + y^2)2 - 2x(2x)}{(x^2 + y^2)^2}$$

$$= \frac{2y^2 - 2x^2}{(x^2 + y^2)^2}$$

Likewise,

$$f_{yy} = \frac{2x^2 - 2y^2}{(x^2 + y^2)^2}$$

and

$$f_{xy} = \frac{\partial}{\partial y}\left[\frac{2x}{(x^2 + y^2)}\right] = \frac{(x^2 + y^2) \cdot 0 - 2x(2y)}{(x^2 + y^2)^2}$$

$$= -\frac{4xy}{(x^2 + y^2)^2}$$

As to be expected, $f_{yx} = f_{xy}$. This can be checked by computing $\frac{\partial}{\partial x}[2y/(x^2 + y^2)]$.

example 41.10   The partial differential equation

$$\frac{\partial z}{\partial t} = \kappa \frac{\partial^2 z}{\partial x^2}$$

is known as the **diffusion equation** and has a wide variety of applications in physics, chemistry, and biology. Find constants a, b, and c so that $z(x, t) = ce^{ax+bt}$ is a solution.

solution 41.10   The first order partial derivatives are $\frac{\partial z}{\partial t} = ce^{ax+bt}(b) = bce^{ax+bt}$ and $\frac{\partial z}{\partial x} = ace^{ax+bt}$. Hence, $\frac{\partial^2 z}{\partial x^2} = a^2 ce^{ax+bt}$. Now $z(x, t)$ will be a solution provided

$$bce^{ax+bt} = \kappa a^2 ce^{ax+bt}$$

Therefore, $b = \kappa a^2$. Set $z(x, t) = ce^{(ax+a^2\kappa t)}$, where a and c are arbitrary constants.

example 41.11   The partial differential equation

$$\frac{\partial^2 z}{\partial r^2} + \frac{1}{r}\frac{\partial z}{\partial r} + \frac{1}{r^2}\frac{\partial^2 z}{\partial \theta^2} = 0$$

is a form of **Laplace's equation**. Verify that any function of the form $z = r^k \cos k\theta$ is a solution.

solution 41.11   The first partial derivative with respect to r is $\frac{\partial z}{\partial r} = kr^{k-1} \cos k\theta$, and so

$\dfrac{\partial^2 z}{\partial r^2} = k(k-1)r^{k-2}\cos k\theta$. Furthermore, $\dfrac{\partial z}{\partial \theta} = r^k(-k\sin k\theta)$, and so $\dfrac{\partial^2 z}{\partial \theta^2} =$ $r^k(-k^2\cos k\theta)$. Substituting into Laplace's equation, we have

$$k(k-1)r^{k-2}\cos k\theta + \frac{1}{r}(kr^{k-1}\cos k\theta) + \frac{1}{r^2}[r^k(-k^2\cos k\theta)]$$

$$= (r^{k-2}\cos k\theta)[k(k-1)+k-k^2] = (r^{k-2}\cos k\theta)(k^2-k+k-k^2)$$

$$= (r^{k-2}\cos k\theta)(0) = 0$$

Hence, Laplace's equation is satisfied.

exercises for section 41

part A

Compute $\dfrac{\partial w}{\partial x}$, $\dfrac{\partial w}{\partial y}$, and $\dfrac{\partial w}{\partial z}$ for each of the following functions w.

1. $w = xyz - x^2z$

2. $w = \dfrac{xy}{z}$

3. $w = 3x^4 - 2y^2 - z + 1$

4. $w = \dfrac{x+z}{y}$

5. $w = xe^{yz}$

6. $w = x\ln(yz)$

Find f_x and f_y for each of the following functions.

7. $f(x, y) = x^2e^{-y}$

8. $f(x, y) = (3x - y + 1)(2x + y + 4)$

9. $f(x, y) = \dfrac{3x+4y}{x-2y}$

10. $f(x, y) = \ln(x+2y)$

11. $f(x, y) = \sin(xy)$

12. $f(x, y) = x\cos(xy)$

Determine $\dfrac{\partial f}{\partial x}$, $\dfrac{\partial f}{\partial y}$, and $\dfrac{\partial f}{\partial z}$ for each of the given functions.

13. $f(x, y, z) = \dfrac{xz}{y+z}$

14. $f(x, y, z) = xyze^{xy}$

15. $f(x, y, z) = \sqrt{x^2+y^2+z^2}$

16. $f(x, y, z) = \ln(xy^2z)$

17. $f(x, y, z) = xye^{yz}$

18. $f(x, y, z) = \tan^{-1}(xyz)$

A *critical point* (a, b) for $z = f(x, y)$ is a point where both partial derivatives f_x and f_y are 0. Find the critical points of each of the following functions.

19. $2x^2 - 4xy + 3y^2 - 8x + 1$

20. $x^3 + 256y - y^4$

21. $xy + 3x - 2y + 7$

22. $e^{-(x^2+y^2)}$

★23. $x^2 - \ln(xy) + y^2$

24. $x^2 + xy + y^2 - 5x + 3$

25. $x^3 - 12xy + y^3$

★26. $x\cos y$

27. $x^2 - 4xy + y^2 + 3$

28. xye^{x+y}

The critical points for a function $w = f(x, y, z)$ can be found by setting all three partials, f_x, f_y, and f_z, equal to 0. Find the critical points of each of the following functions.

29. $x^2 + y^2 + z^2 - 2x + 4y - 3z + 2$ **30.** xye^{yz}

Compute all first and second order partial derivatives of the given functions $f(x, y)$ and verify that $f_{xy} = f_{yx}$.

31. $x^4 - 3x^2y - 2xy^3 + y^3$ **32.** xe^{xy}

33. $(x + y)/(x - y)$ **34.** $3xy - xy^2 - \ln(xy)$

35. $\ln(x^3 + y^3)$

36. The partial differential equation $\dfrac{\partial^2 z}{\partial t^2} = c^2 \dfrac{\partial^2 z}{\partial x^2}$ is the **wave equation**. Verify that

$$z = \sin ax \cos act$$

is a solution for any constant a.

37. Show that the function $z = x^3 - 3xy^2$ satisfies **Laplace's equation**

$$\frac{\partial^2 z}{\partial x^2} + \frac{\partial^2 z}{\partial y^2} = 0$$

38. Verify that any function of the form $u = e^{-ka^2t} \sin ax$ satisfies the **heat equation**

$$k \frac{\partial^2 u}{\partial x^2} = \frac{\partial u}{\partial t}$$

★**39.** Show that the function $C = \dfrac{1}{\sqrt{4\pi at}} e^{-[x^2/(4at)]}$ satisfies the **diffusion equation***

$$\frac{\partial C}{\partial t} = a \frac{\partial^2 C}{\partial x^2}$$

★**40.** Show that any function of the form $w - (\sin ax)(\cos by)(e^{-\sqrt{a^2+b^2}\,z})$ satisfies the three-dimensional Laplace's equation

$$\frac{\partial^2 w}{\partial x^2} + \frac{\partial^2 w}{\partial y^2} + \frac{\partial^2 w}{\partial z^2} = 0$$

part B

In examples 41.10 and 41.11, we verified that a given function satisfied a particular partial differential equation. The following exercises will indicate how such solutions can be found. The technique is known as **separation of variables.**

* E. Pielou, *An introduction to mathematical ecology.* 2nd ed., New York: Wiley-Interscience, p. 166.

41. If $f(x) = g(y)$ for $a < x < b$ and $c < y < d$, show that $f(x)$ and $g(y)$ must be constants. $\left[\textit{Hint: What is } \dfrac{\partial}{\partial y} f(x)? \right]$

42. Suppose that $z = \phi(x)\psi(y)$ and $\dfrac{\partial z}{\partial x} = \dfrac{\partial^2 z}{\partial y^2}$.

    (a) Show that $\phi'(x)/\phi(x) = \psi''(y)/\psi(y)$.

    (b) Use exercise 41 to conclude that $\phi'(x) = c\phi(x)$ and $\psi''(y) = c\psi(y)$ for some constant c.

    (c) When $c = -a^2$, solve the differential equations in (b) to find $\phi(x)$ and $\psi(y)$ and therefore a solution z.

43. Follow the steps in exercise 42 to find solutions to the *wave equation*

$$\frac{\partial^2 z}{\partial x^2} = \frac{\partial^2 z}{\partial y^2}$$

that are of the form $z(x, y) = \phi(x)\psi(y)$.

42 applications of partial derivatives

In this section, we will present three of the many uses for partial derivatives:

1. the partial derivative as a rate of change and related rates problems
2. two-variable optimization
3. the partial derivative as a tool in estimating experimental errors

the partial derivative as a rate of change Since all but one of the variables are fixed in computing $\frac{\partial w}{\partial x}$, $\frac{\partial w}{\partial x}$ represents an instantaneous rate of change with respect to x. This is illustrated in the next example.

example 42.1 The formula $w = .1091x^{.425}y^{.725}$ gives body surface area w (in square feet) in terms of weight x (in pounds) and height y (in inches). We may assume that an adult's height y is fixed, but weight x can fluctuate. To measure this rate of change, we can compute $\frac{\partial w}{\partial x}$:

$$\frac{\partial w}{\partial x} = .1091(.425)x^{-.575}y^{.725} = (.0463675)\frac{y^{.725}}{x^{.575}}$$

The units for $\frac{\partial w}{\partial x}$ will be square feet per pound. If, for example, $y = 76$ inches and $x = 200$ pounds, then $\frac{\partial w}{\partial x} = .051$ square foot per pound. Adding an extra pound will increase w by about $\frac{1}{20}$ square foot. ◧

example 42.2 Most simple models of organism shapes involve combinations of geometric shapes such as spheres, cylinders, and boxes. Shown in figure 42.1 is a model

figure 42.1

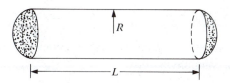

459

for the shape of the *bacillus* bacteria. Two spherical caps have been placed on the ends of a cylinder. Find expressions for the surface area S and volume V. Then compute and interpret $\dfrac{\partial S}{\partial R}$, $\dfrac{\partial S}{\partial L}$, $\dfrac{\partial V}{\partial R}$, and $\dfrac{\partial V}{\partial L}$.

solution 42.2

The volume and surface area are given by the expressions $V = \pi R^2 L + 2(\frac{2}{3}\pi R^3) = \pi R^2 L + \frac{4}{3}\pi R^3$ and $S = 2\pi RL + 2(2\pi R^2) = 2\pi RL + 4\pi R^2$, respectively.

In computing $\dfrac{\partial S}{\partial R}$ and $\dfrac{\partial V}{\partial R}$, L is fixed but R is allowed to vary, as indicated

in figure 42.2. Then $\dfrac{\partial S}{\partial R} = 2\pi L + 8\pi R$ mm²/(mm radius) and $\dfrac{\partial V}{\partial R} =$

figure 42.2

$2\pi RL + 4\pi R^2$ mm³/(mm radius). In computing $\dfrac{\partial S}{\partial L}$ and $\dfrac{\partial V}{\partial L}$, R is fixed and L is

allowed to vary, as depicted in figure 42.3. Then $\dfrac{\partial S}{\partial L} = 2\pi R$ mm²/(mm length)

and $\dfrac{\partial V}{\partial L} = \pi R^2$ mm³/(mm length).

figure 42.3

Since the rate of metabolism in cell growth will depend on S and V, the expressions derived in example 42.2 are helpful in constructing appropriate models. Often, however, in the formula $w = f(x, y)$, both x and y are changing with respect to time t. How can the rate of change $\dfrac{dw}{dt}$ be computed?

First, note that if y were fixed, then $\dfrac{dw}{dt} = \dfrac{\partial f}{\partial x}\dfrac{dx}{dt}$ by the ordinary Chain

Rule of Section 10. Likewise, if x were held constant, then $\dfrac{dw}{dt} = \dfrac{\partial f}{\partial y}\dfrac{dy}{dt}$ by this

same rule. *When x and y both change*, the computation of $\dfrac{dw}{dt}$ takes the form shown in the accompanying box.

> **Chain Rule for partial derivatives**
>
> If $w = f(x, y)$, $x = x(t)$, and $y = y(t)$, then
>
> $$\frac{dw}{dt} = \frac{\partial f}{\partial x}\frac{dx}{dt} + \frac{\partial f}{\partial y}\frac{dy}{dt}$$

This new form of the Chain Rule is illustrated in examples 42.3–42.5.

example 42.3   If $w = x^2 y^3 + 3xy$, $x = t^2 + t$, and $y = 3t - 4$, compute $\dfrac{dw}{dt}$.

solution 42.3   One method is to substitute the expressions for x and y into the equation for w and take the ordinary derivative. *An easier approach is to use the Chain Rule*:

$$\frac{dw}{dt} = \frac{\partial w}{\partial x}\frac{dx}{dt} + \frac{\partial w}{\partial y}\frac{dy}{dt}$$

$$= (2xy^3 + 3y)(2t + 1) + (3x^2 y^2 + 3x)3 \qquad\blacksquare$$

example 42.4   At age 13, a typical boy grows at the rates of 13 pounds per year and 2.4 inches per year, and his height and weight are about 62 inches and 100 pounds. (See example 11.5.) Find the rate at which his body surface area is increasing.

solution 42.4   Although we are not given weight x and height y as explicit functions of t, we are given that $\dfrac{dx}{dt} = 13$ and $\dfrac{dy}{dt} = 2.4$ when $t = 13$. From the body surface area formula in example 42.1, we have

$$\frac{dw}{dt} = \frac{\partial w}{\partial x}\frac{dx}{dt} + \frac{\partial w}{\partial y}\frac{dy}{dt}$$

$$= [(.1091)(.425)x^{-.575}y^{.725}](13) + [(.1091)(.725)y^{-.275}x^{.425}](2.4)$$

Setting $x = 100$ and $y = 62$, we have $\dfrac{dw}{dt} = 1.282$ square feet per year. $\blacksquare$

example 42.5   In example 42.2, suppose R and L increase according to the formulas $R = .1 - .09e^{-t}$ and $L = .8 - .72e^{-2t}$. Find expressions for $\dfrac{dV}{dt}$ and $\dfrac{dS}{dt}$.

solution 42.5

Using the partial derivatives from example 42.2, we have

$$\frac{dV}{dt} = \frac{\partial V}{\partial R}\frac{dR}{dt} + \frac{\partial V}{\partial L}\frac{dL}{dt} = (2\pi RL + 4\pi R^2)(.09e^{-t}) + (\pi R^2)(1.44e^{-2t})$$

and

$$\frac{dS}{dt} = \frac{\partial S}{\partial R}\frac{dR}{dt} + \frac{\partial S}{\partial L}\frac{dL}{dt} = (2\pi L + 8\pi R)(.09e^{-t}) + (2\pi R)(1.44e^{-2t})$$

two-variable optimization   In section 41, we pointed out that the first step in deciding where a relative maximum or relative minimum for $z = f(x, y)$ can occur is to set $\frac{\partial f}{\partial x} = 0$ and $\frac{\partial f}{\partial y} = 0$ and solve this system of equations. If $f(x, y) = y^2 - x^2$, then $\frac{\partial f}{\partial x} = -2x$ and $\frac{\partial f}{\partial y} = 2y$. Thus, both partial derivatives vanish only at $(0, 0)$. As we saw in example 40.5, $(0, 0, 0)$ is a saddle point on the graph, as illustrated in figure 42.4.

figure 42.4

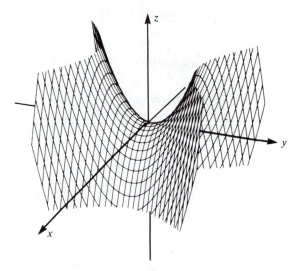

How then can we decide whether a given solution (a, b) to the system $\frac{\partial f}{\partial x} = 0$ and $\frac{\partial f}{\partial y} = 0$ gives a relative maximum, a relative minimum, or a saddle point? The test needed is a generalization of the Second Derivative Test of Section 13, and is summarized in the accompanying box.

> **the second derivative test**
>
> Suppose (a, b) is a point for which $\dfrac{\partial f}{\partial x} = 0$ and $\dfrac{\partial f}{\partial y} = 0$. Let $\Delta(x, y) = f_{xx}f_{yy} - (f_{xy})^2$. Then
>
> **1.** If $\Delta(a, b) > 0$ and $f_{xx}(a, b) > 0$, then $f(x, y)$ has a relative minimum at (a, b).
>
> **2.** If $\Delta(a, b) > 0$ and $f_{xx}(a, b) < 0$, then $f(x, y)$ has a relative maximum at (a, b).
>
> **3.** If $\Delta(a, b) < 0$, then $f(x, y)$ has a saddle point at (a, b).
>
> The function $\Delta(x, y)$ is called the **discriminant**. When $\Delta(a, b) = 0$, the test fails to give any information.

The following examples illustrate the use of this test for the critical points.

example 42.6   Find and classify the critical points of the function $f(x, y) = 3xy - x^3 - y^3 + 1$.

solution 42.6   The first order partial derivatives are $\dfrac{\partial f}{\partial x} = 3y - 3x^2$ and $\dfrac{\partial f}{\partial y} = 3x - 3y^2$. Setting both partials equal to zero, we have $y = x^2$ and $x = y^2$. Hence $x = (x^2)^2 = x^4$ and so $x = 0$ and $x = 1$. The critical points are then $(0, 0)$ and $(1, 1)$. To classify these two critical points we must compute f_{xx} and Δ:

$$f_{xx} = -6x, \qquad f_{yy} = -6y, \qquad f_{xy} = 3 \qquad \text{and } \Delta = 36xy - 9$$

At $(0, 0)$, $\Delta = -9$. Hence, a saddle point occurs there. At $(1, 1)$, $\Delta = 27$ and $f_{xx} = -6$, and so a relative maximum occurs at $(1, 1)$. ▤

example 42.7   In example 14.5 (page 145), we constructed a box with a reinforced base which was twice as thick as the sides or top. If the volume is to be 16 cubic feet, find the dimensions of the box of minimum surface area. *Do not assume, however, that the box has a square base.*

solution 42.7   As in example 14.5, the surface area is given by

$$S = 3xy + 2xz + 2yz$$

Since $xyz = 16$, $z = 16/(xy)$ and we can express S as a function of the two variables x and y:

$$S = 3xy + \frac{32}{y} + \frac{32}{x}$$

Thus, $\dfrac{\partial S}{\partial x} = 3y - \dfrac{32}{x^2}$ and $\dfrac{\partial S}{\partial y} = 3x - \dfrac{32}{y^2}$. Setting both partial derivatives equal to zero gives $3y = \dfrac{32}{x^2}$ and $3x = \dfrac{32}{y^2}$. Hence, $32 = 3yx^2 = 3xy^2$. Since $xy > 0$, it follows that $x = y$ and so $3x = \dfrac{32}{x^2}$ or $x = \sqrt[3]{32/3} \approx 2.201$.

Note that $S_{xx} = 64/x^3$, $S_{yy} = 64/y^3$, $S_{xy} = 3$ and so $\Delta = 64^2/(x^3y^3) - 9$. At the critical point $(2.201\ldots, 2.201\ldots)$, $\Delta = 27$ and $S_{xx} > 0$. Hence, a relative minimum occurs. The optimal design is then $x = 2.201$ feet by $y = 2.201$ feet by $z = 3.302$ feet, and the minimum surface area is 43.61 square feet. *Note that the optimal design necessarily has a square base.* ▤

example 42.8

A small lake is to be stocked with trout and bluegill. If x denotes the number of trout and y the number of bluegill, then the weights of the fish at the end of the year are dependent upon the population densities. The functions $W_1 = 4 - .003x - .002y$ and $W_2 = 3 - .002x - .004y$ give the weights of a single trout and a single bluegill, respectively, at the season's end. Assuming no deaths occur, how many fish of each type should be stocked in the lake in order to maximize the total biomass of fish?

solution 42.8

The total biomass B is given by

$B = (\text{number of trout})(\text{weight/trout}) + (\text{number of bluegill})(\text{weight/bluegill})$

$= x(4 - .003x - .002y) + y(3 - .002x - .004y)$

$= 4x - .003x^2 + 3y - .004y^2 - .004xy$

Thus $\dfrac{\partial B}{\partial x} = 4 - .006x - .004y$ and $\dfrac{\partial B}{\partial y} = 3 - .008y - .004x$. Setting both partial derivatives equal to 0 and multiplying by 1,000, we obtain the system

$$6x + 4y = 4{,}000$$
$$4x + 8y = 3{,}000$$

This system has solution $x = 625$ and $y = 62.5$. Note that $B_{xx} = -.006$, $B_{yy} = -.008$ and $B_{xy} = -.004$. Hence, $\Delta = .000032 > 0$ and $B_{xx} < 0$. A relative maximum occurs at the critical point. Therefore, for a maximum biomass of 1,343.75 pounds, stock the lake with 625 trout and about 63 bluegill. ▤

estimating errors in measurement: the total increment formula
In computing the volume of the box shown in figure 42.5, we must first estimate the length x, width y, and height z, and then use the formula $V = V(x, y, z) = xyz$. Errors in measuring x, y, and z (which are denoted by Δx, Δy, and Δz) are unavoidable. We will now develop a formula for the resulting error ΔV in measuring V.

figure 42.5

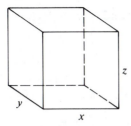

In section 11, we rephrased the limit definition of the derivative in the form

$$f'(x) = \lim_{\Delta x \to 0} \frac{f(x + \Delta x) - f(x)}{\Delta x}$$

$$= \lim_{\Delta x \to 0} \frac{\Delta y}{\Delta x}, \qquad \text{where } \Delta y = f(x + \Delta x) - f(x)$$

Hence, for Δx fairly small, $\frac{\Delta y}{\Delta x} \approx f'(x)$ or $\Delta y \approx f'(x)\,\Delta x$.

When this result is applied to partial derivatives of $w = f(x, y, z, \ldots)$, we obtain $\Delta w \approx \frac{\partial f}{\partial x}\,\Delta x$, $\Delta w \approx \frac{\partial f}{\partial y}\,\Delta y$, $\Delta w \approx \frac{\partial f}{\partial z}\,\Delta z$, etc. The approximation $\Delta w \approx \frac{\partial f}{\partial x}\,\Delta x$ assumes that all variables other than x are fixed. If all variables x, y, $z, \ldots$ are being estimated, this will *not* be the case. The correct approximation for Δw in terms of Δx, Δy, $\Delta z, \ldots$ takes the form given in the box.

Total Increment Formula

If $$w = f(x, y, z, \ldots) \text{ and}$$
$$\Delta w = f(x + \Delta x, y + \Delta y, z + \Delta z, \ldots) - f(x, y, z, \ldots),$$
then $$\Delta w \approx \frac{\partial f}{\partial x}\,\Delta x + \frac{\partial f}{\partial y}\,\Delta y + \frac{\partial f}{\partial z}\,\Delta z + \cdots$$

and the approximation improves as Δx, Δy, $\Delta z, \ldots$ approach 0.

The Total Increment Formula is applied in the following three examples.

example 42.9   The sides of a box have been measured, giving $x = 8$ inches, $y = 6$ inches, and $z = 5$ inches. Each measurement is in error by no more than .1 inch. Estimate the maximum error in computing the volume V of the box.

solution 42.9   Our estimate for the volume is $V = (8)(6)(5) = 240$ cubic inches, and we are given that $|\Delta x| \leq .1$, $|\Delta y| \leq .1$, and $|\Delta z| \leq .1$. Since $\dfrac{\partial V}{\partial x} = yz$, $\dfrac{\partial V}{\partial y} = xz$, and $\dfrac{\partial V}{\partial z} = xy$, we have

$$\Delta V \approx (yz)\,\Delta x + (xz)\,\Delta y + (xy)\,\Delta z$$
$$= 30\,\Delta x + 40\,\Delta y + 48\,\Delta z$$

This last term is no larger than $30(.1) + 40(.1) + 48(.1) = 11.8$. An approximation for the maximum error ΔV is then 11.8 cubic inches.   📚

example 42.10   **measuring blood flow through the lungs**   In a common cardiovascular test, blood flow F through the lungs is estimated by making three separate measurements:

1. $x =$ the concentration of oxygen in the pulmonary artery (measured by catheterization). Typical units are liters of O_2 per liter of blood. (See figure 28.3**(b)** for the anatomical diagram.)
   Typical units are liters of O_2 per liter of blood. (See figure 28.3**(b)** for the anatomical diagram.)

2. $y =$ the concentration of oxygen in the pulmonary vein

3. $z =$ the rate of oxygen consumption by the lungs (measured with a respiratory spirometer). Typical units are liters of O_2 per minute.

It is a consequence of Fick's Principle (see section 37, exercise 36) that

$$F = \frac{z}{y - x}$$

Suppose that the measurements for x, y, and z are $x = .03$, $y = .21$, and $z = 2.7$, and the errors in measurement are at most 1%. Estimate the maximum possible error for F.

solution 42.10   We are given that $|\Delta x| \leq (.01)(.03) = .0003$, $|\Delta y| \leq (.01)(.21) = .0021$, and $|\Delta z| \leq (.01)(2.7) = .027$. The partial derivatives are

$$\frac{\partial F}{\partial x} = \frac{z}{(y - x)^2}, \qquad \frac{\partial F}{\partial y} = \frac{-z}{(y - x)^2}, \qquad \text{and } \frac{\partial F}{\partial z} = \frac{1}{y - x}$$

Hence, by applying the Total Increment Formula and evaluating at $x = .03$, $y = .21$, and $z = 2.7$, we have

$$\Delta F \approx \frac{z}{(y - x)^2}\,\Delta x - \frac{z}{(y - x)^2}\,\Delta y + \frac{1}{y - x}\,\Delta z$$
$$= 83.33\,\Delta x - 83.33\,\Delta y + 5.55\,\Delta z,$$

This last term is no larger than $83.33(.0003) + (83.33)(.0021) + 5.55(.027) = .350.$ Our estimate for F is 15. The error then can be as much as 2.33 percent. ▤

example 42.11

In using mathematical models to make predictions, we must often estimate experimentally the **parameters** that appear in the model. The values for these parameters are then subject to error.

As a simple example, suppose $N(t) = N_0 e^{rt}$ describes the population growth of a species of rodent. To predict the population level at time $t = 1$ year, we must estimate N_0, their present population size, and the intrinsic rate of increase r. These parameters are notoriously hard to estimate. Note that $w = N(1) = N_0 e^r$.

Suppose that our estimates for N_0 and r are $N_0 = 1,200$ and $r = .4$. Then $w = N(1) = 1,790$. To estimate Δw, the error in our prediction, we compute $\dfrac{\partial w}{\partial N_0} = e^r$ and $\dfrac{\partial w}{\partial r} = N_0 e^r$. Hence,

$$\Delta w \approx (e^r) \Delta N_0 + (N_0 e^r) \Delta r$$
$$= e^{.4} \Delta N_0 + 1,200 e^{.4} \Delta r$$

Now if the error in estimating N_0 is no more than 200, and $|\Delta r| \le .1$, then Δw should be no larger than $e^{.4}(200) + (1,200 e^{.4})(.1) = 477$.

We may conclude that $w = N(1)$ is somewhere between $1,790 - 477 = 1,313$ and $1,790 + 477 = 2,267$. These large errors plague the application of population models to wildlife management. ▤

If $w = f(t, \alpha, \beta, \ldots)$, where $\alpha, \beta, \ldots$ are parameters to be estimated, then $\dfrac{\partial w}{\partial \alpha}$ is called the **sensitivity of w to the parameter α**. The larger the value of $\left| \dfrac{\partial w}{\partial \alpha} \right|$, the more the term $\left| \dfrac{\partial w}{\partial \alpha} \Delta \alpha \right|$ can contribute to the overall error for w. Determining those parameters with high sensitivities can indicate to the model builder that very special care must be given to their estimation.

exercises for section 42

part A

The body surface area formula in example 42.1 gives the surface area w (in square feet) of a human in terms of the weight x (in pounds) and the height y (in inches). Compute and interpret $\partial w / \partial x$ for the following values of x and y.

1. $x = 160$, $y = 70$ **2.** $x = 110$, $y = 64$

3. For the body surface area formula of example 42.1, compute $\dfrac{\partial w}{\partial y}$. What is the interpretation of this partial derivative?

4. Compute $\dfrac{\partial S}{\partial r}$, $\dfrac{\partial S}{\partial L}$, $\dfrac{\partial V}{\partial r}$, and $\dfrac{\partial V}{\partial L}$ for a cylinder of radius r, length L, and volume V.

5. Let $T(t, x)$ represent the temperature (in °C) in a lake at time t and depth x. Suppose that

$$T(t, x) = (10 + 5 \sin 2\pi t)e^{-3x}$$

Compute $\dfrac{\partial T}{\partial t}$ and $\dfrac{\partial T}{\partial x}$. What are the interpretations of these two partial derivatives?

6. One form of the ideal gas law asserts that

$$V = k\,\frac{T}{P}$$

for some constant k. Here, V is the volume of a gas (in liters), T is the temperature (in °K), and P is the pressure (in atmospheres). Compute and interpret $\partial V/\partial T$ and $\partial V/\partial P$.

7. Poiseuille's Law gives the rate of flow F of a liquid through cylindrical tubing as a function of the radius R, the pressure gradient $x = \Delta P$, and the length L of the tubing:

$$F = c\,\frac{x}{L}\,R^4, \qquad \text{for some constant } c$$

For the case of arteries in the human body, we may regard L as fixed, but R and x can change through the action of vascular muscle around the artery and by increased pumping of the heart. Compute and interpret $\partial F/\partial R$ and $\partial F/\partial x$.

8. The temperature maintained during the growth of a laboratory population can have profound effects on the population levels. If x is the temperature (in °C) and t is time measured in days, suppose that the predicted number y in the population is given by

$$y = \frac{x(30 - x)}{1 + e^{-2t}}$$

Compute and interpret $\dfrac{\partial y}{\partial t}$.

Using the Chain Rule, compute $\dfrac{dw}{dt}$ in each of the following problems.

9. $w = x^2 y$, $x = \cos t$, and $y = t^2$

10. $w = e^{xy}$, $x = 3t - 1$, and $y = t^2 - 3t + 2$

11. $w = \ln (x^2 + y^2)$, $x = e^{-t}$, and $y = \sin t$

12. $w = \sqrt{xy}$, $x = t^3 + 3t + 1$, and $y = 3e^{-2t}$

13. $w = 3x^2 - 2xy + 9x - 3y + 1$, $x = 4 \sin t$, $y = t \ln t$

14. At age 2, a typical little boy grows at the rate of 3.5 inches per year and 4.4 pounds per year. His height and weight are about 34 inches and 29 pounds, respectively. Find the rate at which his body surface area is increasing.

15. A large pine tree is regarded as a cylinder of radius r and height h. Suppose that $dr/dt = 2$ inches per year $= 1/6$ foot per year, while h (in feet) is given by

$$h = 50 - 4e^{-.5t}$$

Find an expression for the rate at which the volume of the tree is increasing.

16. A particle is moving in the xy-plane. At time t, its position $P = (x, y)$ is given by

$$x = e^{-t} \sin t \qquad \text{and} \qquad y = 3 - 2e^{-t}$$

Find an expression for $\dfrac{dw}{dt}$ if $w = \sqrt{x^2 + y^2}$ is the distance from the particle to $(0, 0)$.

17. The pressure P and temperature T of a gas increase according to the formulas

$$P = 2 - e^{-.5t} \qquad \text{and} \qquad T = 300 - 20e^{-.1t}$$

where t is time in minutes. Find an expression for $\dfrac{dV}{dt}$, assuming the ideal gas law $V = k \dfrac{T}{P}$.

Find the critical points of each of the following functions of two variables and classify each critical point as a relative maximum, a relative minimum, or a saddle point.

18. $f(x, y) = 3x^2 - 6xy + y^2 + 12x - 16y + 1$ [*Hint*: See example 41.6.]

19. $f(x, y) = x^2 + y^3 - 12y$ [*Hint*: See example 41.7.]

20. $f(x, y) = 2x^2 - 4xy + 3y^2 - 8x + 1$

21. $f(x, y) = x^3 - 27x + 256y - y^4$ 22. $f(x, y) = xy + 3x - 2y + 7$

23. $f(x, y) = e^{-(x^2 + y^2)}$ ★24. $f(x, y) = x^2 - \ln (xy) + y^2$

25. $f(x, y) = x^2 - 4xy + y^2 + 3$ 26. $f(x, y) = xye^{x+y}$

27. $f(x, y) = x^3 - 12xy + y^3$

28. A small rectangular aquarium has length x, width y, and height z. If the aquarium is to contain 4 cubic feet, find the design that minimizes the amount of glass needed for the bottom and sides.

29. A large box-like building must contain 36,000,000 cubic feet of space. In an effort to make the building as energy efficient as possible, the designer wishes to find the dimensions that will minimize the daily heat loss through the walls, ceiling, and floor. This heat loss T is given by

$$T = 12xy + 9xz + 9yz$$

where x, y, and z are the length, width, and height, measured in feet. Find the design that minimizes T.

30. Rework the trout-bluegill problem of example 42.8 if the weights of a trout and bluegill, respectively, at the end of the year are now given by

$$W_1 = 5.5 - .005x - .001y \qquad \text{and} \qquad W_2 = 3.5 - .004x - .002y$$

31. When a light is placed at the point $(0, 0, 1)$, the light intensity I at the point (x, y) in the plane below is given by

$$I = \frac{k}{x^2 + y^2 + 1}, \qquad \text{for some constant } k$$

Verify that the intensity is largest at the point $(0, 0)$ in the plane.

32. Common blood type is determined genetically by three alleles A, B, and O. Individuals of genotype AA, BB, or OO are called *homozygous* with respect to this characteristic. *Heterozygous* individuals are AB, AO, and BO. The *Hardy-Weinberg Law* predicts that the proportion P of heterozygous individuals in a given population is given by

$$P = 2pq + 2pr + 2qr$$

where p is the proportion of allele A in the population, q is the proportion of allele B, and r is the proportion of allele O.* Using the fact that $p + q + r = 1$, show that the maximum proportion of heterozygous individuals in any population is 2/3.

★ **33.** Refer to the accompanying figure. Show that of all triangles with a given perimeter c, the triangle with maximum area is an equilateral triangle.

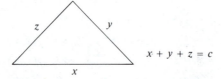

$$x + y + z = c$$

---

* See section 57 for an explanation of the terminology and see section 61 for the Hardy–Weinberg Law.

[*Hint*: If $s = c/2$, then the area A is given by *Heron's Formula*: $A = \sqrt{s(s-x)(s-y)(s-z)}$. Maximize $w = A^2$.]

The sides of a box have been measured, giving $x = 3.0$ meters, $y = 2.4$ meters, and $z = 1.8$ meters. Each of these measurements is in error by no more than 2 centimeters.

34. Estimate the maximum error in computing V, the volume of the box.

35. Estimate the maximum error in computing S, the surface area of the box.

36. The diameter and height of a cylinder are estimated to be 5 inches and 9.5 inches, respectively. These measurements are in error by no more than $\frac{1}{10}$ inch. Estimate the maximum error in computing the volume of the cylinder.

37. The height of a building is to be estimated from the angle θ and distance x shown in the figure. If $|\Delta x| \leq 1$ foot and $|\Delta\theta| \leq 1$ degree $= \pi/180$ radians, estimate h and Δh.

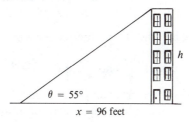

$\theta = 55°$

$x = 96$ feet

38. Compute the blood flow rate F through the lungs if $x = .04$, $y = .23$, and $z = 3.0$. [*Hint*: See example 42.10.] Estimate the maximum error in computing F if measurements for x, y, and z are in error by no more than 2%.

★**39.** *Creatinine* is a waste product from muscular activity that is cleared from the bloodstream by the kidneys. To measure the rate R at which the kidneys clear the blood of creatine, we measure:

1. x, the creatinine concentration in the bloodstream

2. y, the creatinine concentration in the urine

3. z, the urine output per minute

The rate R is measured in units of milliliters of clean blood per minute. Since all the creatinine is eventually removed by the kidneys, Rx, the creatinine per minute flowing into and cleared by the kidneys, is equal to yz, the creatinine per minute in the urine. Hence, $R = yz/x$.

If $x = 1.2$ (mg/100 mL), $y = 40.0$ (mg/100 mL), and $z = 3.0$ (mL/min), estimate R. If these measurements are in error by no more than 10%, estimate the maximum error ΔR.

★**40.** The formula in exercise 39 can also be applied to the removal of *urea* by the kidneys. If $x = 26.0$ (mg/100 mL), $y = 600$ (mg/100 mL), and $z = 3.0$

(mL/min), estimate R, the *urea clearance rate*. If the measurements are in error by no more than 5%, estimate the maximum error in measuring R.

41. A population of rodents is thought to grow exponentially. If

$$N(t) = N_0 e^{rt}$$

we must estimate both N_0 and r in order to predict $w = N(5)$. If $N_0 = 1{,}200 \pm 100$ and $r = .6 \pm .1$, find $w + \Delta w$ using the Total Increment Formula.

★42. A fish population in a large lake is expected to grow in numbers according to the logistic equation

$$N = \frac{K}{1 + ce^{-rt}}$$

Five thousand young fish are stocked in the lake at time $t = 0$, and we wish to predict $w = N(10)$. If $K = 25{,}000 \pm 1{,}000$ and $r = .5 \pm .05$, estimate w and Δw using the Total Increment Formula. $\left[\textit{Hint:} \text{ First show } K = 5{,}000(1 + c) \text{ and find } \dfrac{\partial w}{\partial r} \text{ and } \dfrac{\partial w}{\partial K}. \right]$

★43. The model presented in example 37.7 has been applied by C.S. Gist and F.W. Whicker to predict the effects of radioactive iodine 131 on deer populations (see Part A, exercise 30 of section 37). The maximum amount of iodine in the thyroid gland of a deer occurs at time

$$T = \frac{1}{\lambda_2 - \lambda_1} \ln \left(\frac{\lambda_2}{\lambda_1} \right)$$

In a study of the effect on a Colorado deer population of the 1964 Chinese nuclear test, the estimates for λ_2 and λ_1 were .126 and .107, respectively. If the maximum error for each of these parameters is 5%, estimate the maximum error for T.

44. The vital capacity VC of the lungs is the largest volume (in milliliters) that can be exhaled after a maximum inspiration of air:

$$VC = \begin{cases} (27.63 - .112a)h & \text{for males} \\ (21.78 - .101a)h & \text{for females} \end{cases}$$

where a = age in years and h = height in centimeters. Compute and interpret $\dfrac{\partial VC}{\partial a}$ and $\dfrac{\partial VC}{\partial h}$.

43 linear regression and curve fitting techniques

In section 17, we discussed curve fitting methods that were based on fitting a line through data *by sight*. For the case shown in figure 43.1, most individuals

figure 43.1

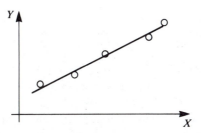

will produce approximately the same line $Y = mX + b$. Shown in figure 43.2 are data which are much more typical of the situations that arise in the biosciences. Here data on egg production Y as related to length X of the spawners are displayed for two populations of whitefish in Alberta, Canada. If

figure 43.2 (Adapted from, G. Hutchinson, *An Introduction to Population Biology.* New Haven: Yale University Press, 1978).

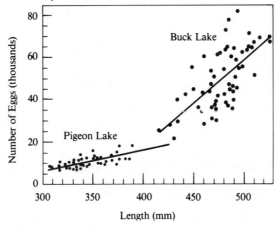

we were to ask ten people to fit lines through the data by sight, chances are that their results would vary quite a bit. How then can we fit the "best line" to the data in a completely objective and mechanical way? As we know from

Section 14 on optimization, we must first establish a criterion for deciding whether one line fits better than another. The commonly used criterion is called the **least squares method**.

figure 43.3

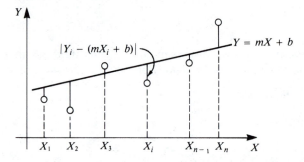

Shown in figure 43.3 are a **trial line** $Y = mX + b$ and the data points $(X_1, Y_1), (X_2, Y_2), \ldots, (X_n, Y_n)$. The discrepancy between the observed value Y_i of Y and the predicted value $mX_i + b$ is $|Y_i - (mX_i + b)|$. As a way of deciding how well this line fits, we will compute the *sum of the squares of the deviations*

$$D = \sum_{i=1}^{n} [Y_i - (mX_i + b)]^2$$

and then attempt to find the particular choices of m and b that minimize D. Since D is a function of m and b, the techniques of Section 42 apply:

$$\frac{\partial D}{\partial m} = \sum_{i=1}^{n} 2[Y_i - (mX_i + b)](-X_i)$$

and

$$\frac{\partial D}{\partial b} = \sum_{i=1}^{n} 2[Y_i - (mX_i + b)](-1)$$

Using the laws of summation (page 210), we can rewrite

$$\frac{\partial D}{\partial m} = 2\left[-\left(\sum_{i=1}^{n} X_i Y_i\right) + m\left(\sum_{i=1}^{n} X_i^2\right) + b\left(\sum_{i=1}^{n} X_i\right)\right]$$

$$\frac{\partial D}{\partial b} = 2\left[-\left(\sum_{i=1}^{n} Y_i\right) + m\left(\sum_{i=1}^{n} X_i\right) + bn\right]$$

Setting both partial derivatives equal to 0, we see that m and b must satisfy

$$m\left(\sum_{i=1}^{n} X_i^2\right) + b\left(\sum_{i=1}^{n} X_i\right) = \sum_{i=1}^{n} X_i Y_i$$

$$m\left(\sum_{i=1}^{n} X_i\right) + bn = \sum_{i=1}^{n} Y_i$$

This system has a unique solution for m and b.* It can be shown (using the Second Derivative Test of section 42) that these two values do indeed minimize D. The resulting line $Y = mX + b$ is called the **regression line of Y versus X**. Let's now give a systematic method for computing the various sums needed to find m and b.

example 43.1

The weights Y (in British tons) and the lengths X (in feet) of five adult blue whales are given in table 43.1. Fit the regression line through the data using the

table 43.1

X	81	75	87	82	78
Y	91	71	110	95	80

previous formulas, and compare the observed values of Y to the predicted values $mX + b$.

solution 43.1

Before we find the proper values of m and b, we first compute ΣX, ΣX^2, ΣY, and ΣXY in table 43.2.

table 43.2

X	Y	X^2	XY
81	91	6,561	7,371
75	71	5,625	5,325
87	110	7,569	9,570
82	95	6,724	7,790
78	80	6,084	6,240
$\Sigma X = 403$	$\Sigma Y = 447$	$\Sigma X^2 = 32,563$	$\Sigma XY = 36,296$

Hence, m and b must satisfy

$$32{,}563m + 403b = 36{,}296$$
$$403m + 5b \quad = 447$$

This system has solution $m = 3.298$ and $b = -176.421$. (Use Cramer's Rule, for example.) The regression line is then $Y = 3.298X - 176.421$. Shown in table 43.3 are the observed values of Y and those predicted by the regression line.

* The general solutions for m and b are $m = \dfrac{n \Sigma XY - (\Sigma X)(\Sigma Y)}{n \Sigma X^2 - (\Sigma X)^2}$ and $b = \dfrac{(\Sigma Y)(\Sigma X^2) - (\Sigma XY)(\Sigma X)}{n \Sigma X^2 - (\Sigma X)^2}$.

table 43.3

X	Y (Observed)	Y (Predicted)
81	91	90.72
75	71	70.93
87	110	110.51
82	95	94.02
78	80	80.83

example 43.2   In a physiological experiment designed to find a relationship between oxygen uptake Y and walking speed X on a treadmill, the data shown in table 43.4 were obtained. Find the regression line, the line of "best fit."

table 43.4

X (mph)	0	1	2	3	4	5
Y (l/min)	.28	.31	.34	.38	.41	.43

solution 43.2   The values of ΣX, ΣY, ΣX^2, and ΣXY are computed in table 43.5.

table 43.5

X	Y	X²	XY
0	.28	0	0
1	.31	1	.31
2	.34	4	.68
3	.38	9	1.14
4	.41	16	1.64
5	.43	25	2.15
$\Sigma X = 15$	$\Sigma Y = 2.15$	$\Sigma X^2 = 55$	$\Sigma XY = 5.92$

Hence, m and b must satisfy the system

$$55m + 15b = 5.92$$
$$15m + 6b = 2.15$$

Solving for m and b yields $m = .0311$ and $b = .2804$. Hence, $Y = .0311X + .2804$ is the line of "best fit." Shown in table 43.6 are the observed and predicted values of Y.

table 43.6

X	Y (Observed)	Y (Predicted)
0	.28	.280
1	.31	.312
2	.34	.343
3	.38	.374
4	.41	.405
5	.43	.436

The method of least squares can be used to fit to data many other commonly occurring functions. Often we will have some idea of the type of function that is appropriate. We may be familiar with a mathematical model that predicts the type of function that should occur. For radioactive decay, for example, our first choice of function would be $y = ce^{-kt}$. In other cases, we may have no idea of what functional relationship to expect. In these cases, we plot the data and try to guess which curves will provide good fits. Such a picture, shown in figure 43.4, is called a **scatter diagram**.

figure 43.4

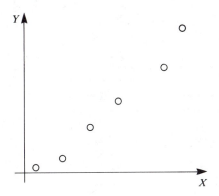

fitting exponential functions   If $y = ce^{kt}$, then

$$\ln y = \ln (ce^{kt}) = \ln c + kt$$

Hence, if the original data are (t_1, y_1), (t_2, y_2), ..., (t_n, y_n), let $X = t$ and $Y = \ln y$. We can then fit a regression line to the **transformed data** (X_1, Y_1), (X_2, Y_2), ..., (X_n, Y_n). If $Y = mX + b$ is the regression line, then $c = e^b$ and $k = m$. This method is illustrated in example 43.3.

example 43.3   In a study of the survival of a species of mountain goat, the data in table 43.7 were obtained. Fit an exponential function to the data, and compare the observed values of y to those predicted by $y = ce^{kt}$.

table 43.7

t (time)	4	6	8	10	12
y (number alive at time t)	387	269	155	73	29

solution 43.3   Let $X = t$ and $Y = \ln y$. The transformed data are shown in table 43.8.

table 43.8

X	4	6	8	10	12
Y	5.9584	5.5947	5.0434	4.2905	3.3673

The regression line is $Y = -.32432X + 7.44542$. (You should check that $\Sigma X = 40$, $\Sigma Y = 24.2543$, $\Sigma X^2 = 360$, and $\Sigma XY = 181.0616$, then solve for m and b.) Hence, $c = e^b = 1712.00$ and $k = m = -.32432$. Hence, $y = 1712.00e^{-.32432t}$ is the final exponential function. Shown in table 43.9 are the observed values of y and those predicted by this function.

table 43.9

t	y (Observed)	y (Predicted)
4	387	467.85
6	269	244.57
8	155	127.85
10	73	66.83
12	29	34.94

fitting power functions   To fit a power function $y = cx^k$ to data (x_1, y_1), $(x_2, y_2), \ldots, (x_n, y_n)$, note that

$$\ln y = \ln (cx^k) = \ln c + k \ln x$$

If we transform the data by letting $X = \ln x$ and $Y = \ln y$, then

$$Y = kX + (\ln c)$$

Once we have found the regression line $Y = mX + b$, we let $k = m$ and $c = e^b$. This is illustrated in example 43.4.

example 43.4   A study of land and freshwater birds in an island group yielded the data shown in table 43.10. Fit a power function $S = cA^k$ to the data (see example 5.11), and display the observed values of S versus the predicted values.

table 43.10

A (area of island in square miles)	50	125	350	700	1,025
S (number of species)	13	17	24	29	32

solution 43.4   We must transform the data before applying the regression formulas. Letting $Y = \ln S$ and $X = \ln A$, we obtain the transformed data shown in table 43.11.

table 43.11

X	Y	X²	XY
3.91202	2.56495	15.30392	10.03414
4.82831	2.83321	23.31261	13.67964
5.85793	3.17805	34.31538	18.61683
6.55108	3.36730	42.91665	22.05943
6.93245	3.46574	48.05883	24.02603
$\Sigma X = 28.08179$	$\Sigma Y = 15.40925$	$\Sigma X^2 = 163.907$	$\Sigma XY = 88.416$

Hence, m and b must satisfy the system

$$163.907m + 28.0818b = 88.416$$
$$28.0818m + 5b \quad = 15.40925$$

whose solution is $m = .302457$ and $b = 1.383140$. Hence, $c = e^b = 3.9874$ and $k = m = .30245$. Thus, $S = 3.9874A^{.30245}$ and we obtain the observed and predicted values shown in table 43.12. As you can see, the fit is excellent.

table 43.12

A	S (Observed)	S (Predicted)
50	13	13.02
125	17	17.18
350	24	23.45
700	29	28.92
1,025	32	32.46

other functional forms   Functions of the form $y = \alpha t/(1 + \beta t)$ arose in example 33.7 on chemical kinetics. To fit this type of function to data, note that

$$\frac{1}{y} = \frac{1 + \beta t}{\alpha t} = \frac{1}{\alpha}\left(\frac{1}{t}\right) + \frac{\beta}{\alpha}$$

If we let $Y = 1/y$ and $X = 1/t$, then the graph of Y versus X is the linear function $Y = (1/\alpha)X + \beta/\alpha$. This observation is applied in example 43.5.

example 43.5   When the gas acetaldehyde CH_3CHO is heated above 518°C, carbon monoxide (CO) is formed as a result of molecular collision. Shown in table 43.13 are the partial data of Hinshelwood and Hutchinson (1926). Fit a function of the form $y = \alpha t/(1 + \beta t)$ to the data.

table 43.13

t (sec)	42	105	840
y (pressure increase in mm due to CO)	34	74	244

solution 43.5   Letting $X = 1/t$ and $Y = 1/y$, we obtain the transformed data shown in table 43.14.

table 43.14

X	.0238	.00952	.00119
Y	.0294	.01351	.004098

The regression line is $Y = 1.1184X + .00280$. Since $m = 1/\alpha$, we have $\alpha = 1/m = .8941$ and $\beta = \alpha b = .0025$. Hence, $y = .8941t/(1 + .0025t)$ and the comparison of observed values to predicted values is shown in table 43.15.

table 43.15

t	y (Observed)	y (Predicted)
42	34	33.98
105	74	74.32
840	244	241.84

Many of the most famous applications of mathematics to problems of growth involve the logistic function $y = K/(1 + ce^{-rt})$. If we solve for ce^{-rt}, we obtain $ce^{-rt} = (K - y)/y$. Taking natural logarithms yields

$$\ln\left(\frac{K-y}{y}\right) = -rt + \ln c$$

If we let $X = t$ and $Y = \ln\left(\frac{K-y}{y}\right)$, then the relationship between Y and X is linear and the regression formulas can be applied to this transformed data. This technique is illustrated in example 43.6.

example 43.6   In a classical experiment by Pearl (1925)*, a population of fruit flies was allowed to grow in a small bottle. Table 43.16 presents partial data from one experiment. Pearl selected $K = 1,035$ for the carrying capacity (see Section 35). Fit a logistic curve to the data.

table 43.16

t (days)	0	9	18	27	36
y (number)	22	39	225	547	877

solution 43.6   Let $X = t$ and $Y = \ln\left(\frac{1,035 - y}{y}\right)$. We then have the information shown in table 43.17.

table 43.17

X	Y	X^2	XY
0	3.8296	0	0
9	3.2402	81	29.1618
18	1.2809	324	23.0562
27	−0.1141	729	−3.0815
36	−1.7139	1,296	−61.7004
$\Sigma X = 90$	$\Sigma Y = 6.5227$	$\Sigma X^2 = 2,430$	$\Sigma XY = -12.5639$

The resulting regression line is $Y = -.16046X + 4.1928$. Hence, $r = .16046$ and $c = e^b = 66.209$. The final logistic curve is then

$$y = \frac{1,035}{1 + 66.209e^{-.16046t}}$$

---

* R. Pearl, *The Biology of Population Growth*, New York: A. Knopf (1925).

More curve fitting methods will be explored in the Part B exercises. In most applications, the data set is large. Fortunately, a computer program for finding regression lines is a standard part of most computer libraries. In the examples and exercises of this text, however, the size of the data sets will be kept small.

exercises for section 43

1. In a study of the growth of the human fetus, the data on length Y versus age X shown in the accompanying table were obtained. Fit a linear function $Y = mX + b$ to the data and display the observed values of Y versus the predicted values.

X (weeks)	12	20	28	40
Y (centimeters)	10	25	38	53

2. Let Y = cardiac output (in l/min) and let X = work output (in kg-m/min). In a study of the relationship between X and Y during different levels of exercise, the data obtained in the table were obtained:

X	0	400	800	1200	1600
Y	2.5	7.4	13	18.8	23.2

   Find the regression line and compute the predicted values of Y.

3. At sea level, slightly less than $\frac{1}{2}$ liter of nitrogen is dissolved in body fluids. When a diver descends into the sea, nitrogen levels will increase, and too rapid an ascent to the surface can cause dangerous nitrogen bubbles to form. The table shows eventual nitrogen levels Y at various depths X. Find the regression line $Y = mX + b$. How well does the line fit the data?

X (m)	33	100	200	300
Y (l)	2	4	7	10

4. The heart rate Y (in beats per minute) changes with age X (in years). Data on a female child are shown in the table. Construct the line of best fit, and compare the observed values of Y to those predicted by $Y = mX + b$.

X	2	4	6	8	10	12	14	16
Y	108	102	92	88	90	86	82	79

5. The genetic disease phenylketonuria results from the lack of a key enzyme that is needed to process phenylalanine. If high concentrations build up in the body, mental retardation results. The method of treatment consists of instituting a diet low in phenylalanine. Let X be the age of the child (in weeks) when treatment is started, and let Y be the resulting IQ. Find the regression line for the data shown in the table.* What is the predicted value of Y when $X = 0$? Can you explain why the answer is not near the normal value of 100?

X	2	4	6	8	10	12
Y	85	84	83	82	80	77

6. The accompanying data on a rodent population (*Microtus arvalis pall*) are cited in Kostitsyn (1939).† Fit an exponential function $y = ce^{kt}$ to the data and compare the observed values of y to those predicted by the function.

t (months)	0	2	4	6	8	10	12	14
y (number of rodents)	2	5	16	20	40	109	200	283

7. Let y be the percentage of light at the ocean's surface that penetrates to a depth of x meters. Fit an exponential function to the data in the table.

x (meters)	1	2	3	4	5	6
y	.34	.14	.05	.015	.006	.002

8. Table 17.2 presented data showing the area A of a wound versus the time t after medication was applied. Fit an exponential function to the data using the regression techniques of this section.

9. For the fiddler crab, let W be the weight of the body minus the claws (in grams), and let C be the weight of the claws (in grams). The data in the table are provided by Thompson (1942). Fit a power function $C = cW^k$ to the data. How well does the curve fit?

W	58	300	536	1080	1449	2233
C	5	78	196	537	773	1380

* For more information see, W.E. Knox, *Pediatrics,* **26**: 1 (1960).

† V.A. Kostitzin, *Mathematical Biology,* London: Harrap (1939).

10. Partial data of Bergmann (1847) show the weight W of particular mammals (in kilograms) and the corresponding amount of heat H produced in a day (in calories). Fit a function of the form $H = cW^k$ to the data.

	W	**H**
Rabbit	2	116
Man	70	2,310
Horse	600	13,200
Elephant	4,000	52,000
Whale	150,000	255,000

11. Thompson (1942) has presented data on the fetal growth of blue whales.[*] Here, time $t = 0$ corresponds to November 1, and t is measured in months. The fetal length y is measured in feet.

t **(months)**	0	1	2	3	4
y **(feet)**	4.2	6.1	8.4	11.2	14.5

(a) Fit a function of the form $y = mt + b$ to the data.

(b) It may be possible to obtain a better fit using an allometric law. Let $z = y - 4.2$. Fit a function of the form $z = ct^k$.

★(c) Which fit is better?

12. It is well known that the rate of oxygen consumption R for fish is related to the body weight W by a function of the form

$$R = cW^k, \qquad \text{for} \quad .7 \le k \le .8\dagger$$

Fit such a function to the data in the table.

W **(grams)**	10	20	25	30	40
R **(ml/hour)**	1.6	2.8	3.3	3.8	4.5

13. The *Michaelis-Menten equation* predicts the relationship between the initial speed V with which an enzyme-induced reaction takes place and the concentration x of the original substance being acted on:

$$V = \frac{\alpha x}{1 + \beta x}$$

[*] d'Arcy Thompson, *On growth and form*. London: Cambridge Univ. Press (1961).

[†] See, D.H. Cushing and J.J. Walsh, *Ecology of the seas*. Philadelphia: W.B. Saunders, (1976), pp. 268 .

for some α and β. (See section 4, part B exercise 31.) Fit this function to the data in the table.

x (molarity)	1.5	2.0	3.0	4.0	8.0
V (mg product/min)	.21	.25	.28	.33	.44

14. Holling (1959)* created an artificial predator-prey situation in which the prey were represented by small sandpaper disks and the predator was a blindfolded subject who searched a square yard table for the disks. When the disks were found by tapping the table with a finger, they were set aside and the search continued. Let y be the number of prey eaten per minute (i.e., the number of disks found in a minute), and let x be the prey density (i.e., the number of disks per square yard). Fit a *Holling's functional response curve*

$$y = \frac{\alpha x}{1 + \beta x}$$

to the data in the accompanying table. Compare the observed values of y to those predicted by the function.

x	5	10	15	25	50	100	250
y	3	5	8	11	14	18	21

15. In a classical experiment on bacterial growth, H.G. Thornton (1933)† measured the area y occupied by the colony at time t. The data are shown in the table

t (days)	0	1	2	3	4	5
y (cm²)	.24	2.78	13.53	36.30	47.50	49.40

Using $K = 50$, fit a logistic function to the data and compare the observed values of y to those predicted by the function.

16. The growth in length y of bamboo as a function of time t can often be summarized by a logistic curve. Using $K = 162$, fit a logistic curve to the partial data of W. Pfeffer (1881).

t (days)	4	8	12	16	20
y (mm)	10.5	42.2	107.4	149.7	161.4

* C.S. Holling, "Some characteristics of simple types of predation and parasitism", *Can. Entomol.*, **91**: 385–398 (1959).

† H.G. Thornton, "On the development of a standardized agar medium for counting soil bacteria", *Ann. Appl. Biol.*, **9**: 265 (1922).

17. Functions of the form $y = y_\infty(1 - ce^{-kt})$, where $y_\infty > 0$ and $k > 0$, have appeared throughout the text. (See page 187, section 34, page 386, and page 403 for examples.)

    (a) When y_∞ is known and $y < y_\infty$, let $Y = \ln[(y_\infty - y)/y_\infty]$ and $X = t$. Show that

    $$Y = -kX + \ln c$$

    (b) When $y > y_\infty$, let $Y = \ln[(y - y_\infty)/y_\infty]$ and $X = t$. Show that $c < 0$ and

    $$Y = -kX + \ln(-c)$$

    (c) Apply (a) and fit a curve $y = y_\infty(1 - ce^{-kt})$ to the blood glucose data shown in the table. Use $y_\infty = 139.2$.

Time t (min)	0	10	20	30	40	50
Concentration y (mg/dl)	85.0	105.4	120.1	127.3	131.9	134.4

18. Often y_∞ is not known in the function $y = y_\infty(1 - ce^{-kt})$ and so the method outlined in part B exercise 17 cannot be used. To find y_∞, c, and k, we will introduce the curve fitting method known in fishery science as the *annual increment method*. Let $\Delta y = y(t + 1) - y(t)$.

    (a) Show that $\Delta y = -(1 - e^{-k})y + y_\infty(1 - e^{-k})$.

    (b) Let $Y = \Delta y$ and $X = y$. Conclude that $Y = mX + b$, where $m = -(1 - e^{-k})$ and $b = -y_\infty m$.

    (c) Apply this curve fitting technique to the tabled data on age t and length y for the Pacific herring.*

t (years)	3	4	5	6	7	8	9	10
y (cm)	25.7	28.4	30.15	31.65	32.85	33.65	34.44	34.97

★19. The method of least squares can be applied to fit a parabola

$$Y = aX^2 + bX + c$$

to data (X_1, Y_1), (X_2, Y_2), (X_3, Y_3), ..., (X_n, Y_n). To judge how well a given parabola fits, we will compute

$$D = \sum_{i=1}^{n} [Y_i - (aX_i^2 + bX_i + c)]^2$$

* H. Kitahama, "Size composition of the spring herring of Hokkaido", *Yoichi: Hokk. Fish. Res. Lab.* (1955).

(a) Compute $\dfrac{\partial D}{\partial a}$, $\dfrac{\partial D}{\partial b}$, and $\dfrac{\partial D}{\partial c}$, and show that a critical point (a, b, c) must satisfy the system

$$\left(\sum X_i^4\right)a + \left(\sum X_i^3\right)b + \left(\sum X_i^2\right)c = \sum X_i^2 Y_i$$

$$\left(\sum X_i^3\right)a + \left(\sum X_i^2\right)b + \left(\sum X_i\right)c = \sum X_i Y_i$$

$$\left(\sum X_i^2\right)a + \left(\sum X_i\right)b + nc = \sum Y_i$$

(b) Apply the formulas in (a) and fit a parabola through the points $(0, 0)$, $(1, 4)$, $(2, 6)$, $(3, 3)$, $(4, 2)$, and $(5, -1)$.

 computing double integrals

Our very first interpretation for $\int_a^b f(x)\, dx$ was the area shown in figure 44.1. When $f(x, y) > 0$ for (x, y) in a region R, a solid is formed and we may be interested in the

figure 44.1

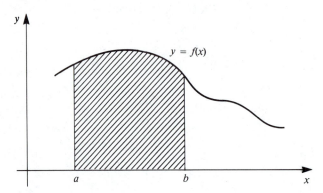

volume that is enclosed (see figure 44.2). The symbol for this volume is $\int \int_R f(x, y)\, dx\, dy$, the so-called *double integral of f over R*. The properties of the

figure 44.2

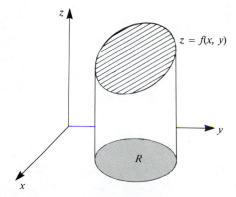

double integral are the same as those of the single integral (see page 222), and are summarized in the accompanying box.

properties of double integrals

1. $\iint_R [f(x, y) + g(x, y)] \, dx \, dy = \iint_R f(x, y) \, dx \, dy + \iint_R g(x, y) \, dx \, dy$

2. $\iint_R cf(x, y) \, dx \, dy = c \iint_R f(x, y) \, dx \, dy$

3. If A and B are disjoint regions in the domain of definition of f, then

$$\iint_{A \cup B} f(x, y) \, dx \, dy = \iint_A f(x, y) \, dx \, dy + \iint_B f(x, y) \, dx \, dy$$

But how is a double integral to be computed? Again, the clue comes from examining sections. First assume that R is the rectangle shown in figure 44.3(a). Let's temporarily think of the solid formed as a loaf of bread. The area

figure 44.3(a)

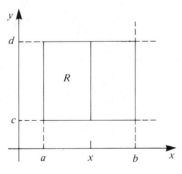

figure 44.3(b)

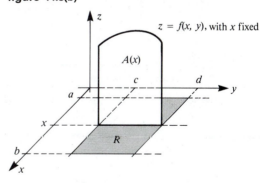

of the cross section $z = f(x, y)$, *for x fixed*, as shown in figure 44.3(b), is given by

$$A(x) = \int_c^d f(x, y) \, dy$$

We may think of $A(x) \Delta x$ as the volume of a thin slice of bread of cross-sectional area $A(x)$ and thickness Δx. The volume $\iint_R f(x, y) \, dx \, dy$ can be obtained by computing:

$$\text{limit}_{\Delta x \to 0} \sum A(x)\,\Delta x$$

that is, by computing

$$\int_a^b A(x)\,dx = \int_a^b \left[\int_c^d f(x, y)\,dy\right] dx$$

In a similar manner, we may fix y and compute the cross-sectional area $A(y) = \int_a^b f(x, y)\,dx$. Then $\int\int_R f(x, y)\,dx\,dy$ is given by

$$\int_c^d A(y)\,dy = \int_c^d \left[\int_a^b f(x, y)\,dx\right] dy$$

This method of computing $\int\int_R f(x, y)\,dx\,dy$, known as the **method of iterated integrals**, is summarized in the box, and is illustrated in examples 44.1–44.3.

---

the method of iterated integrals

If R is the rectangle $\{(x, y): a \le x \le b, c \le y \le d\}$, then $\int\int_R f(x, y)\,dx\,dy$ may be computed by finding either of the following iterated integrals:

$$\int_a^b A(x)\,dx = \int_a^b \left[\int_c^d f(x, y)\,dy\right] dx$$

or

$$\int_c^d A(y)\,dy = \int_c^d \left[\int_a^b f(x, y)\,dx\right] dy$$

---

example 44.1 Compute $\int\int_R (x^2 + y^2)\,dx\,dy$, where R is the rectangle shown in figure 44.4.

figure 44.4

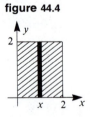

solution 44.1 We can compute either $\int_0^2 [\int_0^2 (x^2 + y^2)\,dx]\,dy$ or $\int_0^2 [\int_0^2 (x^2 + y^2)\,dy]\,dx$ to find the double integral. Let's compute the latter iterated integral:

$$A(x) = \int_0^2 (x^2 + y^2)\, dy = x^2 y + \left. \frac{y^3}{3} \right|_{y=0}^{y=2}, \qquad \text{since } x \text{ is fixed}$$

$$= 2x^2 + \frac{8}{3}$$

Finally,

$$\int_0^2 A(x)\, dx = \int_0^2 \left(2x^2 + \frac{8}{3} \right) dx = \frac{2}{3} x^3 + \left. \frac{8}{3} x \right|_0^2 = \frac{32}{3}$$

Integral tables are often helpful in handling the iterated integrals, as demonstrated in example 44.2.

example 44.2   Find $\iint_R ye^{xy}\, dx\, dy$, where R is the rectangle bounded by the lines $x = 1$, $x = 2$, $y = 0$, and $y = 2$.

solution 44.2   The rectangle R is shown in figure 44.5. This time, we will compute

figure 44.5

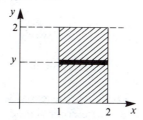

$\int_0^2 \left(\int_1^2 ye^{xy}\, dx \right) dy$ to find $\iint_R ye^{xy}\, dx\, dy$. Fixing y, we have

$$A(y) = \int_1^2 ye^{xy}\, dx$$

$$= y \left. \frac{e^{xy}}{y} \right|_{x=1}^{x=2}, \qquad \text{using integral table formula 7 with } a = y$$

$$= e^{xy} |_{x=1}^{x=2} = e^{2y} - e^y$$

Finally,

$$\int_0^2 A(y)\, dy = \int_0^2 (e^{2y} - e^y)\, dy = \frac{1}{2} e^{2y} - \left. e^y \right|_0^2$$

$$= \frac{1}{2} e^4 - e^2 + \frac{1}{2} = 20.41 \ldots$$

example 44.3   The plane $z = 6 - 2x - 3y$ is shown in figure 44.6. Find the volume of the solid formed by this plane and the rectangle $R = \{(x, y): 0 \le x \le 1, 0 \le y \le 1\}$.

figure 44.6

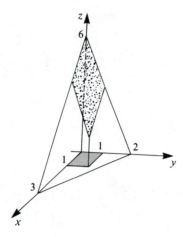

solution 44.3   The desired volume is $\iint_R (6 - 2x - 3y) \, dx \, dy$. This can be found by computing

$$\int_0^1 \left[\int_0^1 (6 - 2x - 3y) \, dx \right] dy = \int_0^1 (6x - x^2 - 3xy \, |_{x=0}^{x=1}) \, dy$$

$$= \int_0^1 (5 - 3y) \, dy = 5y - \frac{3}{2} y^2 \, \Big|_0^1$$

$$= 7/2 \text{ cubic units}$$

The method of iterated integrals can be extended from integrations over rectangles to double integrals over more complicated regions. For example, let R be the region bounded by the lines $x = a$ and $x = b$, and by the graphs of two functions $\psi(x)$ and $\phi(x)$ with $\psi(x) \le \phi(x)$, as illustrated in figure 44.7(a). The cross-sectional area is $A(x) = \int_{\psi(x)}^{\phi(x)} f(x, y) \, dy$, as shown in figure 44.7(b). We

figure 44.7(a)

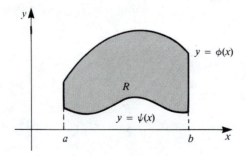

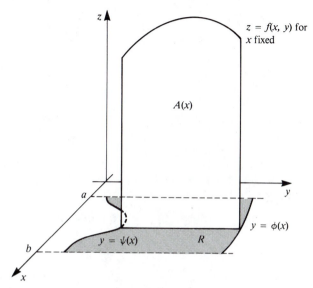

then compute $\int_a^b A(x)\,dx$ to find $\int\int_R f(x, y)\,dx\,dy$. In summary,

---

If $R = \{(x, y): a \le x \le b, \psi(x) \le y \le \phi(x)\}$, then

$$\int\int_R f(x, y)\,dx\,dy = \int_a^b \left[\int_{\psi(x)}^{\phi(x)} f(x, y)\,dy \right] dx$$

This new integration formula is applied in example 44.4.

example 44.4   Compute $\int\int_R (x^2 + y^2)\,dx\,dy$, where R is the triangular region shown in figure 44.8.

figure 44.8

(figure showing triangular region with y = x, base along x-axis up to 1)

y = x

1   x

solution 44.4   Here $\psi(x) = 0$ and $\phi(x) = x$. Hence,

$$\int\int_R (x^2 + y^2)\,dx\,dy = \int_0^1 \left[\int_0^x (x^2 + y^2)\,dy \right] dx$$

Now

$$A(x) = \int_0^x (x^2 + y^2)\, dy = x^2 y + \frac{y^3}{3}\Big|_{y=0}^{y=x}$$

$$= x^3 + \frac{x^3}{3}$$

$$= \frac{4}{3} x^3$$

Finally,

$$\int_0^1 \frac{4}{3} x^3\, dx = \frac{1}{3} x^4\Big|_0^1 = \frac{1}{3}.$$

For the region R shown in figure 44.9**(a)**, $\iint_R f(x, y)\, dx\, dy$ can be found using iterated integrals. For a fixed y, the cross-sectional area is $A(y) =$

figure 44.9(a)   **figure 44.9(b)**

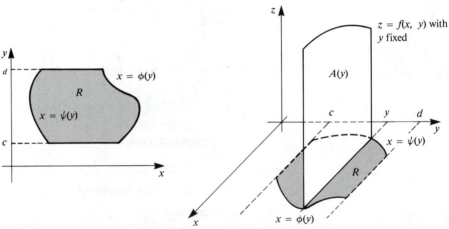

$\int_{\psi(y)}^{\phi(y)} f(x, y)\, dx$, and $\iint_R f(x, y)\, dx\, dy$ is given by $\int_c^d A(y)\, dy$, as illustrated in figure 44.9**(b)**. In summary,

---

If $R = \{(x, y): c \le y \le d,\ \psi(y) \le x \le \phi(y)\}$, then

$$\iint_R f(x, y)\, dx\, dy = \int_c^d \left[\int_{\psi(y)}^{\phi(y)} f(x, y)\, dx\right] dy$$

The formula above is used in our next two examples.

example 44.5   Find $\iint_R 2xy\, dx\, dy$ for the region R shown in figure 44.10.

figure 44.10

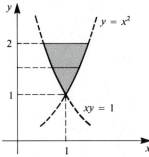

solution 44.5

Since $xy = 1$, $x = 1/y$. Hence, $\psi(y) = 1/y$. The function $y = x^2$ can be rewritten $x = \sqrt{y}$, and so $\phi(y) = +\sqrt{y}$. Hence, we must compute:

$$\int_1^2 \left(\int_{1/y}^{\sqrt{y}} 2xy \, dx \right) dy$$

Now

$$\int_{1/y}^{\sqrt{y}} 2xy \, dx = x^2 y \Big|_{x=1/y}^{x=\sqrt{y}} = (\sqrt{y})^2 y - \left(\frac{1}{y}\right)^2 y = y^2 - 1/y$$

Finally,

$$\int_1^2 (y^2 - 1/y) \, dy = \frac{y^3}{3} - \ln y \Big|_1^2 = \frac{7}{3} - \ln 2 \approx 1.6401$$

example 44.6

Find the volume of the triangular pyramid formed by $z = 6 - 2x - 3y$, the xy-plane, the xz-plane, and the yz-plane.

solution 44.6

The graph of $z = 6 - 2x - 3y$ intersects the xy-plane (where $z = 0$) in the line $2x + 3y = 6$. The region over which we must integrate and the triangular pyramid are shown in figure 44.11. The volume V is given by

figure 44.11(a)

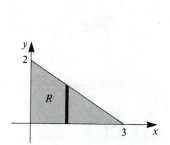

figure 44.11(b)

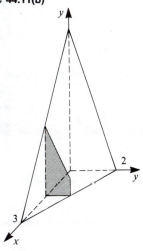

$$\int_0^3 A(x)\, dx = \int_0^3 \left[\int_0^{(6-2x)/3} (6 - 2x - 3y)\, dy \right] dx$$

Now

$$\int_0^{(6-2x)/3} (6 - 2x - 3y)\, dy = 6y - 2xy - \frac{3}{2} y^2 \Big|_{y=0}^{y=(6-2x)/3}$$

$$= 6\left(\frac{6-2x}{3}\right) - 2x\left(\frac{6-2x}{3}\right) - \frac{3}{2}\left(\frac{6-2x}{3}\right)^2 - 0$$

$$= \frac{2}{3} x^2 - 4x + 6$$

Hence,

$$V = \int_0^3 \left(\frac{2}{3} x^2 - 4x + 6\right) dx = \frac{2}{9} x^3 - 2x^2 + 6x \Big|_0^3$$

$$= 6 \text{ cubic units}$$

the double integral as a sum   Like its one-dimensional counterpart, the double integral is *the limit of a sum*. Given the region R shown in figure 44.12, enclose R in a rectangle $\{(x, y): a \le x \le b,\ c \le y \le d\}$. We then perform the following steps, analogous to those on page 212 for the single integral:

figure 44.12

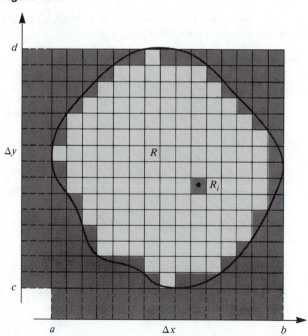

Step 1. Divide each of the intervals $[a, b]$ and $[c, d]$ into n pieces of lengths $\Delta x = (b - a)/n$ and $\Delta y = (d - c)/n$, respectively.

Step 2. Let $R_1, R_2, \ldots, R_m$ be those rectangles that lie entirely *inside* the region R. Let (x_i, y_i) be the midpoint of the rectangle R_i and compute $f(x_i, y_i)$.

Step 3. Form the sum $\sum_{i=1}^{m} f(x_i, y_i) \Delta x \, \Delta y$ or, more simply, $\sum f(x, y) \Delta x \, \Delta y$. Note that each small rectangle has area $\Delta x \, \Delta y$, and, when $f(x, y) > 0$, $f(x_i, y_i) \Delta x \, \Delta y$ may be interpreted as the volume of the slender box shown in figure 44.13.

figure 44.13

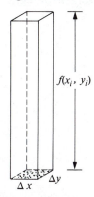

$f(x_i, y_i)$

Δx Δy

Step 4. Let n increase so that $\Delta x \to 0$ and $\Delta y \to 0$. We define

$$\int \int_R f(x, y) \, dx \, dy = \lim_{\Delta x, \Delta y \to 0} \sum f(x_i, y_i) \Delta x \, \Delta y$$

Note that as Δx and Δy approach 0, more and more rectangles R_i will fit into the region R, and, in the case where $f(x, y) > 0$, *more and more slender boxes fill and fit the original volume.*

Although this limit definition of the double integral is rarely used to actually estimate an integral, we will use this definition in Section 46 to discover additional interpretations of the double integral.

If, for example, we set $f(x, y) \equiv 1$, then $\sum_{i=1}^{m} f(x_i, y_i) \Delta x \, \Delta y = \sum_{i=1}^{m} \Delta x \, \Delta y$ is just the area of the small rectangles lying inside R. Hence, as Δx and Δy approach 0, the double integral approaches the area of R:

$$\int \int_R dx \, dy = \text{area } (R)$$

exercises for section 44

part A

Given that $\int\int_R f(x, y)\, dx\, dy = 4$ and $\int\int_R g(x, y)\, dx\, dy = 5$, use the general properties of the double integral to find each of the following.

1. $\displaystyle\int\int_R 3f(x, y)\, dx\, dy$ **2.** $\displaystyle\int\int_R [f(x, y) + g(x, y)]\, dx\, dy$

3. $\displaystyle\int\int_R [f(x, y) - 2g(x, y)]\, dx\, dy$

Let A and B be the rectangles shown in the figure. Make use of the integral tables and the method of iterated integrals to compute each of the following.

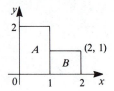

4. $\displaystyle\int\int_A (x + y^2)\, dx\, dy$ **5.** $\displaystyle\int\int_B (x + y^2)\, dx\, dy$

6. $\displaystyle\int\int_B \left(xy - \frac{y}{x}\right) dx\, dy$ **7.** $\displaystyle\int\int_A (xy^2 - x^2 y)\, dx\, dy$

8. $\displaystyle\int\int_B x^2 e^{xy}\, dx\, dy$ [*Hint*: Integrate with respect to y first.]   **★9.** $\displaystyle\int\int_A \sin \pi(x + y)\, dx\, dy$

10. $\displaystyle\int\int_A \frac{y}{x^2 + 1}\, dx\, dy$ **11.** $\displaystyle\int\int_B \frac{x}{x + y}\, dx\, dy$ [*Hint*: Integrate with respect to y first.]

12. $\displaystyle\int\int_{A \cup B} (2x + 3y)\, dx\, dy$ **13.** $\displaystyle\int\int_B y \cos(\pi xy)\, dx\, dy$

14. Find the volume of the solid bounded above by the plane $z = 5 - 2x - y$ over the rectangle $R = \{(x, y): 0 \le x \le 1, 0 \le y \le 1\}$.

15. The graph of $z = x^2 + y^2$ was sketched in example 40.3. Find the volume formed by this graph over the rectangle $R = \{(x, y): 1 \le x \le 2, 1 \le y \le 2\}$.

16. Evaluate the iterated integral $\displaystyle\int_0^1 \left(\int_0^{\sqrt{1-x}} xy\, dy\right) dx$.

17. Evaluate the iterated integral $\displaystyle\int_1^2 \left[\int_{1/y}^{y^2} (x + y)\, dx\right] dy$.

Given the regions A, B, C, and D shown in the figures, compute the double integral. [*Hint*: Often an iterated integral which is difficult to compute when we integrate with respect to y first will be easier when the integration first begins with x (and vice versa).]

(a)

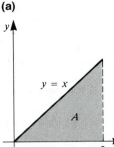

(b)

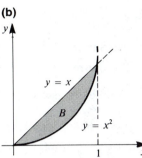

(c)

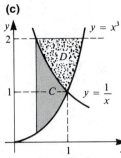

18. $\iint_A x^2 y \, dx \, dy$

19. $\iint_D x^2 y \, dx \, dy$

20. $\iint_D \frac{y}{x} \, dx \, dy$

21. $\iint_C (x^2 + y^2) \, dx \, dy$

22. $\iint_B (5 - 2x + 4y) \, dx \, dy$

23. $\iint_A y^2 e^{xy} \, dx \, dy$

24. $\iint_B \frac{y}{x+1} \, dx \, dy$

25. $\iint_C \frac{y}{x} \, dx \, dy$

26. $\iint_A (3xy^2 - 6yx^2) \, dx \, dy$

27. $\iint_{C \cup D} (2x - 6y) \, dx \, dy$

28. Find the volume of the triangular pyramid formed by the planes $z = 12 - 3x - 4y$, $x = 0$, $y = 0$, and $z = 0$.

★29. The graph of $z = y^2 - x^2$, shown in figure 40.15, intersects the xy-plane in the lines $y = \pm x$. Find the volume formed by this graph and the region R shown in the accompanying figure.

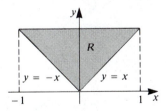

30. The double integral $\displaystyle\int_{-2}^1 \int_{x-1}^{1-x^2} (x^2 + y^2) \, dy \, dx$ is the volume of what solid? Compute the volume.

part B

31. When $h(x, y) = f(x)g(y)$ and R is the rectangle bounded by the lines $x = a$, $x = b$, $y = c$, and $y = d$, show that

$$\iint_R h(x, y) \, dx \, dy = \left[\int_a^b f(x) \, dx \right]\left[\int_c^d g(y) \, dy \right]$$

32. Let $I = \int_0^1 e^{-x^2/2}\, dx$ and $J = \int_0^1 \int_0^1 e^{-(x^2+y^2)/2}\, dx\, dy$.

    (a) Use Simpson's Rule (see section 27) with $n = 10$ to estimate I.

    (b) Use the result of exercise 31 to show $J = I^2$.

33. The double integral $\int_0^1 \int_0^1 f(x, y)\, dx\, dy$ can be estimated by repeated use of Simpson's Rule. Suppose $A(x) = \int_0^1 f(x, y)\, dy$ has been found for x to equal the data in the tables below. Estimate the double integral.

x	0	.1	.2	.3	.4	.5	.6
$A(x)$	.2345	.3457	.5723	.6234	.5673	.2346	.7650

x	.7	.8	.9	1.0
$A(x)$	.8721	.7856	.4523	.3111

45 double integrals using polar coordinates

Many regions in the xy-plane, such as circles and sectors, are not easy to describe using the usual (x, y) coordinates. For example, the 45° sector shown

figure 45.1

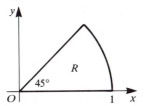

in figure 45.1 is bounded by the line $y = x$, part of the circle $x^2 + y^2 = 1$, and the line $y = 0$. It is difficult to compute a double integral over such a region when the usual (x, y) coordinates are used. The region R has a simple description, however, in **polar coordinates**.

figure 45.2

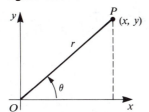

Given a point $P = (x, y)$, let r be the distance from the origin O, and let θ be the angle that the line OP makes with the positive x-axis. Then the *polar coordinates of the point P are (r, θ)*. The angle θ is expressed in radians and is usually chosen to lie between 0 and 2π. The region R can now be described by the inequalities

$$0 \leq r \leq 1 \quad \text{and} \quad 0 \leq \theta \leq \frac{\pi}{4}$$

Note from figure 45.2 that

$$\cos \theta = x/r$$
$$\sin \theta = y/r$$
$$x^2 + y^2 = r^2$$

These relationships hold for all points P in any of the four quadrants. The polar coordinates of $(0, 0)$ are by definition $r = 0$ and $\theta = 0$.

example 45.1   Find the polar coordinates (r, θ) of

$$\textbf{(a)}\ P = (2, 2) \qquad \textbf{(b)}\ Q = (0, 3) \qquad \textbf{(c)}\ S = (-1, 2)$$

solution 45.1   **figure 45.3**

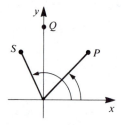

The points P, Q, and S are plotted in figure 45.3. For P, $\theta = 45° = \pi/4$ radians, and $r = \sqrt{2^2 + 2^2} = 2\sqrt{2}$. Thus, P has polar coordinates $(2\sqrt{2}, \pi/4)$. For Q, $\theta = 90° = \pi/2$ and $r = 3$; then Q is specified in polar coordinates by $(3, \pi/2)$. For the point S, however, we must use the formulas derived previously. We have $r = \sqrt{5}$ and $\cos \theta = -1/\sqrt{5}$.   Hence, $\theta = \cos^{-1}(-1/\sqrt{5}) = 116.56° = .6475\pi$ radians. Therefore, S has polar coordinates $(r, \theta) = (\sqrt{5}, .6475\pi)$.   ▤

example 45.2   Describe each of the regions shown in figure 45.4 using polar coordinates.

figure 45.4(a)

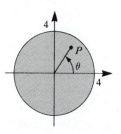

figure 45.4(b)

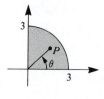

figure 45.4(c)

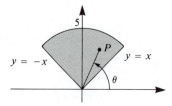

figure 45.4(d)

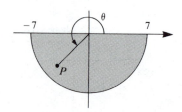

solution 45.2   A typical point $P = (r, \theta)$ is shown in each of the four regions.
For **(a)**, $r \leq 4$, but θ is not restricted. Hence, the region may be described by the inequalities $0 \leq r \leq 4$ and $0 \leq \theta \leq 2\pi$.
In **(b)**, $r \leq 3$ and $0 \leq \theta \leq \pi/2$.
For the sector in **(c)**, θ must lie between $45°$ and $45° + 90° = 135°$. Hence, the region is described by $0 \leq r \leq 5$ and $\pi/4 \leq \theta \leq 3\pi/4$.
Finally, for the region in **(d)**, θ varies between π and 2π and $0 \leq r \leq 7$.

How can $\int \int_R f(x, y) \, dx \, dy$ be computed if polar coordinates are used and R is a region such as one shown in figure 45.4? The formula given in the box provides the answer.

---

formula for changing to polar coordinates

If R is described by the inequalities $0 \leq r \leq r_0$, and $\theta_1 \leq \theta \leq \theta_2$, then

$$\int \int_R f(x, y) \, dx \, dy = \int_{\theta_1}^{\theta_2} \int_0^{r_0} f(r, \theta) r \, dr \, d\theta$$

where $r^2 = x^2 + y^2$, $x = r \cos \theta$, and $y = r \sin \theta$.
Note that an extra factor of r appears in the second integral.

A derivation of the formula will be given at the end of this section. Let's now see how changing a double integral to polar coordinates can simplify the computation.

example 45.3   Compute $\int \int_R (x^2 + y^2) \, dx \, dy$, where R is the region shown in figure 45.4**(b)**.

solution 45.3   Since $r^2 = x^2 + y^2$,

$$\int \int_R (x^2 + y^2) \, dx \, dy = \int_0^{\pi/2} \left(\int_0^3 r^2 r \, dr \right) d\theta = \int_0^{\pi/2} \frac{r^4}{4} \Big|_0^3 \, d\theta$$

$$= \int_0^{\pi/2} \frac{81}{4} \, d\theta = \frac{81}{4} \cdot \frac{\pi}{2} = \frac{81\pi}{8}$$

example 45.4   Compute $\int \int_R x^2 y \, dx \, dy$, where R is the region shown in figure 45.4**(c)**.

solution 45.4   The region R is described by the inequalities $0 \leq r \leq 5$ and $\pi/4 \leq \theta \leq 3\pi/4$. Note that in polar coordinates, $x^2 y = (r \cos \theta)^2 (r \sin \theta) = r^3 \sin \theta \cos^2 \theta$. Hence,

$$\iint_R x^2 y \, dx \, dy = \int_{\pi/4}^{3\pi/4} \left[\int_0^5 (r^3 \sin\theta \cos^2\theta) r \, dr \right] d\theta$$

$$= \int_{\pi/4}^{3\pi/4} \left(\frac{r^5}{5} \sin\theta \cos^2\theta \, \bigg|_0^5 \right) d\theta$$

$$= \int_{\pi/4}^{3\pi/4} \frac{5^5}{5} \sin\theta \cos^2\theta \, d\theta$$

Using the substitution method with $u = \cos\theta$, we compute the last integral to be $-\dfrac{5^5}{5} \cdot \dfrac{\cos^3\theta}{3} \bigg|_{\pi/4}^{3\pi/4} = 147.31 \ldots$.

example 45.5   In example 40.4, we graphed the bell-shaped function $f(x, y) = 10e^{-(x^2+y^2)/2}$. Find the total volume enclosed by the graph and the entire xy-plane, illustrated in figure 45.5.

figure 45.5

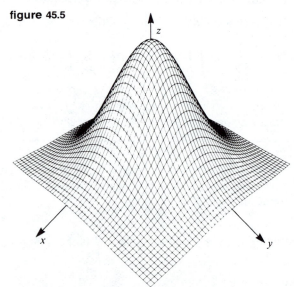

solution 45.5   Our line of attack is to first compute the volume over the disk $r^2 = x^2 + y^2 \le R^2$, and then let $R \to +\infty$. Let D_R denote the disk of radius R. The volume over D_R is then given by

$$\iint_{D_R} 10e^{-(x^2+y^2)/2} \, dx \, dy = \int_0^{2\pi} \left(\int_0^R 10e^{-r^2/2} r \, dr \right) d\theta$$

$$= \int_0^{2\pi} -10e^{-r^2/2} \bigg|_0^R \, d\theta$$

$$= \int_0^{2\pi} (10 - 10e^{-R^2/2}) \, d\theta = 2\pi(10 - 10e^{-R^2/2})$$

Now, as $R \to \infty$, $e^{-R^2/2} \to 0$. Hence, the total volume over the entire xy-plane is 20π cubic units.

We end this section by verifying the formula for the change to polar coordinates. Divide the region R into small circular sectors of length Δr and angle $\Delta\theta$, as shown in figure 45.6(a). The area of a typical sector is

$$\frac{\Delta\theta}{2\pi}(\pi r_2^2 - \pi r_1^2) = \frac{(r_2 - r_1)(r_1 + r_2)}{2}\Delta\theta = \frac{r_1 + r_2}{2}\Delta r\,\Delta\theta$$

figure 45.6   **figure 45.6(b)**

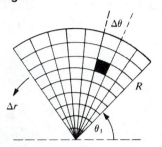
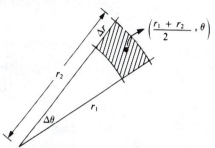

Let $r = (r_1 + r_2)/2$ and let (r, θ) be the midpoint of the sector, as illustrated in figure 45.6(b). The area of the sector is then $r\,\Delta r\,\Delta\theta$. Multiplying by $f(r, \theta)$, we obtain the volume of the slender solid shown in figure 45.7. When we add these

figure 45.7

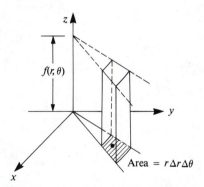

Area $= r\Delta r\Delta\theta$

volumes over all small sectors, we will approximate the true volume $\int\int_R f(x, y)\,dx\,dy$. Thus, $\Sigma f(r, \theta)\,r\,\Delta r\,\Delta\theta$ closely approximates the volume when Δr and $\Delta\theta$ are small.

In forming the sectors, we divided $[0, r_0]$ and $[\theta_1, \theta_2]$ into small subintervals of length Δr and $\Delta\theta$, respectively. Now, by the definition of double integral,

$$\lim_{\Delta r, \Delta\theta \to 0}\sum f(r, \theta)r\,\Delta r\,\Delta\theta = \int_{\theta_1}^{\theta_2}\int_0^{r_0}f(r, \theta)r\,dr\,d\theta$$

The formula for changing to polar coordinates now follows.

exercises for section 45

part A Find the polar coordinates (r, θ) of each of the following points $P = (x, y)$.

1. $(2, 2)$ **2.** $(-3, 3)$ **3.** $(0, 6)$ **4.** $(0, -5)$ **5.** $(-4, 0)$

6. $(1, 0)$ **7.** $(2, 3)$ **8.** $(-1, .2)$ **9.** $(\frac{1}{2}, \frac{\sqrt{3}}{2})$ **10.** $(-\frac{1}{2}, \frac{\sqrt{3}}{2})$

Describe in polar coordinates each of the following regions.

11.

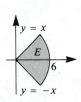

12.

13.

14.

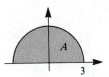

15.

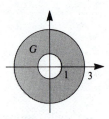

16.

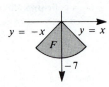

17.

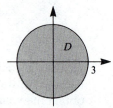

18.

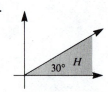

19.

★20.

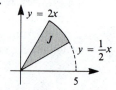

By switching to polar coordinates, compute each of the following double integrals. The regions given refer to those specified in exercises 11–20.

21. $\displaystyle\iint_A (x^2 + y^2)\, dx\, dy$

22. $\displaystyle\iint_B e^{-(x^2+y^2)}\, dx\, dy$

23. $\displaystyle\iint_C x^2 y\, dx\, dy$

24. $\displaystyle\iint_D \frac{1}{x^2 + y^2 + 1}\, dx\, dy$

25. $\displaystyle\iint_E (x + 2y)\, dx\, dy$

26. $\displaystyle\iint_F \sqrt{x^2 + y^2}\, dx\, dy$

27. $\displaystyle\iint_F xy^2\, dx\, dy$

28. $\displaystyle\iint_D \sqrt{9 - x^2 - y^2}\, dx\, dy$

★29. $\displaystyle\iint_G \frac{xy}{x^2 + y^2}\, dx\, dy$

★30. $\displaystyle\iint_H e^{-\sqrt{x^2+y^2}}\, dx\, dy$

★31. $\displaystyle\iint_I \frac{2}{\sqrt{x^2 + y^2}}\, dx\, dy$

★32. $\displaystyle\iint_J (x^2 + y^2)\, dx\, dy$

Following the line of attack in example 45.5, compute the total volume enclosed by each of the following positive functions $f(x, y)$ and the entire xy-plane.

33. $\displaystyle\frac{5}{(x^2 + y^2 + 1)^2}$

34. $e^{-\sqrt{x^2+y^2}}$

★35. $\displaystyle\frac{10}{x^2 + y^2 + 4}$

46 applications of the double integral

When $f(x, y) > 0$, the double integral $\int\int_R f(x, y)\, dx\, dy$ may be interpreted as a volume. In this section, we will use the *limit definition of the double integral* given in Section 44 to discover other common interpretations. In each case, the double integral will perform a *summation over a two-dimensional region*.

the average value of a function   Given a function $z = f(x, y)$ for (x, y) in a region R, we may be interested in the average value that f achieves over R. The appropriate formula, a generalization of the average value formula given in Section 21, is stated in the box.

---

the average value formula

$$\bar{Z} = \frac{1}{\text{area } (R)} \int \int_R f(x, y)\, dx\, dy$$

Let's see why this formula is valid in the special case where R is a rectangle, as shown in figure 46.1. Divide $[a, b]$ and $[c, d]$ into n parts of lengths $\Delta x = (b - a)/n$ and $\Delta y = (d - c)/n$, respectively. Then $m = n^2$ rectangles are formed, each with area $\Delta x\, \Delta y$. If (x_i, y_i) denotes the midpoint of rectangle R_i, then we

figure 46.1

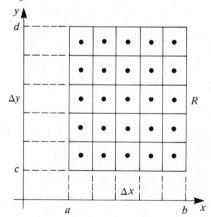

can approximate the average value of $f(x, y)$ by computing the simple average

$$\frac{\sum_{i=1}^{n^2} f(x_i, y_i)}{n^2} = \frac{\sum f(x, y)}{n^2}$$

This last average can be rewritten as $\dfrac{\Delta x \, \Delta y \sum f(x, y)}{n^2 \, \Delta x \, \Delta y} = \dfrac{\sum f(x, y) \, \Delta x \, \Delta y}{\text{area}\,(R)}$, since R consists of n^2 rectangles, each of area $\Delta x \, \Delta y$. As n increases, more and more values of $f(x, y)$ are averaged and

$$\frac{1}{\text{area}\,(R)} \sum f(x, y) \, \Delta x \, \Delta y \rightarrow \frac{1}{\text{area}\,(R)} \int\int_R f(x, y) \, dx \, dy$$

by the limit definition of the double integral. Use of the average value formula is illustrated in the next two examples.

example 46.1   Find the average value of the function $f(x, y) = y/x$ over the rectangle shown in figure 46.2.

figure 46.2

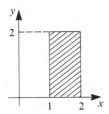

solution 46.1   The rectangle has area 2, and so we must compute $\bar{Z} = \tfrac{1}{2} \int \int_R y/x \, dx \, dy$. Now $\int \int_R y/x \, dx \, dy$ can be computed as follows:

$$\int_1^2 \left(\int_0^2 \frac{y}{x} \, dy \right) dx = \int_1^2 \left(\frac{y^2}{2x} \Big|_{y=0}^{y=2} \right) dx = \int_1^2 \frac{2}{x} \, dx$$

The value of the last integral is $2 \ln 2$. Hence, $\bar{Z} = \tfrac{1}{2}(2 \ln 2) = \ln 2$ ▤

   In some cases, it is more convenient to switch to polar coordinates, as demonstrated in example 46.2.

example 46.2   **Poiseuille's Law for laminar flow**   When a liquid flows through cylindrical tubing of radius r_0 (see figure 46.3), the velocity r units from the center axis is given by $v = k(r_0^2 - r^2)$ for some constant k. Find the average velocity in the tubing.

figure 46.3

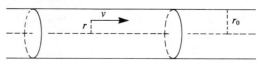

solution 46.2

We must compute the average velocity over the typical cross section shown in figure 46.4. The velocity $v(x, y) = v(r, \theta) = k(r_0^2 - r^2)$, where $r^2 = x^2 + y^2$. Since

figure 46.4(a)

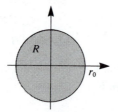

figure 46.4(b)

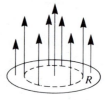

the region R has area πr_0^2, we must compute

$$\bar{V} = \frac{1}{\pi r_0^2} \int \int_R v(x, y) \, dx \, dy$$

Switching to polar coordinates, we obtain

$$\int \int_R v(x, y) \, dx \, dy = \int_0^{2\pi} \left[\int_0^{r_0} k(r_0^2 - r^2) r \, dr \right] d\theta$$

$$= \int_0^{2\pi} \left(k r_0^2 \frac{r^2}{2} - k \frac{r^4}{4} \Big|_0^{r_0} \right) d\theta$$

$$= \int_0^{2\pi} k \frac{r_0^4}{4} \, d\theta$$

Since r_0 is constant, the last integral is equal to $\dfrac{k\pi r_0^4}{2}$. Hence,

$$\bar{V} = \frac{1}{\pi r_0^2} \frac{k\pi r_0^4}{2} = \frac{k r_0^2}{2}$$

area density functions   A population occupies the territory R shown in figure 46.5. The population may consist of humans, animals, or even plants. To

figure 46.5

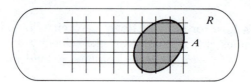

measure how densely populated the region is at various locations, we might divide the region into **quadrats**, or small rectangles, and count the numbers within each quadrat. For example, for a human population, a quadrat might be a city block.

Let $\rho(x, y)$ be the **density** (in numbers per square meter) at position (x, y) in the region. From a practical point of view, we can estimate $\rho(x, y)$ by counting the numbers in a small quadrat centered at (x, y). The following formula shows how the number within a subregion A can be computed in terms of $\rho(x, y)$.

If $\rho(x, y)$ is the density at point (x, y), then

$$\iint_A \rho(x, y)\, dx\, dy = \text{number in the population inside the subregion } A$$

This can be seen as follows. A small quadrat about (x, y) (see figure 46.6) contains $\Delta x\, \Delta y$ square meters, and the density is approximately $\rho(x, y)$ individuals per square meter. Thus, $\rho(x, y)\, \Delta x\, \Delta y$ approximates the number

figure 46.6

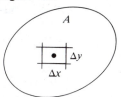

within this quadrat. The corresponding double integral sums and refines these estimates over smaller and smaller quadrats. This formula is applied in the following examples 46.3 and 46.4.

example 46.3   When $\rho(x, y)$ is constant, we say that the population is *uniformly distributed over* R. If 5,000 animals are distributed uniformly over R, then $\iint_R \rho(x, y)\, dx\, dy = 5{,}000$. But

$$\iint_R \rho(x, y)\, dx\, dy = \iint_R c\, dx\, dy = c \iint_R dx\, dy$$

The last integral is just area (R). Hence, $5{,}000 = c\ \text{area}\,(R)$, or $c = 5{,}000/\text{area}\,(R)$. If area $(R) = 25$ square miles, then the expected number of animals in a subregion A is given by

$$\iint_A \rho(x, y)\, dx\, dy = \iint_A 200\, dx\, dy = 200\ \text{area}\,(A)$$

example 46.4   The population densities of many large cities have been estimated, and these studies are an important part of urban planning. Let $(0, 0)$ be placed at the center of the city, as illustrated in figure 46.7. The population density (in

figure 46.7

thousands of individuals per square mile) at a point (x, y) that is $r = \sqrt{x^2 + y^2}$ miles from the center is often represented by a function of the form

$$\rho(x, y) = Ae^{-B\sqrt{x^2+y^2}}$$

The values $A = 120$ and $B = .2$ have been given for New York City in 1940. Estimate the number who lived within 5 miles of the center of the city at that time.

solution 46.4   Let R denote a disk of radius 5 about $(0, 0)$. We must compute

$$\iint_R 120e^{-.2\sqrt{x^2+y^2}}\, dx\, dy$$

Switching to polar coordinates, we have $\int_0^{2\pi} (\int_0^5 120e^{-.2r}r\, dr)\, d\theta$. Using either integration by parts or integral table formula 8 we obtain

$$\int_0^5 120re^{-.2r}\, dr = -600e^{-.2r}(r + 5)\big|_0^5 = 792.72$$

Finally, $\int_0^{2\pi} (792.72)\, d\theta = 2\pi(792.72) = 4{,}980.8$ thousand people, or about 4,980,800 people!

Rather than measuring the density $\rho(x, y)$ in numbers/m², we might measure the *biomass* in a quadrat. Typical units for $\rho(x, y)$ might be kg/m², and then $\iint_A \rho(x, y)\, dx\, dy$ gives the *total biomass within subregion A*.

example 46.5   A large field is bounded on the south by a river and on the west by mountains, as depicted in figure 46.8. The grass biomass density $\rho(x, y)$ (in kg/m²) at a point (x, y) that is x meters from the mountains and y meters from the river is estimated to be $\rho(x, y) = \dfrac{5xe^{-.01y}}{x + 1{,}000}$. Find the total biomass in each of the regions A_1 and A_2 shown in the figure.

figure 46.8

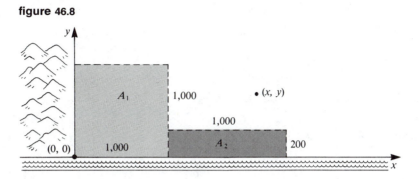

solution 46.5

We are asked to compute $\iint_R \dfrac{5xe^{-.01y}}{x + 1,000}\, dx\, dy$ for $R = A_1$ and A_2. Using formula 66 in the integral tables, we obtain

$$\int \frac{5x}{x + 1,000}\, dx = 5x - 5,000 \ln (x + 1,000)$$

Hence,

$$\iint_{A_1} \frac{5xe^{-.01y}}{x + 1,000}\, dx\, dy = \int_0^{1,000} \int_0^{1,000} \frac{5x}{x + 1,000}\, e^{-.01y}\, dx\, dy$$

$$= \int_0^{1,000} e^{-.01y}[5x - 5,000 \ln (x + 1,000)]\big|_0^{1,000}\, dy$$

$$= \int_0^{1,000} 1534.26 e^{-.01y}\, dy$$

This last integral is easily seen to equal $153,419 \approx 153,400$ kilograms of grass. Now,

$$\iint_{A_2} \frac{5x}{x + 1,000}\, e^{-.01y}\, dx\, dy = \int_0^{200} \int_{1,000}^{2,000} \frac{5x}{x + 1,000}\, e^{-.01y}\, dx\, dy$$

A similar computation shows that this integral yields a biomass of 257,036 kilograms of grass in region A_2.   ▤

probability density functions   When $f(x, y) \geq 0$ on a region R and $\iint_R f(x, y)\, dx\, dy = 1$, we call $f(x, y)$ a **probability density function**. The following example illustrates how a positive function $f(x, y)$ can be converted into a probability density function.

example 46.6

Find the constant c so that $f(x, y) = cxy$ is a probability density function on the rectangle shown in figure 46.9.

figure 46.9

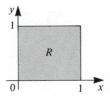

solution 46.6

We must find the constant c so that $\int\int_R cxy\,dx\,dy = 1$. Now

$$\int\int_R cxy\,dx\,dy = \int_0^1 \left(\int_0^1 cxy\,dx\right) dy = \frac{1}{4}c$$

(Check this!) Hence, we must set $c = 4$. Note that $4xy \geq 0$.   ▤

Imagine that a dart is being thrown into the target region R shown in figure 46.10. We may interpret $\int\int_A f(x, y)\,dx\,dy$ as the *probability that the dart*

figure 46.10

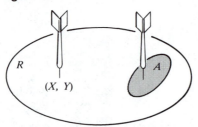

will land in a position (X, Y) *inside the subregion A.* Alternately, if a population is spread over the region R, then $\int\int_A f(x, y)\,dx\,dy$ represents the *proportion of the population inhabiting subregion A.*

In the case of the dart thrower, the form of the probability density function will determine the chances of landing in various subregions. Two common functions $f(x, y)$ are used to model dart throwing:

1. For the completely unskilled thrower, $f(x, y) = c$ for (x, y) in the region R. Here $c = \dfrac{1}{\text{area}\,(R)}$ and the resulting probability density function is known as the *two-dimensional uniform distribution on R.*

2. For skilled throwers, $f(x, y) = \dfrac{1}{2\pi\sigma^2} e^{-(x^2+y^2)/(2\sigma^2)}$. This is a *two-dimensional normal distribution* centered at the bull's-eye $(0, 0)$. The skill of the player will determine the value of σ.

Shown in figure 46.11 is a $10''$ dartboard. The circles have radii $1''$, $2''$, and $3''$, respectively. Let's do some probability computations.

figure 46.11

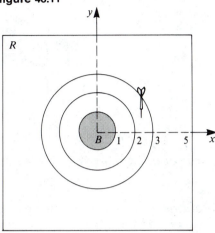

example 46.7   Find the probability that a completely unskilled player will hit the bull's-eye region B.

solution 46.7   Since area $(R) = 100$, the probability density function is $f(x, y) = .01$. The probability of a bull's-eye is just

$$\iint_B .01 \, dx \, dy = .01 \text{ area } (B) = .01\pi = .0314 \ldots$$

In fact, if B is *any region on the board with area* π, then the probability that the dart will land in B is $.0314\ldots$. This explains the designation "unskilled player."

example 46.8   Using the two-dimensional normal distribution, find the probability of a bull's-eye for

    (a) a skilled player with $\sigma = 1$

    (b) a skilled player with $\sigma = 2$

solution 46.8   In **(a)**, we must find $\displaystyle\iint_B \frac{1}{2\pi} e^{-(x^2+y^2)/2} \, dx \, dy$. Switching to polar coordinates, we compute

$$\int_0^{2\pi} \left(\int_0^1 \frac{1}{2\pi} e^{-r^2/2} r \, dr \right) d\theta = \int_0^{2\pi} \left(\frac{-1}{2\pi} e^{-r^2/2} \Big|_0^1 \right) d\theta$$

$$= \int_0^{2\pi} \frac{1}{2\pi} (1 - e^{-.5}) \, d\theta$$

The value of the last integral is $1 - e^{-.5} = .3935$. For the second player, with $\sigma = 2$, we must compute:

$$\iint_B \frac{1}{8\pi} e^{-(x^2+y^2)/8} \, dx \, dy = \int_0^{2\pi} \left(\int_0^1 \frac{1}{8\pi} e^{-r^2/8} r \, dr \right) d\theta$$

$$= \int_0^{2\pi} \left(\frac{-1}{8\pi} 4 e^{-r^2/8} \Big|_0^1 \right) d\theta$$

The last integral is $\int_0^{2\pi} \frac{1}{2\pi} (1 - e^{-.125}) \, d\theta = 1 - e^{-.125} = .1175$. This illustrates that the smaller the value of σ, the more skilled is the dart thrower. ▤

other interpretations   Many times in the physical and biological sciences, double integrals are formed but are not actually computed. In such cases, the purpose of the double integral is *to formulate with mathematical precision a concept that involves summation.* This is illustrated in the two examples that follow.

example 46.9   **measuring primary production in the ocean**   The **euphotic zone** consists of that portion of the seas where plant growth through photosynthesis can take place. The rate P at which photosynthesis takes place depends mainly on the amount of light that reaches depth x at a particular time of the day (see figure 46.12). This light intensity will itself vary with the time of day. If we let $t = 0$

figure 46.12

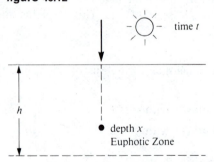

correspond to sunrise and $t = D$ correspond to sunset, then the rate P can be written as a function* of depth x and time t;

$$P = P(x, t), \qquad \text{for } 0 \le x \le h \text{ and } 0 \le t \le D$$

* For a discussion of the actual functional forms that have been proposed for $P(x, t)$, see
  T. Parsons and M. Takahashi, *Biological Oceanographic Processes*, New York: Pergamon Press, (1973). In particular, see pp. 78–81.

The *total daily gross photosynthesis* T over a typical square meter of the water column is given by

$$T = \int_0^D \left(\int_0^h P(x, t)\, dx \right) dt$$

Typical units for T are milligrams of carbon per day per square meter. We have summed production over both time and depth. ▤

example 46.10   **a general formula for laminar flow rate**   A fluid moves through the region R in the upward direction, as shown in figure 46.13. At the point (x, y) in the

figure 46.13

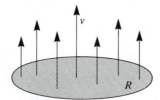

region R, the velocity in the vertical direction is $v = v(x, y)$. If we measure v in cm/sec, then

$$F = \int \int_R v(x, y)\, dx\, dy$$

measures the *flow rate F across R* in cm³/sec. This can be seen by referring to figure 46.14. In one second, the volume that flows through the small rectangle

figure 46.14

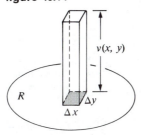

R_i of area $\Delta x\, \Delta y$ is approximately $v(x, y)\, \Delta x\, \Delta y$ cm³, where (x, y) is the center of the rectangle. The double integral has again added and refined the flow estimates across these small rectangles.

Using the formula for average value, we obtain

$$\bar{V} = \frac{1}{[\text{area } (R)] \int \int_R v(x, y)\, dx\, dy} = \frac{F}{[\text{area } (R)]}$$

Hence, we have shown that

$$\text{Flow rate} = (\text{Average velocity})(\text{Area})$$

This is a commonly used formula in fluid dynamics.   ▤

exercises for section 46

part A

Find the average value of each of the following functions $f(x, y)$ over the rectangle shown in the figure.

1. xy^2 **2.** $4x + 3y$ **3.** $y/(x + 1)$

Find the average value of each of the following functions over the disk shown in the figure.

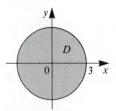

4. xy^2 **5.** $\dfrac{5}{x^2 + y^2 + 1}$ **6.** $e^{-\sqrt{x^2+y^2}}$

★**7.** The density function $\rho(x, y) = Ae^{-B\sqrt{x^2+y^2}}$ (in thousands per square mile) for the population of large cities was introduced in example 46.4. Derive a formula for the *average population density* over an area extending r_0 miles from the center of town.

8. Seventy-five hundred bears are uniformly distributed over a forest of area 500 square miles.

(a) What is the density function $\rho(x, y)$?

(b) Show that the expected number in a region A is 15 area (A).

9. For the density function in example 46.4, the values $A = 120$ and $B = .4$ have been given for Philadelphia in 1940. Estimate the number who lived within 2 miles of the center of the city at that time.

10. Seeds have been dispersed from a large tree onto the ground below. The density of the seeds (in numbers per square meter) is given by

$$\rho(x, y) = 12e^{-r^2/8}$$

where $r = \sqrt{x^2 + y^2}$ is the distance from the tree.

(a) Find the total number of seeds within 5 meters of the tree.

★**(b)** Find the total number of seeds dispersed.

11. Rework exercise 10**(a)**, assuming that $\rho(x, y) = 10/(x^2 + y^2 + 1)$.

12. Rework example 46.5, assuming that the grass biomass density function is

$$\rho(x, y) = 10(1 - e^{-.02x})e^{-.01y}$$

Given the rectangle R shown in the figure, find c so that each of the given functions $f(x, y)$ is a probability density function. Then compute the probability that (X, Y) is in the smaller rectangle A.

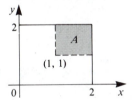

13. $f(x, y) = cx^2y$ **14.** $f(x, y) = c(x^2 + y^2)$

★**15.** Show that the volume bounded by the graph of $f(x, y) = \dfrac{1}{2\pi\sigma^2} e^{-(x^2+y^2)/(2\sigma^2)}$

and the xy-plane is 1. Conclude that $f(x, y)$ is a probability density function.

Shown in the figure is a dart board identical to the one in figure 46.11. Point scores are shown for the various regions.

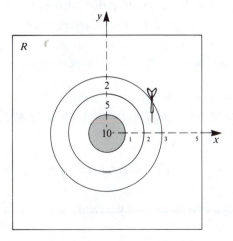

16. Find the probability that a completely unskilled dart thrower will throw the dart:

  (a) into the 5-point region.

  (b) outside the regions that give points.

17. Using the normal distribution with $\sigma = 1$, find the probability that:

  (a) points will be scored.

  (b) the dart will land in the 5-point region.

★**18.** Find σ so that the probability of a bull's-eye is $\frac{1}{2}$ when the two-dimensional normal distribution is used.

19. Water is flowing through the rectangle shown in the figure to the portion of space above the xy-plane. The upward velocity at point (x, y) is $v(x, y) = 4(1 - x^2)(1 - y^2)$ cm/sec.

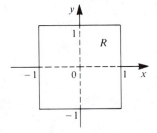 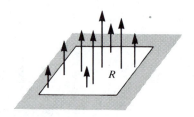

  (a) Find the flow rate F.

  (b) Find the average velocity $\bar{V}$.

20. When the *retina* receives light, it is believed that the excitation of receptors all around the center of the retina contributes to the excitation at the center $(0, 0)$. Let $f(x, y)$ be the excitation per unit area of a receptor at (x, y) and let r_0 be the radius of the circular area R.

  (a) What does $\int \int_R f(x, y)\, dx\, dy$ represent?

  (b) If we assume $f(x, y) = f(r, \theta) \doteq c/r^k$, where $0 < k < 2$, compute $\int \int_R f(x, y)\, dx\, dy$.

part B

21. A population starts at the origin $(0, 0)$ at time $t = 0$ and spreads out over the xy-plane. Let $f(x, y, t)$ be the probability density function for the population at time t. Assume

$$f(x, y, t) = \frac{1}{4\pi Dt}\, e^{-(x^2 + y^2)/(4Dt)}$$

This is a two-dimensional normal distribution with $\sigma^2 = 2Dt$.

(a) Verify that $f(x, y, t)$ satisfies the two-dimensional diffusion equation*

$$\frac{\partial f}{\partial t} = D\left(\frac{\partial^2 f}{\partial x^2} + \frac{\partial^2 f}{\partial y^2}\right)$$

(b) Find a formula for the proportion of the population that lives within r_0 of the origin at time t.

Given the region R shown in the accompanying figures, let $\rho(x, y)$ be the density function (in g/cm², for example). Then

$$T = \int\int_R \rho(x, y) \, dx \, dy$$

gives the total weight of the region. The center of mass $(\bar{x}, \bar{y})$ is that point

(a)   **(b)**

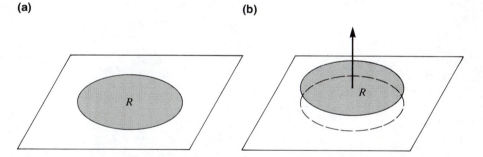

where, if a string were attached and pulled directly upward, the region would remain parallel to the xy-plane. The values of $\bar{x}$ and $\bar{y}$ may be found from the formulas

$$\bar{x} = \frac{\int\int_R x\rho(x, y) \, dx \, dy}{T}$$

$$\bar{y} = \frac{\int\int_R y\rho(x, y) \, dx \, dy}{T}$$

Find the center of mass of the pictured regions given the density function $\rho(x, y)$.

* For more information see, E. Pielou, *An introduction to mathematical ecology*. 2nd ed. New York: Wiley-Interscience, (1969), pp. 172.

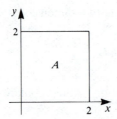 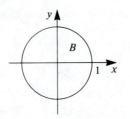 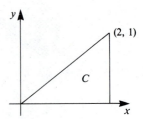

22. A, with $\rho(x, y) = c$ **23.** A, with $\rho(x, y) = x + y$

24. B, with $\rho(x, y) = c$ **25.** B, with $\rho(x, y) = x^2 + y^2$

26. C, with $\rho(x, y) = x^2 - 4y^2$ **27.** C, with $\rho(x, y) = c$

sequences, difference equations, and series

prologue   The models we have introduced thus far in the text have attempted to predict the state $f(t)$ of a given system for values of t that vary over an entire interval. The exponential growth model, for example, produced a function $N(t) = N_0 e^{kt}$ that predicted the number in a population *at any future time*. Such models are given the name **time-continuous models**.

Many systems are able to change only at a discrete set of times $t_1, t_2, \ldots, t_n, \ldots$. For example, a nerve cell or neuron does not fire continuously, but sends its message down the axon only when the total charge on the cell exceeds a certain threshold value. In other cases, we may be able to observe the changes only at selected times. The numbers in a salmon population can be estimated only when the salmon return to spawn in the rivers of their birth. At other times of the year, the population is hidden. A *discrete time-dependent model* would attempt to predict the number of salmon x_{n+1} in year $n + 1$ in terms of the prior year counts $x_1, x_2, \ldots, x_n$.

As a simple example, suppose that we follow the development of a bacteria culture arising from a single cell. Assume that the cell undergoes division every 20 minutes, and let x_n be the number of cells after n divisions have taken place. Then $x_{n+1} = 2x_n$. This equation is called a **difference equation**. Since $x_0 = 1$, it follows that $x_1 = 2x_0 = 2$, $x_2 = 2x_1 = 4, \ldots$, and so a **sequence** is generated. The general expression for x_n is 2^n, and $x_n = 2^n$ is called a **solution** to the difference equation.

We begin Part 6 with a discussion of sequences, the ways they can be generated, and methods for calculating their limits.

sequences and limits

A very popular pastime in Southern California during the summer months is watching the California grey whale. Since the whales return yearly to the warm Pacific ocean waters, their numbers can be estimated. However, it is not possible to keep track of the population at other times of the year. The population then reveals itself only at discrete moments of time. As time goes on, population data on the grey whale will take the form

$$w_1, w_2, w_3, w_4, \ldots, w_n, \ldots$$

where w_n is the population estimate during the summer of year n.

The data above form a **sequence**. From a mathematical point of view, a sequence is a function. For each natural number n, we have associated a unique real number w_n.

definition 47.1 A **real sequence** is a function f from the natural numbers $N = \{1, 2, 3, 4, \ldots\}$ into the real numbers $\mathcal{R}$. We write $a_n = f(n)$ and call a_n the nth *term of the sequence.*

This definition is illustrated in our next example.

example 47.1 Find the first ten terms of each of the following sequences.

$$\textbf{(a)}\ a_n = f(n) = \frac{2n}{n+1} \qquad \textbf{(b)}\ b_n = g(n) = (-1)^n \frac{1}{n}$$

solution 47.1 **(a)** If $a_n = \dfrac{2n}{n+1}$, then $a_1 = \dfrac{2 \cdot 1}{(1+1)} = 1$, $a_2 = \dfrac{2 \cdot 2}{2+1} = \dfrac{4}{3}$, etc.
The first ten terms of the sequence are then

$$1, \quad \frac{4}{3}, \quad \frac{6}{4}, \quad \frac{8}{5}, \quad \frac{10}{6}, \quad \frac{12}{7}, \quad \frac{14}{8}, \quad \frac{16}{9}, \quad \frac{18}{10}, \quad \frac{20}{11}$$

It appears that the terms are approaching 2 as n increases.
(b) In the sequence $b_n = (-1)^n\, 1/n$, the term $(-1)^n$ will alternately be -1 and 1. The first ten terms are then

$$-1, \quad \frac{1}{2}, \quad -\frac{1}{3}, \quad \frac{1}{4}, \quad -\frac{1}{5}, \quad \frac{1}{6}, \quad -\frac{1}{7}, \quad \frac{1}{8}, \quad -\frac{1}{9}, \quad \frac{1}{10}$$

Although the signs are changing, the terms b_n approach 0 as n increases. ▤

definition 47.2   We say that $\text{limit}_{n\to\infty} a_n = L$ if, as n increases, the terms a_n get closer and closer to a **single real number** L. We say that the sequence $\{a_n\}$ **converges** to L.

In the case of the two sequences in example 47.1, we can write:

$$\lim_{n\to\infty} \frac{2n}{n+1} = 2 \quad \text{and} \quad \lim_{n\to\infty} (-1)^n \frac{1}{n} = 0$$

The limit results of Section 6 can be used to deduce the limits of two important sequences. When $f(x)$ is a real-valued function defined for $x > 0$, we can define a sequence by $a_n = f(n)$. If $\text{limit}_{x\to+\infty} f(x) = L$, then the terms of the sequence must also approach L. Letting $f(x) = 1/x^k$ for $k > 0$, we may conclude:

$$\lim_{n\to\infty} \frac{1}{n^k} = 0 \qquad \text{for any } k > 0$$

When $a_n = r^n$ for a fixed real number r, we may write $r^n = e^{(\ln r)n}$. Thus, if $f(x) = e^{(\ln r)x}$, then $a_n = f(n)$. When $0 < r < 1$, $\ln r < 0$, and so $f(x)$ is an *exponential decay function*. Hence, we may conclude:

$$\lim_{n\to\infty} r^n = 0 \qquad \text{when} \quad 0 < r < 1$$

We will use these two basic limits in many of the examples that follow.

The procedures for computing $\text{limit}_{n\to\infty} a_n$ are very similar to the methods for finding $\text{limit}_{x\to+\infty} f(x)$ presented in Section 6. In fact, the **limit principles** summarized in the box are valid.

---

limit principles for sequences

Given that $\text{limit}_{n\to\infty} a_n = L$ and $\text{limit}_{n\to\infty} b_n = M$, then

1. $\lim_{n\to\infty} (a_n + b_n) = L + M$ **2.** $\lim_{n\to\infty} (a_n - b_n) = L - M$

3. $\lim_{n\to\infty} a_n b_n = LM$ **4.** $\lim_{n\to\infty} a_n/b_n = L/M$ provided $M \neq 0$

5. $\lim_{n\to\infty} \sqrt[m]{a_n} = \sqrt[m]{L}$ provided $\sqrt[m]{a_n}$ and $\sqrt[m]{L}$ are defined

example 47.2   Compute (if possible) the limits of each of the following sequences.

$$\textbf{(a)}\ a_n = (-1)^n \qquad \textbf{(b)}\ b_n = \frac{n^2}{n^2+1} \qquad \textbf{(c)}\ c_n = \frac{3n}{n+\sqrt{n}}$$

solution 47.2   **(a)** The terms in the sequence $a_n = (-1)^n$ alternate between -1 and 1, and therefore do not approach a *single* real number. Hence, limit a_n does not exist.
(b) To apply the limit principles to b_n, first multiply numerator and denominator by $1/n^2$:

$$\frac{n^2}{n^2+1} \cdot \frac{1/n^2}{1/n^2} = \frac{1}{1+(1/n^2)}$$

Hence, limit $b_n = \dfrac{1}{1 + \text{limit } 1/n^2} = \dfrac{1}{1+0} = 1$ using limit principles (1) and (4).
(c) For sequence c_n, multiply numerator and denominator by $1/n$:

$$\frac{3n}{n+\sqrt{n}} \cdot \frac{1/n}{1/n} = \frac{3}{1+(1/\sqrt{n})}$$

Since $\displaystyle\lim_{n\to\infty} \frac{1}{\sqrt{n}} = 0$, limit principles (4) and (1) give $\displaystyle\lim_{n\to\infty} c_n = 3$. ▤

example 47.3   Compute $\lim_{n\to\infty} (\sqrt{n^2+n} - n)$ using the limit principles.

solution 47.3   The expression $\sqrt{n^2+n} - n$ must be rewritten before the limit principles can be applied. The algebraic trick known as *rationalizing the numerator* must be used: (You can use your calculator and find terms like a_{100} to anticipate our final answer.)

$$\frac{\sqrt{n^2+n}-n}{1} \cdot \frac{\sqrt{n^2+n}+n}{\sqrt{n^2+n}+n} = \frac{n^2+n-n^2}{\sqrt{n^2+n}+n} = \frac{n}{\sqrt{n^2+n}+n} \cdot \frac{(1/n)}{(1/n)}$$

$$= \frac{1}{1+\sqrt{1+(1/n)}} \qquad \text{since } \sqrt{n^2+n} = n \cdot \sqrt{1+1/n}$$

Since limit $\sqrt{1+(1/n)} = \sqrt{1+\text{limit }(1/n)} = \sqrt{1+0} = 1$, the limit of the original sequence is $1/(1+1) = 1/2$. ▤

A final limit principle, known as the **Sandwich Principle**, is especially useful:

> **6.** If $|a_n| \le b_n$ and $\lim_{n\to\infty} b_n = 0$, then $\lim_{n\to\infty} a_n = 0$.

If $|a_n| \le b_n$, then $-b_n \le a_n \le b_n$. Since limit $b_n = 0$, limit $(-b_n)$ is also 0, and a_n is "sandwiched" between two sequences approaching zero, as illustrated in figure 47.1

figure 47.1

$$-b_n \qquad a_n \qquad 0 \qquad\qquad b_n$$

This principle is illustrated in our next example.

example 47.4   Find (if possible) the limits of the following sequences

(a) $a_n = 4 + \left(\dfrac{-1}{2}\right)^n$ (b) $b_n = \dfrac{\cos(n^3)}{n^4}$ (c) $c_n = (-1)^{n+1}\dfrac{n}{n+1}$

solution 47.4   (a) First note that $(-\frac{1}{2})^n = (-1)^n (\frac{1}{2})^n$. Since $|(-1)^n (\frac{1}{2})^n| = |(\frac{1}{2})^n| = (\frac{1}{2})^n$, and limit $(\frac{1}{2})^n = 0$, it follows that limit $(-\frac{1}{2})^n = 0$, by the Sandwich principle. Hence, limit $[4 + (-\frac{1}{2})^n] = 4 + 0 = 4$.
(b) Since $|\cos(n^3)| \le 1$, we have $|b_n| \le 1/n^4$. Again, by the Sandwich principle, limit $b_n = 0$ since limit $(1/n^4) = 0$.
(c) One is tempted to apply the Sandwich principle to c_n. Since limit $n/(n+1) = 1$, however, the principle does not apply. In fact, the term $(-1)^{n+1}$ causes the sequence to oscillate closer and closer to both -1 and 1, and not a *single* real number. Hence, limit c_n does not exist.   ▤

In our prior examples, we generated terms of a sequence by substituting particular values of n into an **explicit formula** $a_n = f(n)$. A sequence can also be defined **recursively**. In such a case, a_{n+1} is computed in terms of the prior sequence members $a_1, a_2, a_3, \ldots, a_{n-1}, a_n$ that have already been computed, as demonstrated in example 47.5.

example 47.5   Write down the first ten terms of the famous Fibonacci sequence defined by $a_1 = 1$, $a_2 = 1$, and $a_{n+1} = a_n + a_{n-1}$.

solution 47.5   Letting $n = 2$ in the formula $a_{n+1} = a_n + a_{n-1}$, we obtain $a_3 = a_2 + a_1 = 1 + 1 = 2$. Letting $n = 3$ yields $a_4 = a_3 + a_2 = 2 + 1 = 3$. In general, we find the next term by adding the two terms that precede it. Hence, we obtain

$$1, \quad 1, \quad 2, \quad 3, \quad 5, \quad 8, \quad 13, \quad 21, \quad 34, \quad 55 \quad ▤$$

definition 47.3   If a_{n+1} is computed solely in terms of a_n, then we write $a_{n+1} = f(a_n)$ and call this equation a **first order recursion** or a **first order difference equation**. When $a_{n+1} = f(a_n, a_{n-1})$, we have a **second order recursion** or a **second order difference equation**.

The Fibonacci sequence is then defined by a second order recursion. Note that in order to generate the terms of a sequence defined by such a recursion, we must first specify a_1 and a_2. In example 47.5, the conditions $a_1 = 1$ and $a_2 = 1$ are called **initial conditions**.

We now present three common first order difference equations.

example 47.6   **the geometric sequence**   For r a fixed real number, define the sequence a_n recursively by

> **hypothesis** $a_{n+1} = ra_n$

In most applications, it is convenient to regard the first term of the sequence as a_0 rather than a_1. Then a_0 can be thought of as the state of the system at time $t = 0$.

In terms of a_0, we can write $a_1 = ra_0$, $a_2 = ra_1 = r(ra_0) = r^2a_0$, $a_3 = ra_2 = r(r^2a_0) = r^3a_0$, etc. In general, we may conclude:

> **conclusion** $a_n = a_0r^n$

When $0 < r < 1$, the sequence decays to 0. For $r > 1$, the terms of the sequence increase exponentially. If, for example, a seasonally breeding population increases its numbers by 5% each year, then $a_{n+1} = 1.05a_n$. If the population presently has 1,000 members, then $a_0 = 1,000$ and so $a_n = 1,000(1.05)^n$.   ▤

example 47.7   **the arithmetic sequence**   If a population increases its numbers by a fixed number d each time period, then

> **hypothesis** $a_{n+1} - a_n = d$

Hence, $a_{n+1} = a_n + d$, and so, in terms of a_0, $a_1 = a_0 + d$, $a_2 = a_1 + d = (a_0 + d) + d = a_0 + 2d$, $a_3 = a_2 + d = (a_0 + 2d) + d = a_0 + 3d$, etc. In general, we may conclude:

> **conclusion** $a_n = a_0 + nd$

▤

example 47.8   **a discrete logistic equation**   The discrete analogues of population growth differential equations take the form

$$y_{n+1} = f(y_n)y_n$$

If $f(y)$ is a constant r, then $y_{n+1} = ry_n$ and we have the geometric sequence of example 47.6. In 1968, Maynard Smith[*] introduced a **discrete logistic equation**

$$y_{n+1} - y_n = ry_n\left(\frac{K - y_n}{K}\right)$$

or

$$y_{n+1} = y_n\left(1 + r\frac{K - y_n}{K}\right)$$

for positive constants r and K.

Note that if $y_0 = 0$, then $y_1 = 0$, and all subsequent terms are 0. Also, if $y_n = K$, then $y_{n+1} = K[1 + r(K - K)/K] = K$. Again, K is called the **carrying capacity**.

There is no known way to solve this recursion explicitly for y_n. To find terms in the sequence, we must specify y_0 and generate y_1, y_2, etc. Unlike the logistic function of section 35, the population count y_n *can oscillate around the carrying capacity* K in the discrete model. To see this, let $r = 2$, $K = 800$, and $y_0 = 450$. Then

$$y_{n+1} = y_n\left[3 - \left(\frac{y_n}{400}\right)\right]$$

Hence, $y_1 = 450(3 - 1.125) = 843.75$. Subsequent terms are shown in table 47.1.

table 47.1

n	0	1	2	3	4	5	6 ... 10
y_n **(rounded off)**	450	844	751	843	753	842	754 ... 756

This type of oscillation about K has been observed in population growth studies, as depicted in figure 47.2. ▤

In most cases it is difficult to find the actual limit of a sequence a_n that is defined recursively. It is not hard, however, to find the *possible limits* for a_n, as the final example demonstrates.

[*] J. Maynard Smith, *Mathematical ideas in biology.* London: Cambridge University Press, (1968).

figure 47.2 The logistic growth of two laboratory populations of the flour beetle, *Tribolium confusum*; one in 64 gm of flour (upper curve), and one in 16 gm (lower curve). (Adapted: W.C. Allee et al., *Principles of Animal Ecology*, Philadelphia: W.B. Saunders Co., 1949.)

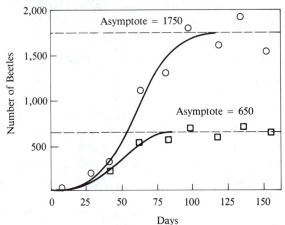

example 47.9   The sequence a_n is defined recursively by $a_1 = 1$ and $a_{n+1} = \sqrt{2 + a_n}$. Determine the possible limits for a_n.

solution 47.9   If $x = \text{limit}_{n \to \infty} a_n$ does indeed exist, then $\text{limit}_{n \to \infty} a_{n+1}$ is also x. Taking the limit of both sides of the recursion $a_{n+1} = \sqrt{2 + a_n}$ and applying the limit principles, we obtain $x = \sqrt{2 + x}$. Since $\sqrt{2 + x}$ denotes the positive square root, $x \geq 0$. Squaring both sides yields $x^2 = 2 + x$ or $x^2 - x - 2 = (x - 2)(x + 1) = 0$. Hence, $x = 2$.

We have *not* shown that $\text{limit}_{n \to \infty} a_n = 2$. It is easy to convince yourself that 2 is the actual limit by examining the first few terms of the sequence:

$$1, \quad 1.73205, \quad 1.93186, \quad 1.98288, \quad 1.99572, \quad 1.99893, \quad \ldots \qquad ■$$

exercises for section 47

part A   Find the first ten terms of each of the following sequences.

1. $a_n = n/(2n + 1)$ **2.** $b_n = n/(n^2 + 1)$ **3.** $c_n = (-1)^n (\tfrac{1}{3})^n$

4. $d_n = 4 - n$ **5.** $t_n = (-2)^{-n}$

Compute (if possible) $\text{limit}_{n \to \infty} a_n$ for each of the following sequences.

6. $a_n = \dfrac{n}{n^2 + 1}$ **7.** $a_n = \dfrac{\sqrt{n + 1}}{n}$ **8.** $a_n = \dfrac{3n^2 - n + 1}{n^2 - 2}$

9. $a_n = \dfrac{4 - n}{2n + 3}$

10. $a_n = \dfrac{n}{n + \sqrt{n}}$

11. $a_n = 3e^{-n}$

12. $a_n = \left(\dfrac{3}{\pi}\right)^n$

13. $a_n = (-3)^n$

14. $a_n = \dfrac{n}{(n + 1)(n + 2)}$

15. $a_n = 5^{1/n}$

16. $a_n = \sqrt{n + 1} - \sqrt{n}$

17. $a_n = \sqrt{n^2 + 1} - n$

18. $a_n = \sqrt{\dfrac{n}{n + 1}}$

19. $a_n = \sqrt[3]{\dfrac{2 - n}{(2 + n)}} - 2$

20. $a_n = \left(\dfrac{1}{n}\right)^n$

21. $a_n = \dfrac{-1^n}{2^n}$

22. $a_n = \dfrac{\sin n}{n}$

23. $a_n = \dfrac{(-1)^n}{n^5}$

24. $a_n = (-1)^n \dfrac{2n}{n + 1}$

25. $a_n = (-1)^n \dfrac{\cos n}{n + 2}$

Find the first five terms of the sequences defined recursively by:

26. $a_{n+1} = a_n + 2a_{n-1}$ with $a_1 = 1$ and $a_2 = 0$

27. $a_{n+1} = 3 - a_n$, with $a_1 = 2$

28. $a_{n+1} = 2a_n$, with $a_1 = 1/2$

29. $a_{n+1} = a_n/(1 + a_n)$, with $a_1 = 1$

30. $a_{n+1} = a_n[2 - (a_n/20)]$, with $a_1 = 20$

31. A herd of elk in a game preserve increases its numbers by 11% each year. Let x_n be the number in the population at the end of year n.

   (a) Find the difference equation relating x_{n+1} to x_n.

   (b) Solve for x_n if $x_0 = 300$.

32. The probability that an adult buffalo survives the year is .9. Suppose that a herd starts with 100 members, and let x_n be the number of original buffalo still alive after n years.

   (a) Write the difference equation relating x_{n+1} and x_n.

   (b) Solve for x_n and compute x_5.

33. A savings account gives an effective annual yield of 13.5%. Let x_n be the amount (in dollars) in the account after n years.

   (a) Relate x_{n+1} to x_n.

   (b) Determine when $x_n \geq 10{,}000$ given that $x_0 = 4{,}000$.

34. The body eliminates 20% of the amount of drug present each hour. Let x_n be the amount of drug (in milligrams) in the body n hours after being administered.

   (a) Relate x_{n+1} to x_n.

   (b) Find x_n if $x_0 = 200$. Then compute x_{12}.

35. A geometric sequence has $a_3 = 25$ and $a_5 = 100$. Find the general term a_n.

36. An arithmetic sequence has $a_3 = 25$ and $a_5 = 100$. Find the general term a_n.

37. A warehouse stocks stereo systems. Four hundred are removed each day, while 150 arrive each day from the factory. Let x_n denote the number in inventory at the beginning of the nth day of the year.

(a) Relate x_{n+1} to x_n.

(b) If $x_1 = 10{,}000$, find x_n and determine when $x_n = 0$.

38. The discrete population growth difference equation

$$y_{n+1} = \lambda y_n e^{-\alpha y_n}$$

where $\alpha > 0$ and $\lambda > 0$, was introduced in 1965 by L.M. Cook.*

(a) If $y_0 = (\ln \lambda)/\alpha$, show that $y_n = y_0$ for all n.

(b) If $\lambda = 3$, $\alpha = .002$, and $y_0 = 500$, compute $y_1, y_2, \ldots, y_5$.

39. The discrete population growth difference equation

$$y_{n+1} = \lambda y_n^{1-b}$$

where $\lambda > 0$ and $b > 0$, is the discrete analogue of the Gompertz growth equation discussed in example 33.8.

(a) If $y_0 = (\lambda)^{b^{-1}}$, show that $y_n = y_0$ for all n.

(b) If $\lambda = 2$, $b = .1$, and $y_0 = 900$, compute $y_1, y_2, \ldots, y_{10}$.

40. In the discrete logistic equation for example 47.8, show that if $y_\infty = \lim_{n \to \infty} y_n$ exists, then $y_\infty = 0$ or K.

41. The sequence x_n is defined recursively by $x_0 = 1$ and $x_{n+1} = (x_n/2) + (1/x_n)$.

(a) Show that if $x_\infty = \lim_{n \to \infty} x_n$ exists, then $x_\infty = \pm\sqrt{2}$.

(b) Compute the first five terms of the sequence.

42. The sequence x_n is defined recursively by $x_{n+1} = \sqrt{3x_n - 2}$.

(a) If $x_\infty = \lim_{n \to \infty} x_n$ exists, show that $x_\infty = 1$ or 2.

(b) If $x_0 = 1.1$, find the first five terms of the sequence.

(c) If $x_0 = 3$, find the first five terms of the sequence.

★**43.** Let a_n denote the Fibonacci sequence of example 47.5, and define $r_n = a_{n+1}/a_n$.

(a) Compute the first five terms of the sequence r_n.

(b) Find $\lim_{n \to \infty} r_n$, assuming that the limit actually exists.

part B   Discrete analogues of differential equations of the form $y' = g(y)$ can be found using the approximation

$$y' = \frac{dy}{dt} \approx \frac{\Delta y}{\Delta t}$$

---

* L.M. Cook, *Nature*, London, **207**, p. 316.

If we let $y_n = y(n)$ and $\Delta y = y_{n+1} - y_n$, then $\Delta y/\Delta t = y_{n+1} - y_n \approx g(y_n)$. The resulting difference equation is then given by

$$y_{n+1} = y_n + g(y_n)$$

44. Find the discrete analogue of the logistic differential equation

$$y' = \frac{r}{K} y(K - y)$$

45. Find the discrete analogue of the limited growth differential equation

$$y' = r(K - y)$$

46. Find a differential equation corresponding to the difference equation

$$y_{n+1} = \beta y_n^2$$

 first order difference equations

A **first order difference equation** takes the form $x_{n+1} = f(x_n)$. Once x_0 is specified, the terms of the sequence can be generated:

$$x_1 = f(x_0), \qquad x_2 = f(x_1), \qquad x_3 = f(x_2), \qquad \ldots$$

Our goal in this section, however, is to present several of the common methods for finding **explicit expressions** for x_n. When $x_{n+1} = rx_n$, we have shown that $x_n = x_0 r^n$. This explicit expression for x_n is called the *general solution* to the difference equation.

Solving difference equations can be a formidable task. There is one class of difference equations whose explicit solutions are known. The **first order linear difference equation** takes the form

$$x_{n+1} = a_n x_n + b_n$$

where a_n and b_n are given sequences. Fortunately, this special difference equation occurs frequently in applications. Our line of attack in solving this equation will closely parallel that found in Section 37 on the first order linear differential equation

$$x' = \alpha(t)x + \beta(t)$$

the homogeneous linear difference equation   When $b_n = 0$ for all n, the difference equation is called **homogeneous** and takes the form

$$x_{n+1} = a_n x_n$$

Given x_0, subsequent terms in the sequence are

$$x_1 = a_0 x_0$$
$$x_2 = a_1 x_1 = a_0 a_1 x_0$$
$$x_3 = a_2 x_2 = a_0 a_1 a_2 x_0$$

etc.

Thus, if we write $A(n) = a_0 a_1 a_2 \cdots a_n$, then the general solution may be written as $x_n = (a_0 a_1 a_2 \cdots a_{n-1})x_0 = A(n-1)x_0$. To summarize,

> The general solution to $x_{n+1} = a_n x_n$ is
>
> $$x_n = A(n-1)x_0$$
>
> where $A(n) = a_0 a_1 a_2 \cdots a_n$.

Note that when $a_n = a$ for all n, the solution takes the form of the familiar geometric sequence $x_n = x_0 a^n$. This formula is applied in our next example.

example 48.1   Solve the difference equation $x_{n+1} = \dfrac{n+1}{n+2} x_n$ subject to the initial condition $x_0 = 2$.

solution 48.1   Note that $a_0 = \frac{1}{2}$, $a_1 = \frac{2}{3}$, $a_2 = \frac{3}{4}$, $a_3 = \frac{4}{5}$, etc. Hence, $A(1) = a_0 a_1 = \frac{1}{2} \cdot \frac{2}{3} = \frac{1}{3}$, $A(2) = a_0 a_1 a_2 = \frac{1}{3} \cdot \frac{3}{4} = \frac{1}{4}$, $A(3) = a_0 a_1 a_2 a_3 = a_3 A(2) = \frac{1}{5}$, and so on. The general formula for $A(n)$ is then $A(n) = 1/(n+2)$. Hence, we can use the formula for the general solution to obtain

$$x_n = A(n-1)x_0 = \frac{2}{n+1}$$

Note that the limit of x_n is 0.

the linear difference equation with constant coefficients   When the sequences a_n and b_n are both constant, the linear difference equation takes the form $x_{n+1} = ax_n + b$. This equation is the analogue of the differential equation $x' = ax + b$ of Section 37. We may find the general solution by performing the following procedure.

Step 1. Solve the homogeneous difference equation $x_{n+1} = ax_n$. The solution takes the form ca^n.

Step 2. Construct one particular solution p_n to the original equation $x_{n+1} = ax_n + b$.

Step 3. Write the general solution as $x_n = ca^n + p_n$.

To see why this procedure is valid, let p_n and q_n be two solutions to $x_{n+1} = ax_n + b$, and let $y_n = q_n - p_n$. Then

$$y_{n+1} = q_{n+1} - p_{n+1} = (aq_n + b) - (ap_n + b)$$
$$= a(q_n - p_n) = ay_n$$

Thus, y_n is a solution to the homogeneous equation, and so $y_n = ca^n$ for some constant c. Hence, $q_n = ca^n + p_n$.

How then can a particular solution be found? Guided by our experience with linear differential equations, we will try a constant particular solution $p_n = A$, where A is to be determined. If p_n is to be a solution, then we must have

$$p_{n+1} = A = ap_n + b = aA + b$$

Hence, provided $a \neq 1$, $A = b/(1 - a)$. Thus, $p_n = b/(1 - a)$ is a particular solution.

If $a = 1$, then $x_{n+1} = x_n + b$. This is an **arithmetic sequence** and we have seen in example 47.7 that $x_n = x_0 + nb$. To summarize,

The general solution to $x_{n+1} = ax_n + b$ is

$$x_n = \begin{cases} ca^n + \dfrac{b}{1 - a}, & \text{for } a \neq 1 \\ x_0 + nb, & \text{for } a = 1 \end{cases}$$

This general formula is illustrated in our next example, and is then used in the construction of three important discrete models.

example 48.2   Find the solution to the difference equation $x_{n+1} = .4x_n + 3$ that satisfies the initial condition $x_0 = 100$. What is the limit of x_n?

solution 48.2   Since $a = .4$ and $b = 3$, $b/(1 - a) = 5$. Hence $x_n = c(.4)^n + 5$. When $n = 0$, $x_0 = 100 = c + 5$. Thus, $c = 95$ and so $x_n = 95(.4)^n + 5$. Then $\lim_{n \to \infty} x_n = 95(0) + 5 = 5$.   ▤

example 48.3   Sergio deposits \$10,000 in a high interest account that pays 12% annually. At the end of each year, however, he withdraws \$1,500 and throws a party for his friends. Find the amount in the account at the end of year n. How long can these parties go on?

solution 48.3   If we let x_n be the amount in the account at time n, then

$$x_{n+1} = x_n + (\text{interest}) - (\text{withdrawal})$$
$$= x_n + .12x_n - 1,500$$

Hence, $a = 1.12$, $b = -1,500$, and so $b/(1 - a) = -1,500/(-.12) = 12,500$. Thus, $x_n = c(1.12)^n + 12,500$. Since $x_0 = 10,000$, it follows that $10,000 = c + 12,500$. The final formula for x_n is $x_n = -2,500(1.12)^n + 12,500$.

One answer to the final question is "all night long."

But, we really want to determine the time n when the money will run out.

Note that $x_n < 0$ when $12{,}500 < 2{,}500\,(1.12)^n$, that is, when $(1.12)^n > 5$. Taking natural logarithms of both sides yields $n > \ln 5/\ln 1.12 = 14.20$. Now, $x_{14} = \$282.22$ and $x_{15} < 0$. Note that $(1.12)x_{14} = \$316.09$ will still be left for the last party. Sergio will buy pretzels and beer! ▤

example 48.4   **a model for harvesting a seasonally breeding population**   An animal population reproduces once a year. We will assume that reproduction takes place at times $n = 0,\ 1,\ 2,\ldots$, and that we count the population before reproduction. In our model, the population changes due to births, deaths, and harvesting by man. Hence, the change in the population over a 1-year period is given by

$$x_{n+1} - x_n = (\text{number of births}) - (\text{number of deaths}) - (\text{number harvested})$$

Next we make two assumptions:
   1. A fixed number h are harvested each year and harvesting takes place shortly after reproduction.
   2. The numbers of births and deaths between times n and $(n+1)$ are directly proportional to x_n.

Hence, $x_{n+1} = bx_n - dx_n - h = (b - d)x_n - h$, where b and d are constants of proportionality. Assuming $b = 2$, $d = .8$, and $h = 2{,}500$, solve the resulting difference equation and predict the long-term effects of this harvesting policy if $x_0 = 10{,}000$.

solution 48.4   The difference equation takes the form

$$x_{n+1} = (1.2)x_n - 2{,}500$$

Hence, $x_n = c(1.2)^n + (2{,}500)/.2 = c(1.2)^n + 12{,}500$. Since $x_0 = 10{,}000$, it follows that $c = -2{,}500$ and so the final form for x_n is $-2{,}500(1.2)^n + 10{,}000$.

The remainder of the solution is similar to that of example 48.3. Note that $x_n < 0$ when $(1.2)^n > 5$, that is, when $n > \ln 5/\ln 1.2 = 8.82$. With this harvesting policy, the population will be extinct in 9 years. Shown in table 48.1 are the values of x_n before extinction.

table 48.1

n	1	2	3	4	5	6	7	8
x_n	9,500	8,900	8,180	7,316	6,279	5,035	3,542	1,750

▤

example 48.5   **designing a drug dosage scheme**   Let $x_0 = b$ be the initial amount of a drug introduced into the body. If no additional doses are added, the drug will gradually be eliminated. We will assume that the amount of drug remaining in the body at time t is given by $x(t) = be^{-kt}$ for some $k > 0$. There is much experimental evidence to support this assumption for many common drugs.

If we decide to administer additional doses of b units every τ hours, we are designing a **drug-dosage scheme**. Let x_n be the amount of drug in the body when the nth dosage has been given. Letting $a = e^{-k\tau}$, we have

$$x_{n+1} = (\text{drug remaining}) + (\text{new dosage})$$
$$= ax_n + b$$

Since $x_0 = b$, it follows that $x_n = \dfrac{b}{1-a}(1 - a^{n+1})$. (Check the details.) Since $0 < a < 1$, we can make two conclusions:

1. $x_\infty = \underset{n \to \infty}{\text{limit }} x_n = \dfrac{b}{1-a} = \dfrac{b}{(1 - e^{-k\tau})}$

2. $x_n = x_\infty(1 - a^{n+1})$ is always less than x_∞. Hence, the *amount of drug in the body never exceeds* $x_\infty = b/(1 - a)$.

A typical graph of $x(t)$ is shown in figure 48.1. Suppose that we measure time

figure 48.1

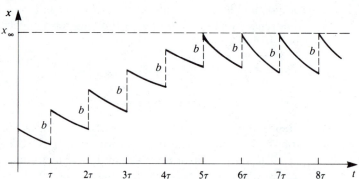

in hours and that $k = .05$. If a level of more than 1,000 milligrams of the drug in the body is considered unsafe and if individual doses are 200 milligrams, determine how frequently the drug can be safely administered.

solution 48.5

Here, $a = e^{-.05\tau}$ and $b = 200$. Since the maximum drug level is $b/(1-a) = 200/(1 - e^{-.05\tau})$, we must guarantee that

$$\frac{200}{1 - e^{-.05\tau}} < 1,000$$

Hence, $e^{-.05\tau} < \frac{4}{5}$. Taking natural logarithms and solving, we obtain $\tau > \ln .8/(-.05) = 4.46$. Hence, administering the drug at time intervals of less than 4.46 hours will eventually result in a dangerously high level of the drug. ∎

The difference equation $x_{n+1} = x_n/(1 + x_n)$ has applications in *population genetics*. Although this equation is not linear, it can easily be transformed into a linear difference equation:

$$\frac{1}{x_{n+1}} = \frac{1 + x_n}{x_n} = 1 + \frac{1}{x_n}$$

If we let $y_n = 1/x_n$, then $y_{n+1} = 1 + y_n$. Using the formula on page 536 we obtain $y_n = y_0 + n$, where $y_0 = 1/x_0$. Thus, we conclude:

$$x_n = \frac{x_0}{1 + nx_0}$$

This sequence is important in applications because it slowly converges to 0. This is illustrated in example 48.6.

example 48.6 Solve the difference equation $x_{n+1} = x_n/(1 + x_n)$ subject to the initial condition $x_0 = 2$. Compute x_n for $n = 10, 20, 30, 40,$ and 50.

solution 48.6 From the previously derived formula, $x_n = 2/(1 + 2n)$. Substituting values of n into this formula, we have the results shown in table 48.2. For an application of this difference equation to genetics, see part B exercise 34.

table 48.2

n	0	10	20	30	40	50
x_n	2	.095	.049	.033	.025	.020

The difference equation $x_{n+1} = x_n + b_n$ is a linear difference equation that we have not yet considered. If $x_0 = 0$, then

$$x_1 = x_0 + b_0 = b_0$$
$$x_2 = x_1 + b_1 = b_0 + b_1$$
$$x_3 = x_2 + b_2 = b_0 + b_1 + b_2$$

etc.

In general, $x_n = b_0 + b_1 + b_2 + \cdots + b_{n-1}$. Now what is $\text{limit}_{n \to \infty} x_n$? To compute this limit, we must determine when the *infinite series*

$$\sum_{n=0}^{\infty} b_n = b_0 + b_1 + b_2 + \cdots + b_n + \cdots$$

converges. We will turn to this topic in Section 50.

exercises for section 48

part A

Solve each of the following homogeneous linear difference equations subject to the given boundary condition.

1. $x_{n+1} = \dfrac{n}{n+1} x_n$, with $x_0 = 3$

2. $x_{n+1} = (n+1)x_n$, with $x_0 = 1$
[*Note*: The solution is commonly written as $n!$ ("n factorial").]

3. $x_{n+1} = \dfrac{1}{n+1} x_n$, with $x_0 = 2$

4. $x_{n+1} - x_n = \frac{1}{2}x_n$, with $x_0 = 1$

5. $x_{n+1} - x_n = \dfrac{1}{n} x_n$, with $x_1 = 2$

Find the general solution to each of the following linear difference equations.

6. $x_{n+1} - 3x_n = 5$

7. $x_{n+1} = 2x_n + 2$

8. $x_{n+1} = x_n - 4$

9. $x_{n+1} + 2x_n = 4 - x_n$

10. $3x_{n+1} - 2x_n + 1 = 0$

Solve each of the following linear difference equations subject to the given boundary condition.

11. $x_{n+1} = .5x_n + 4$, with $x_0 = 10$. What is $\lim_{n \to \infty} x_n$?

12. $x_{n+1} - x_n = .3x_n - 10$, with $x_0 = 20$

13. $x_{n+1} - x_n = 4$, with $x_0 = 100$

14. $x_{n+1} = 3 - x_n$, with $x_0 = 0$

15. $x_{n+1} = (1.1)x_n - 200$, with $x_0 = 1,000$. When is $x_n < 0$?

16. $x_{n+1} - x_n = .1x_n + 10$, with $x_0 = 4$. When is $x_n > 100$?

17. $x_{n+1} = x_n - 300 + .2x_n$, with $x_0 = 2,000$

18. $x_{n+1} = .8x_n + h$, with $x_0 = 1,000$. Find h so that $\lim x_n = 2,000$.

★19. $x_{n+1} = 3x_n^2$, with $x_0 = 1$. [*Hint*: Let $y_n = \ln x_n$.]

★20. $x_{n+1} = 2x_n^{\frac{1}{n}}$, with $x_0 = 900$. What is $\lim_{n \to \infty} x_n$? [*Hint*: Let $y = \ln x_n$.]

21. Every year Jimmy's parents deposit $500 into a special account for his college education. The account pays at an annual interest rate of 8%. If the first $500 was placed in the account on his first birthday, find the expression for x_n, the amount in the account when Jimmy turns n years old. Find x_{18}.

22. To build up a population of trout in a small lake, 200 young trout are added each year. In addition, the population increases its own numbers by 20% each year. Let x_n denote the size of the population after n years.

   (a) If $x_0 = 1,200$, determine when $x_n \geq 2,800$.

   ★**(b)** Once $x_n \geq 2,800$, the lake is no longer stocked and fishermen will catch 600 fish per year. What is the fate of the population?

23. Refer to example 48.3. Now suppose that Sergio withdraws $1,300 at the end of each year for his New Year's Party, and that the account is set up as before.

   (a) Find the amount in the account at time n.

   (b) For how many years can these parties be held?

   (c) Rework (a) and (b), assuming the withdrawal is only $1,200.

24. A population of buffalo can increase its numbers by about 15% each year. Let x_n be the population count after n years, and assume that h buffalo are removed from the herd at the end of each year.

   (a) Find x_n in terms of h if $x_0 = 1,000$.

   (b) Find the largest h so that $x_{10} \geq 2,000$.

25. A population of pheasants lives on a small island. Because of limited resources, the island can sustain no more that K individuals. The population growth during the year, $\Delta x_n = x_{n+1} - x_n$, can be expected to decrease as x_n approaches K. Assuming that Δx_n is directly proportional to $K - x_n$, find x_n given that $x_0 < K$. What is $\text{limit}_{n \to \infty} x_n$?

26. The monthly payment on a mortgage loan of P dollars is denoted by R, and the *interest rate per month* is called i. Let x_n be the amount owed after n payments.

   (a) Show that $x_{n+1} = (1 + i)x_n - R$.

   (b) Solve for x_n given that $x_0 = P$.

   (c) Let N be the total number of payments to be made. Determine R so that $x_N = 0$.

   (d) A 30-year loan ($N = 360$) is taken out at 1% interest per month. Find the monthly payment R if $P = \$50,000$.

27. The body eliminates 10% of a certain drug each hour. Suppose that doses of 200 milligrams are given every 6 hours.

   (a) Find the maximum drug level in the body.

   (b) Find the amount of drug in the body 24 hours after the first dose.

   (c) How frequently should the drug be administered to build up a maximum drug level of 1,000 milligrams?

28. A level of more than 1,500 milligrams of a certain drug in the body is considered unsafe. Individual doses are 250 milligrams, and the drug is removed from the body according to the exponential decay equation

$$x(t) = x_0 e^{-.1t}$$

where t is measured in hours. How frequently can the drug be safely administered?

29. Often, in administering a drug to a patient, an initial *load dose* of $x_0 = A$ is given. This dose is larger than the subsequent doses of size b. Let x_n be the amount of drug in the body after the nth dose has been administered, and, as in example 48.5, let

$$a = e^{-k\tau}$$

   (a) Solve the resulting difference equation for x_n, using the initial condition $x_0 = A$.

   (b) Find $x_\infty = \lim_{n \to \infty} x_n$.

   (c) If $A \geq b/(1 - a)$, show that the maximum amount of drug in the body is $x_0 = A$.

   ★(d) If $A < b/(1 - a)$, show that the maximum drug level is $b/(1 - a)$. [*Hint*: Let $f(t) = ca^t + b/(1 - a)$ and show that $f'(t) > 0$. Conclude that $x_n = f(n) < x_{n+1} = f(n + 1)$.]

30. Approximately 100 new cases of a rare disease arise each year. Through drug therapy, about 25% of all individuals with the affliction are cured each year. Let x_n denote the number with the disease after n years.

   (a) Find and solve the difference equation relating x_{n+1} to x_n given that $x_0 = 700$.

   (b) Compute $\lim_{n \to \infty} x_n$.

★31. The charge on a nerve cell (*neuron*) is increased by 1 millivolt every 2 milliseconds. Individual charges decay exponentially according to the formula

$$x(t) = x_0 e^{-.05t}$$

where t is measured in milliseconds. Thus, if x_0 is the present charge on the cell, the charge remaining after 2 milliseconds is

$$x(2) = x_0 e^{-.1} \text{ millivolts}$$

Let x_n be the charge on the cell after $2n$ milliseconds.

   (a) Show that $x_{n+1} = (e^{-1})x_n + 1$, and solve for x_n if $x_0 = 0$.

   (b) The "all or nothing" law asserts that the neuron will fire as soon as the total charge on the cell exceeds a certain threshold value. If the neuron fires as soon as the charge exceeds 4 millivolts, how frequently will the neuron fire?

part B

32. The discrete Gompertz population growth difference equation

$$y_{n+1} = \lambda y_n^{1-b}$$

where $\lambda > 0$ and $b > 0$, was introduced in exercise 39 of Section 47.

(a) Letting $x_n = \ln y_n$, solve the difference equation for x_n when $b \neq 1$.

(b) Find y_n and determine when $y_\infty = \lim_{n \to \infty} y_n$ exists.

★**(c)** When is the approach to y_∞ oscillatory?

33. **(a)** Find the general solution to the nonhomogeneous difference equation

$$y_{n+1} = ay_n + bc^n$$

when $c \neq a$. [*Hint*: Construct a particular solution of the form $p_n = Ac^n$.]

(b) Solve the equation $x_{n+1} - x_n = \left(\frac{1}{2}\right)^n$ given that $x_0 = 10$.

34. The genetic disease *cystic fibrosis* is a severe respiratory ailment that usually results in death before the individual reaches adulthood. The disease is caused by the presence of two recessive "c" genes. If C denotes the normal dominant form of the gene, then individuals of genotypes CC and Cc are normal, and children of genotype cc result from a match between two Cc individuals. (We will ignore the possibility of a gene mutation.) Let p_n be the frequency of the c gene in generation n, and suppose that there are N individuals in the population.

The Hardy-Weinberg principle (see Section 61) asserts that the number of CC, Cc, and cc individuals is given by

$$N(1 - p_n)^2, \qquad 2Np_n(1 - p_n), \qquad \text{and } Np_n^2$$

respectively. The Np_n^2 individuals of genotype cc are soon lost to the population. The new proportion of Cc individuals is then

$$\frac{2Np_n(1 - p_n)}{N[(1 - p_n)^2 + 2p_n(1 - p_n)]} = \frac{2p_n}{1 + p_n}$$

(a) Conclude that $p_{n+1} = p_n/(1 + p_n)$ and solve for p_n in terms of p_0.

(b) Determine the number of generations needed to reduce p_n from .02 to .002.

 **second order difference equations**

A **second order difference equation** takes the form $x_{n+1} = f(x_n, x_{n-1})$. In order to generate the sequence defined by this recursion, x_0 and x_1 must be specified. Then

$$x_2 = f(x_1, x_0), \qquad x_3 = f(x_2, x_1), \qquad x_4 = f(x_3, x_2), \ldots$$

Hence, *initial conditions* will take the form $x_0 = a$ and $x_1 = b$.

The **second order linear difference equation** is a second order difference equation that can be written in the standard form

$$x_{n+1} + \alpha_1(n)x_n + \alpha_2(n)x_{n-1} = \beta(n)$$

where $\alpha_1(n)$, $\alpha_2(n)$, and $\beta(n)$ are given sequences. The methods for solving this equation are entirely similar to the techniques introduced in Section 38 for the second order linear differential equation

$$y'' + \alpha_1(t)y' + \alpha_2(t)y = \beta(t)$$

In this section we will follow the format of Section 38 and solve various special cases of this second order equation. Applications will be given to population biology.

the homogeneous equation with constant coefficients

Suppose first that $\beta(n) \equiv 0$ and that $\alpha_1(n)$ and $\alpha_2(n)$ are constant sequences. By clearing out fractions if necessary, we can rewrite this difference equation in the form

$$ax_{n+1} + bx_n + cx_{n-1} = 0$$

In example 47.6 we saw that a first order equation such as $bx_n + cx_{n-1} = 0$ will have a solution of the form $x_n = r^n$ where $r = -c/b$. This suggests that we try a geometric sequence as a solution in our present setting:

$$ax_{n+1} + bx_n + cx_{n-1} = ar^{n+1} + br^n + cr^{n-1}$$
$$= r^{n-1}(ar^2 + br + c)$$

Hence, r^n will be a solution provided $ar^2 + br + c = 0$. The polynomial $ar^2 + br + c$ is again called the **characteristic polynomial** of the difference equation.

example 49.1   Find geometric sequence solutions to each of the following difference equations.

$$\text{(a) } x_{n+1} - x_n - 2x_{n-1} = 0 \qquad \text{(b) } 3x_{n+1} - 7x_n + 2x_{n-1} = 0$$

solution 49.1   For **(a)**, r^n is a solution provided $r^2 - r - 2 = (r+1)(r-2) = 0$. Hence, $r = -1$ and $r = 2$, and so the geometric sequences $(-1)^n$ and 2^n are solutions.

For **(b)**, observe that the characteristic polynomial of $3x_{n+1} - 7x_n + 2x_{n-1} = 0$ is $3r^2 - 7r + 2 = (3r - 1)(r - 2)$. The roots are $r = \frac{1}{3}$ and $r = 2$, and hence $(\frac{1}{3})^n$ and 2^n are solutions.

definition 49.1   Two nonzero solutions $x_1(n) = z_n$ and $x_2(n) = w_n$ to the difference equation $ax_{n+1} + bx_n + cx_{n-1} = 0$ are called **linearly independent solutions** provided w_n is not a constant multiple of z_n. We then call z_n and w_n a **fundamental set of solutions**.

Thus, in example 49.1(a), $(-1)^n$ and 2^n are linearly independent and form a fundamental set of solutions. We will now show that any sequence of the form $c_1(-1)^n + c_2 2^n$ is also a solution.

Note that if $x_1(n) = z_n$ and $x_2(n) = w_n$ are solutions, then $x(n) = c_1 x_1(n) + c_2 x_2(n) = c_1 z_n + c_2 w_n$ is again a solution for any choice of constants c_1 and c_2:

$$ax_{n+1} + bx_n + cx_{n-1} = a(c_1 z_{n+1} + c_2 w_{n+1}) + b(c_1 z_n + c_2 w_n) + c(c_1 z_{n-1} + c_2 w_{n-1})$$
$$= c_1(az_{n+1} + bz_n + cz_{n-1}) + c_2(aw_{n+1} + bw_n + cw_{n-1})$$
$$= c_1 \cdot 0 + c_2 \cdot 0 = 0$$

Thus, in example 49.1(a), $x_n = c_1(-1)^n + c_2 2^n$ is a solution. Are there any others? The key theorem of this section asserts that we have found *all solutions* to the difference equation $x_{n+1} - x_n - 2x_{n-1} = 0$.

theorem 49.1   Let $x_1(n) = z_n$ and $x_2(n) = w_n$ be two nonzero linearly independent solutions to $ax_{n+1} + bx_n + cx_{n-1} = 0$. Then the general solution is given by $x_n = c_1 z_n + c_2 w_n$ where c_1 and c_2 are arbitrary constants.

The proof of the theorem is beyond the level of the text. The theorem is illustrated in our next example.

example 49.2   Find the general solution to the difference equation $x_{n+1} - 3x_n + 2x_{n-1} = 0$. Then determine the solution that satisfies the initial conditions $x_0 = 1$ and $x_1 = 3$.

solution 49.2   The characteristic polynomial $r^2 - 3r + 2 = (r-1)(r-2)$ has roots $r = 1$ and

$r = 2$. Therefore, $x_1(n) = (1)^n = 1$ and $x_2(n) = 2^n$ are solutions. Since these solutions are linearly independent, the general solution is $x_n = c_1 + c_2 2^n$.

Now, $x_0 = 1 = c_1 + c_2$ and $x_1 = 3 = c_1 + 2c_2$. It follows that $c_1 = -1$ and $c_2 = 2$. Hence, $x_n = -1 + 2 \cdot 2^n = -1 + 2^{n+1}$. ▤

If the characteristic polynomial $ar^2 + br + c = 0$ has *only one real root* r_1, then $x_1(n) = r_1^n$ is the only solution produced. When this occurred in the second order linear differential equation, we multiplied the single solution $y_1(t)$ by t to obtain a second linearly independent solution $y_2(t) = ty_1(t)$. In a like manner, it is not hard to show that $x_2(n) = nx_1(n) = nr_1^n$ is a second solution to the original second order difference equation. Details are outlined in part B exercise 46.

The results accumulated thus far are summarized in table 49.1.

table 49.1 Solutions to $ax_{n+1} + bx_n + cx_{n-1} = 0$

Roots of $ar^2 + br + c$	Fundamental Solutions	General Solution
1. distinct real roots r_1 and r_2	r_1^n and r_2^n	$c_1 r_1^n + c_2 r_2^n$
2. a single real root r_1	r_1^n and nr_1^n	$c_1 r_1^n + c_2 n r_1^n$

Example 49.3 illustrates the case of a single root r_1.

example 49.3 Solve the difference equation $4x_{n+1} - 4x_n + x_{n-1} = 0$ subject to the initial conditions $x_0 = 1$ and $x_1 = \frac{3}{4}$.

solution 49.3 The characteristic polynomial $4r^2 - 4r + 1 = (2r - 1)^2$ has the single root $r = \frac{1}{2}$. Hence, the general solution takes the form $x_n = c_1(\frac{1}{2})^n + c_2 n(\frac{1}{2})^n$. Applying the initial conditions, we obtain $x_0 = 1 = c_1 + 0 \cdot c_2 = c_1$. Then $x_n = (\frac{1}{2})^n + c_2 n(\frac{1}{2})^n$ and $x_1 = \frac{3}{4} = \frac{1}{2} + \frac{1}{2}c_2$. Hence, $c_2 = \frac{1}{2}$. The final solution is then

$$x_n = \left(\frac{1}{2}\right)^n + \frac{1}{2}n\left(\frac{1}{2}\right)^n = \left(\frac{1}{2}\right)^n \left(1 + \frac{n}{2}\right)$$ ▤

In attempting to solve the difference equation $x_{n+1} + x_{n-1} = 0$, we obtain $r^2 + 1$ as the characteristic polynomial, and hence $r = \pm i$ are the complex roots. As was the case in Section 38, the solutions now involve the trigonometric functions. If we try $x_n = \cos n\theta$ as a solution, and make use of the identity

$$\cos(x + y) = \cos x \cos y - \sin x \sin y$$

we obtain

$$x_{n+1} + x_{n-1} = \cos(n+1)\theta + \cos(n-1)\theta = \cos(n\theta + \theta) + \cos(n\theta - \theta)$$
$$= (\cos n\theta \cos \theta - \sin n\theta \sin \theta) + (\cos n\theta \cos(-\theta) - \sin n\theta \sin(-\theta))$$
$$= 2 \cos n\theta \cos \theta$$

Hence, $x_n = \cos n\theta$ will be a solution provided $\cos \theta = 0$. Setting $\theta = \pi/2$, we see that $x_n = \cos(n\pi/2)$ is one solution. Likewise, we can show that $\sin(n\pi/2)$ is a second solution.

When $r = \alpha + \beta i$ and $r = \alpha - \beta i$ are the complex roots of the characteristic polynomial, it can be shown that $x_1(n) = \rho^n \cos n\theta$ and $x_2(n) = \rho^n \sin n\theta$ will be solutions provided:

$$\rho^2 = \alpha^2 + \beta^2 \qquad \text{and} \qquad \cos \theta = \frac{\alpha}{\rho}$$

The details are outlined in part B exercise 47. These relationships can be remembered from figure 49.1. The results are summarized in table 49.2.

figure 49.1

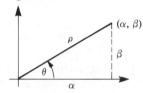

table 49.2   Solutions to $ax_{n+1} + bx_n + cx_{n-1} = 0$

Roots of $ar^2 + br + c$	Fundamental Solutions	General Solution
3. complex roots $\alpha + \beta i$ and $\alpha - \beta i$	$\rho^n \cos n\theta$ and $\rho^n \sin n\theta$ where $\rho^2 = \alpha^2 + \beta^2$ and $\cos \theta = \dfrac{\alpha}{\rho}$	$\rho^n(c_1 \cos n\theta + c_2 \sin n\theta)$

The next example illustrates the case of complex roots.

example 49.4   Find the general solution to the difference equation $x_{n+1} - x_n + \frac{1}{2}x_{n-1} = 0$. Then find the solution that satisfies $x_0 = 1$ and $x_1 = 0$.

solution 49.4   We solve $r^2 - r + \frac{1}{2} = 0$ by applying the quadratic formula:

$$r = \frac{1 \pm \sqrt{1 - 4(1/2)}}{2} = \frac{1 \pm \sqrt{-1}}{2} = \frac{1}{2} \pm \frac{1}{2}i$$

Hence, $\alpha = \beta = \frac{1}{2}$ and so $\rho^2 = \alpha^2 + \beta^2 = \frac{1}{4} + \frac{1}{4} = \frac{1}{2}$ and $\cos \theta = \alpha/\rho = 1/\sqrt{2}$. We may select $\theta - \pi/4$ as a solution. The general solution is given by

$$x_n = \left(\frac{1}{\sqrt{2}}\right)^n \left[c_1 \cos\left(\frac{n\pi}{4}\right) + c_2 \sin\left(\frac{n\pi}{4}\right) \right]$$

We apply the initial conditions to obtain $x_0 = 1 = c_1 \cos 0 + c_2 \sin 0 = c_1$ and $x_1 = 0 = (1/\sqrt{2}) [\cos (\pi/4) + c_2 \sin (\pi/4)]$. Thus, $c_2 = -\tan (\pi/4) = -1$. The final solution is then

$$x_n = \left(\frac{1}{\sqrt{2}}\right)^n \left[\cos \left(\frac{n\pi}{4}\right) - \sin \left(\frac{n\pi}{4}\right)\right]$$

applications to population biology   Second order difference equations have many applications in the areas of population biology and population genetics. The applications to genetics will be presented in Section 61 after we have discussed the relationship between genetics and probability theory. For now we present two models for the growth of populations.

example 49.5

A seasonally breeding population is censused each year. If we let x_n be the population count at time $t = n$, then $\Delta x_n = x_{n+1} - x_n$ is the growth (or decline) during year n. Assuming that $\Delta x_n = b(\Delta x_{n-1})$ for some $b > 0$, solve the resulting difference equation and examine the long-range growth of the population.

solution 49.5

We are given that $x_{n+1} - x_n = b(x_n - x_{n-1})$, which may be written in the standard form $x_{n+1} - (1 + b)x_n + bx_{n-1} = 0$. The characteristic polynomial $r^2 - (1 + b)r + b$ factors as $(r - 1)(r - b)$. Hence, $x_n = c_1 \cdot 1^n + c_2 b^n = c_1 + c_2 b^n$, provided $b \neq 1$. If $b = 1$, the solution takes the form $x_n = c_1 + c_2 n$.

The constant c_2 may be evaluated in terms of Δx_0, the growth during the first year. If $b \neq 1$, then $x_0 = c_1 + c_2$ and $x_1 = c_1 + c_2 b$, from which it follows that $c_2 = \Delta x_0/(b - 1)$. When $b = 1$, $x_0 = c_1$ and $x_1 = c_1 + c_2$. Hence, $c_2 = \Delta x_0$.

Suppose first that $\Delta x_0 > 0$. If $b > 1$, then $c_2 = \Delta x_0/(b - 1) > 0$ and the population increases exponentially. For $0 < b < 1$, $c_2 < 0$ and $\text{limit}_{n \to \infty} x_n = c_1 = x_0 - c_2 > 0$. Thus, the population approaches a finite limit. If $b = 1$, $x_n = x_0 + (\Delta x_0)n$ form an arithmetic sequence and will steadily increase by Δx_0 units each time period.

The case $\Delta x_0 < 0$ will be examined in exercise 40.

example 49.6

an age structure model   The members of an animal population are divided into three categories: newborns, yearlings, and adults. We will assume that once a female reaches the age of two years, she is an adult and will reproduce. The model might be appropriate for certain large mammals.

We will census the population immediately after its annual reproductive period, and define

$$x_n = \text{number of adults at time } n$$
$$y_n = \text{number of yearlings at time } n$$
$$\text{and } z_n = \text{number of newborns at time } n$$

We make the following additional assumptions:

    1. There are equal numbers of males and females in each age category.

    2. Fifty percent of the newborns survive to age one.

    3. Seventy-five percent of the yearlings survive to age two.

    4. Ninety percent of the adults will survive the year.

    5. An adult female gives birth once a year to a single newborn.

Hence, from (2), we have $y_{n+1} = .5z_n$. Now, x_{n+1} is the number of surviving yearlings plus the number of surviving adults, or $x_{n+1} = .75y_n + .90x_n$. Finally, $z_n = .5x_n$, since half of the adults are females who have just given birth. It follows that

$$x_{n+1} = .75y_n + .90x_n = .75(.5z_{n-1}) + .90x_n$$
$$= .75(.5)(.5)x_{n-1} + .90x_n$$

Rearranging into standard form, we obtain:

$$x_{n+1} - .90x_n - .1875x_{n-1} = 0$$

If an animal preserve starts out with 100 adults, find x_n, y_n, and z_n.

solution 49.6

The characteristic polynomial $r^2 - .90r - .1875$ has roots

$$r = \frac{.9 \pm \sqrt{.81 + .75}}{2}$$

or $r = 1.0745$ and $r = -.1745$. Hence, $x_n = c_1(1.0745)^n + c_2(-1)^n (.1745)^n$. Since $x_0 = 100$, we have $x_1 = .90(100) = 90$. Hence, $100 = c_1 + c_2$ and $90 = 1.0745c_1 - .1745c_2$. This system has solution $c_1 = 86.0288$ and $c_2 = 13.9712$. Thus:

$$x_n = 86.0288(1.0745)^n + 13.9712(-1)^n(.1745)^n$$
$$z_n = .5x_n$$

and
$$y_n = .5z_{n-1} = .25x_{n-1}$$

To determine the population levels over the first few years, it is easiest to use the original difference equation with $x_0 = 100$ and $x_1 = 90$ and generate terms. Resulting values of x_n, y_n, and z_n are shown in table 49.3. For $n \geq 5$, the term $|13.9712(-1)^n(.1745)^n|$ is very small. Hence, $x_n \approx 86.0288(1.0745)^n$. Eventually, the population will increase at the rate of about 7.45% per year. Our solution is convenient for evaluating x_n for large n. When $n = 20$, $x_n = 86.0288(1.0745)^{20} \approx 362$.

table 49.3

n	x_n (adults)	y_n (yearlings)	z_n (newborns)
1	100	0	50
1	90	25	45
2	100	23	50
3	107	25	54
4	115	27	57
5	123	29	62

the nonhomogeneous linear difference equation   If p_n and q_n are two solutions to the nonhomogeneous equation

$$ax_{n+1} + bx_n + cx_{n-1} = \beta(n)$$

then $y_n = q_n - p_n$ will satisfy the homogeneous equation:

$$\begin{aligned} ay_{n+1} + by_n + cy_{n-1} &= a(q_{n+1} - p_{n+1}) + b(q_n - p_n) + c(q_{n-1} - p_{n-1}) \\ &= (aq_{n+1} + bq_n + cq_{n-1}) - (ap_{n+1} + bp_n + cq_{n-1}) \\ &= \beta(n) - \beta(n) = 0 \end{aligned}$$

Hence, $q_n = y_n + p_n$, where p_n is a particular solution and $ay_{n+1} + by_n + cy_{n-1} = 0$. To find the general solution to the nonhomogeneous equation, follow the steps given in the accompanying box.

---

procedure for obtaining solutions to $ax_{n+1} + bx_n + cx_{n-1} = \beta(n)$

Step 1. Solve the homogeneous difference equation $ax_{n+1} + bx_n + cx_{n-1} = 0$. Let $x_1(n) = z_n$ and $x_2(n) = w_n$ be the solutions.

Step 2. Construct *one particular solution* p_n to $ax_{n+1} + bx_n + cx_{n-1} = \beta(n)$.

Step 3. Form $x_n = c_1 z_n + c_2 w_n + p_n$. This is the *general solution*.

This step-by-step procedure is practiced in our next example.

example 49.7   Find the general solution to the difference equation $3x_{n+1} - 7x_n + 2x_{n-1} = 4$.

solution 49.7   In example 49.1**(b)**, we showed that the general solution to the homogeneous equation was $c_1(\tfrac{1}{3})^n + c_2 2^n$. For a particular solution, let's try a constant sequence $p_n = A$:

$$3p_{n+1} - 7p_n + 2p_{n-1} = 3A - 7A + 2A = -2A$$

Hence, set $A = -2$ for a particular solution. The general solution is then $x_n = c_1(\frac{1}{3})^n + c_2 2^n - 2$.

Particular solutions can often be constructed by using the **method of undetermined coefficients**. Shown in table 49.4 are two commonly occurring functions $\beta(n)$ and the corresponding trial solutions. As was the case with differential equations, the capital letters in the table denote *constants to be determined*.

table 49.4

The Method of Undetermined Coefficients
If $\beta(n)$ is of the form: Try a solution p_n of the form:

1. a constant d A
2. $dn + e$ $An + B$

If the trial solution fails, multiply the table entry by n and try again.

Table 49.4 is used to construct particular solutions for the difference equations in our final two examples.

example 49.8   Construct particular solutions to each of the following nonhomogeneous difference equations.

$$\text{(a) } 3x_{n+1} - 4x_n + x_{n-1} = 5 \qquad \text{(b) } x_{n+1} - 3x_n + x_{n-1} = 3n - 2$$

solution 49.8   **(a)** If we set $p_n = A$, then $3p_{n+1} - 4p_n + p_{n-1} = 3A - 4A + A = 0$. Hence, $p_n = A$ will not give a particular solution. Following the directions in table 49.4, we try $p_n = An$:

$$3p_{n+1} - 4p_n + p_{n-1} = 3A(n+1) - 4An + A(n-1)$$
$$= 2A$$

Hence, we can set $A = 5/2$. Then $p_n = 5n/2$ is a particular solution.

(b) Setting $p_n = An + B$, we obtain $p_{n+1} - 3p_n + p_{n-1} = [A(n+1) + B] - 3(An + B) + [A(n-1) + B] = -An - B$. For $p_{n+1} - 3p_n + p_{n-1} = 3n - 2$, set $A = -3$ and $B = 2$. Thus, $p_n = -3n + 2$ is a particular solution.

example 49.9   If left undisturbed, a fish population in a small lake will grow according to the difference equation $x_{n+1} = \frac{1}{6}x_n + \frac{1}{3}x_{n-1}$. To build up the stocks for fishermen, 1,000 fish are added each year. Thus,

$$x_{n+1} = \frac{1}{6}x_n + \frac{1}{3}x_{n-1} + 1,000$$

If $x_0 = 0$ and $x_1 = 5,000$, find x_n and compute $\lim_{n \to \infty} x_n$.

solution 49.9   The homogeneous equation $x_{n+1} - \frac{1}{6}x_n - \frac{1}{3}x_{n-1} = 0$ has characteristic polynomial $r^2 - \frac{1}{6}r - \frac{1}{3} = \frac{1}{6}(6r^2 - r - 2) = \frac{1}{6}(3r - 2)(2r + 1)$. Hence, the homogeneous equation has solution $c_1(\frac{2}{3})^n + c_2(-\frac{1}{2})^n = c_1(\frac{2}{3})^n + c_2(-1)^n (\frac{1}{2})^n$.

For a particular solution, let $p_n = A$. Then

$$A = \frac{1}{6}A + \frac{1}{3}A + 1{,}000$$

Hence, $\frac{1}{2}A = 1{,}000$ or $A = 2{,}000$. Thus, $x_n = c_1(\frac{2}{3})^n + c_2(-1)^n (\frac{1}{2})^n + 2{,}000$.

Applying the initial conditions yields $0 = x_0 = c_1 + c_2 + 2{,}000$ and $5{,}000 = x_1 = \frac{2}{3}c_1 - \frac{1}{2}c_2 + 2{,}000$. This system has solution $c_1 = 12{,}000/7$ and $c_2 = -26{,}000/7$. Note that $\lim_{n \to \infty} x_n = 2{,}000$, independent of c_1 and c_2.   ▤

exercises for section 49

part A   Find two fundamental solutions to each of the following second order linear difference equations.

1. $x_{n+1} - x_n - 2x_{n-1} = 0$ **2.** $x_{n+1} - 3x_n - 10x_{n-1} = 0$

3. $6x_{n+1} + x_n - 2x_{n-1} = 0$ **4.** $2x_{n+1} + x_n - x_{n-1} = 0$

5. $4x_{n+1} - x_{n-1} = 0$ **6.** $x_{n+1} - x_{n-1} = 0$

7. $x_{n+1} + 4x_{n-1} = 0$ **8.** $x_{n+1} + 2x_n + 2x_{n-1} = 0$

9. $4x_{n+1} - 4x_n + x_{n-1} = 0$ **10.** $9x_{n+1} + 12x_n + 4x_{n-1} = 0$

Find the general solution to each of the following linear difference equations. Then find the solution that satisfies $x_0 = 1$ and $x_1 = 2$.

11. $x_{n+1} - 2x_n + x_{n-1} = 0$ **12.** $x_{n+1} - 4x_n - 12x_{n-1} = 0$

13. $x_{n+1} - x_{n-1} = 0$ **14.** $x_{n+1} + x_{n-1} = 0$

15. $2x_{n+1} + x_n - x_{n-1} = 0$

Find the solution to each of the following linear difference equations that satisfies the specified boundary conditions.

16. $x_{n+1} - x_n - 2x_{n-1} = 0$, with $x_0 = 1$ and $x_1 = -1$

17. $x_{n+1} - 8x_n + 16x_{n-1} = 0$, with $x_0 = 3$ and $x_1 = 8$

18. $x_{n+1} + 2x_n + 2x_{n-1} = 0$, with $x_0 = 0$ and $x_2 = 1$

19. $x_{n+1} + 2x_n + 5x_{n-1} = 0$, with $x_0 = 1$ and $x_2 = 4$

20. $x_{n+1} - 2x_{n-1}$, with $x_0 = 1$ and $x_1 = 0$

Verify that the given sequence p_n is a particular solution to the given non-homogeneous difference equation. Then find the general solution.

21. $x_{n+1} + 4x_n - 5x_{n-1} = 6$ and $p_n = n$

22. $x_{n+1} - 3x_{n-1} = 2$ and $p_n = -1$

23. $x_{n+1} + 3x_n + 2x_{n-1} = -12n + 8$ and $p_n = -2n + 1$

24. $x_{n+1} + x_{n-1} = 5(2^{n-1})$ and $p_n = 2^n$

25. $4x_{n+1} - 4x_n + x_{n-1} = 16$ and $p_n = 16$

Use the method of undetermined coefficients to construct particular solutions to each of the following nonhomogeneous linear difference equations.

26. $3x_{n+1} - 2x_n - 2x_{n-1} = 4$ **27.** $3x_{n+1} - 2x_n + 2x_{n-1} = 6$

28. $x_{n+1} - x_{n-1} = 1$ **29.** $x_{n+1} - 3x_n + 2x_{n-1} = 6n$

30. $7x_{n+1} - 4x_n - 5x_{n-1} = 2n - 3$

Find the solution to each of the following nonhomogeneous difference equations that satisfies the boundary conditions specified.

31. $x_{n+1} - x_n - 2x_{n-1} = -10$, with $x_0 = 5$ and $x_1 = 2$

32. $4x_{n+1} - x_{n-1} = -18$, with $x_0 = 0$ and $x_1 = -5$

33. $x_{n+1} + x_{n-1} = n$, with $x_0 = 1$ and $x_1 = 2$

34. $x_{n+1} - x_{n-1} = 2n$, with $x_0 = 0$ and $x_1 = 4$

35. $x_{n+1} + 2x_n + 2x_{n-1} = 2$, with $x_0 = 1$ and $x_1 = -2$

36. In a model for weight loss on diet, w_n denotes the weight after n weeks on the diet, and $\Delta w_n = w_{n+1} - w_n$ is the change in weight during the week $(n + 1)$. Assume that $\Delta w_n = .9 \Delta w_{n-1}$, $w_0 = 180$, and $w_1 = 175$.

   (a) Find w_n and compute $w_\infty = \text{limit}_{n \to \infty} w_n$.

   (b) When is $w_n \leq 150$?

37. Trout are to grown in a pond and then harvested. If w_n denotes the total biomass of trout after n months, then $\Delta w_n = w_{n+1} - w_n$ is the growth in weight during month $(n + 1)$. If $\Delta w_n = 1.2 \Delta w_{n-1}$, find w_n and determine w_{12}, the total biomass after one year, given that $w_0 = 200$ pounds and $w_1 = 220$ pounds.

★**38.** The dieter in exercise 36 follows her diet except on Sundays. On this day, she feasts and gains 2 pounds. Find and solve the new difference equation for w_n. How will her eating habits on Sunday affect the eventual outcome of the diet?

★**39.** The manager of the trout farm in exercise 37 takes 30 pounds of trout from the pond at the end of each month to sell in his fish market. Find and solve the new difference equation for w_n and determine w_{12}.

40. In example 49.5, find x_n when $\Delta x_0 = x_1 - x_0 < 0$ and determine when limit $x_n > 0$.

41. A model for the growth of an insect population divides the colony into three age categories: egg larva (0–1 month), pupa (1–2 months), and adult (2–3 months). We will assume the population changes according to the following rules:

(1) Only 1% of the eggs survive to the pupa stage.

(2) About 20% of the pupae survive to the adult stage. At age two months, each adult will reproduce.

(3) Each adult gives rise to 1,000 eggs, and all adults die within the next month.

Let x_n be the number of adults after n months, and assume that the population is counted after reproduction.

(a) Show that x_n satisfies the difference equation $x_{n+1} - 2x_{n-1} = 0$.

(b) Find x_n if a population is started with 100 adults. Describe how the adult population changes.

42. In a model for the growth of a buffalo population, the females in the herd are divided into newborns, yearlings, and adults (2 years of age and older). Sixty percent of the newborns survive their first year, while 75% of the yearlings become adults. An adult female will survive the year with probability .95. Each year about 40 female calves will be born for every 100 adult females.

Let x_n be the number of adult females after n years, and assume that the population is counted after reproduction.

(a) Show that $x_{n+1} = .95x_n + .18x_{n-1}$.

(b) A herd is started with 100 females, 40 newborn female calves, and an appropriate number of adult males. Use the difference equation to find the herd structure over the next 5 years.

(c) Determine x_n for large n by solving the difference equation subject to $x_0 = 100$ and $x_1 = 95$.

43. (a) Find the explicit solution to the Fibonacci sequence defined by

$$x_{n+1} = x_n + x_{n-1} \qquad \text{with } x_0 = x_1 = 1$$

(b) Females in a rabbit population mature after 2 months and will give birth to a single female baby. Assuming that all rabbits survive, show that $x_{n+1} = x_n + x_{n-1}$ where x_n is the number of adult females after n months. Census the population after reproduction.

44. Rework exercise 43, assuming that two female babies are produced each month for each adult female rabbit.

(a) Use the difference equation to find x_n for $n = 1, 2, \ldots, 10$.

(b) If $x_0 = x_1 = 1$, solve the difference equation for x_n and describe the growth of the population for large n.

45. Solve the system of difference equations

$$x_{n+1} = 2x_n + 6y_n$$
$$y_{n+1} = -2x_n - 5y_n$$

subject to $x_0 = 1$ and $y_0 = 2$.

[*Hint*: Follow the line of attack presented in section 39 to find a second order linear difference equation satisfied by x_n.]

part B

46. The linear difference equation $x_{n+1} - 2ax_n + a^2x_{n-1} = 0$ has characteristic polynomial $(r - a)^2$.

(a) Show that $x_n = u_n a^n$ is a solution if and only if

$$u_{n+1} - 2u_n + u_{n-1} = 0$$

(b) Write $u_{n+1} - 2u_n + u_{n-1} = (u_{n+1} - u_n) - (u_n - u_{n-1})$ and let $y_n = u_{n+1} - u_n$. Conclude $y_n = c$, for some constant c.

(c) Solve $y_n = u_{n+1} - u_n = c$ for u_n. Conclude that $x_n = na^n$ is a second solution to the original difference equation.

47. This exercise requires a knowledge of complex number arithmetic and De Moivre's Theorem:

$$[\rho (\cos \theta + i \sin \theta)]^n = \rho^n (\cos n\theta + i \sin n\theta)$$

Suppose $r = \alpha \pm i\beta$ are the complex roots to the characteristic polynomial $ar^2 + br + c$. Write $\alpha \pm i\beta$ in *polar form* (see the figure):

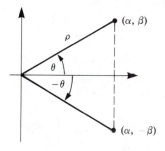

$$\alpha + i\beta = \rho (\cos \theta + i \sin \theta)$$
$$\alpha - i\beta = \rho [\cos (-\theta) + i \sin (-\theta)]$$

(a) If $r_1 = \alpha + i\beta$, show that $r_1^n = \rho^n \cos n\theta + i\rho^n \sin n\theta$.

(b) If $r_2 = \alpha - i\beta$, show that $r_2^n = \rho^n \cos n\theta - i\rho^n \sin n\theta$.

(c) Conclude that $\frac{1}{2}r_1^n + \frac{1}{2}r_2^n = \rho^n \cos n\theta$ is a real-valued solution to $ax_{n+1} + bx_n + cx_{n-1} = 0$.

(d) Conclude that $(1/2i)r_1^n - (1/2i)r_2^n = \rho^n \sin n\theta$ is a second real-valued solution.

50 infinite series and convergence

Given an infinite sequence $a_1, a_2, a_3, \ldots, a_n, \ldots$, we can begin to add more and more terms of the sequence:

$$S_1 = a_1$$
$$S_2 = a_1 + a_2$$
$$S_3 = a_1 + a_2 + a_3$$
$$\cdot$$
$$\cdot$$
$$\cdot$$
$$S_n = a_1 + a_2 + a_3 + \cdots + a_n$$

The term S_n is called the **nth partial sum** of the series. When we let $n \to \infty$, we are attempting to sum the infinite series.

definition 50.1 Given an infinite sequence a_n, we say that the *infinite series* $\sum\limits_{n=1}^{\infty} a_n$ **converges** if the sequence of partial sums $S_n = a_1 + a_2 + \cdots + a_n$ has a finite limit. If $S = \lim\limits_{n \to \infty} S_n$, we write

$$S = \sum_{n=1}^{\infty} a_n$$

and call S the **sum of the infinite series**. If $\lim\limits_{n \to \infty} S_n$ does not exist, we say that the **infinite series** $\sum\limits_{n=1}^{\infty} a_n$ **diverges**.

In the next two examples, we will determine whether the given series converges by first finding an explicit expression for the nth partial sum S_n.

example 50.1 Examine the infinite series $\sum\limits_{n=1}^{\infty} (-1)^{n+1} = 1 - 1 + 1 - 1 + 1 - \cdots$ for convergence.

solution 50.1 Since $S_1 = 1$, $S_2 = 1 - 1 = 0$, $S_3 = 1 - 1 + 1 = 1$, $S_4 = 1 - 1 + 1 - 1 = 0$, etc., the sequence of partial sums S_n is:

$$1, \quad 0, \quad 1, \quad 0, \quad 1, \quad 0, \quad 1, \quad \ldots$$

Hence, $\lim\limits_{n \to \infty} S_n$ does not exist, and so $\sum\limits_{n=1}^{\infty} (-1)^{n+1}$ diverges.

example 50.2   Examine the sequence of partial sums of $\displaystyle\sum_{n=1}^{\infty} \frac{1}{n(n+1)}$ to show that $\displaystyle\sum_{n=1}^{\infty} \frac{1}{n(n+1)} = 1$.

solution 50.2   We must first find a formula for S_n. Since $\dfrac{1}{[n(n+1)]} = \dfrac{1}{n} - \dfrac{1}{(n+1)}$, the original infinite series can be rewritten as:

$$\sum_{n=1}^{\infty} \left(\frac{1}{n} - \frac{1}{n+1}\right) = \left(1 - \frac{1}{2}\right) + \left(\frac{1}{2} - \frac{1}{3}\right) + \left(\frac{1}{3} - \frac{1}{4}\right) + \cdots$$

Hence, $S_1 = 1 - \frac{1}{2}$, $S_2 = 1 - \frac{1}{2} + \frac{1}{2} - \frac{1}{3} = 1 - \frac{1}{3}$, $S_3 = 1 - \frac{1}{4}$, and in general, $S_n = 1 - 1/(n+1)$. Hence, $S = \text{limit}_{n\to\infty} S_n = 1$.

One of the most important series in mathematics is the geometric series

$$a + ar + ar^2 + ar^3 + \cdots + ar^{n-1} + \cdots$$

formed from the geometric sequence $a_n = ar^{n-1}$. To decide when the series converges, we will use the factorization formula

$$x^n - 1 = (x - 1)(1 + x + x^2 + \cdots + x^{n-1})$$

Hence,

$$1 + x + x^2 + \cdots + x^{n-1} = \frac{x^n - 1}{x - 1} = \frac{1 - x^n}{1 - x}$$

Applying this formula to $S_n = a + ar + \cdots + ar^{n-1}$, we obtain

$$S_n = a(1 + r + r^2 + \cdots + r^{n-1}) = \frac{a(1 - r^n)}{(1 - r)}$$

In example 47.4 we established that $\text{limit}_{n\to\infty} r^n = 0$ when $-1 < r < 1$. Hence, we may conclude:

> The geometric series $\displaystyle\sum_{n=1}^{\infty} ar^{n-1}$ converges when $-1 < r < 1$. The sum is given by:
>
> $$S = \frac{a}{1 - r}$$

When $|r| \geq 1$, $\text{limit}_{n\to\infty} S_n$ does not exist, and so $\sum_{n=1}^{\infty} ar^{n-1}$ diverges. This new formula is illustrated in our next example.

example 50.3

Find the sums of each of the following geometric series.

(a) $5 + \dfrac{5}{2} + \dfrac{5}{4} + \dfrac{5}{8} + \cdots + \dfrac{5}{2^{n-1}} + \cdots$ (b) $1 - \dfrac{1}{3} + \dfrac{1}{9} - \dfrac{1}{27} + \cdots + (-1)^{n-1}\dfrac{1}{3^{n-1}} + \cdots$

solution 50.3

(a) Let $a = 5$ and $r = \frac{1}{2}$ in the formula for the sum of a geometric series. Hence $S = 5/(1 - \frac{1}{2}) = 10$.

(b) We can write $(-1)^{n-1}1/(3^{n-1}) = (-\frac{1}{3})^{n-1}$. Setting $a = 1$ and $r = -\frac{1}{3}$, we have $S = 1/[1 - (-\frac{1}{3})] = 1/(\frac{4}{3}) = \frac{3}{4}$.   ▤

Examples 50.1–50.3 are not typical. We determined whether an infinite series converged by finding an *explicit formula* for S_n. In most cases, this is not possible. Instead, we will address ourselves to two basic questions:

Question 1. How can we determine whether a given infinite series converges or diverges?

Question 2. If the infinite series converges, how many terms should be added to obtain the sum S to a desired number of decimal places? (See page 574, section 51.)

a simple test for divergence   If $S_n = a_1 + a_2 + \cdots + a_n$, note that $S_n - S_{n-1} = (a_1 + a_2 + \cdots + a_{n-1} + a_n) - (a_1 + a_2 + \cdots + a_{n-1}) = a_n$. If $\Sigma_{n=1}^{\infty} a_n$ is convergent, then $S = \lim_{n \to \infty} S_n$ exists. Since S_{n-1} will also approach S as n increases, it follows that

$$\lim_{n \to \infty} a_n = \lim_{n \to \infty} S_n - \lim_{n \to \infty} S_{n-1} = S - S = 0$$

Hence, we have a test for divergence, as summarized in the box.

---

test for divergence

If $\displaystyle\sum_{n=1}^{\infty} a_n$ converges, then $\lim_{n \to \infty} a_n = 0$. Hence, if $\lim_{n \to \infty} a_n$ is *not* 0, then $\displaystyle\sum_{n=1}^{\infty} a_n$ must diverge.

---

example 50.4

Show that the infinite series $\displaystyle\sum_{n=1}^{\infty} \dfrac{n}{n + 4}$ diverges.

solution 50.4

We have $a_n = \dfrac{n}{n + 4} = \dfrac{1}{1 + (4/n)}$. Hence, $\lim a_n = 1 \neq 0$. According to the test, the series diverges.   ▤

Note that the test does *not* state that $\sum_{n=1}^{\infty} a_n$ converges when $\lim_{n\to\infty} a_n = 0$. In such cases, the series may diverge, as example 50.5 shows.

example 50.5   Show that the infinite series $\sum_{n=1}^{\infty} \dfrac{1}{\sqrt{n}}$ diverges, although $\lim_{n\to\infty} \dfrac{1}{\sqrt{n}} = 0$.

solution 50.5   Letting $S_n = 1 + 1/\sqrt{2} + 1/\sqrt{3} + \cdots + 1/\sqrt{n}$, note that $1/\sqrt{n}$ is the smallest of all n terms in this sum. Thus,

$$S_n = 1 + \frac{1}{\sqrt{2}} + \frac{1}{\sqrt{3}} + \cdots + \frac{1}{\sqrt{n}} > n\,\frac{1}{\sqrt{n}} = \sqrt{n}$$

As n increases, $\sqrt{n}$ increases and therefore S_n takes on larger and larger values. Thus, although $\lim_{n\to\infty} \dfrac{1}{\sqrt{n}} = 0$, $\lim_{n\to\infty} S_n$ does not exist.

convergence tests for positive series   If each term a_n in the series $\sum_{n=1}^{\infty} a_n$ is positive, then S_n and $\sum_{n=1}^{\infty} a_n$ may be represented as areas, as depicted in figure 50.1. The rectangle based at $x = n$ has length 1, height a_n, and

figure 50.1

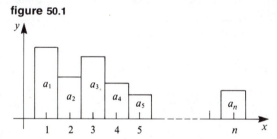

therefore area $1 \cdot a_n = a_n$. Then S_n is the sum of the areas of the first n rectangles, while $S = \lim S_n$ is the *total area* of the rectangles from $n = 1$ to $+\infty$. Hence, $\sum_{n=1}^{\infty} a_n$ will *converge provided that the total area bounded by the rectangles is finite*. Otherwise, an infinite area is formed, and the series will diverge.

Suppose next that $0 < a_n \le b_n$ for each n, as illustrated in figure 50.2. What can be said about the two corresponding series $\sum_{n=1}^{\infty} a_n$ and $\sum_{n=1}^{\infty} b_n$?

figure 50.2

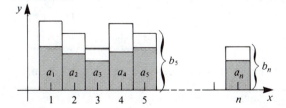

observation 1   If the total area $\sum_{n=1}^{\infty} b_n$ bounded by the larger rectangles is finite, then the rectangles formed from the smaller terms a_n will also bound a finite area.

observation 2   If the total area bounded by the smaller rectangles is infinite, then the larger area $\sum_{n=1}^{\infty} b_n$ must also be infinite.

observation 3   If the area $\sum_{n=1}^{\infty} a_n$ is finite, *no conclusion* can be reached about the larger area. Thus, $\sum_{n=1}^{\infty} b_n$ might converge or diverge.

These observations form the basis of the *comparison test*, which is summarized in the box and illustrated in example 50.6.

---

comparison test for positive series

Suppose that $0 < a_n \le b_n$ for all n.

1. If $\Sigma\, b_n$ converges, then $\Sigma\, a_n$ also converges.

2. If $\Sigma\, a_n$ diverges, then $\Sigma\, b_n$ also diverges.

3. If $\Sigma\, a_n$ converges, the test gives *no information* on $\Sigma\, b_n$.

4. If $\Sigma\, b_n$ diverges, the test gives *no information* on $\Sigma\, a_n$.

---

example 50.6   Examine each of the following infinite series for convergence.

$$\textbf{(a) } \sum_{n=1}^{\infty} \frac{1}{n(2^n)} \qquad \textbf{(b) } \sum_{n=1}^{\infty} \frac{1}{(n+1)^2} \qquad \textbf{(c) } \sum_{n=1}^{\infty} \left(\frac{1}{n}\right)^{1/3}$$

solution 50.6   In setting up the comparisons, we will make use of the **reciprocal rule:** if $0 < a < b$, then $1/a > 1/b$.

(a) Since $n(2^n) \ge 2^n$, $1/[n(2^n)] \le 1/2^n$. Since $\sum_{n=1}^{\infty} \frac{1}{2^n}$ is a convergent geometric series, $\sum_{n=1}^{\infty} \frac{1}{n(2^n)}$ also converges.

(b) In example 50.2, we showed that $\sum_{n=1}^{\infty} \frac{1}{n(n+1)} = 1$, and our original series should be closely related. Since $n(n+1) < (n+1)(n+1) = (n+1)^2$, it follows that $\frac{1}{(n+1)^2} < \frac{1}{n(n+1)}$. Hence, $\sum_{n=1}^{\infty} \frac{1}{(n+1)^2}$ converges.

(c) Since $n \ge 1$, $\sqrt{n} \ge \sqrt[3]{n}$. Hence, $1/\sqrt{n} \le 1/\sqrt[3]{n} = (1/n)^{1/3}$. From example 50.5, we know that $\sum_{n=1}^{\infty} \frac{1}{\sqrt{n}}$ diverges. Hence, from part **(2)** of the comparison test, we conclude that the original series diverges.

Each of the series in example 50.6 can now be used for comparisons with other series. For example, you can now use the result of part **(a)** to show that

$$\sum_{n=1}^{\infty} \frac{1}{[n(2^n)+5]} \text{ converges.}$$

Perhaps the most frequently applied of the many available tests for the convergence or divergence of infinite series is the **ratio test**, given in the box.

---

ratio test for positive series

Suppose $a_n > 0$ for each n and $L = \text{limit}_{n \to \infty}\, a_{n+1}/a_n$ exists.

1. If $L < 1$, then $\Sigma\, a_n$ converges.

2. If $L > 1$, then $\text{limit}\, a_n$ cannot be 0, and so $\Sigma\, a_n$ diverges.

3. If $L = 1$, then the test gives *no information*. Another test must be used.

Although we will not provide the details, the ratio test is actually a consequence of the comparison test. When $L < 1$, a comparison can be set up between a_n and a geometric sequence ar^n where $L < r < 1$ and a is a positive constant. Since $0 < r < 1$, $\Sigma\, ar^n$ is a convergent geometric series, and hence $\sum_{n=1}^{\infty} a_n$ will also converge. The ratio test is illustrated in our next example.

example 50.7   Apply the ratio test to each of the following positive infinite series.

$$\textbf{(a)}\ \sum_{n=1}^{\infty} \frac{n}{2^n} \qquad \textbf{(b)}\ \sum_{n=1}^{\infty} \frac{e^n}{n^3} \qquad \textbf{(c)}\ \sum_{n=1}^{\infty} \frac{1}{n^2}$$

solution 50.7   **(a)** For $a_n = \dfrac{n}{2^n}$, $\dfrac{a_{n+1}}{a_n} = \dfrac{n+1}{2^{n+1}} \cdot \dfrac{2^n}{n} = \dfrac{1}{2}\left(1 + \dfrac{1}{n}\right)$. Hence, $L = \text{limit}\, \dfrac{a_{n+1}}{a_n} = \dfrac{1}{2}$ and so

$\sum_{n=1}^{\infty} \dfrac{n}{2^n}$ converges.

(b) For this series, $\dfrac{a_{n+1}}{a_n} = \dfrac{e^{n+1}}{(n+1)^3} \dfrac{n^3}{e^n} = e\left(\dfrac{n}{n+1}\right)^3 \to e$ as $n \to \infty$. Since $e > 1$, the series diverges.

(c) If $a_n = \dfrac{1}{n^2}$, then $\text{limit}\, \dfrac{a_{n+1}}{a_n} = \text{limit}\left(\dfrac{n+1}{n}\right)^2 = 1$. We receive no information from the ratio test. Note, however, that

$$\sum_{n=1}^{\infty} \frac{1}{n^2} = 1 + \frac{1}{2^2} + \frac{1}{3^2} + \frac{1}{4^2} + \cdots = 1 + \sum_{n=1}^{\infty} \frac{1}{(n+1)^2}$$

and we showed in example 50.6**(b)** that $\sum_{n=1}^{\infty} \dfrac{1}{(n+1)^2}$ converges. Hence, the original series converges.

absolute convergence   If the infinite series $\sum_{n=1}^{\infty} a_n$ contains negative terms, neither the comparison test nor the ratio test can be applied directly. On the other hand, both of these tests could be used on the positive series $\sum_{n=1}^{\infty} |a_n|$. If $\sum_{n=1}^{\infty} |a_n|$ converges, we say that the original series $\sum_{n=1}^{\infty} a_n$ **converges absolutely**. But does $\sum_{n=1}^{\infty} a_n$ itself converge? Theorem 50.1 provides the answer.

theorem 50.1   If the series $\sum_{n=1}^{\infty} |a_n|$ converges, then the series $\sum_{n=1}^{\infty} a_n$ also converges.

A demonstration of this result is not difficult. Given $S_n = a_1 + a_2 + \cdots + a_n$, write $S_n = P_n - N_n$, where P_n is the sum of the positive terms in S_n, and $-N_n$ is the sum of the negative terms in S_n. Since we are given that $\sum_{n=1}^{\infty} |a_n|$ converges, the sum of all the areas shown in figure 50.3 is finite. Hence, the sum of the

figure 50.3

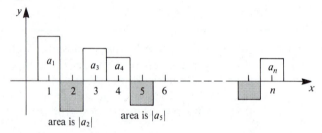

area is $|a_2|$ area is $|a_5|$

areas above the x-axis and the sum of the areas below the x-axis are each finite. The quantities P_n and N_n each have representations in the figure. Observe that P_n is the sum of the areas above the x-axis over to and including the rectangle at $x = n$. Likewise, N_n is the area of the rectangles below the x-axis over to $x = n$. Hence $P = \text{limit}_{n \to \infty} P_n$ and $N = \text{limit}_{n \to \infty} N_n$ both exist and are finite.

Now $\sum_{n=1}^{\infty} |a_n| = P + N$, and $\sum_{n=1}^{\infty} a_n = \text{limit}_{n \to \infty} (P_n - N_n) = P - N$. Thus, the series $\sum a_n$ itself converges.

Therefore one method of showing that $\sum_{n=1}^{\infty} a_n$ converges is to apply tests for positive series to $\sum_{n=1}^{\infty} |a_n|$. This is illustrated in our next example.

example 50.8   The following infinite series contain negative terms. Show that each converges absolutely and therefore converges.

$$\textbf{(a)} \ \sum_{n=1}^{\infty} (-1)^n \frac{n^2}{3^n} \qquad \textbf{(b)} \ \sum_{n=1}^{\infty} \frac{\sin n}{n^3}$$

solution 50.8   **(a)** Note that $\left| (-1)^n \dfrac{n^2}{3^n} \right| = \dfrac{n^2}{3^n}$. We apply the ratio test to $\displaystyle\sum_{n=1}^{\infty} \frac{n^2}{3^n}$:

$$\frac{|a_{n+1}|}{|a_n|} = \frac{(n+1)^2}{3^{n+1}} \cdot \frac{3^n}{n^2} = \frac{1}{3}\left(1 + \frac{1}{n}\right)^2 \to \frac{1}{3}$$

Hence, $\displaystyle\sum_{n=1}^{\infty} (-1)^n \frac{n^2}{3^n}$ converges absolutely and therefore itself converges.

(b) Note that $\left|\dfrac{\sin n}{n^3}\right| \le \dfrac{1}{n^3}$. We will show that the series $\displaystyle\sum_{n=1}^{\infty} \frac{1}{n^3}$ converges using the comparison test. For $n \ge 1$, $n^3 \ge n^2$, and hence $1/n^3 \le 1/n^2$. Since $\Sigma\, 1/n^2$ converges [see example 50.7**(c)**], the series $\Sigma\, 1/n^3$ and therefore $\Sigma |(\sin n)/n^3|$ converge by the comparison test. By theorem 50.1, the original series will also converge. ▤

warning!   Theorem 50.1 is just one of a large number of tests that have been devised by mathematicians for series that contain negative terms. If $\Sigma_{n=1}^{\infty} |a_n|$ diverges, $\Sigma_{n=1}^{\infty} a_n$ *may still converge*. More testing is needed, as demonstrated in examples 50.9 and 50.10.

example 50.9   The infinite series

$$1 - 1 + \frac{1}{\sqrt{2}} - \frac{1}{\sqrt{2}} + \frac{1}{\sqrt{3}} - \frac{1}{\sqrt{3}} + \cdots$$

does not converge absolutely. Show, however, that the series converges to 0.

solution 50.9   The sequence of partial sums S_n follows a clear pattern:

$$1, \quad 0, \quad \frac{1}{\sqrt{2}}, \quad 0, \quad \frac{1}{\sqrt{3}}, \quad 0, \quad \frac{1}{\sqrt{4}}, \quad \cdots$$

and hence limit $S_n = 0$. The series does not converge absolutely since $1 + 1 + 1/\sqrt{2} + 1/\sqrt{2} + \cdots + 1/\sqrt{n} + 1/\sqrt{n} > 2n(1/\sqrt{n}) = 2\sqrt{n}$. ▤

example 50.10   The series $\Sigma_{n=1}^{\infty} n^2 x^n$, called a **power series**, will have negative terms when $x < 0$. Show that this series converges only for $-1 < x < 1$.

solution 50.10   Applying the ratio test to $\Sigma\, n^2 |x|^n$, we obtain

$$\frac{|a_{n+1}|}{|a_n|} = \frac{(n+1)^2 |x|^{n+1}}{n^2 |x|^n} = \left(1 + \frac{1}{n}\right)^2 |x|$$

Hence, the limit L is $|x|$. When $|x| < 1$, the series converges absolutely, and therefore converges. For $|x| \ge 1$, note that $|n^2 x^n| = n^2 |x|^n \ge n^2$. Thus, $\text{limit}_{n\to\infty}\, n^2 x^n$ cannot be 0. The series therefore diverges for $|x| \ge 1$. ▤

exercises for section 50

part A

Decide whether $\Sigma\, a_n$ converges by first finding S_n, the nth partial sum. Then (if possible) compute S, the sum of the infinite series.

1. $\displaystyle\sum_{n=1}^{\infty} (-2)^n$ **2.** $\displaystyle\sum_{n=1}^{\infty} \frac{1+(-1)^n}{2}$

3. $\displaystyle\sum_{n=1}^{\infty} (3/\pi)^n$ **4.** $\displaystyle\sum_{n=1}^{\infty} (\pi/3)^n$

5. $\displaystyle\sum_{n=1}^{\infty} 9/10^n$ **6.** $\displaystyle\sum_{n=1}^{\infty} 2/5^n$

7. $\displaystyle\sum_{n=1}^{\infty} \left(\frac{1}{3^n} - \frac{1}{7^n}\right)$ **8.** $\displaystyle\sum_{n=1}^{\infty} \frac{1+(-1)^n}{2^n}$

9. $\displaystyle\sum_{n=1}^{\infty} \frac{1}{n(n+2)} = \frac{1}{1\cdot 3} + \frac{1}{2\cdot 4} + \frac{1}{3\cdot 5} + \cdots$

10. $\displaystyle\sum_{n=1}^{\infty} \frac{1}{\sqrt{n+1}+\sqrt{n}}$ [*Hint:* Rationalize the denominator.]

★**11.** Use partial fractions (see section 22) to find the sum of $\displaystyle\sum_{n=1}^{\infty} \frac{1}{4n^2-1}$.

★**12.** Show that $\displaystyle\sum_{n=1}^{\infty} \ln\left(\frac{n+1}{n+2}\right)$ diverges by computing S_n.

Show that each of the following infinite series diverges.

13. $\displaystyle\sum_{n=1}^{\infty} \frac{n}{n+2}$ **14.** $\displaystyle\sum_{n=1}^{\infty} \frac{3n^2}{n^2+2}$

15. $\displaystyle\sum_{n=1}^{\infty} \frac{3^n-2^n}{2^n}$ **16.** $\displaystyle\sum_{n=1}^{\infty} (-1)^n\, 10^{-100}$

17. $\displaystyle\sum_{n=1}^{\infty} (1/n)^{1/4}$ **18.** $\displaystyle\sum_{n=1}^{\infty} 1/n^{99}$

The following two classes of infinite series are commonly used in the comparison test.

    (i) The *p-series* $\sum_{n=1}^{\infty} a/n^p$ converges for $p>1$ and diverges for $0<p\le 1$.

    (ii) The *geometric series* $\sum_{n=1}^{\infty} ar^{n-1}$ or $\sum_{n=1}^{\infty} ar^n$ converges if and only if $-1<r<1$.

Use these infinite series for comparison purposes to decide whether each of the following infinite series converges or diverges.

19. $\displaystyle\sum_{n=1}^{\infty} \frac{1}{n^2+4}$ **20.** $\displaystyle\sum_{n=1}^{\infty} \frac{1}{n(3^n)}$ **21.** $\displaystyle\sum_{n=1}^{\infty} \frac{1}{3^n-1}$

22. $\displaystyle\sum_{n=1}^{\infty} \frac{n}{n^2+4}$ **23.** $\displaystyle\sum_{n=1}^{\infty} \frac{2^n}{n(2^n+1)}$ **24.** $\displaystyle\sum_{n=1}^{\infty} \frac{1}{2^n-n}$

25. $\displaystyle\sum_{n=1}^{\infty} \frac{\sin^2 n}{n^2}$ **26.** $\displaystyle\sum_{n=1}^{\infty} \frac{1}{\sqrt{n^2+n}}$ **27.** $\displaystyle\sum_{n=1}^{\infty} \frac{1}{n^3+n}$

28. $\displaystyle\sum_{n=1}^{\infty} e^{-n^3}$

Use the ratio test to decide whether each of the following infinite series converges or diverges.

29. $\displaystyle\sum_{n=1}^{\infty} n/3^n$ **30.** $\displaystyle\sum_{n=1}^{\infty} n^2/3^n$

31. $\displaystyle\sum_{n=1}^{\infty} 2^n/n^4$ **32.** $\displaystyle\sum_{n=1}^{\infty} \frac{2^n}{n!}$, where $n! = n(n-1)\cdots 3\cdot 2\cdot 1$

33. $\displaystyle\sum_{n=1}^{\infty} n^5(.9)^n$ **34.** $\displaystyle\sum_{n=1}^{\infty} n^3 e^{-n}$

35. $\displaystyle\sum_{n=1}^{\infty} \frac{1}{n}\left(\frac{3}{2}\right)^n$ **36.** $\displaystyle\sum_{n=1}^{\infty} \frac{n!}{2^{2n}}$, where $n! = n(n-1)\cdots 3\cdot 2\cdot 1$

★**37.** $\displaystyle\sum_{n=1}^{\infty} \frac{2^n n!}{n^n}$ $\left[\text{Hint: Use } e = \lim_{n\to\infty}\left(1+\frac{1}{n}\right)^n \text{ and } n! = n(n-1)\cdots 3\cdot 2\cdot 1.\right]$

★**38.** $\displaystyle\sum_{n=1}^{\infty} \frac{\pi^n}{n^n}$ $\left[\text{Hint: Use } e = \lim_{n\to\infty}\left(1+\frac{1}{n}\right)^n.\right]$

Each of the following infinite series contains negative terms in its sum. Which of the series converge?

39. $\displaystyle\sum_{n=1}^{\infty} (-1)^n n$ **40.** $\displaystyle\sum_{n=1}^{\infty} (-1)^n \frac{1}{n^2}$

41. $\displaystyle\sum_{n=1}^{\infty} (-1)^n \frac{2^n}{n}$ **42.** $\displaystyle\sum_{n=1}^{\infty} \frac{\sin n}{n^5}$

43. $\displaystyle\sum_{n=1}^{\infty} (-1)^n n e^{-n}$ **44.** $\displaystyle\sum_{n=1}^{\infty} \frac{\cos\left[(2n+1)\pi\right]}{n}$

Find the values of x for which each of the following power series is convergent.

45. $\displaystyle\sum_{n=0}^{\infty} \frac{n}{n+1} x^n$ **46.** $\displaystyle\sum_{n=1}^{\infty} \frac{2^n}{x^n}$

47. $\displaystyle\sum_{n=1}^{\infty} n^3 x^n$ **48.** $\displaystyle\sum_{n=1}^{\infty} n^n x^n$

49. $\displaystyle\sum_{n=1}^{\infty} \frac{(-1)^n}{3^n} x^n$ **50.** $\displaystyle\sum_{n=1}^{\infty} (-1)^n \left(1+\frac{1}{n}\right) x^n$

part B

Let $f(x)$ be a positive decreasing function defined for $x \geq 1$, and let $a_n = f(n)$. The *integral test* is based on the relationship between

$$S = \sum_{n=1}^{\infty} a_n \qquad \text{and} \qquad A = \int_1^{+\infty} f(x)\, dx$$

The series S converges if and only if the integral A is finite.

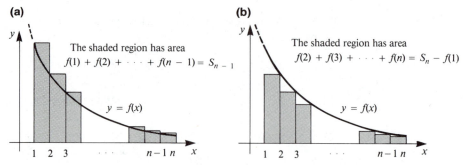

(a)

The shaded region has area

$f(1) + f(2) + \cdots + f(n-1) = S_{n-1}$

$y = f(x)$

(b)

The shaded region has area

$f(2) + f(3) + \cdots + f(n) = S_n - f(1)$

$y = f(x)$

51. Show that if A is finite, then $S < f(1) + A$, and so $S = \Sigma\, a_n$ converges. [*Hint*: See figure **(b)**].

52. If A is infinite, show that S is also infinite and so $\Sigma\, a_n$ diverges. [*Hint*: See figure **(a)**.]

53. Let $f(x) = 1/x^k$. Conclude that $\displaystyle\sum_{n=1}^{\infty} 1/n^k$ converges for $k > 1$.

54. Show that $\displaystyle\sum_{n=1}^{\infty} 1/n$ diverges by using the integral test.

55. Does $\displaystyle\sum_{n=2}^{\infty} \frac{1}{n \ln n}$ converge? Use the integral test.

56. Let $f(x) = x^k e^{-\alpha x}$ where $k > 0$ and $\alpha > 0$.

(a) Show that $f(x)$ is decreasing for $x > k/\alpha$.

(b) Show that $\displaystyle\sum_{n=1}^{\infty} f(n)$ converges by using the ratio test.

(c) Conclude that $\displaystyle\lim_{x \to +\infty} f(x) = 0$ and that $\displaystyle\int_1^{+\infty} f(x)\, dx$ is finite.

The equation $x_{n+1} = ax_n + b_n$ is a first order linear difference equation.

57. If $x_0 = 0$, show that $x_n = a^{n-1}b_0 + a^{n-2}b_1 + \cdots + ab_{n-2} + b_{n-1}$.

58. Find the general solution in terms of x_0, a, and the solution in exercise 57.

59. If $a = 1$, when does $\displaystyle\lim_{n \to \infty} x_n$ exist?

60. If $\Sigma\, b_n$ is convergent and $a \neq 1$, show that $\displaystyle\lim_{n \to \infty} x_n = 0$.

61. If $x_\infty = \displaystyle\lim x_n$ exists, show that $b = \lim b_n$ also exists and $x_\infty = b/(1-a)$.

Taylor polynomial approximations

Of the many functions encountered in calculus, polynomial functions $p(x) = a_0 + a_1x + a_2x^2 + \cdots + a_nx^n$ are the easiest to differentiate and integrate. In addition, computations with polynomials are especially suited to a calculator or computer. Remarkably, most functions can be closely approximated by polynomials. As we will show, this fact can be exploited to approximate the zeros of a differentiable function $f(x)$, and to compute integrals not accessible by standard techniques.

linear approximations and Newton's iteration method   Shown in figure 51.1 is a function $y = f(x)$ with derivative $f'(x)$ that exists at $x = a$. In

figure 51.1(a)

figure 51.1(b)

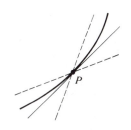

selecting a line that approximates $y = f(x)$ near $x = a$, we must require that this line pass through $P = (a, f(a))$. In addition, it is reasonable to select $f'(a)$ as the slope of this line, because the tangent line would appear to "snuggle up" to the graph of f better than any other line through P. Hence, using the point-slope form for the equation of a line, we have

$$\frac{y - f(a)}{x - a} = f'(a) \quad \text{or} \quad y = f(a) + f'(a)(x - a)$$

definition 51.1   If $f(x)$ has a derivative at $x = a$, we define the **first degree Taylor polynomial approximation** to be

$$P_1(x) = f(a) + f'(a)(x - a)$$

In the next two examples we will construct the polynomial $P_1(x)$ and see how well it approximates the given function $f(x)$.

example 51.1   Find the first degree Taylor polynomial $P_1(x)$ for $f(x) = e^x$ at $a = 0$ and $a = -2$.

solution 51.1   The derivative of $f(x)$ is $f'(x) = e^x$; thus, $f'(0) = 1$ and $f'(-2) = e^{-2}$. Then, when $a = 0$, $P_1(x) = f(0) + f'(0)(x - 0) = 1 + x$.   When $a = -2$, $P_1(x) = f(-2)$ $+ f'(-2)(x + 2) = e^{-2} + e^{-2}(x + 2) = e^{-2}(x + 3)$.

When we use the approximation $e^x \approx 1 + x$, we obtain $e^{.1} \approx 1.1$ and $e^{1.5} \approx$ 2.5. The exact value of $e^{.1}$ is $1.1051709....$ On the other hand, the approximation is poor when applied to $e^{1.5}$, as illustrated in figure 51.2. The approximation $P_1(x)$ does well only for x very close to $a = 0$.

figure 51.2

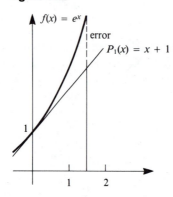

example 51.2   Find the first degree Taylor polynomial $P_1(x)$ for $f(x) = \sin x$ at $a = 0$. Use $P_1(x)$ to estimate $\sin(.01)$

solution 51.2   The derivative of $f(x)$ is $f'(x) = \cos x$, and so $f'(0) = 1$. Hence, $P_1(x) = f(0) + f'(0)(x - 0) = 0 + x = x$. Thus, $\sin(.01) \approx .01$. Using a calculator, we can determine the exact value of $\sin(.01)$ to be $.0099998....$

The first degree Taylor polynomial $P_1(x)$ can be used to approximate a solution $x = c$ to the equation $f(x) = 0$. If $x = a$ is near the solution, then $P_1(x)$ has x-intercept x_1 that should be closer to $x = c$ than is the original point $x = a$, as shown in figure 51.3. This x-intercept is the solution to $f(a) + f'(a)(x - a) = 0$. Thus, $x_1 = a - [f(a)/f'(a)]$, provided $f'(a) \neq 0$. Letting $x_0 = a$, we can express the formula as

$$x_1 = x_0 - \left[\frac{f(x_0)}{f'(x_0)}\right]$$

figure 51.3

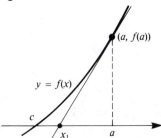

If we repeat the whole process with x_1 playing the role of a, then $x_2 = x_1 - [f(x_1)/f'(x_1)]$ should be even closer to $x = c$ than is x_1, as shown in figure

figure 51.4

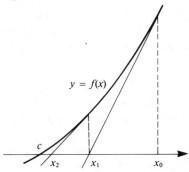

51.4. We can then define a sequence x_n recursively by:

$$x_0 = a \qquad \text{and} \qquad x_{n+1} = x_n - \left[\frac{f(x_n)}{f'(x_n)}\right]$$

This method of approximating c is known as **Newton's Iteration Method**. Our first estimate for the unknown root $x = c$ will be $x_0 = a$. As the next example illustrates, the convergence of x_n to the root $x = c$ is in general *rapid*.

example 51.3   The function $f(x) = x^3 - 5$ has $x = \sqrt[3]{5}$ as a zero. Approximate this zero using Newton's Iteration Method with $x_0 = 2$.

solution 51.3   The derivative of $f(x)$ is $f'(x) = 3x^2$ and so $x_{n+1} = x_n - [f(x_n)/f'(x_n)] = x_n - [(x_n^3 - 5)/(3x_n^2)] = \frac{2}{3}x_n + \frac{5}{3}(1/x_n)^2$. Letting $x_0 = 2$, we have $x_1 = \frac{4}{3} + \frac{5}{12} = 1.75$. Subsequent values of x_n are shown in table 51.1. The actual value of $\sqrt[3]{5}$ is $1.7099759\ldots$, and x_4 gives the zero correct to at least seven decimal places.

table 51.1

n	0	1	2	3	4
x_n	2	1.75	1.710884	1.7099765	1.7099759

The sequence x_n need not converge at all. An example is given in part B exercise 51. Suppose, however, that $\lim_{n\to\infty} x_n = c$ and $f'(c) \neq 0$. If $f(x)$ and $f'(x)$ are both continuous at $x = c$, then $f(x_n) \to f(c)$ and $f'(x_n) \to f'(c)$ since x_n is approaching c. Taking the limit of both sides of the recursion $x_{n+1} = x_n - [f(x_n)/f'(x_n)]$, we obtain

$$c = \lim x_{n+1} = c - [f(c)/f'(c)]$$

Thus $f(c)/f'(c) = 0$ or $f(c) = 0$. We may therefore conclude that *if the sequence does converge, it will converge to a solution of* $f(x) = 0$.

A rough sketch of $y = f(x)$ is helpful in selecting the first approximation x_0, as illustrated in example 51.4.

example 51.4

The graph of the function $f(x) = x^3 - 3x + 1$ is shown in example 12.1. Find the solution to $x^3 - 3x + 1 = 0$ that lies between 0 and 1.

solution 51.4

From the graph on page 123, we see that $x_0 = .4$ may be a good first estimate for the root. Since $f'(x) = 3x^2 - 3$, we have

$$x_{n+1} = x_n - [(x_n^3 - 3x_n + 1)/(3x_n^2 - 3)]$$

Hence,

$$x_1 = .4 - [(.4)^3 - 3(.4) + 1]/[3(.4)^2 - 3] = .34603175\ldots$$

Likewise,

$$x_2 = x_1 - [f(x_1)/f'(x_1)] = .34603175 - (.00333785)/(-2.6401926)$$
$$= .34729571$$

and
$$x_3 = x_2 - [f(x_2)/f'(x_2)] = .34729571 - (.00000167)/(-2.6381571)$$
$$= .34729634$$

The value of x_4 is easily seen to be .34729636. Thus, the zero between 0 and 1 occurs at $x = .347296\ldots$. Note how rapidly the sequence has converged.

second degree Taylor polynomials   The first degree Taylor polynomial $P_1(x)$ has the special properties that $P_1(a) = f(a)$ and $P_1'(a) = f'(a)$. Thus, $P_1(x)$ matches up with $f(a)$ and $f'(a)$ but $P_1''(a) = 0$. The **second degree**

Taylor polynomial $P_2(x)$ satisfies the additional condition that $P_2''(a) = f''(a)$. Thus, $P_2(x)$ can bend at $x = a$ in much the same way as $f(x)$. In particular, if $f(x)$ is concave upward at $x = a$, then $P_2(x)$ will also be concave upward there.

Define $P_2(x) = f(a) + f'(a)(x - a) + c(x - a)^2$, where c is to be determined. Then $P_2(a) = f(a)$, $P_2'(x) = f'(a) + 2c(x - a)$, and hence $P_2'(a) = f'(a)$. Finally, $P_2''(a) = 2c$. In order to have $P_2''(a) = f''(a)$, set $c = f''(a)/2$. Thus:

definition 51.2   Suppose $f(x)$ has first and second derivatives that exist at $x = a$. Then the *second degree Taylor polynomial* for $f(x)$ at $x = a$ is defined to be

$$P_2(x) = f(a) + f'(a)(x - a) + \frac{f''(a)}{2}(x - a)^2$$

The next two examples show that $P_2(x)$ is superior to $P_1(x)$ in approximating $f(x)$ near $x = a$.

example 51.5   Find the second degree Taylor polynomial $P_2(x)$ for $f(x) = e^x$ and $a = 0$.

solution 51.5   The first and second derivatives of $f(x)$ are $f'(x) = f''(x) = e^x$. Thus, $f'(0) = f''(0) = 1$ and so $P_2(x) = f(0) + f'(0)(x - 0) + [f''(0)/2](x - 0)^2 = 1 + x + x^2/2$. A computer-generated graph of e^x, $P_1(x)$, and $P_2(x)$ is shown in figure 51.5. As is usually the case, $P_2(x)$ does a much better job of approximating $f(x) = e^x$ than does $P_1(x)$.

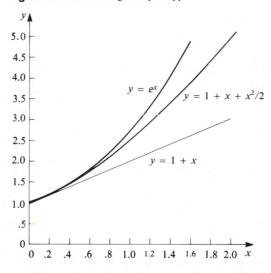

figure 51.5 Second degree Taylor approximation for e^x

example 51.6   Find the second degree Taylor polynomial $P_2(x)$ for $f(x) = \ln x$ at $a = 1$. Use it to estimate $\ln (1.1)$.

solution 51.6   The first and second derivatives of $f(x)$ are $f'(x) = 1/x$ and $f''(x) = -1/x^2$. Hence, $f'(1) = 1$ and $f''(1) = -1$. Then $P_2(x) = \ln 1 + 1 \cdot (x-1) - \frac{1}{2}(x-1)^2 = (x-1) - \frac{1}{2}(x-1)^2$. Thus, $\ln (1.1) \approx .1 - .5(.1)^2 = .095$. The true value of $\ln (1.1)$ is $.09531018. \ldots$. If we were to use $P_1(x) = (x-1)$ to approximate $\ln (1.1)$, we would obtain the less accurate value of $.1$ as the approximation.   ▤

higher order Taylor polynomials   The *nth degree Taylor polynomial* for a function $f(x)$ that is differentiable n times at $x = a$ is defined to be

$$P_n(x) = f(a) + f'(a)(x-a) + \frac{f''(a)}{2}(x-a)^2 + \cdots + \frac{f^{(n)}(a)}{n!}(x-a)^n$$

Here, $f^{(n)}(a)$ is the *nth derivative* of f at $x = a$, and $n!$ is defined to be $n(n-1)(n-2) \cdots 3 \cdot 2 \cdot 1$. Selected values of $n!$ are presented in table 51.2.

table 51.2

Table of factorials $n!$	
$1! = 1$	$6! = 720$
$2! = 2$	$7! = 5,040$
$3! = 3 \cdot 2 \cdot 1 = 6$	$8! = 40,320$
$4! = 4 \cdot 3 \cdot 2 \cdot 1 = 24$	$9! = 362,880$
$5! = 120$	$10! = 3,628,800$

example 51.7   Find $P_5(x)$ for $f(x) = \sin x$ at $a = 0$.

solution 51.7   For $f(x) = \sin x$, $f'(x) = \cos x$, $f''(x) = -\sin x$, $f^{(3)}(x) = -\cos x$, $f^{(4)}(x) = \sin x$, and $f^{(5)}(x) = \cos x$. Letting $x = 0$, we obtain $f(0) = 0$, $f'(0) = 1$, $f''(0) = 0$, $f^{(3)}(0) = -1$, $f^{(4)}(0) = 0$, and $f^{(5)}(0) = 1$. Using table 51.2 and the formula for $P_n(x)$ yields

$$P_5(x) = 0 + 1(x-0) + \frac{0}{2}(x-0)^2 + \frac{-1}{6}(x-0)^3 + \frac{0}{24}(x-0)^4 + \frac{1}{120}(x-0)^5$$

$$= x - \frac{x^3}{6} + \frac{x^5}{120}$$ ▤

The *nth* degree Taylor polynomial $P_n(x)$ satisfies the conditions that $P_n(a) = f(a)$, $P_n'(a) = f'(a)$, $P_n''(a) = f''(a), \ldots, P_n^{(n)}(a) = f^{(n)}(a)$. As a result, $P_n(x)$ will approximate $f(x)$ well over a *larger range of values* than any of the prior polynomials $P_1(x), \ldots, P_{n-1}(x)$. For $f(x) = \sin x$, and $P_n(x) = P_5(x) = x - x^3/6 + x^5/120$, $P_5(.01) = .00999983$. This is the value of $\sin (.01)$ correct to eight decimal places. On the other hand, $P_5(1) = .8416666$ is a fair approximation to $\sin 1 = .84147\ldots$. If we had used $P_1(x) = x$, we would obtain 1 as a very crude approximation for $\sin 1$.

The nth derivatives often follow a nice pattern so that $P_n(x)$ can be found in general, as shown in example 51.8.

example 51.8   Find $P_n(x)$ for each of the following functions.

$\quad$ **(a)** e^x about $a = 0$ $\qquad$ **(b)** $\ln(1 + x)$ about $a = 0$

solution 51.8   **(a)** For $f(x) = e^x$, $f^{(n)}(x) = e^x$ for all n. Hence, $f^{(n)}(0) = 1$. Then $P_n(x)$ is given by

$$1 + x + \frac{1}{2}x^2 + \frac{1}{6}x^3 + \cdots + \frac{1}{n!}x^n$$

(b) When $f(x) = \ln(1 + x)$, $f'(x) = 1/(1 + x) = (1 + x)^{-1}$, $f''(x) = -1(1 + x)^{-2}$, $f^{(3)}(x) = 2(1 + x)^{-3}$, and $f^{(4)}(x) = (-3)2(1 + x)^{-4} = -(3!)(1 + x)^{-4}$. Note the relationship between the coefficient and the power. All terms are of the form $(-1)^n(n - 1)!(1 + x)^{-n}$. Thus, $f^{(n)}(0) = (-1)^n(n - 1)!$ and so

$$\frac{f^{(n)}(0)}{n!} = \frac{(-1)^n(n - 1)(n - 2)\cdots 3 \cdot 2 \cdot 1}{n(n - 1)(n - 2)\cdots 3 \cdot 2 \cdot 1} = \frac{(-1)^n}{n}$$

Hence,

$$P_n(x) = f(0) + f'(0)x + \cdots + \frac{f^{(n)}(0)}{n!}x^n = x - \frac{x^2}{2} + \frac{x^3}{3} + \cdots + (-1)^n\frac{x^n}{n} \quad \blacksquare$$

If we let $E_n(x) = f(x) - P_n(x)$, then $E_n(x)$ measures the *error in approximating* $f(x)$ by $P_n(x)$. The **Lagrange Remainder Formula** allows us to estimate this error.

Lagrange Remainder Formula

Suppose $f(x)$ and its first $(n + 1)$ derivatives are defined on an interval $(a - \delta, a + \delta)$. If

$$E_n(x) = f(x) - P_n(x)$$

then there exists a real number c between x and a with

$$E_n(x) = \frac{f^{(n+1)}(c)}{(n + 1)!}(x - a)^{n+1}$$

It is rarely possible to actually find c. Note however that if $|f^{(n+1)}(t)| \leq M$ on the interval between x and a, then

$$|E_n(x)| \leq \frac{M}{(n+1)!}|x-a|^{n+1}$$

This inequality allows us to obtain an upper bound for the error $E_n(x)$.

example 51.9

Estimate the error when $\sin x$ is approximated by $P_5(x) = x - x^3/6 + x^5/120$ for $0 \leq x \leq .5$.

solution 51.9

In example 51.7, we showed that $f^{(5)}(x) = \cos x$. Hence, $f^{(6)}(x) = -\sin x$ and so $|f^{(6)}(x)| = |-\sin x| \leq 1$. From the inequality obtained previously, we conclude

$$|E_5(x)| \leq \frac{1}{6!}|x|^6 = \frac{|x|^6}{720} \leq \frac{(.5)^6}{720} = .0000217$$

example 51.10

Estimate e using the Taylor polynomial $P_8(x)$ for $f(x) = e^x$ about $a = 0$.

solution 51.10

From example 51.8**(a)**, $P_8(x) = 1 + x + x^2/2 + \cdots + x^8/8!$. Since $f^{(9)}(x) = e^x$, for $0 \leq x \leq 1$, $|f^{(9)}(x)| = |e^x| \leq e$. However, since the goal of the example is to estimate e, we cannot use e in our error formula. We do know, however, that $e < 3$. Hence,

$$|E_8(x)| \leq \frac{e}{9!}|x-0|^9 < \frac{3|x|^9}{362,880}$$

Now let $x = 1$ to obtain $e^1 \approx P_8(1) = 2.7182788$ and $|E_8(1)| \leq .00000827$. Thus, the true value of e is somewhere between $2.7182788 - .00000827 = 2.7182705$ and $2.7182788 + .00000827 = 2.7182871$.

The polynomial $P_n(x)$ can be used to carry out difficult integrations. In example 51.11, we will use the following special property of the integral:

$$\left| \int_a^b f(x)\,dx \right| \leq \int_a^b |f(x)|\,dx$$

This property follows from the limit definition of the integral and the fact that

$$\left| \sum_{i=1}^n f(x_i)\,\Delta x \right| \leq \sum_{i=1}^n |f(x_i)|\,\Delta x.$$

example 51.11   Use $P_5(x) = x - x^3/6 + x^5/120$ as an approximation for $\sin x$ to estimate the integral

$$\int_0^{.5} \frac{\sin x}{x}\, dx$$

How accurate is the computation?

solution 51.11   Since $\sin x = P_5(x) + E_5(x)$, $(\sin x)/x = 1 - x^2/6 + x^4/120 + [E_5(x)/x]$, for $x \neq 0$. By the Fundamental Theorem, we have

$$\int_0^{.5} \left(1 - \frac{x^2}{6} + \frac{x^4}{120}\right) dx = x - \frac{x^3}{18} + \frac{x^5}{600}\Big|_0^{.5}$$

This last expression is computed to be .49310764, and is our approximation for the original integral. From example 51.9, $|E_5(x)/x| \leq |x|^5/720 = x^5/720$ since $x \geq 0$. Hence

$$\left| \int_0^{.5} E_5(x)/x\, dx \right| \leq \int_0^{.5} |E_5(x)/x|\, dx \leq \int_0^{.5} x^5/720\, dx$$

This last integral is equal to .00000362, the largest possible error. ▤

exercises for section 51

part A   In each of the following problems, $P_1(x)$ and $P_2(x)$ refer to the Taylor polynomials of degree 1 and 2, respectively.

1. Find $P_1(x)$ for $f(x) = \sqrt[3]{x}$ around $a = 1$ and $a = 8$.
2. Find $P_1(x)$ for $f(x) = \ln x$ around $a = 1$ and $a = e$.
3. Find $P_1(x)$ for $f(x) = \tan^{-1} x$ around $a = 0$ and $a = 1$.
4. Find $P_1(x)$ for $f(x) = 1/x$ around $a = 1$ and $a = 2$.
5. Find $P_2(x)$ for $f(x) = e^{-x}$ around $a = 0$. Use it to estimate $e^{-.5}$ and compare this estimate to the actual value, found with the aid of a calculator.
6. Find $P_2(x)$ for $f(x) = \sin^{-1} x$ around $a = 0$. Use it to estimate $\sin^{-1}(.15)$ and compare this estimate to the actual value, found with the aid of a calculator.
7. Find $P_2(x)$ for $f(x) = \sqrt[5]{x}$ around $a = 1$. Estimate $\sqrt[5]{.8}$ and compare this estimate to the actual value, found with the aid of a calculator.
8. Find $P_2(x)$ for $f(x) = \cos x$ and $a = \pi$. Estimate $\cos 3$ and compare your approximation to the actual value.
9. For $f(x) = \cos x$ and $a = 0$, graph $f(x)$, $P_1(x)$, and $P_2(x)$ on the same axes, for $-\pi/2 < x < \pi/2$.

10. For $f(x) = \ln x$ and $a = 1$, graph $f(x)$, $P_1(x)$, and $P_2(x)$ on the same axes for $0 < x < 2$.

In each of the following problems, carry out Newton's Iteration Method until the zero is estimated correct to six decimal places.

11. Derive an algorithm for computing square roots by applying Newton's Method to $f(x) = x^2 - c = 0$. Then estimate $\sqrt{3{,}103}$ using $x_0 = 50$.

12. The function $f(x) = x^3 - 3x + 1$ has a zero between 1 and 2. Use the graph in example 12.1 and Newton's Method to find this zero.

13. The function $f(x) = 3x^5 - 5x^3 + 1$ was graphed in example 13.3. A root to $f(x) = 0$ lies between -2 and -1. Find this root using Newton's Method.

14. The function $f(x) = 1 + x^2 - x^6$ was graphed in example 13.5 and has two zeroes, $x = \pm c$. Find the zero that lies between 1 and 1.5 using Newton's Method.

15. Find the root to $x^7 + x^5 - 1 = 0$ using Newton's Method with $x_0 = .9$.

16. We may define π as the first positive solution to $\sin x = 0$. Estimate π by Newton's Method, using the common approximation 22/7 for x_0.

17. The only solution to $\ln x - 1 = 0$ is $x = e$. Estimate e using Newton's Method with $x_0 = 3$.

18. The graphs of $y = x$ and $y = e^{-x}$ will intersect at some point (x, x) where $0 < x < 1$. Find this point of intersection by applying Newton's Method to the equation $f(x) = x - e^{-x} = 0$.

19. The graphs of $y = x$ and $y = \cos x$ intersect at some point (x, x) where $0 < x < 1$. Find the point of intersection by applying Newton's Method to the equation $f(x) = x - \cos x = 0$.

★20. The function $f(x) = x \sin x$ has a relative maximum between 0 and π. Use Newton's Method to find where it occurs.

Find the fourth degree Taylor polynomial $P_4(x)$ for:

21. $\cos x$ around $a = 0$

Estimate $\cos 1° = \cos\left(\dfrac{\pi}{180}\right)$. How accurate is the estimate?

22. e^{-2x} around $a = 0$

Estimate e^{-1}. How accurate is the estimate?

Find the third degree Taylor polynomial $P_3(x)$ for:

23. $\tan x$ around $a = 0$

Estimate $\tan 1° = \tan\left(\dfrac{\pi}{180}\right)$.

24. e^{-x^3} around $a = 0$

Estimate $e^{-.125}$.

Find the fifth degree Taylor polynomial $P_5(x)$ for:

25. $\ln x$ around $a = 1$ **26.** x^5 around $a = -1$

★**27.** The value of $\pi/4$ is given by the formula $\pi/4 = \tan^{-1}(1/2) + \tan^{-1}(1/3)$.

    (a) Find $P_5(x)$ for $f(x) = \tan^{-1} x$ around $a = 0$.

    (b) Estimate $\pi/4$ and π.

Find the general expression for the nth degree Taylor polynomial $P_n(x)$ for each of the following functions.

28. $f(x) = 1/x$ around $a = 1$ **29.** $f(x) = e^{-x}$ around $a = 0$

30. $f(x) = \sqrt{x}$ around $a = 1$

31. Estimate the error when $\cos x$ is approximated by

$$P_2(x) = 1 - \frac{x^2}{2} \text{ for } 0 \le x \le \frac{1}{2}$$

32. Estimate the error when e^x is approximated by

$$P_3(x) = 1 + x + \frac{x^2}{2} + \frac{x^3}{6}$$

for $0 \le x \le 1$.

33. Estimate the error when $\ln(1 + x)$ is approximated by

$$P_4(x) = x - \frac{x^2}{2} + \frac{x^3}{3} - \frac{x^4}{4}$$

for $0 \le x \le 1/2$.

34. Estimate the error when $1/(1 + x)$ is approximated by

$$P_5(x) = 1 - x + x^2 - x^3 + x^4 - x^5$$

for $0 \le x \le .1$.

35. **(a)** Use $P_3(x) = x - \dfrac{x^3}{6}$ to approximate $\sin x$ and estimate

$$\int_0^{1/2} \sqrt{x} \sin x \, dx$$

    ★**(b)** How accurate is the computation?

36. **(a)** Use

$$P_3(x) = (x - 1) - \frac{(x-1)^2}{2} + \frac{(x-1)^3}{3}$$

to approximate $\ln x$ and estimate

$$\int_1^2 \sqrt{x - 1} \ln x \, dx$$

    ★**(b)** How accurate is the computation?

37. (a) Use

$$P_4(x) = 1 - x + \frac{x^2}{2} - \frac{x^3}{6} + \frac{x^4}{24}$$

to approximate e^{-x} and estimate

$$\int_0^{1/2} \sqrt{x} e^{-x} \, dx$$

★**(b)** How accurate is the computation?

38. (a) Use

$$P_4(x) = 1 - \frac{x^2}{2} + \frac{x^4}{24}$$

to approximate $\cos x$ and estimate

$$\int_0^1 \cos(x^3) \, dx$$

[*Hint*: Use $P_4(x^3)$ to approximate $\cos(x^3)$.]

★**(b)** How accurate is the computation?

39. (a) Estimate

$$\int_0^{.5} \frac{x^3}{1-x} \, dx$$

using $P_5(x) = 1 + x + x^2 + x^3 + x^4 + x^5$ for $1/(1-x)$.

★**(b)** How accurate is the computation?

★**40.** If $P_n(x)$ denotes the Taylor polynomial of degree n for e^x around $a = 0$, use $P_n(-x^2)$ to estimate

$$\int_0^1 e^{-x^2} \, dx$$

part B

L'Hospital's rule   Suppose that $f(x)$ and $g(x)$ are defined and have continuous derivatives in the vicinity of a point $x = a$. If $f(a) = g(a) = 0$, then

$$\lim_{x \to a} \frac{f(x)}{g(x)}$$

is called an *indeterminate form*. Note that the substitution $x = a$ leads to "0/0." L'Hospital's Rule asserts that if $g'(a) \neq 0$, then

$$\lim_{x \to a} \frac{f(x)}{g(x)} = \lim_{x \to a} \frac{f'(x)}{g'(x)} = \frac{f'(a)}{g'(a)}$$

A formal proof can be based on the Mean Value Theorem (see section 12). The result is reasonable if we replace $f(x)$ and $g(x)$ by their first degree Taylor polynomials at $x = a$:

$$f(x) \approx f(a) + f'(a)(x - a) = f'(a)(x - a)$$
$$g(x) \approx g(a) + g'(a)(x - a) = g'(a)(x - a)$$

Hence, for x very close to a,

$$\frac{f(x)}{g(x)} \approx \frac{f'(a)(x - a)}{g'(a)(x - a)} = \frac{f'(a)}{g'(a)}$$

As an example,

$$\lim_{x \to 0} \frac{\sin 2x}{x} = \lim_{x \to 0} \frac{2 \cos 2x}{1} = 2 \cos 0 = 2$$

In case $f'(a) = g'(a) = 0$, and $g''(a) \neq 0$, then, assuming f'' and g'' are continuous at a,

$$\lim_{x \to a} \frac{f(x)}{g(x)} = \lim_{x \to a} \frac{f''(x)}{g''(x)} = \frac{f''(a)}{g''(a)}$$

This result can be motivated using second degree Taylor polynomials. Thus,

$$\lim_{x \to 0} \frac{1 - \cos x}{x^2} = \lim_{x \to 0} \frac{\sin x}{2x} = \lim_{x \to 0} \frac{\cos x}{2} = 1/2$$

using the more general form of L'Hospital's Rule.

Find each of the following limits using L'Hospital's Rule.

41. $\lim\limits_{x \to 0} \dfrac{1 - \cos x}{x}$

42. $\lim\limits_{x \to 1} \dfrac{\ln x}{x - 1}$

43. $\lim\limits_{x \to 0} \dfrac{\sin 2x}{\sin 3x}$

44. $\lim\limits_{x \to 0} \dfrac{e^x - 1}{x}$

45. $\lim\limits_{x \to 1} \dfrac{x^8 - 1}{x - 1}$

46. $\lim\limits_{x \to 0} \dfrac{e^x - 1}{\sin x}$

47. $\lim\limits_{x \to 2} \dfrac{x^3 - 8}{x - 2}$

48. $\lim\limits_{x \to 0} \dfrac{x \sin x}{\ln (x + 1)}$

49. $\lim\limits_{x \to \pi} \dfrac{\sin 4x}{\sin 3x}$

50. $\lim\limits_{x \to 0} \dfrac{1 - \cos^2 x}{x^2}$

51. If $f(x) = \sqrt[3]{x}$, show that Newton's Iteration Formula takes the form

$$x_{n+1} = -2x_n$$

Conclude that the only choice of x_0 that makes the sequence converge is $x_0 = 0$.

an introduction to stability theory   The population growth differential equations in part 4 were of the form

$$y' = \frac{dy}{dt} = g(y)$$

If $g(y_1) = 0$, then the constant function $y(t) \equiv y_1$ is a solution to the differential equation. If we imagine that $y(t)$ specifies the position of a particle at time t on a y-axis (see the figure), and $y(t)$ changes position according to $y' = g(y)$, then

if the particle is placed at $y = y_1$, it will remain there. We call such values of y *stationary points*. At all other values of y, the velocity y' is nonzero and so the particle will move.

Find the stationary points of each of the following differential equations.

52. $y' = k(y_\infty - y)$, $k > 0$

53. $y' = ry\left(\dfrac{K - y}{K}\right)$, for r, $K > 0$

54. $y' = -\beta y \ln \dfrac{y}{K}$, for β, $K > 0$

55. $y' = ky(a - y)(b - y)$, where $0 < a < b$

56. $y' = ky\left(\dfrac{I}{y + a} - m\right)$, where k, I, a, and $m > 0$

We call a stationary point y_1 *locally stable* if there is a neighborhood $(y_1 - \delta, y_1 + \delta)$ about y_1 with the property that whenever the particle is at a point y_0 in this neighborhood, $y(t)$ *will return to* y_1 (see the figure). More precisely, if $y' = g(y)$ and $y(0) = y_0$, then

$$\lim_{t \to \infty} y(t) = y_1$$

A stationary point y_1 is *unstable* if slight changes in the position of the particle from y_1 result in the particle moving away from y_1 (see the figure).

How can we determine if a given stationary point is locally stable or unstable? A simple test can be performed, as summarized in the box.

> **test for stability**
>
> Let y_1 be a stationary point of $y' = g(y)$.
> 1. If $g'(y_1) < 0$, then $y = y_1$ is locally stable.
> 2. If $g'(y_1) > 0$, then $y = y_1$ is unstable.

The test can be demonstrated by replacing $g(y)$ by its Taylor polynomial $P_1(y)$ based at $y = y_1$.

Since $g(y_1) = 0$, the first degree Taylor polynomial for $g(y)$ is

$$P_1(y) = g(y_1) + g'(y_1)(y - y_1) = g'(y_1)(y - y_1)$$

Thus, for y_0 close to y_1, $y' = g(y) \approx g'(y_1)(y - y_1)$.

57. Show that $y' = g'(y_1)(y - y_1)$ with $y(0) = y_0$ has solution

$$y = y_1 + (y_0 - y_1)e^{ct}$$

where $c = g'(y_1)$.

The test then asserts that the original differential equation $y' = g(y)$ behaves like $y' = g'(y_1)(y - y_1)$ for $y(0) = y_0$ close to y_1. Note that when $c = g'(y_1) < 0$,

$$\lim_{t \to +\infty} [y_1 + (y_0 - y_1)e^{ct}] = y_1$$

Classify the stationary points in each of the following exercises as locally stable or unstable.

58. Exercise 52

59. Exercise 53

60. Exercise 54

61. Exercise 55

62. Exercise 56

 **functions defined by power series**

All of the functions that have appeared in the applications thus far in the text have been formed from **finite combinations** of the elementary functions x^n, e^{kx}, $\ln x$, $\sin ax$, $\cos ax$, etc. To enlarge the arsenal of functions that can be used for modeling purposes, we can form *functions defined by infinite series*. The simplest examples are the **power series**

$$f(x) = \sum_{n=0}^{\infty} a_n x^n = a_0 + a_1 x + a_2 x^2 + \cdots + a_n x^n + \cdots$$

To specify where such a function $f(x)$ is defined, we must determine the values of x for which the infinite series $\Sigma\, a_n x^n$ converges. Suppose $L = \text{limit}\,|a_{n+1}/a_n|$ exists. If we apply the ratio test to $\Sigma\,|a_n x^n|$, we obtain

$$\frac{|a_{n+1} x^{n+1}|}{|a_n x^n|} = \frac{|a_{n+1}|}{|a_n|}\,|x| \to L|x|$$

as n increases. Hence, if $L|x| < 1$, the power series converges absolutely and therefore converges by Theorem 50.1. If $L = 0$, the series will converge for all x. When $L|x| > 1$, the ratio test tells us that $\Sigma\, a_n x^n$ diverges. Letting $R = 1/L$, we have the conclusion shown in the box.

---

open interval of convergence for $\Sigma\, a_n x^n$

If $L = \text{limit}\,|a_{n+1}/a_n|$ exists, then $f(x) = \Sigma_{n=0}^{\infty} a_n x^n$ is defined for $|x| < R = 1/L$. If $L = 0$, then $f(x)$ is defined for all x. The interval $(-R, R)$ is called the *interval of convergence* and R is called the *radius of convergence*. Convergence at $x = R$ or $x = -R$ is *not guaranteed* and must be examined separately.

---

example 52.1   Find the radius of convergence and hence the domain of definition of each of the following power series.

$$\textbf{(a)}\ f(x) = \sum_{n=0}^{\infty} 2^n x^n \qquad \textbf{(b)}\ g(x) = \sum_{n=0}^{\infty} \frac{(-1)^n}{(n!)^2}\, x^n$$

solution 52.1   For $f(x)$, $|a_{n+1}/a_n| = 2^{n+1}/2^n = 2$. Hence $L = 2$, and so $R = 1/2$ is the radius of convergence. Hence, $f(x)$ is defined for $-1/2 < x < 1/2$.

For $g(x)$, $\left|\dfrac{a_{n+1}}{a_n}\right| = \dfrac{1}{(n+1)!(n+1)!} \cdot \dfrac{n!n!}{1} = \dfrac{1}{(n+1)^2} \to 0$ as n increases. Hence $R = +\infty$ and so $g(x)$ is defined for each real number x. ▤

Power series can also be defined in terms of powers of $(x - a)$. If $f(x) = \sum_{n=0}^{\infty} a_n (x - a)^n$ and $L = \text{limit} \, |a_{n+1}/a_n|$, then the series will converge absolutely provided $L|x - a| < 1$. Again, letting $R = 1/L$ yields $|x - a| < R$ or $a - R < x < a + R$. For $|x - a| > R$, the series diverges, as illustrated in figure 52.1. Note that $\sum_{n=0}^{\infty} a_n (x - a)^n$ and $\sum_{n=0}^{\infty} a_n x^n$ have the same radius of convergence.

figure 52.1

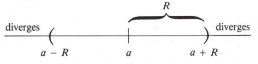

example 52.2

Find the interval of convergence of the power series $f(x) = 1 + \sum_{n=1}^{\infty} \frac{1}{n}(x - 2)^n$.

solution 52.2

Note that $\left|\dfrac{a_{n+1}}{a_n}\right| = \dfrac{1}{n+1} \cdot \dfrac{n}{1} \to 1$ as $n \to \infty$. Thus, $L = 1$ and so $R = 1$. Then $f(x)$ is defined for $|x - 2| < 1$, that is, for $1 < x < 3$.

All of the elementary functions have power series expansions of the form $\sum a_n x^n$. In fact, the nth partial sum is just the nth degree Taylor polynomial $P_n(x) = f(0) + f'(0)x + \cdots + \dfrac{f^{(n)}(0)}{n!} x^n$. Power series expansions of the form $\sum_{n=0}^{\infty} \dfrac{f^{(n)}(0)}{n!} x^n$ are called **Maclaurin series expansions**. Table 52.1 gives the Maclaurin series of the standard elementary functions.

table 52.1

Elementary function	Maclaurin series expansion	Interval of convergence
e^x	$1 + x + \dfrac{x^2}{2!} + \dfrac{x^3}{3!} + \dfrac{x^4}{4!} + \cdots$	$(-\infty, +\infty)$
$\ln(1 + x)$	$x - \dfrac{x^2}{2} + \dfrac{x^3}{3} - \dfrac{x^4}{4} + \dfrac{x^5}{5} - \cdots$	$(-1, 1)$
$\dfrac{1}{1 - x}$	$1 + x + x^2 + x^3 + x^4 + \cdots$	$(-1, 1)$
$\sin x$	$x - \dfrac{x^3}{3!} + \dfrac{x^5}{5!} - \dfrac{x^7}{7!} + \cdots$	$(-\infty, +\infty)$
$\cos x$	$1 - \dfrac{x^2}{2!} + \dfrac{x^4}{4!} - \dfrac{x^6}{6!} + \cdots$	$(-\infty, +\infty)$
$\tan x$	$x + \dfrac{x^3}{3} + \dfrac{2}{15} x^5 + \dfrac{17}{315} x^7 + \cdots$	$\left(-\dfrac{\pi}{2}, \dfrac{\pi}{2}\right)$
$\sin^{-1} x$	$x + \dfrac{1}{2} \dfrac{x^3}{3} + \dfrac{1 \cdot 3}{2 \cdot 4} \dfrac{x^5}{5} + \dfrac{1 \cdot 3 \cdot 5}{2 \cdot 4 \cdot 6} \dfrac{x^7}{7} + \cdots$	$(-1, 1)$
$\cos^{-1} x$	$\dfrac{\pi}{2} - \sin^{-1} x$, where $\sin^{-1} x$ is given above	$(-1, 1)$
$\tan^{-1} x$	$x - \dfrac{x^3}{3} + \dfrac{x^5}{5} - \dfrac{x^7}{7} + \dfrac{x^9}{9} - \cdots$	$(-1, 1)$

Power series look like "long polynomials", and what makes power series so very useful is that they *can actually be manipulated like polynomials within their interval of convergence*. More precisely, the properties in the accompanying box are satisfied.

---

polynomial properties of power series

Let $f(x) = \Sigma\, a_n x^n = a_0 + a_1 x + a_2 x^2 + \cdots + a_n x^n + \cdots$ for $|x| < R$.

1. For $|a| < R$, $\lim\limits_{x \to a} f(x) = f(a) = \sum\limits_{n=0}^{\infty} a_n a^n$. The function $f(x)$ is continuous inside the interval of convergence.

2. $f'(x) = a_1 + 2a_2 x + 3a_3 x^2 + \cdots + na_n x^{n-1} + \cdots$

$$= \sum_{n=1}^{\infty} na_n x^{n-1} \qquad \text{for } |x| < R$$

3. $\int f(x)\,dx = \left(a_0 x + a_1 \dfrac{x^2}{2} + a_2 \dfrac{x^3}{3} + \cdots + a_n \dfrac{x^{n+1}}{n+1} + \cdots \right) + c$

$$= \sum_{n=0}^{\infty} a_n \frac{x^{n+1}}{n+1} + c \qquad \text{for } |x| < R$$

These polynomial properties are illustrated in our next two examples.

example 52.3   The power series expansion for $1/(1-x)$ is the geometric series $1 + x + x^2 + \cdots + x^n + \cdots$, which converges for $|x| < 1$. Obtain the Maclaurin series expansions for

$$\textbf{(a)}\ \frac{1}{(1-x)^2} \qquad \textbf{(b)}\ \ln|1-x|$$

solution 52.3   Note that $1/(1-x)^2 = D_x\, 1/(1-x)$ and $-\ln|1-x| = \int 1/(1-x)\,dx$. Recalling that $1 + x + x^2 + \cdots + x^n + \cdots = 1/(1-x)$, we apply polynomial properties **(2)** and **(3)** to obtain

$$\frac{1}{(1-x)^2} = 1 + 2x + 3x^2 + \cdots + nx^{n-1} + \cdots$$

and

$$-\ln|1-x| = \left(x + \frac{x^2}{2} + \frac{x^3}{3} + \frac{x^4}{4} + \cdots + \frac{x^{n+1}}{n+1} + \cdots \right) + c$$

Letting $x = 0$, we have $0 = -\ln|1-0| = 0 + c$. Then $c = 0$ and so

$$\ln|1-x| = -\left(x + \frac{x^2}{2} + \frac{x^3}{3} + \cdots + \frac{x^{n+1}}{n+1} + \cdots \right)$$

Both expansions are valid provided $|x| < 1$.

example 52.4   Find the power series expansion of $e^{-t^2/2}$ and then construct an antiderivative for $e^{-t^2/2}$.

solution 52.4   For any real number x,

$$e^x = 1 + x + \frac{x^2}{2} + \frac{x^3}{6} + \cdots + \frac{x^n}{n!} + \cdots$$

In particular, letting $x = -t^2/2$, we have

$$e^{-t^2/2} = 1 - \frac{t^2}{2} + \frac{t^4}{2^2 2!} - \frac{t^6}{2^3 3!} + \cdots + \frac{(-1)^n t^{2n}}{2^n n!} + \cdots$$

An antiderivative $F(t)$ for $e^{-t^2/2}$ may be found by integrating this series term by term. Hence:

$$F(t) = t - \frac{1}{2}\frac{t^3}{3} + \frac{1}{2^2 2!}\frac{t^5}{5} + \cdots + \frac{(-1)^n}{2^n n!}\frac{t^{2n+1}}{2n+1} + \cdots$$

The antiderivative $F(t)$ is defined for all numbers t, and provides the tool for estimating the normal probability integral $(1/\sqrt{2\pi})\int_0^z e^{-t^2/2}\,dt$.   ▤

Power series may also be added and multiplied like ordinary polynomials *within a common interval of convergence*, as described in the box.

---

polynomial properties of power series

If $f(x) = \Sigma\, a_n x^n$ and $g(x) = \Sigma\, b_n x^n$ are *both* defined for $|x| < c$, then:

4. $f(x) + g(x) = (a_0 + b_0) + (a_1 + b_1)x + (a_2 + b_2)x^2 + \cdots$

$$= \sum_{n=0}^{\infty} (a_n + b_n)x^n$$

5. $f(x)g(x) = (a_0 + a_1 x + a_2 x^2 + \cdots)(b_0 + b_1 x + b_2 x^2 + \cdots)$

$$= a_0 b_0 + (a_0 b_1 + a_1 b_0)x + (a_0 b_2 + a_1 b_1 + a_2 b_0)x^2 + \cdots$$

We next use these properties to construct power series expansions for more complicated functions. In each case, it would be extremely tedious to compute $[f^{(n)}(0)]/n!$ directly.

example 52.5   For $x \neq 0$, obtain a power series expansion for $(\cos x - 1)/x^2$. Then compute $\lim_{x \to 0} (\cos x - 1)/x^2$.

solution 52.5   Using the power series expansion for $\cos x$ given in table 52.1, we obtain

$$\frac{\cos x - 1}{x^2} = x^{-2}\left[\left(1 - \frac{x^2}{2} + \frac{x^4}{4!} - \frac{x^6}{6!} + \frac{x^8}{8!} - \cdots\right) - 1\right]$$

$$= -\frac{1}{2} + \frac{x^2}{4!} - \frac{x^4}{6!} + \frac{x^6}{8!} - \cdots \qquad \text{for } x \neq 0$$

Hence,

$$\underset{x \to 0}{\text{limit}} \frac{\cos x - 1}{x^2} = \underset{x \to 0}{\text{limit}} \left(-\frac{1}{2} + \frac{x^2}{4!} - \frac{x^4}{6!} + \cdots\right)$$

$$= -1/2$$

example 52.6   Obtain the first five terms of the power series expansion of $f(x) = (\sin x)/(1 + x^2)$ about $x = 0$.

solution 52.6   The power series for $1/(1 - t)$ is $\sum_{n=0}^{\infty} t^n$ and the expansion is valid provided $|t| < 1$. If we let $t = -x^2$, then

$$\frac{1}{1 - (-x^2)} = \sum_{n=0}^{\infty} (-x^2)^n$$

$$= 1 - x^2 + x^4 - x^6 + \cdots + (-1)^n x^{2n} + \cdots$$

This expansion is valid for $|-x^2| = |x|^2 < 1$, that is, for $|x| < 1$. Using the power series for $\sin x$ given in table 52.1, we have

$$\frac{\sin x}{1 + x^2} = \left(x - \frac{x^3}{3!} + \frac{x^5}{5!} - \frac{x^7}{7!} + \cdots\right)(1 - x^2 + x^4 - x^6 + \cdots)$$

$$= x + \left(-1 - \frac{1}{6}\right)x^3 + \left(1 + \frac{1}{6} + \frac{1}{120}\right)x^5 + \cdots$$

[*Note*: that terms involving x^3 are formed from $(-\frac{1}{3!}x^3)(1)$ and $x(-x^2)$. Terms involving x^5 are formed from $x(x^4)$, $(-\frac{1}{3!}x^3)(-x^2)$, and $(\frac{1}{5!}x^5)(1)$.] The interval of convergence is $(-1, 1)$, the interval of convergence for $\sum (-1)^n x^{2n}$.

applications to differential equations   It is easy to produce examples of differential equations that cannot be solved by any of the methods presented in sections 33–39. It is known, for example, that the innocent looking differential equation $y' = t^2 + y^2$ does not have solutions that can be expressed as finite combinations of the elementary functions. Often, however, differential equations have power series solutions $y = \sum_{n=0}^{\infty} a_n(t - a)^n$. For the second order equation

$$y'' + \alpha_1(t)y' + \alpha_2(t)y = 0$$

it is known that solutions can be represented by $y = \sum a_n t^n$ when $\alpha_1(t)$ and $\alpha_2(t)$ are polynomials.

The following examples will illustrate two methods by which the unknown coefficients can be determined. When $y = \sum_{n=0}^{\infty} a_n t^n$,

$$y' = a_1 + 2a_2 t + 3a_3 t^2 + \cdots = \sum_{n=0}^{\infty} (n+1)a_{n+1} t^n$$

Likewise,

$$y'' = 2a_2 + 6a_3 t + 12a_4 t^2 + \cdots = \sum_{n=0}^{\infty} (n+2)(n+1)a_{n+2} t^n$$

These two alternate expressions for y' and y'' are especially useful in finding power series solutions to differential equations.

example 52.7   Find the power series solution to the differential equation $y' = y$ that satisfies the initial condition $y(0) = 1$.

solution 52.7   If $y = \sum_{n=0}^{\infty} a_n t^n$, then $y' = \sum_{n=0}^{\infty} (n+1)a_{n+1} t^n$. By comparing nth powers, we see that $y' = y$ in case $a_n = (n+1)a_{n+1}$, that is, if $a_{n+1} = a_n/(n+1)$. Since $y(0) = a_0$, $a_1 = a_0/1 = 1$, $a_2 = a_1/2 = 1/2$, $a_3 = a_2/3 = 1/6$, and, in general, $a_n = 1/n!$. Thus, $y = \sum_{n=0}^{\infty} \dfrac{t^n}{n!}$. Of course, this is just the Maclaurin series expansion for e^t. ▤

example 52.8   Find power series solutions to the differential equation $y'' - ty' + y = 0$ with:

(a) $y(0) = 1$ and $y'(0) = 0$ (b) $y(0) = 0$ and $y'(0) = 1$.

solution 52.8   If $y = \sum_{n=0}^{\infty} a_n t^n$, then $y' = \sum_{n=0}^{\infty} n a_n t^{n-1}$, and $y'' = \sum_{n=0}^{\infty} n(n-1)a_n t^{n-2}$. Note that $y(0) = a_0$ and $y'(0) = a_1$. Hence, the initial conditions will specify a_0 and a_1. Now, by substitution, we have

$$y'' - ty' + y = \sum_{n=0}^{\infty} n(n-1)a_n t^{n-2} - t\left(\sum_{n=0}^{\infty} n a_n t^{n-1}\right) + \sum_{n=0}^{\infty} a_n t^n$$

$$= \sum_{n=0}^{\infty} n(n-1)a_n t^{n-2} - \sum_{n=0}^{\infty} n a_n t^n + \sum_{n=0}^{\infty} a_n t^n$$

In order to combine nth powers, we need to rewrite y'' in the special form $\sum_{n=0}^{\infty} (n+2)(n+1)a_{n+2} t^n$. Hence:

$$y'' - ty' + y = \sum_{n=0}^{\infty} [(n+2)(n+1)a_{n+2} - (n-1)a_n] t^n$$

To guarantee a solution, we must have $(n+2)(n+1)a_{n+2} = (n-1)a_n$ or

$$a_{n+2} = \frac{n-1}{(n+2)(n+1)} a_n$$

For **(a)**, note that the even coefficients are all given in terms of a_0. If $a_0 = y(0) = 1$, then $a_2 = \dfrac{-1}{2} a_0 = \dfrac{-1}{2}$, $a_4 = \dfrac{1}{4 \cdot 3} a_2 = \dfrac{-1}{4!}$, and $a_6 = \dfrac{3}{6 \cdot 5} a_4 = \dfrac{-1 \cdot 3}{6!}$, etc. The odd coefficients are all given in terms of a_1. If $a_1 = y'(0) = 0$, then all of the odd coefficients are 0. Hence

$$y_1(t) = 1 - \left(\frac{1}{2!} t^2 + \frac{1}{4!} t^4 + \frac{3}{6!} t^6 + \frac{3 \cdot 5}{8!} t^8 + \cdots \right)$$

For **(b)**, $a_0 = 0$ and so all the even coefficients vanish. When $a_1 = y'(0) = 1$, $a_3 = a_1(1-1)/(3 \cdot 2) = 0$. Since a_5, a_7, ... are all computed in terms of $a_3 = 0$, all odd coefficients beyond a_1 vanish. Hence,

$$y_2(t) = 0 + 1t + 0t^2 + 0t^3 + \cdots = t$$

The *general solution* to $y'' - ty' + y = 0$ is given by $c_1 y_1(t) + c_2 y_2(t)$.

If a power series solution appears in many applications, we may give the function a special name, study its special properties, and add it to our list of basic functions. Such an example is the **Bessel function** $J_0(t)$, which appears in the solutions to many of the partial differential equations mentioned in Section 41.

example 52.9   The differential equation $t^2 y'' + ty' + t^2 y = 0$ is *Bessel's equation of order 0*. The solution that satisfies $y(0) = 1$ and $y'(0) = 0$ is called $J_0(t)$. Find the power series representation of J_0.

solution 52.9   If $y = \Sigma_{n=0}^{\infty} a_n t^n$, then $t^2 y = a_0 t^2 + a_1 t^3 + a_2 t^4 + \cdots = \Sigma_{n=2}^{\infty} a_{n-2} t^n$. In addition, $ty' = t(\Sigma \, na_n t^{n-1}) = \Sigma_{n=0}^{\infty} na_n t^n$. Finally,

$$t^2 y'' = t^2 [\Sigma \, n(n-1)a_n t^{n-2}] = \Sigma_{n=0}^{\infty} n(n-1)a_n t^n$$

Hence, $t^2 y'' + ty' + t^2 y = \displaystyle\sum_{n=0}^{\infty} [n(n-1) + n] a_n t^n + \sum_{n=2}^{\infty} a_{n-2} t^n$

We can combine these two power series by writing

$$\sum_{n=0}^{\infty} [n(n-1) + n] a_n t^n = a_1 t + \sum_{n=2}^{\infty} n^2 a_n t^n$$

Hence,

$$t^2 y'' + ty' + t^2 y = a_1 t + \sum_{n=2}^{\infty} (n^2 a_n + a_{n-2}) t^n$$

To construct a solution, set $a_1 = y'(0) = 0$ and, for $n \geq 2$,

$$a_n = \frac{-1}{n^2} a_{n-2}$$

Since all odd coefficients are given in terms of $a_1 = 0$, they will vanish. Finally, $a_2 = \dfrac{-1}{2^2} a_0 = \dfrac{-1}{2^2}$, $a_4 = \dfrac{-1}{4^2} a_2 = \dfrac{1}{2^2 \cdot 4^2}$, and $a_6 = \dfrac{-1}{6^2} a_4 = \dfrac{-1}{2^2 \cdot 4^2 \cdot 6^2}$. The pattern is now apparent and we conclude that

$$J_0(t) = 1 - \frac{1}{2^2} t^2 + \frac{1}{2^2 \cdot 4^2} t^4 - \frac{1}{2^2 \cdot 4^2 \cdot 6^2} t^6 + \cdots$$ 📖

If $y = \sum_{n=0}^{\infty} a_n t^n$ is a solution, then $a_n = f^{(n)}(0)/n!$, and hence the coefficients can be found from the initial conditions and the differential equation itself, as demonstrated in example 52.10.

example 52.10 Find the first five terms of the power series solution to $y' = t^2 + y^2$ that satisfies $y(0) = 1$.

solution 52.10 Since $y' = t^2 + y^2$, $y'(0) = 0^2 + [y(0)]^2 = 1$. To find $y''(0)$, $y^{(3)}(0)$, and $y^{(4)}(0)$, we can apply implicit differentiation repeatedly to $y' = t^2 + y^2$:

$$y'' = 2t + 2yy'$$

and hence $y''(0) = 2(0) + 2(1)(1) = 2$. Similarly,

$$y^{(3)} = 2 + 2yy'' + 2(y')^2$$

Therefore, $y^{(3)}(0) = 2 + 2(1)(2) + 2(1)^2 = 8$. Finally,

$$y^{(4)} = 2yy^{(3)} + 2y'y'' + 4(y')y''$$

Thus $y^{(4)}(0) = 28$.
   Since $a_n = y^{(n)}(0)/n!$, $a_0 = 1$, $a_1 = 1$, $a_2 = 2/2 = 1$, $a_3 = 8/6 = 4/3$, and $a_4 = 28/4! = 28/24 = 7/6$. The first five terms of the solution are then

$$y = 1 + x + x^2 + \frac{4}{3} x^3 + \frac{7}{6} x^4 + \cdots$$ 📖

exercises for section 52

part A Each of the following functions is defined by a power series. Find the radius of convergence and open interval of convergence for each.

1. $f(x) = \displaystyle\sum_{n=0}^{\infty} \frac{x^n}{(n+1)3^n}$

2. $f(x) = \displaystyle\sum_{n=0}^{\infty} (-1)^n \sqrt{n} x^n$

3. $f(x) = \displaystyle\sum_{n=1}^{\infty} \frac{n!}{2^n} (x-1)^n$

4. $f(x) = \displaystyle\sum_{n=1}^{\infty} \frac{2^n}{n^3} (x-2)^n$

5. $f(x) = \sum_{n=0}^{\infty} (-1)^n \dfrac{n}{(n+1)^2} (x+2)^n$ **6.** $f(x) = \sum_{n=0}^{\infty} (-1)^n \dfrac{x^n}{(2n)!}$

7. $f(x) = \sum_{n=0}^{\infty} \dfrac{1}{1 \cdot 3 \cdot 5 \cdots (2n+1)} x^n$ **8.** $f(x) = \sum_{n=1}^{\infty} \dfrac{1}{2 \cdot 4 \cdot 6 \cdots (2n)} x^n$

★9. $f(x) = \sum_{n=0}^{\infty} n(2x-1)^n$ **★10.** $f(x) = \sum_{n=1}^{\infty} \dfrac{(2n)!}{2^{2n}(n!)^2} x^n$

Use the Maclaurin series expansion for $\sin x$ to find power series representations for each of the following functions.

11. $\sin (x^2)$ **12.** $x^4 \sin x$

13. $\displaystyle\int \sin (x^2)\, dx$ **14.** $\sin (x^2) + x \sin x$

Use the Maclaurin series expansion for $\tan^{-1} x$ to find power series representations for each of the following functions:

15. $\dfrac{1}{1+x^2}$ **16.** $\displaystyle\int \tan^{-1} x\, dx$

17. $x \tan^{-1} (x^2)$ **18.** $\dfrac{1}{(1+x^2)^2}$

Use the Maclaurin series expansion for $\ln (1+x)$ to find power series representations for each of the following functions.

19. $\ln (1+x^2)$ **20.** $\displaystyle\int \ln (1+x^2)\, dx$

Compute each of the following limits by first representing the given function $f(x)$ as a power series around $x = 0$.

21. $\displaystyle\lim_{x \to 0} \dfrac{\tan x - x}{x^3}$ **22.** $\displaystyle\lim_{x \to 0} \dfrac{\sin^{-1} x^2}{x^2}$

23. $\displaystyle\lim_{x \to 0} \dfrac{\ln (1+x^4)}{x^3}$ **24.** $\displaystyle\lim_{x \to 0} \dfrac{e^{-x^2} - 1}{x^2}$

★25. $\displaystyle\lim_{x \to 0} \dfrac{1 - e^{-x}}{\sin (4x)}$

Find the first five terms of the power series expansion for $f(x)$ about $x = 0$ for each of the following functions.

26. $f(x) = (\sin x)/(1-x)$ **27.** $f(x) = e^x \tan x$
for $-1 < x < 1$

28. $f(x) = x^3 + 2 \ln (1+x)$ **29.** $f(x) = x \tan x - \sin x$

★30. $f(x) = x \sec^2 2x$

Construct power series solutions to each of the following linear differential equations. Then solve the equation using the standard methods.

31. $y' = -y$ with $y(0) = 1$

32. $y' = ty$ with $y(0) = 2$

33. $y'' + y = 0$ with $y(0) = 0$ and $y'(0) = 1$

34. $y'' - y = 0$ with $y(0) = 1$ and $y'(0) = 0$

35. $y'' - 2y' + y = 0$ with $y(0) = 0$ and $y'(0) = 1$

Construct power series solutions to each of the following second order linear differential equations.

36. $y'' - ty = 0$ with $y(0) = 0$ and $y'(0) = 1$

37. $y'' - ty = 0$ with $y(0) = 1$ and $y'(0) = 0$

38. $y'' - 2ty' + 8y = 0$ with $y(0) = 1$ and $y'(0) = 0$

39. $y'' - 2ty' + 8y = 0$ with $y(0) = 0$ and $y'(0) = 1$

40. $ty'' + (1 - t)y' + 3y = 0$ with $y(0) = 2$ and $y'(0) = -6$

41. The Bessel function $J_1(t)$ satisfies the differential equation

$$t^2 y'' + ty' + (t^2 - 1)y = 0$$

with $y(0) = 0$ and $y'(0) = -\frac{1}{2}$. Find the power series representation for $J_1(t)$.

42. The fourth degree *Legendre polynomial* $p_4(t)$ satisfies the differential equation

$$(1 - t^2)y'' - 2ty' + 20y = 0$$

with $y'(0) = 0$. If $p_4(1) = 1$, use the power series method to find $p_4(t)$.

43. The nth degree *Hermite polynomial* $H_n(x)$ satisfies the differential equation

$$y'' - 2ty' + 2ny = 0$$

Use the power series method to find $H_3(t)$ if $H_3(0) = 0$ and $H_3'(0) = -12$.

44. The first order differential equation $y' = t + y^2$ has a power series solution $y = \Sigma a_n x^n$. If $y(0) = 0$, find the first five terms of the series.

45. Find the first five terms of a power series solution to

$$y'' + 2 \sin y = 0$$

if $y(0) = 1$ and $y'(0) = 0$.

★46. Let $y = 1/\sqrt{1 - t^2}$, for $-1 < t < 1$.

(a) Show that $(1 - t^2)y' = ty$ and $y(0) = 1$.

(b) Find the power series solution to the differential equation in (a) and obtain the Maclaurin series for $1/\sqrt{1 - t^2}$.

(c) Find the Maclaurin series for $\sin^{-1} t$.

7

an introduction to vectors and matrices

prologue   Mathematics in the nineteenth century was distinguished by the creation of many new algebraic systems. Prior to this period, the real number system was thought to be the only number system needed to explain the physical world. In this part of the text, we will study two algebraic systems—vectors and matrices—that have proven to be indispensable tools to the twentieth century scientist.

In 1858, Arthur Cayley invented **matrices**. A **matrix** is a rectangular array of real numbers. Shown below are matrices of various sizes:

$$A = \begin{bmatrix} 3 & 5 & 9 \\ 0 & 1 & -1 \\ 2 & -1 & 3 \end{bmatrix} \qquad B = [3 \quad 4 \quad -2 \quad 1 \quad 5]$$

$$C = \begin{bmatrix} 4 & -1 & .6 \\ 2 & -7 & 1 \end{bmatrix} \qquad D = \begin{bmatrix} 4 \\ 5 \\ 0 \end{bmatrix}$$

A **vector** is a matrix with only one row (such as B above) or one column (such as D). Motivated by his study of geometric transformations, Cayley discovered how these arrays could be added, subtracted, and multiplied so that many (but not all) of the usual laws of arithmetic were still valid. As we will now demonstrate, matrices and vectors can be used to keep track of a large number of variables in an orderly and enlightening fashion. As such, they are especially useful in constructing multivariate models of physical and biological phenomena.

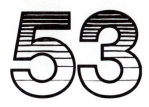

vectors and vector algebra

The mathematical concept of **vector** has a wide variety of uses in the biological and physical sciences. A vector is commonly interpreted as a directed line segment or "arrow" (see figure 53.1) and it is this particular interpretation that

figure 53.1

is so important in geometry and physics. Other uses for vectors involve record-keeping. As we will show in examples 53.1–53.3, vectors may be used to keep track of the age structure of a population or the various components in an ecosystem.

definition 53.1   An **n-dimensional vector** X is an ordered n-tuple $[x_1, x_2, \ldots, x_n]$ of real numbers. The real number x_i is called the **ith component** of the vector X. If $\bar{0} = [0, 0, \ldots, 0]$, we call $\bar{0}$ the **zero vector**.

example 53.1   A population is divided into four age categories: newborns, one-year-olds, two-year-olds, and adults. Let x_1, x_2, x_3, and x_4 be the present numbers in each of these age categories. Then if $X = [300, 200, 150, 75]$, we have 300 newborns, 200 one-year-olds, 150 two-year-olds, and 75 adults. The four-dimensional vector X specifies the **age structure** of the population.   ▤

example 53.2   In demography, a human population is commonly divided into 18 age categories: 0–4 years, 5–9 years, ..., 80–84 years, and 85–89 years. If x_i is the number in the ith age category, then $X = [x_1, x_2, \ldots, x_{18}]$ is the 18-dimensional vector that gives the age structure of the population.   ▤

example 53.3   An aquatic ecosystem is divided into five components: phytoplankton, zooplankton, zooplankton predators, small carnivores, and large carnivores. If $x_i(t)$ denotes the biomass of the ith component at time t, then the 5-dimensional vector $X(t) = [x_1(t), x_2(t), x_3(t), x_4(t), x_5(t)]$ specifies the **state of the ecosystem** at a particular time t.   ▤

The mathematical concept of vector need not involve any geometry. The most common interpretation of vectors, however, is geometric, as demonstrated in example 53.4.

example 53.4   **geometric vectors**   The ordered pair (x, y) is commonly interpreted as the set of coordinates of a point P in a plane. Alternatively, $[x, y]$ may be thought of as a **directed line segment** or **geometric vector** X with initial point $(0, 0)$ and terminal point P. Shown in figure 53.2 are the vectors $X = [2, 1]$, $Y = [1, 4]$, and $Z = [-1, 3]$.

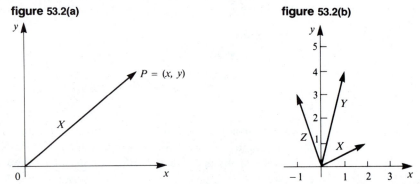

figure 53.2(a)   figure 53.2(b)

Likewise, $X = [x, y, z]$ may be interpreted as a directed line segment $\overrightarrow{OP}$ in three dimensions with initial point $\bar{O} = [0, 0, 0]$ and terminal point $P = [x, y, z]$. Show in figure 53.3 are the geometric vectors $S = [2, 3, 3]$ and $T = [2, -1, 4]$.

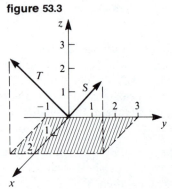

figure 53.3

For now, all our geometric vectors will have initial point at the origin.
We next introduce three algebraic operations on vectors.

definition 53.2   If $X = [x_1, x_2, \ldots, x_n]$ and $Y = [y_1, y_2, \ldots, y_n]$ are two n-dimensional vectors, then their sum is the n-dimensional vector

$$X + Y = [x_1 + y_1, x_2 + y_2, \ldots, x_n + y_n]$$

If a is a real number, the vector aX is defined to be

$$aX = [ax_1, ax_2, \ldots, ax_n]$$

This operation is called **scalar multiplication** of the vector X by the real number (*scalar*) a. Finally, by $X - Y$ we mean the vector $X + (-1)Y$. Hence

$$X - Y = [x_1 - y_1, x_2 - y_2, \ldots, x_n - y_n]$$

Thus, $[2, 3, 3] + [2, -1, 4] = [2 + 2, 3 - 1, 3 + 4] = [4, 2, 7]$ and $-2[2, -1, 4] = [-2(2), -2(-1), -2(4)] = [-4, 2, -8]$. Note, however, that $[0, 1, 2] + [3, -5]$ *is not defined. Both vectors must have the same dimension n in order to perform vector addition or subtraction.*

Vector addition and scalar multiplication obey the usual laws of algebra, as summarized in the accompanying box.

---

laws of vector algebra

Let X, Y, and Z be n-dimensional vectors, and a and b be real numbers. Then:

1. $X + \vec{0} = \vec{0} + X = X$
2. $X + (-X) = \vec{0}$
3. $X + (Y + Z) = (X + Y) + Z$
4. $X + Y = Y + X$
5. $0 \cdot X = \vec{0}$ and $1 \cdot X = X$
6. $(a + b)X = aX + bX$
7. $a(X + Y) = aX + aY$

These laws have simple demonstrations. To prove law **(7)**, for example, note that $a(X + Y) = a[x_1 + y_1, x_2 + y_2, \ldots, x_n + y_n] = [a(x_1 + y_1), \quad a(x_2 + y_2), \ldots, a(x_n + y_n)] = [ax_1 + ay_1, ax_2 + ay_2, \ldots, ax_n + ay_n]$, using the usual distributive law for real numbers. The last vector is just $aX + aY$.

The following two examples illustrate how vector addition and subtraction can be used to perform manipulations on population data.

example 53.5   The vector $F = [300, 200, 150, 80]$ gives the age structure of the females in a population, while $M = [310, 175, 160, 60]$ specifies the age distribution for males. The vector $X = F + M = [610, 375, 310, 140]$ then gives the age structure for the entire population. ▤

example 53.6   A population, divided into four age classes, is subject to harvesting. Let h_1, h_2, h_3, and h_4 denote the numbers harvested from each of the age groups. Then $H = [h_1, h_2, h_3, h_4]$ is called the **harvesting vector**. Last year's harvest was $H = [0, 5, 10, 20]$. If this year's harvest is $2H$ and $X = [450, 220, 150, 120]$, the age structure after last year's harvest, then the age structure after the harvest this year is given by

$$Y = X - 2H = X - [0, 10, 20, 40]$$
$$= [450, 210, 130, 80]$$

When $X = [x_1, y_1]$ and $Y = [x_2, y_2]$ are interpreted as geometric vectors, then $X + Y$ and $X - Y$ are as shown in figures 53.4(a) and (b), respectively.

figure 53.4(a)   **figure 53.3(b)**

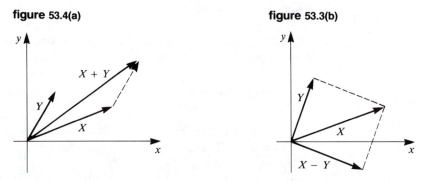

Vector $X + Y$ may be formed geometrically by what is commonly referred to as "tip to tail" addition. Translate the vector Y so that it remains parallel to its original position and so that its initial point is the terminal point of X. If P is the new position of the terminal point of Y, the diagram in figure 53.5 shows that P has coordinates $(x_1 + x_2, y_1 + y_2)$. Hence, $X + Y$ may be interpreted as the directed line segment $\overrightarrow{OP}$.

figure 53.5

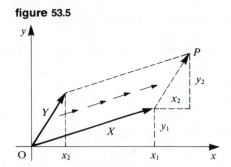

The vector W shown in figure 53.6 is $X - Y$. To see this, note that when W is added "tip to tail" to Y, the vector X results. Hence $W + Y = X$, or $W = X - Y$. Similar interpretations can be given for geometric vectors in three

figure 53.6

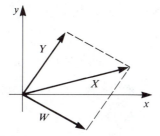

dimensions. All operations are performed in the plane determined by X and Y, as indicated in figure 53.7.

figure 53.7

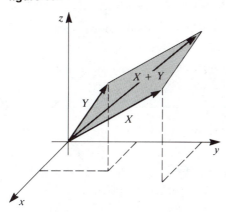

Given two n-dimensional vectors X and Y, we associate with them a special *real number* $X \cdot Y$ called the **dot product**.

definition 53.3   If $X = [x_1, x_2, \ldots, x_n]$ and $Y = [y_1, y_2, \ldots, y_n]$, then the **dot product** $X \cdot Y$ is defined to be the real number

$$x_1 y_1 + x_2 y_2 + \cdots + x_n y_n$$

Thus, if $X = [1, -2, 2]$ and $Y = [3, -4, 5]$, then $X \cdot Y = 1(3) + (-2)(-4) + 2(5) = 21$. An application of the dot product is given in the next example.

example 53.7   An animal population has age structure given by $X = \lfloor 200, 150, 120, 76 \rfloor$. The vector $W = [0, 50, 100, 150]$ specifies the commercial value (in dollars) of an animal in a particular age class. Thus, for example, each animal in the third age class is valued at \$100. The *total commercial value* of the population is given by the dot product

$$X \cdot W = 200(0) + 150(50) + 120(100) + 76(150)$$
$$= \$30{,}900$$

If $H = [0, 5, 60, 50]$ represents this year's harvest, then

$$H \cdot W = 0(0) + 5(50) + 60(100) + 50(150) = \$13{,}750$$

gives the *commercial value of the harvest*. ▤

The dot product has many of the same multiplication properties as real number multiplication. These properties are summarized in the box.

---

properties of the dot product

Let X, Y, and Z be n-dimensional vectors.

1. $X \cdot Y = Y \cdot X$

2. $(aX) \cdot (bY) = (ab)(X \cdot Y)$ for a and b real numbers

3. $X \cdot (Y + W) = X \cdot Y + X \cdot W$

To see that property (3) is valid, note that $Y + W = [y_1 + w_1, \ldots, y_n + w_n]$ and so

$$X \cdot (Y + W) = x_1(y_1 + w_1) + \cdots + x_n(y_n + w_n)$$
$$= (x_1y_1 + x_1w_1) + \cdots + (x_ny_n + x_nw_n)$$
$$= (x_1y_1 + \cdots + x_ny_n) + (x_1w_1 + \cdots + x_nw_n)$$
$$= X \cdot Y + X \cdot W$$

These properties allow us to expand algebraic expressions involving vectors in the usual manner. As an example,

$$(X + Y) \cdot (X + Y) = X \cdot (X + Y) + Y \cdot (X + Y), \qquad \text{using law (3)}$$
$$= X \cdot X + X \cdot Y + Y \cdot X + Y \cdot Y$$
$$= X \cdot X + 2(X \cdot Y) + Y \cdot Y, \qquad \text{using law (1)}$$

This resembles the usual real number expansion $(x + y)^2 = x^2 + 2xy + y^2$. Likewise, the result $(X + Y) \cdot (X - Y) = X \cdot X - Y \cdot Y$ reminds us of the formula for factoring the difference of two squares.

Note that $X \cdot X = x_1^2 + x_2^2 + \cdots + x_n^2$. When $n = 2$, $X \cdot X$ is just the square of the ordinary length of the vector $X = [x_1, x_2]$, as can be seen in figure 53.8. By the Pythagorean Theorem, the length of X is $\sqrt{|x_1|^2 + |x_2|^2} = \sqrt{x_1^2 + x_2^2} = \sqrt{X \cdot X}$.

figure 53.8

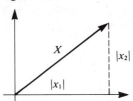

When $n = 3$, $\sqrt{X \cdot X} = \sqrt{x^2 + y^2 + z^2}$ again agrees with the length of X as given by the Pythagorean Theorem, and illustrated in figure 53.9. The length of $\overrightarrow{OP}$ is $\sqrt{d^2 + |z|^2} = \sqrt{x^2 + y^2 + z^2}$. This illustrates the general definition of the length of a vector.

figure 53.9

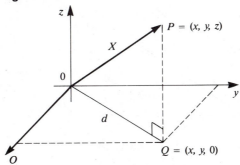

definition 53.4   We define the length $|X|$ of an n-dimensional vector X to be $\sqrt{X \cdot X}$.

example 53.8   Find the lengths of the vectors $X = [1, -1, 4]$ and $Y = [0, 1, -1, 3]$.

solution 53.8   For the vector X, $X \cdot X = 1 + (-1)^2 + 4^2 = 18$, and so $|X| = \sqrt{18} = 3\sqrt{2}$. Finally, $Y \cdot Y = 0 + 1 + (-1)^2 + 3^2 = 11$, and so $|Y| = \sqrt{11}$.   ▤

Note that $|aX|^2 = (aX) \cdot (aX) = a^2(X \cdot X) = a^2|X|^2$. Hence, $|aX| = |a|\,|X|$. Multiplying a vector X by a scalar changes its length by a factor of $|a|$. When $a > 0$, it can be shown that the direction of a geometric vector remains the same. Shown

figure 53.10

in figure 53.10 are the geometric vectors $\frac{1}{2}X$ and $3X$. The geometric vector $-X$ has the same length, but points in the direction opposite to X. Since $-2X = 2(-X)$, the direction of X is reversed and the length is increased by a factor of 2, as illustrated in figure 53.11.

figure 53.11

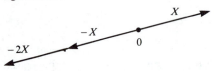

Given two geometric vectors X and Y, how can we decide when the two vectors are perpendicular? The angle AOB shown in figure 53.12 will be a right angle if and only if the Pythagorean Theorem holds. Since the length of

figure 53.12

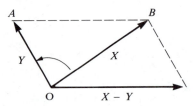

segment AB is $|X - Y|$, we must have

$$|X|^2 + |Y|^2 = |X - Y|^2$$

Writing the lengths in terms of the dot products, we see that vectors X and Y will be perpendicular if and only if

$$X \cdot X + Y \cdot Y = X \cdot X - 2(X \cdot Y) + Y \cdot Y$$

Thus, $-2(X \cdot Y) = 0$ or $X \cdot Y = 0$.

definition 53.5   Let X and Y be n-dimensional vectors. We say that X and Y are **orthogonal** (or perpendicular) in case $X \cdot Y = 0$.

The preceding argument then shows that X and Y are orthogonal precisely when the Pythagorean relationship

$$|X|^2 + |Y|^2 = |X - Y|^2$$

holds. Hence the dot product provides an *algebraic method* for verifying that two vectors are perpendicular. This is illustrated in our final three examples.

example 53.9   If $X = [-1, 2, 3, 4]$ and $Y = [0, -1, 4, k]$, find k so that X and Y are orthogonal.

solution 53.9   We first compute $X \cdot Y = 0 - 2 + 12 + 4k = 4k + 10$. Then X and Y will be orthogonal only if $X \cdot Y = 4k + 10 = 0$, that is, if $k = -5/2$.   ▤

example 53.10   Let $X = [-1, 1, 2]$ and $Y = [0, 1, 4]$ be two vectors in three-dimensional space. Find a vector Z with $|Z| = 1$ that is perpendicular to both X and Y.

solution 53.10   Let $Z = [a, b, c]$. Then Z must satisfy the three equations $Z \cdot X = 0$, $Z \cdot Y = 0$, and $|Z|^2 = 1$. Thus we must solve the system

$$-a + b + 2c = 0$$
$$b + 4c = 0$$
$$a^2 + b^2 + c^2 = 1$$

Hence $b = -4c$, $a = b + 2c = -4c + 2c = -2c$, and $1 = a^2 + b^2 + c^2 = (-2c)^2 + (-4c)^2 + c^2 = 21c^2$. One solution is $c = 1/\sqrt{21}$ and so $a = -2/\sqrt{21}$ and $b = -4/\sqrt{21}$. Then Z is $(1/\sqrt{21})[-2, -4, 1]$. A second solution is $-Z$.   ▤

example 53.11   A rhombus is a quadrilateral with four sides of equal length. Use the vector dot product to show that the diagonals of a rhombus are perpendicular.

solution 53.11   A rhombus is formed from two vectors X and Y with $|X| = |Y|$. One diagonal is $X + Y$, while the other diagonal is parallel to $X - Y$, as shown in figure 53.13. To show that these two vectors are perpendicular, we compute their dot product: $(X + Y) \cdot (X - Y) = X \cdot X - Y \cdot Y = |X|^2 - |Y|^2 = 0$, since $|X| = |Y|$.

figure 53.13

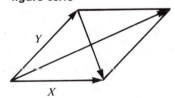

▤

exercises for section 53

part A   In exercises 1–10, let $X = [3, -1, 0]$, $Y = [-2, -1, 3]$, $Z = [0, -1, 4]$, $W = [0, 0, 0, 4]$, and $U = [-1, 2, 3, 0]$. Compute (if possible) each of the following vectors.

1. $X + Y$ 2. $X - Y$ 3. $W + X$

4. $W - U$ **5.** $W + U$ **6.** $2W - 2U$

7. $U + Y$ **8.** $2(Y - 2Z) + 4Z$ **9.** $3W + 3U$

10. $W - (2U - W)$

Given the geometric vectors in the accompanying diagram, use "tip to tail" addition and subtraction to find each of the following.

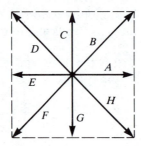

11. $A + C$ **12.** $A - C$ **13.** $F + B$

14. $D - H$ **15.** $E - C$ **16.** $D + B$

17. $B - C$ **18.** $A - B + E + F$ **19.** $2C - B$

20. $D - E + A$

Refer to the geometric vectors in the accompanying diagram.

  (a) Write the vectors in the form $[x, y]$.

  (b) Perform the computation.

  (c) Express your final answer in terms of one of the vectors in the diagram.

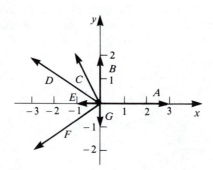

21. $E + B$ **22.** $A - B$ **23.** $D + F$

24. $A + 6G$ **25.** $F + C$

26. A cattle population is divided into six categories: male newborn, female newborn, male yearling, female yearling, adult male, and adult female. If x_i is the number in category i, assume that

$$X = [20, 24, 13, 11, 8, 52]$$

The vector $H = [0, 0, 4, 2, 6, 10]$ specifies the number and type that will soon be sold. Compute and interpret:

    (a) $X - H$ **(b)** $X - .5H$ **(c)** $1.1X$

27. The doctor on a cruise ship must arrange for the insulin needs of three passengers. Three types of insulin are needed. The vector $D_1 = [20, 10, 0]$ indicates that passenger 1 needs 20 units of the first type of insulin and 10 units of the second type each day. Vectors $D_2 = [0, 0, 30]$ and $D_3 = [30, 10, 15]$ specify the daily needs of the other two passengers. Passenger 1 will be on the ship 7 days, while passengers 2 and 3 will each be on board 10 days. Compute and interpret:

    (a) $D_1 + D_2 + D_3$ **(b)** $7D_1 + 10D_2 + 10D_3$

Let $X = [2, -1, 0]$, $Y = [-1, 2, 5]$, and $Z = [0, 3, 4]$, and compute each of the following.

28. $X \cdot Y$ **29.** $X \cdot Z$ **30.** $X \cdot (Y + Z)$

31. $|X|$ **32.** $|Y|$ **33.** $|X + Y|$

34. $(3X) \cdot (2Z)$ **35.** $Z \cdot (2X - 3Y)$

36. Refer to exercise 26. The vector $V = [25, 30, 85, 75, 300, 200]$ specifies the commercial value (in dollars) of an animal in each age category. Compute and interpret $X \cdot V$ and $H \cdot V$.

37. Refer to exercise 27. The costs per unit of the three types of insulin needed are given by $C = [.40, .30, .55]$. Compute and interpret $C \cdot D_1$, $C \cdot (10D_2)$, and $C \cdot (7D_1 + 10D_2 + 10D_3)$.

Show that the following pairs of vectors are orthogonal.

38. $[0, 1, -3, 2]$ and $[1, 0, 2, 3]$ **39.** $[-1, 1, 0, 3]$ and $[2, -1, 1, 1]$

40. Refer to the figure. Find a so that X and Y are perpendicular.   **41.** Refer to the figure. Find a so that X and Y are perpendicular.

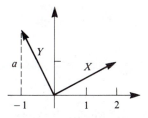

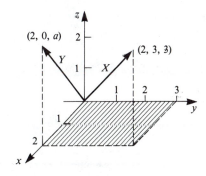

42. If $X = [0, -1, 2]$ and $Y = [1, -2, 3]$, find a vector Z with $|Z| = 1$ that is perpendicular to both X and Y.

43. Use the laws of dot products to verify that

$$(X - 2Y) \cdot (X - 2Y) = X \cdot X - 4(X \cdot Y) + 4(Y \cdot Y)$$

44. Use the laws of dot products to factor

$$|X|^2 - 6(X \cdot Y) + 8|Y|^2$$

45. If Z is perpendicular to both X and Y, show that Z is perpendicular to any vector of the form $aX + bY$.

part B

The box (or parallelepiped) shown in the figure is formed by three non-coplanar vectors X, Y, and Z.

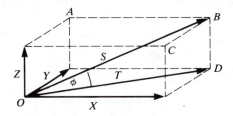

46. Write the vector T joining O to D in terms of X, Y, and Z.

47. Write the vector S joining O to B in terms of X, Y, and Z.

48. If $X \cdot Y = 0$, $X \cdot Z = 0$, and $Y \cdot Z = 0$, compute $|S|$ and $|T|$ in terms of $|X|$, $|Y|$, and $|Z|$.

49. If $X \cdot Y = 0$, $X \cdot Z = 0$, and $Y \cdot Z = 0$, compute $S \cdot T$ in terms of $|X|$, $|Y|$, and $|Z|$.

We will show in section 54 (page 608-609) that the angle ϕ between two vectors X and Y may be computed from the formula

$$\cos \phi = (X \cdot Y)/(|X||Y|)$$

50. Using this formula, compute the angle between S and T if $X \cdot Y = 0$, $Y \cdot Z = 0$, $X \cdot Z = 0$, $|X| = 1$, $|Y| = 1$, and $|Z| = 1$.

free vectors with illustrations from biophysics

In our development of vectors thus far, all geometric vectors have emanated from the origin. In applications to physics, it is convenient to allow vectors to start at different points in space, as depicted in figure 54.1. Such vectors are called **free vectors**.

figure 54.1

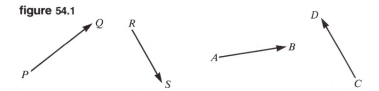

In physics, *vectors* are defined as "quantities with direction and magnitude". Common examples are velocities, accelerations, forces, and torques. In addition, two vectors are called *equal* if they have "the same magnitude and direction." Let's now make these ideas more precise and relate them to the definitions in section 53.

If P and Q are two points in the plane or in space, then $\overrightarrow{PQ}$ denotes the **directed line segment** or *free vector* from P to Q. Two free vectors $\overrightarrow{PQ}$ and $\overrightarrow{RS}$ are called **equivalent** if each of the following conditions holds (see figure 54.2):

1. The length of $\overrightarrow{PQ}$ is equal to the length of $\overrightarrow{RS}$. The length of $\overrightarrow{PQ}$ is denoted by $|\overrightarrow{PQ}|$ and is called the **magnitude of the free vector**.

2. *PRSQ* forms a parallelogram.

figure 54.2

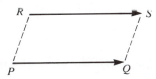

Thus, $\overrightarrow{PQ}$ and $\overrightarrow{RS}$ have the "same direction". In three-dimensional space, it is necessary that P, Q, R, and S be coplanar for a parallelogram to be formed. When conditions **(1)** and **(2)** are satisfied, we write $PQ \cong RS$. It can be shown that the symbol "$\cong$" obeys the same rules as ordinary equality of real numbers. We will use this fact in the examples that follow.

example 54.1   If $P = (1, 1)$, $Q = (3, 2)$, $R = (2, 0)$, and $S = (4, 1)$, show that vectors $\overrightarrow{PQ}$ and $\overrightarrow{RS}$ are equivalent.

solution 54.1   Refer to figure 54.3. The lengths of vectors $\overrightarrow{PQ}$ and $\overrightarrow{RS}$ are $|\overrightarrow{PQ}| = \sqrt{(3-1)^2 + (2-1)^2} = \sqrt{5}$ and $|\overrightarrow{RS}| = \sqrt{(4-2)^2 + (1-0)^2} = \sqrt{5}$, respectively. To show that $PRSQ$ is a parallelogram, note that the slope of PQ is $(2-1)/(3-1) = 1/2$, while the slope of RS is $(1-0)/(4-2) = 1/2$. Thus, PQ is parallel to RS.

figure 54.3

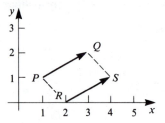

Every free vector in three-dimensional space is equivalent to a geometric vector $[a, b, c]$. To see this, refer to figure 54.4 and let $P = (x_1, y_1, z_1)$ and $Q = (x_2, y_2, z_2)$. If $X = [x_2, y_2, z_2]$ and $Y = [x_1, y_1, z_1]$, then free vector $PQ \cong X - Y = [x_2 - x_1, y_2 - y_1, z_2 - z_1]$. We call $[x_2 - x_1, y_2 - y_1, z_2 - z_1]$ the *standard*

figure 54.4

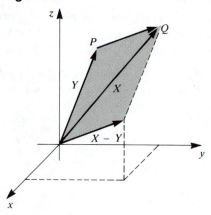

representation of free vector $\overrightarrow{PQ}$, and $x = x_2 - x_1$, $y = y_2 - y_1$, and $z = z_2 - z_1$ the **vector coordinates** of $\overrightarrow{PQ}$. The use of vector coordinates is illustrated in our next two examples.

example 54.2   Let $P = (4, -1, 2)$ and $Q = (1, 3, -4)$.

(a) Find the vector coordinates of $\overrightarrow{PQ}$.

___**(b)** Find a free vector with initial point $R = (2, 1, 3)$ that is equivalent to $\overrightarrow{PQ}$.

solution 54.2   The vector $\overrightarrow{PQ}$ is equivalent to $[1 - 4, 3 - (-1), -4 - 2] = [-3, 4, -6]$. If $\overrightarrow{RS}$ is equivalent to $\overrightarrow{PQ}$, then $\overrightarrow{RS} \cong [-3, 4, -6]$. Let $S = (x, y, z)$. Then $x - 2 = -3$, $y - 1 = 4$, and $z - 3 = -6$. Hence $x = -1$, $y = 5$, and $z = -3$. Then S is $(-1, 5, -3)$. ▤

Two free vectors $\overrightarrow{PQ}$ and $\overrightarrow{RS}$ can be added by either adding their standard representations or by adding geometrically, "tip to tail" as shown in example 54.3.

example 54.3   Consider the free vectors $\overrightarrow{PQ}$ and $\overrightarrow{RS}$ shown in figure 54.5.

figure 54.5

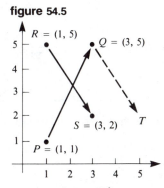

(a) Find $\overrightarrow{PQ} + \overrightarrow{RS}$ by adding their standard representations.
(b) Find $\overrightarrow{PQ} + \overrightarrow{RS}$ by adding geometrically.

solution 54.3   **(a)** We have $\overrightarrow{PQ} \cong [2, 4]$ and $\overrightarrow{RS} \cong [3 - 1, 2 - 5] = [2, -3]$. Thus, $\overrightarrow{PQ} + \overrightarrow{RS} \cong [2, 4] + [2, -3] = [4, 1]$.
(b) To add the vectors geometrically, select T so that $\overrightarrow{QT} \cong \overrightarrow{RS}$. The sum is then represented by the free vector $\overrightarrow{PT}$. Note that since $\overrightarrow{PT} \cong [4, 1]$, T has coordinates (5, 2). ▤

projections: resolving a vector into components   Given geometric vectors X and Y, we can project the vector Y onto the line ℓ determined by the vector X, as illustrated in figure 54.6. Vector Y can be written as $Y_p + Y_n$,

figure 54.6(a)   **figure 54.6(b)**

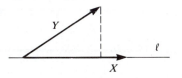   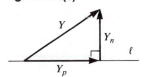

where Y_p is parallel to X and Y_n is perpendicular to X. We say that Y has been *resolved into components with respect to the vector X.* Y_n is called the *component of Y normal to the line ℓ.*

Explicit formulas for Y_p and Y_n can be found if we note that $Y_p = aX$ for some real number a, and that $Y_n = Y - Y_p = Y - aX$. Since $Y_n \cdot X = 0$, it follows that $(Y - aX) \cdot X = (Y \cdot X) - a|X|^2 = 0$. Solving for a, we have $a = (X \cdot Y)/(|X|^2)$. To summarize:

---

components of *Y* on *X*

The projection of vector Y onto the line ℓ determined by vector X is given by

$$Y_p = \left(\frac{X \cdot Y}{|X|^2}\right)X$$

The normal component of Y along ℓ is given by

$$Y_n = Y - Y_p$$

We illustrate these formulas in the next two examples.

example 54.4   Let $X = [1, -1, 2]$ and $Y = [3, 2, 4]$.
  (a) Find the projection Y_p of Y onto X.
  (b) Find the normal component Y_n of Y on X.

solution 54.4   We first compute $X \cdot Y = 3 - 2 + 8 = 9$, and $|X|^2 = 1^2 + (-1)^2 + 2^2 = 6$. Hence, $Y_p = \frac{9}{6}[1, -1, 2] = [\frac{3}{2}, -\frac{3}{2}, 3]$. The normal component Y_n is given by $Y - Y_p = [3, 2, 4] - [\frac{3}{2}, -\frac{3}{2}, 3] = [\frac{3}{2}, \frac{7}{2}, 1]$. Note that $X \cdot Y_n = \frac{3}{2} - \frac{7}{2} + 2 = 0$, as expected.

example 54.5   The vector $U = [\cos\theta, \sin\theta]$ determines the line ℓ through the origin that is shown in figure 54.7.

figure 54.7

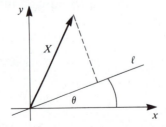

(a) Derive the formula for the projection of a vector $X = [x, y]$ onto ℓ.

(b) If $X = [1, 4]$ and $\theta = 45°$, find X_p and X_n.

solution 54.5

We apply the projection formula to obtain $X_p = (X \cdot U/|U|^2)U = (x \cos \theta + y \sin \theta)[\cos \theta, \sin \theta]$. Thus for $\theta = 45°$ and $X = [1, 4]$, $X_p = (1(\frac{1}{2}\sqrt{2}) + 4(\frac{1}{2}\sqrt{2}))[\frac{1}{2}\sqrt{2}, \frac{1}{2}\sqrt{2}] = [\frac{5}{2}, \frac{5}{2}]$. Finally, the normal component of X along U is given by $X - X_p = [1, 4] - [\frac{5}{2}, \frac{5}{2}] = [-\frac{3}{2}, \frac{3}{2}]$. ▤

When the acute angle θ between Y and the line determined by a vector X is given (as illustrated in figure 54.8), the lengths $|Y_p|$ and $|Y_n|$ of the vector

figure 54.8(a)   **figure 54.8(b)**

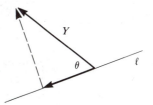

components can be computed in terms of $|Y|$ and θ. Since θ is the acute angle between Y and ℓ,

$$\sin \theta = |Y_n|/|Y| \text{ and } \cos \theta = |Y_p|/|Y|$$

Hence,

$$|Y_p| = |Y| \cos \theta \quad \text{and} \quad |Y_n| = |Y| \sin \theta$$

These two formulas for the lengths of the components are especially useful in physics. They are illustrated in our next two examples.

example 54.6

A sled is pulled by exerting a force F of 15 pounds in the direction shown in figure 54.9. The component of F in the direction of the motion has magnitude

$$|F_p| = |F| \cos 30° = 15 \cos 30° = 12.99 \text{ pounds}$$

figure 54.9

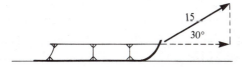

▤

example 54.7   Shown in figure 54.10 is the biceps muscle supporting the forearm and a weight W in the hand. The biceps muscle is attached to the forearm about 4

figure 54.10(a)   **figure 54.10(b)**

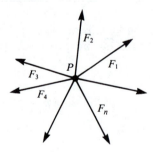

centimeters from the elbow joint and forms an angle ϕ of about $17°$ with the humerus bone of the upper arm. A well-developed biceps muscle can produce a force F of 500 pounds in the direction shown. The component of F normal to the forearm has magnitude $|F_n| = |F| \sin 73° = 500 \sin 73° \approx 478$ pounds. We will return to this example later in the section after we have discussed levers.

forces acting at a point   When forces $F_1, F_2, \ldots, F_n$ act at a single point P (see figure 54.11), then the *net force* is given by $F = F_1 + F_2 + \cdots + F_n$. We say that *the system is in equilibrium at* P in case the net force $F = \vec{0}$.

figure 54.11

Forces in equilibrium are illustrated in examples 54.8–54.10.

example 54.8   Suppose that $F_1 = [2, 3]$, $F_2 = [-1, 3]$, $F_3 = [-2, -1]$, and $F_4 = [1, -3]$, as illustrated in figure 54.12.

figure 54.12

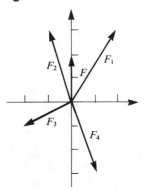

(a) Find the net force F.

(b) Find a fifth force F_5 so that the system is in equilibrium at $P = (0, 0)$.

solution 54.8   The net force $F = [2, 3] + [-1, 3] + [-2, -1] + [1, -3] = [0, 2]$. If $F_5 + (F_1 + F_2 + F_3 + F_4) = F_5 + F$ is to be $\vec{0}$, set $F_5 = [0, -2]$.

example 54.9   **forces acting on an incline**   A body of weight W pounds is placed on the incline shown in figure 54.13. If we resolve the downward force of W into components, we obtain

$$|F_p| = W \sin \theta \qquad \text{and} \qquad |F_n| = W \cos \theta$$

figure 54.13(a)   **figure 54.13(b)**

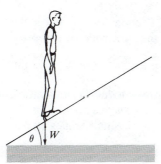   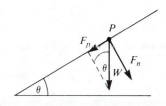

In order to maintain an equilibrium position, the sum of the forces acting at P must be $\vec{0}$. The incline itself must exert a force $X = -F_n$, while the frictional force Y must balance F_p (see figure 54.14). *A common assumption* (which is supported by experimental evidence) *is that the force of friction between the*

figure 54.14

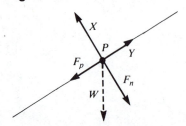

incline and the body is directly proportional to F_n. Thus $|Y| = \mu|F_n|$ for some constant μ, called the **coefficient of friction**.

Since $Y = -F_p$, it follows that $\mu|F_n| = |F_p|$ and hence $\mu W \cos \theta = W \sin \theta$. Therefore, $\mu = \tan \theta$. When $|F_p| = W \sin \theta > |Y| = \mu W \cos \theta$, that is, when $\tan \theta > \mu$, the body will slide down the incline.

example 54.10   When you walk normally, the heel of your shoe will strike the ground at an angle θ and with a downward force of F pounds, as illustrated in figure 54.15.

figure 54.15(a)

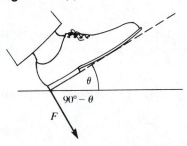

figure 54.15(b)

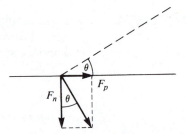

If μ is the coefficient of friction for the particular heel and ground surface, then, as in example 54.9, the heel will not slip provided that

$$\tan \theta \leq \mu \qquad \text{or} \qquad \mu \leq \tan^{-1} \theta$$

For a leather heel against a wooden floor, $\mu \approx .54$ and so θ must not exceed $28°$. For rubber against asphalt, $\mu \approx .60$ and so $\theta \leq 31°$. Finally, for leather soles on ice, $\mu \approx .1$ and $\theta \leq 6°$. Such small angles can be produced by taking short steps.

forces acting along a line: levers   The analysis of the mechanical struct- ure of human limbs can be based on a principle in physics known as the **law of levers**. Shown in figure 54.16 is a line segment resting on a **pivot point** or **fulcrum**. Forces of various magnitudes and directions act along the segment. In biophysical applications, the line segment is typically *bone*, the pivot a *joint*,

figure 54.16(a) Original Forces.   **figure 54.16(b)** Normal Components.

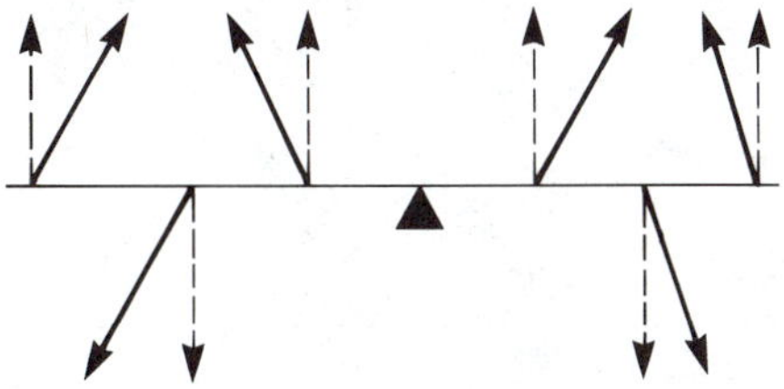

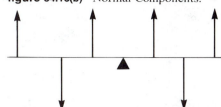

and the forces are produced by *muscles* acting in various directions. When does the lever system balance? The answer is given in terms of the *normal components of the forces and the distances from the pivot.*

We define the **torque** $T_i = x_iF_i$, where F_i is the normal component of a force acting at P_i, and x_i is the **directed distance** or *x-coordinate* with respect to the pivot. Then T_i is a free vector with direction normal to the lever. In figure 54.17, T_1 and T_2 have opposite directions since $T_1 = x_1F_1$ and $x_1 < 0$.

figure 54.17

The **net torque** on the lever system is given by

$$T = T_1 + T_2 + \cdots + T_n$$

The system balances in case the net torque $T = \vec{0}$. Note that we can ignore a force F_i acting at the pivot since $x_i = 0$. Pressure is being exerted on the pivot by the forces $F_1, F_2, \ldots, F_n$. For stability, it is necessary that the pivot itself exert a force P so that

$$P + F_1 + F_2 + \cdots + F_n = \vec{0}$$

In the following examples, we will use these conditions for equilibrium to solve various problems in physics and biophysics.

example 54.11   In order to balance the 500-pound rock shown in figure 54.18, where along the line should a downward force of 100 pounds be applied?

figure 54.18

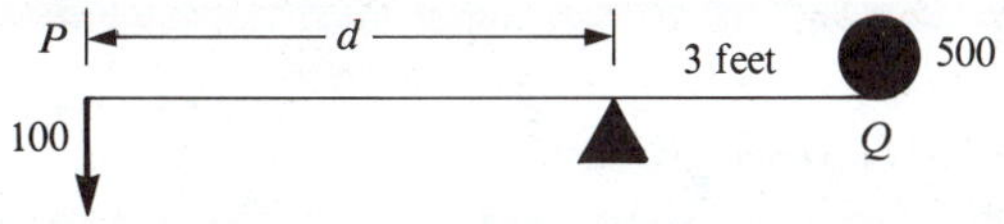

P |← ——— d ——— →| 3 feet   ● 500

100 | ▲ Q

solution 54.11   The point P has x-coordinate $-d$, while Q has x-coordinate 3. Hence, for equilibrium, $-d[0, -100] + 3[0, -500] = \vec{0}$. We must then have $100d - 1{,}500 = 0$ or $d = 15$ feet.

example 54.12   The forces shown in figure 54.19 act at points along the x-axis. The origin $(0, 0)$ is the pivot.

figure 54.19

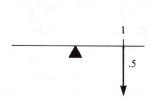

(a) Find the net torque on the lever.
(b) Determine a fourth torque T_4 so that the system balances.

solution 54.12   The normal component of the force acting at -2 is $[0, -1]$, while the force acting at 1 has normal component $[0, 1.5]$. The net torque is then

$$-2[0, -1] + 1[0, 1.5] + 3[0, -1] = [0, .5]$$

For the system to balance, T_4 must be $[0, -.5]$. This torque can be produced in many ways. Shown in figure 54.20 are two possible methods.

figure 54.20(a)   **figure 54.20(b)**

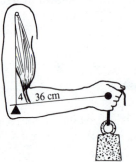

example 54.13   **biceps muscle**   We have mentioned that the biceps is capable of exerting a force of magnitude 500 pounds. However, the actual amount of weight W that the arm can support in the position shown in figure 54.21 is far less. The law of levers can be used to estimate W and the pressure on the elbow joint.

figure 54.21(a)   **figure 54.21(b)**

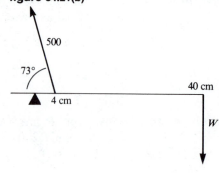

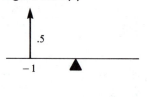

In our analysis, we will ignore the weight of the arm and the action of other smaller muscles. Typical lengths from the elbow joint to the point of attachment of the biceps on the forearm and to the weight in the hand are given in the figure. For equilibrium, we must have

$$4\,(500 \sin 73°) - 40\,W = 0$$

Hence, $W = 50 \sin 73° = 47.8$ pounds. In the position shown, the force exerted on the elbow joint is

$$[-500 \cos 73°,\ 500 \sin 73°] + [0, -47.8] \approx [-146, 430]$$

This force has a magnitude of about 454 pounds.

example 54.14   **spinal muscles**   The lever diagram in figure 54.22 represents the forces acting on the vertebral column when the body is bent over at an angle of $90° - \theta$. In the figure, W_1 is the weight of the trunk, W_2 is the weight of the arms and head, and F is the force produced by the *erector spinae muscles*. The

figure 54.22(a)   **figure 54.22(b)***

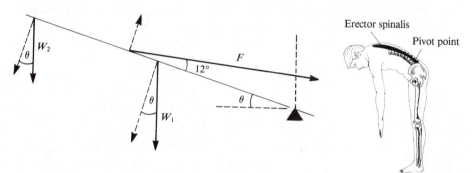

pivot is based at the *fifth lumbar vertebra*. If L denotes the length of the spinal column, then W_1 is concentrated at a point $L/2$ units from the pivot, while F operates $2L/3$ units away. To maintain this position it must be the case that

$$-\frac{2}{3}L(|F| \sin 12°) - \frac{1}{2}L(-W_1 \cos \theta) - L(-W_2 \cos \theta) = 0$$

Solving for $|F|$, we obtain

$$|F| = \frac{3}{2}(.5\,W_1 + W_2)\frac{\cos \theta}{\sin 12°}$$

* *Source*: Paul Davidovits, *Physics in biology and medicine*, Englewood Cliffs, NJ.: Prentice-Hall, (1975).

If W denotes the *total body weight*, then $W_1 \approx .4W$ and $W_2 \approx .2W$. The formula for $|F|$ may then be written

$$|F| \approx .6W \cos \theta \csc 12°$$

Thus, if $W = 200$ pounds and $\theta = 0°$, then $|F| \approx 120 \csc 12° \approx 577$ pounds. If the arms are supporting a 50-pound weight, then W_2 is now $.2(200) + 50 = 90$ pounds, and $|F| \approx 195 \csc 12° = 934$ pounds. In this case the force exerted on the fifth lumbar vertebra is given by

$$P = [934 \cos 12°, \ 934 \sin 12°] + [0, -80] + [0, -90]$$
$$= [913.6, 24.2]$$

The magnitude of P is 913.9 pounds. Thus, tremendous pressure is put on this vertebra.

exercises for section 54

Show that free vectors $\overrightarrow{PQ}$ and $\overrightarrow{RS}$ are equivalent by finding the vector coordinates of each vector.

1. $P = (2, 1)$, $Q = (0, 4)$, $R = (1, 2)$, $S = (-1, 5)$

2. $P = (-1, 1)$, $Q = (1, 3)$, $R = (1, 0)$, $S = (3, 2)$

3. $P = (0, 1, 2)$, $Q = (-1, 1, 4)$, $R = (1, 1, 3)$, $S = (0, 1, 5)$

4. $P = (0, 1, 0)$, $Q = (1, 0, 0)$, $R = (1, 0, 0)$, $S = (2, -1, 0)$

5. If $R = (0, -1, 4)$, find S so that $\overrightarrow{RS} \cong [1, -1, 2]$.

6. If $P = (3, -5, 6)$, find Q so that $\overrightarrow{PQ} \cong [0, 1, 1]$.

7. If $A = (2, 3)$, $B = (-1, 1)$, and $C = (-3, 2)$, find D so that $\overrightarrow{AB} \cong \overrightarrow{CD}$.

8. If $M = (-1, 0, 1)$, $N = (2, 1, 3)$, and $O = (8, -7, 4)$, find P so that $\overrightarrow{MN} \cong \overrightarrow{OP}$.

9. In exercise 7, find E so that $\overrightarrow{AE} \cong \overrightarrow{AB} + \overrightarrow{CD}$. Sketch all three vectors.

10. In exercise 8, find Q so that $\overrightarrow{MQ} \cong \overrightarrow{MN} + \overrightarrow{OP}$.

★11. Let $P = (4, -1, 2)$ and $Q = (2, 3, -1)$. Find $R = (0, -1, z)$ so that $\overrightarrow{PQ}$ and $\overrightarrow{PR}$ are perpendicular.

★12. If $A = (1, 2)$ and $B = (3, 4)$, find C so that $\overrightarrow{AB}$ and $\overrightarrow{AC}$ are perpendicular and $|\overrightarrow{AC}| = 3$.

13. Show that the forces $F_1 = [0, 2]$, $F_2 = [-4, 2]$, and $F_3 = [2, 3]$ are not in equilibrium. Find F_4 so that $F_1 + F_2 + F_3 + F_4 = \vec{0}$.

14. If $F_1 = [2, 1]$, $F_2 = [3, -2]$, $F_3 = [-4, 1]$, and $F_4 = [-1, 0]$, show that the forces are in equilibrium.

Given the vectors X and Y in each of the following situations, find Y_p and Y_n. Verify that $Y = Y_p + Y_n$ and $Y_p \cdot Y_n = 0$.

15. $X = [3, 4]$ and $Y = [1, -1]$

16. $X = [3, -1, 2]$ and $Y = [2, 3, 0]$

Given the vectors A and B in each of the following situations, find A_p and A_n. Verify that $A = A_p + A_n$ and $A_p \cdot A_n = 0$.

17. $A = [-2, 6]$ and $B = [0, 4]$

18. $A = [4, 1, -1]$ and $B = [0, 1, 2]$

19. An airplane wishes to fly from O to D (see the figure). Town D is $45°$ northeast of town O. The plane will fly into a 40 mph wind blowing from west to east.

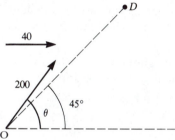

(a) If the plane flies at a speed of 200 mph, find the net velocity V in terms of θ.

(b) If $\theta = 60°$, find $|V|$ and the direction the plane will take.

★(c) Find θ so that the plane will fly directly toward D.

20. Where should a downward force of 50 pounds be applied so that the lever system shown in the figure will balance?

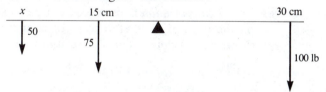

21. In what direction should the force of 40 pounds be applied so that the lever system shown in the figure will balance?

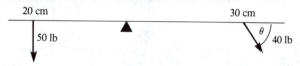

22. Find the largest weight that can be supported given that the force F of 100 pounds acts in the direction shown in the figure.

23. The *deltoid muscle* is the main muscle used in holding the arm in the horizontal position shown in the accompanying figure. Assume that the

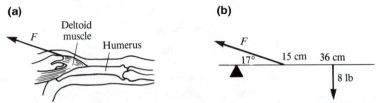

(a)

(b)

arm weighs 8 pounds and that the center of gravity of the arm is about 36 centimeters from the pivot (the shoulder joint). Assume further that the deltoid muscle is attached to the humerus bone about 15 centimeters from the pivot.

(a) Find the magnitude $|F|$ of the force exerted by the deltoid muscle in order to maintain this position.

(b) Find the net force on the joint.

24. Rework exercise 23, assuming a 5-pound weight is being held in the hand and the distance from the hand to the shoulder joint is 72 centimeters.

25. A 250-pound weightlifter bends over at a 45° angle to begin a lift. The weight he will attempt is 300 pounds.

(a) Find the force that must be produced by the erector spinae muscles to begin the lift.

(b) Find the force on the fifth lumbar vertebra.

[*Hint*: Use the equations developed in example 54.14.]

26. The *Achilles tendon* experiences a large tension when an individual stands on the balls of his feet. The pivot is at the *tibia*. Given the dimensions shown in the figure, find the relationship between $|F|$, the magnitude of

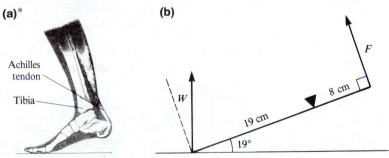

(a)*

(b)

* *Source*: Paul Davidovits, *Physics in biology and medicine*, Englewood Cliffs, NJ.: Prentice-Hall. (1975).

the force exerted by the Achilles tendon, and the weight W being supported.

27. A large sea bass exerts a force of 10 pounds on the 10-foot fishing rod shown in the figure. The fisherman exerts a force of 60 pounds in the

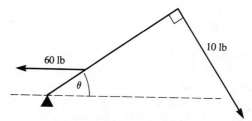

horizontal direction and places his hands 2 feet from the base of the rod, which pivots in a harness. Find θ so that the rod remains stationary.

 **matrices and matrix algebra**

The concept of a *matrix* is a nineteenth century invention due to the English mathematician Arthur Cayley. It represents the natural generalization of an n-dimensional vector $X = [x_1, x_2, \ldots, x_n]$ and has become a useful tool in many areas of science and technology. Matrices are commonly used to store and manipulate data in an orderly fashion, and we will emphasize this particular application in the introductory material of this section. Applications to modeling in the biosciences will be presented in Section 56.

definition 55.1 An $n \times m$ **matrix** A is a rectangular array of numbers arranged in n rows and m columns. We denote the element in the ith row and jth column by a_{ij} and write $A = [a_{ij}]$.

Shown below are matrices of various dimensions:

$$A = \begin{bmatrix} 4 & -2 & 3 \\ 0 & 1 & 6 \end{bmatrix} \qquad B = \begin{bmatrix} 1 & -4 \\ 2 & 3 \\ -7 & 0 \end{bmatrix}$$

$$C = [-3 \quad 5 \quad -9 \quad 6 \quad 2] \qquad D = \begin{bmatrix} 0 & 0 & 1 & -3 \\ 5 & -7 & 2 & 3 \\ -2 & -3 & 0 & 2 \end{bmatrix}$$

Matrix A is a 2×3 matrix, while matrix B has dimension 3×2. Note that a_{13} is the entry in the first row and third column of matrix A. Thus, $a_{13} = 3$. Matrix C is a 1×5 matrix, while D has dimension 3×4. You should verify the following entries in D:

$$d_{12} = 0 \qquad d_{21} = 5 \qquad d_{34} = 2 \qquad d_{22} = -7$$

Note that we may consider an n-dimensional vector $X = [x_1, \ldots, x_n]$ as a $1 \times n$ matrix. Alternatively we may write the vector X vertically and view it as an $n \times 1$ matrix. As was the case with vectors, *two matrices $A = [a_{ij}]$ and $B = [b_{ij}]$ are equal only when all their corresponding entries are equal*, that is, when $a_{ij} = b_{ij}$ for all i and j. Only matrices of the same dimension may be compared. *Thus, no two of the following matrices are equal.*

$$A = \begin{bmatrix} 1 & -1 \\ 2 & 0 \end{bmatrix} \qquad B = \begin{bmatrix} 1 & -1 & 0 \\ 2 & 0 & 1 \end{bmatrix} \qquad C = \begin{bmatrix} 1 & 0 \\ 2 & 0 \end{bmatrix}$$

As the following three examples demonstrate, matrices are especially helpful in storing and displaying data involving many variables.

example 55.1 A company has three warehouses, each of which stocks four types of items, A, B, C, and D. The inventory can be represented by a 4×3 matrix:

	Warehouse number		
Item	1	2	3
A	300	240	420
B	150	200	80
C	560	700	350
D	220	560	280

If we denote the matrix by S, then $s_{32} = 700$ indicates that there are 700 units of item C stored in warehouse 2.

example 55.2 Five different subjects are tested with each of three different drugs, D_1, D_2, and D_3, that are designed to lower blood pressure. If e_{ij} is the effect of drug D_i on individual j, then the results of the experiment can be summarized by a 3×5 matrix E, as shown:

$$E = \begin{bmatrix} -10 & -15 & -13 & -18 & -20 \\ -15 & -16 & -21 & -23 & -10 \\ -7 & -10 & -12 & -17 & -8 \end{bmatrix}$$

The entry $e_{24} = -23$ indicates that a decrease in blood pressure of 23 mm Hg was obtained using drug D_2 on patient 4.

example 55.3 **a competition matrix** Three species compete with one another for the same food resources in an ecosystem. An animal wins the competition for a food item when it attacks and drives off another animal seeking the same item. Let c_{ij} be the percentage of such encounters that an animal of species i wins when in competition with an animal of species j. The matrix $C = [c_{ij}]$ is called a **competition matrix**. If C is given by

$$C = \begin{bmatrix} .5 & .3 & .6 \\ .7 & .5 & .2 \\ .4 & .8 & .5 \end{bmatrix}$$

then $c_{23} = .2$ indicates that species 2 triumphs in only 20% of the encounters with species 3. Can you explain why $c_{ii} = .5$?

When two matrices A and B are of the **same dimension**, they can be added, subtracted, and multiplied by a scalar in much the same manner as n-dimensional vectors.

definition 55.2 Let A and B be $n \times m$ matrices. Then $A + B$ is the $n \times m$ matrix with (i, j)th entry $a_{ij} + b_{ij}$. Likewise, $A - B$ has (i, j)th entry $a_{ij} - b_{ij}$. For any real number c, cA is the matrix $[ca_{ij}]$.

These matrix operations are illustrated in the next three examples.

example 55.4 Consider the matrices A, B, and C:

$$A = \begin{bmatrix} 4 & -2 & 1 \\ 2 & 10 & 4 \end{bmatrix} \qquad B = \begin{bmatrix} 2 & 3 & 0 \\ -1 & 4 & -1 \end{bmatrix} \qquad C = \begin{bmatrix} 4 & -6 \\ 2 & -8 \end{bmatrix}$$

Compute (if possible):

(a) $A + B$ (b) $A - 2B$ (c) $A - C$ (d) $.5C$

solution 55.4 To compute $A + B$, we add corresponding entries of the two matrices:

$$A + B = \begin{bmatrix} 4+2 & -2+3 & 1+0 \\ 2-1 & 10+4 & 4-1 \end{bmatrix} = \begin{bmatrix} 6 & 1 & 1 \\ 1 & 14 & 3 \end{bmatrix}$$

The matrix $2B$ is formed by doubling each entry in B. Hence,

$$A - 2B = \begin{bmatrix} 4-2(2) & -2-2(3) & 1-2(0) \\ 2-2(-1) & 10-2(4) & 4-2(-1) \end{bmatrix} = \begin{bmatrix} 0 & -8 & 1 \\ 4 & 2 & 6 \end{bmatrix}$$

Note that $A - C$ cannot be computed since A and C have different dimensions. Finally,

$$.5C = \begin{bmatrix} .5(4) & .5(-6) \\ .5(2) & .5(-8) \end{bmatrix} = \begin{bmatrix} 2 & -3 \\ 1 & -4 \end{bmatrix}$$

example 55.5 Let S be the inventory matrix of example 55.1. The following matrix T keeps track of the number of items of each type shipped from the various warehouses on a given day:

	1	2	3
A	24	51	34
B	18	24	56
C	28	13	32
D	50	35	100

Item   Warehouse number

Then the *next day's inventory* is given by

$$S - T = \begin{bmatrix} 300-24 & 240-51 & 420-34 \\ 150-18 & 200-24 & 80-56 \\ 560-28 & 700-13 & 350-32 \\ 220-50 & 560-35 & 280-100 \end{bmatrix} = \begin{bmatrix} 276 & 189 & 386 \\ 132 & 176 & 24 \\ 532 & 687 & 318 \\ 170 & 525 & 180 \end{bmatrix}$$

example 55.6   In a study of the age structure and racial makeup in a school district, the population is divided into three age groups: (1) preschool age, (2) elementary school age, and (3) high school age. In addition, the children are categorized as (1) Caucasian, (2) Black, or (3) Hispanic. Let a_{ij} be the number in age category i and racial classification j. The matrices A and B summarize the data obtained for two neighboring school districts:

$$\text{District 1} \qquad\qquad \text{District 2}$$

$$A = \begin{bmatrix} 300 & 500 & 450 \\ 280 & 420 & 360 \\ 200 & 550 & 600 \end{bmatrix} \qquad B = \begin{bmatrix} 200 & 100 & 100 \\ 400 & 600 & 200 \\ 300 & 200 & 400 \end{bmatrix}$$

Note that the first column of each matrix gives the data for the Caucasian population, while the second row gives the data on elementary schools. The *pooled data* for the two school districts are given by matrix $A + B$. If 52% of each age–race category in district 2 is female, then $.52B$ presents the *data for females* in district 2. You can verify that

$$A + B = \begin{bmatrix} 500 & 600 & 550 \\ 680 & 1{,}020 & 560 \\ 500 & 750 & 1{,}000 \end{bmatrix} \quad \text{and} \quad .52B = \begin{bmatrix} 104 & 52 & 52 \\ 208 & 312 & 104 \\ 156 & 104 & 208 \end{bmatrix}$$

A matrix with all its entries equal to 0 is called a **zero matrix** and will be denoted by **0**. Note that $A + \mathbf{0} = A$ for any matrix A that has the dimension as **0**. In fact, matrix addition and scalar multiplication obey the usual laws of algebra, as summarized in the box.

---

laws of matrix algebra

Let A, B, and C be $n \times m$ matrices, and a and b real numbers. Then:

1. $A + \mathbf{0} = \mathbf{0} + A = A$
2. $A + (-A) = (-A) + A = \mathbf{0}$
3. $A + (B + C) = (A + B) + C$
4. $A + B = B + A$
5. $(a + b)A = aA + bA$
6. $a(A + B) = aA + aB$
7. $0A = \mathbf{0}$ and $1A = A$

The derivations of these laws hinge on the fact that the corresponding laws for real numbers hold for the entries in the (i, j)th position.

How are two matrices to be multiplied? The definition we are about to give is a generalization of the vector dot product of Section 53. Writing the second vector Y as a column vector,

$$[x_1 x_2 \ldots x_n]\begin{bmatrix} y_1 \\ y_2 \\ \vdots \\ y_n \end{bmatrix} = [x_1 y_1 + x_2 y_2 + \cdots + x_n y_n]$$

As a first example, let A and B be the two matrices shown. Partition A into rows R_1 and R_2, and B into columns C_1, C_2, and C_3:

$$A = \begin{bmatrix} 2 & 1 \\ \hline -1 & 0 \end{bmatrix}\begin{matrix} R_1 \\ R_2 \end{matrix} \qquad B = \begin{matrix} C_1 & C_2 & C_3 \\ \begin{bmatrix} 1 & 2 & 4 \\ 0 & -3 & 2 \end{bmatrix} \end{matrix}$$

The rows of A and the columns of B may now be considered as 2-dimensional vectors. Hence, all of the dot products $R_i \cdot C_j$ are defined. The **product matrix** AB is the matrix

$$\begin{bmatrix} R_1 \cdot C_1 & R_1 \cdot C_2 & R_1 \cdot C_3 \\ R_2 \cdot C_1 & R_2 \cdot C_2 & R_2 \cdot C_3 \end{bmatrix} = \begin{bmatrix} 2(1)+1(0) & 2(2)+1(-3) & 2(4)+1(2) \\ -1(1)+0(0) & -1(2)+0(-3) & -1(4)+0(2) \end{bmatrix}$$

$$= \begin{bmatrix} 2 & 1 & 10 \\ -1 & -2 & -4 \end{bmatrix}$$

In attempting to form the product BA, we partition B into rows and A into columns:

$$B = \begin{bmatrix} 1 & 2 & 4 \\ \hline 0 & -3 & 2 \end{bmatrix}\begin{matrix} R_1 \\ R_2 \end{matrix} \qquad A = \begin{matrix} C_1 & C_2 \\ \begin{bmatrix} 2 & 1 \\ -1 & 0 \end{bmatrix} \end{matrix}$$

But now it is not possible to form $R_1 \cdot C_1$, for instance, because R_1 is a 3-dimensional vector, while C_1 is 2-dimensional. *The matrix product BA is not defined.* This illustrates that in order to form AB, *the number of columns in A must equal the number of rows in B.*

definition 55.3   Let A be an $n \times r$ matrix, and B an $r \times m$ matrix. Let $R_1, R_2, \ldots, R_n$ denote the rows of A and $C_1, C_2, \ldots, C_m$ the columns of B. Then the **product matrix** AB is the $n \times m$ matrix whose (i, j)th entry is given by the dot product $R_i \cdot C_j$.

The following step-by-step procedure may prove helpful in computing matrix products:

Step 1. Start with the first row R_1 of A and compute $R_1 \cdot C_1$, $R_1 \cdot C_2, \ldots, R_1 \cdot C_m$. This gives the *first row* of AB.

Step 2. Take the second row R_2 of A and form $R_2 \cdot C_1$, $R_2 \cdot C_2$, ..., $R_2 \cdot C_m$. This gives the *second row* of AB.

Step n. The *final row* of AB is formed from $R_n \cdot C_1$, $R_n \cdot C_2$, ..., $R_n \cdot C_m$.

example 55.7

Compute (if possible) AB, BA, CA, and AC, where the matrices A, B, and C are as shown:

$$A = \begin{bmatrix} -3 & 0 \\ 2 & 1 \end{bmatrix} \qquad B = \begin{bmatrix} 3 & -1 \\ 4 & 2 \end{bmatrix} \qquad C = \begin{bmatrix} 2 & 0 \\ 0 & -1 \\ 1 & 5 \end{bmatrix}$$

solution 55.7

To compute AB, we start with $R_1 = [-3 \;\; 0]$ and form dot products with the columns of B:

$$\begin{bmatrix} -3 & 0 \\ 2 & 1 \end{bmatrix} \begin{bmatrix} 3 & -1 \\ 4 & 2 \end{bmatrix} = \begin{bmatrix} -9 & 3 \\ - & - \end{bmatrix}$$

For the second row of AB, we take $R_2 = [2 \;\; 1]$ and repeat the process:

$$\begin{bmatrix} -3 & 0 \\ 2 & 1 \end{bmatrix} \begin{bmatrix} 3 & -1 \\ 4 & 2 \end{bmatrix} = \begin{bmatrix} -9 & 3 \\ 10 & 0 \end{bmatrix}$$

Likewise, BA and CA are given by

$$BA = \begin{bmatrix} 3 & -1 \\ 4 & 2 \end{bmatrix} \begin{bmatrix} -3 & 0 \\ 2 & 1 \end{bmatrix} = \begin{bmatrix} -11 & -1 \\ -8 & 2 \end{bmatrix}$$

and

$$CA = \begin{bmatrix} 2 & 0 \\ 0 & -1 \\ 1 & 5 \end{bmatrix} \begin{bmatrix} -3 & 0 \\ 2 & 1 \end{bmatrix} = \begin{bmatrix} -6 & 0 \\ -2 & -1 \\ 7 & 5 \end{bmatrix}$$

The product AC is *not defined* since the matrix A has two columns, and the matrix C has three rows.   ▤

Note that $AB \neq BA$ for the matrices A and B in example 55.7. *The commutative law of multiplication does not hold for matrices.* There is, however, one important class of matrices that commute with any other matrix. We call the matrix D a **diagonal matrix** in case $D = [d_{ij}]$ with $d_{ij} = 0$ for $i \neq j$. Thus, the only possible nonzero entries in D are the **diagonal entries** $d_{11}, d_{22}, \ldots, d_{nn}$. Shown below are diagonal matrices of various dimensions:

$$\begin{bmatrix} 4 & 0 \\ 0 & -2 \end{bmatrix} \qquad \begin{bmatrix} 1 & 0 & 0 \\ 0 & 2 & 0 \\ 0 & 0 & 3 \end{bmatrix} \qquad \begin{bmatrix} 1 & 0 & 0 & 0 \\ 0 & 0 & 0 & 0 \\ 0 & 0 & 2 & 0 \\ 0 & 0 & 0 & 3 \end{bmatrix}$$

To determine when $DA = AD$ for a diagonal matrix D and arbitrary A, let's examine the 2×2 case in detail:

$$DA = \begin{bmatrix} d_{11} & 0 \\ 0 & d_{22} \end{bmatrix} \begin{bmatrix} a & b \\ c & d \end{bmatrix} = \begin{bmatrix} d_{11}a & d_{11}b \\ d_{22}c & d_{22}d \end{bmatrix}$$

and

$$AD = \begin{bmatrix} a & b \\ c & d \end{bmatrix} \begin{bmatrix} d_{11} & 0 \\ 0 & d_{22} \end{bmatrix} = \begin{bmatrix} d_{11}a & d_{22}b \\ d_{11}c & d_{22}d \end{bmatrix}$$

Thus, d_{11} must equal d_{22} in order to guarantee $DA = AD$. In particular, if we set all diagonal entries equal to 1, we obtain a matrix I for which $IA = AI = A$. The matrix I is called an **identity matrix**. Shown below are identity matrices of various dimensions:

$$\begin{bmatrix} 1 & 0 \\ 0 & 1 \end{bmatrix} \qquad \begin{bmatrix} 1 & 0 & 0 \\ 0 & 1 & 0 \\ 0 & 0 & 1 \end{bmatrix} \qquad \begin{bmatrix} 1 & 0 & 0 & 0 \\ 0 & 1 & 0 & 0 \\ 0 & 0 & 1 & 0 \\ 0 & 0 & 0 & 1 \end{bmatrix}$$

Except for the commutative law of multiplication, the other common laws of multiplication are valid for matrices, provided *all matrix multiplications and additions are defined*. In particular, if all matrices are $n \times n$ matrices, then the laws summarized in the box are valid.

---

laws of matrix algebra

Let A, B, and C be matrices, and a and b real numbers. Then, *provided all operations are defined*,

 8. $A(BC) = (AB)C$

 9. $A(B + C) = AB + AC$

 10. $(B + C)A = BA + CA$

 11. If A is an $n \times n$ matrix and I is the $n \times n$ identity matrix, then

$$AI = IA = A$$

 12. $(aA)(bB) = (ab)AB$

The following diagram shows an easy way of testing whether matrix multiplication is defined:

$$\underbrace{\underset{n \times k}{A} \cdot \underset{p \times m}{B}}_{\substack{\text{these must be} \\ \text{equal}}} = \underset{\substack{\text{dimension of } C \text{ when} \\ k = p}}{\underset{n \times m}{C}}$$

Thus, if A is a 3×4 matrix, B is a 4×2 matrix, and C is a 2×5 matrix, then AB has dimension 3×2, and $(AB)C$ has dimension 3×5. Note, however, that $C(AB)$, CA, and CB are not defined.

Note that two distributive laws **(9)** and **(10)** are needed since we cannot guarantee that multiplication is commutative. Illustrations of the matrix algebra laws are presented in the final examples.

example 55.8

Expand each of the following matrix expressions.

$$\textbf{(a)} \ (A+B)^2 \qquad \textbf{(b)} \ (A+B)(A-B)$$

solution 55.8

Since $(A+B)^2 = (A+B)(A+B)$, we may apply distributive law **(9)** to obtain $(A+B)^2 = (A+B)A + (A+B)B$. Finally, using law **(10)**, we have $(A+B)A + (A+B)B = A^2 + BA + AB + B^2$. Note that since we cannot assume that A and B commute, this is the final form for the expansion.

Likewise for **(b)** $(A+B)(A-B) = (A+B)A + (A+B)(-B) = A^2 + BA + A(-B) + B(-B) = A^2 + BA - AB - B^2$, using **(9)**, **(10)**, and **(12)**.   ▤

example 55.9

Factor each of the following matrix expressions.

$$\textbf{(a)} \ XC - X \qquad \textbf{(b)} \ A^2B - AB^2$$

Assume all matrices are of dimension $n \times n$.

solution 55.9

For **(a)**, $XC - X = XC - XI$, where I is the appropriate identity matrix. Finally, using law **(9)**, $XC - XI = X(C - I)$.

For **(b)**, first note that $A^2B - AB^2 = (AA)B - AB^2 = A(AB) - AB^2 = A(AB - B^2)$. Note that we have used both laws **(8)** and **(9)** in the manipulations. Finally,

$$A(AB - B^2) = A(AB - BB) = A(A - B)B \qquad ▤$$

example 55.10

Use law **(12)** as an aid in computing AB, where

$$A = \begin{bmatrix} .01 & -.2 \\ .24 & .05 \end{bmatrix} \quad \text{and} \quad B = \begin{bmatrix} 1/2 & 1/4 \\ 3/2 & -1/3 \end{bmatrix}.$$

solution 55.10

We can avoid working with decimals and fractions by noting that

$$AB = \left(.01 \begin{bmatrix} 1 & -20 \\ 24 & 5 \end{bmatrix}\right)\left(\frac{1}{12} \begin{bmatrix} 6 & 3 \\ 18 & -4 \end{bmatrix}\right) = \frac{1}{1,200} \begin{bmatrix} -354 & 83 \\ 234 & 52 \end{bmatrix} \qquad ▤$$

·exercises for section 55

Find the values of x and y that make each of the following pairs of matrices equal.

1. $\begin{bmatrix} 3 & -1 \\ 2x & y \end{bmatrix}$ and $\begin{bmatrix} 3 & -1 \\ 4 & 1-y \end{bmatrix}$ **2.** $\begin{bmatrix} 3 \\ 2x \\ -1 \end{bmatrix}$ and $\begin{bmatrix} 3 \\ -4 \\ 0 \end{bmatrix}$

3. $\begin{bmatrix} 2 & x-3 & -5 \\ 0 & 1 & 3-y \end{bmatrix}$ and $\begin{bmatrix} 2 & 1 & -5 \\ 0 & 1 & 4 \end{bmatrix}$

4. $\begin{bmatrix} -2 & 3 & x \\ -1 & 2y-1 & 3 \end{bmatrix}$ and $\begin{bmatrix} -2 & 3 & 2x \\ -1 & \frac{1}{2} & 3 \end{bmatrix}$

5.   Given the matrix

$$A = \begin{bmatrix} 4 & -3 & 2 & 0 & 1 \\ 5 & -2 & 0 & 6 & 4 \\ \frac{1}{2} & 3 & 7 & -\frac{5}{2} & 0 \\ 0 & 4 & -5 & 6 & -3 \end{bmatrix}$$

identify a_{31}, a_{13}, a_{45}, a_{34}, and a_{25}.

Use the following matrices and the laws of matrix algebra to perform each of the following computations (if possible).

$$A = \begin{bmatrix} 2 & -1 \\ 0 & 3 \end{bmatrix} \qquad B = \begin{bmatrix} 4 & 0 & 2 \\ -3 & 1 & -2 \end{bmatrix} \qquad C = \begin{bmatrix} 0 & -1 & 1 \\ 1 & 2 & 0 \end{bmatrix}$$

$$D = \begin{bmatrix} 4 & 2 \\ -1 & 3 \\ 0 & 6 \end{bmatrix} \qquad E = \begin{bmatrix} 3 & 0 & -1 \end{bmatrix} \qquad F = \begin{bmatrix} 4 & -1 & 6 \\ 2 & 0 & 3 \\ -1 & 0 & 0 \end{bmatrix}$$

6. $.2A$ **7.** CD **8.** DC

9. $B+C$ **10.** $AB+AC$ **11.** DA

12. AB **13.** $AB-D$ **14.** DF

15. $DF+3B$ **16.** $(AC)F$ **17.** $A(CF)$

18. ED **19.** $2A-BD$ **20.** FE

21. If $A = \begin{bmatrix} 2 & 1 \\ 1 & 1 \end{bmatrix}$ and $B = \begin{bmatrix} 1 & -1 \\ -1 & 2 \end{bmatrix}$, verify that $AB = BA = I$, where I is the 2×2 identity matrix. The matrix B is the **multiplicative inverse** of A.

22. If $A = \begin{bmatrix} 1 & 0 \\ 1 & 0 \end{bmatrix}$ and $B = \begin{bmatrix} 0 & 0 \\ 0 & 1 \end{bmatrix}$, verify that $AB = 0$. The matrices A and B are called **zero divisors**.

23. If $A = \begin{bmatrix} 1 & 1 \\ 0 & 1 \end{bmatrix}$, find all matrices $B = \begin{bmatrix} a & b \\ c & d \end{bmatrix}$ for which $AB = BA$.

24. If $A = \begin{bmatrix} 1 & 1 \\ 1 & 1 \end{bmatrix}$, find a general formula for A^n.

25. If $A = \begin{bmatrix} a & b \\ c & d \end{bmatrix}$ and $ad - bc \neq 0$, verify that

$$B = \frac{1}{ad - bc} \begin{bmatrix} d & -b \\ -c & a \end{bmatrix}$$

satisfies $AB = BA = I$, where I is the 2×2 identity matrix.

26. Let S be the inventory matrix in example 55.1 and assume that the matrix $W = \begin{bmatrix} 30 & 20 & 25 & 50 \end{bmatrix}$ specifies the selling price for each of the items A, B, C, and D sold by the company. Compute WS and give its interpretation.

27. For the blood pressure experiment described in example 55.2, the reductions in blood pressure could be in error by as much as 5%. Compute $E \pm .05E$.

28. The matrices A and B in example 55.6 conveniently summarized the racial makeup of two neighboring school districts. Define

$$X = \begin{bmatrix} 1 & 1 & 1 \end{bmatrix} \quad \text{and} \quad Y = \begin{bmatrix} 1 \\ 1 \\ 1 \end{bmatrix}$$

Compute and interpret XA, XB, $X(A + B)$, AY, BY, and $(A + B)Y$.

29. Show that if $AB = BA$, then

    (a) $(A + B)A = A(A + B)$ **(b)** $A^2B = BA^2$

Factor each of the following matrix expressions.

30. $AX - 2X$ **31.** $A + BA$ **32.** $A^2 + AB$

33. $AB - CB$ **34.** $A^2 + 2A + I$, where $IA = AI = A$

35. Let X and Y be $n \times 1$ column vectors and A be an $n \times n$ matrix.

    (a) If $AX = X$ and $AY = Y$, show that $A(X + Y) = X + Y$.

    (b) If $AX = C$ and $AY = C$, show that $A(X - Y) = 0$.

 **applications of matrices**

In this section we present two important applications of matrices to population biology and ecology. In each case we will use the fact that a system of equations

$$y_1 = a_{11}x_1 + a_{12}x_2 + \cdots + a_{1n}x_n$$
$$x_2 = a_{21}x_1 + a_{22}x_2 + \cdots + a_{2n}x_n$$
$$\vdots$$
$$y_n = a_{n1}x_1 + a_{n2}x_2 + \cdots + a_{nn}x_n$$

can be rewritten in the matrix form $Y = AX$, where

$$Y = \begin{bmatrix} y_1 \\ y_2 \\ \vdots \\ y_n \end{bmatrix}, \quad A = \begin{bmatrix} a_{11} & a_{12} \cdots a_{1n} \\ a_{21} & a_{22} \cdots a_{2n} \\ \vdots & \vdots & \vdots \\ a_{n1} & a_{n2} \cdots a_{nn} \end{bmatrix}, \quad \text{and} \quad X = \begin{bmatrix} x_1 \\ x_2 \\ \vdots \\ x_n \end{bmatrix}$$

Instead of manipulating the cumbersome system involving variables x_1, $x_2, \ldots, x_n$ and $y_1, y_2, \ldots, y_n$, we will use the laws of matrix algebra on $Y = AX$ to perform the calculations, as illustrated in example 56.1.

example 56.1 Write each of the following systems of equations in matrix form.

$$\textbf{(a)} \ z_1 = 3y_1 - y_2 \qquad \textbf{(b)} \ y_1 = x_1 - x_2$$
$$z_2 = -2y_1 + 4y_2 \qquad \qquad y_2 = x_1 - 2x_2$$

Then use the laws of matrix algebra to express z_1 and z_2 in terms of x_1 and x_2.

solution 56.1 Let $X = \begin{bmatrix} x_1 \\ x_2 \end{bmatrix}$, $Y = \begin{bmatrix} y_1 \\ y_2 \end{bmatrix}$, and $Z = \begin{bmatrix} z_1 \\ z_2 \end{bmatrix}$. Then the systems in **(a)** and **(b)** can be written as $Z = AY$ and $Y = BX$, where

$$A = \begin{bmatrix} 3 & -1 \\ -2 & 4 \end{bmatrix} \qquad \text{and} \qquad B = \begin{bmatrix} 1 & -1 \\ 1 & -2 \end{bmatrix}$$

Hence, $Z = AY = A(BX) = (AB)X$, and $AB = \begin{bmatrix} 2 & -1 \\ 2 & -6 \end{bmatrix}$. It follows that $z_1 = 2x_1 - x_2$ and $z_2 = 2x_1 - 6x_2$.

the Leslie matrix model   In the models for the growth of populations introduced thus far, we have kept track only of the *total number* $N(t)$ in the population. In human demography, for example, it is essential to follow also the numbers in various *age categories*. We may want to predict the number of college age students or the number at retirement age several years in the future. For animal populations subject to harvesting, the harvesting of too many mature animals can harm reproduction and affect future harvests. It is therefore important to keep track of age classes.

In 1945, P.H. Leslie[*] introduced a matrix algebra model that does take into account the *age structure* of the population. Earlier versions of the model were presented by E. Lewis (1941)[†] and H. Bernadelli (1942),[‡] but it was Leslie who developed the model in detail and popularized its use. In the version of the model we will discuss, *only females are considered*.

Assume that the population is divided into n age categories, each of length k years. By the **age** of an animal, we mean *age on its last birthday*. Thus, the age categories are

$$C_1 = [0, k), C_2 = [k, 2k), \ldots, C_n = [(n-1)k, nk)$$

Let x_i be the number in the population who fall in age category C_i, and form the vector $X = [x_1, x_2, \ldots, x_n]$. The vector X specifies the age structure of the population. For human populations, k is usually 5 and $n = 18$. Thus, the age groups are 0–4 years, 5–9 years, . . . , 85–89 years. For large whales, k is usually taken to be 2.

If Y specifies the age structure k years later, the purpose of the model is to relate Y to X. We will show that[**]

$$Y = AX$$

where A is an $n \times n$ matrix whose entries are given in terms of *survival* and *fecundity rates*. Define

S_1 = percentage of individuals in age category C_1 that survive to age class
   C_2, k years later.

---

[*] P.H. Leslie, "On the Use of Matrices in Certain Population Mathematics", *Biometrika*, **33**: 183–212, (1945).

[†] E.G. Lewis, "On the Generation and Growth of a Population," *Sankya*, **6**: 93–96, (1943).

[‡] H. Bernadelli, "Population Waves," *J. Burma Res. Soc.*, **31**: 1–18, (1941).

[**] In order for the multiplication to be defined, X must be the **column vector** formed from $[x_1, x_2, \ldots, x_n]$. *For simplicity of notation, we will regard X as either a row or column vector, whichever is appropriate.*

S_2 = percentage of individuals in age category C_2 that survive to age class C_3,
⋮ k years later.

S_{n-1} = percentage of individuals in age class C_{n-1} that survive to the final age class C_n, k years later.

Note that each S_k is assumed to be constant over time and independent of the total number in the population. Thus, the model does not take crowding into account. In addition, note that the model does not keep track of those that live more than kn years.

figure 56.1 Leslie Matrix Model.

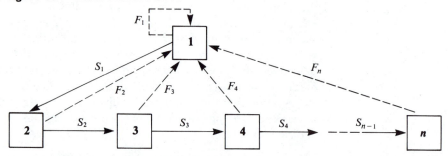

Let F_k be the average number of female offspring, born to an individual in age class C_k, that survive to the next census. Then F_k is determined by the average number of female births per mother in age category C_k and the infant survival rates. As illustrated in figure 56.1, note that

$$y_1 = F_1x_1 + F_2x_2 + \cdots + F_nx_n$$
$$y_2 = S_1x_1$$
$$y_3 = \quad\quad S_2x_2$$
$$\cdot$$
$$\cdot$$
$$\cdot$$
$$y_n = \quad\quad\quad\quad S_{n-1}x_{n-1}$$

Thus, in matrix form, $Y = AX$, where A is the *Leslie matrix*

$$A = \begin{bmatrix} F_1 & F_2 & F_3 & \cdots & F_n \\ S_1 & 0 & 0 & \cdots & 0 \\ 0 & S_2 & 0 & \cdots & 0 \\ \cdot & \cdot & \cdot & & \cdot \\ \cdot & \cdot & \cdot & & \cdot \\ \cdot & \cdot & \cdot & & \cdot \\ 0 & 0 & 0 & S_{n-1} & 0 \end{bmatrix} \quad \text{and} \quad X = \begin{bmatrix} x_1 \\ x_2 \\ x_3 \\ \cdot \\ \cdot \\ \cdot \\ x_n \end{bmatrix}$$

To project the population $2k$ years, let $Z = AY = A(AX) = (AA)X = A^2X$.

example 56.2   **Bernadelli's beetles**   In his 1941 paper, H. Bernadelli gave the following Leslie matrix A for a fictitious beetle population:

$$A = \begin{bmatrix} 0 & 0 & 6 \\ 1/2 & 0 & 0 \\ 0 & 1/3 & 0 \end{bmatrix}$$

Assume that time is measured in months and that $k = 1$ month. If $X = [0, 0, 18]$, project the population forward 6 months.

solution 56.2   The population structure one month later is

$$Y = \begin{bmatrix} 0 & 0 & 6 \\ 1/2 & 0 & 0 \\ 0 & 1/3 & 0 \end{bmatrix} \begin{bmatrix} 0 \\ 0 \\ 18 \end{bmatrix} = \begin{bmatrix} 108 \\ 0 \\ 0 \end{bmatrix}$$

After 2 months, the population structure is $Z = AY = [0, 54, 0]$. The population structure after 3 months is $AZ = [0, 0, 18]$. If we repeat the process, we obtain the population projections summarized in table 56.1.

table 56.1

Month	Age class 1	Age class 2	Age class 3
0	0	0	18
1	108	0	0
2	0	54	0
3	0	0	18
4	108	0	0
5	0	54	0
6	0	0	18

Note that the population cycles, and that only one age class is represented at each census. ▤

example 56.3   In a 1971 paper,* A.L. Jensen applied the Leslie matrix model to a population of brook trout in Hunt Creek, Michigan. The population was divided into five-year classes (fingerlings, yearlings, etc.). The Leslie matrix A, with entries rounded off to two significant digits, is given by

$$A = \begin{bmatrix} 0 & 0 & 37 & 64 & 82 \\ .06 & 0 & 0 & 0 & 0 \\ 0 & .34 & 0 & 0 & 0 \\ 0 & 0 & .16 & 0 & 0 \\ 0 & 0 & 0 & .08 & 0 \end{bmatrix}$$

* A.L. Jenson, *Trans. Amer. Fish Soc. 100* (**1**): pp. 456–9.

If $X = [10{,}000, 583, 200, 34, 3]$, then the population projection for the next year is $Y = AX = [9{,}822, 600, 198, 32, 3]$. Another matrix multiplication gives $Z = AY = [9{,}620, 589, 204, 32, 3]$. This is the population projection for year two. ▤

The basic Leslie matrix model has been generalized in many ways. One simple generalization applies to animals who, after reaching a certain age, survive subsequent time periods with the same probability (see figure 56.2). In

figure 56.2

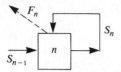

addition, they continue to reproduce with the same regularity. For such populations, the last equation in the Leslie model, $y_n = S_{n-1}x_{n-1}$, is replaced by

$$y_n = S_{n-1}x_{n-1} + S_n x_n$$

where S_n is the proportion of animals in the last age class that survive until the next census. The last age class therefore consists of animals at least $(n-1)k$ years of age.

This new model can often be applied to populations of large mammals, as described in example 56.4.

example 56.4   The Leslie matrix for the American bison is given by

$$A = \begin{bmatrix} 0 & 0 & .42 \\ .60 & 0 & 0 \\ 0 & .75 & .95 \end{bmatrix}$$

The population is divided into calves, yearlings, and adults (age two or more). Thus, females who reach the age of 2 years survive an additional year with probability .95 and reproduce with the same regularity. If we start a herd with 100 adult females (and an appropriate number of adult males), then next year's herd structure is given by

$$Y = \begin{bmatrix} 0 & 0 & .42 \\ .6 & 0 & 0 \\ 0 & .75 & .95 \end{bmatrix} \begin{bmatrix} 0 \\ 0 \\ 100 \end{bmatrix} = \begin{bmatrix} 42 \\ 0 \\ 95 \end{bmatrix}$$

Multiplying the prior year's age structure vector by A results in the population predictions given in table 56.2.

table 56.2

Year	Calves	Yearlings	Adults
0	0	0	100
1	42	0	95
2	40	25	90
3	38	24	104
4	44	23	117
5	49	26	128

Under mild conditions on the Leslie matrix, it can be shown that, for t large, the population age structure $X(t)$ at time t is given by

$$X(t) = (\lambda_0)^t X_0$$

for some $\lambda_0 > 0$ and n-dimensional vector X_0. Thus, exponential growth or decay results. It is highly unlikely, however, that the survival and fecundity rates will remain constant for long periods of time.

discrete compartmental models   In section 39, we introduced the general two-compartment model, whose purpose was to keep track of the amount of tracer material that flowed between the compartments. In that model, material was *continually interchanged*. Here, we present a model that keeps track of compartmental contents every Δt units of time and assumes that the system changes only at times $\Delta t, 2\Delta t, \ldots, n\Delta t, \ldots$ Of course, by selecting Δt very small, we can approximate the continuous case.

In constructing a compartmental model of a physical system, we conceptually separate the system into a distinct number of smaller components between which material is transported. Compartments need not be spatially distinct but must be distinguishable on some basis. The following are a few examples:

1. Acid rain (containing strontium 90, for example) is deposited onto pasture land. Compartments might be grasses, soil, streams, and litter.

2. In studying the flow of energy through an aquatic ecosystem, we might separate the system into phytoplankton, zooplankton, plankton predators, seaweed, small carnivores, large carnivores, and decay organisms.

3. A tracer is infused into the bloodstream and is lost to the body by the metabolism of a particular organ and by excretion. Appropriate compartments might be arterial blood, venous blood, the organ, and urine.

Suppose then that a system is divided into n compartments and, after each Δt units of time, material is interchanged between compartments. We will assume that a fixed fraction τ_{ij} of the contents of compartment J are passed to

figure 56.3 τ_{ij} indicates a transfer to I from J

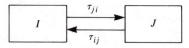

compartment I every Δt units of time, as depicted in figure 56.3. This hypo-thesis is known as the **linear donor-controlled hypothesis**.

Let $X = [x_1, x_2, \ldots, x_n]$, where x_i is the amount of tracer in compartment I. We say that X specifies the **state of the system**. If $Y = [y_1, y_2, \ldots, y_n]$ is the state Δt units of time later, we will show that

$$Y = TX$$

for some $n \times n$ matrix T determined by the **transfer coefficients** τ_{ij}. To find T, note that

$$y_1 = x_1 + (\text{amount of tracer entering 1}) - (\text{amount of tracer leaving 1})$$
$$= x_1 + (\tau_{12}x_2 + \tau_{13}x_3 + \cdots + \tau_{1n}x_n) - (\tau_{21} + \tau_{31} + \cdots + \tau_{n1})x_1$$
$$= (1 - \tau_{21} - \tau_{31} - \cdots - \tau_{n1})x_1 + \tau_{12}x_2 + \cdots + \tau_{1n}x_n$$

If we let $\tau_{11} = 1 - \tau_{21} - \tau_{31} - \cdots - \tau_{n1}$, then τ_{11} is just the fraction of compart-ment 1 that remains in 1.

Letting $\tau_{ii} = 1 - \sum_{j \neq i} \tau_{ji}$, it follows that

$$Y = TX, \qquad \text{where } T = [\tau_{ij}]$$

The matrix T is called the **transfer matrix**. Note that the *sum of the transfer coefficients in any column is 1.*

Discrete compartmental models are illustrated in our final two examples.

example 56.5   In the compartmental diagram given in figure 56.4, the transfer coefficients are shown and the compartment contents at a particular time t are shown in the boxes.

figure 56.4

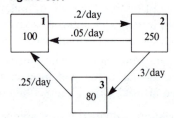

(a) Find the transfer matrix T.

(b) If $\Delta t = 1$ day, find the state of the system Y one day later.

solution 56.5 $X = [100, 250, 80]$ gives the state of the system at time $t = 0$. Remember that τ_{ij} specifies the rate of transfer to compartment i from compartment j. Hence, we are given that $\tau_{21} = .2$, $\tau_{12} = .05$, $\tau_{32} = .3$, $\tau_{23} = 0$, $\tau_{13} = .25$, and $\tau_{31} = 0$. The matrix T is then

$$T = \begin{bmatrix} - & .05 & .25 \\ .2 & - & 0 \\ 0 & .3 & - \end{bmatrix}$$

Since the column entries must sum to 1, we conclude that

$$T = \begin{bmatrix} .8 & .05 & .25 \\ .2 & .65 & 0 \\ 0 & .3 & .75 \end{bmatrix}$$

Hence

$$Y = TX = \begin{bmatrix} .8 & .05 & .25 \\ .2 & .65 & 0 \\ 0 & .3 & .75 \end{bmatrix} \begin{bmatrix} 100 \\ 250 \\ 80 \end{bmatrix} = \begin{bmatrix} 112.5 \\ 182.5 \\ 135 \end{bmatrix}$$

If X_0 denotes the initial state of the system, and X_n is the state after $n(\Delta t)$ units of time, then

$$X_1 = TX_0, \ X_2 = TX_1, \ X_3 = TX_2, \ldots, X_{n+1} = TX_n$$

Note that $X_2 = T(TX_0) = T^2 X_0$, $X_3 = T(T^2 X_0) = T^3 X_0$, and in general,

$$X_n = T^n X_0$$

It is usually easier to apply the recursion formula $X_{n+1} = TX_n$ repeatedly than to compute the matrix T^n, as illustrated in our final example.

example 56.6   Strontium 90 is deposited into pasture land by rainfall. To study how this material is cycled through the ecosystem, we divide the system into the compartments shown in figure 56.5. Suppose that $\Delta t = 1$ month and the transfer coefficients (which have been estimated experimentally) are measured in fraction/month.

(a) Ignoring the losses due to radioactive decay, form the transfer matrix T.

(b) Suppose that rainfall has quickly deposited the strontium 90 into the compartments so that $X_0 = [20, 60, 15, 20]$. (Units might be grams per hectare.) Compute $X_1, X_2, \ldots, X_{12}$, the states of the ecosystem over the next 12 months.

figure 56.5

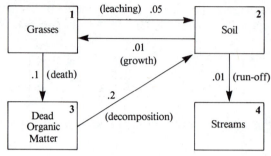

solution 56.6

The matrix T of transfer coefficients is

$$T = \begin{bmatrix} .85 & .01 & 0 & 0 \\ .05 & .98 & .2 & 0 \\ .1 & 0 & .8 & 0 \\ 0 & .01 & 0 & 1 \end{bmatrix}$$

Hence,

$$X_1 = TX_0 = \begin{bmatrix} .85 & .01 & 0 & 0 \\ .05 & .98 & .2 & 0 \\ .1 & 0 & .8 & 0 \\ 0 & .01 & 0 & 1 \end{bmatrix} \begin{bmatrix} 20 \\ 60 \\ 15 \\ 20 \end{bmatrix} = \begin{bmatrix} 17.6 \\ 62.8 \\ 14.0 \\ 20.6 \end{bmatrix}$$

Subsequent states, computed using the recursion $X_{k+1} = TX_k$, are shown in table 56.3.

table 56.3

Month	Grasses	Soil	Dead organic matter	Streams
0	20.00	60.00	15.00	20.00
1	17.60	62.80	14.00	20.60
2	15.59	65.22	12.96	21.23
3	13.90	67.29	11.93	21.88
4	12.49	69.03	10.93	22.55
5	11.31	70.46	9.99	23.24
6	10.32	71.61	9.13	23.95
7	9.48	72.52	8.33	24.66
8	8.79	73.21	7.61	25.39
9	8.20	73.71	6.97	26.12
10	7.71	74.04	6.40	26.86
11	7.29	74.22	5.89	27.60
12	6.94	74.28	5.44	28.34
.	.	.	.	.
.	.	.	.	.
60	3.60	50.76	1.89	58.75

exercises for section 56

part A

Write each of the following linear systems in matrix form. Then use the laws of matrix algebra to find Z in terms of X.

1. $\begin{aligned} z_1 &= 4y_1 - 2y_2 \\ z_2 &= -y_1 + y_2 \end{aligned}$ and $\begin{aligned} y_1 &= 3x_1 - 2x_2 + x_3 \\ y_2 &= x_1 + x_2 - 2x_3 \end{aligned}$

2. $\begin{aligned} z_1 &= y_1 - y_2 \\ z_2 &= 3y_1 + y_2 \end{aligned}$ and $\begin{aligned} y_1 &= x_2 \\ y_2 &= x_1 + x_2 \end{aligned}$

★3. $\begin{aligned} z_1 &= y_1 - y_2 + 3 \\ z_2 &= 3y_1 + 4y_2 - 1 \end{aligned}$ and $\begin{aligned} y_1 &= 3x_1 - x_2 - 1 \\ y_2 &= 4x_1 + 2x_2 + 6 \end{aligned}$

4. The Leslie matrix for an insect population is

$$A = \begin{bmatrix} 0 & 20 & 0 \\ .1 & 0 & 0 \\ 0 & .2 & 0 \end{bmatrix}$$

The age classes are 0–1 weeks, 2–3 weeks, and 4–5 weeks, and only insects 2–3 weeks old can reproduce.

(a) If $X = [0, 100, 0]$ specifies the initial age structure, project the population forward 2 months (or 4 time periods).

(b) Show that $A^3 = 2A$ and $A^4 = 2A^2$. Conclude that the population in each age class doubles every month.

5. The growth of many fish populations, such as salmon or striped bass, is characterized by very high fecundities and extremely small survival rates among the newborns. Suppose that such a population is divided into 0-year-olds (eggs), 1-year-olds, 2-year-olds, and 3-year-olds. The Leslie matrix is given by

$$A = \begin{bmatrix} 0 & 0 & 0 & 1{,}000 \\ .005 & 0 & 0 & 0 \\ 0 & .6 & 0 & 0 \\ 0 & 0 & .7 & 0 \end{bmatrix}$$

(a) Project the population forward 5 years given that $X_0 = [10{,}000, 0, 0, 0]$.

(b) Show that $A^4 = 2.1I$ where I is the 4×4 identity matrix. Conclude that the population in each age class increases by 210% every 4 years.

★6. Cattle on a large ranch are divided into calves, yearlings, and adults. A calf survives the first year with probability .7, while 80% of the yearlings mature into adults. In addition, adults survive a given year with probability .9, and an adult female produces a single calf each year.

(a) Construct the Leslie matrix, assuming that we census the population after reproduction and count females only.

(b) Construct the Leslie matrix, assuming that we census the population before reproduction and count both males and females.

7. Based on vital statistics from the 1930s, M.B. Usher[*] constructed the following Leslie matrix for the female blue whale population:

$$
\begin{bmatrix}
0 & 0 & .19 & .44 & .50 & .50 & .45 \\
.87 & 0 & 0 & 0 & 0 & 0 & 0 \\
0 & .87 & 0 & 0 & 0 & 0 & 0 \\
0 & 0 & .87 & 0 & 0 & 0 & 0 \\
0 & 0 & 0 & .87 & 0 & 0 & 0 \\
0 & 0 & 0 & 0 & .87 & 0 & 0 \\
0 & 0 & 0 & 0 & 0 & .87 & .8
\end{bmatrix}
$$

The age classes are 0–1, 2–3, 4–5, 6–7, 8–9, 10–11, and 12 years or over. A typical 10-year-old female gives birth once every 2 years, with a 50% chance of producing a female. If $X = [800, 700, 650, 500, 400, 300, 280]$ specifies the present population level, predict the population structure 2, 4, and 6 years from now.

8. If A is the Leslie matrix

$$
\begin{bmatrix}
0 & F_2 & 0 \\
S_1 & 0 & 0 \\
0 & S_2 & 0
\end{bmatrix}
$$

show that $A^3 = (S_1 F_2)A$ and $A^4 = (S_1 F_2)A^2$. If X_n denotes the population structure after n time periods, find X_n in terms of X_1 and X_2.

9. Let A be the Leslie matrix

$$
\begin{bmatrix}
0 & 0 & F_3 \\
S_1 & 0 & 0 \\
0 & S_2 & 0
\end{bmatrix}
$$

(a) Show that $A^3 = (S_1 S_2 F_3)I$, where I is the 3×3 identity matrix.

(b) If $S_1 S_2 F_3 < 1$, what can you predict about future populations?

(c) If $F_3 = 4$ and $S_2 = .7$, find S_1 so that the eventual population growth is 50% every three time periods.

In each of the compartmental diagrams shown in the figures, the transfer coefficients are shown and the contents at a particular time t are given.

10.

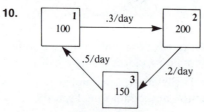

[*] M.B. Usher, *Biological Management and Conservation*, London: Chapman and Hall, (1973), pp. 176–179.

(a) Form the transfer matrix T and the present state X.

(b) Find the state of the system 1 day later.

11.

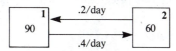

(a) Form the transfer matrix T and the present state X.

(b) Find the state of the system 1 day later.

(c) Eventually, the system will reach an equilibrium state $\hat{X} = [x_1, x_2]$ that satisfies $TX = \hat{X}$. Find $\hat{X}$. [*Hint*: $x_1 + x_2 = 150$.]

12.

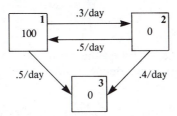

(a) Form the transfer matrix T and the present state X.

(b) Find the state of the system 1 day later.

(c) Find the equilibrium state $\hat{X} = [x_1, x_2, x_3]$ by solving $T\hat{X} = \hat{X}$ with $x_1 + x_2 + x_3 = 100$.

13. Radioisotopes (such as phosphorus 32 and carbon 14) have been used to study the transfer of nutrients in food chains. Shown in the figure is a compartmental representation of a simple aquatic food chain. One hundred units (e.g., microcuries) of tracer are dissolved in the water of an aquarium containing a species of phytoplankton and a species of zooplankton.

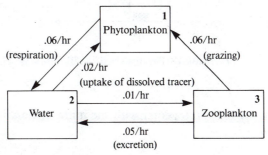

(a) Find the transfer matrix T.

(b) Predict the state of the system over the next 6 hours.

14. A field has been completely devastated by fire. Two types of vegetation, grasses and small shrubs, will first begin to grow, but the small shrubs can take over an area only if preceded by the grasses. In the accompanying figure, the transfer coefficient of .3 indicates that, by the end of the summer, 30% of the prior bare space in the field becomes occupied by grasses.

(a) Find the transfer matrix T.

(b) If $X = [10, 0, 0]$, with area measured in acres, predict the ground cover over the next 6 years.

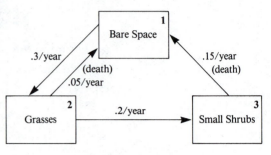

part B   **harvesting matrices**   When the Leslie matrix model is used to model the growth of a population, we define a *harvesting matrix* to be a diagonal matrix of the form

$$H = \begin{bmatrix} h_1 & 0 & 0 & \cdots & 0 \\ 0 & h_2 & 0 & \cdots & 0 \\ 0 & 0 & h_3 & \cdots & 0 \\ 0 & 0 & 0 & \cdots & h_n \end{bmatrix}$$

where h_i is the percentage of age class i *that remains after harvesting.*

We imagine that the population is harvested once a year, soon after the census takes place, as indicated in the figure. The population structure after harvesting is given by

$$Z_1 = HY = H(AX) = (HA)X$$

X $Y = AX$ ▬ = harvesting period

├─────────────────┼──▬────────────▶

t $t + \Delta t$

If this harvesting procedure is used at the end of the next time period, then

$$Z_2 = (HA)Z_1$$

We say that H, together with the initial population X, determines a *harvesting policy.*

15. If $A = \begin{bmatrix} 0 & 0 & .5 \\ .6 & 0 & 0 \\ 0 & .7 & .9 \end{bmatrix}$, $H = \begin{bmatrix} 1 & 0 & 0 \\ 0 & 1 & 0 \\ 0 & 0 & .4 \end{bmatrix}$, and $X = [100, 50, 40]$,

compute and interpret $Z_k = (HA)^k X$ for $k = 1, 2, \ldots, 5$.

an introduction to discrete probability

prologue   Probability theory, the mathematics of chance, has become an essential tool to the scientist, and has a wide variety of applications. It is that branch of mathematics used to make election predictions, to control traffic signals, and to aid in counseling couples concerned about the health of their potential offspring. The subject had its origins, however, in games of chance played in the Middle Ages. Cardano's *Book on Games of Chance* (published in 1663) and later books such as De Moivre's *Doctrine of Chance* (1716) analyzed the popular card and dice games of their day and often served as gamblers' manuals.

The development of this subject was hampered not only by the superstitions of gamblers but also by the errors of professional mathematicians. Even simple computations involving the roll of two dice were not well understood by the scientific community in the seventeenth and eighteenth centuries. In the 1920s, however, a systematic method for analyzing chance phenomena was introduced, based on set theory and the **sample space concept**. This is where we begin our study.

Serious applications of probability theory are a more recent phenomenon. The first and most famous application to biology is due to the Austrian monk Gregor Mendel (1822–1884). Through his experiments with garden peas, Mendel discovered that the laws of heredity could be explained using those same rules that were to dice rolls and coin flips by Cardano and De Moivre. Simple games of chance still provide excellent settings for introducing probability concepts. In this final part of the text, we will emphasize games of chance including nature's own drawing of lots—i.e., genetics.

57 sample spaces, events, and probability

The first step in constructing a model for a chance phenomenon is to list all observable outcomes of the experiment. We confine our attention to experiments that can be repeated again and again under essentially the same conditions. Of course, coin flips and dice rolls are simple first examples. The complete list of the physical outcomes of such an experiment $\mathcal{E}$ is known as the **sample space** for $\mathcal{E}$.

definition 57.1 Let $\mathcal{E}$ be an experiment with outcomes $e_1, e_2, \ldots, e_n$. If S is the set $\{e_1, e_2, \ldots, e_n\}$, we call S the *sample space* for $\mathcal{E}$. Elements of the set S are called **elementary events**.

The concept of a sample space is illustrated in the next four examples.

example 57.1 When a single coin is flipped, $S = \{H(\text{Heads}),\ T(\text{Tails})\}$ serves as a sample space for the experiment. If the coin is flipped twice, the elementary events can be found with the aid of a **tree diagram**, as illustrated in figure 57.1. Thus, $S = \{HH,\ HT,\ TH,\ TT\}$ is a way of listing all observable outcomes.

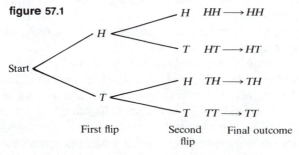

figure 57.1

$$
\begin{array}{llll}
 & H & HH \longrightarrow HH \\
H & & \\
 & T & HT \longrightarrow HT \\
\text{Start} & & \\
 & H & TH \longrightarrow TH \\
T & & \\
 & T & TT \longrightarrow TT
\end{array}
$$

First flip   Second flip   Final outcome

example 57.2 When a single die is rolled, $S = \{1, 2, 3, 4, 5, 6\}$ is a suitable sample space. The sample space for the roll of two dice has in the past caused some confusion even among mathematicians. Leibniz, one of the coinventors of calculus, asserted that a sum of 11 could be achieved in only one way—namely, by rolling a 5 and a 6. Actually, there are two physically distinct ways of rolling a sum of 11. If one die is red, and the other is green, then physical outcomes can be recorded as (x, y) where x is the outcome on the red die and y is the outcome on the green die. Shown below are the 36 members of the sample space S:

$$S = \left\{ \begin{array}{llllll} (1,1), & (1,2), & (1,3), & (1,4), & (1,5), & (1,6), \\ (2,1), & (2,2), & (2,3), & (2,4), & (2,5), & (2,6), \\ (3,1), & (3,2), & (3,3), & (3,4), & (3,5), & (3,6), \\ (4,1), & (4,2), & (4,3), & (4,4), & (4,5), & (4,6), \\ (5,1), & (5,2), & (5,3), & (5,4), & (5,5), & (5,6), \\ (6,1), & (6,2), & (6,3), & (6,4), & (6,5), & (6,6) \end{array} \right\}$$

Note that $(5, 6)$ and $(6, 5)$ each give a sum of 11. When we form the subset $\{(5, 6), (6, 5)\}$ we have specified the event "sum of 11." Likewise, the event "sum of 7" can be summarized by the subset

$$\{(1,6), \quad (2,5), \quad (3,4), \quad (4,3), \quad (5,2), \quad (6,1)\}$$

definition 57.2   If $\mathscr{E}$ is an experiment with sample space S, we call a subset E of S an **event.**

example 57.3   Shown in figure 57.2 is the betting layout for the American version of roulette.

figure 57.2

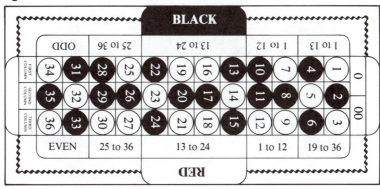

A small ball is spun around the wheel and then settles into one of 38 slots marked $0, 00, 1, 2, \ldots, 36$. Thus, $S = \{0, 00, 1, 2, \ldots, 36\}$ serves as an appropriate sample space. A common bet is "black." As you can see from the layout, the event "black" is given by

$$B = \{2, 4, 6, 8, 10, 11, 13, 15, 17, 20, 22, 24, 26, 28, 29, 31, 33, 35\}$$

The event "second column" is specified by

$$C_2 = \{2, 5, 8, 11, 14, 17, 20, 23, 26, 29, 32, 35\}$$

example 57.4   When each of two children flips a penny, $S = \{HH, HT, TH, TT\}$ is an appropriate sample space. The event "the coins match" is given by the subset $M = \{HH, TT\}$.

The next example illustrates a fundamental counting technique called the Multiplication Principle.

example 57.5   Four balls numbered 1, 2, 3, and 4 are placed in a bag. A single ball is removed and its number is noted. A second ball is then selected from the remaining three. The tree diagram of figure 57.3 shows the possible outcomes of the experiments. There are $4 \cdot 3 = 12$ ways of performing the draws.

figure 57.3

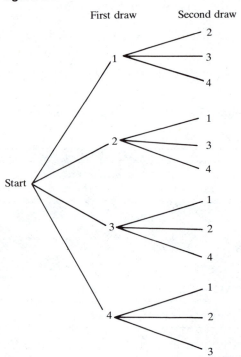

> **Multiplication Principle**   If act 1 can be performed m ways and act 2 can be performed n ways, then act 1 followed by act 2 can be accomplished in mn ways.

It is not hard to see why the principle is true. Shown in figure 57.4 are m roads from town A to town B, and n roads connecting B to C. Given a choice of m roads from A to B, there are then n ways of completing the trip. Thus, A is connected to C by mn routes. If there were k roads from C to a fourth town D, then the number of routes from A to D would be given by $(mn)k = mnk$. The Multiplication Principle is illustrated in example 57.6.

figure 57.4

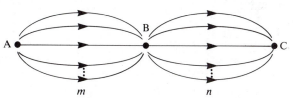

example 57.6   A young couple plans to have three children. In terms of the sex of the children, there are $2 \cdot 2 \cdot 2 = 8$ ways the family can be formed, as illustrated in the tree diagram of figure 57.5. The event "exactly two boys" is given by {*GBB, BGB, BBG*}.

figure 57.5

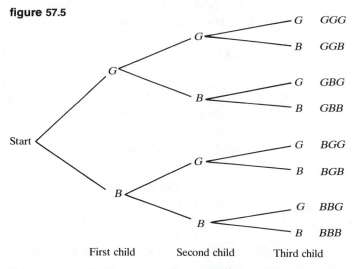

How does one assign "chance" or "probability" to events? We will discuss two commonly employed methods. In the case of a single die, the symmetry of the cube suggests that the same probability should be assigned to each of the six elementary events 1, 2, 3, 4, 5, and 6. Thus, the probability $\frac{1}{6}$ is assigned to each outcome. We would expect the number 3 to occur about $\frac{1}{6}$ of the time when the die is rolled repeatedly. Likewise, when two dice are tossed, it is as difficult to achieve $(3, 4)$ as $(1, 1)$, because we are specifying exactly what must occur on each die. Consequently, the probability $\frac{1}{36}$ is assigned to each of the 36 outcomes. This method of assigning probabilities judges the relative difficulty of performing the elementary events and then *weighs the elementary events* accordingly. This is illustrated in the following famous example.

example 57.7   Amy is trying to get "a head." More precisely, she flips a penny and will stop if the coin lands *H*. Otherwise, she will try one more time. Thus, the sample

figure 57.6

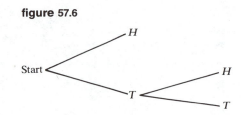

space for Amy's experiment is $S = \{H, TH, TT\}$, as depicted in figure 57.6. The event "a head" is then $E = \{H, TH\}$. The French mathematician Roberval (1602–1675) analyzed this game and assigned the probability $\frac{2}{3}$ to E based on the fact that there are three elementary events. However, Roberval's assessment was incorrect. When the experiment is performed repeatedly, E actually occurs about 75% of the time. This can be explained as follows. Obtaining the outcome TH should be twice as difficult as obtaining a single H on the first toss. If we define $P(TH) = P(TT) = \frac{1}{4}$ and $P(H) = \frac{1}{2}$, then $P(H) = 2P(TH)$ and $P(E) = P(TH) + P(H) = \frac{1}{4} + \frac{1}{2} = \frac{3}{4}$. ▤

When a doctor tells her patient that there is a 90% chance of full recovery, she is not basing this probability on a sample space analysis. Rather, based on a large number of similar cases, she knows that approximately 90% of all such patients fully recovered. This is the **empirical method** of assigning probabilities, and is illustrated in the following example.

example 57.8 In a study of 190,177 British airmen, Race and Sanger* found that 79,334 had blood type A. The probability that a given Englishman will have blood type A is estimated empirically to be $79{,}334/190{,}177 \approx .417$. ▤

In general, we must assign probabilities to elementary events according to the following rules:

definition 57.3 Given a sample space $S = \{e_1, e_2, \ldots, e_n\}$, a **probability function** ω is a rule that assigns to each elementary event e_i a real number $p_i = \omega(e_i)$ satisfying:

  1. $0 \le p_i \le 1$
  2. $p_1 + p_2 + \cdots + p_n = 1$

If $E = \{s_1, s_2, \ldots, s_k\}$ is an event, we define $P(E)$, the **probability of E**, to be $\omega(s_1) + \omega(s_2) + \cdots + \omega(s_k)$, the sum of the probabilities of the elementary events constituting E.

Note that the definition does not tell us how to assign the weights $\omega(e_i)$. When we assign weights to elementary events, we are constructing a

* R.R. Race and R. Sanger, *Blood Groups in Man*, 6th ed., Oxford: Blackwell (1973).

probability model, and the predictions of the model must then be checked empirically.

example 57.9   In the experiment of tossing two dice, considered in example 57.2, we are inclined to assign probability $\frac{1}{36}$ to each of the 36 outcomes. With this assignment,

$$P(\text{"sum of 11"}) = P(\{(5, 6), (6, 5)\}) = \frac{1}{36} + \frac{1}{36} = \frac{1}{18}$$

and

$$P(\text{"sum of 7"}) = P(\{1, 6), (2, 5), (3, 4), (4, 3), (5, 2), (6, 1)\})$$
$$= \frac{6}{36} = \frac{1}{16}$$

When equal weights are assigned to the elementary events $e_1, \ldots, e_n$, then $\omega(e_i) = 1/n$ in order to satisfy $P(S) = 1$, and it follows that

$$P(E) = \frac{\text{number of elements of } E}{n}$$

The sample space is then called **uniform** or **equiprobable**. Uniform sample spaces are presented in the next three examples.

example 57.10   Refer to example 57.3. The symmetry of the roulette wheel suggests that we should assign probability $\frac{1}{38}$ to each of the 38 elementary events. Thus

$$P(\text{"black"}) = \frac{\text{number of black outcomes}}{38} = \frac{18}{38} = \frac{9}{19}$$

and

$$P(\text{"column 2"}) = \frac{\text{number of outcomes in column 2}}{38} = \frac{12}{38} = \frac{6}{19}$$

example 57.11   If we are willing to assume that births of girls and boys are equiprobable, then the probability of each of the elementary events in example 57.6 should be $\frac{1}{8}$. Thus

$$P(\text{"2 boys"}) = P(\{GBB, BGB, BBG\}) = \frac{3}{8}$$

example 57.12   If three dice are thrown, find the probability of a sum of 6.

solution 57.12   The Multiplication Principle tells us that there are $6 \cdot 6 \cdot 6 = 216$ outcomes to the experiment. If we assume that the outcomes are equiprobable, then we assign weight $\frac{1}{216}$ to each elementary event. Then all we need to do is to count the number of ways of forming a sum of 6. This can be done with the aid of the portion of the tree diagram shown in figure 57.7. Since 10 of the 216 outcomes yield a sum of 6, we conclude that $P(\text{"sum of 6"}) = 10/216 \approx .046$.

figure 57.7

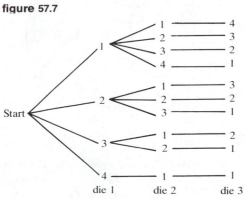

die 1   die 2   die 3

Before presenting examples from genetics, let's first review the physical basis for heredity and some of the standard terminology.

Mendel's First Law asserts that the determinants of heredity are **genes** which occur in pairs. In the case of mammals, one gene originates in the sperm of the father while the other comes from the egg of the mother. The actual carriers of this genetic information are **chromosomes**, which occur in pairs called **homologues**. Again, one chromosome in the pair is paternal and the other is maternal. Shown in figure 57.8 are the standard 23 chromosome pairs for humans.

figure 57.8   (Adapted from Barbara R. Landau, *Essential anatomy and physiology*. 2nd ed. Glenview, Ill.: Scott,, Foresman, and Co., 1980.

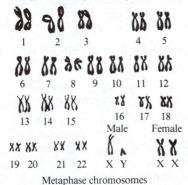

Metaphase chromosomes

We may consider a gene as the genetic material DNA in a special position on a chromosome, and will depict a chromosome and gene as shown in figure 57.9. Forms of the gene that can occupy a given position (*locus*) are called

figure 57.9

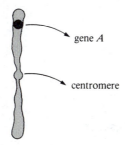

We may consider a gene as the genetic material DNA in a special position

gene *A*

centromere

alleles. If $A_1, A_2, \ldots, A_n$ are the complete set of alleles for a given locus, then, by examining the alleles on a given chromosome pair, we can say that an individual is of **genotype** $A_i A_j$ (see figure 57.10). The actual physical expression

figure 57.10

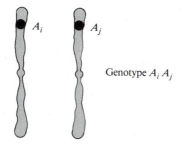

A_i A_j

Genotype $A_i A_j$

of these two genes is called the **phenotype**. This terminology in illustrated in the following examples which present two genetically determined human characteristics.

example 57.13 The genetically determined condition of albinism results from the inability to produce the dark skin pigment *melanin*. The result is extremely light hair and skin and often a pinkish tint to the eyes. Although this defect is relatively rare in humans, it is quite common in rabbits and mice. Two alleles *A* and *a* are responsible. Genotypes *AA* and *Aa* are normal in appearance and so have the same phenotype. Individuals of genotype *aa* are albinos. The allele *A* is **dominant** over *a*.

example 57.14 Common blood type in humans is determined by three alleles, *A*, *B*, and *O*. Thus, the possible genotypes are

$$AA, BB, OO, AB, AO, \text{ and } BO$$

Alleles A and B each result in the production of their own antigens, while allele O is inactive. The phenotypes are determined by the antigens produced. Shown in table 57.1 are the phenotypes together with their possible genotypes.

table 57.1

Phenotype (blood group)	A	B	AB	O
Genotypes	AA, AO	BB, BO	AB	OO

An individual of genotype A_iA_j will produce sex cells or *gametes* of type A_i or A_j by the process of *meiosis*. This process selects one chromosome from each chromosome pair. Thus, each gamete contains exactly half the usual number of chromosomes found in a body cell, as illustrated in figure 57.11.

figure 57.11

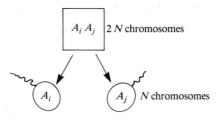

Notice that for $i \neq j$, an individual of genotype A_iA_j can be formed in two different ways (see figure 57.12):

figure 57.12(a)   An A_i sperm unites with an A_j egg.

figure 57.12(b)   An A_j sperm unites with an A_i egg.

These facts are used in the final examples.

example 57.15

Potential parents are each of genotype Aa with respect to albinism, and so each carry the defective allele a. Find the probability that their child will be

(a) an albino   **(b)** a carrier (Aa)

solution 57.15

Both A and a gametes are produced by each parent. The appropriate sample space may be determined from either the tree diagram of figure 57.13 or the so-called *Punnett square* shown in table 57.2. Assuming each physical event is equiprobable, we assign probabilities of $\frac{1}{4}$ to genotype aa and $\frac{1}{4}+\frac{1}{4}=\frac{1}{2}$ to

figure 57.13   **table 57.2**

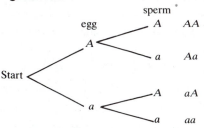

genotype *Aa*. While the outcome *aa* is an elementary event, *Aa* is an event that can occur in two distinct ways. Thus the probability that the child will be an albino is $\frac{1}{4}$ while the probability that the child will carry a single *a* allele is $\frac{1}{2}$. ▤

example 57.16   A husband and wife are of known blood types *AO* and *AB*, respectively. Using either a tree diagram or a Punnett square, it is easy to see that each of the genotypes *AA*, *AB*, *BO*, and *AO* are equiprobable for their offspring. Thus $P(\text{"blood type }A\text{"}) = P(\{AA, AO\}) = \frac{1}{2}$. It is not possible to produce a child of blood type *O* since the mother cannot produce an *O* gamete. ▤

exercises for section 57

part A   Form appropriate sample spaces for each of the following experiments.

1. A coin is flipped and then a single die is tossed.

2. A card is selected from a standard deck of fifty-two cards.

3. A red ball, a white ball, and a blue ball are placed in a box. Then the balls are removed, one at a time.

4. In the playoffs of the National Basketball Association, two teams play in a "best two out of three" series.

5. A couple will have no more than four children, but will complete their family if they have a son and a daughter.

6. The spinner shown in the figure is spun twice and the numbers selected are noted.

In each of the following exercises, assume that all elementary events are equiprobable.

7. For the American version of roulette in example 57.3, find the probability of the events:

    (a) "odd"   (b) "13 to 24"   (c) "0 or 00"

8. When two dice are rolled, find the probability of

    (a) "doubles"   (b) "a sum of 6"   (c) "a sum of 9"

9. For the experiment described in example 57.5, find the probability that:

    (a) a sum of 5 is formed.

    (b) the second ball selected has a higher number than the first.

10. The following "system" was once proposed for playing roulette: bet simultaneously on "black" and "third column." Find the probability of:

    (a) winning both bets   (b) losing both bets

11. A single card is selected from a standard deck of fifty-two cards. Find the probability of:

    (a) "a diamond"   (b) "an ace"
    (c) "a face card" (jack, queen, or king)

12. For the experiment described in exercise 3, find the probability that the red ball is withdrawn from the box before the white ball.

13. A husband and wife each have blood type *AB*. Find the probability that their child will have:

    (a) blood type *AB*   (b) blood type *A*

14. A very bouncy rubber ball is thrown onto the grid shown here and settles into one of the slots. The contestant wins if the ball settles into one of the black squares. Find the probability of a win.

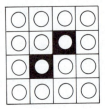

15. When a single die is viewed from a given angle, only three of the six faces are visible. This fact has been used to construct trick dice. Shown in the figure are two dice. The same numbers of spots are on the opposite faces.

    Find the probability of:

    (a) a sum of 6   (b) a sum of 7

16. For the trait of albinism discussed in examples 57.13, find the probability that a child will be normal if one parent is an albino and the other has genotype *Aa*.

17. Among all families with four children, find the probability of two boys and two girls.

18. The menu of the day at a children's hospital allows the patient to select from three meat dishes, two vegetables, two desserts, and three beverages. A single choice of each type must be made. Find the total number of different meals that could be prepared.

★19. *Meiosis* is the process by which sex cells or gametes are formed. As mentioned in the text, this process selects one of the two chromosomes from each chromosome pair for the gamete. Find the number of different kinds of gametes that can be formed by a given human.

20. Shown here are the characters on the three reels of a small slot machine. Each reel spins independently of the others.

(a) Find the total number of elementary events.

(b) What is the probability of winning the jackpot (i.e., obtaining the outcome 7–7–7)?

(c) Find the number of ways of obtaining two cherries.

In the following problems, you must first decide what weights should be attached to the elementary events. Then compute the probability of the event specified in the problem.

21. Given the spinner shown here, find the probability of obtaining an even number.

22. Given the spinner shown here, find the probability that an odd number occurs.

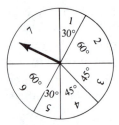

23. Five red balls, three white balls, and four blue balls are placed into an urn, and a single ball is selected. If we write

$$S = \{\text{red, white, blue}\}$$

as the sample space, assign appropriate weights to the elementary events.

★ 24. Two evenly matched teams play in a "best two out of three game" series. Find the probability that the series goes three games.

★ 25. Amy will flip a coin at most three times, but will stop once she gets "a head." Find the probability of obtaining a head.

26. When a thumbtack was tossed 200 times, it landed upright on its head 75 times. If we express the sample space as shown in the figure, assign weights to the elementary events.

$$S = \left\{ \quad , \quad \right\}$$

27. A survey is conducted in Bay City to determine the average number of children in a family. The results are given in the table.

Number of children	0	1	2	3	4	5	6
Number of families	21	40	38	12	9	6	4

Estimate the probability that a Bay City family selected at random will have more than two children.

28. Of all two-children families with at least one boy, find the probability that both children are boys.

29. An experiment has four outcomes, *a*, *b*, *c*, and *d*. The outcomes *a* and *b* are equiprobable, *c* is twice as likely to occur as *b*, and *d* is twice as likely to occur as *c*. Find the probability that either *a* or *b* occurs.

★**30.** Shown here is a die that is not quite cubical. Assuming that probability is proportional to surface area, find the probability of observing a 3 when the die is tossed.

31. *Achondroplasia* is a type of dwarfism caused by a dominant allele *D*. Thus, genotypes *Dd* and *DD* are dwarfs, although genotype *DD* is not viable and usually dies before birth.

   (a) If a dwarf marries a person of normal size, what is the probability that their first child will be of normal size?

  ★**(b)** Among marriages between two such dwarfs, what percentage of the children are of normal size?

32. Color in flowers is often determined by two alleles *R* and *r*. In particular, *RR* flowers are red, *Rr* flowers are pink, and *rr* flowers are white.

   (a) If red flowers are crossed with white flowers, find the proportion of flowers in the next generation that are pink.

   (b) If pink flowers are self-fertilized, find the probability that a flower in the next generation will be pink.

33. Two individuals of blood types *A* and *B* have just produced a baby daughter with blood type *O*.

   (a) What are the genotypes of the parents?

   (b) Find the probability that the next child is of blood type *O*.

34. One hypothesis for eye color in humans assumes that eye color is controlled by two alleles *B* and *b*. The brown allele *B* is dominant over the blue allele *b*, and so genotypes *BB* and *Bb* have brown eyes. Jack has brown eyes, but his mother had blue eyes.

   (a) What is Jack's genotype?

   (b) If Jack marries a blue-eyed woman, find the probability that their first child will be a blue-eyed girl.

part B

permutations and combinations   If five cards are dealt from a standard deck and the *order* in which the cards are dealt is noted, then, by the Multiplication Principle, this deal can be made in $52 \cdot 51 \cdot 50 \cdot 49 \cdot 48$ ways. In terms of factorials, the number can be written as $52!/(52 - 5)!$

In general, if k objects are selected from n objects with the *order of selection noted*, this act can be accomplished in

$$P[n, k] = \frac{n!}{(n-k)!}$$

ways. The expression $P[n, k]$ gives the number of *permutations* of k objects selected from n objects.

35. Find the number of ways of arranging six books on a shelf.

36. Ten horses run in a race at Santa Anita. Find the number of ways the race could end in "win," "place," and "show." Assume there are no ties.

37. How many "words" of two or three letters can be formed from the word "dog"?

In most card games, the order in which the cards are dealt is not relevant. Given a five-card poker hand, there are $P[5, 5] = 5!$ ways that these five cards could have been dealt. Hence

$$(\text{number of five-card hands})5! = P[52, 5]$$

The number of five-card hands is therefore $P[52, 5]/5! = 52!/(5! \, 47!) = 2,598,960$.

When a subset of k objects is selected from n objects *without regard to the order of their selection*, we let $C[n, k]$ denote the number of such subsets. Then $C[n, k]$ is given by

$$C[n, k] = \frac{P[n, k]}{k!} = \frac{n!}{k!(n-k)!}$$

The expression $C[n, k]$ gives the number of *combinations* of k objects selected from n objects.

38. Find the number of five-card hands consisting only of jacks, queens, kings, or aces. Find the probability of obtaining such a hand.

★39. Find the probability of obtaining a seven-card hand containing all four aces.

58 compound events and the laws of probability

In the genetics examples of section 57, we concentrated on a single characteristic determined by a gene pair. Often, however, we are interested in estimating the probability that a given organism has several different genetically determined characteristics. A good example is the Rh classification of blood type, which is determined by two alleles R and r. An individual of genotype rr is Rh-negative (lacking the Rh-antigen), while Rh-positive individuals are either RR or Rr. Given the blood types of both parents, we may ask:

1. What is the probability that a child will be Rh-positive *and* type B?
2. What is the probability that a child will be Rh-positive *or* type B?
3. What is the probability that a child will be *neither* Rh-positive *nor* type B?

Since we are representing *events as subsets* of a sample space S, the set-theoretic representation of these more complicated or **compound events** helps to clarify how their probabilities should be computed.

If $S = \{e_1, e_2, \ldots, e_n\}$ and E and F are events, then the event "E or F" will occur if the outcome e of experiment $\mathscr{E}$ is in at least one of the two subsets E or F. Thus,

> The event "E or F" is the subset $E \cup F$.

If we represent the sample space S by a box in a standard Venn diagram, then the event "E or F" $= E \cup F$ may be depicted as shown in figure 58.1. Although the sample space and events are not necessarily planar areas, the Venn diagram is an excellent tool that can be used to clarify relationships among events.

figure 58.1 *E or F*

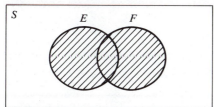

The event "*E* and *F*" will occur when the outcome *e* lies in both subsets *E* and *F*. Thus,

The event "*E* and *F*" is the subset $E \cap F$.

The Venn diagram representation of "*E* and *F*" is shown in figure 58.2.

figure 58.2 *E* and *F*

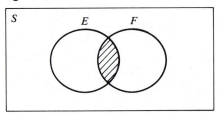

If we let Ø represent the *null set* or *impossible event*, we call events *E* and *F* **mutually exclusive** in case $E \cap F = \emptyset$. It is not possible for *E* and *F* to occur simultaneously, as illustrated in figure 58.3.

figure 58.3

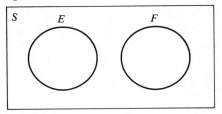

Finally, the event "not *E*" can be represented as the *complement* of the set *E*, as depicted in figure 58.4. Our symbol for the event complementary to *E* will be $\bar{E}$. Thus,

The event "not *E*" is the subset $\bar{E}$

figure 58.4 not *E*

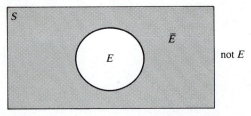

not *E*

Compound events are constructed in our next three examples.

example 58.1   Two dice are tossed. Let A be the event "doubles," B the event "a sum 6," and C the event "at least one 6." Then

$$A = \{(1, 1), (2, 2), (3, 3), (4, 4), (5, 5), (6, 6)\}$$
$$B = \{(1, 5), (2, 4), (3, 3), (4, 2), (5, 1)\}$$

and $\quad C = \{(6, 1), (6, 2), (6, 3), (6, 4), (6, 5), (6, 6), (1, 6), (2, 6),$
$$(3, 6), (4, 6), (5, 6)\}$$

The event "A and B" is the event "doubles and a sum of 6." We write $A \cap B = \{(3, 3)\}$. Note that the events B and C are mutually exclusive. If one die shows a 6, the sum will be at least 7. Finally, note that $\bar{C}$ is the event "no 6."

example 58.2   When a single card is selected from a standard deck of fifty-two cards, the sample space is as shown in figure 58.5. If we define H = "heart," D =

figure 58.5

$$S = \begin{cases} A\clubsuit, K\clubsuit, Q\clubsuit, J\clubsuit, 10\clubsuit, \ldots, 2\clubsuit & \text{(clubs, black)} \\ A\diamondsuit, K\diamondsuit, Q\diamondsuit, J\diamondsuit, 10\diamondsuit, \ldots, 2\diamondsuit & \text{(diamonds, red)} \\ A\heartsuit, K\heartsuit, Q\heartsuit, J\heartsuit, 10\heartsuit, \ldots, 2\heartsuit & \text{(hearts, red)} \\ A\spadesuit, K\spadesuit, Q\spadesuit, J\spadesuit, 10\spadesuit, \ldots, 2\spadesuit & \text{(spades, black)} \end{cases}$$

"diamond," F = "face card," and R = "red card," then the event "heart or diamond," $H \cup D$, is just R, the collection of red cards. The event $H \cap F = \{K\heartsuit, Q\heartsuit, J\heartsuit\}$. Finally, $\bar{R}$ is just the collection of spades and clubs shown in rows 1 and 4 of S.

example 58.3   If an individual has blood type AB and is Rh-negative, the genotype is denoted by AB/rr. Alleles A and B occur on chromosome pair 9, while r is on chromosome pair 1. Thus, through meiosis, gametes of types A/r and B/r are produced in equal quantities. If the second parent is OO/Rr, gametes of type O/R and O/r are formed. Possible genotypes for the offspring are shown in the Punnett square of table 58.1. The event "blood type A" = $\{AO/Rr, AO/rr\}$,

table 58.1

	O/R	O/r
A/r	AO/Rr	AO/rr
B/r	BO/Rr	BO/rr

while the event "Rh-positive" = $\{AO/Rr, BO/Rr\}$. Denoting the first event by A and the second by R^+, we have

"Blood type A and Rh-positive" = $A \cap R^+ = \{AO/Rr\}$

and   "Type A or Rh-positive" = $A \cup R^+ = \{AO/Rr, AO/rr, BO/Rr\}$

In section 57 we assigned probability $P(E)$ to an event E by adding the probabilities or weights of each elementary event in E. If $A = \{a_1, \ldots, a_k\}$ and $B = \{b_1, b_2, \ldots, b_r\}$ are mutually exclusive events, then $A \cap B = \emptyset$ and $A \cup B$ consists of the $k + r$ elementary events

$$\{a_1, a_2, \ldots, a_k, b_1, b_2, \ldots, b_r\}$$

Hence,

$$P(A \cup B) = \omega(a_1) + \omega(a_2) + \cdots + \omega(a_k) + \omega(b_1) + \omega(b_2) + \cdots + \omega(b_r)$$
$$= P(A) + P(B)$$

This assignment of probability thus satisfies the three basic *probability axioms* given in the box.

---

probability axioms

1. $P(S) = 1$

2. If E is an event, $0 \le P(E) \le 1$.

3. If A and B are mutually exclusive events, then

$$P(A \cup B) = P(A) + P(B)$$

Axion **(3)** is used in our next example.

example 58.4   In the dice game of craps, the object is to throw "7 or 11" on the first roll. We have already seen in example 57.2 that

$$P(\text{"sum of 7"}) = \frac{1}{6} \text{ and } P(\text{"sum of 11"}) = \frac{1}{18}$$

Since these two events are mutually exclusive,

$$P(\text{"7 or 11"}) = P(\text{"7"}) + P(\text{"11"})$$
$$= \frac{1}{6} + \frac{1}{18} = \frac{2}{9}$$

Probabilities are often assigned to compound events without first assigning weights to elementary events. We must make sure that such assignments obey the probability axioms, as demonstrated in the following example.

example 58.5   Mr. Big owns two horses that will run in a race. He believes that each horse has a 60% chance of winning. Such an assignment of probability is *not consistent* with the axioms. If A is the event "horse 1 wins" and B the event "horse 2 wins," then

$$P(A \cup B) = P(A) + P(B) = .6 + .6 = 1.2$$

and axiom 2 is violated. We have assumed in applying axiom 3 that it is not possible for both horses to finish in a dead heat. ▤

All other laws of probability for compound events can be derived from the three basic axioms. As a first example, note that events A and $\bar{A} =$ "not A" are mutually exclusive. Since $A \cup \bar{A} = S$, axioms 1 and 3 give

$$1 = P(S) = P(A \cup \bar{A}) = P(A) + P(\bar{A})$$

We have therefore established:

> **Law 1** $P(\bar{A}) = 1 - P(A)$

Law 1 is illustrated in the following two examples.

example 58.6   In a family with three children, the probability that all three are boys was shown in example 57.6 to be $\frac{1}{8}$. If A is the event "all boys," then $\bar{A}$ is the event "at least one girl." Hence, using Law 1, we have

$$P(\text{"at least one girl"}) = 1 - P(A) = \frac{7}{8}$$

In a family of five children, the probability that all are boys is $1/2^5 = 1/32$. Again, by Law 1, $P(\text{"at least one girl"}) = 1 - 1/32 = 31/32$. ▤

It is often considerably easier to perform the probability computation involving $\bar{A}$ than to compute $P(A)$ directly. This is illustrated in the following incident reported in John Scarne's *A Complete Guide to Gambling*.

example 58.7   In 1952, "Fat the Butch" lost $49,000 by betting that he could throw a double 6 at least once in twenty-one rolls of two dice. Fat reasoned as follows: "The chances of rolling a double 6 are 1 in 36, and so in eighteen rolls the chances should be even. I'll give myself the advantage by rolling twenty-one times." After 12 hours of continuous dice rolling, he realized there must be something wrong with his logic. His opponent, a big-time gambler known as "The Brain," netted $49,000. Find the true probability of $A =$ "at least one double 6 in twenty-one rolls."

solution 58.7   Estimating $P(A)$ directly is a difficult task, because there are so many ways at least one double 6 could be achieved. However, we will compute $P(\bar{A})$, where $\bar{A}$ is the event "(6, 6) does not occur in twenty-one rolls."

Since each roll has 36 outcomes, the Multiplication Principle tells us that

$$\underbrace{36 \cdot 36 \cdots \cdots 36}_{\text{(21 times)}} = 36^{21}$$

is the number of elementary events for the twenty-one rolls. For each roll, there are thirty-five ways for double 6 *not* to occur, and therefore 35^{21} ways to observe no double 6 in twenty-one rolls. Since each elementary event should be equiprobable, we have

$$P(\bar{A}) = \frac{35^{21}}{36^{21}} = \left(\frac{35}{36}\right)^{21} = .5535$$

Hence, $P(A) = 1 - P(\bar{A}) = .4465.$ ▤

The event "*A* but not *B*" can be expressed in set notation as $A - B$ or $A \cap \bar{B}$. Note that events $A - B$ and $A \cap B$ are mutually exclusive, and their union is A, as can be seen in figure 58.6. Again, using axiom 3,

figure 58.6

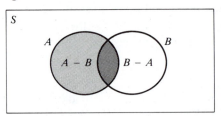

$P(A) = P(A - B) + P(A \cap B)$. We have therefore established:

> **Law 2** $P(\text{"}A \text{ but not } B\text{"}) = P(A) - P(A \cap B)$

Since $A \cup B$ is the union of the mutually exclusive events B and $A - B$, we have

$$P(A \cup B) = P(B) + P(A - B)$$
$$= P(B) + P(A) - P(A \cap B)$$

using Law 2. This demonstrates the following very useful probability law:

> **Law 3** $P(A \cup B) = P(A) + P(B) - P(A \cap B)$

Laws 2 and 3 are illustrated in the next two examples.

example 58.8   In rolling two dice, let A be the event "a sum of 6" and B the event "doubles." We have shown that $P(A) = \frac{5}{36}$ and $P(B) = \frac{6}{36}$. Now $A \cap B$ is just the elementary event $(3, 3)$, which has probability $\frac{1}{36}$. Hence, by Law 3,

$$P(\text{"sum of 6 or doubles"}) = P(A) + P(B) - P(A \cap B)$$
$$= \frac{11}{36} - \frac{1}{36} = \frac{10}{36} = \frac{5}{18}$$

Using Law 2,

$$P(\text{"sum of 6 but not doubles"}) = P(A - B)$$
$$= P(A) - P(A \cap B) = \frac{4}{36} = \frac{1}{9} \qquad \blacksquare$$

example 58.9   A study of a tribe in South America revealed that 75% were of blood type A and the rest were of blood type O. Sixty percent of the general population were Rh-negative, while 30% were Rh-positive and of blood type A. Using this information, find the probability that a member of the tribe is:

    (a) Type A or Rh-positive   **(b)** Type A and Rh-negative

    (c) Rh-positive but not type A **(d)** Type O and Rh-negative

solution 58.9   Let A be the event "type A blood" and R^+ the event "Rh-positive." We are given that $P(A) = .75$, $P(A \cap R^+) = .30$, and $P(\text{"Rh-negative"}) = P(\overline{R^+}) = .6$. Hence, by Law 1, $P(R^+) = .4$. The desired events can be located easily in the Venn diagram shown in figure 58.7.

figure 58.7

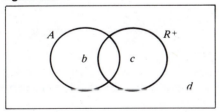

Now

$$P(\text{"Type } A \text{ or Rh-positive"}) = P(A \cup R^+)$$
$$= P(A) + P(R^+) - P(A \cap R^+)$$
$$= .75 + .4 - .30 = .85$$

$P(\text{"Type } A$ and Rh-negative"$) = P(\text{region } b) = P(A) - P(A \cap R^+) = .75$
$- .30 = .45$, and $P(\text{"Rh-positive but not Type } A") = P(\text{region } c) = P(R^+) -$
$P(A \cap R^+) = .4 - .3 = .1$. Finally, the event "type O and Rh-negative" is just the

event labeled d in the diagram. This event is the complement of $A \cup R^+$. Hence, $P(\text{"Type } O \text{ and Rh-negative"}) = 1 - P(A \cup R^+) = 1 - .85 = .15.$ ◼

When three events A, B, and C are involved, the generalization of Law 3 takes the form:

Law 4   the Inclusion-Exclusion Principle

$$P(A \cup B \cup C) = [P(A) + P(B) + P(C)] - [P(A \cap B)$$
$$+ P(A \cap C) + P(B \cap C)] + P(A \cap B \cap C)$$

A derivation of this law is outlined in exercise 15. When *all three events are mutually exclusive*, we have $A \cap B = A \cap C = B \cap C = A \cap B \cap C = \emptyset$ and the law simplifies to

$$P(A \cup B \cup C) = P(A) + P(B) + P(C)$$

The Inclusion-Exclusion Principle is put to work in our final example.

example 58.10   The dice game called "Chuck-a-Luck" was a popular carnival game around the turn of the century. It is played with three large dice that are tumbled in an hourglass-shaped cage. The player bets on one of the numbers from 1 to 6 and wins if the number comes up at least once. Suppose we bet on the outcome 1 (see figure 58.8). The following "con" was sometimes used to entice people to

figure 58.8

1	2	3	4	5	6	Betting layout
●						

play. "Each die has a 1 in 6 chance of landing on 1. Since there are three dice, your chances of winning are $\frac{1}{6} + \frac{1}{6} + \frac{1}{6} = \frac{1}{2}$." Explain the error and compute the true probability of winning.

solution 58.10   Let A be the event "1 on the first die." Likewise, let B and C be the corresponding events on the other two dice. Now $P(A) = P(B) = P(C) = \frac{1}{6}$ and we win if $A \cup B \cup C$ occurs. The law

$$P(A \cup B \cup C) = P(A) + P(B) + P(C)$$

holds *only* when the events A, B, and C are mutually exclusive. Clearly, this is *not* the case in the present experiment. Instead, the event $A \cap B$ (1 on each of the first two dice) occurs with probability $\frac{1}{36}$, and the event $A \cap B \cap C$ (1 on all three dice) occurs with probability $\frac{1}{216}$. Using Law 4, we then conclude

$$P(A \cup B \cup C) = P(A) + P(B) + P(C) - [P(A \cap B) + P(A \cap C) + P(B \cap C)] + P(A \cap B \cap C)$$

$$= \frac{1}{6} + \frac{1}{6} + \frac{1}{6} - \left[\frac{1}{36} + \frac{1}{36} + \frac{1}{36}\right] + \frac{1}{216}$$

$$= \frac{91}{216} = .421$$

The probability of any event formed from A, B, and C can be computed once we know $P(A)$, $P(B)$, $P(C)$ and the probabilities of the intersections given in Law 4. For example, region e in figure 58.9 is the event "A and C but

figure 58.9

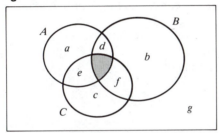

not B," and its probability is given by $P(A \cap C) - P(A \cap B \cap C)$. Region a is the event "A but neither B nor C" and has probability

$$P(A) - [P(A \cap B) + P(\text{region } e)]$$
$$= P(A) - [P(A \cap B) + P(A \cap C) - P(A \cap B \cap C)]$$

We will discover in section 59 when the probabilities of intersections can be computed in terms of $P(A)$, $P(B)$, and $P(C)$.

exercises for section 58

part A

1. In the Venn diagram shown, A is the event "male" and B is the event "right-handed." Express in words the events labeled a, b, and c in the diagram.

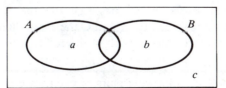

2. In the Venn diagram shown, M is the event "male," H is the event "a heavy smoker," and E is the event "developed emphysema."

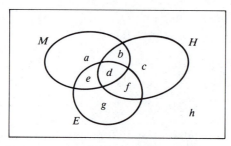

(a) Express in words the event labeled e.

(b) Locate the event "a female with emphysema who is not a heavy smoker."

(c) Locate the event $\bar{M} \cap \bar{H} \cap \bar{E}$ in the Venn diagram.

In the circuits shown in the figures, let A_i be the event "the ith fuse is operative," and let C be the event "current flows from left to right." In each case, write C in terms of A_1, A_2, and A_3.

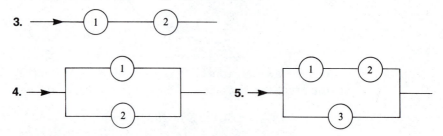

3.

4.

5.

6. For the game of roulette (see example 57.3), let R be the event "red," E the event "even," and C_1 the event "first column." Write each of the following events as subsets, and list all elementary events.

(a) "red and even"   (b) "red but not even"

(c) "column one but neither red nor even"

7. For a roll of two dice, let A be the event "a sum of 10," B the event "at least one 5," and C the event "doubles." Express each of the following events verbally and list all the elementary events.

(a) $A \cup C$ (b) $A \cap C$ (c) $A \cap \bar{B}$

8. A coin is tossed three times. Let A be the event "at least one head," B the event "at least one tail," and C the event "a head on the third toss." List all elementary events in each of the following events.

(a) $\bar{A}$ (b) $B \cap C$ (c) $A - B$

9. Which of the following pairs of events are mutually exclusive?

(a) For a roll of two dice: "a sum of 9" and "at least one 2"

(b) For a coin flipped twice: "at least one head" and "at least one tail"

(c) On a particular day: "rain" and "sunshine"

(d) For a single card selected from a standard deck: "red" and "a spade"

10. If B_1 and B_2 are mutually exclusive and A is a third event, draw a Venn diagram to illustrate that $A \cap B_1$ and $A \cap B_2$ are mutually exclusive.

11. If A and B are mutually exclusive, when are events $\bar{A}$ and $\bar{B}$ mutually exclusive?

12. If event A occurs only when event B occurs, use the probability laws to conclude that $P(A) \le P(B)$.

★13. For any two events A and B, use Laws 1–4 to conclude that

$$P(A \cap B) \le P(A) \le P(A \cup B) \le P(A) + P(B)$$

14. If $P(A) = P(B)$, show that $P(A \text{ but not } B) = P(B \text{ but not } A)$.

★15. Prove the Inclusion-Exclusion Principle (Law 4) by applying Law 3 first to $P[(A \cup B) \cup C]$ and then to $P[(A \cap C) \cup (B \cap C)]$.

In the following problems, use the probability laws to find the probability of the event in question.

16. When two dice are rolled, $P(\text{"doubles"}) = \frac{1}{6}$ and $P(\text{"sum of 5"}) = \frac{1}{9}$. Find $P(\text{"a sum of 5 or doubles"})$.

17. A single die is rolled six times. Find $P(\text{"at least one 6"})$.

18. In roulette, $P(\text{"black"}) = 18/38$, $P(\text{"column 3"}) = 12/38$, and $P(\text{"black and column 3"}) = 4/38$. Find $P(\text{"black or column 3"})$ and $P(\text{"column 3 but not black"})$.

19. If a single card is selected from a deck, then $P(\text{"queen"}) = \frac{1}{13}$ and $P(\text{"red"}) = \frac{1}{2}$. Find $P(\text{"queen or red"})$ and $P(\text{"queen but not red"})$.

20. If three dice are tossed, use the Multiplication Principle to establish that $P(\text{"all three numbers are different"}) = 120/216$. Next find $P(\text{"at least two numbers are identical"})$.

In the following problems, use a Venn diagram as an aid in computing the probability of the given event.

21. Given that $P(A) = .3$, $P(B) = .4$, and $P(B \text{ but not } A) = .2$, find:

(a) $P(A \cap B)$ **(b)** $P(A \cap \bar{B})$ **(c)** $P(A \cup B)$

22. In the population of Dry Gulch, 75% of the population are cowboys and 90% are beer drinkers. Only 5% stay away from beer and horses. Find the proportion of the population that are beer-drinking cowboys.

23. Last week's television survey revealed that 60% of all soap opera fans watch "Search for Yesterday," 50% view "Four Lives to Live," and 40%

watch both shows. Your mother-in-law is a devoted soap opera fan but you have no idea which shows she watches. Find the probability that she watches:

(a) at least one of the two shows

(b) only one of the shows

(c) neither of the two programs

24. Given events A, B, and C with $P(A) = .5$, $P(B) = .4$, $P(C) = .3$, and $P(A \cap C) = .1 = P(B \cap C) = .1 = P(B \cap C)$. If A and B are mutually exclusive, find the probability of:

(a) A but not C **(b)** A or B

(c) C but neither A nor B

(d) neither A nor B nor C

25. Determine whether each of the following assignments of probability is consistent with the probability axioms and laws.

(a) $P(A) = .8$, $P(B) = .7$, and $P(A \cap B) = .4$

(b) $P(A) = .6$, $P(\bar{B}) = .6$, and $P(A \cap B) = .2$

★ 26. For the game of "Chuck-a-Luck" in example 58.10, let $A =$ "1 on the red die," $B =$ "1 on the white die," and $C =$ "1 on the green die." As we point out in the example, $P(A) = P(B) = P(C) = \frac{1}{6}$, $P(A \cap B) = P(A \cap C) = P(B \cap C) = \frac{1}{36}$, and $P(A \cap B \cap C) = \frac{1}{216}$.

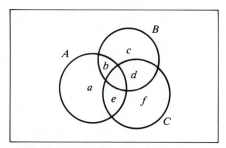

(a) Compute the probability of the events labeled a through f in the figure.

(b) Find $P($"exactly two 1s"$)$ and $P($"single 1"$)$ using the probabilities computed in **(a)**.

part B   The solutions to the following exercises use the permutation and combination formulas developed in the part B exercises of section 57.

27. One hundred tickets have been sold for a raffle and you have purchased six tickets. If five prizes are to be given, find the probability that you will win at least one prize.

28. In a five-card poker hand, let A be the event "no aces."

(a) Show that A can occur in $C[48, 5]$ ways.

(b) What is the event $\bar{A}$? Find $P(\bar{A})$.

29. At a gathering of n people, each individual is asked his or her birthday.

(a) Show that there are 365^n events in the sample space.

(b) If A is the event "no two people have the same birthday," show that there are $P[365, n]$ ways A could occur.

(c) Assuming each elementary event is equiprobable, compute $P(A)$. When is it reasonable to make this assumption?

(d) What is the event $\bar{A}$? Compute $P(\bar{A})$ for $n = 23$.

★**30.** An experiment is to be conducted in which k objects ($k \geq 3$) are to be placed in three boxes. For each object, a box is selected at random using the spinner shown in the figure and the object is then placed in that box.

(a)   **(b)**

Let A_i be the event "box i is unoccupied."

(a) Show that $P(A_i) = 2^k/3^k$.

(b) Show that $P(A_1 \cap A_2) = P(A_2 \cap A_3) = P(A_1 \cap A_3) = 1/3^k$.

(c) Find the probability that all boxes are occupied.

59 conditional probability and independence

When a patient with certain symptoms is first examined, the doctor formulates a set of possible diagnoses, to which he or she may attach certain subjective probabilities. With the results of a medical test, these probabilities might change. Some possibilities may be completely eliminated, while other diagnoses may become more probable. These new or **conditional probabilities** will again change as more information is added. Suppose then that a certain event B has just occurred. How do we compute the new probability of an event A? The following examples provide some insight.

example 59.1

A new shooter is about to roll the dice at a craps table. You have bet on "Big 8" and will win immediately if a sum of 8 occurs on the next roll. It is not difficult to see that if A = "sum of 8," then $P(A) = \frac{5}{36}$. As the dice settle at the far end of the table, one of the bystanders calls out "doubles." At this point in time, the collection of possible outcomes has been reduced to

$$B = \{(1, 1), (2, 2), (3, 3), (4, 4), (5, 5), (6, 6)\}$$

and of these possibilities, only $(4, 4)$ is favorable. Thus, your probability of obtaining a sum of 8 has increased to $\frac{1}{6}$. We will write $P(A$ given $B) = P(A|B) = \frac{1}{6}$. Note that in terms of the original probabilities,

$$P(A|B) = \frac{1/36}{6/36} = \frac{P(A \cap B)}{P(B)}$$

example 59.2

Tay-Sachs disease is a serious disorder of the nervous system that usually results in death by age two or three. Affected individuals are of genotype tt. The normal allele T is dominant and so genotypes Tt and TT are normal. Judy has a little brother with the disease and is concerned that she might carry the allele. Since both of Judy's parents must be of genotype Tt, the Punnett square (table 59.1) indicates the possibilities for their offspring. Since Judy is normal,

table 59.1

	T	t
T	TT	Tt
t	tT	tt

672

the possibility *tt* must be eliminated from the sample space. Hence,

$$P(\text{``Judy is } Tt\text{''}|\text{``Judy is normal''}) = \frac{2}{3}$$

Again, note that if A is the event "Judy is Tt" and B is the event "Judy is normal," then

$$P(A|B) = \frac{2/4}{3/4} = \frac{P(A \cap B)}{P(B)}$$

When we know that a certain event B has occurred, the sample space S shrinks to B, as shown in figure 59.1. Although there may be a strong

figure 59.1

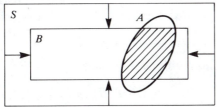

temptation to define $P(A|B)$ as $P(A \cap B)$, note that $P(B|B)$ has to be 1. This definition then will not suffice. If, however, we define $P(A|B) = P(A \cap B)/P(B)$, then $P(B|B)$ will be 1. Of course, this definition agrees with the probability considerations in examples 59.1 and 59.2.

definition 59.1   Let B be an event with $P(B) > 0$. Then we define the *conditional probability of A given B* by

$$P(A|B) = \frac{P(A \cap B)}{P(B)}$$

In the following example, we compute conditional probabilities from empirical data.

example 59.3   A test of the new sleeping pill "Bye-Bye" involved 200 individuals. One hundred were given "Bye-Bye" and the other 100 were given a sugar tablet that they thought was the drug. The next morning, all subjects were asked whether they thought they had slept better. The results are recorded in table 59.2. If B is the event "took Bye-Bye" and I is the event "improved sleep," note that $P(B \cap I) = 71/200 = .355$, $P(B) = 100/200 = .5$, and $P(I) =$

table 59.2

	Improved sleep	Did not sleep better
Took "Bye-Bye"	71	29
Took sugar pill	58	42

$(71 + 58)/200 = 129/200 = .645$. Hence,

$$P(I|B) = \frac{P(I \cap B)}{P(B)} = \frac{.355}{.5} = .71$$

and

$$P(B|I) = \frac{P(I \cap B)}{P(I)} = \frac{.355}{.645} = .55$$

By solving the equation $P(A|B) = P(A \cap B)/P(B)$ for $P(A \cap B)$, we obtain the important Multiplication Law for probabilities.

Law 5   the Multiplication Law

$$P(A \cap B) = P(A|B)P(B) \text{ and } P(A \cap B) = P(B|A)P(A)$$

The Multiplication Law is illustrated in the next two examples.

example 59.4

The probability that a basketball player will make his first free throw is .70. Once the first free throw is made, his confidence increases and the probability of his making the second shot is .80. Define A = "first free throw made" and B = "second free throw made." We are given that $P(A) = .7$ and $P(B|A) = .8$. Hence, by Law 5,

$$P(\text{"making both shots"}) = P(A \cap B) = P(B|A)P(A)$$
$$= (.8)(.7) = .56$$

example 59.5

An urn contains six white balls and four black balls, as illustrated in figure 59.2.

figure 59.2

The balls are mixed and then one ball is selected. Then, *without returning the first ball to the urn*, a second ball is selected. Let W_1 be the event "the first ball selected is white," and let W_2 be the event "the second ball selected is white." We will compute $P(W_2)$ using Law 5. We first note $P(W_1) = \frac{6}{10}$ and $P(\bar{W}_1) = P(\text{"black on the first draw"}) = .4$. Now, $P(W_2|W_1) = \frac{5}{9}$ since, given W_1, only five of the remaining nine balls are white. Likewise, $P(W_2|\bar{W}_1) = \frac{6}{9}$, since all six original white balls still remain. Hence, by the Multiplication Law,

$$P(W_1 \text{ then } W_2) = P(W_1 \cap W_2) = P(W_2|W_1)P(W_1) = \frac{5}{9} \cdot \frac{6}{10} = \frac{1}{3}$$

and

$$P(\bar{W}_1 \text{ then } W_2) = P(\bar{W}_1 \cap W_2) = P(W_2|\bar{W}_1)P(\bar{W}_1) = \frac{6}{9} \cdot \frac{4}{10} = \frac{4}{15}$$

The event W_2 occurs only when one of the two events W_1 or $\bar{W}_1$ precedes it. Hence, by axiom 3,

$$P(W_2) = P(W_1 \text{ then } W_2) + P(\bar{W}_1 \text{ then } W_2) = \frac{1}{3} + \frac{4}{15} = \frac{9}{15} \cdots$$

When the occurrence of an event B does not change the original probability that A will occur, we call events A and B **independent**.

definition 59.2   We say that events A and B are *independent* if $P(A|B) = P(A)$ and $P(B|A) = P(B)$.*

Note that when events A and B are independent, the Multiplication Law takes a special form.

> **Law 6**   If A and B are independent, then
>
> $$P(A \cap B) = P(A)P(B)$$

If $P(A|B) = P(A)$, we can automatically deduce that

$$P(B|A) = \frac{P(A \cap B)}{P(A)} = \frac{P(A|B)P(B)}{P(A)} = \frac{P(A)P(B)}{P(A)}$$
$$= P(B)$$

Hence, to show that events A and B are independent, we need verify only *one* of the two conditional probability equations in definition 59.2. This fact is used in our next example.

---

* We confine our attention to events with positive probabilities.

example 59.6

Two dice are tossed, one at a time. Let A be the event "6 on the first die," B the event "sum of 7," and C the event "sum of 8." We have shown that $P(A) = \frac{1}{6}$, $P(B) = \frac{1}{6}$, and $P(C) = \frac{5}{36}$. Which pair of events are independent?

solution 59.6

For events A and B, $P(B|A) = \frac{1}{6}$, since if 6 occurs on the first die, 1 must occur on the second for a sum of 7 to occur. Hence, $P(B|A) = P(B)$ and so events A and B are independent.

For events A and C, $P(C|A)$ is again $\frac{1}{6}$, since now 2 must occur on the second die. Since $P(C) = \frac{5}{36}$, $P(C|A) \neq P(C)$ and so events A and C are not independent.

Events B and C are mutually exclusive. Hence, $P(B|C) = [P(B \cap C)]/[P(C)] = 0/P(C) = 0$. Events B and C are therefore dependent. ▤

There is sometimes a tendency to confuse the notions of independent and mutually exclusive events. Note that in terms of conditional probabilities,

$$P(A|B) = \begin{cases} P(A) & \text{when } A \text{ and } B \text{ are independent} \\ 0 & \text{when } A \text{ and } B \text{ are mutually exclusive} \end{cases}$$

In most cases, independence is not something that is proven, but is perceived as being true (or at least reasonable) from the very nature of the experiment. If, in example 59.5, we had *returned* the first ball chosen to the urn before selecting the second ball, then the result of the second draw should be independent of the result of the first draw. Likewise, it is reasonable to assume that the sex of a second child is independent of that of the first child. Once such assumptions are made, Law 6 can be applied. This is demonstrated in our next two examples.

example 59.7

A craps shooter automatically wins if "7 or 11" is rolled first. Find the probability of two consecutive "7 or 11" throws.

solution 59.7

We have shown in example 58.4 that $P(\text{"7 or 11"}) = \frac{2}{9}$. It is sometimes hard for a dice shooter to believe that once he or she has thrown a "7 or 11," the probability of throwing "7 or 11" on the next toss is still $\frac{2}{9}$. Nevertheless, assuming independence, it follows that

$$P(\text{two consecutive "7 or 11" throws}) = \frac{2}{9} \cdot \frac{2}{9} = \frac{4}{81}$$

It is hard to explain why pit bosses will retire dice when a player has a run of good luck. Perhaps they are just as superstitious as the shooters. ▤

example 59.8

Judy and Jack Smith each had brothers who were afflicted with Tay-Sachs disease. As we saw in example 59.2, the probability that Judy is Tt is $\frac{2}{3}$, and the same holds for Jack.* Let A be the event "Judy is Tt," B the event "Jack is Tt," and let C be the event "their child is tt." If Judy and Jack are not related, then events A and B should be independent. Therefore, $P(A \cap B) = P(A)P(B) = \frac{4}{9}$. Given that Judy and Jack are both Tt, the probability is $\frac{1}{4}$ that their child will have Tay-Sachs disease. Thus, $P(C|A \cap B) = \frac{1}{4}$. Since the event C will occur only when event $A \cap B$ happens, we have

$$P(C) = P[C \cap (A \cap B)] = P(C|A \cap B)P(A \cap B)$$
$$= \frac{1}{4} \cdot \frac{4}{9} = \frac{1}{9}$$

More examples of independent events are provided by **Mendel's Second Law of Independent Assortment**: When the gene loci are on different chromosome pairs, $P(A_iA_j/B_rB_s) = P(A_iA_j)P(B_rB_s)$. Mendel's Second Law also holds for genes on the same chromosome, provided the gene loci are far enough apart. Such genes are called **unlinked**. The following example is one first presented by Mendel.

example 59.9   Yellow garden peas have genotype YY or Yy, while green pea plants are of genotype yy. A second characteristic studied by Mendel was the length of the plants. Tall plants are either TT or Tt, while short plants are tt. Plants with genotype combination Yy/Tt are first formed from pure stocks and then self-fertilized. Assuming the genes are unlinked, we can use Mendel's Second Law to predict the proportions of phenotypes in the next generation.

Note that $P(YY) = P(yy) = \frac{1}{4}$ and $P(Yy) = \frac{1}{2}$. Likewise, $P(TT) = P(tt) = \frac{1}{4}$ and $P(Tt) = \frac{1}{2}$. Hence,

$$P(\text{"yellow and tall"}) = P(YY/TT) + P(YY/Tt) + P(Yy/TT) + P(Yy/Tt)$$
$$= P(YY)P(TT) + P(YY)P(Tt) + P(Yy)P(TT) + P(Yy)P(Tt)$$
$$= \frac{1}{4} \cdot \frac{1}{4} + \frac{1}{4} \cdot \frac{1}{2} + \frac{1}{2} \cdot \frac{1}{4} + \frac{1}{2} \cdot \frac{1}{2}$$
$$= \frac{9}{16}$$

This result was verified experimentally by Mendel.

For a fixed event B, $P(\_\_|B)$ satisfies the three basic probability axioms of section 58. To see that axiom 3 is satisfied, note that if events A_1 and A_2 are mutually exclusive, then $A_1 \cap B$ and $A_2 \cap B$ are also mutually exclusive events,

---

*It should be remarked that there is now a standard biochemical test that can determine whether an individual carries the t allele.

figure 59.3

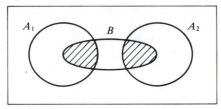

as illustrated in figure 59.3. Hence,

$$P(A_1 \cup A_2|B) = \frac{P[(A_1 \cup A_2) \cap B]}{P(B)} = \frac{P(A_1 \cap B) + P(A_2 \cap B)}{P(B)}$$
$$= P(A_1|B) + P(A_2|B)$$

It follows that P(__|B) also satisfies all of the probability laws that were deduced from the axioms. In particular,

$$P(\bar{A}|B) = 1 - P(A|B)$$

and

$$P(A_1 \cup A_2|B) = P(A_1|B) + P(A_2|B) - P(A_1 \cap A_2|B)$$

When three events A, B, and C are involved, the definition of independence takes the following form.

definition 59.3   Three events A, B, and C are called *independent* in case each pair of the events are independent and $P(C|A \cap B) = P(C)$.

Thus, if events A, B, and C are independent, and both A and B occur, then the probability of C remains unchanged. It is also not hard to prove that $P(A|B \cap C) = P(A)$ and $P(B|A \cap C) = P(B)$ when events A, B, and C are independent.

Using the Multiplication Law repeatedly yields

$$P(A \cap B \cap C) = P[C \cap (A \cap B)] = P(C|A \cap B)P(A \cap B)$$
$$= P(C|A \cap B)P(B|A)P(A)$$

When all three events are independent, the Multiplication Law reduces to a special form:

Law 7   When A, B, and C are independent,

$$P(A \cap B \cap C) = P(A)P(B)P(C)$$

Law 7 is illustrated in the following example.

example 59.10   Let A be the event "tall," B the event "dark," and C the event "handsome." Assume that $P(A) = .2$, $P(B) = .3$, and $P(C) = .1$. Assuming that the events A, B, and C are independent, find the probability of the events "tall, dark, and handsome," "tall and dark, but not handsome," and "tall, but neither handsome nor dark."

solution 59.10   Since all three events have been assumed to be independent, we can compute the probability of intersections by multiplication. To keep track of the events, we will use the Venn diagram in figure 59.4.

figure 59.4

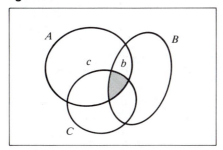

$$P(\text{"tall, dark, and handsome"}) = P(A \cap B \cap C) = P(A)P(B)P(C)$$
$$= (.2)(.3)(.1) = .006$$

Thus, the shaded area in the figure has probability .006. The event "tall and dark, but not handsome" is represented by region b in the Venn diagram, and

$$P(\text{region } b) = P(A \cap B) - P(A \cap B \cap C)$$
$$= (.2)(.3) - .006 = .054$$

Finally, the event "tall, but neither handsome nor dark" is represented by region c, and

$$P(\text{region } c) = P(A) - P(A \cap C) - P(\text{region } b)$$
$$= .2 - (.2)(.1) - .054 = .126$$

The definition of independence for more than three events can take many forms. Receiving our cue from Laws 6 and 7, however, we will define a collection of events to be independent if, given any finite subcollection E_1, $E_2, \ldots, E_n$ of the events,

$$P(E_1 \cap E_2 \cap \cdots \cap E_n) = P(E_1)P(E_2) \cdots P(E_n)$$

Again, independence is something that is usually assumed reasonable from the nature of the chance phenomena as the following example demonstrates.

example 59.11   If we assume that births of girls and boys are equiprobable and independent, then the probability that in a family with six children, all will be the same sex is given by

$$P(BBBBBB) + P(GGGGGG) = \left(\frac{1}{2}\right)^6 + \left(\frac{1}{2}\right)^6 = \frac{1}{32}$$

Mr. and Mrs. Emory Landon Harrison of Johnson City, Tennessee, have a family of thirteen sons and no daughters. How rare is such an event? Again, assuming independence, we have

$$P(13 \text{ sons in a row}) = \left(\frac{1}{2}\right)^{13} = \frac{1}{8,192} = .00012$$

exercises for section 59

part A   In each of the following problems, given that the event B has occurred, compute the conditional probability $P(A|B)$. Then decide if events A and B are independent.

1. In tossing two dice, let B = "1 on the red die" and A = "sum of 7."

2. In tossing two dice, let B = "both numbers are odd" and A = "sum of 6."

3. In selecting a card from a standard deck, let B = "an ace" and A = "a club."

4. For the slot machine in exercise 20 of section 57, let B = "a lemon on the first reel" and A = "three lemons."

5. For the slot machine in exercise 20 of section 57, let B = "plums on the first two reels" and A = "three plums."

6. In the game of roulette (see example 57.3), let B = "black" and A = "even."

7. In the game of roulette (see example 57.3), let B = "black" and A = "third column."

★ 8. For the spinner in exercise 22 of section 57, let B = "an even number is selected" and A = "2 has been selected."

9. A couple have known blood genotypes BO and AB. Let B = "their child has blood type B" and A = "their child has genotype BO."

10. For two individuals each flipping a coin, let B = "the coins match" and A = "the first person's coin shows head".

11. A study of 150 couples concerning their attitudes toward a piece of legislation produced the results shown in the table. If W = "wife favors legislation" and H = "husband favors legislation," estimate $P(W|H)$ and $P(H|W)$.

	Husband favors legislation	Husband is against legislation
Wife favors legislation	50	25
Wife is against legislation	35	40

12. One thousand entering college freshmen were polled to study the relationship between hair and eye color, with the results shown in the table. If A is the event "light hair" and B is the event "light eyes," estimate $P(A|B)$, $P(B|A)$, $P(\bar{A}|B)$, and $P(\bar{B}|\bar{A})$.

	Light hair	Dark hair
Light eyes	350	120
Dark eyes	230	300

13. In a study of the survival rates of elk on a game preserve, a large group of young elk were followed over a 10-year period. Eighty percent were alive after 5 years, while 50% were still living after 10 years. Estimate the probability that a 5-year-old elk will live an additional 5 years.

14. If $P(A) = P(B)$, show that $P(A|B) = P(B|A)$.

15. If event A occurs only when event B occurs, find $P(A|B)$ and $P(B|A)$.

16. Joan feels there is an 80% chance that she will make the volleyball team, and, if she makes the team, there is a 50% chance that she will be on the starting team. Find the probability that she will make the team and start.

17. The probability that it will rain tomorrow is .7, and, if it does rain tomorrow, there is a 90% chance that the rain will continue another day. Find the probability that it will rain the next two days.

18. Welcome to that new game show "Urn a Fortune." Two $10,000 bills are placed in an urn together with ninety-eight $1 bills. You are allowed to select two bills from the urn. Find the probability that you will win $20,000.

19. Ten blue socks and eight red socks are thoroughly mixed up in a drawer. Waking up at 5:00 A.M. for work, Cosmo does not want to wake up his wife by turning on the lights. He selects two socks at random from the drawer. Find the probability that the socks will match.

20. A ball is selected at random from the first urn shown in the figure and is

Urn 1   Urn 2

transferred to the second urn. The contents of urn 2 are thoroughly mixed and then a single ball is selected and transferred back to urn 1. Find the probability that urn 2 will contain 4 black balls.

21. Let A and B be events with $P(A) = P(B) = .5$ and $P(A \cap B) = .25$. Are events A and B independent?

22. Find the probability that the event "red" will occur three times in a row in the game of roulette (refer to example 57.3).

23. Assume that births of girls and boys are equiprobable. Find the probability of:

    (a) all girls in a family with six children.

    (b) at least one boy in a family with six children.

24. Two brothers run a family business. The chances that the brothers will be alive 20 years from now are .8 and .7, respectively. Assuming that the events in question are independent, find the probability that:

    (a) both brothers will be alive 20 years from now.

    (b) at least one of the two brothers will be alive to run the family business 20 years from now.

25. The cure rate for a certain disease is 85%. If a couple have both contracted the disease, find the probability that:

    (a) both will be cured.

    (b) neither will be cured.

    (c) only one of the two will be cured.
    What must you assume in order to carry out the calculations?

26. In example 59.10, page 679, find the probability of the event

    (a) "handsome, but neither tall nor dark."

    (b) "tall and handsome, but not dark."

27. In example 59.9 page 677, find the probability that a garden peak plant is:

    (a) "green and short."

    (b) "yellow and short."

28. By examining Dick's family history, it has been determined that the probability that he carries the recessive allele c responsible for cystic fibrosis is $\frac{1}{6}$. Similarly, Jane's chance of being Cc is $\frac{1}{3}$.

    (a) Find the probability that both Dick and Jane are carriers of the c allele.

    (b) Find the probability of producing a child with cystic fibrosis.

29. Mary is a 25-year-old woman with two children. She has just learned that her father has Huntington's Chorea, a serious degenerative disease of the nervous system that often does not manifest itself until late in life. This

affliction is caused by the presence of a single dominant allele D. Thus, Mary's father is now known to have genotype Dd. Assuming that Mary's mother and husband are both of genotype dd (which is justified by the rareness of the disease), find the probability that:

(a) Mary has Huntington's Chorea.

(b) neither of Mary's children has Huntington's Chorea, given that Mary has the disease.

30. In cattle, coat color is determined by two alleles R and r, and the possible coat colors are red (RR), white (rr), and roan (Rr). In addition, horned cattle have genotype hh, while hornless cattle are either HH or Hh. A hornless roan cow is mated to a horned roan bull and a horned white calf results. Assuming the genes are unlinked, find the probability that the next calf from this mating will be horned and roan.

★31. In parakeets, the color of the feathers is determined by the combined effect of two unlinked sets of genes. Let A and a be the alleles at one locus and B and b the alleles at the second locus. Shown in the table are the colors resulting from various genotypes. A yellow female and a blue male have produced white offspring.

(a) What are the genotypes of the parents?

(b) If this pair were to have many offspring, what proportions would you expect to be green, yellow, blue, and white?

(c) Answer question **(b)**, assuming that both parents are Aa/Bb.

Phenotype	green	yellow	blue	white
Genotypes	$A\_/B\_$	$aa/B\_$	AA/bb	$\_a/bb$

part B

binomial probabilities   In many of our examples and exercises, we have assumed that births of girls and boys are equiprobable and then we proceeded to compute the probability of a certain number of boys and girls in a particular family. In the United States, the proportion of male births, is actually between .51 and .52. If $p = P(\text{"male birth"})$, we may ask, for example, what is the probability of k boys in a family of five children?

32. Find $P(BBBBB)$ in terms of p.

33. Show that $P(BBBBG) = p^4(1-p)$. Conclude that the probability of exactly one boy in five children is $5p^4(1-p)$.

34. Compute $P(BBBGG)$ and $P(BGBGB)$.

35. Show that there are $C[5, 2]$ different ways that a family with two boys in five children can be formed. Conclude that

$$P(\text{"2 boys in 5 children"}) = C[5, 2]p^2(1-p)^3$$

The general expression for k boys among n children is the *binomial probability*

$$C[n, k] p^k (1 - p)^{n-k}$$

The p^k comes from the k boys, $(1 - p)^{n-k}$ from the $(n - k)$ girls, and $C[n, k]$ reflects the number of ways of forming such a family.

36. Find the probability of three boys and three girls in a family with six children if $P(\text{"male birth"}) = .52$.

sequential experiments and Bayes' theorem

Many of the more interesting applications of probability models involve a sequence of experiments $\mathscr{E}_1, \mathscr{E}_2, \ldots, \mathscr{E}_n$, where experiment $\mathscr{E}_2$ is performed after $\mathscr{E}_1$ and the probabilities associated with $\mathscr{E}_2$ may depend on the outcome of $\mathscr{E}_1$. A simple sequential experiment considered in section 59 was the repeated rolling of dice, as depicted in figure 60.1. Since in this example the results of

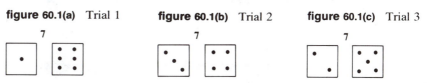

figure 60.1(a) Trial 1   **figure 60.1(b)** Trial 2   **figure 60.1(c)** Trial 3

each experiment are independent of one another, we have

$$P(\text{``three 7s in a row''}) = \frac{1}{6} \cdot \frac{1}{6} \cdot \frac{1}{6} = \frac{1}{216}$$

The sequential experiment involving the urns illustrated in figure 60.2 does not submit to such an easy analysis. What is the probability that the final ball

figure 60.2

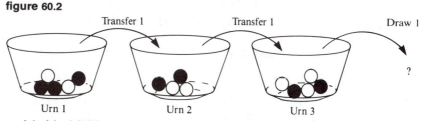

selected is black? Here the result of experiment $\mathscr{E}_3$ depends heavily on what occurred in experiments $\mathscr{E}_2$ and $\mathscr{E}_1$.

A third example of a sequential experiment comes from an analysis of human pedigrees. Shown in figure 60.3 are the standard symbols used in

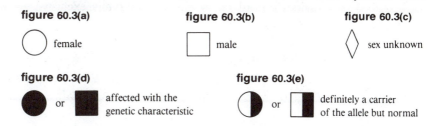

figure 60.3(a)   **figure 60.3(b)**   **figure 60.3(c)**

female   male   sex unknown

figure 60.3(d)   **figure 60.3(e)**

or affected with the genetic characteristic   or definitely a carrier of the allele but normal

figure 60.3(f)

a mating

figure 60.3(g)

siblings

recording a family's genetic history. From a mathematical point of view, a family pedigree can be considered as a sequential experiment in which genetic information is transferred from generation to generation, as depicted in figure 60.4. We might ask, "What is the chance that granddaughter (III-1) carries the

figure 60.4

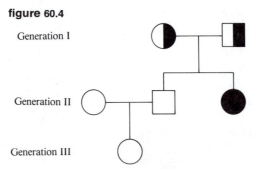

Generation I

Generation II

Generation III

allele in question?" Again, the calculation depends on exactly what has occurred in the second generation.

The main tool we will use in analyzing sequential experiments is the **Partition Theorem** or **Law of Total Probability**. Suppose first that the sample space S has been partitioned into two mutually exclusive events B_1 and B_2, as shown in figure 60.5. Hence, $B_1 \cup B_2 = S$ and $B_1 \cap B_2 = \emptyset$. This partition

figure 60.5

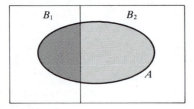

divides a third set A into two mutually exclusive events $A \cap B_1$ and $A \cap B_2$. Hence, by axiom 3,

$$P(A) = P(A \cap B_1) + P(A \cap B_2)$$

By the Multiplication Law $P(A \cap B_1) = P(A|B_1)P(B_1)$ and $P(A \cap B_2) = P(A|B_2)P(B_2)$. We have therefore established:

Law 8   Partition Theorem
If B_1 and B_2 form a partition of the sample space S, then

$$P(A) = P(A|B_1)P(B_1) + P(A|B_2)P(B_2)$$

When S is partitioned into n disjoint events, $B_1, B_2, \ldots, B_n$, as shown in figure 60.6, then $P(A) = \sum_{k=1}^{n} P(A \cap B_k) = \sum_{k=1}^{n} P(A|B_k)P(B_k)$, using Law 5. In summary:

Law 9   Generalized Partition Theorem
If mutually exclusive events $B_1, B_2, \ldots, B_n$ partition the sample space S, then

$$P(A) = \sum_{k=1}^{n} P(A|B_k)P(B_k)$$

figure 60.6

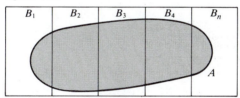

In the next example, we will use the Partition Theorem to analyze the urn experiment presented at the beginning of this section.

example 60.1   A single ball is selected at random from urn 1 and transferred to urn 2, as shown in figure 60.7. Finally, a ball is randomly selected from urn 2. Find the

figure 60.7

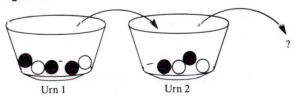

Urn 1   Urn 2

probability that the final ball chosen is black. What is the probability that the final ball chosen will be white?

solution 60.1   Let $B_1 =$ "the ball transferred is white" and $B_2 =$ "the ball transferred is black." Since exactly one of the two mutually exclusive events B_1 and B_2 will occur, B_1 and B_2 partition the sample space. If A is the event "the final ball selected is black," note that $P(A|B_1) = 2/5 = .4$ and $P(A|B_2) = 3/5 = .6$. Hence, by Law 8,

$$P(A) = P(A|B_1)P(B_1) + P(A|B_2)P(B_2)$$
$$= (.4)(.4) + (.6)(.6) = .52$$

It follows that $P(\bar{A}) = 1 - P(A) = .48$. This is the probability of selecting a white ball.　　　　　　　　　　　　　　　　　　　　　　　　　　　　　　▤

example 60.2   Again, we have the same question as shown in example 60.1 but we show an alternative solution.

solution 60.2   A second, more mechanical way of solving this problem involves the use of the **probability tree** shown in figure 60.8. On the diagram we place **transition**

figure 60.8

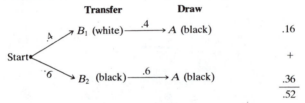

probabilities. For example, $P(B_1) = .4$ is the probability of reaching B_1 from the start of the experiment. Given that we have reached $B_1 =$ "white," $P(A|B_1)$ is the transition probability from B_1 to A. If we multiply along the paths to A and then add, we obtain, as in solution 1, $P(A) = .52$.　　　　　　　　　　　▤

The General Partition Theorem can be neatly expressed in tree diagram form, as shown in figure 60.9. Note that $P(B_1)$ is the probability of progressing from the start of the experiment to event or *state* B_1. Having arrived at B_1, we

figure 60.9

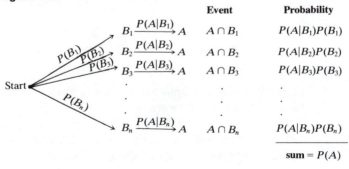

then insert $P(A|B_1)$, the transition probability to event A. By multiplying probabilities along the various paths to A, and then adding, we obtain $P(A)$. We will use probability trees in the examples to follow.

example 60.3   We are now ready to solve the urn transfer problem posed at the beginning of this section. In example 60.1 we showed that the probability of selecting a black ball from urn 2 is .52. If we have transferred a black ball to urn 3, then the probability of drawing a black ball is now $\frac{3}{6} = .5$. By forming the probability tree shown in figure 60.10, we find that $P(\text{"final ball drawn is black"}) = .42$. The probability that the final ball selected is white is then $1 - .42 = .58$.

figure 60.10

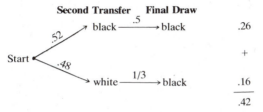

example 60.4   Shown in figure 60.11 is a family pedigree for albinism. As in example 59.2, $P(\text{"Jay is } Aa\text{"}) = \frac{2}{3}$. Assuming that Mary's family has no history for the trait,

figure 60.11

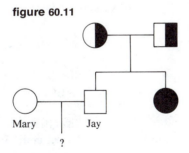

we will suppose that she is genotype AA. If Jay is Aa, then there ia 50% chance that he will pass the allele a to the child. Then $P(\text{"child is } Aa\text{"}) - 1/3$, as shown in figure 60.12.

figure 60.12

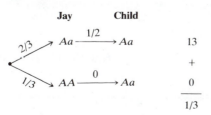

On the other hand, the incidence rate of albinism in the general population is roughly 1 in 20,000. As we will show later in the text, this implies that there is a slight chance of about .02 that Mary is *Aa*. Taking this into account, we obtain the revised tree diagram shown in figure 60.13. Note that in computing

figure 60.13

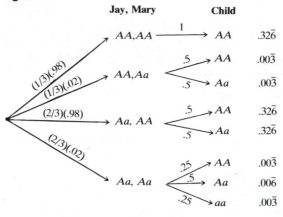

P("Jay is *AA* and Mary is *AA*"), for example, we have assume independence of the two events in question. Hence

$$P(\text{``Child is } aa\text{''}) = .00\overline{3} = \frac{1}{300}$$

and

$$P(\text{``Child is } Aa\text{''}) = .00\overline{3} + .32\overline{6} + .00\overline{6} = .3366$$

Although the probability of having an albino child has increased from 1 in 20,000 to 1 in 300, the probability is still quite small.

The final mathematical tool of this section is **Bayes' Theorem**, which gives a way of computing $P(B|A)$ in terms of **reversed conditional probabilities** of the form $P(A|\_)$. Suppose that S is partitioned by mutually exclusive events B_1 and B_2. By the definition of conditional probability, we have

$$P(B_1|A) = \frac{P(A \cap B_1)}{P(A)}$$

But by the Multiplication Law, $P(A \cap B_1)$ can also be written as $P(A|B_1)P(B_1)$. Hence,

$$P(B_1|A) = \frac{P(A|B_1)P(B_1)}{P(A)}$$

and by Law 8, the denominator can be calculated as:

$$P(A) = P(A|B_1)P(B_1) + P(A|B_2)P(B_2)$$

The more general form of Bayes' Theorem is as follows:

Law 10   Bayes' Theorem
  If S is partitioned by mutually exclusive events $B_1, B_2, \ldots, B_n$, then

$$P(B_k|A) = \frac{P(A|B_k)P(B_k)}{P(A)}$$

where $P(A) = \sum_{i=1}^{n} P(A|B_i)P(B_i)$.

As we will show in the examples, the terms in the numerator and denominator of Bayes' Theorem can be read directly from the probability tree.

example 60.5   Footprints indicating a shoe size of 7 or 8 have been found at the scene of the crime. Fifty-five percent of the population in this area are female, and it is estimated that 25% of all women wear this size, whereas only 5% of all men wear a shoe this small. Find the probability that the footprints belong to a female.

solution 60.5   If A is the event "size 7–8 shoe," we are given that $P(A|\text{"female"}) = .25$, while $P(A|\text{"male"}) = .05$. We wish to compute the reversed conditional probability $P(\text{"female"}|A)$. The probability tree for this problem is shown in figure 60.14.

figure 60.14

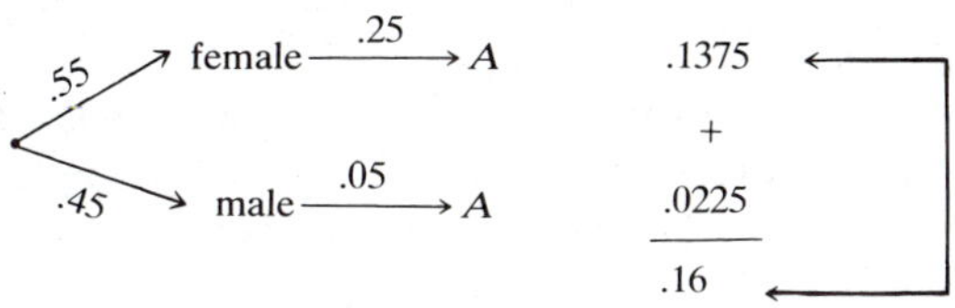

We see that $P(A) = .16$, and by applying Bayes' Theorem, we obtain

$$P(\text{"female"}|A) = \frac{P(A|\text{"female"})\, P(\text{"female"})}{P(A)}$$

$$= .1375/.16 = .86$$

Note that the numerator appears directly in the probability tree.

  Bayes' Theorem has proven to be a useful tool in assessing probable diagnoses from symptoms of clinical tests, as discussed in example 60.6.

example 60.6   Among individuals who have sustained head injuries, X rays reveal that only about 6% have skull fractures. One of the standard symptoms of a skull fracture is nausea, which occurs in 98% of all skull fracture cases. With other types of head injuries, nausea is present in about 70% of all cases. An individual who has just suffered a head injury is not nauseous. Find the probability that he has a skull fracture.

solution 60.6   Let N be the event "nausea" and F the event "skull fracture." We are given that $P(F) = .06$, $P(N|F) = .98$, and $P(N|\bar{F}) = .70$. We are asked to compute the reversed conditional probability $P(F|\bar{N})$. Note that $P(\bar{N}|F) = 1 - P(N|F) = .02$ and $P(\bar{N}|\bar{F}) = 1 - P(N|\bar{F}) = .30$. From the probability tree shown in figure

figure 60.15

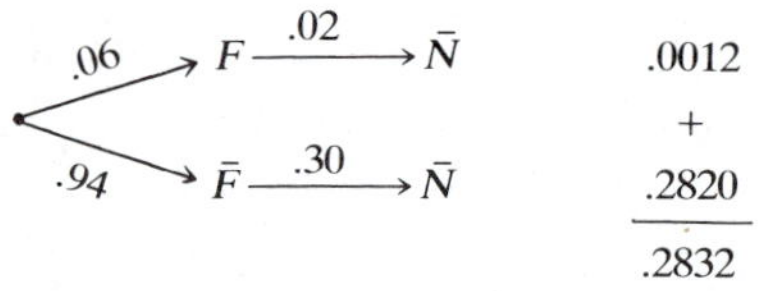

60.15, it follows that $P(\bar{N}) = .2832$. Therefore, by Bayes' Theorem, we conclude

$$P(F|\bar{N}) = \frac{P(\bar{N}|F)P(F)}{P(\bar{N})} = \frac{.0012}{.2832} = .004$$

Hence, it is highly unlikely that an individual without the symptoms actually has a fracture. It has therefore been suggested that such an individual not be subjected to the risks of X rays.

You may have noticed that the major difficulty in examining human pedigrees comes from the uncertainty in identifying a normal person as *Aa* or *AA*. There are many tests for identifying *heterozygotes* (individuals of genotype *Aa*). Unfortunately, some tests are not completely reliable, and Bayes' Theorem must be used to compute the new probabilities, as illustrated in example 60.7.

example 60.7   *Phenylketonuria* (PKU) is a genetic disease caused by two recessive alleles *aa*. Affected individuals lack the ability to produce the key enzyme needed to process *phenylalanine*. As a result, this substance accumulates in the body and causes severe mental retardation. Since the normal allele *A* codes for the key enzyme, there is a possibility that enzyme levels in *AA* individuals can be made to differ from *Aa* genotypes. Although this fact has been used to develop tests for heterozygosity, the tests are not perfect. Suppose that, *among known Aa individuals*, a test correctly identifies the presence of the *a* allele in 80% of all cases. On the other hand, among known AA individuals, the test indicates the possible presence of the a allele in 20% of all cases.

Let $+$ be the event "test indicates that the a allele is present." We are therefore given that $P(+|Aa) = .8$ and $P(+|AA) = .2$. For the pedigree shown in figure 60.16, P(individual II-1 is Aa) $= \frac{2}{3}$. What is $P(Aa|+)$ in this case?

figure 60.16

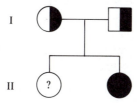

figure 60.17

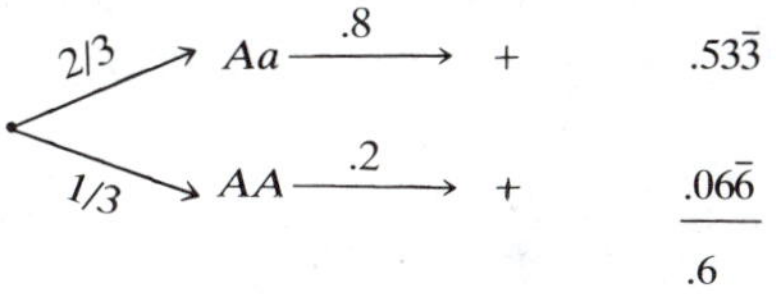

$$
\begin{array}{llll}
2/3 \nearrow \ Aa \xrightarrow{\ .8\ } & + & & .53\overline{3} \\
\searrow \ AA \xrightarrow{\ .2\ } & + & & \underline{.06\overline{6}} \\
1/3 & & & .6
\end{array}
$$

From the probability tree shown in figure 60.17, it can be seen that $P(+) = .6$ and so

$$P(Aa|+) = \frac{P(+|Aa)\,P(Aa)}{P(+)} = \frac{.53\overline{3}}{.6} \approx .89$$

If the test results are positive, then the probability that individual II-1 is Aa increases from $\frac{2}{3}$ to .89.

When an individual has no family history for PKU, it can be shown $P(Aa)$ is about .02. In such cases, the probability tree is as shown in figure 60.18.

figure 60.18

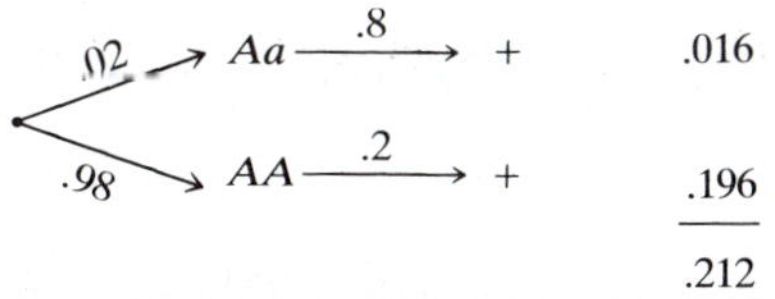

$$
\begin{array}{llll}
.02 \nearrow \ Aa \xrightarrow{\ .8\ } & + & & .016 \\
\searrow \ AA \xrightarrow{\ .2\ } & + & & \underline{.196} \\
.98 & & & .212
\end{array}
$$

Thus, $P(Aa|+) = \dfrac{.016}{.212} \approx .075$, indicating that the test result is of little value.

example 60.8   The genetic disease *hemophilia*, which is characterized by the inability of blood to clot properly, is the classic example of a **sex-linked trait**. The allele h responsible for this trait resides on the X sex chromosome. Since males have chromosome pair XY, affected males are of genotype X^hY. The X^h notation emphasizes that the h allele is on the X chromosome. The normal allele H is dominant over h. Therefore, affected females are of genotype X^hX^h and their fathers must be hemophiliacs. Such cases are extremely rare. Male offspring with hemophilia generally result from the mating of an X^HX^h female and a normal X^HY male, as shown in table 60.1. Hence, $P(\text{"affected child"}) = \frac{1}{4}$ and the probability of a normal son is $\frac{1}{4}$.

table 60.1

Egg \ Sperm	X^H	Y
X^h	X^HX^h	X^hY
X^H	X^HX^H	X^HY

Shown in figure 60.19 is a family pedigree for hemophilia. Except for Jean, the genotypes of all individuals in the pedigree are known. Normal males must

figure 60.19

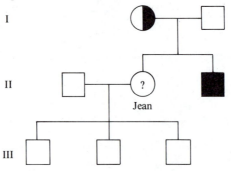

be X^HY. Since the hemophiliac received his X chromosome from his mother, the mother (I-1) is X^HX^h. There is a 50% chance that Jean received the h allele from her mother. The fact that Jean has three normal sons leads us to believe that she is much more likely to be X^HX^H than X^HX^h. If Jean is X^HX^h, then the probability of three normal sons in a row is, by independence, $\frac{1}{4} \cdot \frac{1}{4} \cdot \frac{1}{4} = \left(\frac{1}{4}\right)^3$. From Bayes' Theorem and the tree diagram of figure 60.20, we see

figure 60.20

Jean

$.5 \rightarrow X^HX^h \xrightarrow{(1/4)^3} 3$ normal sons   .0078

                                                                        +

$.5 \rightarrow X^HX^H \xrightarrow{(1/2)^3} 3$ normal sons   .0625
                                                                        _____

                                                                        .0703

that

$$P(\text{``Jean is } X^H X^H \text{''}|\text{``3 normal sons''}) = \frac{.0625}{.0703} = .889$$

There is now about a 90% chance that Jean does not carry the h allele.

exercises for section 60

part A

[*Note*: In the following exercises, use a probability tree as an aid in implementing either the Partition Theorem or Bayes' Theorem.] Shown in the figure are two urns containing black and white balls.

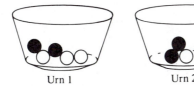

Urn 1   Urn 2

1. If two balls are selected from urn 1, find the probability that the second ball chosen is white.

2. A ball is transferred from urn 1 to urn 2, and then a single ball is selected from urn 2. Find the probability that the ball chosen is white.

3. One of the two urns is selected at random (by a coin flip) and a single ball is withdrawn. Find the probability that the ball chosen is black.

4. Seventy percent of the registered voters in a town are Democrats and 30% are Republicans. Eighty percent of the Republicans favor Proposition X, while only 30% of the Democrats will vote for X.

   (a) If all registered voters show at the polls, predict the percentage of the voters that will support Proposition X.

   ★**(b)** If all registered Republicans vote, how low must the Democratic voter turnout be for Proposition X to win?

5. Two evenly matched teams play in a "best two out of three game" series. Find the probability that the series goes to three games.

6. Twenty percent of the rabbits in a population are weak and their chances of being captured by a fox are .5. Among stronger rabbits, there is a 10% chance of being caught. Predict the proportion of rabbits that will be caught.

7. The San Diego Chargers and the Cincinnati Bengals are set to play in Cincinnati. If it does not snow, the probability that the Chargers will win is .6. In snow, the Bengals' chances of winning increase to 70%. If the weatherman predicts a 60% chance of snow, compute the probability, at this point in time, that the Chargers will win.

8. Suppose you are in Las Vegas with $72 and you need about twice this much to buy a bus ticket home. You decide to play roulette (see example 57.3) and adopt the following strategy. You will first bet $36 on one of the columns. If you win, the payoff is 2 to 1, and so you will take the additional $72 and head for home. If you lose, you will take the remaining $36 and bet $4 on each of nine numbers. Since the payoff on a single number is 35 to 1, you will have $4(35) + 4 = 144 for bus fare.

   (a) Find the probability that this strategy will succeed.

   (b) Is the suggested strategy any better than simply making one bet of $72 on "red"?

9. The accompanying pedigree shows the mating of two half-siblings. If the

   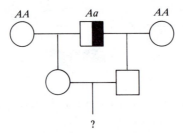

   allele a is recessive, find the probability that an offspring will be

   (a) Aa

   (b) aa

10. Pattern baldness is a trait caused by the presence of a dominant B allele. The trait, however, does not express itself in Bb females who appear perfectly normal. Shown here is a pedigree for this trait.

   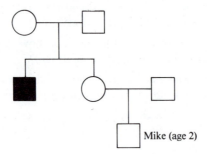

   Mike (age 2)

   (a) What are the genotypes of Mike's grandparents?

   (b) What is the probability Mike will not be bald?

11. Refer to the accompanying figure. A single ball is transferred from urn 1 to urn 2. After mixing the balls in urn 2, a ball is passed to urn 3. A final ball is selected from the contents of urn 3. Find the probability that the final ball selected is white.

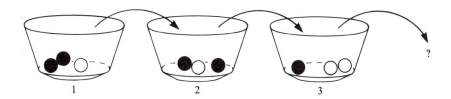

12. In a certain city, 30% of the registered voters are Republican, 60% are Democrats, and the remainder are Independents. A poll reveals that Proposition A is favored by 80% of all Republicans. Among Democrats and Independents, the figures are 50% and 30%, respectively.

(a) Find the probability that a voter favors Proposition A.

(b) If an individual supports Proposition A, what is the chance that he or she is a Democrat?

13. Voice analyzers are used to detect tiny modulations in a person's voice when that person is lying. To test the effectiveness of this device, an individual was instructed to lie on half the responses. When a lie was told, the voice analyzer indicated "lie" in 90% of all cases. When the truth was told, the device indicated "lie" in only 5% of the cases. Based on these results, find the probability that the person is lying, given that the device indicates "lie".

14. In example 60.6, compute the probability of a skull fracture given that the patient is nauseous.

15. In example 60.7, find the probability that individual II-1 is of genotype *AA*, given that the test is negative.

16. You are leaving on a long summer vacation and have no choice but to leave your pet parrot Polly with the next door neighbor Mrs. Kravitts. Mrs. Kravitts is rather forgetful and there is a 50% chance that she will forget to feed Polly. Without food, Polly is certain to find eternal rest. If Polly is fed, there is a slight chance of 5% that Polly will not survive. Upon returning, you find Polly lying in her cage—stiff as a board. "Polly is only sleeping," insists Mrs. Kravitts. Find the probability that Polly was not fed.

17. Tests for the presence of the *h* hemophilia allele in normal women are not completely reliable.* Assume that in known *Hh* women, a certain test correctly identifies the presence of the *h* allele in 85% of all cases. Among known *HH* women, the test fails in only 1% of all cases. In example 60.8, we pointed out that the probability that Jean is *Hh* is $\frac{1}{2}$. In her case, find:

(a) P(Jean is *Hh*|test is positive)

(b) P(Jean is *HH*|test is negative)

* See *Genetic Counseling* by W. Fuhrmann and F. Vogel, Springer-Verlag, 1976, Chapter 6.

18. *Cystinosis* is a rare hereditary disease that leads to the accumulation of cystine crystals in bone marrow, the cornea, and internal organs. In patients with cystinosis (*cc*) the concentration of free cystine in white blood cells is eighty times higher than in normal (*CC*) individuals. Among known *Cc* individuals, the concentration levels are in general much higher than normal. Among heterozygotes, only one in nine show normal concentration levels, while among *CC* genotypes, all results are in the normal range.*

From a family pedigree, it has been determined that the probability that Jack carries the *c* allele is $\frac{2}{3}$. With this information, find:

   (a) *P*(Jack is *Cc*|high concentration level)

   (b) *P*(Jack is *CC*|normal concentration level)

19. Shown here is a family pedigree for Huntington's Chorea. This genetic disease is caused by the presence of a single dominant allele *D* and has the unusual property that it does not manifest itself until age 35 or more.

   (a) Find the probability that Mary has Huntington's Chorea.

   (b) Find the probability that neither of Mary's daughters is affected.

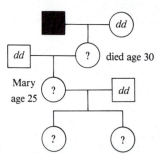

20. The accompanying probability tree shows three possible diagnoses for thyroid disease: *A* (hypothyroidism), *B* (euthyroidism), and *C* (hyperthyroidism). *S* represents the symptoms "increase in nervousness, heat sensitivity, and sweating."† Given that these symptoms are present in a patient, find the most probable diagnosis by computing *P*(*A*|*S*), *P*(*B*|*S*), and *P*(*C*|*S*).

$$
\begin{array}{c}
\overset{.15}{\nearrow} A \xrightarrow{\ 0\ } S \\
\overset{.65}{\rightarrow} B \xrightarrow{.015} S \\
\underset{.20}{\searrow} C \xrightarrow{.27} S
\end{array}
$$

* (Based on J. Schneider et al., *Science* (1967), vol. 157, pp. 1321–22.)

† (Based on "Conditional Probability Program for Diagnosis of Thyroid Function" by J.E. Overall et al, *Journal of American Medical Association,* **183** V, (1963) pp. 307–313.)

21. In the family pedigree for albinism shown in example 60.4, suppose that Mary and Jay have two normal children. Assuming that Mary is *AA*, compute the new probability that Jay is *Aa*.

In the following two problems, use the pedigree shown here.

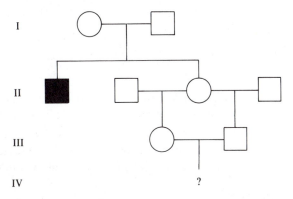

22. If the trait is caused by the presence of two recessive alleles, what are the genotypes of individuals I-1 and I-2? Assuming that males II-2 and II-4 are *AA*, find the probability that individual II-1 will have this trait.

23. If the trait is caused by a sex-linked recessive allele *r*, what are the genotypes of individuals I-1 and I-2? Find the probability that individual IV-1 will have the trait.

24. There are known cases of traits caused by a dominant allele *D*, in which the trait does not always appear in individuals of genotype *Dd*. An example is *retinoblastoma*. Suppose that 80% of *Dd* individuals show the trait.

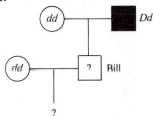

(a) Refer to the accompanying pedigree. Given that Bill does not show the trait, find the probability that Bill is *Dd*.

(b) Find the probability that Bill's child will show the trait.

25. Shown here is a family pedigree for PKU (see example 60.7) showing a marriage between first cousins. Assume that individuals II-1 and II-5 are *AA*.

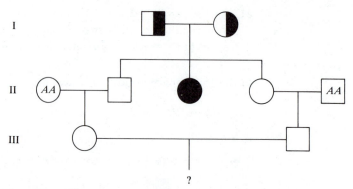

(a) Find the probability that both first cousins are carriers. What is the probability that their first child will be affected?

(b) Suppose that individual III-1 married someone from the general population where the proportion of Aa individuals is about .02. Compute the probability that their first child will have PKU.

 # probability models in population genetics

We have emphasized thus far the transfer of genetic information from generation to generation within a given family. **Population genetics**, however, studies this same phenomenon on a grander scale. It attempts to discover the mechanisms responsible for changes in the genetic constitution of an entire population. How can a given allele maintain itself over time? What are the sources of evolutionary change? It is quite remarkable that elementary probability theory can provide much insight into such issues. One of the fundamental tools of the population geneticist is the famous Hardy-Weinberg model. This is where we begin our brief introduction to population genetics.

the Hardy-Weinberg model (1908)   Suppose that we are following the prevalence of a certain genetic trait from generation to generation and that this trait is determined by two alleles A and a. When the alleles are not sex-linked (see example 60.8), it is reasonable to assume that the frequency of various genotypes is independent of sex.

Let $P_0 = P(AA)$ be the present frequency of genotype AA in the general population. Likewise, let $Q_0 = P(Aa)$ and $R_0 = P(aa)$ denote the frequencies of Aa and aa individuals. In population genetics models it is customary to speak of the "frequency p of allele A in the population." Imagine that an individual has been selected at random from the population and that one of his or her gametes is chosen. *Then p is the probability that the selected gamete is a type A gamete.* If q denotes the frequency of the a allele, then p and q may be expressed in terms of P_0, Q_0, and R_0, as shown in figure 61.1 and on page 702.

figure 61.1

Genotype	Gamete	
$AA \xrightarrow{\ 1\ } A$	P_0	
$Aa \xrightarrow{.5} A$	$.5Q_0$	
$ \xrightarrow{.5} a$	$.5Q_0$	
$aa \xrightarrow{\ 1\ } a$	R_0	

P_0, Q_0, R_0

$$p = P(A) = P_0 + .5Q_0 \quad \text{and} \quad q = P(a) = R_0 + .5Q_0$$

As illustrated in the following example, these formulas can often be used to estimate gene frequencies. There must, however, be a way of distinguishing between AA and Aa genotypes.

example 61.1   A screening of school-age children in Ghana for sickle cell anemia produced the results shown in table 61.1. The normal dominant allele is A and its frequency in this group is $P_0 + .5Q_0 = .834 + .5(.161) = .9145$. The frequency of the sickle cell allele a is then $1 - .915 = .0855$.

table 61.1

Genotype	AA	Aa	aa (sickle cell)
Frequency	.834	.161	.005

The Hardy-Weinberg model assumes that individuals, when selecting their mate, make that choice without regard to genotype AA, Aa, or aa. When this occurs, we say that *mating is at random* with respect to the given set of alleles. This hypothesis is hardly satisfied in the case of such visible traits as height, complexion, or level of intelligence. On the other hand, it is difficult to imagine that an individual would base his or her choice of mate on blood type or the ability to taste PTC*! In fact, for the vast majority of gene loci in man, the **random mating hypothesis** is reasonable.

What are the proportions of genotypes in the next generation? Shown in figure 61.2 are hypothetical *gene pools*, which represent the collection of

figure 61.2(a)

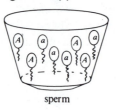

sperm

figure 61.2(b)

eggs

gametes for the entire population. The probability of selecting an A gamete from either pool is just p, the overall gene frequency of A. Imagine that we can form an individual by selecting a gamete from each collection, as illustrated in figure 61.3. Then $P(AA) = p^2$, while $P(Aa) = 2pq$ and $P(aa) = q^2$. It is possible

---

* PTC is phenylthiocarbamide. For some people, the taste is extremely bitter. Other people they can't taste a thing.

figure 61.3

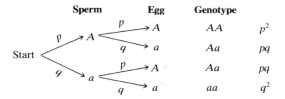

to show that this rather strange way of visualizing the formation of the next generation is actually equivalent to the random mating hypothesis. Hence, we have the *Hardy-Weinberg proportions* shown in the box.

$$P_1 = P(AA) = p^2$$
$$Q_1 = P(Aa) = 2pq$$
$$R_1 = P(aa) = q^2$$

The subscript 1 in P_1, Q_1, and R_1 refers to the genotype frequencies in the next generation. When the population is fairly large, we would expect the actual proportion of Aa individuals, for example, to be close to $2pq$.

The surprise comes when we compute the new gene frequencies p_1 and q_1:

$$p_1 = P_1 + .5Q_1 = p^2 + pq = p(p + q) = p$$

and

$$q_1 = R_1 + .5Q_1 = q^2 + pq = q(p + q) = q$$

When the second generation is formed, $P_2 = P(AA)$ will again be p^2, $Q_2 = 2pq$ and $R_2 = q^2$. (We have assumed that the first generation mates only with itself.) *Thus, the Hardy-Weinberg proportions stabilize after a single generation.* This fact is used in our next two examples.

example 61.2   A study of Navaho Indians in New Mexico revealed that 77.7% had blood type O, while the remaining 22.3% had blood type A. If q is the frequency of allele O, then, according to the Hardy-Weinberg model, $q^2 = .777$. Hence, $q = .881$ and so $p = 1 - q = .119$. The proportion of those with type A blood (and genotype AA or AO) is predicted to be $p^2 + 2pq = .223$. This is in agreement with the data produced by the study.

example 61.3   **estimating the proportion of carriers**   In the case of a very rare genetic disease such as albinism, the random mating hypothesis, although not strictly valid, is very close to being satisfied. The Hardy-Weinberg model can be used to estimate the proportion of Aa individuals in the population. In the United

States, the incidence rate of albinism at birth is about 1 in 20,000. Then $q^2 = 1/20,000$ and so $q \approx .007$. The proportion of carriers of albinism is then estimated to be $2pq = .014$. 🗏

Notice that we have implicitly assumed in our calculations that once a new generation is formed, genotypes survive to the reproductive stage with the same probability. In other words, we have assumed that the proportions of AA, Aa, and aa genotypes in the population do not vary over time. When we alter our basic model and take into account these possibly different survival rates, we are developing a **selection model**.

a Hardy-Weinberg selection model   Suppose that the children of the next generation have been formed according to the random mating hypothesis, and let p be the frequency of the allele A among the children. If R is the event "survival to the reproductive age," we will let $s_1 = P(R|AA)$, $s_2 = P(R|Aa)$, and will assume that $P(R|aa) = 0$. Thus, the model is appropriate for a fatal genetic disease (such as Tay-Sachs) caused by two a recessive alleles. The probability tree in the Hardy-Weinberg model now takes the form shown in figure 61.4. From Bayes' Theorem, we have $P(AA|R) = \dfrac{P(R|AA)P(AA)}{P(R)} = (p^2s_1)/(p^2s_1 + 2pqs_2)$. Likewise, $P(Aa|R) = 2pqs_2/(p^2s_1 + 2pqs_2)$ and $P(aa|R) = 0$.

figure 61.4

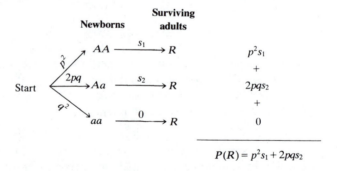

$$P(R) = p^2s_1 + 2pqs_2$$

These expressions give the the proportions of AA, Aa, and aa genotypes *among the survivors*. These proportions now play the same role as P_0, Q_0, and R_0 in the original Hardy-Weinberg model.

The new gene frequencies among the survivors and their children are

$$p_1 = P(AA|R) + .5P(Aa|R) = \frac{ps_1 + qs_2}{ps_1 + 2qs_2}$$

and

$$q_1 = P(aa|R) + .5P(Aa|R) = q\left(\frac{s_2}{ps_1 + 2qs_2}\right)$$

In general, the difference equation for the frequency of allele a in generation n is given by

$$q_{n+1} = q_n\left(\frac{s_2}{p_n s_1 + 2q_n s_2}\right)$$

If we let $s = s_1/s_2$, then this difference equation can be written as

$$q_{n+1} = q_n\left[\frac{1}{(1 - q_n)s + 2q_n}\right]$$

This model is illustrated in our next two examples.

example 61.4

A screening of young adults in Ghana for sickle cell anemia showed that 75.6% were of genotype AA, while 24.4% were of genotype Aa. Thus, the frequency of allele a among young adults was .122, a significant increase from the frequency of .0895 among children (see example 61.1). Using the data from example 61.1, we compute

$$s_1 = P(R|AA) = \frac{P(AA|R)P(R)}{P(AA)} = \frac{.756P(R)}{.834}$$

and

$$s_2 = P(R|Aa) = \frac{P(Aa|R)P(R)}{P(Aa)} = \frac{.244P(R)}{.161}$$

It follows that $s = s_1/s_2 \approx .6$. Thus, the survival rate among Aa individuals is significantly higher than among AA genotypes. (It is known that the allele a in Aa individuals provides some protection against malaria.) If we use the difference equation with $s = .6$ and $q_1 = .122$, we obtain the predictions shown in table 61.2 for the frequency of the sickle cell allele in subsequent generations.

table 61.2

Generation n	2	3	4	5	6	7
Frequency of a	.158	.193	.222	.243	.259	.269

example 61.5

When $s_1 = s_2$ in the Hardy-Weinberg selection model, the difference equation for q_n simplifies to $q_{n+1} = q_n/(1 + q_n)$. In Section 48, we showed that this difference equation has solution

$$q_n = \frac{q_0}{1 + nq_0}$$

Hence, $\bar{q} = \text{limit}_{n\to\infty} q_n = 0$.* This special case of the model is appropriate for Tay-Sachs disease (see example 59.2). If the present frequency of the allele t in a population is .01, then the number of generations needed to reduce this frequency to .005 is easily determined:

$$q_n = \frac{.01}{1 + .01n} = .005 \qquad \text{or } .01n + 1 = 2$$

Hence, $n = 100$ generations are required. The model then indicates how very slowly the defective allele will be eliminated from the population.   ▤

The equilibrium frequency for allele a is $\bar{q} = \text{limit } q_n$. Assuming that this limit actually exists (see part B exercise 21), we have

$$\lim_{n\to\infty} q_{n+1} = \lim_{n\to\infty} q_n\left[\frac{1}{(1 - q_n)s + 2q_n}\right]$$

Hence, $\bar{q}$ satisfies $\bar{q} = \bar{q}/[(1 - \bar{q})s + 2\bar{q}]$. One obvious solution is $\bar{q} = 0$. If $\bar{q} \neq 0$, then $(1 - \bar{q})s + 2\bar{q} = 1$. Hence, $\bar{q} = (s - 1)/(s - 2)$. It is not difficult to show that $0 < \bar{q} < 1$ if and only if $s < 1$. Hence, in order for the allele a to maintain itself, we must have $s_1 < s_2$.   This condition, illustrated by the classical example of sickle cell anemia in example 61.4, is known as **heterozygote superiority**.

When the alleles are located on the X sex chromosome, we cannot assume that the proportions of various genotypes are independent of sex. A more delicate analysis is required. In our final population genetics model, we will make use of the techniques for solving difference equations discussed in Section 49.

Hardy-Weinberg model for a sex-linked trait   Suppose that alleles A and a reside on the X chromosome and so the possible genotypes in the population are $X^A Y$, $X^a Y$, $X^A X^A$, $X^A X^a$, and $X^a X^a$. Let p be the present frequency of allele A in males, and let f denote the frequency in females. Note that p is just the proportion of $X^A Y$ males in the population. We will again assume the *random mating hypothesis* and imagine that the next generation is formed by selecting from gamete pools, as illustrated in figure 61.5. Since an

figure 61.5(a)

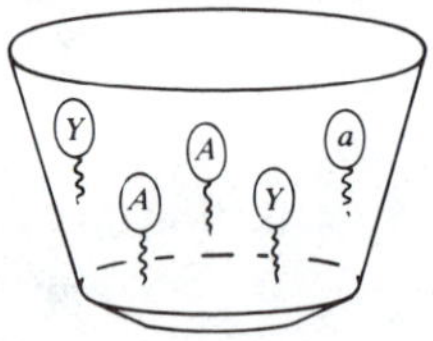

sperm

figure 61.5(b)

eggs

* Although we have used q_∞ to denote the limit of a sequence q_n in our prior work, it is customary to use $\bar{q}$ or $\hat{q}$ in population genetics.

$X^A Y$ male received the X chromosome from his mother,

$$p_1 = P(X^A Y) = f$$

Given that a female is formed, the probability tree in figure 61.6 shows that

figure 61.6

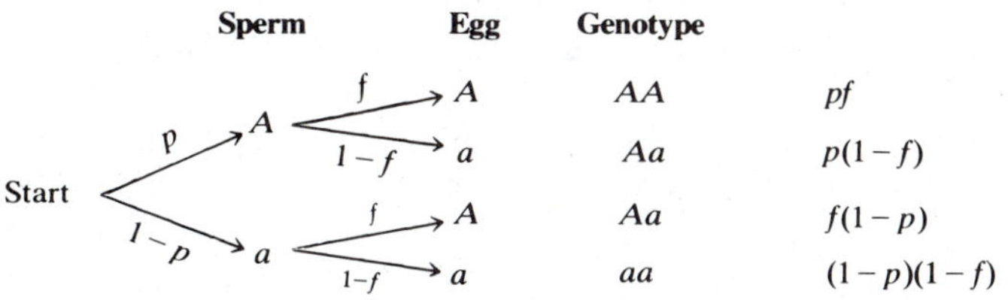

	Sperm	Egg	Genotype	
	A	A	AA	pf
	A	a	Aa	$p(1-f)$
	a	A	Aa	$f(1-p)$
	a	a	aa	$(1-p)(1-f)$

among females, $P(AA) = pf$ and $P(Aa) = p(1-f) + f(1-p) = p + f - 2pf$. The frequency of the A allele in the female population is now

$$f_1 = P(AA) + .5P(Aa)$$

$$= .5(p + f)$$

The difference equations for the frequency of allele A in generation n take the form shown in the box.

$$p_{n+1} = f_n$$
$$f_{n+1} = .5p_n + .5f_n$$

example 61.6 Coat color in cats is a sex-linked trait determined by two alleles A and a. Each genotype gives rise to a distinct phenotype, as summarized in table 61.3. The

table 61.3

Genotype	$X^A X^A$	$X^A X^a$	$X^a X^a$	$X^A Y$	$X^a Y$
Phenotype	black	tortoise shell	yellow	black	yellow
Frequency	.4	.5	.1	.5	.5

frequency of the allele A in male cats is $p_0 = .5$, while the allele frequency for females is $f_0 = .4 + .5(.5) = .65$. The allele frequencies in the next generation are $p_1 = f_0 = .65$ and $f_1 = .5(p_0 + f_0) = .575$. Repeating the process yields the allele frequencies for subsequent generations, as shown in table 61.4. Notice how

table 61.4

Generation n	1	2	3	4	5	6	7
Allele frequency for males	.650	.575	.613	.594	.603	.598	.601
Allele frequency for females	.575	.613	.594	.603	.598	.601	.600

very quickly the two sequences are converging. The equilibrium frequency of A in both males and females appears to be $\bar{p} = .6$. ▤

Formulas for the equilibrium gene frequencies $\bar{p}$ and $\bar{f}$ can be found by solving the second order difference equation satisfied by f_n:

$$f_{n+1} = .5p_n + .5f_n = .5f_{n-1} + .5f_n$$

and

$$2f_{n+1} - f_n - f_{n-1} = 0$$

The characteristic polynomial $2r^2 - r - 1 = (2r+1)(r-1)$ has roots $r = -.5$ and $r = 1$. Hence, $f_n = c_1 + c_2(-.5)^n$ and $p_n = c_1 + c_2(-.5)^{n-1}$. It follows that $\bar{f} = \bar{p} = c_1$ and, since $(-\frac{1}{2})^n$ converges to 0 very quickly, the *convergence of sequences f_n and p_n is rapid*.

The constant c_1 can be expressed in terms of the original gene frequencies p and f:

$$f_0 = f = c_1 + c_2$$
$$f_1 = .5p + .5f = c_1 - .5c_2$$

Solving for c_1, we obtain $c_1 = \frac{1}{3}p + \frac{2}{3}f$. The conclusions of the model are summarized in the accompanying box.

Hardy-Weinberg model:
conclusions for a sex-linked trait

1. The equilibrium frequencies of allele A are identical in males and females and

$$\bar{p} = \bar{f} = \frac{1}{3}p + \frac{2}{3}f$$

where p and f are the present frequencies.

2. The convergence of sequences p_n and f_n is rapid.

3. The equilibrium proportions of the various genotypes are as shown in the table (where $\bar{q} = 1 - \bar{p}$).

Genotype	$X^A X^A$	$X^A X^a$	$X^a X^a$	$X^A Y$	$X^a Y$
Equilibrium frequency	$\bar{p}^2$	$2\bar{p}\bar{q}$	$\bar{q}^2$	$\bar{p}$	$\bar{q}$

The conclusions of the model are applied in our final two examples.

example 61.7   Find the equilibrium frequencies for each phenotype in the coat color example 61.6.

solution 61.7   We have determined that $p = .5$ and $f = .65$ are the present frequencies of allele A in male and female cats, respectively. Hence, $\bar{p} = \frac{1}{3}(.5) + \frac{2}{3}(.65) = .6$ is the equilibrium frequency. It follows that the equilibrium frequencies of the various phenotypes are as shown in table 61.5.

table 61.5

Phenotype	Females			Males	
	black	tortoise shell	yellow	black	yellow
Equilibrium frequency	.36	.48	.16	.6	.4

example 61.8   A type of color blindness in humans is due to a recessive allele a residing on the X chromosome. The incidence rate in men is about 1 in 10. The rapid convergence of the allele frequencies in our model makes it reasonable to assume that equilibrium has been reached in a human population. Our estimate for the proportion of color-blind females is therefore $(.1)^2 = .01$. The fraction of women who carry the allele is $P(X^A X^a) = 2(.1)(.9) = .18$.

exercises for section 61

part A   Shown in the following three exercises are actual genotype frequencies of *M-N* blood types in various populations. In each case,

(a) find the frequencies of alleles M and N.

(b) use the Hardy-Weinberg model to find the predicted genotype frequencies.

1. Eskimos: *MM* (.83), *MN* (.16), *NN* (.01)

2. Germans: *MM* (.306), *MN* (.491), *NN* (.203)

3. Japanese: *MM* (.274), *MN* (.502), *NN* (.224)

4. Determine whether the genotype frequencies presented in example 61.1 for sickle cell anemia are in agreement with the Hardy-Weinberg model.

5. Allele A is dominant over allele a and so genotypes AA and Aa are identical in appearance. Suppose there are four times as many $A\_\_$ individuals as aa genotypes. Use the Hardy-Weinberg model to estimate the frequency of allele a, and predict the proportions of various genotypes in the population.

6. Find the maximum possible proportion of heterozygotes (Aa) in a randomly mating population.

In the following three problems, use the Hardy-Weinberg model to estimate the proportion of carriers in the given population. Each genetic disease is due to the presence of two recessive alleles.

7. Cystic fibrosis occurs in approximately 1 out of every 1,600 births in the United States.

8. In the Jewish population of northern Europe, Tay-Sachs disease has an incidence rate of 1 in 6,000.

9. Albinism among the Indians on the San Blas Islands off Panama occurs with probability 1/130.

10. If R_0 is the frequency of a rare recessive trait, show that the proportion Q_0 of carriers can be approximated by $2\sqrt{R_0}$.

★11. Let p, q, and r denote the frequencies of alleles A, B, and O, respectively, in a given population for the standard blood groups. Assuming the random mating hypothesis, use a probability tree to establish that:

(a) $P(\text{"Blood type } O\text{"}) = r^2$

(b) $P(\text{"blood type } A\text{"}) = p^2 + 2pr$

(c) $P(\text{"Blood type } B\text{"}) = q^2 + 2qr$

(d) $P(\text{"Blood type } AB\text{"}) = 2pq$

★12. A study of the blood types in France produced the information shown in the table. Use the results of exercise 11 to find the frequencies of alleles

Blood type	O	A	B	AB
Frequency	.441	.435	.090	.034

A, B, and O. What fraction of those with blood type A carry the O allele?

13. In the Hardy-Weinberg model developed in the text, we assumed that the frequency of allele A was the same in males as in females. Suppose this hypothesis is dropped, and let p be the frequency of A in males and f be the frequency of A in females. Show that after a single generation, the frequency of allele A in both sexes is

$$\bar{p} = \frac{p+f}{2}$$

What occurs in subsequent generations?

14. Consider the special case of the selection model discussed in example 61.5.

   (a) Show that $n = 1/q_n - 1/q_0$.

   (b) Conclude that the number of generations required to reduce q_0 to $.5q_0$ is $n = 1/q_0$.

   (c) Apply part **(b)** to cystic fibrosis, which has a present c allele frequency of about .025 in the United States.

15. For Huntington's Chorea (see exercise 29, section 59), it has been estimated that $s_2 \approx .75s_1$, where s_1 is the survival rate for dd genotypes.

   (a) If q_n is the frequency of the rare D allele, let $A = d$ and $a = D$ in the selection model to conclude

   $$q_{n+1} = \frac{3q_n}{4 + 2q_n}$$

   (b) If $q_0 = .006$, find $q_1, q_2, \ldots, q_5$.

★16. In the Hardy-Weinberg selection model presented in the text, assume that $P(R|AA) = P(R|Aa) = s_1$ and $P(R|aa) = s_2$.

   (a) Show that

   $$p_{n+1} = p_n\left[\frac{s_1}{s_1 + (s_2 - s_1)q_n^2}\right]$$

   (b) If $\bar{p} = \text{limit } p_n$ exists and $s_1 \neq s_2$, show that $\bar{p} = 0$ or 1.

17. For a certain sex-linked recessive trait, the proportion of $X'X'$ females is .2. Estimate the proportion of $X'Y$ males. What are you assuming in order to carry out the computations?

18. A rare sex-linked recessive allele is fatal in males. If the incidence rate in males is 1 in 50,000, estimate the proportion of women who carry the allele.

19. If Q is the incidence rate of a rare sex-linked recessive allele, show that the proportion of mothers who carry the allele can be approximated by $2Q$.

20. Eye color in *Drosophila* is a sex-linked trait. The dominant allele R codes for red eyes, while the r allele is responsible for white eyes. A laboratory population is started with $X'Y$ white-eyed males and $X^R X^R$ females. Assuming the random mating hypothesis,

   (a) find the frequency of the R allele over the next five generations.

   (b) find the equilibrium proportions of the various genotypes.

part B   **21.** The difference equation

$$q_{n+1} = q_n\left[\frac{1}{(1 - q_n)s + 2q_n}\right]$$

in the Hardy-Weinberg selection model can be solved using the techniques of Section 48.

(a) Let $x_n = 1/q_n$ and show that $x_{n+1} = sx_n + (2 - s)$.

(b) Conclude that $x_n = cs^n + (2 - s)/(1 - s)$ for some constant c.

(c) Finally, show that

$$\bar{q} = \text{limit } q_n = \begin{cases} 0, & \text{for } s > 1 \\ (1 - s)/(2 - s), & \text{for } 0 < s < 1 \end{cases}$$

★**22.** Let p denote the present frequency of allele A in adult women and let R be the event "survival to reproductive age." We will assume that

$$P(R|X^AX^A) = s_1, \quad P(R|X^AX^a) = s_2, \quad \text{and } P(R|X^aY) = 0$$

Thus, only allele A is represented in adult males.

(a) Show that q_1, the frequency of allele a in adult women in the next generation, is given by

$$q_1 = \frac{1}{2} q/(sp + q)$$

where $s = s_1/s_2$.

(b) When $s_1 = s_2$, solve the difference equation $q_{n+1} = \frac{1}{2}q_n$ and show that $\bar{q} = \text{limit } q_n = 0$.

(c) For $s \neq 1$, let $x_n = 1/q_n$ and show that $x_{n+1} = 2sx_n + (2 - 2s)$.

(d) Conclude that $x_n = c(2s)^n + (2 - 2s)/(1 - 2s)$ when $s \neq 1/2$.

(e) When is $\bar{q} \neq 0$?

references

references

Alexander, R.M. *Animal mechanics.* Seattle: Univ. of Washington Press, 1968.

Allee, W.C. et al, *Principles of animal ecology,* Philadelphia: W.B. Saunders, 1949.

Atkins, G.L. *Multicompartmental models for biological systems.* London: Methuen Publishers, 1969.

Batschelet, E. *Introduction to mathematics for life scientists.* New York: Springer-Verlag, 1973.

Benedek, G.B., and Willars, F.M.H. *Physics with illustrative examples from medicine and biology.* Vol. 1. Reading, Mass.: Addison-Wesley, 1973.

Beverton, R.J.H. and Holt, S.J. *On the dynamics of exploited fish populations.* Fishery Investigations, Series II, Vol. XIX. London: H.M.S.O., 1957.

Bodner, W.F. and Cavalli-Sforza, L.L. *Genetics, evolution, and man.* San Francisco: W.H. Freeman and Co., 1976.

Clark, C. "Urban population densities." *Journal of the Royal Statistical Society,* Series A, 1951, 114, pp. 490–96.

Clark, C.W. *Mathematical bioeconomics*: *The optimal management of renewable resources.* New York: John Wiley and Sons, 1976.

Comroe, J.H. *Physiology of respiration.* 2d ed. Chicago: Yearbook Medical Publishers, 1974.

Crow, J.F. *Genetics notes.* 7th ed. Minneapolis: Burgess Publishing Co., 1976.

Crow, J.F. and Kimura, M. *An introduction to population genetics.* New York: Harper & Row, 1970.

Cushing, D.H. *Science and the fisheries.* London: Edward Arnold Publishers, 1977.

Davidovits, P. *Physics in biology and medicine.* Englewood Cliffs, N.J.: Prentice-Hall, 1975.

Defares, J.G. and Sneddon, I.N. *An introduction to the mathematics of medicine and biology.* Amsterdam: North-Holland, 1961.

De Sapio, R. *Calculus for the life sciences.* San Francisco: W.H. Freeman, 1976.

Emlen, J.M. *Ecology*: *An evolutionary approach.* Reading, Mass.: Addison-Wesley, 1973.

Ford, J.M. and Munroe, J.E. *Living systems.* 2d ed. San Francisco: Canfield Press, 1974.

Funderlic, R.E. and Heath, M.T. *Linear compartment analysis of ecosystems.* Oak Ridge, Tenn.: Oak Ridge National Laboratory, 1971.

Fuhrmann, W. and Vogel, F. *Genetic counseling.* 2d ed. New York: Springer-Verlag, 1976.

Gause, G.F. *The struggle for existence.* New York: Hafner, 1969.

Goldberg, S. *Introduction to difference equations.* New York: John Wiley and Sons, 1958.

Goldsby, R.A. *Race and races.* New York: Macmillan, 1971.

Gray, J. *How animals move.* London: Cambrdige University Press, 1960.

Guyton, A.C. *Textbook of medical physiology.* Philadelphia: W.B. Saunders, 1971.

Haberman, R. *Mathematical models.* Englewood Cliffs, N.J.: Prentice-Hall, 1977.

Hoppensteadt, F.C. *Mathematical methods of population biology.* New York: Courant Institute of Mathematical Sciences, New York Univ., 1977.

Hughes, W. *Aspects of biophysics.* New York: John Wiley and Sons, 1979.

Jacquez, J.A. *Compartmental analysis in biology and medicine.* Amsterdam: Elsevier Publishing Co., 1972.

Jacquez, J.A. *Respiratory physiology.* New York: Hemisphere Publishing Co., McGraw-Hill, 1979.

Keyfitz, N. *Introduction to the mathematics of population.* Reading, Mass.: Addison-Wesley, 1977.

Kostitzin, V.A. *Mathematical biology.* London: Harrap, 1939.

Lackey, R.T. and Nielson, L.A. eds. *Fisheries management.* New York: John Wiley and Sons, 1980.

Landau, B.R. *Essential human anatomy and physiology.* 2d ed. Glenview, Ill.: Scott, Foresman, and Co., 1980.

Lenz, W. *Medical genetics.* Chicago: The University of Chicago Press, 1963.

Li, C.C. *Population Genetics.* Chicago: The University of Chicago Press, 1955.

Mac Arthur, R.H. and Wilson, E.O. *The theory of island biogeography.* Princeton, N.J.: Princeton Univ. Press, 1967.

Metcalf, H.J. *Topics in classical biophysics.* Englewood Cliffs, N.J.: Prentice-Hall, 1980.

Mettler, L.E. and Gregg, T.G. *Population genetics and evolution.* Englewood Cliffs, N.J.: Prentice-Hall, 1969.

Myhre, R.J. "Minimum size and optimum age of entry for Pacific halibut." Scientific Report No. 55. Seattle, Washington: International Pacific Halibut Commission, 1974.

Novitski, E. *Human genetics.* New York: Macmillan, 1977.

Parsons, T.R. and Takahashi, M. *Biological oceanographic processes.* New York: Pergamon Press, 1973.

Pielou, E.C. *An introduction to mathematical ecology,* New York: Wiley-Interscience, 1969.

Poole, R.W. *An introduction to quantitative ecology.* New York: McGraw-Hill, 1974.

Riggs, D.S. *The mathematical approach to physiological problems.* Baltimore: Williams and Wilkins, 1963.

Rosen, R. *Optimality principles in biology.* London: Butterworth, 1967.

Rubinow, W.I. *Introduction to mathematical biology.* New York: John Wiley and Sons, 1975.

Ruch, T.C. and Patton, H.D. *Physiology and biophysics,* Vol. II (*Circulation, respiration, and fluid balance*). 20th ed. Philadelphia: W.B. Saunders, 1974.

Savage, J.M. *Evolution.* 2d ed. New York: Holt, Rinehart and Winston, 1969.

Simon, W. *Mathematical techniques for physiology and medicine.* New York: Academic Press, 1972.

Smith, C.A.B. *Biomathematics.* Vol. 1, 4th ed. London: Griffin, 1969.

Smith, J.M. *Mathematical ideas in biology.* London: Cambridge University Press, 1968.

Smith, J.M. *Models in ecology.* London: Cambridge University Press, 1974.

Strahler, A.N. *The earth sciences,* New York: Harper and Row, 1963.

Strickberger, M.W. *Genetics.* 2d ed. New York: Macmillan, 1976.

Tallarida, R.J. and Jacob, L.S. *The dose-response relation in pharmacology.* New York: Springer-Verlag, 1979.

Thorson, G. *Life in the sea.* New York: McGraw-Hill, 1971.

Thrall, R.M., et al. "Some mathematical models in biology." Report No. 40241-R-7. Ann Arbor: University of Michigan, 1967.

Usher, M.B. *Biological management and conservation.* London: Chapman and Hall, 1973.

Watt, K. *Ecology and resource management.* New York: McGraw-Hill, 1968.

Whittaker, R.H. *Communities and ecosystems.* 2d ed. New York: Macmillan, 1975.

Wilson, E.O. and Bossert, W.H. *A primer of population biology.* Sunderland, Mass.: Sinauer Associates, 1971.

answers
to selected
exercises

answers
to selected
exercises

answers to section 1 (page 8)

1. $\{y: y \geq 2\}$ **3.** $\{y: y \geq 0\}$ **5.** $\{y: y \geq 0\}$ **7.** $\{y: y > 2\}$ **9.** $\{y: 0 < y < \frac{1}{2}\}$

11. $\{x: x \neq 1\}$ **13.** $\{x: x \neq 0, 1\}$ **15.** $\{x: x \geq -3\}$ **17.** $\{x: x \geq -\frac{4}{3}\}$ **19.** $\{x: x \geq 0\}$

21. $4a^2, x^2 + 2xh + h^2, 2x + h$ **23.** $6a + 2, 3x + 3h + 2, 3$ **25.** $4, 4, 0$

27. $\sqrt{2a}, \sqrt{x+h}, 1/(\sqrt{x+h} + \sqrt{x})$ **29.** $\dfrac{1}{2a+1}, \dfrac{1}{x+h+1}, \dfrac{-1}{(x+h+1)(x+1)}$

31. $f(0) = 1, f(1) = 1, f(2) = 4$ **33.** $f(0) = -7, f(1) = -4, f(2) = -1$ **35.** $f(0) = -3, f(1) = -1, f(2) = 4$

37.
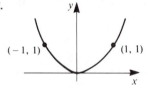
$(-1, 1)$ $(1, 1)$

39.
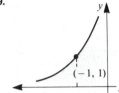
$(-1, 1)$

41.
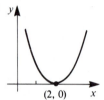
$(2, 0)$

43.
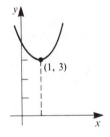
$(1, 3)$

45.
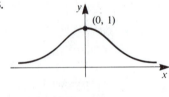
$(0, 1)$

47. $x = 0, -3$

49. $x = 0, 2, -1$

51. $x = 1, -3/2, 5/4$

53. $x = (3 \pm \sqrt{21})/2$

55. $x = 1, -3, -4$

part B   **56.** -1 **58.** 3 **60.** $.25 = \frac{1}{4}$ **62.** $.5$ **64.** $x = 0, 1$ **66.** $x = \frac{5}{3}$

68. $x = 2$ **70.** $f(f(x)) = x$

answers to section 2 (page 11)

1. $[8, +\infty)$ **3.** $[2, +\infty) \cup (-\infty, -2]$ **5.** $(2, +\infty) \cup (-\infty, -2)$ **7.** $(3, 4)$ **9.** do not intersect

11. $x > 1$ **13.** $x > 1$ or $x < 0$ **15.** $(-1, \frac{3}{2})$ **17.** $x < 13$ **19.** $(-\infty, 1) \cup (1, 2)$

21. $(0, \frac{1}{2})$ **23.** $(-\infty, -2) \cup (-1, 1)$ **25.** all real numbers   **27.** $[0, 1)$ **29.** $(-2, -\frac{1}{3})$

31. $(-\infty, 0] \cup [1, +\infty)$

answers to section 3 (page 23)

1.

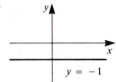

$y = -1$

3.

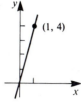

(1, 4)

5.

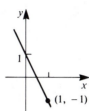

1

(1, −1)

7.

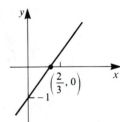

$\left(\frac{2}{3}, 0\right)$

−1

9.

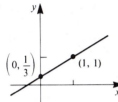

$\left(0, \frac{1}{3}\right)$ (1, 1)

11. $y = -2x + 4$ **13.** $y = 4x - 14$ **15.** $y = 4x - 5$
17. $y = -x + 3$ **19.** $y = 10x + 5$

21. $y = \frac{1}{4}x$ **23.** $W = 4x$ **25.** \$1.56/gallon **27.** $C = \frac{1}{5}N$ **29. (a)** V_0 is the volume of the gas a 0°C **(b)** At $T = 273°C$, $V = 2V_0$ **31.** 60 **part B 32.** $x + 1$ **34.** $-1/x$
36. $L = kS$, reasonable **38.** $E = kS$, reasonable **40.** $W = kV$, reasonable if, e.g., all objects are made of the same material

42.

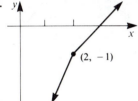

(2, −1)

answers to section 4 (page 37)

1.

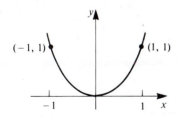

(−1, 1) (1, 1)

−1 1

3.

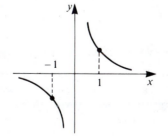

−1

1

5.

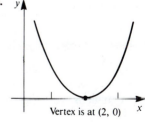

Vertex is at (2, 0)

7.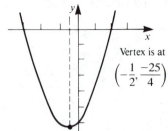

Vertex is at
$$\left(-\frac{1}{2}, \frac{-25}{4}\right)$$

9.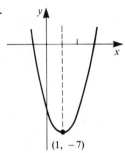

$(1, -7)$

23.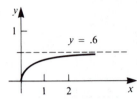

$y = .6$

11. $y = 8/x$ **13.** $\frac{200}{3}$ per day   **15.** $\frac{20}{3}$ cubic feet   **17.** 30   **19.** all reals   **21.** $x \neq 1, 2$
23. see above   **part B** **29.** $50' \times 50'$

answers to section 5 (page 51)

1. $x/3$ **3.** $1/x$ **5.** $(x-1)/2$ **7.** $\sqrt{x}$, for $x \geq 0$ **9.** x^2, for $x \geq 0$ **11.** $f(1) = f(-1) = 1$
13. $f(1) = f(-1) = \frac{1}{2}$ **15.** $f(0) = f(-2) = 1$ **17.** $f(0) = f(\pi) = 0$ **19.** $(x+3)/4$
21. $1/x$ **23.** $(-2x-1)/(x-1)$ **25.** $f^{-1}(x) = (x-1)^2$ for $x \geq 1$ **27.** $\sqrt[3]{x+1}$ **29.** $f^{-1}(x) =$
$-\sqrt[4]{x-1}$, for $x \geq 1$ **31.** $f^{-1}(x) = (x+3)/4$ **33.** $f^{-1}(x) = \sqrt[3]{x}$ **37.** $x^{1/8} = [(x^{1/2})^{1/2}]^{1/2}$
39. $.95639\ldots$ **41.** does not exist   **43.** $1.58740\ldots$ **45.** $1.15996\ldots$ **47.** $.757858\ldots$
49. $2\sqrt{2}$ **51.** $1/25$ **53.** $1/2$ **55.** 50.17 feet   **57.** increase number of species by factor of 2
part B **62.** 28.49 square feet   **66.** 161,080 miles/sec

answers to section 6 (page 63)

1. 7.000 **3.** $.203$ **5.** 5.000 **7.** does not exist   **9.** $.693$ **11.** 2   **13.** 1
15. -1 **17.** $\frac{1}{4}$ **19.** $-\frac{3}{2}$ **21.** 3   **23.** 3   **25.** 1   **27.** 4   **29.** $\frac{1}{3}$
31. (a) $\frac{1}{6}$ **(b)** 3   **33.** 3   **35.** 1   **37.** 0   **39.** -3 **41.** 2

answers to section 7 (page 75)

1. 2   **3.** $\sin \pi = 0$ **5.** $\sqrt{3} = 1.732\ldots$ **7.** $\log_{10} 2 = .301\ldots$ **9.** $1/4$ **11.** continuous
everywhere except $x = 0$, where a jump discontinuity exists   **13.** continuous everywhere except $x = 2$,
where a removable discontinuity exists   **15.** continuous everywhere except $x = 2$, where a jump discontinuity exists   **17.** continuous everywhere except $x = 0$, where a vertical asymptote exists
19. continuous everywhere   **21.** continuous everywhere   **23.** continuous for $x \neq 0$
25. $3x^2$ **27.** 2   **29.** $-\frac{4}{3}$ **31.** 1   **33.** $\frac{1}{5}$ **35.** 3   **37.** $-\frac{1}{4}$ **39.** $1/(2\sqrt{2})$
41. 1   **43.** -1 **45.** $.3678\ldots$ **47.** $(-\infty, -1) \cup (-1, 1) \cup (1, 3)$ **49.** $(-1, 1)$ **51.** $(-1 - \sqrt{3},$
$-1 + \sqrt{3}) \cup (1, +\infty)$ **53.** all intervals of the form $((2n+1)\pi, (2n+2)\pi)$, where $n = 0, \pm 1, \pm 2, \ldots$

answers to section 8 (page 87)

1. The correct formula is $m(x) = 1/x$. When $P = (1, 0)$, $Q = (2, 1)$ and $m(1) = 1$. When $P = (2, .693)$, $Q = (0, -.3)$ and so $m(2) \approx \frac{1}{2}$.   **3.** 3   **5.** 3   **7.** 6   **9.** .434...   For **11–17**, $\dfrac{f(x) - f(a)}{x - a}$ and $f'(x)$ are given: **11.** $3(x + a)$, $6x$ **13.** $-(x + a)/(x^2 a^2)$, $-2/x^3$ **15.** $\frac{1}{2}(x + a) + 1$, $x + 1$

17. $1/(\sqrt{x} + \sqrt{a})$, $1/(2\sqrt{x})$ For **19–23**, $[f(x + h) - f(x)]/h$ and $f'(x)$ are given: **19.** $3x^2 + 3xh + h^2 - 1$, $3x^2 - 1$ **21.** $1/(\sqrt{x + h + 1} + \sqrt{x + 1})$, $1/(2\sqrt{x + 1})$ **23.** $1/[(x + 1)(x + h + 1)]$, $1/(x + 1)^2$

25. 12   **27.** 3   **29.** 7   **31.** no solution   **33.** no solution   **35.** $-\frac{1}{2}$ **37.** $4x^3$

39. $-\dfrac{1}{x^2}$ **41.** $\dfrac{1}{3} x^{-2/3}$ **43.** $\dfrac{7}{3} x^{4/3}$ **45.** $-\dfrac{4}{5} x^{-9/5}$ **47.** $\dfrac{2}{5} x^{-3/5}$ **49.** $\dfrac{f(1 + h) - f(1)}{h} = |h|/h$

and $\text{limit}_{h \to 0} |h|/h$ does not exist   **51.** $(f(h) - f(0))/h = |h| \to 0$ as $h \to 0$

answers to section 9 (page 98)

1. 0   **3.** 12   **5.** $\cos 2$ **7.** $-\frac{1}{4}$ **9.** $6x^5$ **11.** $y' = ky$ where $k = \ln a$

13. $2/\sqrt{3}$ **15.** $2x - 8$ **17.** $6x - 1/x^2$ **19.** $9x^2$ **21.** $\frac{3}{2}\sqrt{x} + \frac{3}{4}x^{-3/4} + \frac{2}{3}x^{-4/3}$

23. $4x^3 - 2x - 1$ **25.** $1/x$ **27.** $1/(1 + x^2) - 2x$ **29.** $3 \cos x - 4/\sqrt{1 - x^2}$ **31.** $3x^2 - 6x + 3$

33. $3 \, 3^x (\ln 3) = (\ln 3)(3^{x+1})$ **35.** $3e^x - 4x^3$ **37.** $\xrightarrow{\quad - \quad 0 \quad + \quad} f' = 4x^3 - 4$

39. $\xrightarrow[\quad 0 \quad \sqrt[3]{2} \quad]{\quad - \quad u \quad - \quad 0 \quad + \quad} f' = 2x - \dfrac{4}{x^2}$ **41.** $(x^2 + 1)3x^2 + (x^3 - 1)2x$ **43.** $(x^3 + 3x + 2)(2x - 4) +$

$(x^2 - 4x + 3)(3x^2 + 3)$ **45.** $(x + 1/x)(2x - 1) + (x^2 - x)(1 - 1/x^2)$ **47.** $x \cos x + \sin x$

49. $(1 - x^2)/(x^2 + 1)^2$ **51.** $-3x^2/(x^3 - 1)^2$ **53.** $5 - 4/(x + 4)^2$ **55.** $9 - \ln x$

57. $\dfrac{(x^5 - 1)(3x^2 - 2x - 1) - (x^3 - x^2 - x)(5x^4)}{(x^5 - 1)^2}$ **59.** $-\dfrac{(x^2 + 1)\,2}{(x - 1)^2} + 2x\,\dfrac{x + 1}{x - 1}$

61. $xe^x \cos x + xe^x \sin x + e^x \sin x$ **63.** 12   **65.** $-e^{-x}$ **part B**   **66.** **(a)** -18 and 6

(b) $3 - \sqrt{3}$ **(c)** $3 + \sqrt{3}$

answers to section 10 (page 106)

1. $f(x) = 2x + 3$, $g(x) = x^5$,   and $f(g(x)) = 2x^5 + 3$ **3** $f(x) = (x + 2)/(x - 2)$, $g(x) = x^3$,   and $f[g(x)] = (x^3 + 2)/(x^3 - 2)$ **5.** $f(x) = 3 - 4x$, $g(x) = 1/x^4$,   and $f(g(x)) = 3 - 4/x^4$ **7.** $x^8 + 3$ **9.** $1/(x - 3)$

11. **(a)** $5(x - 1)^4$ **(b)** $6(2x + 1)^2$ **(c)** $(x - 1)^5 6(2x + 1)^2 + 5(x - 1)^4 (2x + 1)^3$ **13.** **(a)** $(2x + 3)^{-1/2}$

(b) $-3(4 - x)^2$ **(c)** $(4 - x)^3 (2x + 3)^{-1/2} - \sqrt{2x + 3} \, 3(4 - x)^2$ **15.** **(a)** $\frac{1}{2}(x + 1)^{-1/2}$ **(b)** $-10(2 - 5x)$

(c) $\dfrac{\sqrt{x + 1}[-10(2 - 5x)] - (2 - 5x)^2(1/2)(x + 1)^{-1/2}}{x + 1}$ **17.** **(a)** $-2e^{-2x}$ **(b)** $2 \cos 2x$ **(c)** $2e^{-2x} \cos 2x - 2e^{-2x} \sin 2x$

19. **(a)** $-4 \sin 4x$ **(b)** $4e^{4x}$ **(c)** $-4e^{4x} \sin 4x + 4e^{4x} \cos 4x$ **21.** $-12\left(\dfrac{x + 2}{x - 2}\right)^2 \dfrac{1}{(x - 2)^2}$ **23.** $\dfrac{5}{2}(2x - 3)^{1/4}$

25. $x^4[\frac{1}{2}(x + 1)^{-1/2}] + 4x^3 \sqrt{x + 1}$ **27.** $(2x^2 + 1)^2 5(x + 1)^4 + (x + 1)^5 8x(2x^2 + 1)$ **29.** $15x^2(x^3 + 1)^4$

31. $\dfrac{(3x + 1)^2 3(3x - 1)^2 - (x - 1)^3 6(3x + 1)}{(3x + 1)^4}$ **33.** $-6x(x^2 + 1)^{-4}$ **35.** $2[x - (x + 1)^5][1 - 5(x + 1)^4]$

37. $-3e^{-3x}$ **39.** $\dfrac{2x}{x^2 + 2}$ **41.** $10(\ln x)^9 \dfrac{1}{x}$ **43.** $\dfrac{(x + 2)^5(-4e^{-4x}) - e^{-4x} 5(x + 2)^4}{(x + 2)^{10}}$

45. $3[\sin(x+1)]^2 \cos(x+1)$ **47.** $3(e^{3x}+1)^2 3e^{3x}$ **49.** $1/(x \ln x)$ **51.** $y' = 0$ for $x = 2$, $-\frac{1}{2}$, and $\frac{1}{2}$
53. $x = 30$ **55.** 0   **part B 57.** $(0,0)$; won't hit the ground (since we have ignored the influence of gravity)   **59.** $(1/3, 0)$

answers to section 11 (page 114)

1. $v = 4$ and $a = 0$ **3.** $v = 4/(t+4)^2$ and $a = -8/(t+4)^3$ **5.** $v = -5e^{-t}$ and $a = 5e^{-t}$ **7.** Each runner finishes in 10 seconds   **9.** Runner 1 finishes at 12 m/sec while runner 2 comes in at $100/11 \approx$ 9.09 m/sec   **11.** 15.5 ft/sec   **13.** 13.754 ft   **15.** $v = 0$ **17.** 62.3 mph   **19.** 12.5, .37, and .05 lbs/yr   **21.** 18.39 and .0023 lbs/yr   **23.** At age 2, 9 cm/yr $\approx$ 3.5 in/yr; at age 13, 7 cm/yr $\approx$ 2.8 in/yr; at age 18, 0 cm/yr (growth has ceased)   **25.** 8.56, 7.01, 5.74, and 3.15 cm/yr   **27.** 4.6 cm/yr
29. $S(1) = 17$ beats/cc and $S(5) = 25$ beats/cc   **31.** $S(1)/S(2) = 2.1924/.9184 = 2.387 \approx 2.4$
33. $S = \frac{1}{3}k^3/R^2 = c/R^2$ **part B 34.** 3,000/year   **36.** 0, $-1,000$, -30, and -4 per year
38. 36,000 ft³/month, 81,000 ft³/month, 144,000 ft³/month, and 225,000 ft³/month   **40. (a)** $C'(1) =$ 3200.01 \$/ton, $C'(10) = 32,000.01$ \$/ton, and $C'(25) = 80,000$ \$/ton   **(b)** Since $C'(x) \approx C(x+1) - C(x)$, $C'(10)$, for example, approximates the cost of producing the next ton of sugar when present production is 10 tons.
42. $I' = 10(\ln .5)(.5)^x$. Hence $I'(1) = -3.4657$ candles/m²/m, $I'(2) = -1.73286$ candles/m²/m, and $I'(5) = -.2166$ candles/m²/m.

answers to section 12 (page 129)

1.   **3.**

All graphs referred to in Answers 5–33 can be found on pages a11–a12.
5. relative max at $x = 2$, see graph A – 5   **7.** relative min at $x = 3$, see graph A – 7   **9.** relative max at $-\sqrt{2}$, relative min at $\sqrt{2}$; see graph A– 9   **11.** relative min at $x = 1$, other critical point is $x = 0$; see graph A – 11   **13.** relative minima at $x = \pm2$, relative max at $x = 0$; see graph A – 13   **15.** relative max at $x = 1$; see graph A – 15   **17.** relative minima at $x = \pm1$, relative max at $x = 0$; see graph A – 17   **19.** relative min at $2 + \frac{1}{3}\sqrt{3}$, relative max at $2 - \frac{1}{3}\sqrt{3}$; see graph A – 19   **21.** relative min at $x = 1$, relative max at $x = -1$; see graph A– 21   **23.** no critical points; see graph A – 23   **25.** no critical points; graph has same general shape as A – 23   **27.** relative max at $x = -3$, relative min at $x = 3$; see graph A– 27   **29.** relative min at $x = -\sqrt[3]{13.5}$; see graph A – 29   **31.** relative max at $3 - \sqrt{11}$, relative min at $3 + \sqrt{11}$; see graph A – 31   **33.** no critical points; see graph A – 33

35.   **37.**

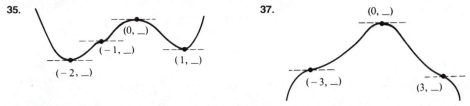

39. stationary points are $(0, 1)$ and $(2, -63)$; only $(2, -63)$ is locally stable   **41.** $(2, 0)$ and $(\frac{4}{3}, \frac{4}{27})$; only $(2, 0)$ is locally stable   **43.** $(.5, -.5)$ and $(-\frac{1}{6}, \frac{23}{18})$; only $(.5, -.5)$ is locally stable   **45.** $(-4, 228)$ and $(3, -115)$; only $(3, -115)$ is locally stable   **47.** $(.8, 4.89257\ldots)$ is locally stable   **part B 49.** $a = -\frac{3}{2}$, $b = 0$, $c = \frac{9}{2}$, $d = -1$ **51.** $a = \frac{8}{3}$, $b = \frac{1}{9}$

These graphs pertain to questions **5–33** in Sections 12 and 13. In the following graphs, inflection points are labeled with "I".

A-5.
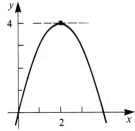

A-7.
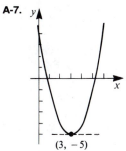
$(3, -5)$

A-9.
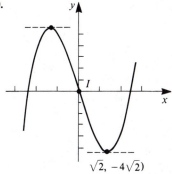
$\sqrt{2}, -4\sqrt{2})$

A-11.
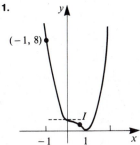
$(-1, 8)$

A-13.
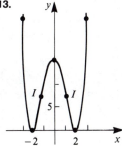

A-15.
$(1, 3)$

A-17.

A-19.
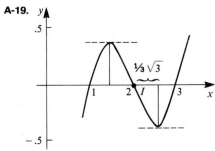
$\frac{1}{3}\sqrt{3}$

A-21.
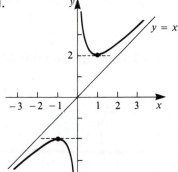
$y = x$

A-22.

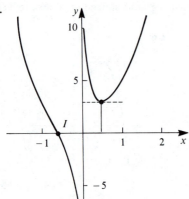

A-23.

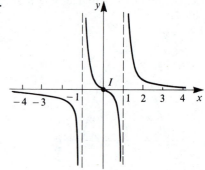

A-27.

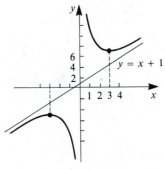

A-29.

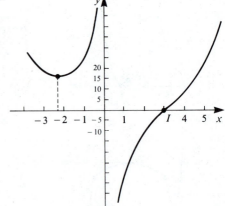

A-31.

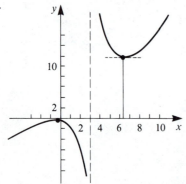

A-33.

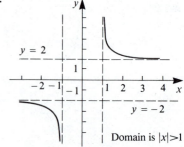

Domain is $|x| > 1$

answers to section 13 (page 138)

1.
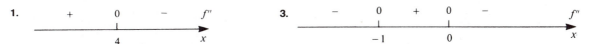

+ 0 − f''

4 x

3.

− 0 + 0 − f''

−1 0 x

All graphs referred to in Answers 15-33 can be found on pages a11-a12.

5. inflection point at $(0, -4)$, concave upward for $x > 0$, concave downward for $x < 0$ **7.** inflection points at $x = 0$ and $x = \sqrt[3]{3.2}$, concave upward for $x < 0$ and $x > \sqrt[3]{3.2}$, concave downward for $0 < x < \sqrt[3]{3.2}$ **9.** no inflection points, concave upward for $x > 0$, concave downward for $x < 0$ **11.** concave upward for all x **13.** inflection points at $x = 0, .7101\ldots$; and $1.6898\ldots$; concave upward for $0 < x < .7101$ and $x > 1.6898$ **15.** concave downward for all x; see graph $A-5$ **17.** concave upward for all x; see graph $A-7$ **19.** inflection point at $(0, 0)$; see graph $A-9$ **21.** inflection points at $(0, 1)$ and $(2/3, .4074\ldots)$; see graph $A-11$ **23.** inflection points at $(\pm\frac{2}{3}\sqrt{3}, 64/9)$; see graph $A-13$ **25.** no inflection points; see graph $A-15$ **27.** inflection points at $(\pm\sqrt{.6}, -.29548\ldots)$; see graph $A-17$ **29.** inflection point at $(2, 0)$; see graph $A-19$ **31.** inflection point at $(\sqrt[3]{-.25}, 0)$; see graph $A-22$ **33.** inflection point at $(0, 0)$; graph looks like $A-23$ **35.** relative min at $x = \frac{4}{3}$ **37.** relative min at $x = 7$, relative max at $x = -1$ **39.** relative min at $x = 0$; point of horizontal inflection at -3 using the first Derivative Test **41.** relative min at $x = -1$, relative max at $x = 1$ **43.** relative min at $x = 1$, point of horizontal inflection at $x = 0$ **45.** relative max at $x = 3$; point of horizontal inflection at $x = 0$, by first Derivative Test **47.** $q''(x) = 2a > 0$, since $a > 0$

answers to section 14 (page 150)

1. $M = 1$, $m = 0$ **3.** $M = 24$, $m = -20$ **5.** $M = 9$, $m = 0$ **7.** $M = 0$, $m = -16$ **9.** $M = 5$, $m = 4$ **11.** $M = 38$, $m = -38$ **13.** $M = 2.0099$ and $m = 1.1732$ **15.** $M = 2 + \sqrt{5}$, $m = 1$ **17.** $M = -3$ at $x = -1$ **19.** $m = 4$ at $x = 1$ **21.** $m = \frac{3}{4}$ at $x = \frac{5}{4}$ **23.** $m = -1$ at $x = 0$ **25.** $m = -3.047\ldots$ at $x = 5$ **27.** $x = 105$, $y = 105$ **29.** $x = 625$, $y = 178.57$ **31.** maximum area $= 6.158$ **33.** $x = 400/\pi$, $y = 0$ (i.e., a circular garden) maximizes the area **35.** $8'' \times 8'' \times 2''$ **37.** $V = 403.066$ cubic inches and $r = 8.1649\ldots$, $h = 5.7735\ldots$ **39.** $r = h$ in the optimal design **41.** head for a point 6.508 miles downshore; time $= 1.37$ **43.** at $x = 0$ **45.** at $x = -\sqrt{.5}$ **part B** **46.** $x = 18$ and $y = 18$ **48.** There is no maximum; $x = 18$ gives the absolute minimum **50.** Minimum brightness occurs at a distance $x = .44249d$ from the single light **52.** Wait $\frac{5}{12}$ year ≈ 150 days and a maximum profit of \$5,062.5 will result. (Is putting up with the turkeys for 150 days worth the extra \$62.50?) **54.** A price of \$15 results in a maximum monthly revenue of \$112,500. **56.** 17.43 feet **part C** **57.** radius $= \frac{7}{12}$ mile **59. (a)** 50 thousand **(b)** .6 thousand **(c)** 56.25 thousand **(d)** 4.327 thousand **(e)** 5.071 thousand **(f)** No harvest is possible; the population is naturally declining.

answers to section 15 (page 165)

1.

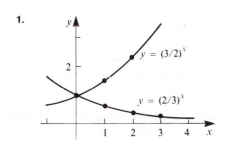

$y = (3/2)^x$

$y = (2/3)^x$

3.

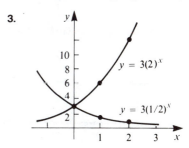

$y = 3(2)^x$

$y = 3(1/2)^x$

5.
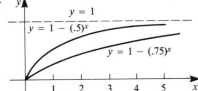

7.
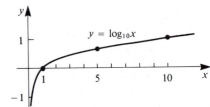

9.
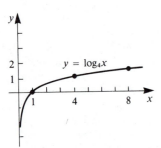

11. $\log_4 64 = 3$

13. $\log_{10} 10{,}000 = 4$

15. $\log_3 5 = x$

17. $\log_2 5 = 3x$

19. $\log_2 x = 8$

21. 7   **23.** 3   **25.** 6   **27.** 0   **29.** 1   **31.** $\ln 1.31 = .270027\ldots$, $e^{1.31} = 3.70617\ldots$
33. $\ln(-3.2)$ is not defined, $e^{-3.2} = .04076\ldots$ **35.** $\ln e = 1$, $e^e = 15.15426\ldots$ **37.** $.356207\ldots$
39. $.946394\ldots$ **41.** $x = 2$ **43.** $x = -\ln\left(\frac{4}{3}\right)$ **45.** $\frac{1}{3}$ **47.** $x = 35.835189\ldots$ **49.** $x = .736965\ldots$ **51.** $x = \left(\frac{1}{2}\right)10^{50}$ **53.** 252   **55.** $3^{16}\left(\frac{1}{2}\right)$ **57.** $1.23\overline{3}$ **59.** $x = \sqrt{8}$
61. Set $N = .5N_0$ and solve.   **63.** 130   **65. (a)** $t = 11.38895\ldots \approx 11.4$ years   **(b)** 6.30 years
67. $x = 86.37 \approx 86$ feet   **69.** $A = A_0(10^R)$ **71. (a)** 158.5   **(b)** at least 4 times more acidic

answers to section 16 (page 176)

1. $.693$ **3.** 1.0050167, 1.0005, 1.0005 **5.** $(-\ln 3)(\frac{1}{3})^x$ **7.** $-3e^{-3x+1}$ **9.** $-e \cdot e^{-ex}$
11. $(\ln 2)x2^x + 2^x$ **13.** $-e^{-x}\cos(e^{-x})$ **15.** $(2x+1)/(x^2+x-4)$ **17.** $xe^x + e^x$
19. $75e^{-5x}$ **21.** $2xe^{-2x}(1-x)$ **23.** $[2e^{2x}(x^2-x+1)]/(x^2+1)^2$ **25.** $(7/x)e^{7x-11} + e^{7x-11}(-1/x^2)$ **27.** $x^4 + 5x^4\ln x$ **29.** $e^{-2x}(\cos x - 2\sin x)$ **31.** 11.05/year, 12.21/year,
10/year   **33.** 603/year, 716/year, and 819/year   **35.** $v(0) = -1$ unit/sec, $v(\pi/2) = -.208$ unit/sec,
$v(\pi) = .043$ unit/sec   **37. (a)** $M = e^{-1}$ at $x = 1$ **(b)** inflection point at $x = 2$ **39.** relative min at
0, inflection points at $x = \pm 1$; see graph below   **41.** Absolute max occurs at $x = n+1$. The function is
increasing the most rapidly at $x = (n+1) - \sqrt{n+1}$ **43.** relative min at $x = \ln 2$, relative max at $x = \ln 3$;
see graph below   see graph below

39.
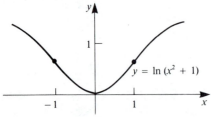

43.

$$y = e^{-x} - \frac{5}{2}e^{-2x} + 2e^{-3x}$$

45. $t = 1.393$ years   **part B 47.** $f'(x) = \dfrac{\sqrt{x-1}}{x^2+1}\left(\dfrac{1}{2(x-1)} - \dfrac{2x}{x^2+1}\right)$ **49.** $f'(x) = \dfrac{x^2 e^x}{(x+4)^4}\left[\dfrac{2}{x} + 1 - \dfrac{4}{x+4}\right]$

51. (b) K **(c)** Show $\dfrac{dR}{dN} = r - \dfrac{2r}{K}N$ **part C** **52. (a)** 0   **(b)** $x = e^{-.5}$ **(c) (a)** $M = \dfrac{1}{1 - e^{-\kappa \tau}}$

answers to section 17 (page 191)

1. $y = 3e^{-2x}$ **3.** $y = 10e^{(3/2)x}$ **5.** $y = 2e^{x}$ **7.** $y = \frac{40}{9}(1.5)^{x}$ **9.** $y = 5e^{.4x}$ **11.** $.03465 \approx$ $.035$/min   **13.** ≈ 99 years   **15.** 49, 72.4 years   **17.** $y' = (.462098\ldots)y$ **19.** ≈ 46 years **21.** age is between 30,655 and 34,900 years   **23.** Death occurred 1.79 hours before the body was discovered   **25.** $N = 1.6e^{.027t}$ **27.** $C = .011B^{1.54}$ **29.** $W = 10.38L^{3}$ [*Note:* Answers to 25–29 may vary a bit.]   **part B** **31. (a)** 30.9 percent   **(b)** Assuming that the number of rabbits eaten is directly proportional to the number of foxes present, about 340 more foxes are needed.

answers to section 18 (page 205)

1. $x^{7}/7$ **3.** $\ln x$, for $x > 0$ **5.** $\frac{1}{2}x^{6}$ **7.** $-1/3\, e^{-3x}$ **9.** $3x$ **11.** $3/2\ x^{2/3}$ **13.** $x^{3}/3 + x^{2}$ **15.** $x^{4}/4 - 2x^{2} + 3x$ **17.** $x^{3}/3 + 2x - 1/x$ **19.** $x^{3} - x^{2}$ **21.** $x^{3} + 3/x$ **23.** $9x - 3x^{2} + x^{3}/3$ **25.** $e^{2x}/2 - e^{2}x$ **27.** $\frac{1}{10}(x^{2} - 1)^{5}$ **29.** Compute $D_{x} - \sqrt{9 - x^{2}}$ **31.** $\frac{1}{3}e^{x^{3}}$ **33.** e^{x2-1} **35.** $e^{\sin x}$ **37.** $\frac{2}{15}(1 + x^{3})^{5}$ **39.** $\frac{1}{3}(1 + x^{2})^{3/2}$ **41.** $\ln(x + 1)$, for $x > -1$ **43.** $\frac{1}{2}\ln(x^{2} - 4)$ for $x^{2} > 4$ **45.** $-\frac{1}{2}\ln(2 - x^{2})$, for $-\sqrt{2} < x < \sqrt{2}$ **47.** $-\frac{1}{2}e^{-2x} + c$ **49.** $-\ln x +$ $c_{1}x + c_{2}$ **51.** $x^{2}/2 - 2x + 9/2$ **53.** $x^{2} + 2x + 1$ **55.** $y = x^{3}/3 - x^{2}/2 + x - 5/6$ **57.** $R(x) =$ $(\alpha/k)(1 - e^{-kx})$ **59.** $R(x) = \alpha x^{2}(b/2 - x/3)$ **61.** 44.0625 feet   **63.** 305.0508 ft/sec ≈ 208 mph **65.** $322.6\bar{6}$ ft   **part B** **67.** $y = 3x - x^{2}/2 - 1/2$ **69.** $y = (x - 1)^{3}/3 + 4/3$ **71.** No limit to growth   **73.** No limit to growth   **75.** $6,666.\bar{6}$ cubic feet

answers to section 19 (page 216)

1. $\sum x = 384$, $\bar{x} = 76.8$, $\sum x^{2} = 29,536$, $\sigma^{2} = 8.96$ **3.** $\sum x = 576$, $\bar{x} = 57.6$, $\sum x^{2} = 35,392$, $\sigma^{2} = 221.44$ **5.** 853   **7.** 42.32   **9.** $7.10509\ldots$ **11.** 1.1   **13.** $1.16909\ldots$ **15.** $.66877\ldots$ **17.** $7.3939\ldots$ **part B** **18.** 8.22152, 8.60349, 8.64232   **20.** .901466, .914793, .916138 **22.** $-.147701$, $-.148939$, $-.148953$

answers to section 20 (page 225)

1. $\frac{7}{3}$ **3.** $\frac{1}{4}$ **5.** 4   **7.** $1 - e^{-1}$ **9.** $-\frac{2}{3}$ **11.** $26 - \ln 3$ **13.** 6   **15.** $\frac{5}{2}$ **17.** $\frac{8}{3}$ **19.** $\frac{4}{3}R^{3}$ **21.** $1 - 2e^{-1}$ **23.** $1/x^{3}$ is not continuous on $[-1, 1]$ **25.** $\frac{1}{2}(1 - 1/b^{2})$, $\frac{1}{2}$ **27.** $k > 1$ **29.** area is infinite   **31.** $\frac{16}{3}$ **33.** $\frac{3}{2}$ **35.** 6   **37.** $\frac{7}{12}$ **39.** 15.2   **part B** **40.** $W(b) - W(a) = \Delta W$, the change in weight from time $t = a$ to $t = b$ **42.** the change in volume from $t = a$ to $t = b$—i.e., the amount of new fluid introduced

answers to section 21 (page 236)

1. $\frac{13}{3}$ **3.** $\frac{8}{3}$ **5.** 14.03   **7.** \$1,065.19   **9.** Compute $\dfrac{1}{t_{0}}\displaystyle\int_{0}^{t_{0}} - 32t\, dt$, where $t_{0} = \dfrac{\sqrt{s_{0}}}{4}$

11. 1   **13.** $\frac{1}{6}$ **15.** .05813...   **17.** .51324...   **19.** 32π **21.** $\pi(1-e^{-2})$
23. $6\pi/5$ **25.** $16\pi/3$ **27.** $32\pi/3$ **29.** Compute $\int_{r-h}^{r}\pi(r^2-x^2)\,dx$

answers to section 22 (page 246)

[*Note*: Your answers may differ from the following answers *by a constant*.]   **1.** $-(\frac{1}{4})\,1/(2x-5)^2$
3. $\frac{2}{9}(3x-2)^{3/2}$ **5.** $-\frac{1}{5}/(1+x^5)$ **7.** $-\frac{3}{8}(3-2x)^{4/3}$ **9.** $\ln(1+e^x)$ **11.** $\frac{1}{27}(3x-2)^9$
13. $(\ln x)^2/2$ **15.** $\frac{1}{6}(1+x^4)^{3/2}$ **17.** $\frac{1}{2}\sin x^2$ **19.** $3-x-3\ln(3-x)$ **21.** $7/576\approx.01215...$
23. $\frac{8}{3}$ **25.** 5   **27.** $\frac{2}{3}(x-2)^{3/2}+2(x-2)^{1/2}$ **29.** Try $u=1+e^{2x}$ **31.** $\ln(\ln x)$
33. $\frac{1}{2}\ln 26 = 1.629048...$ **35.** $\frac{1}{2}\ln|x|-\frac{1}{2}\ln|x+2|$ **37.** $-\frac{1}{2}[\ln|x+1|+\ln|x-1|]=-\frac{1}{2}\ln|x^2-1|$
39. $\ln|x+2|+2/(x+2)$ **41.** $6\ln|x+1|+3/(x+1)-6\ln|x+2|$ **43.** $x-\frac{1}{2}\ln|x+1|+\frac{1}{2}\ln|x-1|$
45. $x^2/2+3x-\ln|x-1|+8\ln|x-2|$

answers to section 23 (page 251)

1. $c(b-a)$ **3.** $k(1-1/b);\quad k$ **5.** 90 ft-lbs   **7.** **(a)** $k=25$ **(b)** $k=\frac{125}{8}\sqrt{2}$
9. $15\pi(\ln 3)$ ft-lbs   **11.** [*Hint*: First show $W=c\ln(x_2/x_1)$]   **13.** 1,310.4 ft-lbs
15. 62,500 ft-lbs   **part B   16.** $9kt$ ft-lb/sec

answers to section 24 (page 258)

[*Note*: Your answers may differ from the ones given *by a constant*.]   **1.** $-x\cos x+\sin x$ and $x^2\sin x+$
$2x\cos x-2\sin x$ **3.** $(-\frac{1}{2}x^3-\frac{3}{4}x^2-\frac{3}{4}x-\frac{3}{8})e^{-2x}$ **5.** $2x^{1/2}\ln x-4x^{1/2}$ **7.** $(x^2+3x)\ln x-$
$(x^2/2+3x)$ **9.** **(a)** $\frac{1}{3}x^2(x^2+1)^{3/2}-\frac{2}{15}(x^2+1)^{5/2}$ **(b)** $x^2\sqrt{1+x^2}-\frac{2}{3}(1+x^2)^{3/2}$
11. $2(-\sqrt{x}\cos\sqrt{x}+\sin\sqrt{x})$ **13.** $\frac{1}{2}(-x^2e^{-x^2}-e^{-x^2})$ **15.** $2e^{\sqrt{x}}(x-2\sqrt{x}+2)$ **17.** $\frac{1}{3}(x^3e^{x^3}-e^{x^3})$
19. Formula 89 with $a=4$ **21.** Formula 15 with $m=3$ **23.** 14.944 (Formula 60 with $a=1$)
25. Formula 66 with $a=4$ and $b=3$ **27.** Formula 9 with $a=3$ **29.** Formula 137 with $a=4$ and $b=-3$
31. .05969... (Formula 76 with $a=1$ and $b=2$)   **33.** $\frac{1}{4}\ln 3$ (Formula 78 after factoring)
35. $\sqrt{3-2x^2}-\sqrt{3}\ln\left(\dfrac{\sqrt{3}+\sqrt{3-2x^2}}{\sqrt{2}x}\right)\left(\text{Formula 48 with }a=\dfrac{\sqrt{3}}{\sqrt{2}}\right)$
37. Formula 28 with $a=3$ **39.** $\dfrac{3x+1}{9}\left\{\dfrac{3x+1}{2}\left[\ln(3x+1)-\dfrac{1}{2}\right]-\ln(3x+1)+1\right\}$
41. $\sqrt{e^{2x}+1}-\ln\left[\dfrac{1+\sqrt{e^{2x}+1}}{e^x}\right]$ **43.** $2[\sqrt{x}-4\ln(\sqrt{x}+4)]$

answers to section 25 (page 272)

1. 5 organisms/m^3 **3.** $150x-3x^2$ organisms/m^3 **5.** $xe^{-.1x}$ organisms/m^3 **7.** **(a)** 2,083 cod
(b) 25 m   **(c)** 617 cod   **(d)** .296   **9.** **(a)** 100 cod   **(b)** 30 m   **(c)** 24.4 cod
(d) .244   **11.** **(a)** 159 cod   **(b)** 25 m   **(c)** 49 cod   **(d)** .309   **13.** [*Hint*: Show that
$s'(0)=as_x(c-b)$ and $s'(x)=0$ for $x=(c-b)/bc$.]   **15.** **(b)** $\beta=1/35$ and $\alpha=8.93...$ **(c)** 4,573 organisms

answers to section 26 (page 283)

1. (a) $f(x) = 1/50$ and $F(x) = x/50$ for $0 \leq x \leq 50$ **(b)** .20   **3. (a)** $f(x) = (3/62,500)\, x(50 - x)$ and $F(x) = (3/62,500)(25x^2 - x^3/3)$, for $0 \leq x \leq 50$ **(b)** .248   **5. (a)** $c = \frac{1}{2}$ **(b)** $F(x) = x/2$, for $0 \leq x \leq 2$
(c) $\frac{1}{2}$ **7. (a)** $c = 1/\ln 3$ **(b)** $F(x) = \ln(x + 1)/\ln 3$, for $0 \leq x \leq 2$ **(c)** .369   **9. (a)** $c = 2$
(b) $F(x) = 1 - e^{-2x}$, for $x > 0$ **(c)** $e^{-2} = .135\ldots$ **11. (a)** $f(x) = 1$, for $0 \leq x \leq 1$
(b) $\frac{1}{10}$ **13.** $e^{-2} = .135\ldots$ **15.** $\mu = 1$, $\sigma^2 = \frac{1}{3}$ **17.** $\mu = (2/\ln 3) - 1 = .82047\ldots$, $\sigma^2 = 1 - \mu^2 = .32681\ldots$ **19.** $\mu = \frac{1}{2}$, $\sigma^2 = \frac{1}{4}$ **21.** 5 years   **23.** $\alpha = 4$ **part B   27.** $c = \alpha^{n+1}/n!$

answers to section 27 (page 294)

1. (a) $\frac{1}{3}$ **(b)** .335   **(c)** .333333   **3. (a)** $\frac{2}{3}$ **(b)** .6730347   **(c)** .66685404
5. (a) 6   **(b)** 6.000000   **(c)** 6.000000   **7. (a)** $\frac{14}{3}$ **(b)** 4.6647956   **(c)** 4.666666
9. 4.10   **11.** 4.01   **13.** 1.12   **15.** 4.066666   **17. (a)** 79.55   **(b)** 80.166
19. .1151   **21.** .6554   **23.** $z = 1.645$ **part B   25. (a)** .002777   **(b)** .083282
27. (a) .188   **(b)** .944

answers to section 28 (page 300)

1. $F = .1264 \text{ l/sec} = 7.584 \text{ l/min}$ **3.** $F = 7.04 \text{ l/min}$ **5.** $F = 3.32 \text{ l/min}$ **7.** $F = \kappa A_0/a$

answers to section 29 (page 316)

1. $\cos \pi/2 = 0$, $\sin(\pi/2) = 1$ **3.** $\cos(-2\pi) = 1$, $\sin(-2\pi) = 0$ **5.** $\cos(-5\pi/2) = 0$, $\sin(-5\pi/2) = -1$
7. $\sin(7\pi/4) = y$-coordinate of $P_4 = -1/\sqrt{2}$ **9.** $\sin(5\pi/4) = y$-coordinate of $P_3 = -1/\sqrt{2}$, $\cos(5\pi/4) = -1/\sqrt{2}$
11. $\cos(.5) = .8776$, $\sin(.5) = .4794$ **13.** $\cos(-2.1) = -.5048$, $\sin(-2.1) = -.8632$
15. $\cos(2.7) = -.9041$, $\sin(2.7) = .4274$ **17.** [*Hint:* What races will end at $(-1, 0)$?]   **19.** $t = \pm n \, \pi/2$
21. $t = \pm(4n + 2)\pi$ **23.** $t = -\pi/2 \pm 2n\pi$ **25.** $f(t) = 4 \sin \pi t$ **27.** $f(t) = 4 \sin \pi(t - 1)$
29.

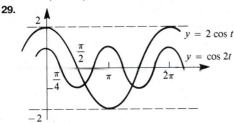

31.

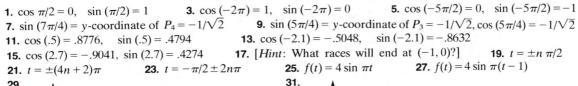

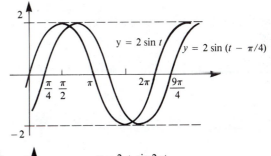

33.   **35.**

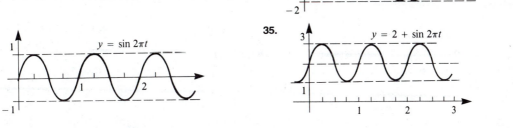

37.

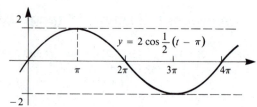

$y = 2 \cos \frac{1}{2}(t - \pi)$

39. $N = 275 + 75 \sin \frac{\pi}{6}(t - 3)$ where t is in months

41. $T = 80 + 20 \sin \frac{\pi}{12}(t - 9)$ where t is in hours after midnight

43. $\theta = 60 \sin 40\pi t$ **45.** $t = \pm.83548 \pm n\pi$

47. $t = \pi/6 \pm 2n\pi,\ -\pi/6 \pm 2n\pi,\ 5\pi/6 \pm 2n\pi,\ 7\pi/6 \pm 2n\pi$

49. $t = \pm\pi/3 \pm 2n\pi,\ \pm(2n+1)\pi$ [*Note:* $\pi/3 = 1.04719$]

51. Between 11 A.M. and 7 P.M.

answers to section 30 (page 331)

1. $-2 \sin(2t - 1)$ **3.** $3 \sin^2 t \cos t$ **5.** $(\cos t)e^{\sin t}$ **7.** $-2t^2 \sin 2t + 2t \cos 2t$ **9.** $2/(t+2)^2 \cos[t/(t+2)]$ **11.** $t = \pm n\pi$. Starting at $t = 0$, the critical points alternate between relative minima and relative maxima.   **13.** relative max at $\pi/3$, point of horizontal inflection at π, relative min at $5\pi/3$
15. relative max at $\ln(\pi/2)$ **17.** $y = \frac{1}{3} \sin 3t$ **19.** $y = 3 \sin 4(t - \pi/8)$ **21.** $t = .785399 \pm n\pi = \pi/4 \pm n\pi$ **23.** $t = .6629 \pm n(\pi/2)$ **25.** $t = \pm n\pi,\ \pi/4 \pm n\pi$ **27.** $12 + K/\pi$ **29.** $12 - K/\pi$
31. Flow $= (2An)/\omega$ cubic units per minute   **33.** $e^{-t} \sec^2 t - e^{-t} \tan t$ **35.** $-\sec t \csc t$
37. $2 \sec^2 t \tan t$ **39.** $2 \sec^2 t \tan t$

41.

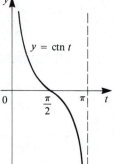

$y = \text{ctn } t$

43. Critical points are at $\pi/4 \pm n\pi$.

$y = e^{-t} \sin t$

45. [*Hint:* First show that the critical points are $t_0 + (1/\omega) \tan^{-1}(\omega/\alpha) \pm n(\pi/\omega)$.]   **47.** $t/(\sqrt{1-t^2}) + \sin^{-1} t$
49. $[1 - 2t \tan^{-1} t]/(t^2 + 1)^2$ **51.** $3t^2/(1 + t^6)$ **53.** Show that $\sin^{-1} t + \cos^{-1} t = c$ and let $t = 0$ to find c.
55. $\dfrac{x\sqrt{9 - x^2}}{2} + \dfrac{9}{2} \sin^{-1} \dfrac{x}{3}$ **57.** $\dfrac{x}{8(x^2 + 4)} + \dfrac{1}{16} \tan^{-1} \dfrac{x}{2}$ **59.** $\dfrac{-x\sqrt{2 - x^2}}{2} + \sin^{-1} \dfrac{x}{\sqrt{2}}$

answers to section 31 (page 341)

1. $\pi/4$ **3.** $4\pi/9$ **5.** $8\pi/9$ **7.** $114.59°$ **9.** $85.94°$ **11.** $\alpha = 55°,\ b = 3.501,\ c = 6.1038$ **13.** $\beta = 70°,\ b = 8.242,\ c = 8.771$ **15.** $\gamma = 110°,\ c = 13.1565$ **17. (a)** 10.5 feet
(b) 36.66 ft/sec ≈ 25 mph   **19.** 62.75 feet   **21.** 109.5 feet up the tree   **part B   23.** [*Hint:* First show $A = \frac{1}{2}ab \sin \theta$ and express a and b in terms of θ.]   **25.** $\theta = \sin^{-1} .8 \approx 53.1°$; the minimum distance is 3.75 miles

answers to section 32 (page 357)

1. $2y \dfrac{dy}{dt}$ **3.** $t \dfrac{dy}{dt} + y$ **5.** $e^x \dfrac{dx}{dt}$ **7.** $x \dfrac{dy}{dt} + y \dfrac{dx}{dt}$ **9.** $\dfrac{1}{t} + \dfrac{1}{x}\dfrac{dx}{dt}$ **11.** $\dfrac{dy}{dt} = -\dfrac{t}{y}$

13. $\dfrac{dy}{dt} = \dfrac{-2ty}{t^2 + 1}$ **15.** $\dfrac{dy}{dt} = \dfrac{\sec(ty) - y}{t}$ **17.** $\dfrac{dy}{dt} = \dfrac{e^y}{1 - te^y}$ **19.** $\dfrac{dy}{dt} = \dfrac{y}{3y^2 - t}$ **21.** $\dfrac{dy}{dt} = 20x^3 \dfrac{dx}{dt}$

23. $\dfrac{dy}{dt} = -\dfrac{y}{x}\dfrac{dx}{dt}$ **25.** $x\dfrac{dy}{dt} = \sec(xy) - y\dfrac{dx}{dt}$ **27.** $\dfrac{1}{2}$ **29.** $\dfrac{dA}{dt} = 60\,\text{ft}^2/\text{sec}$ and $\dfrac{dr}{dt} = \dfrac{30}{\pi r}\,\text{ft/sec}$

31. $\dfrac{dh}{dt} = \dfrac{5}{4\pi} \approx .398\,\text{ft/min}$ **33.** $\dfrac{d\theta}{dt} = -50 \sin^2 \theta$ radians/hour   **35.** $\dfrac{1}{W}\dfrac{dW}{dt} = 3.18\left(\dfrac{1}{L}\dfrac{dL}{dt}\right)$

37. .97 cc/sec   **39.** Maximum $= \dfrac{2}{9}\sqrt{3}$ **41.** Maximum $= \dfrac{1}{4}$ **Part B:** [*Hint:* Compute the two distances in terms of θ.]

answers to section 33 (page 368)

1. $y^2 = t^2 + c$ **3.** $y \equiv 0$ and $y = -1/(t + c)$ **5.** $y = c(1 + t)$ **7.** $y = \ln(e^t + c)$ **9.** $(y + 1)^3 = \frac{3}{4}t^4 + c$ or $y = \sqrt[3]{\frac{3}{4}t^4 + c} - 1$ **11.** $t^2 + y^2 = 25$ or $y = \sqrt{25 - t^2}$ **13.** $y = [(t^2 + 2)/3]^{3/2}$

15. $y = \sqrt[4]{t^3 + 56}$ **17.** $y = 1/(1 - \frac{1}{2}e^{-t})$ **19.** $y = \sin t^2/2$ **21. (b)** $V = (kt/3 + V_0^{1/3})^3$

23. (a) $N = N_0 \exp[-(k/2\pi)(\cos 2\pi t + 1)]$ **(b)** At time $t = 1/2$ **25.** $y^2 = 2x + c$. Graphs are parabolas with vertices on the x-axis   **27. (b)** $A = (I/k)(1 - e^{-kt})$ **(c)** I/k **29. (a)** [*Hint:* Use $A' = c\sqrt{A}(K - A)$ and take the derivative with respect to t.]   **(b)** $\dfrac{-1}{\sqrt{K}} \ln\left(\dfrac{\sqrt{K} - \sqrt{y}}{\sqrt{K} + \sqrt{y}}\right)$ **(c)** For c_1 a constant depending on $A(0)$, $A = K\left(\dfrac{e^{-c\sqrt{K}t} - c_1}{e^{-c\sqrt{K}t} + c_1}\right)^2$.   **(d)** K **part B 30. (a)** $y = [(1 - \alpha)kt + c]^{1/(1-\alpha)}$

(b) $t_1 = \dfrac{c}{k(\alpha - 1)}$ **32. (a)** $x = \dfrac{a^2 kt}{akt + 1}$ and $x_\infty = a$ **(b)** $x = ab\left[\dfrac{1 - e^{(b-a)kt}}{a - be^{(b-a)kt}}\right]$ and $x_\infty = a$

answers to section 34 (page 379)

1. 4.86 cm/yr   **3.** 130.37 g at time $t = 2.83$ yr   **5.** $L(15) = 53.25$ cm, $W(15) = 1,348$ g, $L'(15) = 1.45$ cm/yr   **7.** $L(0) = 5.10$ cm, $W(0) = 1.19$ g, $L'(0) = 6.02$ cm/yr, $W'(0) = 3.68$ g/yr

answers to section 35 (page 387)

1. $N_1(0) = 3$, $N_2(0) = 4$, $K_1 = 105$, $K_2 = 64$ **3.** For population (1): 3.14 days; for population (2): 3.41 days   **5.** $r = .05$ **7.** $N = (rkt + N_0^k)^{1/k}$, $\lim_{t\to\infty} N(t) = +\infty$ for any $k > 0$ **9. (a)** $N(t) = N_0 e^{rt}$ for $0 \le t \le t_1$, where $N(t_1) = M$ **(b)** For $t \le t_1$, $N = K/(1 - ce^{r_1 t})$, where $r_1 = r[K/(K - M)]$ **(c)** $0 < N_0 < K/2$

answers to section 36 (page 397)

1. MSY $= 14.3 \times 10^6$ kg with $F = .355$ **3. (a)** 53.5 years   **(b)** $F = .03\overline{3}$ **5.** $t = 2.4$ years   **7.** 5.9 years   **9. (a)** $F = .182$ **(b)** 3,344 trout

answers to section 37 (page 406)

1. $y = 4e^{t^3/3}$ **3.** $y = -2t^5$ **5.** $y = \frac{3}{2}e^2 t^2 t e^{-2t}$ **7.** $y = ce^{-t^2} + \frac{1}{2}$ **9.** $y = ct - t^2$ **11.** $y = -te^t$
13. $y = -\frac{5}{2}e^{-3t} + \frac{1}{2}e^{-t}$ **15.** $y = 3t^2 \ln t + 2t^2$ **17.** $y = 2t \cos t + \cos t$ **19.** $y = (\sin t)/t^2$
21. 69.3 minutes   **23.** $y = ce^{-\lambda t} + I/\lambda$ **25. (a)** $y = ce^{-.05t} + 1,000$ and the limit is 1,000
(b) $y = ce^{-.05t} + (1,000 - 20h)$ and the limit is $(1,000 - 20h)$ **(c)** 35   **27.** [*Hint*: Factor $e^{-\lambda t}$ out of
the derivative.]   **29. (b)** $c(t) = c_0 e^{-(F/V)} + (I/F)(1 - e^{-(F/V)})$ **(c)** 6 years   **(d)** 432 years
part B   31. $y = y_0 e^{-Bt} e^{(A/\omega)(\cos \omega t - 1)}$ **33.** $w' = (1 - n)y^{-n}y'$. Now substitute for y' and simplify.

answers to section 38 (page 418)

1. e^{2t} and e^{-t} **3.** 1 and e^{5t} **5.** $\sin 2t$ and $\cos 2t$ **7.** e^{-3t} and te^{-3t} **9.** $e^{2t} \cos 3t$ and
$e^{2t} \sin 3t$ **11.** $y = \cos 3t$ **13.** $y = 1$ **15.** $y = \frac{8}{9}e^{-5t} + \frac{1}{9}e^{4t}$ **17.** $y = -\frac{2}{3}e^{.5t} + \frac{2}{3}e^{2t}$
19. $y = 3e^t - 2te^t$ **21.** $y = 2e^{-t} \sin t$ **23.** $y = -4 + 4e^t$ **25.** $y = 2 \sin 2t$
27. $y = c_1 e^{4t} + c_2 e^{-3t} - \frac{1}{10}e^{-t}$ **29.** $y = c_1 \sin(2\sqrt{2}t) + c_2 \cos(2\sqrt{2}t) + \sin 2t$ **31.** $y_p = \frac{7}{4}$
33. $y_p = -3 - 4t$ **35.** $y_p = -\cos 2t$ **37.** $y_p = \frac{3}{2}x^2 + \frac{3}{2}x - \frac{3}{4}$ **39.** $y_p = -2t$
41. $y = -\frac{5}{3}e^t + \frac{11}{12}e^{4t} + \frac{7}{4}$ **43.** $y = -5e^{-t} + e^{-2t} + t^2 - 3t + 4$ **45.** $y = -\frac{1}{2}\sin t + \frac{\pi}{2}\cos t - \frac{1}{2}t \cos t$
part B   46. $y = \frac{1}{3}e^{5(t-1)} + \frac{5}{3}e^{-(t-1)}$ **48.** $y_p = -\frac{1}{36} - \frac{1}{6}t + (\frac{1}{50})\sin t - (\frac{7}{50})\cos t$ **50.** $\sin \omega T \neq 0$

answers to section 39 (page 430)

1. $x = c_1 e^{-t} + c_2 e^{-2t}$
$y = -\frac{1}{2}c_1 e^{-t} - \frac{2}{3}c_2 e^{-2t}$
5. $x = c_1 \sin 2t + c_2 \cos 2t$
$y = -2c_2 \sin 2t + 2c_1 \cos 2t$
9. $x = e^{-2t}(-6 \sin t + 4 \cos t)$
$y = e^{-2t}(10 \sin t + 2 \cos t)$
13. $x = c_1 e^{3t} + c_2 e^{4t} - \frac{2}{3}$
$y = c_1 e^{3t} + \frac{2}{3}c_2 e^{4t} + \frac{4}{3}$
17. $x = c_1 + c_2 e^{-2t} + \frac{1}{5}\sin t - \frac{3}{5}\cos t$
$y = c_1 - c_2 e^{-2t} - \frac{1}{5}\sin t - \frac{2}{5}\cos t$

21. $x = 400 + 50e^{-3t}$
$y = 200 - 50e^{-3t}$
Neither wins. The two populations stabilize at
(400, 200).

25. $x = 25e^{-.059t} + 25e^{-.341t}$
$y = 35.355e^{-.059t} - 35.355e^{-.341t}$

3. $x = c_1 \sin 4t + c_2 \cos 4t$
$y = \frac{-3c_1 + 4c_2}{5}\sin 4t + \frac{-3c_2 - 4c_1}{5}\cos 4t$
7. $x = -3e^{2t} + 7e^{4t}$
$y = -9e^{2t} + 7e^{4t}$
11. $x = (2 + 4t)e^{-3t}$
$y = (1 + 4t)e^{-3t}$
15. $x = c_1 e^{3t} + c_2 e^t + (\frac{5}{3} - 2t)$
$y = 5c_1 e^{3t} + 3c_2 e^t + (\frac{4}{3} - t)$
19. $x = 1,500e^{-t} - 500e^{3t}$
$y = 1,200e^{-t} + 400e^{3t}$
The McCoys win; the Hatfields become extinct at
time $t = .275$.

23. If x and y denote *population densities*, then $x' = \frac{k}{2}(y - x)$ and $y' = k(x - y)$. The solution is $x = 10,000 + 5,000e^{-\alpha t}$ and $y = 10,000 - 10,000e^{-\alpha t}$ where $\alpha = 3k/2$.

27. (a) $x = 432.185 - 425.185e^{-.2025t}$
$y = 34,574.815 + 425.185e^{-.2025t}$

27. (b) $x = 466.326 - 459.326e^{-.2025t} + \frac{7}{81}t$

$y = 37{,}306.106 + 459.326e^{-.2025t} + (6.914t - 2{,}765.432)$

29. (from hip)

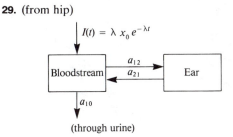

$I(t) = \lambda\, x_0\, e^{-\lambda t}$

(through urine)

part B   31. (a) $x(t) = \dfrac{x_0}{a_{12} + a_{21}}\left[a_{21} + a_{12}e^{-(a_{12}+a_{21})t}\right]$ and $y(t) = \dfrac{x_0 a_{12}}{a_{12} + a_{21}}\left[1 - e^{-(a_{12}+a_{21})t}\right]$

(b) $x_\infty = \dfrac{a_{21}x_0}{a_{12} + a_{21}}$ **(c)** $a_{12} = .282,\ a_{21} = .018$

answers to section 40 (page 447)

1. 70°F   **3.** 80°F   **5.** $f(900, 20) \approx 4{,}400$ cubic feet per acre, $f(950, 15) \approx 3{,}500$ **7.** $f(0, 0)$ and $f(-2, 1)$ are not defined, $f(1, -2) = \ln 4$ **9.** $f(0, 0) = 1,\ f(1, -2) = e^3,\ f(-2, 1) = e^{-3}$
11. $A = 2xz + 2yz$ **13.** $V = \frac{4}{3}\pi x^3 + \pi x^2 y$ **15.** $h = (\tan \alpha \tan \beta)/(\tan \beta - \tan \alpha)$ **17.** $(2, -2, 4)$
19. $(0, 5, -1)$

21.

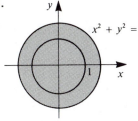

$x^2 + y^2 = 2$

23.

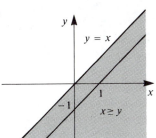

$y = x$

$x \geq y$

25.

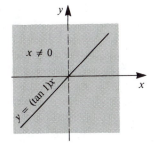

$x \neq 0$

$y = (\tan 1)x$

27.

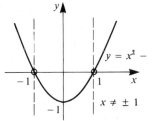

$y = x^2 - 1$

$x \neq \pm 1$

29.

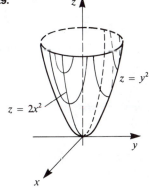

$z = y^2$

$z = 2x^2$

31. The graph is identical to the graph in example 40.5, except that the saddle rises up along the x-axis.

33.

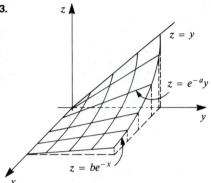

35.

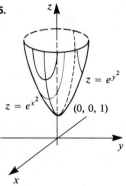

37.

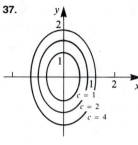

39.

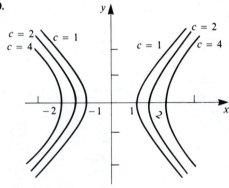

41.

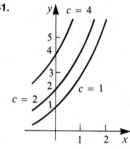

43.

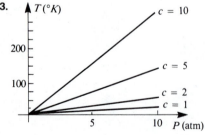

45.

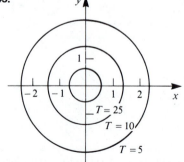

answers to section 41 (page 456)

part A 1. $\dfrac{\partial w}{\partial x} = yz - 2xz$, $\dfrac{\partial w}{\partial y} = xz$, $\dfrac{\partial w}{\partial z} = xy - x^2$ 3. $\dfrac{\partial w}{\partial x} = 12x^3$, $\dfrac{\partial w}{\partial y} = -4y$, $\dfrac{\partial w}{\partial z} = -1$

5. $\dfrac{\partial w}{\partial x} = e^{yz}$, $\dfrac{\partial w}{\partial y} = xze^{yz}$, $\dfrac{\partial w}{\partial z} = xye^{yz}$ 7. $f_x = 2xe^{-y}$, $f_y = -x^2e^{-y}$ 9. $f_x = \dfrac{-10y}{(x-2y)^2}$,

$f_y = \dfrac{10x}{(x-2y)^2}$ 11. $\dfrac{\partial f}{\partial x} = y\cos(xy)$, $\dfrac{\partial f}{\partial y} = x\cos(xy)$ 13. $f_x = z/(y+z)$, $f_y = -xz/(y+z)^2$,

$f_z = xy/(y+z)^2$ 15. $f_x = x/\sqrt{x^2+y^2+z^2}$, $f_y = y/\sqrt{x^2+y^2+z^2}$, $f_z = z/\sqrt{x^2+y^2+z^2}$ 17. $f_x = ye^{zy}$,

$f_y = e^{yz}(xyz + x)$, $f_z = xy^2 e^{yz}$ 19. $(6, 4)$ 21. $(2, -3)$ 23. $(\sqrt{\tfrac12}, \sqrt{\tfrac12})$ and $(-\sqrt{\tfrac12}, -\sqrt{\tfrac12})$

25. $(0, 0)$ and $(4, 4)$ 27. $(0, 0)$ 29. $(1, -2, \tfrac32)$ 31. $f_x = 4x^3 - 6xy - 2y^3$, $f_y = -3x^2 - 6xy^2 + 3y^2$,

$f_{xx} = 12x^2 - 6y$, $f_{yy} = -12xy + 6y$, $f_{xy} = f_{yx} = -6x - 6y^2$ 33. $f_x = \dfrac{-2y}{(x-y)^2}$, $f_y = \dfrac{2x}{(x-y)^2}$, $f_{xx} = \dfrac{4y}{(x-y)^3}$,

$f_{yy} = \dfrac{4x}{(x-y)^3}$, $f_{xy} = f_{yx} = \dfrac{-2x-2y}{(x-y)^3}$ 35. $f_x = \dfrac{3x^2}{x^3+y^3}$, $f_y = \dfrac{3y^2}{x^3+y^3}$, $f_{xx} = \dfrac{-3x^4+6xy^3}{(x^3+y^3)^2}$, $f_{yy} = \dfrac{-3y^4+6x^3y}{(x^3+y^3)^2}$, $f_{xy} =$

$f_{yx} = \dfrac{-9x^2y^2}{(x^3+y^3)^2}$ 37. $\dfrac{\partial^2 z}{\partial x^2} = 6x$ and $\dfrac{\partial^2 z}{\partial y^2} = -6x$ 39. $\Big[$Hint: Show $\dfrac{\partial C}{\partial t} = C\dfrac{x^2-2at}{4at^2}$ and $\dfrac{\partial^2 C}{\partial x^2} =$

$C\dfrac{x^2-2at}{4a^2t^2}.\Big]$ **part B** 43. $z = (c_1 \sin ax + c_2 \cos ax)(c_3 \sin ay + c_4 \cos ay)$

answers to section 42 (page 467)

1. When $x = 160$ and $y = 70$, $\dfrac{\partial w}{\partial x} = .0545$ ft^2/lb   3. $\dfrac{\partial w}{\partial y} = (.0790975)\dfrac{x^{.425}}{y^{.275}}$ ft^2/in. For a fixed weight, this is the

rate of change of surface area with respect to height.   5. If $[x] =$ meters and $[t] =$ hours, $\dfrac{\partial T}{\partial t} =$

$10\pi(\cos 2\pi t)e^{-3x}$ °C/hour and $\dfrac{\partial T}{\partial x} = (10 + 5\sin 2\pi t)(-3e^{-3x})$ °C/meter. The derivative $\dfrac{\partial T}{\partial x}$ measures the rate at

which temperature changes with respect to depth x at a *fixed time of the day*.   7. If $[F] =$ cc/sec and

$[R] =$ mm, $\dfrac{\partial F}{\partial R} = c\dfrac{x}{L}(4R^3)$ cc/sec per mm and $\dfrac{\partial F}{\partial x} = \dfrac{c}{L}R^4$ cc/sec per unit pressure. $\dfrac{\partial F}{\partial R}$ measures the rate at which

flow rate changes with respect to the arterial radius for a *fixed x*, that is, when the heart is pumping at a fixed

rate.   9. $\dfrac{dw}{dt} = 2xy(-\sin t) + x^2(2t)$ 11. $\dfrac{dw}{dt} = \dfrac{2x}{x^2+y^2}(-e^{-t}) + \dfrac{2y}{x^2+y^2}\cos t$

13. $\dfrac{dw}{dt} = (6x - 2y + 9)(4\cos t) + (-2x - 3)(1 + \ln t)$ 15. $\dfrac{dV}{dt} = (2\pi rh)\dfrac16 + \pi r^2(2e^{-.5t})$

17. $\dfrac{dV}{dt} = \dfrac{-kT}{P^2}(.5e^{-.5t}) + \dfrac{k}{P}(2e^{-.1t})$, where $P = 2 - e^{-.5t}$ and $T = 300 - 20e^{-.1t}$ 19. relative minimum at

$(0, 2)$, saddle point at $(0, -2)$ 21. saddle point at $(3, 4)$, relative maximum at $(-3, 4)$ 23. relative

maximum at $(0, 0)$ 25. saddle point at $(0, 0)$ 27. saddle point at $(0, 0)$, relative minimum at $(4, 4)$

29. $x = 300'$, $y = 300'$, and $z = 400'$ 31. Show $I_{xx} = -2$ and $\Delta = 4$ 33. $\Big[$Hint: If $w = A^2$, show that

$\dfrac{\partial w}{\partial x} = s(s - y)(2s - 2x - y)$ and $\dfrac{\partial w}{\partial y} = s(s - x)(2s - x - 2y).\Big]$ 35. $|\Delta S| \le .576$ 37. $|\Delta h| \le 6.52$

feet   39. $R = 100.0$ mL clean blood/min, $|\Delta R| \le 30$ 41. $N(5) \approx 24{,}103 \pm 14{,}060$ 43. $T = 8.6026$

days, and $|\Delta T| \le .403$

answers to section 43 (page 482)

part A **1.** $Y = 1.52804X - 6.70093$

X	Y (Observed)	Y (Predicted)
12	10	11.6355
20	25	23.8598
28	38	36.0841
40	53	54.4206

3. $Y = .0300X + 1.0095$

X	Y (Observed)	Y (Predicted)
33	2	1.99764
100	4	4.00394
200	7	6.99842
300	10	9.9929

5. $Y = -.757143X + 87.1333$
9. $C = .010537W^{1.5445}$

W	C (Observed)	C (Predicted)
58	5	5.577
300	78	70.5863
536	196	172.984
1,080	537	510.44
1,449	773	803.707
2,233	1,380	1,567.42

7. $y = 1.03884e^{-1.03808x}$

11. (a) $y = 2.57t + 3.74$ **(b)** $z = 1.86648t^{1.21419}$ **(c)** The second fit is better. Compute $D = \Sigma\, [y_i - f(t_i)]^2$ for each function.

13. $V = \dfrac{.227655x}{1 + .42800x}$

15. $y = 50/(1 + 147.941e^{-1.94348t})$

t	y (Observed)	y (Predicted)
0	.24	.3357
1	2.78	2.2537
2	13.53	12.3949
3	36.30	34.8561
4	47.50	47.0713
5	49.40	49.5584

part B **17. (c)** $y = 139.2(1 - .3834e^{-.049t})$ **19. (b)** $Y = (-.83928)X^2 + 3.7964X + .53573$

answers to section 44 (page 488)

part A **1.** 12 **3.** -6 **5.** $\frac{11}{6}$ **7.** $\frac{2}{3}$ **9.** 0 **11.** $\frac{3}{2}\ln 3 - 2\ln 2 + \frac{1}{2}$ **13.** $-2/\pi^2$
15. $\frac{14}{3}$ cubic units **17.** 5.6 **19.** $\frac{11}{18}$ **21.** .6776 **23.** $\frac{1}{4}e^4 + \frac{3}{4}$ **25.** .667968 . . .
27. -5.545 **29.** $\frac{1}{3}$ cubic unit **part B** **33.** .5346

answers to section 45 (page 506)

part A **1.** $(2\sqrt{2}, \pi/4)$ **3.** $(6, \pi/2)$ **5.** $(4, \pi)$ **7.** $(\sqrt{13}, .3128\pi)$ **9.** $(1, \pi/3)$
11. $0 \le r \le 3, 0 \le \theta \le \pi$ **13.** $0 \le r \le 4, 0 \le \theta \le \pi/2$ **15.** $0 \le r \le 6, -\pi/4 \le \theta \le \pi/4$ **17.** $1 \le r \le 3,$
$0 \le \theta \le 2\pi$ **19.** $2 \le r \le 3, \ 0 \le \theta \le \pi/4$ **21.** $81\pi/4$ **23.** 1,024/15 **25.** $72\sqrt{2}$ **27.** 0
29. 0 **31.** $\pi/2$ **33.** 5π **35.** $10\pi \ln [(R^2 + 4)/4] \to +\infty$, as $R \to +\infty$

answers to section 46 (page 518)

part A 1. $\frac{4}{3}$ **3.** $(\ln 3)/2$ **5.** $\frac{5}{9}\ln 10$ **7.** $2A/(B^2 r_0^2)[1 - e^{-Br_0}(r_0 B + 1)]$ **9.** 901,040
11. about 102 seeds **13.** $c = \frac{3}{16}$, A has probability $\frac{21}{32}$ **15.** [*Hint*: Show that the integral over
$0 \le r \le R$, $0 \le \theta \le 2\pi$ is $1 - e^{-R^2/(2\sigma^2)}$.] **17. (a)** $1 - e^{-4.5} = .9889$ **(b)** $e^{-.5} - e^{-2} = .4712$
19. (a) $\frac{64}{9}$ cm^3/sec **(b)** $\frac{16}{9}$ cm/sec **part B 21. (b)** $1 - e^{-r_0^2/(4Dt)}$ **23.** $\bar{x} = \bar{y} = \frac{7}{6}$ **25.** $\bar{x} = \bar{y} = 0$
27. $\bar{x} = \frac{4}{3}$, $\bar{y} = \frac{1}{3}$

answers to section 47 (page 530)

Part A 1. $\frac{1}{3}$, $\frac{2}{5}$, $\frac{3}{7}$, $\frac{4}{9}$, $\frac{5}{11}$, $\frac{6}{13}$, $\frac{7}{15}$, $\frac{8}{17}$, $\frac{9}{19}$, $\frac{10}{21}$ **3.** $-\frac{1}{3}$, $\frac{1}{9}$, $-\frac{1}{27}$, $\frac{1}{81}$, $-\frac{1}{243}$, $\frac{1}{729}$,
$-\frac{1}{2,187}$, $\frac{1}{6,561}$, $-\frac{1}{19,683}$, $\frac{1}{59,049}$ **5.** $-\frac{1}{2}$, $\frac{1}{4}$, $-\frac{1}{8}$, $\frac{1}{16}$, $-\frac{1}{32}$, $\frac{1}{64}$, $-\frac{1}{128}$, $\frac{1}{256}$, $-\frac{1}{512}$, $\frac{1}{1,024}$
7. 0 **9.** $-\frac{1}{2}$ **11.** 0 **13.** does not exist **15.** 1 **17.** 0 **19.** -3
21. 0 **23.** 0 **25.** 0 **27.** 2, 1, 2, 1, 2 **29.** 1, $\frac{1}{2}$, $\frac{1}{3}$, $\frac{1}{4}$, $\frac{1}{5}$ **31. (a)** $x_{n+1} = (1.11)x_n$
(b) $x_n = 300 \ (1.11)^n$ **33. (a)** $x_{n+1} = 1.135x_n$ **(b)** $n \ge 8$ **35.** $a_n = (3.125) 2^n$ or $(-3.125)(-2)^n$
37. (a) $x_{n+1} = x_n - 250$ **(b)** $n = 41$ **39.** Table for y_n is shown here:

n	1	2	3	4	5	6	7	8	9	10
y_n	911.69	922.35	932.04	940.85	948.85	956.11	962.69	968.65	974.05	978.93

41. (b) 1, 1.5, 1.4166, 1.4142157, 1.4142136, **43. (a)** 1, .5, .66, .6, .625 **(b)** $(1 + \sqrt{5})/2$
part B 44. $y_{n+1} = y_n[1 + r(K - y_n)/K]$ **46.** $y' = \beta y^2 - y$

answers to section 48 (page 540)

part A 1. $x_n = 0$ for $n \ge 1$ **3.** $x_n = 2/n!$ **5.** $x_n = 2n$ **7.** $x_n = c(2^n) - 2$ **9.** $x_n =$
$c(-3)^n + 1$ **11.** $x_n = 2(.5)^n + 8$, limit is 8 **13.** $x_n = 100 + 4n$ **15.** $x_n = -1,000(1.1)^n + 2,000$, $x_n < 0$
for $n \ge 8$ **17.** $x_n = 500(1.2)^n + 1,500$ **19.** $x_n = 3^{2n-1}$ **21.** $x_n = 6,250(1.08)^n - 6,250$, $x_{18} =$
$\$18,725.12$ **23. (a)** $x_n = -833.33(1.12)^n + 10,833.33$ **(b)** 22 years **(c)** $x_n = 10,000$, and the parties
can continue forever! **25.** If r is the constant of proportionality and $r > 0$, then $x_n = (K - x_0)(1 - r)^n + K$. If
$0 < r < 2$, limit $x_n = K$. Otherwise the limit does not exist. **27. (a)** 426.84 mg **(b)** 425.98 mg
(c) 2.1 hours (or about every 2 hours) **29. (a)** $x_n = Aa^n + \dfrac{b}{1 - a}(1 - a^n)$ **(b)** $\dfrac{b}{1 - a}$ **(c)** [*Hint*:
Use the explicit solution to show $x_n \le b/(1 - a) \le A$.] **31. (a)** $x_n = 10.508 (1 - e^{-.1n})$ **(b)** $n = 5$, that is,
every 10 milliseconds **part B 32. (a)** $x_n = c(1 - b)^n + (\ln \lambda)/b$ **(b)** $(\lambda)^{b-1}e^{c(1-b)n}$; if $0 < b < 2$,
limit $y_n = \lambda^{b-1}$ **(c)** When $1 < b < 2$, the approach to y_∞ is oscillatory. **34. (b)** 450 generations

answers to section 49 (page 552)

part A 1. 2^n and $(-1)^n$ **3.** $(\frac{1}{2})^n$ and $(-\frac{2}{3})^n$ **5.** $(\frac{1}{2})^n$ and $(-\frac{1}{2})^n$ **7.** $2^n \sin (n\pi/2)$ and
$2^n \cos (n\pi/2)$ **9.** $(\frac{1}{2})^n$ and $n(\frac{1}{2})^n$ **11.** $x_n = 1 + n$ **13.** $x_n = \frac{3}{2} - \frac{1}{2}(-1)^n$ **15.** $x_n =$
$2(\frac{1}{2})^n - (-1)^n$ **17.** $x_n = 3(4^n) - n(4^n)$ **19.** $x_n = (\sqrt{5})^n(\cos n\theta - \frac{7}{4}\sin n\theta)$ where $\theta = \pi - \tan^{-1} 2$
21. $x_n = c_1 + c_2(-5)^n + n$ **23.** $x_n = c_1(-2)^n + c_2(-1)^n - 2n + 1$ **25.** $x_n = c_1(\frac{1}{2})^n + c_2 n(\frac{1}{2})^n + 16$

27. $p_n = 2$ **29.** $p_n = -3n^2 - 9n$ **31.** $x_n = -(2)^n + (-1)^n + 5$
33. $x_n = \frac{3}{2}\sin(n\pi/2) + \cos(n\pi/2) + \frac{1}{2}n$ **35.** $x_n = (\sqrt{2})^n[.6\cos(3\pi n/4) - 1.8\sin(3\pi n/4)] + .4$
37. $w_n = 100 + 100(1.2)^n$; $w_{12} = 991.61$ lbs   **39.** If w_n denotes the total biomass *before* the 30 pounds are
taken, then $[w_{n+1} - (w_n - 30)] = 1.2[w_n - (w_{n-1} - 30)]$. Then $w_n = 130 + 100(1.2)^n - 30n$ and $w_{12} = 661.61$ lbs.
41. $x_n = 50(\sqrt{2})^n[1 + (-1)^n]$. Thus, $x_n = 0$ for n odd.
43. $x_n = \left(\frac{1}{2} - \frac{1}{10}\sqrt{5}\right)\left(\frac{1+\sqrt{5}}{2}\right)^n + \left(\frac{1}{2} + \frac{1}{10}\sqrt{5}\right)\left(\frac{1-\sqrt{5}}{2}\right)^n$ **45.** $x_n = 16(-1)^n - 15(-2)^n$, $y_n = -8(-1)^n + 10(-2)^n$

answers to section 50 (page 564)

part A   **1.** diverges   **3.** converges to $3/(\pi - 3)$ **5.** converges to 1   **7.** converges to $\frac{1}{3}$
9. converges to $\frac{3}{4}$ **11.** converges to $\frac{1}{2}$ **13.** limit $a_n = 1 \neq 0$ **15.** limit a_n does not exist
17. [*Hint*: Show $S_n > n^{3/4}$.]   **19.** converges   **21.** converges   **23.** diverges   **25.** converges
27. converges   **29.** converges   **31.** diverges   **33.** converges   **35.** diverges
37. converges   **39.** diverges   **41.** diverges   **43.** converges   **45.** converges for $-1 < x < 1$
47. converges for $-1 < x < 1$ **49.** converges for $-3 < x < 3$ **part B**   **53.** $\left[\text{\textit{Hint}: Show}\right.$
$\left.\int_1^{+\infty}\frac{1}{x^k}\,dx = \frac{1}{k-1}\text{ for } k > 1.\right]$ **55.** diverges   **59.** when $\sum_{n=0}^{\infty} b_n$ converges

answers to section 51 (page 575)

part A   **1.** For $a = 1$, $P_1(x) = 1 + \frac{1}{3}(x - 1)$. For $a = 8$, $P_1(x) = 2 + \frac{1}{12}(x - 8)$.   **3.** For $a = 0$, $P_1(x) = x$. For
$a = 1$, $P_1(x) = \pi/4 + 1/2(x - 1)$.   **5.** $P_2(x) = 1 - x + \frac{1}{2}x^2$ and $P_2(.5) = .625$.   **7.** $P_2(x) = 1 + \frac{1}{5}(x - 1)$
$- \frac{2}{25}(x - 1)^2$ and $P_2(.8) = .9568$.   **9.** $P_1(x) = 1$ and $P_2(x) = 1 - x^2/2$ **11.** $x_{n+1} + \frac{1}{2}(x_n + c/x_n)$;
55.704578   **13.** -1.343210 **15.** .889891   **17.** 2.718282   **19.** .739085   **21.** $P_4(\pi/180) = $
.9998477 and all digits shown are correct; $|E_4(x)| \leq 7.5 \times 10^{-14}$.   **23.** $P_3(x) = x + x^3/3$ and $P_3(\pi/180) = $
.01745506.   **25.** $P_5(x) = (x - 1) - (x - 1)^2/2 + (x - 1)^3/3 - (x - 1)^4/4 + (x - 1)^5/5$ **27. (a)** $P_5(x) = $
$x - x^3/3 + x^5/5$ **(b)** $\pi/4 \approx .78639403$ **29.** $P_n(x) = 1 - x + x^2/2? - x^3/3! + \cdots + (-1)^n x^n/n!$
31. $|E_2(x)| \leq .0208\overline{3}$ **33.** $|E_4(x)| \leq .000823$ **35. (a)** .06907386   **(b)** error $\leq .0001674$
37. (a) .176149   **(b)** error $\leq .000006$ **39. (a)** .4500699   **(b)** error $\leq .055633$ **part B**   **41.** 0
43. $\frac{2}{3}$ **45.** 8   **47.** 12   **49.** $-\frac{4}{3}$ **part C**   **52.** y_∞ **54.** $y = 0$ and $y = K$
56. $y = 0$ and $y = (I/m) - a$ **58.** stable   **60.** $y = 0$ is unstable; $y = K$ is stable   **62.** If $(I/m) - a$
> 0, it is stable and $y = 0$ is unstable. If $(I/m) - a < 0$, it is unstable and $y = 0$ is stable.

answers to section 52 (page 589)

part A   **1.** $R = 3$, $(-3, 3)$ **3.** $R = 0$, converges only for $x = 0$ **5.** $R = 1$, $(-3, -1)$
7. $R = +\infty$, $(-\infty, +\infty)$ **9.** $R = \frac{1}{2}$, $(0, 1)$ **11.** $x^2 - \frac{x^6}{3!} + \frac{x^{10}}{5!} - \frac{x^{14}}{7!} + \cdots$ **13.** $\frac{x^3}{3} - \frac{1}{7}\frac{x^7}{3!} + \frac{1}{11}\frac{x^{11}}{5!} - \frac{1}{15}\frac{x^{15}}{7!} + \cdots$
15. $1 - x^2 + x^4 - x^6 + x^8 - \cdots$ **17.** $x^3 - \frac{x^7}{3} + \frac{x^{11}}{5} - \frac{x^{15}}{7} + \frac{x^{19}}{9} - \cdots$ **19.** $x^2 - \frac{x^4}{2} + \frac{x^6}{3} - \frac{x^8}{4} + \frac{x^{10}}{5} + \cdots$
21. $\frac{1}{3}$ **23.** 0   **25.** $\frac{1}{4}$ **27.** $x + x^2 + \frac{5}{6}x^3 + \frac{1}{2}x^4 + \cdots$ **29.** $-x + x^2 + \frac{1}{6}x^3 + \frac{1}{3}x^4 - \frac{1}{120}x^5 + \cdots$

31. $y = 1 - t + t^2/2 - t^3/6 + \cdots = e^{-t}$

33. $y = t - \dfrac{t^3}{3!} + \dfrac{t^5}{5!} - \dfrac{t^7}{7!} + \cdots = \sin t$

35. $y = t + t^2 + t^3/2 +$

$t^4/6 + \cdots = te^t$

37. $y = 1 + \dfrac{1}{2 \cdot 3} t^3 + \dfrac{1}{2 \cdot 3 \cdot 5 \cdot 6} t^6 + \dfrac{1}{2 \cdot 3 \cdot 5 \cdot 6 \cdot 8 \cdot 9} t^9 + \cdots$

39. $y = t + \dfrac{(-6)}{3!} t^3 +$

$\dfrac{(-6)(-2)}{5!} t^5 + \dfrac{(-6)(-2)(2)}{7!} t^7 + \dfrac{(-6)(-2)(2)(6)}{9!} t^9 + \cdots$

41. $J_1(t) = \dfrac{t}{2} - \dfrac{1}{2^2 \cdot 4} t^3 + \dfrac{1}{2^2 \cdot 4^2 \cdot 6} t^5 - \dfrac{1}{2^2 \cdot 4^2 \cdot 6^2 \cdot 8} t^7 + \cdots$

43. $H_3(t) = -12t + 8t^3$

45. $y = 1 - (\sin 1)t^2 + \dfrac{\sin 1 \cos 1}{6} t^4 + \cdots$

answers to section 53 (page 601)

part A **1.** $[1, -2, 3]$ **3.** not possible **5.** $[-1, 2, 3, 4]$ **7.** not possible **9.** $[-3, 6, 9, 12]$
11. B **13.** $\vec{0}$ **15.** F **17.** A **19.** D **21.** C **23.** $6E$ or $-2A$ **25.** $4E$
27. (a) $[50, 20, 45]$ specifies the total number of units of each type needed on a particular day
(b) $[440, 170, 450]$ specifies the total number of units needed for the three passengers for the entire cruise.
29. -3 **31.** $\sqrt{5}$ **33.** $3\sqrt{3}$ **35.** -84 **37.** $C \cdot D_1 = \$11 =$ daily cost of insulin for passenger 1;
$C \cdot (10D_2) = \$165 =$ total cost of insulin for passenger 2; $C \cdot (7D_1 + 10D_2 + 10D_3) = \$474.50 =$ total cost of insulin
for all three passengers **41.** $a = -\frac{4}{3}$ **part B** **46.** $T = X + Y$ **48.** $|T| = \sqrt{|X|^2 + |Y|^2}$ and $|S| =$
$\sqrt{|X|^2 + |Y|^2 + |Z|^2}$ **50.** $\phi = 35.26°$

answers to section 54 (page 616)

part A **1.** $[-2, 3]$ **3.** $[-1, 0, 2]$ **5.** $S = (1, -2, 6)$ **7.** $D = (-6, 0)$ **9.** $E = (-4, -1)$
11. $z = \frac{14}{3}$ **13.** $F_4 = [2, -7]$ **15.** $Y_p = -\frac{1}{25} [3, 4]$, $Y_n = \frac{1}{25} [28, -21]$ **17.** $A_p = [0, 6]$, $A_n =$
$[-2, 0]$ **19. (a)** $[200 \cos \theta + 40, \ 200 \sin \theta]$ **(b)** $|V| = 222.7$ mph in the direction $51.05°$ north-
west **(c)** $\theta = 53.13°$ **21.** $\theta = \sin^{-1} \frac{5}{6} = 56.44°$ **23. (a)** 65.67 lbs **(b)** $[-62.8, 11.20]$
25. (a) $2,040.6$ lbs **(b)** $[1,996, -25.74]$, which has a magnitude of $1,996.17$ lbs **27.** $\theta = \sin^{-1} \frac{5}{6} = 56.44°$

answers to section 55 (page 628)

part A **1.** $x = 2$, $y = \frac{1}{2}$ **3.** $x = 4$, $y = -1$ **5.** $a_{31} = \frac{1}{2}$, $a_{13} = 2$, $a_{45} = -3$, $a_{34} = -\frac{5}{2}$, $a_{25} = 4$

7. $\begin{bmatrix} 1 & 3 \\ 2 & 8 \end{bmatrix}$ **9.** $\begin{bmatrix} 4 & -1 & 3 \\ -2 & 3 & -2 \end{bmatrix}$ **11.** $\begin{bmatrix} 8 & 2 \\ -2 & 10 \\ 0 & 18 \end{bmatrix}$ **13.** not defined **15.** not defined

17. $\begin{bmatrix} -14 & 1 & -18 \\ 24 & -3 & 36 \end{bmatrix}$ **19.** $\begin{bmatrix} -12 & -22 \\ 13 & 21 \end{bmatrix}$ **23.** $c = 0$ and $a = d$

27. $\begin{bmatrix} -10 \pm .5 & -15 \pm .75 & -13 \pm .65 & -18 \pm .9 & -20 \pm 1.0 \\ -15 \pm .75 & -16 \pm .8 & -21 \pm 1.05 & -23 \pm 1.15 & -10 \pm .5 \\ -7 \pm .35 & -10 \pm .5 & -12 \pm .6 & -17 \pm .85 & -8 \pm .4 \end{bmatrix}$ **31.** $(B + I)A$ **33.** $(A - C)B$

answers to section 56 (page 639)

part A **1.** $Z = CX$ where $C = \begin{bmatrix} 10 & -10 & 8 \\ -2 & 3 & -3 \end{bmatrix}$ **3.** $Z = \begin{bmatrix} -1 & -3 \\ 25 & 5 \end{bmatrix} X + \begin{bmatrix} -4 \\ 20 \end{bmatrix}$ **5. (a)** $X_1 = [0, 50, 0, 0]$,
$X_2 = [0, 0, 30, 0]$, $X_3 = [0, 0, 0, 21]$, $X_4 = [21000, 0, 0, 0]$, $X_5 = [0, 105, 0, 0]$
7. (*Note*: Population projections were rounded off after each computation.)   **9. (c)** 15/28

t	0–1	2–3	4–5	6–7	8–9	10–11	≥ 12
0	800	700	650	500	400	300	280
2	820	696	609	566	435	348	485
4	975	713	606	530	492	378	691
6	1,094	848	620	527	461	428	882

11. (a) $T = \begin{bmatrix} .8 & .4 \\ .2 & .6 \end{bmatrix}$, $X = \begin{bmatrix} 90 \\ 60 \end{bmatrix}$ **(b)** $Y = \begin{bmatrix} 96 \\ 54 \end{bmatrix}$ **(c)** $\hat{X} = [100, 50]$

13. (a) $T = \begin{bmatrix} .88 & .02 & 0 \\ .06 & .97 & .05 \\ .06 & .01 & .95 \end{bmatrix}$ **(b)**

t	Phytoplankton	Water	Zooplankton
0	0	100	0
1	2	97	1
2	3.7	94.26	2.04
3	5.14	91.76	3.10
4	6.36	89.47	4.17
5	7.39	87.38	5.24
6	8.25	85.45	6.30

part B **15.** (*Note*: Projections are rounded off to the nearest whole number after each computation.)

t	age class 1	age class 2	age class 3
0	100	50	40
1	20	60	28
2	14	12	27
3	14	8	13
4	7	5	7
5	4	4	4

answers to section 57 (page 653)

part A **1.** $\{H1, H2, H3, H4, H5, H6, T1, T2, T3, T4, T5, T6\}$ **3.** $\{RWB, RBW, WRB, WBR, BWR, BRW\}$ **5.** $\{BG, GB, BBG, GGB, BBBG, GGGB, BBBB, GGGG\}$ **7. (a)** $\frac{9}{19}$ **(b)** $\frac{6}{19}$
(c) $\frac{1}{19}$ **9. (a)** $\frac{1}{3}$ **(b)** $\frac{1}{2}$ **11. (a)** $\frac{1}{4}$ **(b)** $\frac{1}{13}$ **(c)** $\frac{3}{13}$ **13. (a)** $\frac{1}{2}$ **(b)** $\frac{1}{4}$
15. (a) $\frac{1}{9}$ **(b)** 0   **17.** $\frac{3}{8}$ **19.** $2^{23} = 8,388,608$ **21.** $\frac{1}{2}$ **23.** $P(\text{red}) = \frac{5}{12}$, $P(\text{white}) = \frac{3}{12}$,
$P(\text{blue}) = \frac{4}{12}$ **25.** $\frac{7}{8}$ **27.** .238   **29.** $\frac{1}{4}$ **31. (a)** $\frac{1}{2}$ **(b)** $\frac{1}{3}$ **33. (a)** AO and BO
(b) $\frac{1}{4}$ **Part B** **35.** 720   **37.** 12   **39.** .000129

answers to section 58 (page 667)

part A **1.** a = a left-handed male, b = a right-handed female, c = a left-handed female **3.** $C = A_1 \cap A_2$ **5.** $C = (A_1 \cap A_2) \cup A_3$ **7. (a)** "a sum of 10 or doubles" = {(1, 1), (2, 2), (3, 3), (4, 4), (5, 5), (6, 6), (6, 4), (4, 6)} **(b)** "a sum of 10 and doubles" = {(5, 5)} **(c)** "a sum of 10 but no 5s" = {(6, 4), (4, 6)} **9.** Only the events in **(a)** and **(d)** **11.** When $A = \bar{B}$ **17.** $1 - (\frac{5}{6})^6$ **19.** P("queen or red") = $\frac{7}{13}$ and P("queen but not red") = $\frac{1}{26}$ **21. (a)** .2 **(b)** .1 **(c)** .5 **23. (a)** .7 **(b)** .3 **(c)** .3 **25.** Only assignment **(b)** is consistent with the axioms. **part B** **27.** .2709 **29. (c)** $P[365, n]/365^n$ **(d)** $\bar{A}$ is the event "at least two people have the same birthday." For $n = 23$, $P(\bar{A})$ = .507.

answers to section 59 (page 680)

part A **1.** $\frac{1}{6}$; independent **3.** $\frac{1}{4}$; independent **5.** $\frac{3}{8}$; dependent **7.** $\frac{2}{9}$; dependent **9.** $\frac{1}{2}$; dependent **11.** $P(W|H) = 50/85$ and $P(H|W) = 50/75$ **13.** $\frac{5}{8}$ **15.** $P(A|B) = P(A)/P(B)$ and $P(B|A) = 1$ **17.** .63 **19.** .477 **21.** Yes **23. (a)** $\frac{1}{64}$ **(b)** $\frac{63}{64}$ **25. (a)** .7225 **(b)** .0225 **(c)** .255 **27. (a)** $\frac{1}{16}$ **(b)** $\frac{3}{16}$ **29. (a)** $\frac{1}{2}$ **(b)** $\frac{1}{4}$ **31. (a)** aa/Bb and AA/bb **(b)** P("green") = $\frac{1}{2}$ = P("white"), P("yellow") = P("blue") = 0 **(c)** P("green") = $\frac{9}{16}$, P("yellow") = $\frac{3}{16}$, P("blue") = $\frac{1}{16}$, P("white") = $\frac{3}{16}$ **part B** **32.** p^5 **34.** $p^3(1-p)^2$ **36.** .311

answers to section 60 (page 695)

part A **1.** $\frac{3}{5}$ **3.** $\frac{3}{5}$ **5.** $\frac{1}{2}$ **7.** .42 **9. (a)** $\frac{3}{8}$ **(b)** $\frac{1}{16}$ **11.** $\frac{7}{12}$ **13.** .945 **15.** $\frac{2}{3}$ **17. (a)** .988 **(b)** .868 **19. (a)** $\frac{1}{4}$ **(b)** $\frac{13}{16}$ **21.** New probability is still $\frac{2}{3}$. **23.** $\frac{1}{16}$ **25. (a)** The probability that both cousins are carriers is $\frac{1}{9}$. The probability of a PKU child is $\frac{1}{36}$ **(b)** $\frac{1}{600}$

answers to section 61 (page 709)

part A **1. (a)** M: .91, N: .09 **(b)** MM: .828, MN: .164, NN: .008 **3. (a)** M: .525, N: .475 **(b)** MM: .276, MN: .499, NN: .226 **5.** AA: .306, Aa: .494, aa: .2 **7.** .049 **9.** .16 **15. (b)** .0045, .0034, .0025, .0019, .0014 **17.** .447, assuming random mating and equilibrium

index

index

a

absolute convergence for infinite series, 562–563

absolute maximum and minimum, 141–142
  procedures for finding, 141, 143

acceleration, 109, 114, 227

acetaldehyde and the formation of CO, 480

Achilles tendon, 618

Achondroplasia, 657

acid rain, 168, 192, 194, 637–638

age structure models, 548–549, 553–554, 593, 595, 602, 631–635

air temperature as a function of latitude and longitude, 436–437

albinism, 651, 655, 689–691, 699
  in the San Blas Indians, 710

alleles, 651–652
  dominant, 651
  recessive, 651

Allen's problem, 396

allometric growth model, 352, 364–365

allometric laws, 190, 193, 358, 479, 483–484

Amphiporous, 149

annual increment method, 486

antiderivative
  definition, 199
  of elementary functions, 200
  rules, 201–202, (See also integration methods)

aquatic ecosystem, 593

area
  as a double integral, 497
  between two curves, 229–231
  of a wound, 483
  under a curve, 197, 219–223

area density functions, 510 513

area-species curve, 47–48, 53–54, 193, 479

arithmetic sequence, 528, 536

arterial blood flow, 236
  to a limb, 352–354

arterial bifurcation, 339–340

artic cod, 196

asymptote
  horizontal, 32
  linear, 124
  vertical, 32, 66

average value, 209
  of a function, 210–211, 228–229
  of a function of two variables, 508–509

b

bacillus bacteria, 448, 460

bacterial growth, 173, 485

baldness, 696

bathtub models, 427–428

Bayes' Theorem, 690–695

Beer-Lambert Law, 164, 167, 237, 483

Bernadelli, H. 631, 633

Bessel functions, 588–589, 591

Beverton-Holt year class model, 392–398

beta distributions, 278

biceps muscle, 610, 614–615

binomial probabilities, 683–684

biomass, 176, 390, 464, 512–513

blood flow
  through the aorta, 404–406
  through the lungs, 466–467
  to a limb, 452–454

blood type
  ABO, 652–654, 661, 665
  MN, 710
  in Navaho Indians, 703
  Rh, 659, 661, 665

blue whale, 22, 24, 191, 271, 397, 475, 484, 640

body surface area, 54, 445, 459, 467

boundary conditions, 182, 414

Boyle's Law, 34, 250

breathing, 311, 410–411

brightness, 35, 113

c

d

QH 323.5 .C84

Cullen, Michael R.

Mathematics for the
    biosciences

Date Due

DEMCO